CAMBRIDGE LIBRARY COLLECTION

Books of enduring scholarly value

Life Sciences

Until the nineteenth century, the various subjects now known as the life sciences were regarded either as arcane studies which had little impact on ordinary daily life, or as a genteel hobby for the leisured classes. The increasing academic rigour and systematisation brought to the study of botany, zoology and other disciplines, and their adoption in university curricula, are reflected in the books reissued in this series.

On the Anatomy of Vertebrates

Richard Owen F.R.S. (1804–92) was a controversial and influential palaeontologist and anatomist. Owen studied medicine at the University of Edinburgh and at London's St Bartholomew's Hospital. He grew interested in anatomical research, and after qualifying he became assistant conservator in the museum of the Royal College of Surgeons, and then superintendent of natural history in the British Museum. He quickly became an authority on comparative anatomy and palaeontology, coining the term 'dinosaur' and founding the Natural History Museum. He was also a fierce critic of Darwin's theory of evolution by natural selection, and engaged in a long and bitter argument with Darwin's 'Bulldog', Thomas Huxley. Published in 1866, this is the second book in a highly illustrated three-volume set that comprises a thorough overview of vertebrate anatomy. This volume focuses on the anatomy of birds, and includes the first part of the analysis of mammalian anatomy.

Cambridge University Press has long been a pioneer in the reissuing of out-of-print titles from its own backlist, producing digital reprints of books that are still sought after by scholars and students but could not be reprinted economically using traditional technology. The Cambridge Library Collection extends this activity to a wider range of books which are still of importance to researchers and professionals, either for the source material they contain, or as landmarks in the history of their academic discipline.

Drawing from the world-renowned collections in the Cambridge University Library, and guided by the advice of experts in each subject area, Cambridge University Press is using state-of-the-art scanning machines in its own Printing House to capture the content of each book selected for inclusion. The files are processed to give a consistently clear, crisp image, and the books finished to the high quality standard for which the Press is recognised around the world. The latest print-on-demand technology ensures that the books will remain available indefinitely, and that orders for single or multiple copies can quickly be supplied.

The Cambridge Library Collection will bring back to life books of enduring scholarly value (including out-of-copyright works originally issued by other publishers) across a wide range of disciplines in the humanities and social sciences and in science and technology.

On the Anatomy of Vertebrates

Volume 2: Birds and Mammals

Richard Owen

CAMBRIDGE UNIVERSITY PRESS

Cambridge, New York, Melbourne, Madrid, Cape Town,
Singapore, São Paolo, Delhi, Tokyo, Mexico City

Published in the United States of America by Cambridge University Press, New York

www.cambridge.org
Information on this title: www.cambridge.org/9781108038263

This edition first published 1866
This digitally printed version 2011

ISBN 978-1-108-03826-3 Paperback

THE

ANATOMY OF VERTEBRATES.

VOL. II.

ON THE

ANATOMY OF VERTEBRATES.

VOL. II.

BIRDS AND MAMMALS.

BY

RICHARD OWEN, F.R.S.

SUPERINTENDENT OF THE NATURAL HISTORY DEPARTMENTS OF THE BRITISH MUSEUM,
FOREIGN ASSOCIATE OF THE INSTITUTE OF FRANCE, ETC.

LONDON:
LONGMANS, GREEN, AND CO.
1866.

LONDON
PRINTED BY SPOTTISWOODE AND CO.
NEW-STREET SQUARE

CONTENTS, OR SYSTEMATIC INDEX.

CHAPTER XIII.

CHARACTERS OF HÆMATOTHERMA.

CHAPTER XIV.

OSSEOUS SYSTEM OF AVES.

CHAPTER XV.

MUSCULAR SYSTEM OF AVES.

CHAPTER XVI.

NERVOUS SYSTEM OF AVES.

CHAPTER XXII.

TEGUMENTARY SYSTEM OF AVES.

CHAPTER XXIII.

GENERATIVE SYSTEM OF AVES.

CHAPTER XXIV.

DEVELOPMENT OF AVES.

CHAPTER XXV.

CHARACTERS AND PRIMARY GROUPS OF THE CLASS MAMMALIA.

CHAPTER XXVI.

OSSEOUS SYSTEM OF MAMMALIA.

ERRATA.

Page 133, note [2] *for* 'vol. i. p. 208,' *read* 'CCXXXVI, vol. i. p. 173.'
" 258, note [4] *for* 'old' *read* 'odd.'
" 260, note [5] *for* 'LXIX.' *read* 'LIX.'
" 333, twenty-three lines from top, *for* 'met-apophyses,' *read* 'par-apophyses.'
" 452, two lines from bottom, *for* 'fig. 303,' *read* 'fig. 304.'
" 479, fifteen lines from bottom, *for* 'promixal,' *read* 'proximal.'

THE

ANATOMY OF VERTEBRATES.

CHAPTER XIII.

CHARACTERS OF HÆMATOTHERMA.

§ 121. *Thermogenous Conditions.*—Life is attended with constant molecular change; and such vital motion becomes converted directly, or through intermediate modes of chemical or electrical force, into that of heat; the chief chemical action preceding or producing the calorific force being due to the introduction of oxygen by air or food into the body, where it operates in a manner analogous to combustion.

The evolution of animal heat more directly relates to the amount of air inspired in a given period, and to the rapidity with which the oxygenated blood is conveyed to the tissues.

In these the molecular changes are governed by the nervous system, and whatever tends to paralyse the nervous force operates in the same degree in arresting those molecular movements on which more directly depends the evolution of heat. In this act the nervous system is accordingly concerned, in so far as it influences the exercise of the muscular movements.

In the *Hæmatotherma*, or Warm-blooded Vertebrates, the atmosphere is directly inspired and applied to a vascular surface which, in proportion to the bulk of the body, is much more extensive than in any of the *Hæmatocrya*. For mechanical convenience the respiratory surface is closely packed, in small compass, by subdivision of the pulmonary cavity into countless minute cells, giving to the lung a spongy texture, obliterating all trace of a visible or conspicuous cavity.

The whole of the venous blood is propelled over this extensive but compactly disposed capillary area by successive contractions of a special ventricle, receiving it from a distinct auricle, and the blood, changed by the respiratory action, is conveyed to another distinct auricle and propelled by a second distinct ventricle over the entire system.

Thus a four-chambered heart and spongy lungs are the chief anatomical characteristics by which the 'warm-blooded' are distinguished from the 'cold-blooded' Vertebrates, although respiration and circulation are subsidiary or auxiliary, not immediate, thermogenous functions.

Whatever tends to obstruct the flow of blood to a part of the body, as ligature of an artery, e. g., lowers in a certain degree the heat of that part; and whatever augments such flow of blood, as, exercise, e. g., or increases the quantity of blood in a part, as where the capillaries dilate through paralysis of the vaso-motory filaments from a ganglion of the sympathetic nerve, raises the heat of such part; temporarily, at least, in the latter case.[1]

In all *Hæmatotherma* the mass of nervous matter constituting the cerebral portions of the prosencephalon is relatively larger both to the rest of the brain and to the bulk of the body than in *Hæmatocrya*, although the degrees of this predominance in the warm-blooded series relate to other functions than the evolution of heat.

Concomitantly with the advance of the circulating and respiratory organs in *Hæmatotherma* is that of the blood itself, in quantity, in the proportion of organic (*proteine*) principles to the water in it, and in depth of colour due to the more abundant blood-discs. The voluntary muscular fibre shows, in most *Hæmatotherma*, by its deeper colour than in *Hæmatocrya*, the influence of this more abundant, richer, and redder blood; and the longer duration and greater energy of the contractions have relation to the hæmatothermal conditions of the nervous, respiratory, and circulating systems.

In every muscular contraction some molecules of the fibre may be said to be burnt, and heat is evolved. Needles of a delicate thermo-electrical apparatus, thrust into a living muscle, indicate a rise of temperature at each act of contraction.[2] The heat-producing results of the sum of such actions is a matter of common experience, and a loss of animal heat results from the cessation of such actions. So, Hunter writes: 'When a man is asleep he is colder than when he is awake; and I find, in general, that the difference is about one degree and a half' (of Fahr.)[3]

[1] v. p. 377.
[2] As in the 'biceps flexor cubiti' of the man so experimented on, in I. p. 402.
[3] xciv. p. 144. See also ii'.

The molecular movements and changes in the organs of vegetative life constitute a more unintermitting source of caloric. The blood which returns from the extensive seat of such operations afforded by the mucous intestinal tract is warmer than before it enters that tract: the blood of the hepatic vein after its passage through the portal circulation, and its work in the liver, shows a more marked rise of temperature. Urine in mammals, before its escape, is hotter than blood;[1] and the rich supply of nerves to the adrenals may relate to the calorific functions of the kidneys.

The production of heat from the actions of organic life depends on the amount of material for the support of such actions—on the quantity of oxidizable substance introduced as aliment into the body. The greater vigour, activity, waste, or wear and tear, in the warm-blooded machinery necessitate, while they enable, a greater energy, and more regular and rapidly recurring performance of the digestive functions; and the warm-blooded differ from the cold-blooded vertebrates in the greater amount of food which they consume, and the shorter intervals between the times of eating. Warm-blooded animals exemplify this influence:—'I weakened,' says Hunter, 'a mouse by fasting, and then introduced the ball of the thermometer into its belly: the ball being at the diaphragm, the quicksilver rose to 97°; in the pelvis to 95°, being two degrees colder than in the strong mouse.'[2] The difference of being 'full' or 'fasting' in resisting cold is a matter of common experience.

§ 122. *Thermogenous Results.*—The more active and unremitting vital combustion, due to the above-defined advanced conditions of the nervous, respiratory, circulating, digestive, and muscular systems, keeps up a constant temperature, as a general rule, in the *Hæmatotherma*, which is usually so much higher than that of the surrounding medium as to cause the sensation of warmth to the hand touching the body. In man the mean temperature of the interior of the body is 100° Fahr.; in the dog, 101°; in the ox, 100°; in the mouse, 99°; in the whale, 105°. In Birds[3] the mean temperature ranges in different species from 106° to 112°.

The heat-producing powers in healthy *Hæmatotherma* are more active as the surrounding medium is cooler; and cold, much below freezing, is long resisted, and habitually, by the warm-

[1] III'. 100° or 101° Fahr. as against 97° Fahr.; (39.5, as against 37 or 30 Cent.)

[2] XCIV. p. 145. See also p. 16, for a similar illustration of loss of heat through starvation in ducks.

[3] IV.

blooded denizens of arctic and antarctic zones. The nature of the external covering has much influence in this resistance, whether it be the thick layer of subcutaneous fat in the whale-tribe, the fur and hair of the quadruped, or the down and feathers of the bird. Save in the case of mankind and the whalekind, the warm-blooded Vertebrate may be distinguished at a glance from the cold-blooded one by the non-conducting, heat-preserving, nature of its clothing, which is 'hair,' as a general rule in Mammals, and 'feathers' in Birds.

There are, however, gradations of the heat-maintaining power in the *Hæmatotherma*. Some Mammals, e. g. the Alpine Marmot, the Hamster, the Squirrel, the Dormice (*Myoxus*), the Porcupine, the Virginian Opossum, at the approach of winter-cold, seek a retreat, fall into a deep sleep, and lose from 10° to 20° Fahr. of heat. In the Squirrel, e. g., the heat of the body has been found to sink from 98° to 78°. Respiration is continued, though slowly, in these winter-sleepers. The Hedgehog (*Erinaceus*) and the Bat (*Vespertilio*, Linn.) fall into a deeper and more lasting torpor; in which breathing is suspended, and a slow and languid circulation is the sole sign of animation. In the Bat, the heart's pulsations fall from 200 in a minute, as when in active wakefulness, to 30 in a minute, during torpidity; the blood being then in a dark venous state, and the temperature of the body down to 40°. In this condition these Insectivora survive the season during which their allotted food is unattainable. In the tropics some allied species, e. g. the Tenrecs (*Centetes*) fall into a similar torpidity, without the excitement of a freezing cold, during the season unfavourable to the presence of their food.

The feeble and inactive young *Hæmatotherma* use up less oxygen than adults; and, when exposed to cold, lose their heat, and also their sensibility, differing in this latter respect from the hybernators. The least touch to a spine of a torpid Hedgehog rouses it to draw a deep sonorous inspiration: the merest shake induces respiration in the torpid Bat.

In all these instances of loss of power to preserve the average mammalian temperature, the physiological conditions of the species approximate more or less to those of the cold-blooded animals; and it is interesting to observe that the winter-sleeping and torpid Mammals are those which most resemble reptiles in their cerebral organisation: they are also of small size. Whether the Edentata and Monotremata would become torpid, and so accommodate themselves to other than their native climates is a question well worthy of experimental determination.

No approach to torpidity with loss of animal temperature has been determined to take place in any bird. The insectivorous kinds migrate—Swifts and Swallows, e. g., to and fro between England and Africa; and migration is performed by numerous other birds in relation to localities furnishing the food most appropriate for the nourishment of their newly-hatched young. Experiments have failed to induce torpidity in birds through artificial cold.

§ 123. *Characters and Orders of Birds.*—The two Hæmatothermal classes *Aves* and *Mammalia,* are defined in vol. i. p. 6; and I here proceed to a fuller exposition of the avian characteristics, and of the modifications on which the class has been divided into orders or other primary groups.

Birds constitute a class of oviparous vertebrate animals, with warm blood, a double circulation, and a covering of feathers. They are organised for flight, and as this, the fleetest and most vigorous kind of locomotion, demands the greatest energy in the contractility of the muscular fibre, so the respiratory function

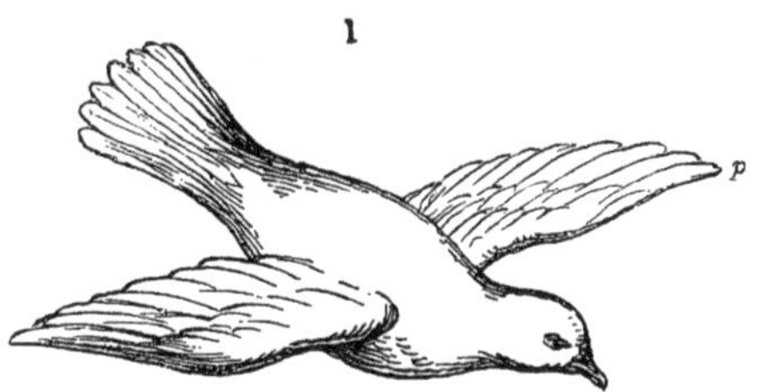

finds its highest developement in the present class. Not only the ramifications of the pulmonary artery, but many of the capillaries of the systemic circulation, from the singular extension of the air-cells through the body, are submitted to the influence of the atmosphere, and hence Birds may be said to enjoy a double respiration.

Although the heart resembles in some particulars that of the *Reptilia,* the four cavities are as distinct as in the *Mammalia,* but they are relatively stronger, their valvular mechanism is more perfect, and the contractions of this organ are more forcible and frequent in Birds in accordance with their more extended respiration and their more energetic muscular actions.

As Birds exceed Mammals in the activity of those functions on which the waste and renovation of the general system more immediately depend, so they possess, as has been shown, a higher standard of animal heat.

The modification of the tegumentary covering characteristic of the present class is to be regarded rather as dependent upon, than occasioning, this high degree of internal temperature, which requires for its due maintenance against the agency of external cold an adequate protection of the surface of the body by means of non-conducting down and imbricated feathers; and this warm clothing is more especially required to meet the sudden variations of temperature to which the bird is exposed, when soaring in the higher regions of air and stooping to the earth, during rapid and extensive flights.

The generative product is excluded from the oviduct in an undeveloped state, inclosed, in a liquid form, within a calcareous case or shell. Collision of two brittle eggs *in transitu* is obviated by the female organs being developed only on the left side of the body. The ovum is subsequently perfected by means of *incubation*, for which action the bird is especially adapted by its high degree of animal heat.

Birds form the best characterised, most distinct, and natural class in the whole animal kingdom, perhaps even in organic nature. They present a constancy in their mode of generation and in their tegumentary covering, which is not met with in any other of the vertebrate classes. No species of Bird ever deviates, like the whales among Mammals, the serpents among Reptiles, and the eels among Fishes, from the tetrapodous type characterising the vertebrate division of animals.

The anterior extremities are constructed according to that plan which best adapts them for the actions of flight; and although, in some few instances, the developement of the wings proceeds not so far as to enable them to act upon the surrounding atmosphere with sufficient power to overcome the counteracting force of gravity; yet, in these cases they assist, by analogous motions, the posterior extremities: either, as in the ostrich, by beating the air while the body is carried swiftly forward by the action of the powerful legs; or, as in the penguin, by striking the water after the manner of fins, and by the resistance of the denser medium carrying the body through the water in a manner analogous to that by which the birds of flight are borne through the air. In a few exceptions, as the cassowary and apteryx, the wings are outwardly represented by a few quills or a small claw In no instance do the anterior extremities take any share in stationary support or in prehension.

Birds are therefore biped, and the operations of taking the food, cleansing the plumage, &c., are almost exclusively performed by means of the mouth, which consists of two lipless and toothless jaws, sheathed with horn. To facilitate the prehensile and other actions thus transferred to the head, the neck is elongated, and the body generally inclined forward and downward from the hip-joints. The thighs are accordingly extended forward at an acute angle from the pelvis toward the centre of the trunk, and the toes are lengthened and spread out to form an adequate base of support. The actions of perching, walking, hopping, running, scratching, burrowing, wading, and swimming, require for their perfect performance different modifications of the posterior extremities. The mandibles, again, present as many varieties of form, each corresponding to the nature of the food, and in some degree indicative of the organisation necessary for its due assimilation. Ornithologists have, therefore, founded their divisions of the class chiefly on the modifications of the bill and feet. Since, however, Birds in general are associated together by characters so peculiar, definite, and unvarying, it becomes in consequence more difficult to separate them into subordinate groups, and these are necessarily more arbitrary and artificial than are those of the other vertebrate classes.

A *binary* division of the class[1] may be founded on the condition of the newly-hatched young, which in some orders are able to run about and provide food for themselves the moment they quit the shell (*Aves præcoces*); while in others the young are excluded feeble, naked, blind, and dependent on their parents for support (*Aves altrices*).

Nitzsch[2] grouped together the feathered tribes under *three* series, according to the great divisions of the terraqueous globe which form respectively the principal theatres of their actions. The first order consists of the birds of the air, *Aves aereæ* (Luftvögeln); the second embraces the birds of the land, *Aves terrestres* (Erdvögeln); the third includes the birds which frequent the waters, *Aves aquaticæ* (Wasser-vögeln). The eagle and lark exemplify the first; the ostrich and common fowl the second; the heron and the gull the third, of these extensive divisions of the class.

Vigors[3] proposed a more definite system upon a similar principle, distributing Birds into five orders. The first includes those which soar in the upper regions of the air, which build their nests and rear their young on high cliffs or lofty trees; they are the chief of aerial birds and form the order termed *Raptores*,

[1] VII'. p. 265. [2] VIII'. [3] IX'.

from the rapacious habits and animal food of the species so grouped together.

The second order affects the lower regions of the air: the birds composing it are peculiarly arboreal in their habits, and are, therefore, termed 'Perchers,' *Insessores.*

The third order corresponds with Nitzsch's *Aves terrestres,* and is denominated *Rasores,* from their general habit of scratching up the soil in quest of food.

By dividing his *Aves aquaticæ* into those which wade to obtain their food, and into those which swim, we get the two remaining orders of the quinary arrangement—viz. the *Grallatores* and *Natatores.* The merit of this system mainly lies in the endeavour to trace the natural affinities of the several families, and show how they pass one into another to form a connected circular whole.

The *Raptores* of Vigors answers to the *Accipitres* of Linnæus and Cuvier; the *Insessores* to the *Passeres* and *Pici* of Linnæus, and to the *Passeres* and *Scansores* of Cuvier; the *Rasores* to the *Gallinæ* of Linnæus, *plus* the *Columbæ,* and to the *Gallinaceæ* of Cuvier; the *Grallatores* to the *Grallæ* of Linnæus and Cuvier; the *Natatores* to the *Anseres* of Linnæus, and the *Palmipedes* of Cuvier.

AVES (Birds).

Class-characters.

Animal, vertebrated, oviparous, biped.

Pectoral[1] *limbs* organised for flight.

Integument, plumose.

Blood, red, warm.

Respiration and *circulation,* double.

Lungs, fixed, perforated.

Negative Characters, no ear-conchs, lips, teeth, epiglottis, diaphragm, fornix, corpus callosum, scrotum.

The following are the orders, with their characters and sample families, adopted as most convenient for the purpose of the present work:—

[1] CCCXLVIII.

Order I. NATATORES.

Swimming Birds. Toes united by a membrane, fig. 2. Legs placed behind the equilibrium, and body covered with a thick coat of down beneath the feathers.

Fam. 1. *Brevipennatæ*. *Ex*. Penguin, Auk, Guillemot, Grebe.
2. *Longipennatæ*. *Ex*. Skimmer, Tern, Mew, Gull, Petrel, Albatross.
3. *Totipalmatæ*. *Ex*. Pelican, Gannet, Cormorant, Anhinga, Frigate Bird, Tropic Bird.
4. *Lamellirostratæ*. *Ex*. Duck, Goose, Swan, Flamingo.

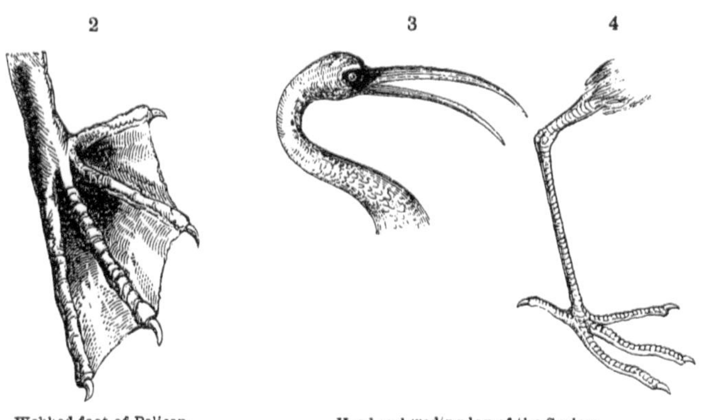

Webbed foot of Pelican. Head and wading leg of the Curlew.

Order II. GRALLATORES.

Wading Birds. Legs long, naked from above the distal extremity of the tibia downwards, fig. 4.

Fam. 1. *Macrodactyli*. *Ex*. Coot, Rail, Crake, Screamer, Jacana.
2. *Cultrirostres*. *Ex*. Boatbill, Crane, Heron, Ibis, Stork, Tantalus, Spoonbill.
3. *Longirostres*. *Ex*. Gambet, Avocet, Snipe, Ruff, Turnstone, Sandpiper, Godwit, Curlew, fig. 3.
4. *Pressirostres*. *Ex*. Oystercatcher, Thicknee, Plover, Lapwing, Bustard, Courser.

Order III. RASORES.

5

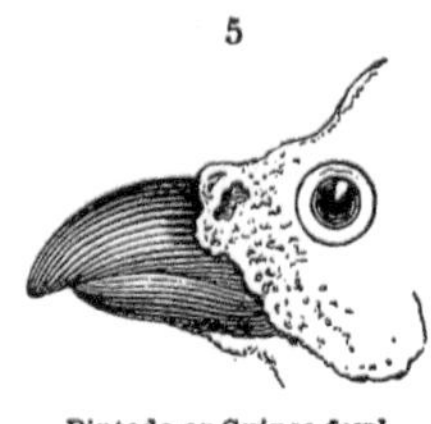

Pintado or Guinea-fowl.

Scratching Birds. Feet strong, provided with obtuse claws for scratching up grains, etc. Upper mandible vaulted; nostrils pierced in a membranous space at the base, and covered by a cartilaginous scale, fig. 5. Nest rude. Sternum with four, rarely two, deep fissures.

Suborders.

Gallinacei or *Clamatores*; Polygamous. *Ex.* Megapode, Peafowl, Partridge, Quail, Pheasant, Ganga, Grouse, Pintado, Tinamú, Turkey, Curassow, Guan.

Columbacei or *Gemitores;* Monogamous. *Ex.* Dove, Goura, Vinago.

Order IV. CANTORES (Oscines).

Singing Birds. Legs short and slender, with three toes before and one behind, the two external toes being united by a very short membrane, fig. 6. Sternum with one hind-notch on each side, manubrium bifurcate, fig. 15; larynx 5—muscular. The brain arrives in this order at its greatest proportional size, and the organ of voice here attains its utmost complexity. Nests complex; eggs usually coloured. Monogamous.

6

Foot of Percher.

Fam. 1. *Dentirostres.* *Ex.* Manakin, Shrike, Wren, Wagtail, Warbler, Thrush.

2. *Conirostres.* *Ex.* Paradise Bird, Crow, Starling, Bunting, Tit, Lark, Finch, Grosbeak.

3. *Tenuirostres.* *Ex.* Sunbird, Nuthatch, Creeper.

4. *Fissirostres.* *Ex.* Swallow, Martin.

Order V. VOLITORES.

Moving solely by flight. Skeleton light and highly pneumatic; sternum with a simple manubrium, and a deep keel; in some entire, fig. 18, in most with two hind-notches on each side, fig. 20;

larynx trimuscular; intestinal cæca usually absent, or large; wings powerful, in some long and pointed; legs small and weak, with few exceptions not used in locomotion; with the back toe *i* short, sometimes turned forward (*Cypselus*), or wanting (*Ceyx*); the outer toe *iv* is reversible in some (*Trogon*), in others united to the mid-toe *iii*, as far as the penultimate joint, fig. 7. Many nidificate in holes of trees, or in the earth; the eggs are white and subspherical. They are monogamous. The

7

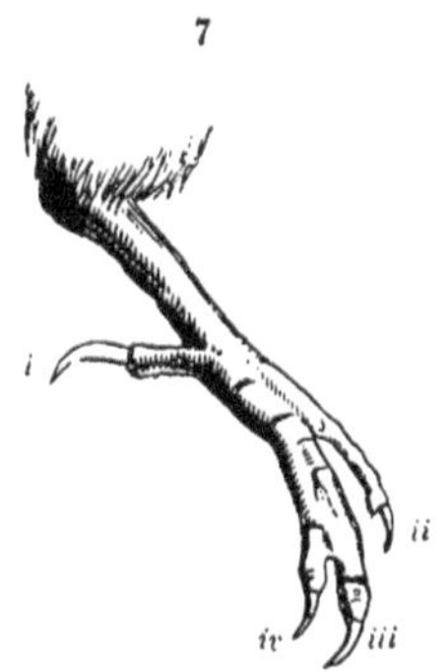

Syndactylous foot of Kingfisher.

8

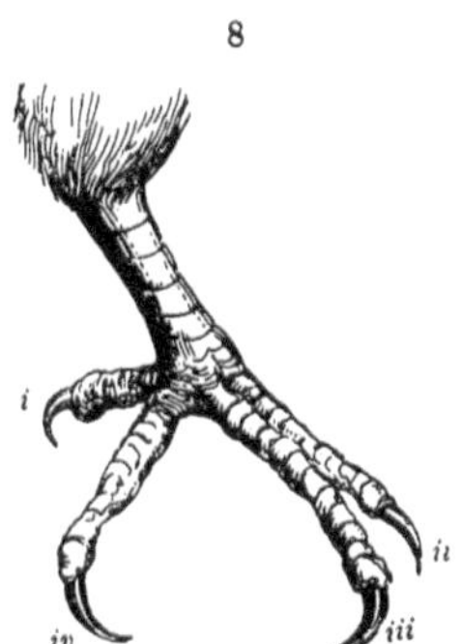

Scansorial foot of Woodpecker.

head is large, and in most the beak is remarkable for its length or width, or both. The gape is wide; the food taken on the wing.

Fam. 1. *Cypselidæ*. *Ex.* Swift.
2. *Trochilidæ*. *Ex.* Humming-bird.
3. *Caprimulgidæ*. *Ex.* Nightjar.
4. *Trogonidæ*. *Ex.* Trogon.
5. *Prionitidæ*. *Ex.* Mot-mot.
6. *Meropidæ*. *Ex.* Bee-eater.
7. *Galbulidæ*. *Ex.* Jacamar.
8. *Coraciadæ*. *Ex.* Roller.
9. *Capitonidæ*. *Ex.* Puff-bird.
10. *Alcedinidæ*. *Ex.* Kingfisher.
11. *Bucerotidæ*. *Ex.* Hornbill.

Order. VI. SCANSORES.

Climbing Birds. Toes arranged in pairs, two before and two behind, fig. 8. Most oviposit in holes of decayed trees. Larynx trimuscular. Monogamous.

Fam. 1. *Ramphastidæ.* *Ex.* Túcan.
2. *Bucconidæ.* *Ex.* Barbet.
3. *Cuculidæ.* *Ex.* Cuckoo.
4. *Picidæ.* *Ex.* Woodpecker.
5. *Musophagidæ.* *Ex.* Touraco or Plantain-eater.
6. *Coliidæ.* *Ex.* Coly.
7. *Psittacidæ.* *Ex.* Parrot.

Order VII. RAPTORES.

Rapacious Birds. Beak, strong, curved, sharp-edged, and sharp-pointed, fig. 9; legs short and robust, with three toes before and one behind, armed with long, strong, crooked talons, fig. 10.

Head of Eagle.

Raptorial foot of Eagle.

Fam. 1. *Nocturnes.* *Ex.* Owl.
2. *Diurnes.* *Ex.* Hawk, Eagle, Vulture.

An eighth group of birds has been characterised under the name CURSORES, Coursers, or 'Running-birds,'[1] by the arrested developement of the wings unfitting them for flight, and by the compensating size and strength of the legs, by which they are enabled to run swiftly on the ground. This is not, however, a natural order; some of its exponents have demonstrably closer affinities to other groups of which they are wingless members, just as the Penguins and Auks bear relation to families of the Natatorial order. Thus the *Notornis* is a modified Coot. The Ostrich bears the same relation to the Bustards. The extinct *Didus* and *Pezophaps* are most nearly allied to the Columbaceous group of *Rasores*. *Apteryx* and the allied extinct *Dinornis* and *Palapteryx*, bear affinity to the Megapodial family of *Gallinæ*.

[1] *Proceri*, Illig.; *Platysternæ*, Nitzsch; *Struthionidæ*, Vigors.

In all the Cursorial genera the sternum is devoid of keel.

Struthio is the only genus of birds in which the toes are reduced to two, fig. 11.

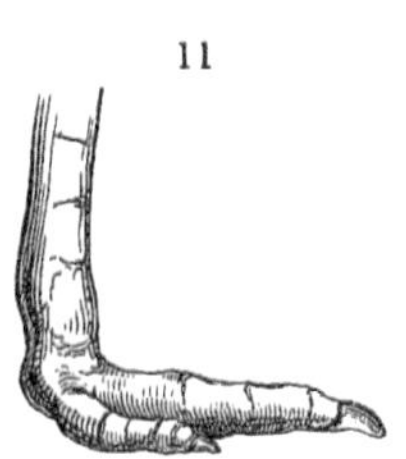

Cursorial foot of Ostrich.

In like manner the web-footed order is an artificial one, including derivatives from different natural groups or types; and the same may be said of the order including the birds that have the legs long and naked above the tarsal joint.

Derivatively the class of Birds is most closely connected with the *Pterosaurian* order of cold-blooded air-breathers. In equivalency it is comparable rather with such a group than with the *Reptilia* in totality, or with the *Mammalia;* and, hence, the corresponding inferiority of value of the avian 'orders' to the subdivisions so called of those larger classes.

In relation to time, indications of *Aves* date as far back as those of *Pterosauria,* in the 'ornithichnites' or foot prints of the New-Red Sandstones, for example.[1] The lithographic slates of a later mezozoic period have revealed a true feathered bird,[2] wanting only the adaptive modification of the caudal vertebræ characteristic of all neozoic birds, even those of the oldest tertiary strata, in which fossil remains of representatives of nearly all the present orders of *Aves* have been found.[3] The most recent instances of extinction of species are of the birds that have lost the power of flight; as, e.g., the gigantic Moas (*Dinornis, Palapteryx, Aptornis, Cnemiornis*)[4] of New Zealand; the equally gigantic *Epyornis* of Madagascar; the Dodo (*Didus*) of the Mauritius; the Solitaire (*Pezophaps*) of Rodriguez; the Gare-fowl (*Alca impennis*) of Northern shores or islands.

Notwithstanding the characteristic powers of locomotion of the class generally, it is amenable, most suggestively, to laws of geographical distribution and limitation.

[1] XVII'. pp. 5 and 324. [2] XV'. [3] CLI'. and XVIII'. p. 549.
[4] XVI'.

CHAPTER XIV.

OSSEOUS SYSTEM OF AVES.

§ 124. *General Characters.*—The skeleton of Birds is remarkable for the rapidity of its ossification and the light and elegant mechanism displayed in the adaptation of its several parts. The osseous substance is compact, and exhibits more of the laminated and less of the fibrous disposition than in the other vertebrate classes. This is more especially the case in those parts of the skeleton which are permeated by the air. The bones which present this singular modification have a greater proportion of the phosphate of lime in their composition than is found in the osseous system of the mammalia, and they are whiter than the bones of any other animal. In the bones where the medulla is not displaced by the extension of the air-cells into their interior, the colour is of a duller white. In the Silk- or 'black-boned' fowl of the Tropics (*Gallus Morio,* Temminck), the periosteal covering of the bones is of a dark colour; but this is a peculiarity of the cellular rather than of the osseous texture, which does not differ in colour from that of other birds; indeed the thin aponeurosis covering the lateral tendons of the gizzard of the Silk-fowl has the same dark hue as the membrane which invests the bones.

§ 125. *Dorsal Vertebræ.*—The modifications of the common vertebrate type of skeleton required by the exigencies of the present class are extreme. Anchylosis so fetters the vertebral column that from no part can a single segment with all the elements be detached without using the saw. The skull includes four, the sacrum a greater number, of vertebræ, of more or less of which the hæmal portions alone retain freedom. The remaining segments may be classified as 'cervical,' 'dorsal,' and 'caudal': in the first and last the pleurapophysis, if present, is confluent with the neural arch: in the dorsal series, the pleur- and hæm-apophyses are flexibly articulated, but the hæmal spines are connate, and represented by a single bony plate.

In fig. 12, is given a sketch of three dorsal segments, 1, 2, 3, with the hæmal arches, 52, 58, of two others. In the first and second dorsals the pleurapophyses (1 and 2) terminate in a free pointed end, like the 'false floating ribs,' of Anthropotomy; in the third, the pleurapophysis, *pl*, 3, articulates with the hæmapophysis, *h*; and this with the expanded spine, *h s*, which, in connation with its homotypes, constitutes the bone called 'sternum,' *f*. Every succeeding dorsal segment has the hæmal arch completed by bone.

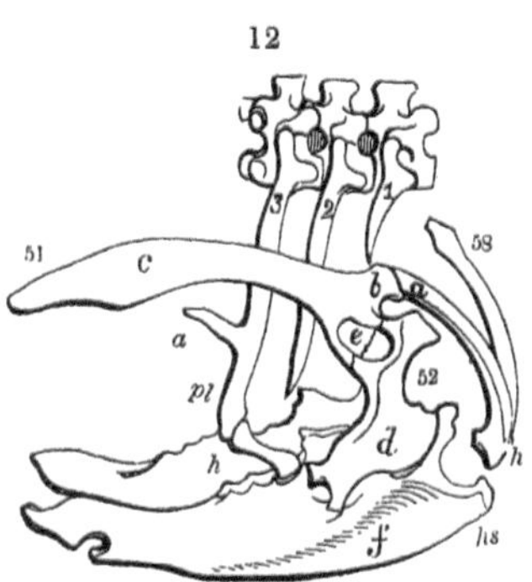

First three dorsal vertebræ and scapular arch of a bird, in diagrammatic side view.

Fig. 13, gives a diagrammatic front view of the connate dorsal or thoracic hæmal spines, *c*, *s*; the hæmapophyses, *d*, of five corresponding segments, and also a modified pair, *h*, *b*, of the hæmapophyses of an antecedent segment.

The pleurapophyses, *pl*, *a*, of the dorsal segment are shown in connection with the centrum, *c*, and neural arch, *n*; it is to this part of the segment that the term 'vertebra,' is commonly restricted.

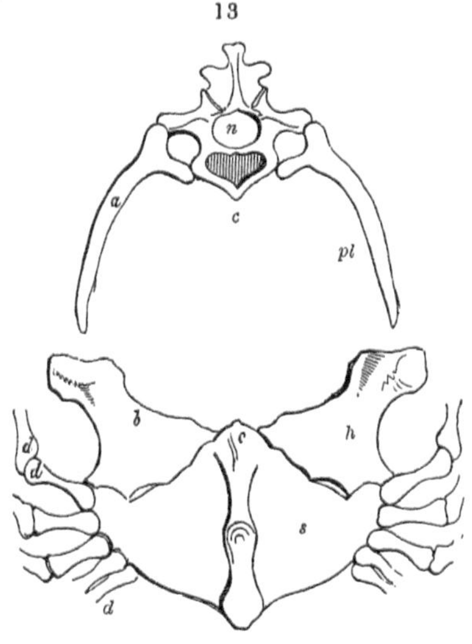

Anterior vertebra, with other hæmal arches of dorsal or thoracic region, in diagrammatic front view.

The *dorsal* vertebræ, thus defined, rarely form more than a fourth part of the entire column, and in some of the long-necked *Grallatores*, as the Stork and Flamingo, fig. 14, form only an eighth part; they have not been observed to be fewer than four (in some Vultures), nor more than nine throughout the class; the latter number obtains in the Apteryx: the most common numbers are six or seven.

The dorsal vertebræ are shorter than most of the cervicals, and with broader neural arches, in consequence of the greater developement of the transverse processes; but their bodies become much compressed, and in some Birds are reduced almost to the form of vertical laminæ towards the sacral region (*Aptenodytes, Catarrhactes*); but, in the Ostrich, the bodies of the dorsal vertebræ retain their breadth throughout the series.

The bodies are united by capsular ligaments and synovial membranes; the anterior articular cartilaginous surface is convex in

14

Skeleton of Flamingo. A. Corax Mitis. B. Tinamou (*T. robustus*). C. Phœnicopterus.

the vertical direction, and concave in the transverse; the posterior surface is the reverse. The Penguins and Auks, however, present an exception to this rule: the posterior surface of the second or third dorsal vertebra is concave, to which the opposed end of the succeeding vertebra presents a corresponding convexity; the 'opisthocœlian' ball-and-socket-joint is continued between the centrums to the last dorsal.[1] In many Birds the bodies of some of the middle dorsal vertebræ are anchylosed together; and in general those which are nearest the sacrum. In the Flamingo, fig.

[1] VII. p. 270, and X. pl. 52, figs. 50, 51.

14, the anchylosis extends from the second to the fifth dorsal vertebra. In the Sparrow-hawk, the same vertebræ are consolidated into one piece, while the sixth enjoys considerable lateral motion, both upon the fifth and seventh, which last is anchylosed to the sacrum; so that the body can be rapidly and extensively inflected toward either side during the pursuit of prey.

From some or most of the dorsal centrums inferior processes (hypapophyses) are sent down, for extensive and favourable origin of the flexor muscles, *longi colli* and *recti antici*, of the neck. In a vulture (*Gyps fulvus*) the hypapophysis is a low median ridge in the first and second dorsals; to this, in the third dorsal, is added a pair of outstanding depressed plates: in the fourth the pair of plates are smaller, and, with the medial ridge, are supported on a common stem: in the fifth dorsal, the hypapophysis is again reduced to a median compressed plate, but it is expanded at the end; the vertebra, which by anchylosis has become the foremost sacral, has a similar but stronger and slightly bifurcate hypapophysis. In both Vultures and Eagles the parial hypapophyses are seen to be due to modified parapophyses, which descend and are progressively lost in the median hypapophysis of the fourth and fifth dorsals (*Harpeya*, Cuv.); the sixth and seventh have only the low median ridge. The parapophysial pairs of inferior processes are broad divergent plates in the anterior dorsals of *Aptenodytes*[1] and *Alca*[2], and subside upon the large and long compressed median hypapophysis which characterises the posterior dorsals. The unusual developement of these inferior processes relates to the size and strength of the subvertebral muscles, which combine with other muscles of the trunk in the shuffling movement by which the Penguin, like the seal, makes progress, prone, upon dry land. In the anterior dorsals the parapophysis, besides forming the articulation for the head of the rib, sends off a muscular process subject to the modifications above mentioned: the diapophysis is larger and more constant in character; it is extended from before backward, is horizontally flattened, and forms the surface for the joint of the tubercle of the rib at a small part of its outer border: a metapophysial ridge is developed from the upper surface, and is frequently produced into filaments coalescing with those of contiguous dorsals. The pneumatic foramina are at the back part of the base of the diapophyses. The zygapophyses are small, the front pair look upward and inward; the back pair outward and downward; the latter often support anapophysial ridges. The neural spine is a

[1] x'. pl. 51, fig. 48, *h*, *h*. [2] xii'.

compressed quadrate plate, its truncate summit is often thickened, sometimes produced forward and backward to fix the vertebræ from their highest points; ossified tendons of spinal muscles, also, aid the coalesced spinous and transverse processes in fixing part of the dorsal region, but only in birds of powerful flight, and not in all such. The partial anchylosis of the dorsal region is associated in Falcons with their 'hovering' action. The pleurapophyses or 'vertebral ribs' articulate moveably to the dorsal vertebræ, as also to the anterior sacral, when developed there to form part of the compages of the 'chest.' In the first, and usually the second dorsal, they are free, pointed, floating ribs, fig. 12, 1, 2, fig. 13, *pl.*; they articulate with bony 'hæmapophyses' or 'sternal ribs,' ib., *h*, *d*, in the remaining dorsals. As the vertebral ribs are placed more backward, the neck or pedicle supporting the head elongates, and this articulates with the parapophysial surface or tubercle, close to the anterior border of the centrum; rarely, as in the Penguin[1] and Ostrich, encroaching upon the intervertebral space. The tubercle of the rib is, in most, supported on an elongate compressed base, and articulates by a synovial joint with the diapophysis. The body of the rib, where formed by the union of the two articular processes, is compressed, or thin from side to side, but broad from within outward; but the outer margin soon expands both forward and backward beyond the compressed part of the body of the rib; this part, as the rib extends down, subsides, the outer margin maintaining or increasing its breadth, and forming the rest of the rib, giving to it a flattened surface externally. This is the common but not constant character of the dorsal pleurapophyses. These ribs are broadest in proportion to their length in the *Apteryx*,[2] narrowest and also longest in the Guillemots and Auks[3]; they are slender in most *Insessores*; broad and strong in *Raptores*. The second, third, and fourth ribs are partially and remarkably expanded in Wood-peckers. In all birds the end of the vertebral rib articulating with a sternal one is thickened to form the subconvex surface of the synovial joint. There may be several minute pneumatic foramina, but the most constant and conspicuous is below the tubercle.

An 'epipleural' appendage, fig. 12, *a*, is attached to most, if not all, the moveable pleurapophyses between the first and last, and consequently may be found in the pair of which the centrum has become part of the sacrum. These appendages are oblong flat bones, varying in the proportions of length and breadth in different species, and also in their mode of union to their rib: they

[1] x'. pl. 52, fig. 48. [2] xi'. pl. 54. [3] xii'.

are directed upward and backward, usually overlapping the succeeding rib. In the Apteryx they occur in the second to the eighth pair of ribs inclusive, and are articulated by a broad base to a fissure in the hind border a little below the middle of the rib: those belonging to the third—sixth ribs are the largest and overlap. The articulations of the appendages persist in other wingless birds, including the Penguins and Auks; also in some birds of flight: the *Raptores* well exemplify the coalescence of the appendage with its rib. The appendages to some of the ribs in *Picus* are broader than they are long: the length much exceeds the breadth in some *Natatores* (*Uria*, *Larus*) and *Grallatores* (*Hæmatopus*, *Phœnicopterus*, *pl.* 14).

The moveable hæmapophyses, or sternal ribs, usually begin at the third, sometimes the second, rarely, as in the Emeu, at the fourth pair, more rarely, as in the Cassowary, at the fifth pair, of the moveable pleurapophyses; a pair of sternal ribs may also exist answering to the segment succeeding the last of those which have the long and moveable vertebral ribs (*Vultur*). The common number of such hæmapophyses is six pairs, of which the first five articulate with the sternum; the last usually having its sternal end attached to the antecedent one. The hæmapophyses are longest, most slender and most numerous in the Guillemots and Auks. There are eight pairs in *Phalerus*, Temm.; seven pairs in *Uria*. In *Rhea* and *Dinornis elephantopus* but three pairs of hæmapophyses articulate with the sternum. The sternal ribs progressively increase in length from the first to the penultimate, and converge towards the costal border of the sternum, where they articulate with transverse elevations divided by narrow depressions. Their upper end is but little, if at all, expanded, and its articular surface is subconcave; their lower or sternal end is expanded from within outward, subcompressed from before backward, and here is usually found the pneumatic foramen. In the ostrich the sternal end supports two distinct articular surfaces, each having its own capsular and synovial articulation with part of the costal eminence.[1] The joint between the pleur- and hæmapophyses is also synovial and capsular. This is the main centre upon which the respiratory movements hinge, the angles between the vertebral and sternal ribs and between these and the sternum, becoming more open in inspiration when the sternum is depressed, and the contrary when the sternum is approximated to the dorsal region in expiration. In some birds, chiefly of the terrestrial or aquatic kinds, the vertebral and sternal portions of one or

[1] xx'. i. p. 54, no. 254.

more of the last pairs of thoracic ribs are unconnected with each other, in the skeleton; such sternal ribs resembling the abdominal hæmapophyses in Saurians, or the 'intersections,' ossified, of the rectus abdominis muscle in Mammals.

The modifications of the sternum in Birds relate to their faculty of flight; more directly, to the adequate origin of the muscles acting upon the pectoral limb, less directly to the mechanism of respiration needed by the conditions of the lungs; also, in Perchers, to sustaining the body in sleep.

The sternum of the Bird is the bony ventral wall of the trunk, fig. 18, 60, *s*, *r*; it is not, however, the homologue of the plastron of the Tortoise, fig. 53, p. 63, vol. i., but of the series of hæmal spines forming the episternum and sternum of the Crocodile (fig. 56, p. 68, *ib.*); it is developed, in most Birds, from one pair of ossific centres, which, coalescing in the midline, usually consolidate the cartilaginous basis of the keel by continuous ossification therein.

15

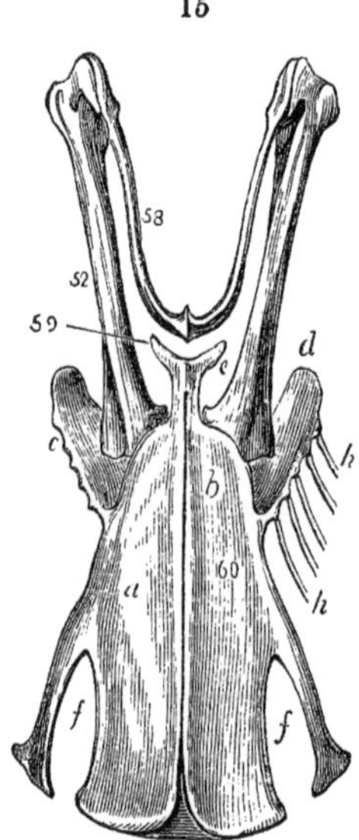

Cantorial sternum, Shrike.

The chief parts to be noted in the Bird's sternum are the 'body,' fig. 15, *a*, with its notches or holes, *f*, *f*; the 'keel,' fig. 18, *s*; the 'costal processes,' fig. 15, *d*; the 'costal borders,' with their articular surfaces, fig. 15, *c*; the 'coracoid grooves,' figs. 15 and 16, *b*; and the 'manubrium,' fig. 15, *e*, 59.

The body may be almost flat, as in *Apteryx* and *Dinornis*; or very concave, the sides being bent upward at an acute angle, as in *Aquila*; it is commonly less concave toward the trunk. It varies greatly in the proportions of length and breadth: the latter dimension is in excess in *Apteryx* and *Dinornis*; the breadth nearly equals the length in other *Struthionidæ*, the Albatross, and the Pelican. The length progressively gains in other birds, until it becomes four times the breadth of the sternum in *Tinamus*, fig. 14, B. Extreme length is associated with ordinary breadth in the sternum of the Anserines, Auks, Guillemots, many Waders, diurnal *Raptores*, and some *Volitores*, reaching to the pelvis, and occasionally to the pubic bones, fig. 18, *s*, *p'*, and requiring removal for exposure of the abdominal cavity.

Examples of an 'entire' sternal body, i. e., neither notched nor perforated, are afforded by Swifts, Humming-birds, fig. 18, some Eagles (*Aquila, Pandion, Haliæetus*) and Petrels (*Thalassidroma*), in connection with excessive developement of the bone and great powers of flight; also by the Emeu, Rhea, Cassowary, and Notornis, fig. 16, where the bone stops short of the parts exhibiting the notches or grooves in birds of flight, and retains almost the small lacertian proportions, fig. 17, *a*. The oblong and not very large sternum of some Cockatoos (*Calyptorhynchus*) is entire; in the

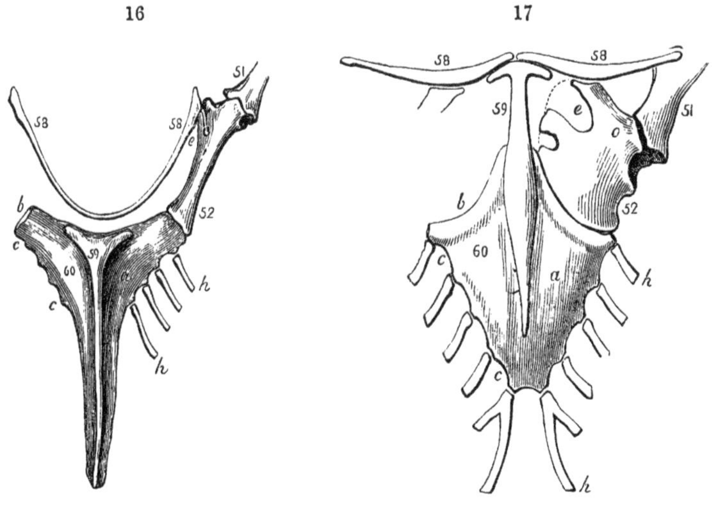

Cursorial sternum, Notornis.

Lacertian sternum, Iguana.

Balearic and Demoiselle Cranes, in the Ibis (*Tantalus Ibis*); in the Agami (*Psophia*), the sternum is long, narrow, and entire. The broad sternum of the Frigate-bird (*Tachypetes*) and the long sternum of the wingless Auk (*Alca impennis*) are also entire.[1]

The anterior margin of the sternal body is impressed by the articular cavities for the coracoids. In *Notornis*, fig. 16, *b*, *Dinornis* and *Apteryx* they are small shallow depressions, near the outer angles: they are similarly situated, but longer and deeper, in the Rhea and Cassowary; are more extended and with a shorter interspace in the Ostrich. In birds of flight they are deep grooves, with the upper or hinder border thickened and convex in many, affording a concavo-convex surface for the broad end of the coracoid. They mostly meet at the midline; they are continuous, perforating the base of the manubrium, in some *Gallinæ* (*Perdix*); and have their medial ends decussating, extending one in advance of the

[1] xii

other across the midline in some *Grallæ* (*Ardea*); and in a slight degree in some diurnal *Raptores.* The borders of the coracoid grooves show modifications characteristic of genera and species.[1] In the Albatross the coracoid grooves extend to the outer angles of the sternum, between *h* and *s*, fig. 13. In most Birds with a like extent of groove the upper or inner border is developed behind and beyond it into a 'costal process:' but the coracoid grooves do not reach the outer angle in many Birds, and the angle itself is then produced to form the process, figs. 15, 19 and 20, *d*. It is long and slender in some *Rasores* (*Perdix*); short and broad in most *Raptores*: but, in many birds it is represented, as in the Eagles, merely by the angle between the anterior and costal borders. On an average about half of the lateral margin of the sternum is adapted for articulating with the dorsal hæmapophyses, figs. 13, *h d* and 15, *h*: but, when the sternum is long, the 'costal border,' fig. 15, *c*, is shorter; and when the sternum is short it occupies a larger extent of the lateral margin. The part of the bird's sternum answering to that of Mammals is included between the costal borders, fig. 16, *c*, *c*: the rest corresponds with the 'xiphoid' prolongation. Thus the Apteryx, Emeu, and Ostrich most resemble Mammals in the proportions of the costal and non-costal parts of the sternum; whilst in most birds of flight the non-costal part, fig. 15, *a*, *f*, extends along that part of the great visceral cavity, which would be similarly defended were the xiphoid car-

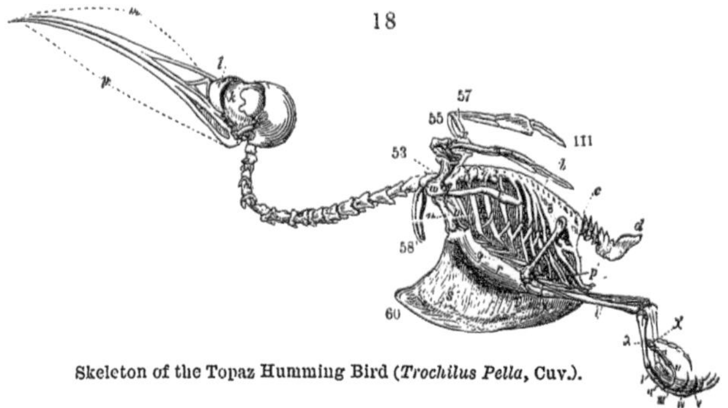

Skeleton of the Topaz Humming Bird (*Trochilus Pella*, Cuv.).

tilage to be produced and expanded in the same degree in Man. In the Crocodile, where it is so produced, without expanding,

[1] The value of these and other sternal characters in Palæontology may be estimated by reference to my 'British Fossil Mammals and Birds,' p. 549 (*Lithornis*), and p. 236 (*Calobates*).

the costal borders are co-extended therewith, fig. 56, p. 68, vol. i. In most *Gallinæ* the lateral margins of the sternum are deeply concave: in the Guan (*Penelope*) almost angularly incised, with the costal border on the anterior slope. In the Tinamou, fig. 14, B, the long margin beyond the short costal border is convex: in many Waders (*Platalea*, *Phœnicopterus*, fig. 14, C) and Swimmers (*Procellaria*, *Diomedæa*) the lateral borders are straight and parallel, or nearly so: in *Rhea*, *Casuarius*, *Dromaius*, *Notornis*, fig. 16, they converge to the hind border: in most birds the lateral borders are moderately concave and diverge, figs. 15 and 20. The costal border is thickened, and divided by the transverse articular ridges for the hæmapophyses into hollows, which usually show pneumatic foramina. The modifications of the posterior border will be noticed in connection with the sternal characteristics of orders, or other groups, of Birds.

The part of the sternum bearing the most direct relation to the force with which the pectoral limbs are worked is the 'keel,' figs. 18 and 19, *s*. In order to afford origin to the accumulated fasciculi of the pectoral muscles, which otherwise would become blended together over the middle of the sternum, this osseous crest is extended downward, analogous to the cranial crest which intervenes to the temporal muscles in the carnivorous mammalia; and which, in like manner, indicates the power of the bite.

The keel varies in depth, length, contour of the front and lower borders, and degree of production, freedom, or otherwise of the angle between those borders. The keel is long and deep in the wingless Auk and Penguin, relating to the mass of muscle working the fore limbs as fins in these excellent and habitual divers: in the Penguin both the free borders are straight, and meet at rather an acute angle, fig. 19. The keel is deep, descending anteriorly far below the furculum in most *Gallinæ*: it is remote from the furculum in *Limosa*, *Ibis*, *Scolopax*; but touches it in many other *Grallatores* (*Otis*, *Psophia*, *Ciconia*). It coalesces with the furculum in *Grus Virgo* and *Grus Antigone*; in the stilted Vulture (*Gypogeranus*); in the Frigate-bird (*Tachypetes*); also in the Pelican (*Onocrotalus*), Gannet (*Sula*), and in old Cormorants (*Carbo*), the fore part of the keel being much produced in these Totipalmates. The keel is thick in the few birds in which a fold of the windpipe penetrates it; the anterior border being excavated to admit the fold. In the larger *Raptores* the front border of the sternum is rather thick and subcarinate. The outer surface of the sternum shows in many birds a 'carinal' ridge, a 'subcostal' ridge and a 'pectoral' ridge, the latter defining the origin of the 'pectoralis secundus.' The subcostal ridge varies in its distance from the costal border, being more remote in *Aquila*, e. g. than in *Uria*: the pectoral ridge varies in position, direction, and extent. In the Eagle it reaches from the outer end of the coracoid groove to the middle of the base of the keel. In the Razor-bill (*Alca torda*) it extends from the costal border to the posterior sternal notch; these differences relate to the form and proportion of the *pectoralis secundus*. The 'manubrium' forms but a small portion of the sternum, and is often absent or rudimentary: it may be compressed, spatulate, long and simple, or bifurcate; the latter is its character in all *Cantores*, fig. 15, *e*.

The parts of the sternum of the Lizard, fig. 17, homologous with parts of the sternum of the Bird are those forming the 'coracoid groove,' ib. *b*, the 'costal border,' ib. *c*, *c*, and the median bone, 59, passing forward to join the clavicles, ib. 58. The broad flat bone, including the first two parts, exists in all birds; the third, or 'episternal' part, is wanting as a distinct element, but its positions and connections are repeated by the exogenous keel and 'manubrium.' The episternum, moreover, is not present in all Lizards: it is wanting in the Chameleons, e. g., in which the sternum partakes of the simplicity of that in the Notornis, fig. 16, the Apteryx and Emeu.

In the Apteryx the anterior border of the sternum between the coracoid grooves is concave, and the posterior border has a deep and wide emargination on each side. In the Emeu the coracoid grooves meet at the middle of the anterior border; and the sternum contracts posteriorly to an obtuse point. The sternum is rhomboid, also, in the Cassowary: it is broader in proportion to its length, and subquadrate in the Ostrich. In Notornis, fig. 16, the costal borders converge posteriorly, as in Lizards, and the narrow breast-bone is continued as a 'xiphoid' part, gradually contracting to a blunt point. The depressions, 60, *a*, for the pectoral muscles are separated by a narrow median tract, expanding anteriorly, 59, and showing the beginning of the 'keel.' In *Brachypteryx*[1] the keel is rather more prominent: two obtuse ridges diverge from its fore-part to the coracoid grooves, between which the fore margin is deeply concave, as in the Apteryx. There is no distinct ossific centre for the keel in *Brachypteryx*, any more than in its feebler rudiment in *Notornis*. In all these keel-less sternums ossification begins, as in the Ostrich[2], by a pair of centres expanding until they meet and coalesce in the

[1] XLIV'. p. 238, no. 1280. [2] Ib. p. 264, no. 1366.

middle line, and thence, according to the stimulus of the growth and pressure of the pectoral muscles, extending, as a keel, into the interspace. A separate ossification answering to the episternum in Lizards and Crocodiles is not formed: but the body of the sternum with the keel has a centre distinct from that of the long bifurcate side-processes, exceptionally, in *Gallinæ*.

In the Penguins, fig. 19, the sternum is long and narrow, with a deep fissure, *f*, on each side of the posterior border: the free borders of the well-developed keel are straight, and meet at an acute angle, which almost touches the furculum. There is a short manubrium, *e*, behind which the coracoid grooves, *b*, meet.

19

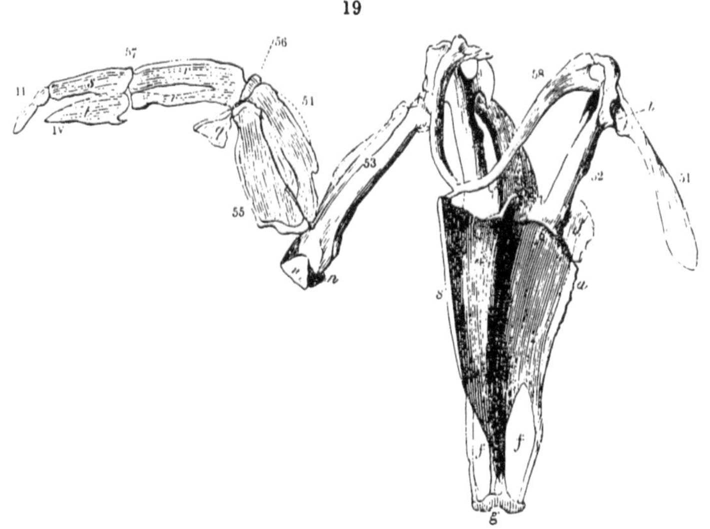

Sternum, scapular arch, and limb, Penguin (*Aptenodytes*).

In the Auks and Guillemots the sternum[1] is very long and narrow; the lower border of the keel is convex, the front one concave; the manubrium is short and wedge-shaped; the sternum is entire in *Alca impennis*; but has a narrow notch on each side the posterior border in *Alca torda*, to which the pectoral ridge extends from the costal border; the notches are converted into foramina in *Uria* and *Phaleris pygmæa*. The Loons (*Colymbus*) have a similar two-notched sternum, but with larger costal borders. In the Grebes (*Podiceps*), the sternum is broader; and there is a median notch between the two lateral posterior ones. In the

[1] xii*.

Skimmers, Gulls, and Terns, the sternum has two shallow notches on each side the posterior margin. In the Petrels and Albatross the posterior border is feebly incised or entire and the sternum acquires great breadth, especially in the Albatross. The keel reaches the furculum in all the Longipennate family. In the *Pelecanidæ*, confluence of the two bones usually here occurs; there is a pair of shallow posterior emarginations. In the Lamellirostrals the sternum is both large and long, boat-shaped, with extensive costal borders; the keel is of moderate depth, with almost straight free borders, excavated for tracheal folds in some swans; there is a short notch or small foramen on each side the broad posterior margin in all the Sifters; the manubrium curves downward in many. The Flamingo's sternum is given in fig. 14, C. The foregoing diversities of sternal structure in the web-footed birds indicate from how many types they have been derived, and shows the artificial character of the webbed-foot.

The same testimony is borne by the breast-bone of the long-legged birds, from which, in some instances, the species have been detached when the truer affinities were sufficiently strongly marked, as, e. g., the Flamingo to the Sifters or Lamellirostrals; the Secretary Bird to the Vultures; and the Couas to the Cuckoos. In the long and narrow sternum of the Coots and Rails the two posterior notches are deep, with the outer boundary the longest, and *Brachypteryx* shows a third intermediate shallow notch.

The Ibis and Spoonbill have a four-notched sternum; the Adjutants and Herons have a two-notched one; the notches are short in both. Peculiarities in the breast-bone of certain Cranes have already been noticed. The Woodcock (*Scolopax*) has a pair of notches, with the outer boundary slender and shorter than the broad intermediate tract; the Gambets (*Totanus*), Avocets, Sandpipers (*Tringa*), Curlews (*Numenius*), Pratincoles (*Glareola*), have the four-notched sternum. In the Godwits (*Limosa, Helias*), the medial notches are almost obsolete, and the lateral ones wide. The 'Thick-knees' (*Œdicnemus*) and Bustards (*Otis*) have the four-notched sternum, the notches being small.

In the Gallinaceous group of Rasores, the four posterior notches are so wide and deep as to reduce the bony parts of the sternum almost to five slender processes, diverging from a short and broad anterior stem, and the points of ossification are multiplied accordingly. The middle process is the broadest, and from it is developed the keel, of which, in some (*Ortyx, Perdix*), it seems to be almost wholly composed. As the median pair of notches is usually deepest, the processes on each side the mid one appear as unequal, styliform, terminally expanded, prongs of a fork, the outer prong being the shortest. The costal border is very short, and is continued upon the costal process, which is long: the manubrium is compressed and terminally dilated and deflected, often perforated transversely by confluence of the short coracoid grooves. The fowls (*Gallus*), pheasants (*Phasianus*), partridges (*Perdix, Francolinus*), quails (*Coturnix, Ortyx, Lophortyx*), grouse (*Tetrao*), exemplify the gallinaceous type of the sternum.

In the Turkeys (*Meleagris*), Pea-fowl (*Pavo*), and Kaleeges (*Polyplectron, Lophophorus, Oreophasis*), the sternum is more ossified, and the lateral processes are shorter and broader; in the Curassows (*Crax, Ourax*), they present the proportions shown in fig. 14, A. In the Gangas or Sand-grouse (*Pterocles, Syrrhaptes*), the outer pair of notches are chiefly present, the inner pair nearly obsolete[1]; in the Tinamous[2] they are wanting, the outer notches are of extreme length, and the whole sternum is reduced to a trifid form, as in fig. 14, B. The sternum of *Columba coronata* resembles that of the Curassow, with the median pair of notches shorter and narrower. In *Columba magnifica*[3], the four notches are more equal in size, and the whole sternum is broader. In the *Columba livia* the median pair of notches are often converted into small foramina.

The transitional steps in the foregoing series from the type-sternum of *Gallinæ* to that of the swiftest of the doves indicate the natural character of the order *Rasores.*

In diurnal *Raptores* the sternum is a large elongate parallelogram, convex outwardly both transversely and longitudinally. The manubrium is short and trihedral; the lower border of the keel is convex; the front border concave; their angle of union rounded off. The instances where the sternum is entire have been cited: in other birds of prey the arrest of ossification is limited to very small parts of the hind border; usually a foramen, rarely a notch (*Sarcoramphus*), on each side; one of which may be filled up, wholly or partially. Eyton[4] figures two small notches on each side the posterior border in *Hierax bengalensis*; and both hole and notch on each side in *Cathartes aura.* In the Nocturnal Raptores the sternum is relatively shorter, the keel less deep, its lower margin less convex and not thickened, the costal border is shorter. The posterior margin usually presents two notches on each side, the outer one the

[1] XIII'. p. 228, fig. 109. [2] Ib. p. 230, fig. 110. [3] XIV'. pl. 2 a. [4] Ib. pl. 16.

deepest; but the Barn-owl (*Strix flammea*) has but one on each side; while *Strix praticola* shows a third intervening notch.[1]

In the *Cantores* the sternum, fig. 15, is broadest behind, with the lateral margins slightly concave, the costal, *c*, usually meeting the rest of the margin at a very open angle. The keel has a convex lower border meeting the concave front border at a sharp angle: the manubrium *e*, is bifurcate: the costal processes, *d*, are broad and flat: the posterior border has a notch, *f*, usually of angular form, on each side, near the lateral margin, and with this outer boundary terminally dilated.

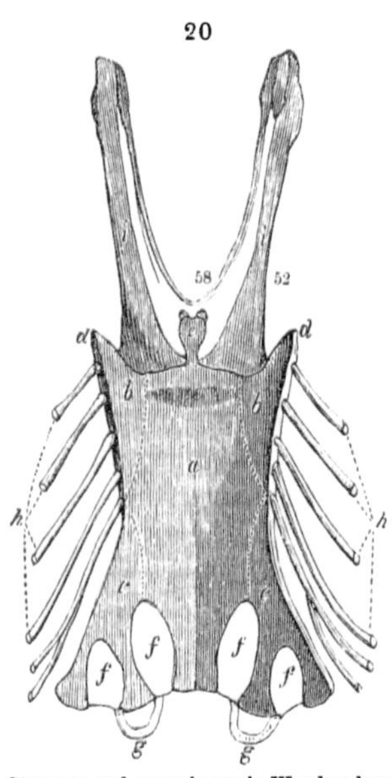

Sternum and scapular arch, Woodpecker, (*Picus*).

Among the *Scansores* the Toucans, Barbets, Touracos, and Woodpeckers, fig. 20, have a four-notched sternum: the Cuckoos have but one pair of short notches; many Parrots (*Psittacus* proper, *Pezoporus*) have one pair of small foramina, and *Calyptorhynchus* has the sternum entire: it is keel-less in *Strigops*. In most parrots the costal border is extensive; the manubrium is trihedral and truncate. None of the *Scansores* have the manubrium bifurcate; it may be notched; in most it is small; in some (*Cuculus*, *Ramphastos*) obsolete.

In the *Volitores*, as a rule, the posterior border of the sternum has a pair of notches on each side: the Eurylaim and Hoopoe have one notch or foramen on each side. The Hornbills, Swifts, and Humming-birds have the sternum entire. In none of this group is the manubrium bifurcate: it is wanting in *Podargus*, *Harpactes*, *Todus*: the costal process is wanting in some. In the Swifts (*Cypselus*) the sternum corresponds in its proportional magnitude with the superior length and power of wing which characterizes the genus: the depth of the keel equals the breadth of the entire bone. The manubrial process is wanting, but the costal processes are moderately long and pointed.

In the Humming-birds, which sustain themselves on the wing during the greater part of the day, and hover above the plant while extracting its juices, the sternum, *r*, *s*, fig. 18, is still

[1] xiv. pl. 4, fig. 5.

further developed as compared with the body; it approaches to a triangular form, expanding posteriorly, where the margin is entire, and convex. The depth of the keel exceeds the entire breadth of the sternum. The coracoid depressions are deeply trochlear: the manubrial process is small, and directed upward; the costal processes are also present, but of small size: the costal border is short. In these pre-eminently volant Vertebrates, the breast-bone reaches the maximum of developement.

21

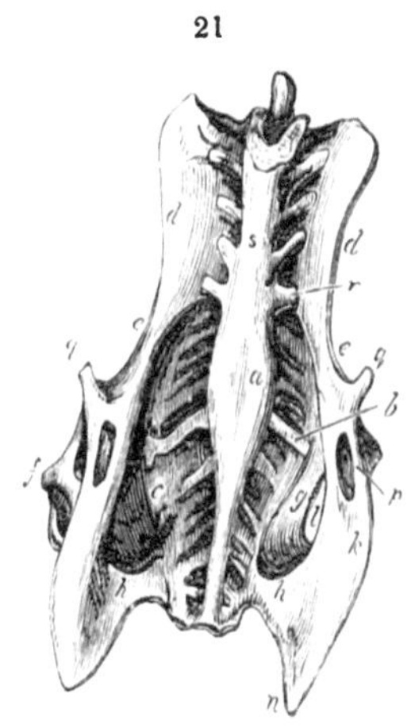

Front view of pelvis, Partridge.

§ 126. *Sacral Vertebræ.*—In vertebrate anatomy the term 'sacrum' is applied to the centrum and neural arch of the vertebra, having its hæmal arch complete, as in the thorax, but with its appendage developed into a hind-limb (vol. i., figs. 101, D and 114). If two or more vertebræ coalesce beyond the thorax, they are likewise said to form 'a sacrum,' although but one may be typically complete, and the rest support only stunted pleurapophyses. In all warm-blooded Vertebrates the sacrum, when present, is so characterized, and confluence is carried to an extreme in birds, converting a large proportion of the vertebral column into a 'sacrum,' fig. 21, *s*, *a*, *c*, which in the Ostrich may include seventeen or more vertebræ. Thirteen is the average number in *Natatores*, twelve in *Grallatores* and *Rasores*, eleven in *Altrices* or the higher birds of flight.[1]

In analyzing this most complex of all compound bones, in a young Ostrich[2], I find the centrum of the first sacral vertebra distinct, although its neural arch and spine have coalesced with those of the second vertebra and with the ilia. Traces of the articulation between the centrum of the second and third sacral vertebræ remain: they are obliterated in the remaining vertebræ, and the bodies of all are cellular and permeated by air.

The pleurapophysis of the first sacral retains its moveable articulation to the par- and di-apophyses of its vertebra; it is long, slender, and terminates in a free point. That of the second sacral vertebra is styliform, half the length of the preceding, and terminates in a free point projecting downward and backward; its head and tubercle, free in the young bird, become confluent

[1] See the Table in VII. p. 273.

[2] Nos. 1835 and 1837, Osteol. Series, Mus. Coll. Chir.

in the full-grown. The third sacral has no pleurapophysis: its parapophysis is a stumpy process; its diapophysis is longer and abuts against the ilium. In the fourth sacral both par- and di-apophyses abut against the ilium.

The neural arch of the fifth sacral vertebra has advanced and rests over the interspace between its own and the preceding centrum: this interlocking relation continues to the eleventh vertebra, where the arch resumes its normal position and connections. The pleurapophyses of the fifth to the eleventh sacral vertebræ inclusive have undergone a corresponding change of position, and are synchondrosed by an expanded head to a rough flat surface formed by the base of the neurapophysis and by a portion of their own and of the preceding centrum: their distal extremities expand and coalesce, forming a broad abutment applied to the iliac bones. The diapophyses are directed upward and outward against the same part, and are of considerable length, especially in the ninth to the fifteenth sacral vertebræ. The dilated part of the neural canal is formed by the increased breadth and flatness of the centrums, and by the wide expanse of the neural arches at the middle of the sacrum. In the seventh to the ninth of these arches there is a wide aperture in each between the diapophysis and the base of the spine. The outlets for the nerves are single and at the interspace of the neural arches, but those at the middle of the canal show two grooves for the separate exit of the motor and sensory roots.

The spines of all the vertebræ are lofty, and already confluent with each other at the middle of the sacrum. They are compressed from before backward, consist of little more than a lace-work of osseous tissue, and diverge in curves from the neural arches, through the interspace between the iliac bones, with both of which their lateral margins are confluent, and which they thus serve to bind firmly together. By the peculiar cellular and pneumatic structure of the parts, not more osseous texture is expended in performing the office of tie-beams across the elongated roof of the pelvis than is absolutely required. The last seven vertebræ are seen between the narrow parts of the ilia produced backward beyond the acetabula, until full-growth, when ossification extends from the summits of the spines bridging over the interval, leaving only a linear fissure on each side, fig. 24. In the Cassowary a few pairs of foramina similarly indicate the last three or four sacral-vertebræ.

In the Apteryx the first four sacral vertebræ send outwards parapophyses which abut against the ilia, and progressively increase in length and thickness. The breadth of these vertebræ also gradually increases; but it diminishes in the four succeeding vertebræ, in which the parapophyses are wanting: then the ninth and tenth sacral vertebræ send outward each a pair of strong parapophyses to abut against the inner surface of the ossa innominata immediately behind the acetabulum: the anchylosis of the bodies is continued through the four succeeding vertebræ, which are of a very simple structure, devoid of transverse or oblique processes, becoming gradually more compressed and more extended vertically, so as to appear like mere bony laminæ; the line of the articulation between the bodies of these posterior sacral vertebræ is obvious, but their spines coalesce to form a continuous bony ridge, which is closely embraced by the posterior extremities of the ilia. The foramina for the nerves are pierced in the sides of the bodies of the sacral vertebræ; they are double in the anterior ones, but single in the posterior compressed vertebræ, where they are seen close to the posterior margin.

The species of *Dinornis* show from 17 to 20 sacral vertebræ. In *D. robustus* the pleurapophyses of the first retain their moveable articulations: those of the second and third are anchylosed, but project freely beyond the ilia: those of the fourth to the eighth abut as parapophyses against the ilia, the last, which is opposite the acetabula, being the thickest: those and the four following sacrals, which have no parapophyses, are very short: from the thirteenth to the twentieth sacral the parapophysial buttresses reappear, and the vertebræ increase in length. A continuous bony roof of the pelvis extends from the sacral spines to the ilia. When vertically and longitudinally bisected, the sacrum shows the great expanse of the canal for that part of the myelon in connection with the nerves of the large and strong hinder extremities. All traces of the original joints between the bodies of the vertebræ, with the exception of the last, are obliterated. The primitive distinction of the neural arches is indicated by undulating transverse folds of the roof of the spinal canal: the motor and sensitive roots issue separately, as in other birds.

In the Penguins (*Aptenodytes*) the sacrum forms the middle third of the upper surface of the pelvis: in *Podiceps* and *Colymbus* the ilia converge to the summits of the posterior sacral spines: in *Uria, Diomedea, Procellaria,* and the *Anatidæ* they converge to the anterior ones, fig. 22, *b*. Pairs of foramina usually indicate the sacral vertebræ, forming a broader posterior sacral roof (ib. *a*, *c*), of the pelvis: but in the Petrels ossification obliterates them. In most *Grallatores* the ilia come near to the neurospinal ridge, ib. *b*, of the anterior sacral vertebræ; whilst the posterior ones form a broad middle tract of that part of the pelvic roof: usually perforated by pairs of foramina, as in the Duck, but becoming obliterated more or less in *Scolopax, Ptilinopus, Psophia, Scops, Ibis,* and perhaps in others with age. In most *Gallinacea,* including the Doves, the ilia converge to a less proportion of the anterior sacral spines, and the space at the middle and posterior part of the pelvis formed by the sacrum, fig. 21, *a, c,* is both broad and long. In the *Tinamus brasiliensis,* figured by Eyton, this part of the roof is almost wholly ossified, as it likewise is in *Oreophasis derbianus,* where a pair of oblique grooves lead forwards, deepening, to 'ilio-neural' canals beneath the anterior sacro-iliac bony roof on each side the neurospinal ridge. In *Hemipodius, Columba,* and *Goura,* the pairs of foramina in the sacral part of the pelvic roof are very small; in *Crax Mitu* they continue large to a late period. The ilio-neural grooves and canals are seen in most *Gallinæ* as in *Oreophasis.*

22

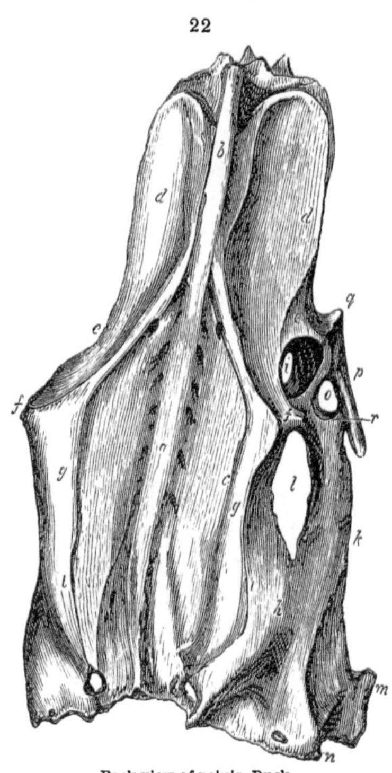

Back view of pelvis, Duck.

In *Cantores, Volitores, Scansores,* and *Raptores,* the proportion of the hind-part of the pelvic roof formed by a neural expanse of the sacrum is less than in *Gallinæ*: the ilioneural grooves are commonly wanting. The bony roof is entire in *Neomorpha, Centropus, Psittacus, Falco, Aquila*: and the parial foramina are very small in *Cypselus, Trochilus, Cassicus, Fregilus,* and most *Cantores*: the ileoneural grooves are present in *Turacus gigas,* and are open canals in *Cypsirhina* and some others. In the diurnal *Raptores* the pelvic roof, of which the sacrum contributes a broad medial tract to about a third of the hinder portion, is strongly and very completely ossified, fig. 23. The ribs of the first two vertebræ retain their moveable joints: in the third to the sixth vertebræ they abut as parapophyses against the lower border of the ilia; the seventh to the tenth vertebræ have no parapophyses; the eleventh to the fourteenth have them long and strong, thickest in the last. All these abutments, with the expansions from the neural spines, coalesce with the innominata and convert the pelvis into one complex mass of bone.

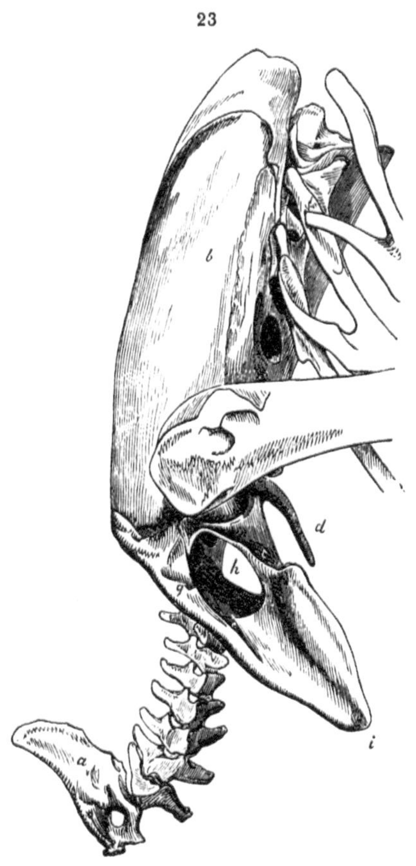

Side view of pelvis, Eagle.

The iliac, ischial, and pubic elements are developed as distinct bones, but speedily coalesce at their point of junction around the acetabulum and usually elsewhere: their independence is longest maintained in the *Cursores*.[1] Ossification begins in each from a single point, even in the much elongated ilium of *Struthio* and *Dromaius*. This bone is, in fact, a single vertebral element, or rather part of one; it is homologous with the pelvic bone, 62, in figs. 43 and 101, D (vol. i.), and with 62 in fig. 28, p. 159, of my work on the 'Archetype Skeleton' (CXL. vol. i.), where it is shown to complete the pleurapophysial element of the pelvic hæmal arch; the ischium being the hæmapophysis of the same arch. The ilium in Birds, figs. 21 and 22, *d*, *h*, fig. 23, *b*, fig. 24, *b*, *c*, *c′*, is remarkable for its developement in the direction of the axis of the vertebral column, extending its connections with many more segments than its own: it is accordingly long and narrow, thickest midway, fig. 22, *f*, where it contributes the upper wall of the acetabulum, ib. *i*, in front of which, *d*, it is outwardly concave; behind the acetabulum, ib. *g*, *h*, it is convex. It differs

[1] XLIV'. vol. i. p. 267. Nc. 1386

in the proportions of the pre-acetabular and post-acetabular extensions, and in the degree of divergence of the latter from the sacrum. The longest and narrowest ilia are seen in certain *Natatores* (*Podiceps, Colymbus*, fig. 34, *a, d, Uria*) and in *Cursores* (*Struthio*, fig. 24, *b, c′, Dromaius*): the shortest and broadest ilia are seen in certain *Volitores, Scansores*, and *Insessores*. In the Grebe and Loon the ilia unite with the summits of the sacral spines behind the acetabula, and diverge for a broader interposed neural expansion anteriorly: in most birds the divergence is shown at the post-acetabular portions, as in fig. 22, *g, g*, the pre-acetabular plates *d, d*, converging to the summits of the sacral spines, ib. *b*. A few birds (*Podargus, Tachypetes*) retain the extent of sacral interposition which obtains at an early stage of pelvic developement in all birds.[1] In the old *Apteryx* the ilia almost meet along the summit of the sacral ridge to within a short distance of their hind end, where an epiphysial piece of bone is sometimes found wedged between this end and the anterior caudal vertebræ. The anterior border of the ilium is usually more or less convex: in *Tinamus, Crax, Onocrotalus*, it is almost straight: in *Geococcyx, Corythaix, Scolephagus*, it is emarginate or concave, the external angle being produced outward: in *Limosa* it is angular; the point being formed by the commencement of the 'gluteal ridge:' this, which is well-marked in most birds, describes a curve, concave downward, and terminates above or behind the acetabulum, as at *f*, fig. 22, marking off the post-acetabular convex part of the ilium, *g, h*. This part is the longest in Grebes, Loons, fig. 34, *d*, and the Ostrich, fig. 24, *c′*: it is the shorter division in Petrels, Gulls, Cranes, and most smaller *Grallatores*, in the *Apteryx*, in most *Insessores*, and especially in diurnal *Raptores*, fig. 23, *g, i*: in many birds it forms half the length of the ilium. In some birds (*Cursores*) it is narrower than the fore part of the ilium; in others, especially *Geococcyx*, it is broader: in most the breadth is about equal, although the ilium may seem broadest behind from its coalescence with the horizontal expansions from the sacral spines. The upper is divided from the outer surface of the post-acetabular part of the ilium by a prominent ridge in most birds, fig. 22, *g*, which generally overhangs the outer surface; in *Geococcyx* to a remarkable extent, like a wide pent-house, producing a deep concavity in the outer and back part of the ilium where it coalesces with the ischium. This coalescence, converting the ischiadic notch into a foramen, fig. 22, *l*, fig. 23, *h*, is common

[1] xv′. p. 45, pl. iii. fig. 6.

to most birds. It does not take place in *Apteryx*, *Dinornis*, *Struthio*, fig. 24, *c'*. The ilium forms an angular projection above the posterior ischial junction in the Albatross, Skimmer, Duck, fig. 22, *n*, Ibis, Spoon-bill, Woodcock, Pigeon, most *Volitores* and *Insessores*. The principal pneumatic foramen of the ilium is on the outer and under part of the post-acetabular division. The ilium developes an oblong articular surface on the prominence extending from the upper and back part of the acetabulum.

The ischium, fig. 24, 63, fig. 22, *k*, *m*, fig. 23, *c*, *i*, is a long, narrow, flattened bone; thickest where it forms the back part of the acetabulum, becoming thinner and broader as it extends backward, with the lower border turned slightly outward; generally placed parallel with its fellow, but diverging in the Ostrich; of nearly uniform breadth in this wingless bird, fig. 24, and in the Apteryx, but usually expanding to its hinder end, and there coalescing with the ilium. Just beyond the acetabular part the ischium contracts, presenting a smooth and thick upper border to the ischiadic notch or foramen, fig. 22, *l*, fig. 23, *h*, and a similar lower border to the foramen or notch, ib. *o*, fig. 24, *l*, which transmits the tendon of the obturator internus muscle; it then becomes lamelliform, with thin margins, usually increasing in depth, and often bent down at its termination to join the pubis, and circumscribe, as in fig. 24, the obturator foramen, *o*. In the Ostrich, the ischium does not join the ilium posteriorly, and the ischiadic notch remains open; its coalescence with the ilium, beyond the ischiadic foramen, is usually extensive, as in figs. 22 and 23.

The most singular modification of the ischia is seen in the *Rhea*, in which they meet below the sacrum and coalesce with each other for some extent, almost obliterating here the bodies of the sacral vertebræ.

The pubic bones, fig. 21, *p*, fig. 24, 64, present an analogous exceptional condition in another member of the *Cursores*, viz. the Ostrich, in which they unite together at their hind ends, forming a 'symphysis,' which is curved downward and forward, fig. 24, *h*; in *Gyps fulvus* the same ends curve toward and almost touch each other. In other birds the pubic bones are directed backward, with usually a curve convex outward, and terminate freely, or are united to the ischium above, as in fig. 34, *b*, the pelvis being thus an open one, as a rule, in Birds. The pubis forms the lower and front portion of the acetabulum, beyond which it quickly contracts, exchanging its trihedral for a subcompressed form, and is more slender than the ischium. The shortest pubis is seen in certain Eagles, in which it terminates after forming the lower boundary of the obturator foramen; its extremity there projecting freely, as in fig. 23, *d*, or being joined by ligament to the ischium, as in the Harpy Eagle, in which it is an inch in length, whilst the ilium is six inches long. The opposite extreme may be seen in the Guillemots and Grebes; and in the latter the pubic styles diverge from the acetabula with a slight outward bend, the interspace of their extremities being twice the breadth of the fore part of the pelvis: they are usually longer than the ischia, figs. 24 and 34; but in the Apteryx they equal that bone in length, and in the Emeu they are shorter. The pubis coalesces with the ilium and ischium at the acetabulum; usually again with the ischium, as at *k*, fig. 22, to close the tendinal foramen, and, in some birds, a third time with the end of the ischium, as in fig. 24, to circumscribe the obturator vacuity, *o*. In Doves, the pubis after uniting with the ischium to close the tendinal foramen, extends backward parallel with and close to its lower margin, sometimes contracting a bony union therewith and obliterating the 'obturator' interspace. The pubic bones as they extend backward in the *Apteryx* are nearly parallel; in the *Emeu*, *Neomorpha*, *Cassicus*, *Podiceps*, they diverge. In most birds the fore part of the acetabular portion of the pubis forms a ridge or tuberosity, figs. 24, *m*, and 22, *q*; in some it is produced to a greater extent (*Geococcyx*, *Corythaix*, *Tinamus*, *Oreophasis*).

24

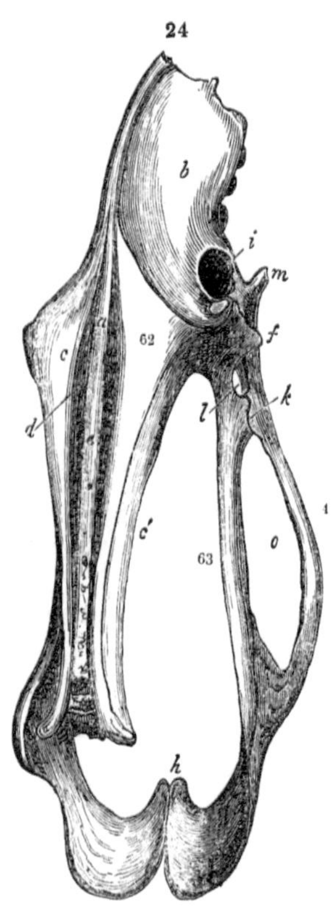

Pelvis of Ostrich.

In accordance with the above-stated differences in the form and proportions of sacrum, ilium, ischium, and pubis, the pelvis of the Bird varies in its general form and proportions. From that of all cold-blooded Vertebrates it differs in the greater number of vertebral segments entering into its composition, and in their bony confluence; from that of Mammals by being unclosed, and by the widely perforate acetabulum, fig. 22, *i*.

The large size and brittle shell of the egg are the teleological conditions of the open pelvis, and the transference of the weight of a horizontal trunk upon a single pair of legs necessitates an extensive grasp of its segments. When the legs require to be pulled far and strongly back, as in diving and cursorial motions, the origins of the requisite muscles are extended far behind the limbs' centre of motion, as in the pelvis of the Grebes, Loons, Guillemots, Ostriches, Emeus; when the bird slowly stalks, or hops, or climbs, or uses the legs chiefly in grasping and perching, the pelvis is short and broad, especially behind, and its breadth may exceed its length (*Cyclarius guanensis*).

The caudal vertebræ are few, short, not produced into a conspicuous appendage, the so-called 'tail' of birds being due to the feathers attached to the terminal vertebræ; these, in birds of flight, coalesce to the number of two or more, and form a compressed vertically extended bone, like a plough-share,[1] fig. 23, *a*, presenting a concave surface to the antecedent centrum; rising above as a sharp crest, anteriorly perforated by the termination of the neural canal; expanding below and there perforated by the hæmal canal which terminates by one inferior and two lateral orifices. This compound bone, 'os en soc de charrue,' supports the coccygeal oil-glands, and gives attachment to the 'rectrices' or rudder-quill-feathers, which are disposed fan-wise. In the Woodpeckers the hæmal part extends far in advance of the articular surface of the centrum, and expands into a broad subquadrate plate concave below; the neural part forms as large a vertical plate; this relates to the use of the stiff tail-feathers in climbing. The horizontal developement prevails in the Peacock. Reckoning the terminal bone as one, the common number of caudal vertebræ is nine. The anterior ones have vertically extended transverse processes including di- par- and pleur-apophysial elements; the neural arch has prezygapophyses, very small postzygapophyses, and a short and thick neural spine. In the third or fourth vertebræ caudal hæmapophyses appear, increase in length, and in the fifth or sixth inclose a hæmal canal. The transverse processes in many birds increase in length to the antepenultimate; in a few (*Ibis*, *Uria*) they gradually shorten to the last; the caudal centrums are joined by ligament and are procœlian. In the Toucan the joint between the sixth and seventh vertebræ

[1] XII. vol. i. p. 208.

has a capsule and synovial fluid, and the neural spine is shortest in the sixth, where the tail has the greatest extent of motion vertically, the transverse bend being checked by the size and length of the transverse processes. The neural spines can be brought by dorsal inflection into contact with the sacrum; and in this motion the side-muscles, which at first tend rather to oppose the elevators, become, as the motion proceeds, themselves elevators, and complete it by a jerk; this throwing up the tail upon the back, as if operated on by a spring, is a conspicuous characteristic of the living bird.[1] In most birds of flight the caudal series are habitually curved upward, as in fig. 23; in the few birds that cannot fly the tail is straight, and the terminal centrum is not expanded. In the *Apteryx* there are nine caudal vertebræ, which are deeper, and project farther below the posterior portions of the iliac bones than in the other birds: as they recede, they increase in lateral and diminish in vertical extent; the spinal canal is continued through the first five, and they are all moveable upon each other, excepting the last two, homologous with the expanded terminal mass in other birds, but which here exceeds the rest only in its greater length, and gradually diminishes to an obtuse point. In the Ostrich the corresponding vertebra is expanded for the support of the caudal plumes, but in the *Apteryx* it offers the same inconspicuous developement as in the Rhea and Emeu. In the 'rumpless' breed of domestic fowl the coccyx is reduced to a single stumpy bone. In some brevipennate sea-birds I have found as many as eleven free caudal vertebræ; only in the extinct *Archeopteryx* of the upper oolitic period was the tail a conspicuous appendage to the trunk, formed by about twenty elongate vertebræ, each of which supported a pair of small and slender quill-feathers.

The terminal vertebræ, ungrasped by the pelvis in the embryo bird, may equal in number those of the ancient feathered fossil; and if such vertebræ participated in the ratio of growth of other parts of the skeleton, without subsequent stunting and confluence, they would more or less repeat the strange and unique feature in the skeleton of *Archeopteryx*; but the metamorphosis of the tail which has taken place in the bird's skeleton in the transition from the mesozoic to the neozoic life-periods of the class, is analogous to that from the protocercal to the homocercal type of tail, which marks the progress in fishes from the palæozoic to the mesozoic periods.[2]

[1] xx'. [2] xv'. p. 45, pls. i. and iii.

§ 127. *Cervical Vertebræ.*—As the prehensile functions of the hand are transferred to the beak, so those of the arm are performed by the neck of the bird; this portion of the spine is therefore composed of numerous, elongated, and freely moveable vertebræ, and is never so short or so rigid but that it can be made to apply the beak to the coccygeal oil-gland, and to every part of the body for the purpose of oiling and cleansing the plumage. In birds that seek their food in water it is in general remarkably elongated, whether they support themselves on the surface by means of short and strong natatory feet, as in the Swan, or wade into rivers and marshes on elevated stilts, as in the Flamingo, fig. 14.

The articular surfaces of the bodies of the cervical vertebræ, like those of the dorsal series, are concave in one direction and convex in the other, so as to lock into each other, and in such a manner that the superior vertebræ move more freely forward, the middle ones backward, while the inferior ones again bend forward; producing the ordinary sigmoid curve observable in the neck of the bird.

This mechanism is most readily seen in the long-necked waders which live on fish and seize their prey by darting the bill with sudden velocity into the water. In the common Heron, for example (*Ardea cinerea*), the head can be bent forward on the atlas or first vertebra, the first upon the second in the same direction, and so on to the sixth, between which and the fifth the forward inflection is the greatest; while in the opposite direction these vertebræ can only be brought into a straight line. From the sixth cervical vertebra to the thirteenth the neck can only be bent backward; while in the opposite direction it is also arrested at a straight line: from the fourteenth to the eighteenth the articular surfaces again allow of the forward inflection, but also limit the opposite motion to the straight line.

An inter-articular cartilage is inclosed between reduplications of the synovial membrane in most of the joints between the bodies of the cervicals, as in the joint of the lower jaw in mammalia. The zygapophysial articulations are simply synovial. The par- and di- apophyses are at the fore part of the vertebræ, and, usually at the third cervical, coalesce with a styliform pleurapophysis projecting backward. The vertebrarterial canal, thus formed, is large, and gives passage to both the vertebral artery and the sympathetic nerve.

The inferior processes from the cervical centrums are of two kinds; one single, developed from the mid-line, usually toward the back part, and answering to the 'hypapophysis' in lizards; the other parial, developed from the under part of the vertebrarterial canal, answering to parapophyses, bent or directed downward, after coalescing with the pleurapophysis. In a Vulture (*Gyps fulvus*), the latter inferior processes begin at the sixth cervical, and are continued to the thirteenth; the hypapophysis begins at the second, and is continued to the fifth, where it is reduced to a low ridge. In the Guillemot the hypapophysis exists as a ridge or process in all the cervicals. In the Apteryx the single hypapophysis for the attachment of the longus colli anticus is present in the last three vertebræ, as in the contiguous dorsals. The parapophysial arch for the protection of the carotid arteries is most complete in the twelfth cervical, but the two sides of the arch are not anchylosed together; the interspace progressively increases in the eleventh, tenth, and ninth vertebræ, and the groove widens and is lost at the fifth vertebra. In many birds the parapophyses after forming the sides of a wide canal in the middle cervicals, converge and unite to inclose a hæmal canal, as in the lower cervical vertebra of the Pelican, fig. 25.

25

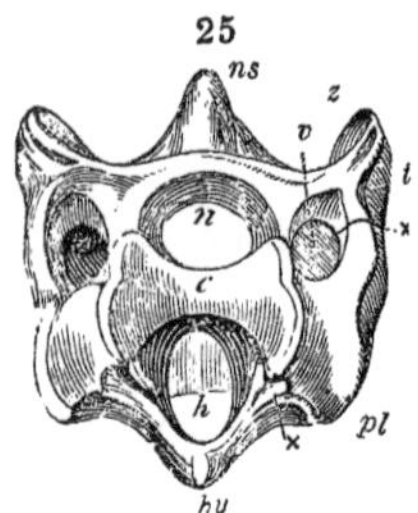

Cervical vertebra, Pelican.

The neural arch in most of the cervicals developes, in addition to the spine, ib. *ns*, zygapophyses, ib. *z*, and diapophyses ib. *t*, also anapophyses, or tubercles, above the posterior zygapophyses. The arch is, in some vertebra, strengthened by bony plates, one of which may be specified as 'interzygapophysial,' and this may be perforated vertically, as in the second, third, and fourth cervicals of the Hornbill (*Buceros*), and in the third and fourth cervicals of the Vulture, and many other birds; the neural arch thus becomes remarkable for its breadth, and the square or quadrate platform of bone from which the small and short neural spine rises. In one or more succeeding vertebræ the incomplete 'interzygapophysial' bar projects backward as a process or tubercle from the prezygapophysis. The prezygapophyses look upward and inward; the postzygapophyses downward and outward; the neural spine is feebly developed, if at all, in the middle of the cervical region; it is most conspicuous in the second to the fifth, and again in the last two or three cervicals. In the Apteryx it is thick and strong in the second, but progressively diminishes to the seventh, cervical, where it is reduced to a mere tubercle; from the eleventh it progressively increases to the last cervical, in which it presents the strong quadrate figure which characterizes the same process in the dorsal vertebræ. The neural canal, ib. *n*, varies in form and diameter in the same vertebræ. If, e.g., the sixth cervical of a Stork be sawed lengthwise vertically, the diameter is greatest in the middle, least at the ends; but if it be sawed lengthwise horizontally, the transverse diameter is the reverse, being narrowest at the centre and widest at the ends. In the Ostrich, the Swan, and many other birds, the canal widens in every direction at its extremities; and on the dorsal or posterior aspect of the spine, the canal remains open for some extent in the intervals of the vertebræ, the myelon being there protected only by membrane and the elastic ligaments which connect the neural spines together. This modification subserves the prevention of compression of the myelon during the frequent, varied, and extensive inflections of the neck in birds.

The atlas and axis speedily effect a partial coalescence; the body of the first, e.g., as an 'odontoid process' to that of the second, and usually presenting a pair of small facets to articulate with its own neurapophyses, which are mainly supported by the 'hypapophysis' simulating the entire centrum of the atlas. The back part of the hypapophysis offers a flat surface to the centrum of the axis, beneath which it is slightly produced, being here wedged into a notch between the true bodies of the atlas and axis. The fore part of the hypapophysis combines with the neurapophyses to form the major part of the cup for the condyle of the occiput, which is completed by the 'odontoid.' The atlantal neurapophyses usually diverge as they rise, and are joined together above by a broad plate slightly arching across from one to the other; in some (*Aptenodytes, Dinornis*) they do not meet: rarely is a neural spine developed. The centrum of the axis is sometimes carinate below with a slight posterior production (*Alca impennis*), sometimes produced into a hypapophysis, as long as the neural spine above (*Aptenodytes*, most *Raptores.*) Postzygapophyses of the atlas articulate with the prezygapophyses of the axis. In a Hornbill (*Buceros*) I have seen complete coalescence of the atlas and axis.

§ 128. *The Skull.*—The neural and hæmal plates of the embryonic trace become modified in the head of the chick by the early expansion of the cerebral part of the neural axis, and by the almost contemporary appearance of the capsules of the organ of hearing, which are speedily followed by the rudiments of the eyeballs. The neural plates are dilated by the primitive vesicles of the ep- mes- and pros-encephalon, the latter speedily showing its greater size and bending down. The notochord extends into the hind part of the future basis cranii, its gelatinous axis terminating at the bend; but the blastemal capsule—the true seat of the histological changes resulting in vertebral structure—is continued forward, expanding, dividing below the part of the vertical cerebral canal called 'infundibulum,' and again uniting anteriorly to form a vertical plate extending between the eye-capsules and becoming lost in the deflected fore part of the cephalic blastema. At this stage neural segments are not shown. The hæmal ones appear as the so-called 'visceral arches' of the head. The foremost is incomplete below at the 'blastemic' stage, and is represented by a pair of obtuse lobes or buds beneath the eyes; the next is larger and becomes closed below; a third, a fourth, and a feeble indication of a fifth, correspond with the primitive vascular arches, and are more truly 'visceral' than 'vertebral.' Of the latter significance is that which descends on each side the heart itself, and is soon indicated by the buds of the appendages which become articulated with such 'scapular' arch.

The cartilage formed round the fore part of the notochord, extends neurad, and attains great thickness at the sides of the cranium in connection with that of the acoustic capsules; it becomes thinner as it rises, and the primitive tissue closes the expanded cranial cavity. The cartilage behind the ear-capsule is of the hindmost neurapophysis: that in front of the capsule is of the next; that which is formed at the optic foramen is the third in advance: these latter neurapophysial cartilages are formed in the blastemal walls of the cranium distinct from the notochordal cartilage. This, advancing along the base of the skull, follows the disposition of the extensions of the notochordal capsule, and bifurcates into the so-called 'trabeculæ' (vol. i. figs. 58–60, 5, 5), which again unite to form the basis of the neural arch and apex of the hæmal arch of the foremost segment of the skull; it becomes compressed between the eyes, and expands in advance of them, the end of the hæmal closing up to that of the neural arch in a way which reminds one of the modification of the vertebral axis at the opposite end of the column. Here, however, the nature of the Bird overrides that of the Vertebrate, and every subsequent step in cranial developement relates to adaptive conditions of vertebral elements and appendages. Distinct cartilages in the buds or piers of the foremost hæmal arch form the basis of the palato-maxillary bones. The palatine cartilage arches outward and backward, like that marked 24 in fig. 60, vol. i., and in it is developed the pterygoid. The cartilage in the second arch forms the basis of the tympanic and mandible; that of the third arch forms the stylo-hyal, rarely ossified in birds, and in connection with it is developed the 'stapes.' The product of the fourth is homologous with a branchial arch in the fish: but further evidence of such conformity with the segmental structure of the trunk-skeleton as is discernible in the much modified anterior termination of the body is given by the ossific centres established in the primordial cartilages; and by the special homologies, determinable prior to confluence, of the bones developed therefrom, with the skull-bones of the lower cold-blooded Vertebrates, which retain their distinctness and depart less from the archetypal arrangement.

Although, as a general rule in the class *Aves*, the separate cranial bones can be discerned only at an early period, yet in those birds in which the power of flight is abrogated, the indications of the primitive centres of ossification endure longer; and in the species here selected for the illustration of the cranial

26

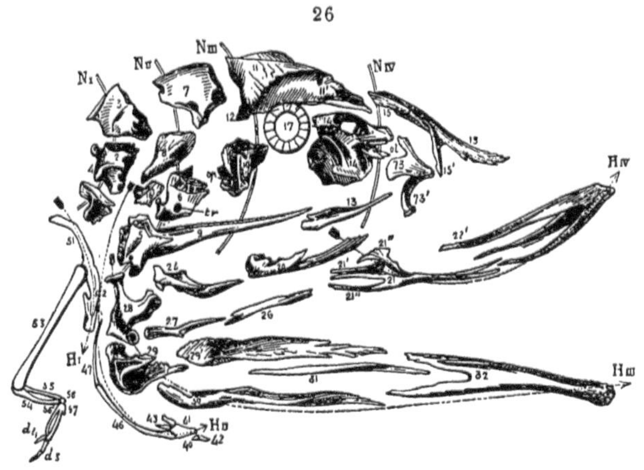

Side view of disarticulated cranial vertebræ and sense-capsules, Ostrich.

segments, the constituent bones of the skull, with the exception of the basioccipital, 1, the basi-pre-sphenoid, 5, 9, and the bones 2, 6, and 8, which coalesce with the petrosal, 16, have been separated by maceration merely in the half-grown bird.

The basioccipital, figs. 26, and 27, 1, developes the major part of the single articular condyle, and sends down a process, more marked in the Struthious genera, and especially in *Aptornis*, than in most other birds: in all respects this primitively distinct bone retains the character of the centrum of its vertebra.

The exoccipitals, figs. 26 and 27, 2, contributing somewhat more to the occipital condyle than in the Crocodile, develope, as in that reptile, the paroccipital, figs. 27 and 28, 4, as an outstanding exogenous ridge or process: but it is lower in position than in the Crocodile (vol. i. p. 135, fig. 93). The superoccipital, figs. 26, 27, 28, 3, as compared with that of the Crocodile, ib., manifests more strongly the flattening and developement in breadth, by which the spinous elements lose the formal character from which their name originated, and are converted from long into flat bones. It always protects the cerebellum; is absent in the Frog, where this organ is a mere rudiment; and is present in the Crocodile in the ratio of the superior size of the cerebellum. The further developement of the cerebellum is the condition of the superior breadth of the spine or crown of the epencephalic arch, fig. 26, N I, in the Bird.

27

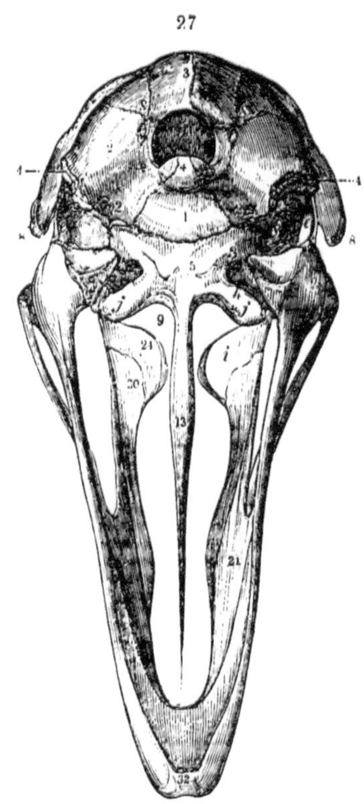

Base of skull, Ostrich.

Of the three bones above defined, 2 is developed in the back part of the cartilage inclosing the ear-capsule, and all bear the same relation thereto, in the primordial cranium, as Nos. 1, 2, 3, 4, in *Chelonia* (p. 131, fig. 92), and as Nos. 1, 2, 3, in the Crocodile (p. 135, fig. 93). No. 2, in the Bird, as in the Crocodile, includes, connately developed therewith, the bone 4, in the Emys. A basal view of the epencephalic arch is given in the young Ostrich, fig. 27, showing the proportions in which the centrum, 1*x*, the neurapophyses 2, 2, and the neural spine, 3, enter into the formation of the neural canal or 'foramen magnum,' 1. The connate element, 4, stands out, like 4 in figs. 81 and 92, as the transverse process of the neural arch.

The second segment of the skull has for its central element a bone, figs. 26, 27, 5 (basisphenoid), ossified, like some trunk-centrums, from three points,[1] and in the Bird, as in other Ovipara, becoming confluent with that, 9, which stands in the same relation to the third cranial segment; the pit for the pituitary body marks the boundary; but the essential distinction of these centrums is given by the neural and hæmal arches. The neural arch of the parietal vertebra retains the same characters which it first manifested in Fishes. Besides the neurapophyses, 6 (alisphenoids), impressed by the mesencephalic ganglia and transmitting the chief part of the trigeminal nerves, besides the vastly expanded and again, as in Fishes, divided neural spine, 7 (parietal bones), the parapophysis, 8 (mastoid) is originally distinct. It has a similar proportional size to that in the Crocodile (vol. i. figs. 93, 95, 8); but owing to the raised dome of the neural arch, is relatively lower in position; it extends apophysially downward and outward, is ossified with the petrosal, and forms a large proportion of the outer wall of the otocrane. Owing to the breadth and shortness of the bird's brain, and the displacement of the optic lobes, the neurapophyses of the mesencephalon, 6, converge toward each other anteriorly, and support part of the neural spine of the prosencephalon, 11, as well as their own, 7. On comparing a side

28

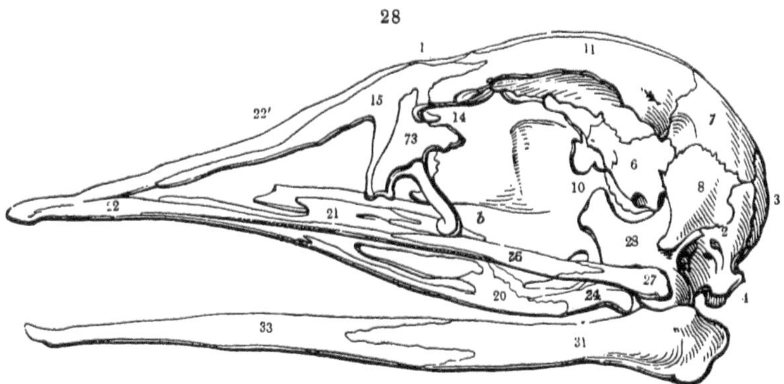

Skull of young Ostrich.

view of the cranium of the Bird, fig. 28, with that of the Tortoise (vol. i. fig. 91), and Crocodile (fig. 95), the greater developement of the epencephalon brings its neural arch, 2, 3, into view, which is obscured by the growth of the apophysial part of 8, in the cold-blooded Ovipara: but the connections of 8 with 2 behind, with 7 above, and with the ear-capsule within, are the same; the latter, however, being ossified in the Bird, but retaining its gristly state in the Tortoise.

[1] CCCXXX. and XXIII.

The hæmal arch of the parietal vertebra, fig. 26, 40, 43, is more reduced than in the Crocodile, and owes much of its apparently typical character to the retention of the thyrohyals, 46, 47, borrowed from a branchial arch of the visceral system, which arches are transitorily manifested in the embryo bird. These spurious cornua project freely or are freely suspended.

The bones, 10 (orbitosphenoids), of the third neural arch coalesce with each other, and with the centrum below, protect a smaller proportion of the prosencephalon than in the Crocodile, but maintain their neurapophysial relation to it and to the optic nerves, below the exit of which they begin to ossify. The neural spines, 11 (frontal), cover a larger proportion of the hemispheres, and, with their homotypes, 7, exhibit a marked increase of developement in conformity with that of the cerebral centres protected by their respective arches. The parapophysis of the frontal vertebra, 12 (postfrontal), is relatively smaller in the Bird than in the cold-blooded Vertebrates, and is rarely ossified from an independent centre, as it is in the Emeu. The hæmal arch of the frontal vertebra, receding from its typical position as the *Hæmatocrya* advanced in time and in developement, is now wholly transferred to the parietal one; its pleurapophysis (28, the ' tympanic '), which is simple, as in the Crocodile, articulates with the parietal parapophysis, 8 (mastoid), though this in some Birds unites with that of the frontal vertebra, 12. The bone, 28, is the chief and most direct osseous developement from the proximal portion of the cartilage of the tympano-mandibular visceral arch: the special appendages of the acoustic organ are developed, as in the Lizards and Snakes (vol. i. fig. 444, B, *e*), in connection with, but not in or from that cartilage. In the young Ostrich and many other birds traces of the composite character of the hæmapophysis (mandibula) are long extant; and bear obviously a homological relation to the teleologically compound character of the same element in the Crocodile: the pieces, Nos. 29, 29′, 30′ and 31, first coalesce with each other, and then with the hæmal spine (32, 'dentary element'), the halves of which are confluent at the symphysis.

The centrum, (13, ' vomer ') of the nasal vertebra is single, and usually coalesces with the neurapophyses (prefrontals), 14, and pleurapophyses (palatines), 20, of its own segment, and with the rostral production of the frontal centrum, 9: it is elongated and pointed at its free termination, and deeply grooved above where it receives the above-named rostrum; indicating both by its form and position that it owes its existence, as bone, to the ossification of the under and outer part of the anterior production of the notochordal capsule. In the young Ostrich the presphenoidal rostrum intervenes between the vomer, 13, and prefrontals, 14. These latter bones manifest the essential neurapophysial relations to the rhinencephalon and olfactory nerves: but they early coalesce together and with the rhinal capsules, as in the tailless Batrachians. The anterior contraction of the cranial cavity, which affects the orbito-sphenoids, influences still more the prefrontals, and, in connection with the large relative size of the eye-capsules, becomes the condition of the extreme modification of the neurapophyses of the foremost cranial vertebra. The neural spine (nasals), 15, is divided along the middle line; but in most Birds the suture becomes obliterated and the spine coalesces with its neurapophyses, with the frontal spine, and with those parts of the hæmal arch of the nasal vertebra with which it comes in contact.

The pleurapophyses (palatines), 20, of this inverted arch retain their typical connections with the nasal centrum and neurapophyses at one end, and with the hæmapophysis (maxillary), 21, at the other end, and they also support the constant element of the diverging appendage of the arch (pterygoid), 24. The hæmapophysis (maxillary), 21, resumes in birds more of its normal proportions and elongated slender form, as such: but the hæmal spine (premaxillary), 22, is largely developed though undivided, and sends upward and backward from the part corresponding to the symphysis of the spine, a long pointed process, 22′, which joins and usually coalesces with the neural spine, 15, and divides the anterior outlet of the hæmal canal into two apertures called the nostrils. The modification of the hæmal arch of the nasal vertebra in the Lizard tribe is here repeated. The pleurapophysial appendage (pterygoid), 24, connects the palato-maxillary arch with the tympanic, and in the Ostrich and some other birds, also with the basisphenoid, 5, and fig. 27, *f*: the second or hæmapophysial ray of the diverging appendage (malar and squamosal) is developed in all Birds, as in the squamate Saurians, combining the movements of the hæmal arch of the nasal vertebra with that of the frontal vertebra, and consisting of the two styliform ossicles (malar, 26, and squamosal, 27), which extend from the hæmapophysis, fig. 28, 21, 21″, to the pleurapophysis, 28: the essential relationship of the compound ray, 26 and 27, with the nasal vertebra, is indicated by their becoming confluent with its hæmapophysis, at 21″, whilst they maintain an arthrodial articulation with the pleurapophysis, 28, of the succeeding vertebra.

The bones of the splanchno-skeleton intercalated with the

segments of the endoskeleton in the bird's skull are the petrosal, 16, between the neural arches of the occipital and parietal vertebræ, connate or co-ossifying with the elements of those vertebræ with which it comes in contact; the sclerotals, 17, interposed between the frontal and nasal neural arches; and the thyrohyals, 47, retained in connection with the debris of the hæmal arch of the parietal vertebra. The olfactory capsule may be represented by ethmoturbinal and turbinal processes in the skull; but chiefly remains cartilaginous. The dermal bone (lacrymal), 73, is well developed and constant: one or more superorbital dermal bones are occasionally present.

As the characters of the occipital segment of the bird's cranium are so obviously those of the vertebral neural arch as to compel acceptance of the interpretation of its elements according to the terms of general homology, there is *à priori* probability in the segmental type being continued to the front end, as it is to the hind end, of the vertebral axis, notwithstanding the modifying influences of the large intercalated sense-capsules, and of the special uses to which certain of the lower bony arches are destined in the head. Developement, obedient to these demands, gives such evidence as it can in favour of the presumption, while relative position and connections afford the proof. In the foregoing description I have, therefore, explained the chief constitution of the bird's skull in the terms of general homology, and I proceed to point out some of its principal modifications in those of special homology.

The occipital condyle is single in Birds, varying from the hemispheroid to the transversely elliptic form, with sometimes a median notch or pit, and in the extent to which it projects; being pedunculate in *Dinornis*,[1] but sessile as a rule. The foramen magnum varies from a subcircular to a full transverse ellipse, and to a vertically oval (*Dinornis giganteus*) form, with lateral encroachments as in *Aptornis*[2] and *Didus*;[3] its plane may be vertical (ib.), but as a rule is oblique from above downward and forward, thus departing further from the reptilian character. The basioccipital in the extinct *Aptornis* sends down, as in the Crocodile, a deep plate below the condyle: it descends in a less degree in *Dinornis*, in both swelling out laterally into a pair of tuberosities, completed by ossifications in the basisphenoid cartilage which afterwards coalesce.[4] As a rule the condyle is on a level with the basis cranii. The paroccipitals in the low flat cra-

[1] XVI vol. iv. pl. 24, fig. 2.
[2] Ib. vol. iii. pl. 52, fig. 4.
[3] Ib. vol. iii. p. 350.
[4] Ib. fig. 1, 5′.

nium of *Dinornis* retain much of their crocodilian position, but they hold a lower one in the loftier domed crania of other birds: they vary in the developement of their apophysial part, standing further out, e.g., in *Rhea* and *Struthio*, than in *Dromaius*. The occipital region is bounded above by the arched ridge formed by the insertion of the muscles *longus colli posticus* and *complexus*, in large and powerful birds; and is bisected, as, e.g., in the Eagle, by a median-vertical ridge dividing the transverse one into a pair of arches: in *Dinornis* a prominence between the insertions of the *longus colli posticus* and *complexus* subdivides the transverse ridge into four arches; and, here, a lower transverse ridge, bounding the insertions of the *recti cap. postici* and *trachelo-mastoidei*, overarches the foramen magnum.[1] In smaller and less robust birds a cerebellar prominence marks the middle of the occipital region: in some species the pressure of the brain from within, and the muscles from without, reduces the thin, bony wall in some places to its membranous lining, leaving openings in the dry skull commonly on each side of the cerebellar prominence. These have been termed 'fontanelles,' as if they were due to original arrest of cranial ossification, but the latter explanation applies to openings, usually reduced to a venous outlet, between the exoccipital and mastoid. In certain Doves, Owls, Parrots, and the Dodo, there is a median 'superoccipital' foramen, usually accompanied by a pair of venous foramina.

The basisphenoid chiefly differs in the presence or absence of 'pterapophyses.'[2] They are longest in the *Struthionidæ*, fig. 27, *j*,[3] are short and thick in *Dinornis* and *Apteryx*, and abut against the tympanic end of the pterygoids; they are shorter in *Grallatores* (*Vanellus*) and *Rasores* (*Columba*, *Tinamus*, *Syrrhaptes*), and their abutment is nearer the middle of the pterygoids; they are absent, or are too short to reach the pterygoids, in the Dodo, Owls, Diurnal Raptores, and most other Birds. In the Emeu, Apteryx, and Dinornis, the basisphenoid shows a median perforation. The sides of the basisphenoid, obliquely grooved by the Eustachian canals (*Dinornis*) and excavated to form the base of the tympanic cavity, in some birds extend outward to the tympanic process of the mastoid, and with it grasp the hinder condyle of the tympanic.

The mesencephalic fossa and the 'foramen ovale' for the transit of the fifth or trigeminal nerve indicate the alisphenoid, fig. 28, 6,

[1] xvi'. vol. iii. pl. 52, fig. 4.

[2] First indicated as such in xvi'. vol. iii. p. 351 (January, 1848); see also xliv. p. 303, no. 1601.

[3] xliv'. p. 259.

and its general homology as a 'neurapophysis;' its extent and connections are shown in fig. 8, p. 22, of CXL, and in the specimen, No. 1363, of XLIV; it articulates below with the basisphenoid, behind with the mastoid, 8, and petrosal, above with the parietal, 7, and frontal, 11, in front with the orbitosphenoid, 10, combining with it to form the *foramen lacerum anterius,* through which orbital portions of the fifth nerve pass; and which usually blends with the common *foramina optica,* encompassed in great part by the orbitosphenoid.

The homology of 8, figs. 25, 28, 31, with the bone so numbered, and called 'mastoid' in vol. i. figs. 75, 81, 91, 92, 93, 95, and 97, is plain; it forms part of the cavity for the otic capsule, as in Fishes, and part of the tympanic cavity, as in the air-breathing Hæmatocrya. As in Reptiles, it offers the articular cavity to 28, a relation partially fulfilled in Fishes; it sends off the second, counting forward, of the great outstanding processes for the insertion of muscles from the trunk and neck; it is developed in and from the thick lateral cartilaginous mass of the primordial cranium, connately with the otic capsule (petrosal). Besides the 'mastoid process,' figs. 27, 29, 8, a second so-called 'tympanic process' is developed in some Birds.[1] The articular cavity for the bone, 28, is single in *Apteryx, Dinornis, Struthio,* and a few other birds, but double in most; in *Aptornis* there are three articular surfaces.[2] The mastoid process varies in shape and size; in a few birds, as in *Calyptorhynchus* and *Aptornis,* it repeats the secondary character more commonly seen in Reptiles by uniting with the third cranial diapophysis, 12, and forming an upper zygomatic arch across the temporal fossa.[3] The optic foramina indicate the orbitosphenoids, which, like neurapophyses of the trunk, in certain instances, coalesce below and divide their segment of the neural axis from the vertebral centrum. In Birds they are uplifted, as in the Perch (vol. i. fig. 85, 10), far above the representative, 9, of their centrum. The divergence of the neural laminæ above the median confluence, to support the prosencephalon, is well shown in CXL. fig. 8, and in the specimen, No. 1363, XLIV., p. 262. In the antecedent neurapophyses, 14, the tendency to median confluence increases, and they become confluent not only below, but above their segment, encompassing the canal, foramen, or foramina, for the olfactory nerves (rhi-

[1] Bustard, XVIII'. vol. iii. p. 352, pl. 52, fig. 9, 8'. [2] Ib. p. 352.

[3] Ib. p. 352, pl. 52, fig. 1, 8, 12; CXL'. pl. 1, fig. 1, 8, 12. In the memoir above cited this skull is described as belonging probably to *D. Casuarinus,* ib. p. 376; it is, however, of *D. didiformis,* since generically separated as *Aptornis.*

nencephalic prolongations of the neural axis), and in some birds expanding above, and appearing, as in *Batrachia,* on the exterior of the cranium, as between 11 and 15, in fig. 29, and at 14, fig. 31. The confluent prefrontals, fig. 28, 14, support the fore part of the frontals, 11, and a greater proportion of the nasals, 15; they are in most birds raised far above the (hypapophysial) ossification of the lower cortical part of their centrum, called the 'vomer,' fig. 32, 13, the large eyeballs, and their thin but deep interorbital septum, being interposed; this septum is ossified in different degrees in different birds. The typical position of the prefrontals is retained in *Apteryx* and *Dinornis.* The condition of the prefrontals in some fishes (*Xiphias*)[1] will aid the comprehension of the developement of the prefrontals between the orbits in these wingless birds. The frontals are large, triangular (fig. 29, 11), or subrhomboid, plates. The post-frontal appears as a distinct bone in some birds (Emeu)[2] fig. 31, 12; it varies much in length; it bends down, and is the longest of the three cranial diapophyses in the Eagles; it curves forward, meets, and coalesces, with a backward production of the lacrymal in some Parrots, fig. 30, *o g.*

29

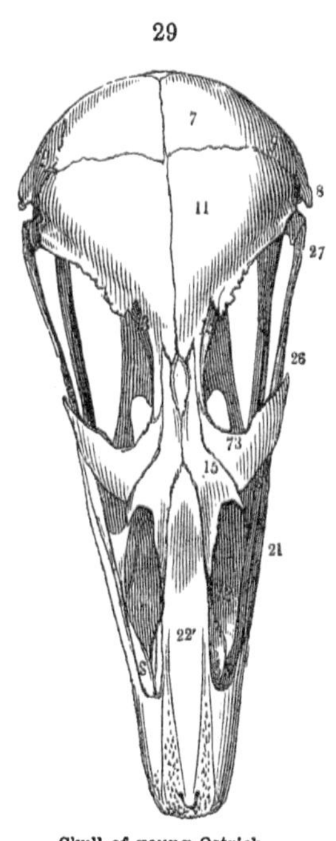

Skull of young Ostrich.

The nasal bones rarely unite with each other in any proportion, fig. 29, 15; in most birds they are separated by the union of the nasal process of the premaxillary, 22′, with the frontal or prefrontal: save in the Emeu, fig. 31, 15,[3] and some other *Struthionidæ,* the nasal bifurcates anteriorly to form the hind boundary of the nostril, and the hinder prong descends to join the maxillary, dividing the nostril from the antorbital vacuity.

30

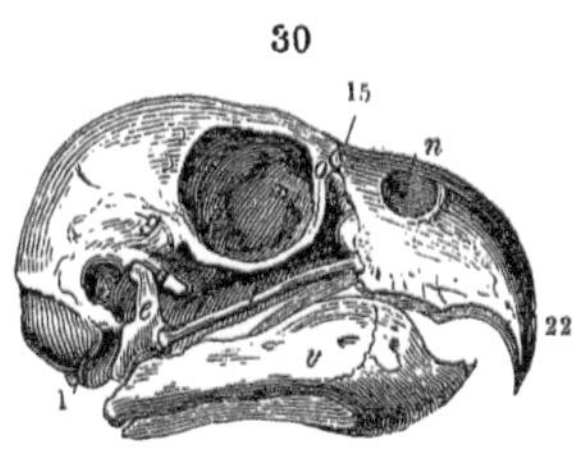

Skull of Parrot.

The premaxillary, figs. 28-32, 22, upon which the upper mandible is moulded, follows, in a minor degree, all the varieties of that much diversified part in Birds: it

[1] CXL′. p. 52, pl. 1, fig. 5, 14. [2] XVIII′. vol. iii. pl. 39, figs. 1 and 2, 12.
[3] XVIII′. vol. iii. pl. 39, 14.

is a single bone, expands from before backward, and divides into a superior and medial nasal process, ib. 22′, and a pair of infero-lateral maxillary processes, ib. 22; the interval between the nasal and maxillary processes being the fore part of the outer bony nostril. The nasal process indicates by a median slit or groove the typical duality of the bone in the *Gallinæ* and a few other birds: the maxillary process usually sends off a 'palatal' plate, fig. 32, 22.

31

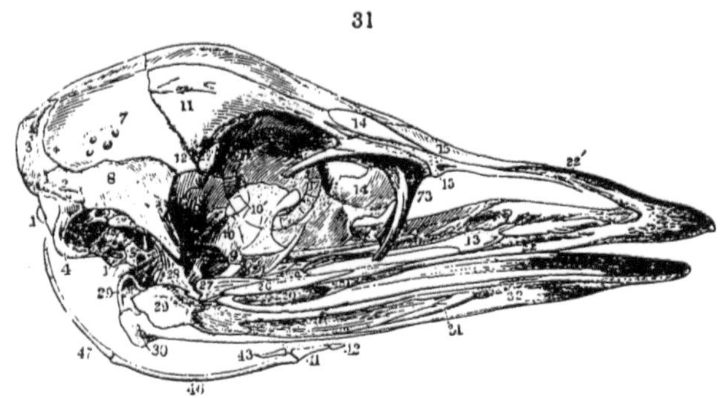

Skull of Emeu (*Dromaius*).

The maxillary, figs. 27–32, 21, is a small and usually slender bone, anteriorly expanded and engrained into the notch between the maxillary and palatal processes of the premaxillary: usually uniting also with the nasal, and in some birds with the lacrymal, 73, and vomer, 13; it then bifurcates posteriorly on the horizontal plane to join the malar, 26, and the palatine, 20. The palatine process sometimes developes from its upper surface a turbinal structure. In the Rhea the palatine plate of the maxillary is perforated: in the Emeu, Ostrich, Apteryx, and in most birds, it is entire: it is of great breadth in the Night-jars (*Caprimulgus, Podargus*), and is both long and broad in the *Apteryx*. In *Struthio* and *Rhea* the maxillary sends upward a process towards the nasal, the descending maxillary process of which is wanting.

The palatines, fig. 32, 20, articulate by a longitudinally grooved mesial surface to the vomer, 13 (*Emeu*), more commonly to both this bone and the presphenoid, or to the latter only: they give attachment posteriorly to the pterygoids, 24: they diverge as they extend forward, developing a 'meatal' plate mesially, which partially bounds the posterior nostril, and a 'muscular' plate externally for the attachment of the entopterygoideus; they then extend forward, parallel or converging, to join the maxillaries, and sometimes also the vomer, and there complete the roof of the mouth. In the *Struthionidæ* the 'palatal' part is short and broad; articulates laterally with the maxillary, and, as it retrogrades, expands, mesiad, to abut upon the presphenoid (*Struthio*) or hind part of the vomer (*Dromaius*, fig. 32, 13), and to articulate with the pterygoids: the palatines nowhere meet in the median line, and the meatal process is wanting. In the *Apteryx* the palatines coalesce anteriorly with the maxillaries, posteriorly with the pterygoids, have a straight outer and a concave inner border, from which, posteriorly, is continued the 'meatal' plate curving inward

32

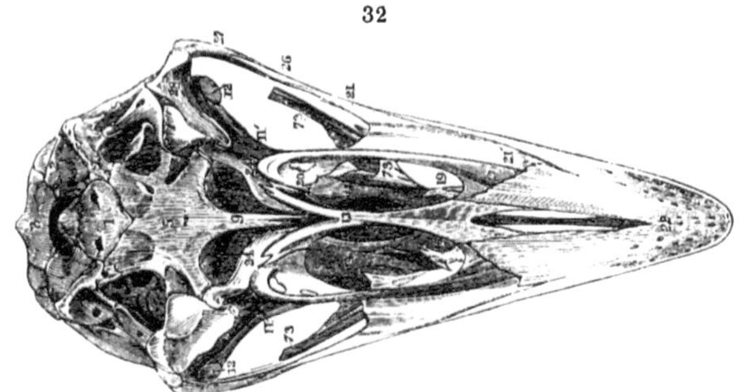

Skull of Emeu (*Dromaius*).

and forward, obliquely, about the hind part of the meatus, and applying itself mesiad to the vomer and presphenoid. In *Natatores*, and most *Grallatores*, the palatines meet each other posteriorly, for a short extent, before diverging as they advance to bound posteriorly the palatine nostril. In the *Pelecanidæ* the mesial union is extensive beneath the presphenoid and vomer. In *Tinamus* the palatines join behind; they there touch each other for a shorter extent in *Columbidæ*; but in most *Gallinacea* they are kept apart by the presphenoid. In *Podargus* and some other Fissirostrals the palatines are short and broad, and extensively joined together behind the small palatal nostril. In *Raptores* the palatines meet behind beneath the presphenoid and vomer: the meatal plate developes a ridge, descending, to increase the concavity of the entopterygoideal surface, the outer border of which also descends; and these carinal boundaries of that surface are found in many birds. The extent of the palatine plate varies in different birds; it is largest in *Struthionidæ*, large in *Raptores*, and least in the *Gallinæ*: in the Tinamous and Pigeons it is of moderate size. In the Emeu it has a large vacuity;[1] it is perforate in *Tinamus*, and in some *Grallatores*.

The pterygoids, figs. 27, 32, 24, articulate with the outer and hinder angles of the palatines by a squamous overlapping

[1] xvi'. vol. iii. pl. 39, fig. 2, 20.

suture in *Struthio*, where they are wide apart, and pass almost parallel backward to join the pterapophyses of the basisphenoid and the tympanics; developing a broad plate mesially to abut upon the presphenoid: their palatine ends are nearer each other, and their course to the tympanics more divergent, in the Emeu, Apteryx, Hemipode: they touch each other anteriorly in some birds; and, in a few (*Podiceps, Sula, Ibis, Argala, Scolopax*), have a short extent of mutual junction before diverging. In *Rhea* the fore part of the pterygoid is slender, is attached to the vomer and presphenoid, traverses obliquely the upper surface of the palatine plate, with which it ultimately coalesces, and becomes engrained between the pterapophysis and orbital process of the tympanic before abutting upon the inner and lower condyle of that bone. In general the pterygoids are straight and slender; they diverge at an open angle in *Raptores*, at an acute angle in *Colymbus*, are nearly parallel and longitudinal in *Struthio*, with intermediate relative positions in other birds. In *Raptores* their connections are limited to the essential terminal ones with the palatines and tympanics: in some other birds they articulate, occasionally by a distinct process, with pterapophysial extensions of the basisphenoid, limiting the movement of the tympanic, and adding strength and fixity to the upper mandible: in most birds there is a prominence or process from the hinder border for ligamentous attachment to the basisphenoid. In *Rhea* there is an articular process for the orbital plate of the tympanic. The tympanic joint is double in Pheasants and Plovers.

The bone, figs. 28–32, 26, answering to that so numbered in figs. 91, 92, 93, 95, vol. i., is a straight, slender, usually triedral, style in birds, articulating with 20 by one end and with 27 by the other, and in some birds being joined by a descending process from the lacrymal. It combines all the essential homological characters of the 'malar,' those, viz., derived from relative position and connections; and exemplifies the unimportance of configuration: showing the opposite extreme to the scale-like shape of the bone in the Turtle, fig. 91, 26, vol. i. The malar of the Crocodile, fig. 95, 26, offers an intermediate modification of form. The malar in the bird overlaps the maxillary by an oblique suture, as in the Reptiles.

The zygomatic connection between the maxillary and tympanic is completed by the bone, figs. 28–32, 27, which has the same slender figure as 26, with which it early coalesces; but preserves a moveable articulation with 28 by a convex condyle adapted to the acetabulum on the outer side of the tympanic. The two bones, 26 and 27, which become blended together in young *Struthionidæ* before the confluence with the maxillary is complete, extend backward in all adult birds, usually in a straight line, from the maxillary to the tympanic. In the Cassowary the zygomatic arch presents a slight expanse and outward bend of the squamosal; in *Didus* it shows a slight downward as well as outward bend. In some Parrots and in Hornbills (*Buceros*), the malo-squamosal zygomatic style, fig. 30, *l*, has a moveable cotyloid joint at both ends; in some *Caprimulgi* it is anchylosed at both ends.

In the birds in which the upper mandible is moveable, either, as in Parrots, by articulation, or as in many other birds by flexibility of the nasal process of the premaxillary, the movements of the tympanic, to and fro, upon its proximal joint, are transferred by the zygoma to the maxillary, and by the pterygoid to the palatine: and thus by the forward rotation of the tympanic the upper jaw is raised, at the same time that the lower jaw by the action of the digastricus may be depressed.

Before anatomy had reached its homological phase, ornithotomists called the zygomatic styles 'ossa communicantia,' and the pterygoids 'ossa homoidea, seu interarticularia;' the following bone was termed 'os quadratum.'

In the tympanic, figs. 28, 31, 28, are to be noticed the 'mastoid' and 'mandibular' ends, and the intermediate body giving attachment at its back part to the ear-drum, and sending from its fore part the 'orbital' process. The mastoid articular end is obliquely extended from behind forward and outward: the body slightly contracts below; then expands and becomes triedral at the setting off of the broad compressed angular orbital process: below this process the mandibular end is much expanded, chiefly transversely: it presents two articular surfaces; the outer one, elongate or reniform, partly concave, partly convex; the inner one a shorter elliptic or oblong convexity; the intermediate non-articular tract varies in different birds. On the outer side of the mandibular end is a hemispheric articular cavity for the 'squamosal.' In most birds the mastoid condyle is divided into two, the inner and posterior encroaching upon the paroccipital, and showing, in an interesting way, the course of retrogression of the tympano-mandibular arch from the fish to the warmblooded ovipara. Most *Cursores* and *Rasores*, *Apteryx*,[1] *Pezus*, *Rhynchotis*, have but one condyle, as in Lizards. The ear-drum is attached to the back part of the pedicle obliquely from its outer margin above to

[1] XVI'. tom. iii. pl. 39, figs. 8, 9, and XXIV'.

the inner one below, whence the membrane is continued to the basisphenoid, paroccipital, and round by the mastoid to the tympanic again. A part of the periphery of the drum may show an epiphysial bony rim. In some birds there is a well-defined flat oval surface on the outer side of the pedicle for a corresponding surface on the mastoid process:[1] most show a distinct articular surface[2] on the inner side of the lower part of the base of the orbital plate for the pterygoid: thus, including the squamosal pit and two mandibular condyles, there may be not fewer than seven articular surfaces in the tympanic bone of the Bird. Its orbital process is a greater developement of the anterior lamina of the Crocodile's tympanic, fig. 93, 28, vol. i.; the size of the process is one of the chief characteristics of the tympanic in the Bird, and shows much variety of shape and proportion in the class.[3] Its apex may be truncate (*Didus*),[4] or rounded (*Dinornis*), or pointed (*Aquila*). A large pneumatic foramen may be situated on the inner side of the pedicle; or on the hinder facet below and between the upper condyles, or in both situations.

The mandible or lower jawbone is ossified usually from nine centres; the anterior being the first to appear, forming the chief and characteristic part, fig. 25, H. iii, of the bone. It bifurcates as it extends backward to form the homologues of the dentary elements, fig. 31, 32, which are thus 'connate' at their symphysis. The Pelicans are an exception, and exemplify the normal separate ossification of each dentary, becoming subsequently confluent for a small extent anteriorly.[5] The 'surangular,' ib. 29′, speedily unites, if it be not connate, with the 'articular,' 29: the angular, 30, remains longer distinct, but coalesces first with the articular: the splenial element, 31, coalesces first with the dentary, and retains longest its primitive independence posteriorly.

In the Garefowl (*Alca impennis*), each dentary retains its bifurcate hind end distinct, the upper prong overlapping the surangular, the lower one the angular; and these two latter elements are divided by an oblong space partly closed within by the splenial: there is also a foramen at the back part of the surangular. The splenial retains its distinctness posteriorly, and a groove on the lower margin of the ramus indicates the extent of its forward production to its confluence with the dentary. A vacuity between the angular and surangular remains in many birds,

[1] XVI′. vol. iii. p. 356, pl. 53, fig. 9, *f*.
[2] Ib. fig. 10, *g*.
[3] Compare ib. pl. 39, figs. 7, 8, 9, *a*, and pl. 53, figs. 8, 9, *k*.
[4] Ib. vol. iii. p. 35.
[5] XXIII.

e.g. *Cracticus, Anthochæra, Lanius,* amongst *Cantores,* but chiefly in the aquatic, wading, and terrestrial orders (*Tetrao, Dinornis, Didus, Notornis,*[1] *Porphyrio, Tantalus, Rhyncops, Uria*). The Coots show a second elliptical vacuity at the base of the coronoid rise of the surangular.[2] In *Rhyncops,* the long compressed symphysial part of the mandible descends below the level of the angular, the lower border of the mandible having a deep notch there. The symphysial part partakes in a minor degree of all the various modifications of the lower mandible. The angular is chiefly extended transversely, and to the inner side of the ramal axis, to form the surfaces adapted to the tympanic condyles, a deep and smooth depression usually dividing them: the inner joint, in Parrots, is a longitudinal groove; the outer one is a longitudinal convexity. An angular process extends from the inner or medial side of the articular expansion; there is also, in some birds, a similar process from its back part, and this, in the Grouse tribe, especially the male *Urogallus,* is much elongated and bent up. The temporal muscles are inserted into an elongate rough tract, or slight elevation of the upper border of the surangular: it is rarely raised into a 'coronoid' process: but this is conspicuous in the conirostral *Cantores,* and especially in the Grosbeak and Crossbill. The latter bird shows a want of symmetry in the mandibular rami; and there is a large sesamoid, wedged into the back and inner part of the joint of the lower jaw.[3]

Through the arrested developement of the hyoid arch (cerato- and stylo-hyals), the tongue of Birds is not suspended by attached inverted piers, but is slung to the cranium, when its branches are sufficiently long, by 'thyro-hyals,' usually including the hypo-, figs. 26, 31, 46, and cerato-, ib. 47, branchial elements; they are long and slender. The basihyal, figs. 26, 31, 33, 41, *bh,* is subcylindrical and expanded at the ends; the front end usually presenting a trochlear articular surface, convex transversely, concave vertically, for the glossohyal, or for the ceratohyal, or for both elements. The ceratohyal, ib. 40, *ch,* is always short, usually extending forward from its attachment as well as backward, and the forward production often unites with its fellow, so as to form the basal part of the direct support of the tongue. In this case the glossohyal, ib. 42, articulates with the ceratohyals; rarely also with the basi-hyal, *bh,* as in the Crane, fig. 33, c. The basihyal is of

[1] xvi'. vol iii. pl. 53, fig. 1. [2] Ib. figs. 1 and 7, *w.*
[3] vii'. p. 277.

extreme length and tenuity in the Woodpeckers, ib. *bh*, D, and supports at its fore end the barbed glossohyal, ib. *gh*. Sometimes the urohyal, fig. 31, 43, is confluent with the basihyal (*Alca impennis, Vanellus, Columba*); more often articulated therewith, or with both basi- and thyro-hyals, fig. 33, C, *uh* (*Grus cinerea*). The 'thyro-hyals' usually retain the two elements of the branchial arch above cited, and shown in figs. 26 and 31, 46 and 47, fig. 33, *hb, cb*.

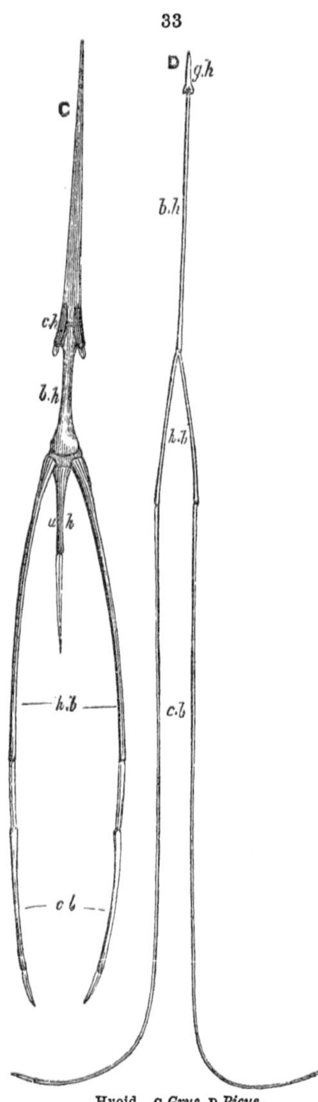

Hyoid. C *Grus*, D *Picus*.

Only in a skull of the extinct *Aptornis* have I seen an ossified 'stylohyal:' it was anchylosed as a styloid process to the side of the inferiorly produced basisphenoid.[1]

The bone, figs. 26, 28, 29, 31, 73, situated at the fore part of the orbit and pierced or grooved by the lacrymal duct, articulates, when not anchylosed, to the frontal, nasal, and prefrontal; it usually sends one process from its upper part arching over the upper and fore part of the orbit; and a second process, from its under part, downward to abut upon the maxillary, dividing the orbital from the antorbital vacuity.

Like the lacrymal in Fishes, this bone in some birds is connected with a suborbital frame extending to the post-frontal (*Macrocercus, Strigops*), fig. 30, *o, g*, and even with the mastoid (*Calyptorhynchus, Plyctolophus, Licmetes, Microglossus aterrimus, Lathamus*).

The superorbital part of the lacrymal is broad and flat, in *Aquila*, and articulates with a similar superorbital derm-bone: in *Vultur*

[1] XVI'. vol. iii. pl. 52, fig. 3, 38.

it is longer and more slender. In *Casuarius* the lacrymal coalesces with the frontal, prefrontal, and nasal, but retains its freedom in other *Struthionidæ*. The lacrymal is very large in *Dacelo* and *Trochilus*, fig. 18, *l*.

The skull in the *Raptores*, especially in the nocturnal division, is short, broad, and high, in proportion to its length, and the cranium is large compared with the face. The occipital foramen is almost horizontal. The superoccipital muscular depressions are well defined. The temporal fossæ are not very deep, and are wide apart superiorly. The cerebral convexities are not strongly marked; the frontal region is flat. A longitudinal furrow extends along the whole upper surface of the cranium in some Owls.

The cranium of the *Warblers* presents a more regular sphericity, but the interorbital space is very concave. The anterior parietes of the orbits are large, from the size of the lacrymal bone and of the transverse lamina of the prefrontal; the internal and posterior orbital parietes are defective; the optic foramina are commonly blended into one, and continuous with the larger fissures above.

In the Parrots the upper surface of the cranium is flattened or slightly convex, and greatly extended in breadth between the orbits for the articulation of the naso-premaxillary bone, fig. 30, *n*.

In the Toucans the cranium slightly increases in breadth to the anterior part where it is joined to the enormous bill. Its superior surface presents an equable convexity. The temporal fossæ, like those of the Parrots, are small, and wholly confined to the lateral aspects of the cranium. The posterior surface, which is absolutely concave in the Macaws, from the backward extension of the paroccipitals, is slightly convex in the Toucans, where it is separated from the upper surface by a regularly arched ridge. The cerebellar prominence extends over the occipital foramen, the plane of which inclines forward and downward from the horizontal line at an angle of 45°. The circumference of the orbit is uninclosed by bone at the posterior part, the postorbital processes of the frontal not being developed as in most Parrots. The septum of the orbits is very incomplete. The nostrils open on the posterior part of the upper mandible, which presents a smooth entire surface formed by the thin parietes of the dilated cellular osseous tissue.

In the Helmeted Hornbill (*Buceros galeatus*) the outer surface of the skull is sculptured with irregular furrows and risings, recalling the surface of the skull in the Crocodiles. The occiput is concave, and separated by a strongly developed ridge from the temporal fossæ, which almost meet at the vertex. The bony septum of the orbits is complete, and formed by two strong plates, separated by an intermediate cellular diploë, except at the posterior part. The optic foramina are distinct; each is directed transversely outwards.

In the Woodpeckers the cranium is rounded, the temporal fossæ shallow, the internal wall or septum of the orbits incomplete, but the anterior boundary is well developed. The posterior facet of the cranium is raised. The superior surface is traversed by a wide furrow extending longitudinally forward, generally to the right, but sometimes also to the left, as far as the lacrymal bone. In some of the larger species of Woodpecker, as the *Picus major*, L., the cranial furrow is more symmetrical. In the Humming-birds it is double, the hyoidean furrows being separated at first by the cerebellic protuberance, and afterwards by a mesial longitudinal ridge.

The skull is remarkable for its length in the majority of the *Waders*. In the Herons and Bitterns the occipital region is low, and inclines from below upward and forward; it is separated from the upper and lateral regions by a well-developed, sharp, lambdoidal crest; and it is divided into two lateral moieties by a slight longitudinal ridge. The temporal fossæ are deep and wide, and extend upward to the sagittal line, along which an osseous crest is developed. The cranium is expanded anteriorly to the above fossæ, for the lodgment of the cerebral hemispheres, the interspace of which is indicated by a deep longitudinal furrow. The roof of the orbits is expanded laterally, which gives great breadth to this part of the head, but the posterior orbital walls are very imperfect, and the internal walls or septum almost wholly wanting. The optic foramina are blended with each other and with the smaller foramina, which in other birds represent the *foramen lacerum orbitale*.

Woodcocks, Snipes, Curlews, and Lapwings resemble Herons in their defective bony orbits; but they want the extended superior parietes of those cavities, and differ much in the almost spherical form of the cranium, which is smooth and devoid of the muscular ridges characteristic of the fish-feeding *Grallæ*. In this order the premaxillary bones present some of their most eccentric forms. They are narrow, elongated, and curved downward in the Ibises and Curlews; bent upward in the contrary direction in the Avosets; extended in a straight line in the Snipes; singularly widened and hollowed out in the Boatbills (*Cancroma*, *Balæniceps*); widened, flattened, and dilated at the extremity in the Spoonbill; thickened, rounded, and bent downwards at an obtuse angle in the Flamingo, fig. 14.

Among the *Natatores*, the Divers (*Colymbus*), Grebes (*Podiceps*), and Cormorants (*Carbo*) show a defective condition of the bony orbits, and of the anterior parietes of the cranium; the septum of the orbits is almost entirely wanting; in place of the posterior orbital parietes, there are two lacunæ leading directly into the cranial cavity, one superior, of large size, and one inferior, smaller; they are, in general, separated by a narrow osseous bar, but in the Coulterneb (*Fratercula arctica*) this is also wanting, so that all the orbital and optic nerves escape by a common opening. In the Petrels and Albatrosses, the internal and posterior walls of the orbits are more complete. In the *Diomedea exulans* the optic foramina are separated both from each other and from the neighbouring outlet. The occipital region is low, and divided into a superior and an inferior facet, the latter being concave from side to side. The plane of the occipital foramen is almost vertical. The occipital or lambdoidal crista is well-marked, and the temporal fossæ nearly approximate in the middle line. In these Sea-birds and in the Gulls, the lateral lacunæ in the bony parietes of the face are very considerable.

A most remarkable characteristic of the cranium of both the Brachypterous and Macropterous Sea-birds is the presence of the two deep, elongated, semilunar glandular depressions extending along the roof of the orbits. In the aquatic birds which frequent the marshes and fresh waters, as the *Anatidæ* or *Lamellirostres*, these glandular pits are wanting, or very feebly marked, as in the Swans. They are, however, again met with of large size, though shallow, in the Curlews (*Numenius*) and Avosets (*Recurvirostra*); and are also found, though of smaller size, in the Flamingo.

The cranial cavity has but a limited range of size in the class of Birds, although an extreme one in relation to the bulk of the body: that of the smallest Humming-bird is proportionally greater than in any other animal, while that of the great *Dinornis* is almost crocodilian in its contracted area: the size of the cranium, small as it is in relation to the trunk and legs in the giant bird, being expanded to the requisite extent for muscular and other attachments by a thick pneumatic cellular diploë between the outer and inner tables.[1] The owls have a similar developement of diploë: in most birds the free cranial wall is thin and compact. The cavity is closely moulded to the brain, and shows well-marked fossæ for the cerebellum, medulla oblongata, optic lobes, hypophysis, cerebral hemispheres, and, in *Dinornis* and *Apteryx*, for the olfactory lobes. Some birds show also a depression upon the petrosal, which is deep in the Heron. In *Dinornis* an upper transverse ridge divides the pros- from the ep-encephalic compartment, and a lower one divides the pros- from the mes-encephalic compartment, which 'tentorial' ridges, being on nearly the same vertical parallel, almost equally bisect the cranial cavity into a wider front and narrower hind division. The roof of the prosencephalic compartment sinks a little into the interspace of the hemispheres, and is here usually grooved by the longitudinal sinus: but in a few birds it developes a bony 'falciform' ridge, which, in *Buceros galeatus*, e. g., bisects the fore part of the prosencephalic compartment.

The principal foramina observed in the cranium are, in the epencephalic fossa, one or more minute 'precondyloid,' the large foramen for the 'vagus' and internal jugular vein, the meatus auditorius internus; in the mesencephalic fossa the 'foramen ovale' for the third and second division of the 'fifth,' the 'carotid,' which opens into the deep 'sella,' the 'foramina,' which transmit nerves to the orbit, not always distinct from the wide foramen opticum; this also being blended with its fellow in many birds; in the prosencephalic compartment, are the rhinencephalic foramina, which, in *Apteryx* and *Dinornis*, from the backward extension and interorbital position of the enormous olfactory chambers, become 'rhinencephalic fossæ,' distributing thereto olfactory nerves by a 'cribriform' plate.

The tympanic cavity is formed by the paroccipital, basi- and ali-sphenoids, petrosal, mastoid, and tympanic. It presents the stapedial canal leading to the 'fenestra ovalis;' and pneumatic apertures by which the air from the Eustachian tube is conducted to the pericranial diploë. The 'petrosal' as the osseous capsule of the acoustic organ, and the 'stapes,' with the cartilaginous 'incus' and 'malleus,' as appendages thereof, will be noticed in connection with the sense-organs.

The orbits are large and lateral, but encroach upon the anterior wall of the cranium, the eyeballs moulding it into a pair of concavities looking forward and usually a little downward and out-

[1] XVI. vol. iv. pl. 24, fig. 4.

ward, with extreme thinning and sometimes partial loss of bone. The roof of the orbit is formed by the frontal, prefrontal, and lacrymal; the hind wall by the frontal, ali- and orbito-sphenoids; there is no bony floor; but the eyeball rotates on a sort of air-cushion resting upon the palatal, the pterygoid, and the orbital process of the tympanic. The bony septum is usually more or less incomplete, and the orbital freely communicates with the temporal vacuity. Only in a few species is the periphery of the orbit completed by bone, as in certain Maccaws and Cockatoos (*Macrocercus*, fig. 30, *Plyctolophus*, *Calyptorhynchus*); the lacrymal extending to the postfrontal as a continuous suborbital bar. In the Woodcock the large lacrymal so extends the front wall of the orbit as to cause it to look a little backward as well as outward: and the orbits are so large as to push the brain-case to the lower and back part of the cranium. In the Owls the postfrontals have the form of broad thin plates, compressed from before backward, and unusually produced downward to increase the wall of the large orbit and give it a more anterior aspect. In most diurnal *Raptores* the upper wall of the orbit is supplemented by a dermal oblong flat superorbital bone, ligamentously connected with the lacrymal. The orbits are smallest and worst defined in the nocturnal small-eyed *Apteryx*: there are no superorbital ridges, no antorbital or postorbital processes, and the interorbital septum is complete and thick, the optic foramina being wide apart. In *Dinornis* the orbits are small, and also divided by the rhinal chamber: but the superorbital ridge is present and developes a strong postorbital process. The interorbital septum, as a rule, is very thin, even when entire, as in *Tachypetes*, *Coracias*, *Eurystomus*: it may have a small vacuity (*Aquila*) or a very large one (*Buceros*), or two or three as in most birds.

The olfactory cerebral crura emerge from the cranium at the upper angle between the hind wall, roof, and septum of the orbit; groove the upper part of the septum as they pass forward to penetrate the prefrontal and expand into the rhinencephalon, dispersing the olfactory nerves to the turbinal membranes. The frontal olfactory foramen, in the *Raptores*, is smaller than the prefrontal one. Between the Vulture and the Crocodile the difference is that the rhinencephalic crura extend along a common canal above the interorbital space in the Reptile, while in the Bird the ossification of the septum divides the rhinencephalic fossa into two: but many birds resemble the Crocodile in this respect. The bones which hold the neurapophysial relation to the rhinencephala, anterior to the frontals, are the same, or homologous, in both *Ovipara*: but in the Bird the secondary peripheral developments of the prefrontals are suppressed as in Batrachians and some fishes (*Xiphias*),[1] in which they form the anterior wall of the orbit, occupying the anterior part of the interorbital space, joining each other at the median line by an extensive vertical cellular surface, and dividing the orbital from the rhinal cavities. In the *Apteryx* and *Dinornis* the latter cavities are so developed as to extend backward between the orbits to the cranium, the front wall of which forms the back wall of the rhinal, instead of the orbital, cavities.

In most birds, however, the orbits intervene: the rhinal chambers are small, and communicate with the upper and back part of the nasal passages on each side of the prefrontal septum. The passages are partly divided by bone developed from the vomer. They usually extend obliquely backward from the outer to the inner or palatal nostrils: but in the Toucans and Hornbills the nasal passage descends vertically at the base of the huge bill. The outer nostril is formed in front by the premaxillary, behind by the nasal—each bone bifurcating to include the area, into the lower part of the circumference of which the maxillary usually enters. In the *Rhea* and *Emeu*, fig. 31, the outer bony nostril is incomplete behind, the maxillary process or prong of the nasal not being developed. In the Ostrich it does not reach the maxillary. The external nostril is near the apex of the bill in the Cassowary. In the *Apteryx* the external nostrils are minute and subterminal; but a linear groove extends back and widens into a large triangular vacuity, on each side the base of the upper mandible in the skull. In the Petrels the nostrils are pierced at the end of a tube upon the upper mandible. In the *Pelecanidæ* there is no outer nostril. The bony septum between the nostrils is rarely entire. The nasal passage is continued backward between the vomer and palatine, or between the presphenoid and palatine, to open, usually by a single median foramen or fissure, or by a pair of such, divided, as in *Dromaius*, by the vomer, fig. 32, 13, or, as in *Struthio*, by the vomer and presphenoid, upon the palate.

Amongst the cranial peculiarities in Birds may be noticed the bony style attached to the occiput in the Cormorant: the light cellular bony core or support of the thick horn or horny crest, in *Casuarius galeatus*; which is expanded and flattened behind in *Casuarius Mooruk*: the longer and narrower horn-core, re-

[1] xvi'. p. 52, pl. 1, fig. 5, 14.

stricted to the space above the orbits, in *Oreophasis Derbyanus*: the bony extensions of the upper part of the premaxillary in certain Hornbills, especially *Buceros galeatus*: the elevated base of the short and thick upper mandible in *Ourax Pauxi*: the multiplied superorbitals in *Tinamus.* Nor, perhaps, should the spherical bony cyst above the fore part of the cranium in a variety of common fowl be omitted, though this, like the stunted mandibles of some varieties of pigeon, may rather rank among the phenomena of pathology.

§ 129. *Scapular Arch and Appendage.*—The simplest condition of this arch is manifested in *Apteryx* and *Dinornis.* It consists of scapula and coracoid, uncomplicated by connection with the hæmapophysis of any other segment: moreover, the pleur- and hæm-apophyses of the occipital rib have coalesced. A man must shut his eyes, and with a tight squeeze, to escape recognising the significance of the propinquity of the scapular arch to the hyoidean one in the embryo bird. As developement proceeds, segment after segment is added to the cervical series, and the occipital ribs, with the myelonal centres supplying their appendages recede far back from the typical position they maintain in the Fish (vol. i. figs. 34, 85, 51, 52). In *Dinornis,* as in *Murœna* and *Anguis,* the arch has no appendages. The scapula is rib-like, compressed, slightly bent, measuring but 4½ inches long in a species (*D. robustus*) with a tibia a yard long; it is barely an inch across its broadest end where it coalesces with the coracoid, and the breadth of the opposite free end is but 5 lines. The coracoid is straight, 2 inches 10 lines long, ½ inch broad, becomes thicker to its sternal end, which is convex and adapted to the small 'coracoid' fossa at the angle of the sternum. There is no trace of glenoid cavity at the confluence of the two bones, but the confluent part is here produced into a ridge, showing that there was no humerus, and that the fore-limb, or appendage of the scapular arch, was wholly absent in *Dinornis.*

In *Apteryx* the scapula is relatively more expanded where it coalesces with the coracoid, and the bone is broader in proportion to its length, and shows a vascular perforation near the humeral articulation, as in the Monitor (vol. i. p. 174). The glenoid cavity is very small, but of the usual shape in Birds. In these the scapular arch includes on each side a *scapula,* fig. 19, 51, a *coracoid* bone, ib. 52, and a *clavicle,* ib. 58—the clavicles, coalescing in most birds at their mesial extremities, constitute a single bone, which, from its peculiar form, is termed the *os furcatorium* or *furculum.* In the Ostrich the two clavicles are distinct from each other, but are severally anchylosed with the coracoid and scapula, so as to form with them one bone on either side. In the Frigate-bird the clavicles coalesce with the coracoids, as well as with each other and with the sternum. In almost every other species of bird the scapula, coracoid, and clavicle are moveably articulated to each other throughout life. In *Rhea* and *Casuarius* the acromial element or clavicle is anchylosed with, or rather is a continuous ossification from, the scapula; but the coracoid bone is free, a condition which the bones of the shoulder present in the Chelonian Reptiles (vol. i. p. 172, fig. 106).

In the Emeu (*Dromaius*) it is interesting to observe that each clavicle commences by a distinct ossification, and long continues separate from the scapula; it does not reach the sternum, but holds the same relative situation as the continuous acromial or clavicular process of the scapula in the other Struthious birds. The clavicles are distinct from each other and from the coracoid in some Ground Parrots and carpophagous Doves (*Columba galeata*, e. g.).

The *scapula*, fig. 19, 51, is broader and flatter in the Penguins (*Aptenodytes*) than in other birds. In the rest of the class it is a long and narrow sabre-shaped bone, increasing in thickness as it approaches the joint of the shoulder; there it is extended in the transverse direction, forming externally the posterior half of the glenoid cavity, and being internally more or less produced, acromially, to meet the clavicle, while it is strongly attached in the remainder of its anterior surface to the coracoid. The blade of the scapula may expand towards the free end (*Gallinæ*); and this may be obliquely truncate (*Gallinæ*), or taper to a point (most *Aves*), which point may be decurved (*Columba*); it is rarely obtuse (*Tetrao, Apteryx, Dinornis*). The position of the scapula is longitudinal, being extended backward from the shoulder, parallel to the vertebral column, towards which, however, it presents a slight convexity. In some birds it extends over the ribs to, or even above, the fore part of the ilium; while in the Emeu and *Apteryx* it crosses over two ribs only. In the Humming-bird (*Trochilus*), fig. 18, *t*, its posterior third is bent downward at a slight angle. In birds where the scapula is pneumatic, the perforations are at the base of the acromial process.

The *coracoid*, fig. 18, *u*, figs. 16, 19, 20, 52, is the strongest of the bones composing the scapular arch: its expanded extremity is securely lodged below in the transverse groove at the anterior part of the sternum, from which it extends upward, outward, and forward, but sometimes almost vertically, to the shoulder-joint, where it is articulated usually at an acute angle with the scapula and commonly also with the clavicle. It thus forms the main support to the wing, and point of resistance to the humeri during the downward stroke of the aerial oar. The humeral end of the bone is commonly bifurcate; the outer process is the strongest, and forms the fore part of the glenoid cavity (*l*, fig. 19), above which it rises, to a greater or less extent, and usually affords, on its inner side, an articular surface for the clavicle: the inner process is short and compressed, articulates with the scapula, and is also joined by ligament to the end of the clavicle. The coracoid is perforated at the base of the inner process. The coracoid is of great breadth in the Albatross, fig. 13, *h*, *b*; and is both long and strong in the Penguin, fig. 19, 52. It is pneumatic in *Aves aereæ* and in *Rasores*; in some *Grallæ* (*Psophia*), and in most longipennate Palmipeds. The sternal ends of the coracoids join each other in *Tachypetes*, decussate in Herons, send up a process above the mesial end in *Aptenodytes* and above the lateral or outer end in *Tachypetes*: the outer angle of the sternal end is produced in *Raptores*. The glenoid cavity resulting from the union of the coracoid and scapula is not equal to the reception of the entire head of the humerus. In *Raptores, Scansores,* and *Cantores,* an ossicle (*Os humero-scapulare*) lies between the scapula and humerus at the upper and back part of the glenoid cavity. In *Rasores, Grallatores,* and *Natatores,* there is, in place of this bone, a strong elastic ligament or fibro-cartilage extended between the scapula and coracoid, against which that part of the head of the humerus rests, which is not in contact with the glenoid cavity.

The *clavicles,* figs. 15, 16, 18, 19, 58, are the most variable elements of the scapular apparatus. In the Ground Parrots of Australia (*Pezophorus,* Illiger) they are rudimentary or wholly deficient; they are slender styles in *Columba galeata*; they are represented by short processes in the Emeu, Rhea, and Cassowary; they do not come in contact inferiorly in the Ostrich, although they reach the sternum. In the Toucans they are separate, and do not reach the sternum. In the Hornbills and Screech Owl (*Strix Ulula*) they are united at their inferior extremities by cartilage. In the rest of the class they are anchylosed together inferiorly, and so constitute one bone, the *furculum* or 'merry-thought.' From the point of confluence a compressed process extends downward in the *Diurnal Raptores,* the *Conirostral Cantores,* the *Rasores,* most of the *Grallatores,* and *Natatores,* in which a ligament extends from its extremity to the ento-sternum. The process itself reaches the sternum, and is anchylosed therewith in the Pelicans, Cormorants, Grebes, Petrels, Frigate-bird, and Tropic-bird; also in the Gigantic Crane, and Storks in general. In the Humming-birds, where the sternum is so disproportionately developed, the furculum terminates almost opposite the commencement of the keel, but at some distance before it; it is of equal length with the coracoid. As the principal use of this elastic bony arch is to oppose the forces which tend to press the humeri toward the mesial plane during the downward stroke of the wing, and restore them to their former position, the piers of the arch are stronger, and the angle of their union is more open, as the powers of flight are enjoyed in greater perfection: of this adjustment the Swifts, Goat-suckers, and Diurnal Birds of Prey afford the best examples. In the Eagle the clavicles are arched both forward and outward, much expanded above, with an articular surface for the fore part of the outer prong of the coracoid. The arch becomes narrower, and the bone itself weaker, as flight is feebler or less sustained; in the *Gallinæ* the U- is changed to the V-shape; and at the point of confluence of the straight and slender piers a process is continued, usually compressed, sometimes styliform (*Crax*), becoming almost obsolete in *Hemipodius*; in the Lapwing the process is at right angles to the arch. In *Tachypetes* the upper ends of the clavicles coalesce with the coracoids: and the lower confluent ends expand into a triangular plate coalesced with the sternal keel. In the crested Pintado the apex of the furculum is dilated and hollowed into a cup opening forward and receiving a fold of the windpipe.[1]

In Birds the *humerus* has a smooth shaft, sub-elliptic in transverse section, with expanded ends, the proximal[2] one being the broadest. Lengthwise the bone is gently sigmoid, the proximal half being convex palmad, the distal half concave, with the plane of the terminal expansions vertical, as the bone extends along the side of the trunk from its scapulo-coracoid articulation backward, in its position of rest.

The head of the humerus is an elongate, semi-oval convexity with the long axis transverse from the radial to the ulnar sides (vertical, as naturally articulated), and with the ends continued

[1] XLIV'. no. 1411, p. 271.

[2] I here avail myself of the terms indicative of aspect and position proposed by Dr. Barclay, in his 'Anatomical Nomenclature.'

Proximal signifies the upper, *distal* the lower, end of the bone, as it hangs in Man; *anconal* is the posterior, *palmar* the anterior, surface, as when the palm of the hand is directed forward; *radial* is the outer, *ulnar* is the inner, side, according to the same position of the human arm and hand. *Proximad*, *palmad*, are adverbial inflections, meaning towards the proximal (upper) end, and towards the palmar (anterior) side.

into the upper and lower crests. Of these, the upper one, in the natural position of the bone, is on the same side as the radius, the lower more tuberous one is on the same side as the ulna; the one marks the 'radial' side, the other the 'ulnar' side, of the bone. The side of the humerus next the trunk answers to that called 'anconal,' the opposite side to that called 'palmar.' The expanded, proximal part of the shaft on the palmar side, fig. 6, is concave across, convex lengthwise: on the anconal side it is convex across to where the ulnar ridge bends anconad near the pneumatic orifice. The radial crest answers to the 'greater tuberosity,' and to the 'pectoral' and 'deltoidal ridges' in mammals; the 'ulnar' crest to the 'lesser tuberosity' and to the ridge for the 'latissimus dorsi,' in mammals. In a few exceptions the shaft of the humerus is almost cylindrical; in still fewer (*Aptenodytes*) it is flat; in the Albatross it becomes triedral toward the distal end.

In the Vulture (*V. monachus*) the ulnar crest forms a thick tuberosity at its proximal end, projecting anconad, and overarching the 'pneumatic' foramen; it descends a short way obliquely palmad, decreasing in breadth, but still thick, convex, and terminating obtusely. The radial crest better merits the name; it extends twice the length of the ulnar one, down the shaft, to the palmar side, towards which the whole crest is slightly bent; its margin describes a very open or low, obtuse, angle at its middle part. A ridge upon the palmar side of its distal half indicates the boundary of the insertion of the pectoralis major into the crest. At the middle of the anconal surface of the proximal part of the shaft there is a low, longitudinal ridge. The tuberosity at the proximal part of the ridge gives insertion to the middle pectoral.

At the distal part of the humerus a ridge on the radial side of the palmar surface, and a rising of the bone on the ulnar side of the same surface, diverge to the opposite angles or tuberosities of the expanded end of the bone; they include a shallow, subtriangular concavity above the articular surfaces. These are two, and are convex. The radial surface is a narrow, subelongate convexity, extending from near the middle of the palmar surface obliquely to the lower part of the radial tuberosity, where the convexity subsides; it is very prominent at its palmar end, with a groove on each side, the deeper one dividing it from the ulnar articular convexity. This is of a transversely oval or elliptical shape, most prominent palmad; all the part of the end of the humerus forming the two articular convexities is as if bent toward the palmar aspect. The ulnar end of the ulnar convexity is continued anconad to that end of the ulnar tuberosity. An oblique, longitudinal channel divides the anconal end of the radial tuberosity from an almost longitudinal ridge, which is nearer the middle of the anconal side of the distal end of the humerus; a similar, but shorter, longitudinal ridge or rising of bone, terminates in the anconal part of the ulnar tuberosity. Between the above almost parallel ridges the anconal surface is nearly flat transversely; it is traversed along the middle by a low, narrow, longitudinal ridge. Lengthwise the bone is here convex.

The differences in the humerus of different birds are seen chiefly in the forms and proportions of the proximal crests; the radial one in the *Columbidæ*, e.g., is shorter and more produced than in most birds of flight. The humerus in the Swift is very short and thick, with strong pectoral and deltoid processes, and with a trochlear groove on the back of the outer condyle: the proximal processes are still further developed in the humerus of the Humming-bird: a distinct ulnar sesamoid plays upon the anconal trochlea in both. The pectoral process is much produced and is angular in *Tachypetes*: also in *Sterna* where it is deflected. The bone maintains its general ornithic character when the processes have subsided, with the abrogation of the power of flight. In the *Apteryx* the humerus is a slender, cylindrical, styliform bone, 1 inch 5 lines in length; slightly expanded at the two extremities, most so at the proximal end, which supports a transverse oval articular convexity, covered with smooth cartilage, and joined by a synovial and capsular membrane to the scapulo-coracoid articulation. A small tuberosity projects beyond each end of the humeral articular surface. The distal end of the humerus is articulated by a true but shallow ginglymoid joint with the rudimental bones of the antibrachium, and both the external and internal condyles are feebly marked.

The humerus is not always developed in length in proportion to the powers of flight; for, although it be shortest in the Struthious Birds and Penguins, it is also very short in the Swifts and Humming-birds. In the latter, however, it is characterised by its thickness and strength, the size of its muscular processes, and the consequent transverse extension of its extremities; while in the *Cursores* it is as attenuated as it is short, and in the Penguins the shaft is reduced to a mere lamina of bone resembling the corresponding part in the paddle of the turtle. In the *Rasores* it rarely equals half the length of the trunk (thorax and pelvis); it equals it in the Argala and Pelican; exceeds it in the

Frigate-bird and Albatross. In this and some other sea-birds, as the Gulls, Awks, and Petrels, the humerus presents a notable 'ectocondyloid' process on the radial side, near its distal extremity.

A sesamoid bone is found attached (like the fabella of Marsupials) to the capsular ligament and the tendons of the extensor muscles, in many of the *Raptores*, in the Swifts and Humming-birds; it is double in the Guillemots and Penguins, fig. 19, *n*, *n*.

There is a deep depression beneath the tuberosity at the ulnar side of the proximal expansion of the humerus in the Penguins, Ostrich, Awks, and other birds which have no air in that bone; and it is here that the air-cells are continued into the bone in the majority of the class which have the humerus pneumatic.

A section of the humerus of the Penguin[1] shows it to be solid; that of the Awk (*Alca*) shows a small medullary cavity with dense and rather thick walls; that of a pneumatic humerus, as of the Argala,[2] e.g., exposes the extreme thinness of the compact wall of a very large cavity, and the loose cancellous lacework at the extremities of the bone: osseous filaments shoot more or less obliquely across different parts of the cavity, serving to strengthen, like tie-beams, the thin walls, and also, being hollow, to convey minute blood-vessels. The proximal half of the bone is divided longitudinally by a loose cancellous partition; the decussation or anastomoses of the delicate hollow columns give an open reticulate structure to the inner surface of the air-cavity at the two extremities of the bone, which is highly characteristic of the long bones of birds.

The radius, fig. 19, *p*, and ulna, ib. *o*, are present in all birds, and co-extended between the joints of the elbow and wrist. The antibrachium, so formed, is short where flight is abrogated; it is but one-third the length of the humerus, e.g., in the Ostrich: it is rather shorter in Guillemots, Divers (*Colymbus*), and Gannets (*Sula*); is about equal in length in *Gallinæ*, Psophia, Cariama; but exceeds the length in most birds of flight. However, in some of the best flyers, e.g. the Albatross where the humerus is extremely long, and in the Swifts and Humming-birds where it is very short, the antibrachium is of the same length therewith. Its two bones are so articulated to each other and to the humerus as to be restricted to flexion and extension: scarcely any degree of rotation is admitted, and this adds to the firmness and resistance required for the action of flight. Owing to the obliquity of the

[1] XLIV'. p. 219, no. 1137. See also, no. 1373, Ostrich. [2] Ib. no. 1108.

humeral tubercle for the radius, the fore-arm moves in a plane not quite perpendicular to the palmar surface of the humerus. When the fore-arm is flexed and the wing is folded, the distal end of the antibrachium is near or in advance of the proximal end of the humerus; the radius being superior and the ulna a little external as well as inferior.

The *radius* is always the more slender bone of the two, sometimes in a remarkable degree: its proximal end is expanded, subelliptic, with a concavity for the oblique tubercle, and a thickened convex border next the ulna for articulation with that bone: a little beyond that articular expansion is the tubercle for the insertion of the biceps. The shaft here becomes slender, usually subcompressed, with a slight bend, convex upward from the ulna; the rest of the shaft, which becomes subtriedral, showing an opposite flexure toward the ulna, though very slightly marked. The distal end is rather more, though less equally, expanded, from the radial to the ulnar side: rather flattened with one or two tendinal grooves on the anconal side, with a terminal transverse convexity for the scaphoid, produced palmad to articulate with the ulna; with a tuberosity (*Aquila*) or ridge (*Tachypetes*) on the radial side of the expansion. The orifice of the medullary artery in the non-pneumatic radius is on the ulnar side of the shaft about one-fourth from the proximal end.

The *ulna* is straight or with a single and slight curve, more marked in the shorter antibrachium of *Gallinæ* than in the long one of long-winged waders and swimmers. The proximal end is most expanded, and is obliquely truncate for the articular excavation adapted to the ulnar tubercle of the humerus: the obtuse angular production of the ulna, behind or anconad of the cavity, represents in different degrees in different birds the olecranon, but is always short: an extension of the bone radiad is obliquely excavated for the head of the radius. The shaft of the ulna gradually decreases to near the distal end, where the subtriedral is exchanged for the subcylindric shape. A ridge is developed below the head on the ulnar side in *Raptores*. In birds (*Tachypetes*, e.g.) in which the ulna is pneumatic, the foramen is on the palmar surface a little below the head. On the ulnar and anconal sides of the shaft are the two rows of quill-knobs (in *Raptores*) for the 'secondaries;' the anconal row is most marked in longipennate *Natatores*; and is the only row in many birds. But this character of the bird's ulna is wanting in the flightless and some other birds. The distal end of the ulna slightly expands into a trochlear joint very convex from the radial to the ulnar side, rather concave from the anconal to the palmar side, and this chiefly at the ulnar part of the trochlea. On the radial side of this trochlea, supported by a tuberosity, is the small surface for the radius. The interosseous space, owing to the greater bend of the ulna, is widest in *Gallinæ*; it is narrower and chiefly seen at the proximal half of the antibrachium in most other birds. The ulnar trochlea articulates with the two free carpal bones, one—the 'scapho-lunar'—being wedged into the radial part, the other—'cuneiforme'—into the ulnar part, leaving a small intermediate tract for the 'magnum' which is confluent with the base of the mid-metacarpal.

In the young Ostrich the *metacarpus* consists of three bones. The one on the radial side answers to that of the index-finger; it is very short, and supports a digit of two phalanges, the second phalanx being armed with a long curved and pointed claw. The second metacarpal is the longest and largest, its base being increased by the confluence therewith of the 'magnum,' which presents a trochlear surface to the two proximal carpals and to the part of the ulnar joint not occupied by them. The third metacarpal, answering to that of the digitus annularis, is bent, its extremity resting against that of the large and straight middle metacarpal, with which it subsequently becomes anchylosed. The middle digit consists of three phalanges; the outer one of two phalanges. In all birds the three metacarpals, here seen to be distinct, coalesce with one another and form a single bone, having an interesting analogy to the metatarsus, which likewise consists in all birds of a coalescence of the three bones supporting the corresponding toes, namely, those answering to the second, third, and fourth in the pentadactyle foot.

The bones of the hand are developed in length, but contracted in breadth. The wedge-like adjustment of the free carpals is such as to restrict the movements of the hand upon the arm to abduction and adduction, or flexion in the ulno-radial plane, requisite for the outspreading and folding up of the wing. The hand of the bird moves thus in a state of pronation, without the power of rotation or of proper flexion or extension, i.e. in the ancono-palmar direction; so that the wing strikes firmly and with the full force of the depressor muscles upon the air.

The following state of anchylosis commonly exists in the metacarpus:—The short 'index' metacarpal coalesces with the base of the 'medius': the slender 'annularis' metacarpal anchyloses by its two ends with those of the medius which it equals in length. The 'index' supports one phalanx, usually terminating in a point about the middle of the 'medius' metacarpal. This supports two phalanges, fig. 19, *s*, *s*: the proximal one singularly expanded by a lamelliform growth from its whole ulnar side, excavated outwardly for the attachment of primaries: the next phalanx is smaller and ends in a point. The 'annularis' metacarpal supports a short and slender pointed phalanx, which in the Frigate-bird is closely joined, lengthwise, to the contiguous expanded phalanx of the mid-digit.

The hand-segment is the longest of those of the pectoral limb in Swifts and Humming-birds: exceeding by three times the length of the humerus: and the bones have a proportionate thickness. The mid-metacarpal shows a series of large impressions for the distal 'primaries' in *Raptores*, and also a longitudinal tendinal groove on the anconal side. The metacarpal, fig. 19, *r*, and phalangeal, ib. *s*, *t*, bones, in the Penguin are flattened, like the antibrachial bones.

The index digit in *Struthio* and the medius digit in *Apteryx*, support each their claw. The claw or spur, when present in other birds, e. g. Syrian Blackbird (*Merula dactyloptera*), Spur-winged Goose (*Anser Gambensis*), Knob-winged Dove (*Didunculus*), Jacana (*Parra Jacana*), Mound-bird (*Megapodius*), Screamer (*Palamedea*), is developed from the radial side of the metacarpus or from the index digit. The Screamer has two spurs, the homotypes of the metatarsal ones in *Pavo bicalcaratus*. The claw upon the index digit of *Archeopteryx* was curved and sharp; and the remains of the unique example of this ancient fossil bird make it probable that the hand had a second free unguiculate digit, perhaps the homologue of the pollex.[1]

Although the instances of these weapons and the occasional use of the wings in Birds not so armed, e. g. the Swan, show them in the light of means of attack, the bones of the pectoral limb in Birds are modified mainly for volant action; the articulations restrict the movements of the several segments to the service of wings, and the processes for muscular attachments relate to such development and disposition of the moving forces as flight requires.

The larger feathers which overlie, in a series, the humerus, are termed, in ornithology, 'scapularies:' those still larger which overlie or are attached to the ulna are the 'secondaries' or 'wing-coverts;' those which are attached to the manus are the 'primaries,' they are the longest: a group of feathers attached to the stunted index digit are the 'spurious' or 'bastard' feathers.

[1] xv'. p. 39, pl. 2, fig. 1.

The primary quill-feathers being the chief direct mechanical instrument in the displacement of the air, the segment of the limb supporting them is the longest and strongest in the most powerful flyers, e. g. Swifts and Hummers, in which the primaries are proportionally longer and stronger than in other birds: but the various habits, habitats, and food of the feathered tribes are associated with different kinds of aerial motion and call for corresponding modifications of the instrument: thus the Frigate and Tropic birds, Albatrosses, Terns, and other ablest flyers among the *Natatores*, contrast strangely with the above-cited *Volitores* in the proportionate length of the brachial and antibrachial segments of the pectoral limb: whilst the powerful Raptorial flyers show an intermediate more harmoniously balanced proportion of the several segments. All these are relatively short and feeble in the heavier land birds which take but brief and occasional flights; and, as circumstances have rendered this exertion less and less necessary, so the wings and their framework have wasted away to the diminutive rudiments in the *Apteryx*, and to zero in *Dinornis*.

§ 130. *Bones of the pelvic limb.*—The segments of this limb do not wholly correspond with those of the pectoral one, the tarsus being absent or blended with the tibia or the metatarsus, which immediately succeeds it.

The *femur*, fig. 34, 65, has a cylindrical shaft, which, when not straight, is slightly bent forward: it nearly equals the pelvis in length in the *Apteryx* and some Ground-cuckoos (*Geococcyx*), but is usually shorter; it is very short in *Dinornis elephantopus*; shortest of all and most bent in *Colymbus*: it is always shorter than the tibia, but in a minor degree in most *Rasores, Scansores, Volitores, Cantores*, some *Natatores* (*Tachypetes*), and the *Apteryx*. The head is hemispherical, proportionally small, and largely scooped out above for the round ligament which fills up the vacuity in the acetabular wall: it is sessile, with its axis nearly at right angles to that of the shaft: the articular surface is continued upon the upper end of the bone which expands as it recedes from the head, and usually rises above its level to form a trochanterian ridge extending from behind forward and there produced and continued a short way down the shaft. The outer (fibular) side coextensive with this ridge is rather flattened and impressed by insertion-marks of muscles. Rarely is there, as in *Aptornis*, a trochanter minor, situated a little below the head on the inner (tibial) side of the bone, or represented by a round rough surface, more anterior, as in *Dinornis*. Assuming its cylindrical or subcylindrical form below the great trochanterian ridge, the shaft at its lower half expands transversely, and, in forming the distal condyles, also from before backwards, with a bend in that direction. The inner condyle begins anteriorly as a ridge, expanding into a convexity which attains its greatest breadth posteriorly, where it becomes more flattened. The outer condyle, commencing in the same manner, is indented at its broad lower end by an angular groove, which, widening, divides the back part of the condyle into two convexities. The inner of these is the broadest and most produced, is applied to the outer facet of the tibia, and represents the ordinary outer condyle: the more external convex ridge and the groove dividing it from the outer condyle are adapted to the head of the fibula. This is the most characteristic part of the bird's femur. The space between the anterior beginnings of the condyles is the 'rotular' channel: it is usually broad and moderately concave transversely, convex lengthwise; sometimes divided from, commonly continued into, the intercondyloid fossa which is marked with pits for ligamentous attachment. The inner side of the inner condyle is flattened, with a tuberosity at its mid-part, and sometimes a second just above the hind part of the condyle. There is usually a tuberosity above the hind end of the fibular ridge, exterior to which the surface is sometimes flattened, sometimes prominent, in *Dinornis* impressed by a deep fossa.[1] At the lower part of the outer condyle before the 'fibular' groove begins, there is usually a small pit. The popliteal depression is divided by a ridge from the intercondyloid one. The shaft shows intermuscular linear ridges: in *Aquila* one extends from the fore and outer angle of the epitrochanterian articular surface to near the beginning of the inner condyle; a second extends from the inner and back part of the upper third of the shaft to the tubercle above the back part of the inner condyle: the third shorter 'linea aspera' is at the back part of the middle third of the shaft near its outer side. In *Dinornis* a ridge continued from the anterior trochanterian one bifurcates at the middle of the fore part of the shaft diverging to the beginnings of the two condyles. In *Apteryx* the two posterior lineæ asperæ approximate at the middle of the shaft and then diverge to the condyles: in *Dinornis* they expand into tuberosities, or the inner one alone is continued as a ridge, but interrupted above the condyle: the inner ridge is strongly marked and continued to the condyle in *Aptornis*. The orifice of the medullary artery is at the back part of the shaft above its middle.

[1] In *Dinornis maximus* the femur is 16 inches in length, and 6½ inches across the distal end.

When the femur is pneumatic the proximal orifice is commonly anterior, near the trochanterian ridge: but in the Ostrich it is behind: the distal orifices when present are in the popliteal fossa. Of the two condyles the outer one is most prominent posteriorly, and, when the femur is held vertically, descends the lowest. In the flexion of the leg on the thigh this puts the ligaments on the stretch; and, as they are partly elastic, the fibula enters its fossa at the conclusion of the bend with somewhat of a jerk.

The *tibia*, fig. 34, 66, is the chief or longest bone of the hind limb, showing its extreme character in this respect in most Stilt-birds, especially the Argala and Flamingo, fig. 14, and its smallest proportions in *Volitores*, fig. 18, and the Frigate-bird, in which the tibia is not half the length of the skull. The shaft is straight and mainly subtriedral, expanded at both ends and most so at the upper one. This presents a semi-oval surface not quite transverse to the shaft, but with the truncate margin raised toward the fore part of the bone, and more or less developed above the level of the undulating articular part. Of this the least marked is the almost flat reniform 'entocondylar' surface for the inner condyle, feebly hollowed near its back part which projects in that direction over the shaft; it is divided from the smaller and less defined 'ectocondylar' surface for the outer femoral condyle by an 'intercondylar' convexity. In advance of these the head of the tibia extends into a 'rotular' process, usually extended transversely and truncate. From the fore or outer part of this process there descend two vertical ridges or plates: the one from near the inner or tibial angle of the rotular process is the 'procnemial ridge,' the other from the outer or fibular angle is the 'ectocnemial ridge;' they subside more or less gradually upon the shaft, and intercept a deep triangular concavity. On the outer side of the intercondylar tuberosity is a single surface for ligamentous union with the head of the fibula, and a little way below this there is a vertical ridge for close attachment to part of the shaft of that bone: below and behind the 'fibular' ridge is the orifice of the medullary artery. From this point the tibia maintains a uniform size usually to its lower third, where along a rough tract, in a line with the above ridge, the styliform end of the fibula terminates by close union or anchylosis. There the tibia begins to gain in transverse breadth, exchanging the triedral for a transversely oval section, and it gradually expands in both directions to the distal condyles, which are most developed from behind forward, in advance of the shaft. The inner condyle is the largest, usually in fore-and-aft, sometimes (*Aquila*) in transverse, extent. A groove commencing near the lower end of the fore part of the shaft leads, deepening, toward the intercondyloid space: it is bridged over by a strong ligament, which becomes ossified in most birds: Parrots, Hornbills, and existing *Cursores* are exceptions. The position—median or submedian and direction—straight or oblique—of the precondylar groove, the presence and direction—transverse or oblique—of its bony bridge, the relative breadths, anteroposterior and transverse, of the distal end, the relative size of the intercondyloid space to the anterior parts of condyles,—help in the determination of bird-affinities, when they have to be deduced from a fossil tibia.[1] The distal condyles commencing behind as ridges bounding an articular surface con-

34

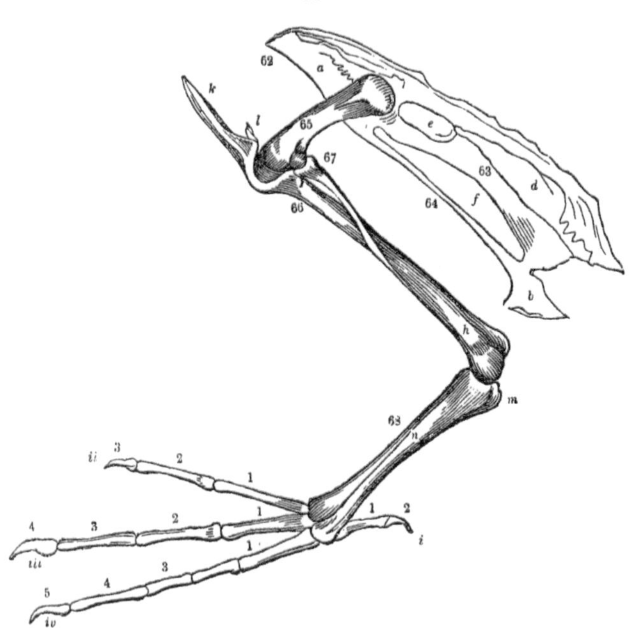

Pelvis and bones of the leg, Loon (*Colymbus*).

cave across, increase in breadth and convexity as they curve to their anterior ends: these are more prominent than their posterior ridged beginnings, but in different degrees in different birds; and the inner condyle is usually the most prominent anteriorly:

[1] See XIX'. p. 204, pl. 3, for the illustration and application of the characters of the tibia in the determination of the affinities of birds indicated by that bone fossil.

thus the distal end of the tibia is like that of the femur with the back of the condyles turned forward, and without the notch in either.

Among the modifications of the proximal end of the tibia may be noted the production of the rotular process in the axis of the shaft two inches above the knee-joint in the Divers (*Colymbus*), fig. 34, *k*; both pro- and ecto-cnemial ridges descend from the fore part of the base of this process, the former extending half-way down the shaft of the tibia. In the Albatross the pro- and ecto-cnemial ridges are much developed; but are still more so in the extinct *Cnemiornis*, without corresponding production of the rotular or 'epicnemial' process. In the Ostrich this process extends forward, without rising above the level of the proximal surface, and contracting to its termination there divides into small pro- and ecto-cnemial processes; the latter the shortest and tuberous. The distal condyles are less produced anteriorly, commence more abruptly and are more produced posteriorly, than in other birds: their articular surfaces are so continuous as to leave no 'intercondylar' space; there is no tendinal groove or bridge: but a tuberosity above the middle of the confluent condyles. The articular surface of these being concave in one direction, convex in the other, forms a 'trochlea,' and the same in the conjoined parts of the distal condyles in other birds. It limits the movements of the next segment of the limb to one plane.

The *fibula*, fig. 34, 67, is a styliform bone ending in a point below at various distances down the tibia in different birds. The articular head is subcompressed, convex in the longer axis, slightly curved backward, hollowed on the inner (tibial) side: rather convex externally: the shaft shows the rough linear tract for attachment to the tibia: and there are sometimes tuberosities for tendinal insertions on the opposite side.

The femur is ossified from one centre: the tibia has an epiphysis for the distal condyles; the proximal end of the metatarsus is ossified from one centre, forming an epiphysis which caps the ends of the three metatarsals that coalesce, first with each other, then with the epiphysis, to form the single compound bone.

The trochlear epiphysis of the tibia most resembles the astragalus in those mammals (Ruminants, e.g.) in which the metatarsals coalesce. The term 'tarso-metatarse' applied by some ornithotomists to the present segment, fig. 34, 68, implies the tarsal homology of the epiphysis; the same might, more probably, be predicable of the distal one of the tibia; but neither being demonstrated, I prefer to call the present segment the 'metatarse.' It consists of the foot-bones of three digits coalesced, and often of a fourth ligamentously joined thereto. This always small and short seeming appendage is the distal end of the metatarsal of the 'hallux' or innermost digit of the pentadactyle foot. The three coalesced bones are the metatarsals of the second (ii), third (iii), and fourth (iv) toes, fig. 34. In their original position the proximal end of the third metatarsal is behind those of the second and fourth which meet in front of it. A fossa below the meeting shows, afterwards, two fore-and-aft canals which diverge to outlets at the back part indicative of the breadth there of the middle metatarsal. When, as in *Aquila*, there are two foramina in front of, as well as two behind, the upper part of the metatarse, the interspace of the former shows the extent to which the mid-metatarsal intervened between the others anteriorly, and this structure is concomitant with a great excess of the transverse over the fore-and-aft diameter of the proximal end of the metatarse. In most birds a fore-and-aft canal also remains to indicate the primitive distinction of the outer (iv) from the middle (iii) metatarsal near their distal ends.

The metatarse, fig. 34, 68, presents a proximal end with two articular cavities ('ento-' and 'ecto-condylar') and the intercondylar space, a shaft with its processes, grooves, and perforations, and a distal end divided (save in *Struthio*) into three trochlear condyles for the three principal (ii, iii, iv) toes: in most birds, also, there is a rough depression on the distal half of the inner metatarsal, for that end of the innermost or first (i) metatarsal. The proximal end varies in the proportions of its transverse and antero-posterior diameters, in the depth of its articular surfaces, and configuration of the intercondylar surface. As a rule the entocondylar surface is largest and deepest; the ectocondylar surface is nearly flat in the Eagle. A tuberosity rises from the fore part of the intercondylar space in birds which sleep standing on one leg (*Grallæ* and some others): it passes into the corresponding space of the tibia, the bar anterior to which affords so much resistance to flexion of the leg as counteracts the effect of oscillations of the body: it requires a muscular effort to bring the tuberosity over that bar, and the elastic lateral ligaments are then put on the stretch; but as soon as the bar is passed the tuberosity slips into the depression above with a snap or jerk. One or more longitudinal ridges at the back of the upper end of the metatarsal are called 'calcaneal;' they intercept or bound tendinal grooves which, in some instances, are bridged over by bone and converted into canals: the ridges may be expanded and flattened.[1] In the Birds of Prey the metatarsal is most modified by the muscles and tendons operating upon the raptorial toes. There are three calcaneal processes, the innermost large, the two outer ones small and approximate. The fore part shows a tuberosity for the insertion of the strong 'tibialis anticus' (fig. 35, 48): below this is the process on the inner margin extending the surface of attachment for the metatarsal of the back toe (*i*). The trochlear ends of the three confluent metatarsals are nearly on the same level, the inner one is the broadest, the outer one the narrowest: each is produced, at an opposite angle, so as to bound the wide concavity behind this end of the metatarsal. In the King-Vulture (*Sarcoramphus*), the mid-trochlea is broadest and most produced: in the Snake-Vulture (*Gypogeranus*) with a metatarsal of stilt-like length, the inner trochlea is shorter than the others and further apart. In most Owls the metatarsal is shorter in proportion to its breadth than in diurnal *Raptores*; a bony bridge overspans the beginning of the tendinal canal on the fore part: the outer trochlea is the shortest and is bent backward and inward. In most *Cantores* and *Volitores* the distal end of the metatarsal is little expanded, and the three trochleæ are of nearly equal length: in *Podargus* and *Dacelo* the outer trochlea is the shortest. In *Cypselus* the trochleæ terminate on the same transverse line: in *Trochilus* the middle one is a little more prominent. In the short and strong metatarsal of the Parrot-tribe, the middle trochlea extends wholly below the others, which are oblique and twisted, especially the outer one, backward and inward: a like twist is noticeable in most *Scansores*, especially the Woodpeckers and Cuckoos. In the spurred *Gallinæ* the weapon is supported on a conical process from the back part of the metatarsal; sometimes there are two, as in *Pavo bicalcaratus*. In all *Rasores* the mid-trochlea is longest; in Pigeons and Curassows the outer (*iv*) is shorter or higher than the inner (*ii*) trochlea: in the Tinamou and Syrrhaptes it is longer. In the Apteryx and tridactyle *Cursores* the mid-trochlea is largest, and extends by almost its whole length beyond the other two which are nearly on a level. In *Struthio* the inner metatarsal (*ii*) terminates in a point near the base of the great trochlea (*iii*): the outer trochlea (*iv*) is comparatively small and short. In *Grallatores* the mid-trochlea is longest, the other two of equal or nearly equal length in most: the inner (*ii*) trochlea is the shorter in the Demoiselle Crane (*Scops Virgo*); the outer one

[1] XVI'. vol. iii. part iv. (1846), pp. 321, 322; XLIV'. p. 274, &c.

(*iv*) in the Woodcock (*Scolopax*). In the Spoonbill (*Platalcea*) and Flamingo (*Phœnicopterus*) the mid-trochlea is but little produced beyond the others. The surface for the attachment of the innermost metatarsal (*i*) is raised well above the trochlear end of the next (*ii*) in most of those Waders that have the back-toe. Amongst *Natatores* the Albatross has the three trochleæ nearly on the same level: in others the mid one is usually most produced: in the Gannet and Pelican the outer trochlea is the shortest and furthest from the middle one. In the Guillemot (*Uria*) the inner trochlea (*ii*) ends at the base of the mid one, whilst the outer trochlea is of nearly equal length with the mid one: in the Grebe (*Podiceps*) the inner trochlea is the longest of the three: in the Loon (*Colymbus*) it is the shortest: the metatarsal in this bird is much compressed, and the outer and inner trochlea are bent backward. The Penguins show the most instructive modification of the metatarsal: it is very short and broad; but the primitive divisions are to a great degree retained, especially on the fore part of the compound bone. The stunted or abortive metatarsal supporting the backwardly directed toe consists of a trochlear articulation supported on a compressed stem, twisted, and obtusely pointed above, with one margin thickened and rough for syndesmosis with the next anchylosed metatarsal. This bone is best developed in the Raptorial birds and the Dodo,[1] in which its length may exceed a fourth part of the length of the metatarsal segment: in Pigeons it is about the fifth or sixth part that length: by the twist the bone forms a pulley upon which the flexor tendon of the back toe plays. In the Dodo the distal expansion is increased by an obtuse process from the outer side of the trochlea; and this character is repeated in the *Columbidæ.*[2] In *Gallinæ* the trochlea is less expanded and the twist is feebly shown: it disappears in the still smaller loose metatarsal of *Apteryx*, *Palapteryx*,[3] and in many Natatores, in some of which it is represented by the ligamentous matter tying the short back toe to the metatarse.

Both this (*i*) and its metatarsal are undeveloped in the larger existing *Cursores*, the Bustards (*Otis*), the Plovers (*Charadicus*), the Thick-knees (*Œdicnemus*), the Oystercatchers (*Hæmatopus*), the Coursers (*Tachydromus*), and the Albatrosses (*Diomedæa* and *Haladroma*).

The toes of birds never exceed four in number, and of these, three are usually elongated and directed forward, diverging, while

[1] XXVI'. [2] XXVII'. pl. 11. [3] XVI'. vol. iii. p. 307.

one is short and directed backward. The hind toe articulates on the same plane as the others in grasping and perching birds, but on a higher level in terrestrial and aquatic kinds. By the analogy of the number of the phalanges of these toes with those in Lizards (vol. i. p. 192, fig. 122) the back toe, fig. 34, *i*, is the innermost, answering to 'hallux;' the inner of the front toes, ib. *ii*, is the second; the middle one, ib. *iii*, is the third: the outer one, ib. *iv*, is the fourth: it will be seen that the number of phalanges progressively increases from two to five. The fifth toe of the four phalanges in the Monitor is not developed in any bird. The constancy of the number of phalanges in each toe is such that the toes retained in a tridactyle bird, e. g., Emeu, are seen to be the second, third, and fourth; those in a didactyle bird, e. g., Ostrich, to be the third and fourth: and, although the latter is much the smaller and shorter toe, it retains its superior number of joints. Among the very few exceptions to this rule may be cited the outer toe of the Caprimulgus and of the Swift, which has but four phalanges; in the Swift, also, the innermost toe is directed forward like the rest. The last phalanx in each toe is pointed, and usually curved, corresponding in some measure with the shape of the claw it supports: the articular ends of the phalanges are slightly expanded and coadapted with trochlear joints limiting motion to one plane.

The chief of the sesamoid bones in the hind limb is the patella: it is of unusual size in the Penguin, is ossified from two centres, and articulates with the procnemial process of the tibia: it coexists with the long rotular process in the Loon, fig. 34, *l*; it is large and of an angular form in the Musk-duck (*Biziura*): in the Merganser the patella is largest and deeply notched; in the Coot it is elongate. In most aerial birds a patella is wanting. A calcaneal sesamoid is wedged into the outer and back part of the ankle-joint in the Apteryx, and plays upon the back part of the tibial trochlea in the Turkey, Guan, Curassow, and some other *Rasores*.

Ossification normally extends into the tendons of some of the muscles in most birds: e. g., of the deep seated spinal ones of the back (*Uria Troile* and many others); of the muscles of the foot and toes (*Gallinæ*). The bony plates at the corneal margin of the sclerotic tunic of the eye, and the columelliform stapes of the ear, are appendages to sense-organs. Mr. William Home Clift discovered small ossifications at the attachments of the semilunar valves of the aorta and pulmonary artery in some Birds.[1]

[1] vii'. p. 331.

CHAPTER XV.

MUSCULAR SYSTEM OF AVES.

§ 131. *General Characters.*—The muscular system of Birds is remarkable for the distinctness and density of the fasciculi or visible fibres, the deep red colour of those chiefly employed in vigorous action, and their marked separation from the tendons, which are of a pearly shining colour, and have a peculiar tendency to ossification. This high degree of developement results from the rapid circulation of very warm and rich blood, highly oxygenated through the extent of the respiratory system. The energy of the muscular contraction in this class is in the ratio of the activity of the vital functions, but the irritability of the fibre rapidly goes after death. The elementary fibres are much smaller and less sharply angular than in Reptiles; the blood-vessels being more abundant and occupying more space in their intervals.

These characteristic properties are manifested in the greatest degree in the muscles of the *Volitores,* and of those *Cantores* that take their food on the wing, as the *Hirundinidæ*; in those of the *Diurnal Raptores* and the long-winged *Palmipedes,* as the Albatross, Tropic Bird, &c. In the more heavy and slow-moving Herbivorous families, the muscles resemble those of the Reptilia in their softness and pale colour. In birds of flight the mechanical disposition of the muscular system is admirably adapted to the aerial locomotion of this class; the principal masses being collected below the centre of gravity, beneath the sternum, beneath the pelvis, and upon the thighs, they act like the ballast of a vessel and assist in maintaining the steadiness of the body during flight, while at the same time the extremities require only long and thin tendons for the communication of the muscular influence to them, and are thereby rendered light and slender.

§ 132. *Muscles of the vertebræ.*—The muscles of the cervical region are the most developed, as might be expected from the size and mobility of this part of the spine; the muscles which are situated on the dorsal and lumbar regions are, on the other hand, very indistinct, feeble, and but slightly carneous; they are not, however, entirely wanting. In the Struthious and short-winged sea birds, in which the dorsal vertebræ are unfettered by anchylosis, these muscles are more fleshy and distinct, most so in the *Apteryx*, and will here be described as seen in that bird.[1]

The *sacro-lumbalis* is the most external or lateral of the muscles of the back, and extends from the anterior border of the ilium to the penultimate cervical vertebra. It arises by short tendinous and carneous fibres from the outer half of the anterior margin of the ilium, and by a succession of long, strong, and flattened tendons from the angles of the fifth and fourth ribs, and from the diapophyses of the third, second, and first dorsal vertebræ; also by a shorter tendon from that of the last cervical vertebra; these latter origins represent the *musculi accessorii ad sacro-lumbalem*; to bring them into view, the external margin of the *sacro-lumbalis* must be raised. These accessory tendons run obliquely forward, expanding as they proceed, and are lost in the under surface of the muscle. It is inserted by a fleshy fasciculus with very short tendinous fibres into the angle of the sixth rib, and by a series of corresponding fasciculi, which become progressively longer and more tendinous, into the angles of the fifth, fourth, third and second ribs, and into the parapophyses of the first dorsal and last two cervical vertebræ: the last insertion is fleshy and strong; the four anterior of these insertions are concealed by the upper and outer fleshy portions of the *sacro-lumbalis*, which divides into five elongated fleshy bundles, inserted successively into the diapophyses of the first three dorsal and last two cervical vertebræ.

35

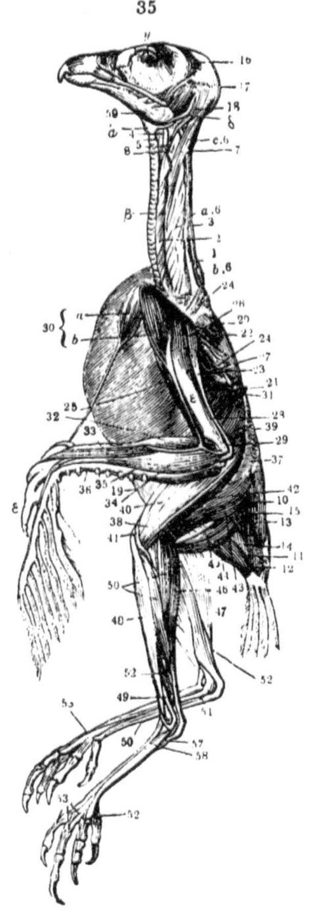

Muscles of a Hawk.

[1] xxiv'. vol. iii. p. 280, pls. 32 and 33.

These last insertions seem to represent the continuation of the *sacro-lumbalis* in Man, which is termed the *cervicalis descendens* or *ascendens.*

The *longissimus dorsi* is blended posteriorly both with the *sacro-lumbalis* and the *multifidus spinæ,* and anteriorly with the outer portion of the *spinalis dorsi.* It extends as far forward as the thirteenth cervical vertebra. It arises from the inner or mesial half of the anterior margin of the ilium; from a strong aponeurosis attached to the spines of the eighth, seventh and sixth dorsal vertebræ; and from the diapophyses of the sixth, fifth, fourth and third dorsal vertebræ. The carneous fibres continued from the second origin, or series of origins from the spinous processes, incline slightly outward as they pass forward, and are inserted into the anapophyses of the first three dorsal vertebræ, receiving accessory fibres from the *spinalis dorsi.* The fasciculi from the diapophyses incline inward, and are also inserted into the anapophyses of the vertebræ anterior to them; they receive fibres from the iliac origin, and soon begin to form a series of oblique carneous fasciculi, which become more distinct as they are situated more anteriorly; they are at first implanted in the vertebra next in front of that from which they rise, and then into the vertebra next but one in front: the most anterior of these tendons of insertions, to which can be traced any of the fibres of the main body of the *longissimus dorsi* is that which is implanted into the thirteenth cervical vertebra; it is this fasciculus which is joined by the first or most posterior of the *fasciculi obliqui* of the *longus colli posticus.*

Obliquus colli, a series of oblique carneous fasciculi, evidently a continuation of, or part of the same system with those in which the *longissimus dorsi* terminates anteriorly, is continued between the diapophysis of one cervical vertebra to the anapophysis or posterior zygapophysis of the next vertebra but one in advance, as far forward as the fourth cervical vertebra. This series of muscles seems to represent the *transversalis colli,* which is the anterior continuation of the *longissimus dorsi* in Mammalia, but it differs in being inserted into the oblique, instead of the transverse processes. In the direction of their fibres these fasciculi resemble the *semispinalis colli,* but they are inserted into the oblique processes instead of the spines of the vertebræ. There are no other muscles with which they can be compared in the Mammalia than these two, with neither of which, however, do they precisely correspond; they seem to represent the second series of oblique muscular fasciculi in the trunk of Fishes.

The *fasciculi obliqui* which rise from the first two dorsal and five lower cervical vertebræ are joined near their tendinous terminations by corresponding oblique fasciculi of the *longus colli posticus*, and the strong round tendons continued from the points of convergence of these fascicles are inserted successively into the posterior oblique processes of the twelfth to the sixth cervical vertebra inclusive; the two fasciculi next in succession receive no accessory fibres from the *longus colli posticus*; the anterior one derives an extensive origin from the upper transverse processes of the eighth, seventh, and sixth cervical vertebræ. It must be observed, however, that the whole of each fasciculus is not expended in the strong round tendinous insertion above described; the portion which arises from the anterior ridge of the diapophysis passes more directly inwards than the rest, and is attached to the tendon which terminates the fasciculus immediately behind; at the middle of the neck these accessory fibres approach to the character of distinct origins. The tendons of insertion, moreover, severally receive accessory fleshy fibres from the base of the zygapophysis of the two vertebræ next behind; and thus they become the medium of muscular forces acting from not less than five distinct points, the power of which is augmented by each tendon being braced down by the oblique converging series of muscles immediately anterior to it. The fasciculus from the eighth cervical vertebra, besides its insertion by the ordinary tendon, sends off externally a small pyramidal bundle of muscular fibres which soon terminates in a long and slender tendon which is inserted into the oblique process of the third cervical vertebra. Corresponding portions of muscle are detached from the two anterior fasciculi, which converge and terminate in a common slender tendon inserted into the posterior oblique process of the fourth cervical vertebra; and thus terminates this complex muscle or series of muscles. It is partially represented by the muscle 3, in fig. 35 (Hawk).

The *longus colli posticus* is most internal or medial of the superficial muscles of the dorsal aspect of the thoracic and cervical regions. At its posterior part it seems to be a continuation of the *longissimus dorsi*; its medial and anterior part offers a strong analogy with the *biventer cervicis*; it is the homologue of the first, or medio-dorsal series of oblique fibres of the muscular system in Fishes. It commences by long and slender, but strong, subcompressed tendons from the spines of the sixth, fifth and fourth dorsal vertebræ: these tendons gradually expand as they proceed forward and downward, and send off from their under surface muscular fibres which continue in the same course, and begin to be grouped into distinct fasciculi at the base of the neck: the first of these bundles joins a fasciculus of the *longissimus dorsi*, which is inserted into the anapophysis of the thirteenth cervical vertebra; the succeeding fasciculi derive their origins from a broad and strong aponeurotic sheet attached to the spines of the fourth, third and second dorsal vertebræ: the second to the eighth fasciculi inclusive are compressed, broad, and fleshy, and are inserted in the strong round tendons described in the preceding muscle, and attached to the zygapophyses of the twelfth to the sixth cervical vertebræ inclusive: the ninth fasciculus, which forms the main anterior continuation of the *longus colli posticus*, is larger than the rest, and receives, as it advances, accessory fibres from the spinous processes of the seventh to the third cervical vertebræ inclusive, and is inserted, partly fleshy, partly by a strong tendon, into the side of the broad spine of the *vertebra dentata*. A slender fasciculus is detached from the mesial and dorsal margin of the *longus colli posticus*, near the base of the neck, which soon terminates in a long round tendon, fig. 35, *a* 6: this tendon is braced down by short aponeurotic fibres to the spine of the fifth, fourth, third and second cervical vertebræ inclusive, immediately beyond which it again becomes fleshy, and expands to be inserted into the occipital ridge: this portion is the *digastrique* or *biventer capitis* of Cuvier, ib. *c*, 6.

In *Raptores* the carneous exceeds the tendinous part of this muscle. The displacement of the dorsal portion of the preceding muscle and the *longissimus dorsi* brings into view the *spinalis dorsi*, which is a well-developed and distinct muscle in the *Apteryx*. It arises by two long, narrow, flattened tendons from the spines of the eighth and seventh dorsal vertebræ: these pass obliquely downward and forward, expanding as they proceed, and terminate in two fasciculi of muscular fibres: the posterior bundle passes forward beneath the anterior one, and inclining inward and upward, divides into two portions, inserted by long tendons into the spines of the second and first dorsal vertebræ; it then sends a few fibres forward to join the outer and anterior fasciculus, which is partly inserted by a slender tendon into the spine of the last cervical vertebra: the rest of the fibres of the second fasciculus join the portion of the *longissimus dorsi* which is implanted into the posterior oblique process of the last cervical vertebra. The three inserted tendons of the *spinalis dorsi* are also the medium of attachment of fibres continued from the *multifidus spinæ*, beneath them.

The series of muscles called *multifidus spinæ* arises by fleshy fibres from the diapophyses of the five last dorsal vertebræ, which pass upward, forward, and inward, to be inserted by four flat tendons into the spines of the seventh to the third dorsal vertebræ inclusive, and by the tendons of the *spinalis dorsi* into the two anterior dorsal spines.

Obliquo-spinales. The removal of the *multifidus spinæ* brings into view a series of long, narrow, flat tendons, coming off from the spines of all the dorsal vertebræ, and slightly expanding as they proceed forwards and obliquely downwards and outwards; they become fleshy half-way from their origin, and are inserted into the posterior oblique and transverse processes of the six anterior dorsal vertebræ, and into the posterior oblique processes of the three last cervical vertebræ.

The *interspinales* muscles do not exist in the region of the back, unless we regard the preceding oblique fibres as a modified representation of them. The most posterior fasciculus of muscular fibres, which is directly extended between the spinous processes, commences at the interspace of the spines of the two last cervical vertebræ, and the series is continued as far as the *vertebra dentata.*

Interarticulares. The muscles which form the more direct continuation of the *obliquo-spinales* are continued from the posterior zygapophysis of one vertebra to that of the next in front.

Obliquo-transversales. A third series of deep-seated intervertebral muscles is situated external to the preceding, and passes obliquely between the diapophysis and the posterior zygapophysis of the vertebra in front. These fasciculi appear to be a continuation of the *multifidus spinæ* in the neck.

The *intertransversales* are two series of short carneous fasciculi passing the one between the diapophyses, and the other between the parapophyses.

Levatores costarum. The first or most anterior of this series of muscles seems to represent the *scalenus medius*; it arises from both the di- and pleur-apophysis of the last cervical vertebra, and expands to be inserted into the first rib, and into the upper and outer part of the second rib. The remaining *levatores* successively diminish in size as they are placed backwards; they come off from the diapophyses of the first six dorsal vertebræ; those from the first and second expand to be inserted into the rib attached to the same transverse process and to the one next behind; the rest have a single insertion: the angle and the part of the rib immediately beneath are the situations of their attachments.

Complexus, fig. 35, 7. This strong triangular fleshy muscle arises from the met- and di-apophyses of the fourth, third and second cervical vertebræ, and gradually expands as it advances forward to be inserted into the occipital ridge, from the outer side of the insertion of the *biventer cervicis* to the mastoid process.

Recti capitis postici. These small muscles are concealed by the preceding; they rise successively from the spines of the third, second and first cervical vertebræ, and expand as they advance to be inserted into the occiput.

Trachelo-mastoideus. This strong, subdepressed carneous muscle arises from the diapophyses of the fifth, fourth, third and second cervical vertebræ, and is inserted into the paroccipital.

Longus colli. This large and long muscle, which appears simple when first exposed, is found to consist, when unravelled by further dissection, of a series of closely succeeding long, narrow fasciculi, arising from the hypapophyses of the sixth dorsal to the first dorsal, and from the ten posterior cervical vertebræ; and sending narrow tendons which increase in length as they are given off more anteriorly, obliquely forward and outward, to be inserted into the pleurapophyses of all the cervical vertebræ save the first two: the highest or foremost tendon is attached to the tubercle at the under part of the ring of the atlas; but this tendon is also the medium of insertion of five small fasciculi of muscular fibres arising from the diapophyses of the sixth, fifth, fourth, third and second cervical vertebræ.

The *rectus capitis anticus major* is continued, or arises by as many distinct tendons, from the five superior tendons of insertion of the preceding muscle; these origins soon become fleshy, converge, and coalesce previous to their insertion into the base of the skull.

The *rectus capitis anticus minor* is a strong fleshy compressed triangular muscle arising from the anterior part of the body of the first four cervical vertebræ, and inserted into the basioccipital.

The *rectus capitis lateralis* arises from the diapophyses of the sixth to the second cervical vertebræ inclusive; it is inserted into the lateral ridge or tubercle of the basioccipital.

The *obliquus externus abdominis* arises, fleshy, from the second and third ribs, and by a strong aponeurosis from the succeeding ribs near the attachment of the costal processes, and from those processes. The fleshy fibres are continued from this aponeurotic origin to nearly opposite the ends of the vertebral ribs; they run almost transversely, very slightly inclined towards the pubis, to within half an inch of the linea alba, and there terminate, by an almost straight, parallel line, in their aponeurosis of insertion. The fibres of this aponeurosis decussate those of the opposite side, and adhere to the tendinous intersections of the *rectus* beneath. The aponeurosis from the last rib passes to be inserted into a strong ligament extending between the free extremities of the pubic bones, leaving the abdomen, behind the last rib, defended only by the *internal oblique* and *transversalis.*

The *obliquus internus abdominis* arises from the whole of the anterior and outer surface of the pubis; aponeurotic from the upper part, fleshy for half an inch from the lower or ventral extremity: the carneous fibres run longitudinally, and cannot be distinctly defined from the *intercostales* on their outer border, or from the *rectus abdominis* on their inner or mesial border, which forms the medium of the insertion of the internal oblique.

The *rectus abdominis* is the medial continuation of the preceding muscle, which arises by a strong, flat, triangular tendon from the lower or ventral extremity of the pubis and from the inter-pubic ligament: it soon becomes fleshy; the carneous portion is interrupted by three broad, oblique, but distinct aponeurotic intersections, and is finally inserted into the sternum.

Transversalis abdominis. A layer of loose, dark-coloured cellular tissue divides the internal oblique from the transverse abdominal, except at its origin from the pubis, and for half an inch anterior to that part. The *transversalis* then proceeds to derive carneous fibres from the inner surface of the ribs near their lower third; they pass obliquely upward and forward, and terminate by a regular, slightly concave line midway between their origins and the extremities of the ribs; a strong aponeurosis passes thence to the linea alba, but becomes thin at the pubic region, where a mass of fat is interposed between it and the peritoneum.

The *diaphragm* presents more of its mammalian character in the Apteryx[1] than in any other known bird. It is perforated by vessels only, in consequence of the non-developement of the abdominal air-cells. The origin corresponding to that of the lesser muscle in Mammals is by two strong and distinct, short tendinous pillars from the sides of the body of the last costal vertebra; they are united by a strong tendon or fascia, forming the anterior boundary of the aortic passage. The tendinous pillars may be traced forward for some way in the central aponeurosis, expanding without crossing; they are then lost in that aponeurosis, which is perforated by the gastric arteries and veins, divides anteriorly to

[1] xɪ'. vol. ii. pl. 52; vol. iii. pl. 36, p. 287.

give passage to the gullet and the apex of the heart, expands over the anterior part of the thoracic air-cells, and becomes, at its lateral circumference, the point of attachment of muscular fibres arising from the inner surface of the anterior ribs, and forming apparently a continuation of the *transversalis abdominis.*

The *appendico-costales*[1] arise from the posterior edge and extremity of the costal processes, and run down to be inserted severally into the rib posterior to that to which the process affording them origin is attached. These processes are supported by strong triangular aponeuroses continued from their anterior and upper margins, severally, to the rib anterior to them.

The *levator caudæ* arises from the posterior and superior extremity of the ischium; it is inserted into the spines of the caudal vertebræ. In birds with a posteriorly expanded sacrum, that bone affords the chief origin to this muscle, fig. 35, 10.

The *adductor caudæ superior* is smaller than the preceding, with which it runs parallel; it rises below from the posterior extremity or tuber of the ischium, and is inserted into the pleurapophyses of the caudal vertebræ.

The *adductor caudæ inferior* arises from the tuber ischii and the ligament connecting this with the posterior extremity of the pubis. It is inserted into the diapophyses of the caudal vertebræ.

The *depressor caudæ* arises, ib. 15, from the under part of the middle line of pelvis; it is inserted into the inferior spines of the caudal vertebræ.

In birds of flight the 'rectrices,' or rudder-quills attached to the coalesced and modified terminal vertebræ call for moving powers not developed in the *Apteryx.*

The *quadratus coccygis,* fig. 35, 11, arises from the diapophyses of the coccygeal vertebræ, and is inserted into the shafts of the tail-quills, which it separates and raises. On the lateral aspect the *pubo-coccygeus,* ib. 12, arises from the posterior margin of the pubis, and inserted also into the shafts of the exterior *rectrices*; it is by means of these muscles in conjunction with the quadratus and levator caudæ, that the Peacock raises the gorgeous plumes overlying the true tail-feathers.

The *ilio-coccygeus,* ib. 13, extends from the posterior margin of the ilium to the last coccygeal vertebra, and to the small inferior tail-feathers.

On the ventral or inferior aspect of the tail, the muscles are in general more feebly developed than on the opposite side, except

[1] xi'. vol. iii. pl. 32, *h*, p. 287.

in the Woodpeckers, where the tail, by means of its stiff and pointed quill-feathers, serves as a prop to support the bird on the perpendicular trunks of trees on which it seeks its food. In these the *ischio-coccygeus*, ib. 14, is of large size, extending from the lower edge of the ischiadic tuberosity, and from the diapophyses of the anterior caudal vertebræ to the hæmapophyses of the posterior ones, and to the sides of the ploughshare bone.

Of the *Muscles of the head*, those which are attached to it for its general motions have already been described; the remaining muscles of this part are devoted to the movements of the jaws, the tongue, the eye, and the ear.

The muscles of the jaws are chiefly modified in relation to the moveable condition of the upper mandible and tympanic bone, and the subserviency of the latter to the actions of these parts.

The *temporalis*, fig. 35, 17, fills the temporal fossa, which consequently indicates the bulk of that muscle in the dry skull. It arises from a greater or less extent of the temporal and parietal bones, and, as it passes within the zygoma, becomes closely blended with the *masseter;* the united muscles derive an accession of fibres from the lower part of the orbit, and are inserted into the raised superior margin, representing the coronoid process; and into the sides of the lower jaw from the articulation as far forward as the commencement of the horny bill. In the Cormorant, the osseous style, moveably articulated to the superoccipital, affords to the temporal muscles a more extensive origin. This, indeed, is its essential use,[1] for the muscles of the upper part of the neck are inserted into the occipital bone, and glide beneath the posterior or superadded fasciculi of the temporalis.

The *biventer maxillæ*, ib. 18, arises by two portions, the one from the lateral depression, the other from the lower part of the paroccipital; they are inserted into the back part and angle of the lower jaw.

The openers and closers of the mandibles present very slight differences of bulk in relation to the developement of the parts they are destined to move; their disproportion to the bill is, on the contrary, truly remarkable in the Horn-bills, Toucans, and Pelican, and the bill is but weakly closed in these in comparison with the shorter-billed birds.

The upper mandible when moveable is acted on by three muscles on either side. The first is of a radiated form, arises from the septum of the orbits, the fibres converging to be inserted

[1] xxiv'. p. 234.

into the pterygoid near its articulation with the tympanic. It draws forward the pterygoid bone, which pushes against and raises the upper jaw.

The *entotympanicus*, or levator tympanicus, arises from the side of the basisphenoid and is inserted into the fossa, on the inner surface of the tympanic bone: in adducting that bone it pushes forward the pterygoid, and, consequently, the upper mandible in the same way as the preceding muscle, and assists in opening the bill.

The *pterygoideus externus* arises from the outer side of the orbital process of the tympanic, and is inserted into the mandible in front of the outer articular cavity. The *pterygoideus internus* arises by a tendon from the fore part, and by fleshy fibres from the rest, of the depression upon the palatine bone, and is inserted into the inner part of the inflected angle of the mandible. This muscle draws forward the lower jaw.

In the Cross-bill (*Loxia curvirostra*) there is a remarkable want of symmetry in the muscles of the jaws on the two sides of the head corresponding to their peculiar position. Those of the side towards which the lower jaw is drawn in a state of rest (which varies in different individuals) are most developed, and act upon the mandibles with a force that enables the bird to dislodge the seeds of the fir-cones, which constitute its food.[1]

The articulation of the lower jaw is strengthened and its movements restrained by two strong ligaments; one of these is extended from the squamosal to the outer protuberance near the joint of the lower jaw. The second ligament extends from the hind end of the squamosal directly backward to the posterior part of the inner articular depression of the lower jaw, and guards against the backward dislocation of the lower jaw.

§ 133. *Muscles of the wings.*—Some of those inserted into the humerus, are prodigiously developed, and form the most characteristic part of the myology of the Bird. The muscles of the shoulder, however, are but small, and those of the distal segments of the wing still more feeble.

The *Trapezius*, fig. 35, 20, arises from the spines of the lower cervical, and a varying number of the contiguous dorsal vertebræ, and is inserted into the dorsal margin of the scapula and the corresponding extremity of the coracoid.

The *rhomboideus* lies immediately beneath the preceding, and is always single; it passes in a direction contrary to the trapezius from the spines of the anterior dorsal vertebræ to the dorsal edge of the scapula. It has no representative in the Apteryx.

[1] xxvii'. p. 459.

The *levator scapulæ* arises by digitations from the pleurapophyses of the last cervical, and the first two dorsal vertebræ; it is inserted into the posterior part of the dorsal edge of the scapula, which it pulls forwards. In the Apteryx it seems to be the most anterior portion of the series of fasciculi composing the *serratus magnus anticus*. This muscle, fig. 35, 21, is most developed in birds of prey; it arises by large digitations from three or four of the middle ribs, and converges to be inserted into the extremity of the scapula.

The *serratus parvus anticus* arises by digitations from the first and second ribs, and is inserted into the commencement of the inferior margin of the scapula. This is the largest of the muscles of the scapula in the Penguins.

A muscle, which may be regarded either as a portion of the *pectoralis minor* or as the analogue of the *subclavius* muscle, arises from the anterior angle of the sternum, and is inserted into the external margin of the sternal extremity of the coracoid bone.

The *supra-spinatus*, ib. 22, arises from the anterior and outer part of the humeral end of the scapula, and is inserted behind the largely developed radial crest of the humerus.

The muscle which seems to represent both the *infra-spinatus* and *teres major*, ib. 23, has a more extensive origin from the scapula, and is inserted into the ulnar tuberosity of the humerus, where it is closely attached to the capsule of the shoulder-joint.

The *subscapularis* arises from the anterior part of the inner surface of the scapula, and is inserted into the ulnar humeral tuberosity. It is divided into two portions by the *pectoralis minor*.

The *latissimus dorsi*, ib. 24, is but a feeble muscle in Birds, and is constantly divided into two distinct slips. The anterior portion arises, more superficial than the trapezius, from the spines of the four or five anterior dorsal vertebræ, and is inserted near the tendon of the deltoid into the outer side of the humerus. The posterior slip comes from the spines of the dorsal vertebræ above the origin of the broad *abductor femoris*, ib. 40, and sometimes from the anterior margin of the same muscle, and is inserted by a broad and thin tendon immediately in front of the preceding portion.

The *deltoides*, ib. 26, arises from the anterior part of the scapula; also in *Volitores* and *Cantores* from the acromial end of the furculum and the coraco-furcular ligament: a distinct fasciculus from the inner angle of the humeral end of the scapula passes over the os humero-scapulare, or the humero-scapular ligament, to be inserted into the angle of the pectoral ridge; this portion is large and distinct in Doves;[1] where the humero-scapular ossicle exists, a fasciculus therefrom, large in Owls, appears as a distinct origin of the deltoid, the main mass of which muscle is inserted into the pectoral ridge from its angle distad. The deltoid raises and retracts the wing.

Birds have the *pectoralis* muscle divided, as in many Mammals, into three portions, so distinct as to be regarded as separate muscles; they all arise from the enormous sternum, and act upon the proximal extremity of the humerus.

The *first* or *great pectoral muscle*, ib. 25, is extraordinarily developed, and is in general the largest muscle of the body. In birds of flight it often equals in weight all the other muscles of the body put together. It arises from the anterior part of the outer surface of the clavicle or furculum, from the keel of the sternum, and from the posterior and external part of the lower surface of that bone; it is inserted by an extended fleshy margin into the palmar surface of the pectoral crest of the humerus. It forcibly depresses the humerus, and, consequently, forms the principal instrument in flight.

This muscle is very long and wide in the *Natatores* generally, but in the Penguin, its origin is limited to the external margin of the subjacent pectoral muscle, which is here remarkably developed. The great pectoral is very long, but not very thick in the *Rasores*. In the *Herons* it is shorter, but much stronger and thicker. Its size is most remarkable in the Humming-birds, Swallows, and diurnal Birds of Prey, where it is attached to almost the whole outer surface of the sternum and its crest, and has an extended insertion. In the Ostrich its origin is limited to the anterior and external eighth part of the sternum, and it is inserted by a feeble tendon into the commencement of the pectoral crest of the humerus, to which it gives a strong rotatory motion forwards. In the Apteryx the *pectoralis major*[2] is represented by two thin triangular layers of muscular fibres attached to the under and lateral part of the sternum, and converging to be inserted into the proximal third of the minute humerus.

The *second pectoral muscle* is situated in birds of flight beneath the great pectoral; it has the form of an elongated triangle: it arises from the base of the crest of the sternum and from the mesial part of the inferior surface of that bone; it increases in size as it ascends, then again becomes suddenly contracted, passes upward and backward round the coracoid, between that bone and

[1] XLIII'. p. 400. [2] XI'. vol. iii. p. 289, pls. 31, 32, *p*.

the clavicle, then turns downward and outward, and is inserted, fleshy, above and in front of the great pectoral, into the upper extremity of the humeral crest.

The interspace between the clavicle, coracoid, and scapula, through which its tendon passes, serves as a pulley, by means of which the direction of the force of the carneous fibres is changed, and although these fibres ascend from below toward their insertion, yet they forcibly raise the humerus, and thus a *levator* of the wing is placed without inconvenience on the lower part of the trunk, and the centre of gravity proportionally depressed.

In the Penguins, Guillemots, and Gulls, this muscle is almost the largest of the three, occupying the whole length of the sternum. It is remarkable for the length and strength of its tendon, which is inserted so as to draw forwards the humerus with great force. It is proportionally the smallest in the *Raptores*; and is very small and slender in the Struthious birds.

We have already alluded to the use which the Penguin makes of its diminutive anterior extremities as water-wings, or fins; to raise these after making the down-stroke obviously requires a greater effort in water than a bird of flight makes in raising its wings in air: hence the necessity for a stronger developement of the second pectoral muscle in this and other diving birds, in all of which the wings are the chief organs of locomotion, in that action, and consequently require as powerful a developement of the pectoral muscles as the generality of birds of flight.

The *third pectoral muscle,* which is in general the smallest of the three, arises from the anterior part of the sternum at the angle between the body and keel, and also by a more extended origin, from the posterior moiety of the inferior surface of the coracoid and the coraco-clavicular membrane; it is directed forward, rising, to pass through the scapulo-coracoid trochlea; its tendon glides through a sheath, attached to the capsule of the shoulder-joint, and in some birds to the os humero-scapulare; and is inserted into the radial tuberosity of the humerus which it helps to raise.

It is proportionally large in the Penguins and Gulls, but attains its greatest developement in the Gallinaceous order.

Above the preceding muscle there is another longer and more slender one, analogous to the *coraco-brachialis,* which arises from the middle of the posterior surface of the coracoid; its direction upward is less vertical than that of the third pectoral, along the outer side of which it is attached to the anterior tuberosity of the humerus. This muscle is wanting in the *Struthionidæ,* is of small size in the Heron and Goose, is much more developed in the *Raptores* and many *Natatores*, especially the Penguins, and attains its greatest relative size in the *Rasores*, where it arises from almost the whole of the coracoideum.

Birds in general possess two *flexors* 31 and one *extensor*, fig. 35, 27, of the fore-arm. They have also muscles corresponding to *pronators* and *supinators*, but their action is limited in the feathered tribes to *inflexion* and *extension* of the fore-arm, and to *adduction* and *abduction* of the hand.

A remarkable muscle, partly analogous in its origin to the clavicular portion of the deltoid, but differently inserted, is the *extensor plicæ alaris*, ib. 30, *a*, *b*, and forms one of the most powerful flexors of the cubit. It is divided into two portions, of which the anterior and shorter arises from the internal tuberosity of the humerus; the posterior and longer from the clavicular extremity of the coracoid bone. In the Ostrich and Rhea, however, both portions arise from the coracoid. The posterior muscle, *b*, sends down a long and thin tendon which runs parallel with the humerus, and is inserted, generally by a bifurcate extremity, into both the radius and ulna. The anterior muscle, *a*, terminates in a small tendon which runs along the edge of the aponeurotic expansion of the wing. In this situation it becomes elastic; it then resumes its ordinary tendinous structure, passes over the end of the radius, and is inserted into the short confluent metacarpal, *u*. It combines with the preceding muscle in bending the fore-arm; and further, in consequence of the elasticity of its tendon, puckers up the soft part of the fold of the wing.

A lesser flexor of the fore-arm, and stretcher of the alar membrane, ib. 31, arises, as a portion of the serratus magnus from the ribs, and terminates in an aponeurosis inserted into the alar membrane and fascia of the fore-arm; it is represented in fig. 35 as turned aside.

The *extensor metacarpi radialis longus*, ib. 32, is the first muscle which detaches itself from the external condyle of the humerus, *ε*, and it forms the radial border of the muscular mass of the fore-arm; it terminates in a large tendon about the middle of the fore-arm, and this tendon passes along a groove of the radius, over the carpus, to the phalanx of the metacarpal, *ε*, into the radial margin of which it is inserted. It raises the hand, draws it forward toward the radial margin of the fore-arm, and retains it in the same plane. In the Penguin this muscle is extremely feeble, and the tendon is lost in that of the *tensor plicæ alaris*.

The *extensor metacarpi radialis brevis*, ib. 33, arises below the preceding from the ulnar edge of the radius, and is inserted into the phalanx of the thumb immediately beyond the tendon of the preceding muscle. The two tendons are quite distinct from one another in the birds of prey, the Ostrich and Parrots, but unite at the lower end of the fore-arm in the *Anatidæ*, *Phasianidæ*, and *Gruidæ*.

The *extensor carpi ulnaris*, ib. 34, comes off from the inferior extremity of the outer condyle of the humerus, passes along the middle of the exterior surface of the fore-arm, and its tendon, after passing through a pulley at the distal end of the ulna, is inserted into the ulnar phalanx. It draws the hand toward the ulnar edge of the fore-arm, and is the principal abductor or folder of the pinion.

The *flexor metacarpi radialis*, ib. 35, is a short and weak muscle, which arises from the inferior part of the ulna, descends along the internal side of that bone, winds round its lower extremity and the radial edge of the carpus, passes beneath the tendon of the radial extensors, and is inserted, external to the latter, high up into the dorsal aspect of the radial phalanx of the metacarpus. In the Ostrich it arises from the lower third of the ulna. In the Penguin it is wanting.

The *flexor metacarpi ulnaris*, ib. 36, arises beneath the fore-arm from the internal pulley of the ulna, continues fleshy to the pinion, and is inserted, first into the ulnar carpal bone, then into the ulnar phalanx. The latter insertion is wanting both in the Ostrich and Penguin.

The *muscles* of the pinion or hand are few, and very distinct from one another; the index or spurious wing is moved by four small muscles, viz. two *extensors*, an *abductor*, which draws the digit forward, and an *adductor*. The middle digit receives three short muscles, two of which are *extensors*, and the third an *abductor;* in this action it is aided by one and opposed by another of the extensors. The outer digit receives an *abductor*, which comes from the ulnar edge of the preceding phalanx.

§ 134. *Muscles of the Legs.*—The muscles of the pelvic limb are here described chiefly as they exist in the Apteryx, in which they present their full developement. The most superficial of the muscles on the outer side of the leg is that very broad one which combines the functions of the *tensor vaginæ* and *rectus femoris*, but which, in the opinion of Cuvier[1] and Meckel,[2] is the homologue of the *tensor vaginæ* and *glutæus maximus* (*seu externus*): since,

[1] XII. tom. i. p. 502. [2] XLVI. th. iii. p. 361.

however, it is exclusively inserted into the leg, it is here described with the other muscles moving that segment of the pelvic extremity. The removal of this muscle, of the *sartorius,* and the *biceps cruris,* is requisite to bring into view the true *glutæi.*

Glutæus externus[1] is smaller than the middle *glutæus,* but is relatively larger in the *Apteryx* than in birds of flight. Besides its origin from the outside of the pelvis, it overlaps part of the *glutæus medius,* and has its insertion into the femur at some distance below the great trochanter, all of which are marked characteristics of the *glutæus magnus.* It takes its origin from the superior margin of the os innominatum, extends along an inch and a quarter of that margin, directly above the hip-joint, and is chiefly attached by distinct short tendinous threads, which run down upon the external surface of the muscle: it rises also by carneous fibres from the external surface of the os innominatum for three lines below the superior margin. The fibres converge and pass into a tendinous sheet, beginning on the external surface of the muscle half-way down its course, which ends in a broad, flat, strong tendon, inserted into a rising on the outer side of the femur nearly an inch below the great trochanter. It abducts and raises the femur.

The *glutæus medius*[2] is a large, triangular, strong and thick muscle, which has an origin of three inches' extent from the rounded anterior and superior margin of the ilium, and from the contiguous outer surface of the bone for an extent varying from an inch to eight lines. Its fibres converge to a strong, short, broad and flat tendon, implanted in the external depression of the great trochanter, having a bursa mucosa interposed between the tendon and the bony elevation anterior to the depression.

The *glutæus minimus*[3] rises below and internal to the preceding muscle from the anterior and inferior extremity, and from one inch and three-fourths of the inferior and outer margin of the ilium, and contiguous external surface, as far as the origin of the *glutæus medius*; also by some fleshy fibres from the outside of the last rib. These fibres slightly converge as they pass backward to terminate in a broad flat tendon which bends over the outer surface of the femur, to be inserted into the elevation anterior to the attachment of the *glutæus magnus.*

A muscle[4] which may be regarded either as a distinct accessory to, or a strip of, the preceding one, arises immediately behind it from half an inch of the outer and inferior part of the ilium; its

[1] xi. vol. iii. p. 290, pl. 32, a [2] Ib. pl. 32, b. [3] Ib. pl. 32, c. [4] Ib. pl. 32, d.

fibres run nearly parallel with those of the *glutæus minimus*, and terminate in a thin flat tendon, which similarly bends round the outer part of the femur, to be inserted into the outer and under part of the trochanter immediately below the tendon of the *glutæus medius*. This muscle and the preceding portion, or *glutæus minimus*, are described by Prof. Mayer[1] under the names of *glutæus quartus* and *glutæus quintus*, in the Cassowary; one of them is absent in most Birds.

Use. All the preceding muscles combine to draw the femur forward, and to abduct and rotate it inward.

Iliacus internus. This is a somewhat short thick muscle, of a parallelogrammic form, fleshy throughout; rising from the tuberosity of the innominatum in front of the acetabulum immediately below the *glutæus minimus*, and inserted at a point corresponding to the inner trochanter, into the inner side of the femur near the head of that bone, which it thus adducts and rotates outwards. This muscle is present both in the Ostrich and Bustard.

Pyramidalis. The same kind of modification which affects the *iliacus internus*, viz. the displacement of its origin from the inner surface of the ilium to a situation nearly external, affects this muscle. It arises fleshy from the outer surface of the ischium for the extent of an inch, and converges to a broad flat tendon which is inserted into the *trochanter femoris* opposite, but close to, the tendon of the *glutæus minimus*, which it opposes, abducting and rotating the femur outwards.

The *adductor brevis femoris*[2] arises from the innominatum immediately behind the acetabulum, passes over the back part of the great trochanter, becomes partially tendinous, and is inserted into the back part of the femur in common with the following muscle.

The *adductor longus*[3] is a long, broad and thin muscle, separated from the preceding by the ischiadic nerve and artery. The origin of this muscle extends one inch and a quarter from near the upper margin of the innominatum which is behind the acetabulum; it is joined by the preceding strip, and is inserted into the whole of the lower two-thirds of the back part of the femur.

The *adductor magnus*[4] is a broad and flat muscle, which has an extensive origin (two inches) from the outer edge of the ischium and the obturator fascia; its fibres slightly diverge as they pass downward to be inserted into the back part of the lower half of the femur, and into the upper and back part of the tibia.

[1] XXIX. p. 12. [2] XI. vol. iii. pl. 32, E. [3] Ib. pls. 32, 35, F.
[4] Ib. pl. 35, G.

The *obdurator internus* arises from the inner side of the opposite margins of the pubis and ischium, where they form the posterior boundary of the *obturator foramen*, and from the corresponding part of the *obturator fascia*; the fleshy fibres converge in a slightly penniform manner to the strong round tendon which glides through the notch, separated from the rest of the foramen by a short, strong, transverse, unossified ligament, and is inserted into the posterior part of the base of the trochanter. In its length and size this muscle resembles the corresponding one in the Ostrich and other Struthious birds.

The *gemellus* is represented by a single small fleshy strip arising from the margin of the *obturator foramen*, close to the emergence of the tendon of the *obturator internus*, with which it is joined, and co-inserted into the femur.

The *quadratus* is a broad fleshy muscle which arises from the pubis, below the *obturator foramen*, and which increases in breadth to be inserted into the femur internal and posterior to the obturator tendon.

Abductor magnus.[1] The largest and most remarkable of the muscles which act upon the bones of the leg is that already alluded to as the most superficial of those on the outer side of the thigh. It has a broad, thin, triangular form, and arises from the spines of the sacrum by a strong but short aponeurosis which soon becomes fleshy; the carneous fibres converge as they descend,[2] and pass into a thin aponeurosis at the lower third of the thigh: this is closely attached to the muscles beneath (*vastus externus* and *cruræus*), then spreads over the outer and anterior part of the knee-joint, is inserted into the patella, and into the anterior process of the head of the tibia.

Owing to the great antero-posterior extent of the origin of this muscle, its anterior fibres are calculated to act as a flexor, its posterior ones as an extensor, of the femur: all together combine to abduct the thigh and extend the leg, unless when this is in a state of extreme flexion, when a few of the posterior fibres glide behind the centre of motion of the knee-joint.

Sartorius.[3] The origin of this muscle is characterised by an

[1] xi'. vol. iii. pl. 31, h.

[2] They are not divided into a superficial and deep layer, as in the Ostrich, but form a simple stratum, as in the Cassowary. Meckel regards the *rectus femoris* as entirely wanting in the Cassowary, supposing, with Cuvier, the present muscle to be the analogue of the *glutæus maximus* and *tensor vaginæ* united. He says that Professor Nitzsch observed a like absence of the *rectus femoris* in the Emeu. Cuvier calls that muscle *rectus anticus femoris*, which is here described as the '*pectineus.*'

[3] xi. vol. iii. pls. 31 and 35, i.

unusual extension, like that of the preceding, with which it is posteriorly continuous: it comes off aponeurotic, from the anterior and superior margin or labrum of the ilium; the fibres soon become fleshy, and the muscle diminishes in breadth and increases in thickness as it descends; it is inserted by short and strong tendinous filaments obliquely into the anterior part of the tendon of the broad rectus, and into the anterior and inner part of the head of the tibia. Its insertion is partly covered by the internal head of the *gastrocnemius.* It bends and adducts the thigh, and extends the leg.

The homologue of the *biceps flexor cruris*[1] is a unicipital muscle, corresponding with the *abductor magnus,* by the removal of which it is exposed, in the characteristic modification of its extended origin, in relation to the great antero-posterior developement of the pelvic bones. *Orig.* By a broad and thin aponeurotic tendon, which at first is confluent with that of the *abductor*, but soon becomes distinct, from the posterior prolongation of the ilium: there is no second head from the femur. *Ins.* The fleshy fibres converge as they descend along the back and outer part of the thigh, and finally terminate in a strong round tendon, which glides through a loop formed, as in the common Fowl, Ostrich, &c., by a ligament extended from the back of the outer condyle of the femur to the head of the tibia, and is inserted into the process on the outside of the fibula. By means of the loop the weight of the hinder parts of the body is partially transferred, when the leg is bent, to the distal end of the femur; and the biceps is enabled, by the same beautiful and simple mechanism, to effect a more rapid and extensive inflection of the leg than it otherwise could have produced by the simple contraction of its fibres.

The *semimembranosus*[2] arises from the side of the caudal vertebræ, and from the posterior end of the ischium; it crosses the superficial or internal side of the *semitendinosus.* It is inserted into the fascia covering the *gastrocnemius* and the inside of the tibia: through the medium of the fascia it acts upon the tendon of the internal *gastrocnemius.*

The *semitendinosus*[3] arises from the posterior and outer part of the sacrum and the aponeurosis connecting it with the ischium: it is a flattened triangular muscle, which receives the square *accessorius* muscle from the lower and posterior part of the femur. It gradually diminishes as it descends, and having passed the

[1] xi. vol. iii. pls. 31, 32, K. [2] Ib. pls. 32, 35, L. [3] Ib. pls. 32, 35, M.

knee-joint, sends off at right angles a broad and square sheet of aponeurosis, which glides between the two origins of the *gastrocnemius internus*, and is inserted into the lower part of the angular ridge continued from the inside of the head of the tibia. The terminal tendon, continued from the apex of the muscle, then runs along the outer or fibular margin of the internal head of the *gastrocnemius*, and becomes confluent with the tendon of that muscle.

The *cruræus*[1] is a simple but strong muscle: it commences at the upper and anterior part of the thigh by two extremities, of which the outer and upper one, representing the *vastus externus*, has its origin extended to the base of the trochanter; the inner and inferior comes off from the inner side of the femur, beneath the insertion of the *glutæus magnus*; the two portions blend into one muscle much earlier than in the Ostrich. It is inserted by the ligamentum patellæ into the fore-part of the head of the tibia.

The *gracilis*[2] lies on the inner side of the *cruræus*, but more superficially; it rises by two heads, one from the anterior and upper part of the femur, the other from the os pubis; both soon become blended together and transmit a broad thin tendon to be inserted into the lower and lateral part of the patella with the *cruræus*.

Two other muscles succeed the preceding, and rise beneath it from the inner and anterior part of the femur; they have a similar insertion, and obviously represent the *vastus internus*.[3] The fibres converge to a middle aponeurosis, which increases to a strong short tendon, inserted into the upper and anterior projection of the tibia.

Popliteus. This small muscle is brought into view when the superficial muscles of the leg which are inserted into the foot are removed. Its carneous fibres extend from the fibula inward and downward to the tibia. It is of relatively smaller extent than in the Cassowary.

Gastrocnemius. This complex and powerful muscle consists, as in other Birds, of several distinct portions, the chief of which correspond with the external and internal origins of the same muscle in Mammals. The *gastrocnemius externus*[4] arises by a strong, narrow, rather flattened tendon from the ridge above the external condyle of the femur, which, about an inch below its

[1] XI. vol. iii. pls. 32, 35, O. [2] Ib. pl. 35, P. [3] Ib. pl. 35, Q.
[4] Ib. pls. 31, 32, R.

origin, becomes firmly attached to the strong ligamentous loop attached by one end to the femur above the preceding tendon, and by the other to the outer ridge of the fibula. This trochlear loop is lined by synovial membrane, and supports the tendon of the *biceps cruris*, which glides through it. The carneous fibres of the external *gastrocnemius* come off from the outer side of the tendon, and from the fascia covering the outer surface of the muscles of the leg: they are continued in a somewhat penniform arrangement two-thirds down the leg, upon the inner surface of the muscle, where they end in a strong subcompressed tendon. This joins its fellow-tendon, from the internal *gastrocnemius*, behind the ankle-joint, and both expand into a thick, strong ligamentous aponeurosis, which extends over three-fourths of the posterior part of the tarso-metatarsal bone. The lateral margins of this fascia are bent down under the flexor tendons behind the joint, and become continuous with a strong ligamentous layer gliding upon the posterior surface of the distal condyles of the tibia, and attached to the tendons of the *peroneus* and *tibialis anticus*: the conjunction of the thickened tendons of the *gastrocnemii* with this deeper-seated layer of ligamento-tendinous substance constitutes a trochlear sheath lined by synovial membrane, through which the flexor tendons of the toes glide. The synovial membrane of the ankle-joint is continued upward, half an inch above the articular surface of the bone, between it and the fibro-cartilaginous pulley. Below the joint the margins are inserted into the lateral ridges of the tarso-metatarsal bone, becoming gradually thinner as they descend, and ending below in a thin semilunar edge directed downward.

The *gastrocnemius internus*[1] has two powerful heads, one from the femur, the other from the tibia; the first arises fleshy from the internal condyle of the femur, expands as it descends, and receives additional fibres from the lower edge of the *accessorius semitendinosi*. About one-fifth down the tibia this muscular origin in the right leg terminated in a flattened tendon which became attached to the inner side of the tibial portion of the *gastrocnemius internus*. The second head, which is separated from the preceding by the insertion of the *semitendinosus*, arises partly from the internal and anterior part of the strong fascia of the knee-joint by short tendinous fibres, which almost immediately become fleshy, and partly from a well-defined triangular surface on the inner and anterior aspect of the head of the tibia: the fleshy fibres

[1] XI'. vol. iii. pl. 35, R.

converge, receive the tendinous slip from the femoral portion, and end on the inner side of the muscle in a strong flattened tendon, about two-thirds down the leg: this joins the tendon of the *gastrocnemius externus* and is inserted as described above.

The *soleus*[1] is a slender flattened muscle arising from the posterior part of the head of the tibia, the tendon of which joins that of the *gastrocnemius internus*, behind the tarsal joint.

The *flexor perforans digitorum*[2] lies immediately anterior to the external *gastrocnemius*; it arises fleshy from the outer condyle of the femur, below the tendinous origin of that muscle, and terminates in a slender flat tendon half-way down the leg. Its tendon, fig. 35, 51, glides behind the tarsal joint through the sheath of the *gastrocnemius*, expands beneath the metatarsus and bifurcates, sending its smallest division to the inner toe, ib. 52, and its larger one to blend with the tendon of the *peroneus medius*.

Flexor perforatus of the outer toe.[3] This arises by very short tendons from the proximal end of the fibula, and from the ligament forming the bicipital pulley; it continues to derive a thin stratum of fleshy fibres from the fascia covering the anterior surface of the muscles of the leg: the fleshy fibres terminate half-way down the leg in a flattened tendon, which, after entering the gastrocnemial sheath, pierces the tendon of the first *perforatus* of the middle toe, then runs forward to the outer toe, expands into a thick ligamentous substance beneath the proximal phalanx, and sends off two tendinous attachments on each side, one to the proximal, the other to the second phalanx, and is continued to be finally inserted into both sides of the third phalanx.

Flexor perforatus digitorum[4] is the strongest of the three; it arises fleshy from the posterior part of the distal extremity of the femur, above the external condyle, and also by a distinct flattened tendon, one inch in length, from the proximal end of the tibia, fig. 35, 50: this tendon, moreover, receives the long slender tendon, ib. 41, sent off obliquely across the front of the knee-joint from the *pectineus*, by which its origin is extended to the pelvis. This accessory tendon perforates the inner fleshy surface of the muscle, and is finally lost about half-way down the carneous part. Before the *flexor perforatus* is joined by the tendon of the *pectineus*, it subdivides posteriorly into four muscular fasciculi. The anterior division receives principally the above tendon, and this division of the muscle becomes wholly tendinous two-thirds down the leg; its tendon passes through the posterior

[1] xi'. vol. iii. pl. 35, s. [2] Ib. pls. 31, 32, 35, 1. [3] Ib. pls. 31, 32, 35, 2.
[4] Ib. pls. 32, 35, 3, 4, 5, 6.

part of the pulley of the *gastrocnemius*, and expands as it passes along the metatarsus: a thick ligamentous substance is developed in it opposite the joint of the proximal phalanx of the second toe, into the sides of which it is inserted, dividing for that purpose, and giving passage to the two other flexor tendons of that toe. The second portion of the present muscle terminates in a tendon situated behind the preceding, which passes through a distinct sheath behind the tarsal joint, expands into a sesamoid fibro-cartilage beneath the corresponding expansion of the previous tendon, which it perforates, and then becomes itself the perforated tendon of the second phalanx of the second toe, in the sides of which it is inserted. The third portion of this muscle ends in a somewhat smaller tendon than the preceding, which forms the second *perforatus flexor* of the third or middle toe. The fourth and most posterior portion soon becomes a distinct muscle; its fleshy fibres cease on the inner side, one-fourth down the leg, but on the outside they are continued three-fourths down the leg; its tendon passes through the gastrocnemial pulley behind the ankle-joint, and divides to form a sheath for the *flexor perforatus* of the fourth toe; it is then joined by the tendon of the *peroneus*, which passes through a pulley across the external malleolus, and finally becomes the perforated tendon of the first phalanx of the middle or third toe.

Pectineus (*rectus anticus femoris* of Cuvier[1] and Meckel[2]).—This is a long, thin, narrow strip of muscle arising from the spine of the pubis, anterior to the acetabulum, and passing straight down the inner side of the thigh; it degenerates into a small round tendon near the knee, which tendon, fig. 35, 41, traverses a pulley, formed by an oblique perforation in the strong rotular tendon of the extensors of the leg, and thus passing across the knee-joint to the outer side of the leg, finally expands, and is lost in the *flexor perforatus digitorum* last described. It is this muscle which causes the toes to be bent when the knee is bent, as in the act of perching.

Peroneus longus[3] arises, tendinous from the head of the tibia, and by carneous fibres from the upper half of the anterior margin of the tibia; these fibres pass obliquely to a marginal tendon, which becomes stronger and of a rounded form where it leaves the muscle. The tendon gives off a broad, thin, aponeurotic sheath to be inserted into the capsule of the tarsal joint; it is then continued through a synovial pulley on the side of the outer

[1] XII'. p. 523. [2] XLVI'. th. iii. p. 365. [3] XI'. vol. iii. pls. 32, 35, 7.

malleolus, and is finally inserted or continued into the perforated tendon of the middle toe.

The *tibialis anticus*,[1] fig. 35, 48, is overlapped and concealed by the peroneus; it arises partly in common with that muscle, and partly by separate short tendinous threads from the outer part of the head of the tibia; it gradually becomes narrower, and finally tendinous two-thirds of the way down the leg; its strong tendon glides through the oblique pulley[2] in front of the distal end of the tibia, expands as it passes over the ankle-joint, and is inserted into the anterior part of the proximal end of the tarso-metatarsal bone, sending off a small tendinous slip to the aponeurosis covering the extensor tendons of the toes, and a strong tendon which joins the fibular side of the tendon of the following muscle.

Extensor longus digitorum.[3] This lies between the *tibialis anticus* and the front and outer facet of the tibia, from which it derives an extensive origin; its tendon commences half-way down the leg, runs along the anterior part of the bone, first under the broad ligamentous band representing the anterior part of the annular ligament, then through a ligamentous pulley, and inclines to the inner or tibial side of the anterior surface of the metatarsal bone, where it expands and divides into three tendons. Of these the innermost is given off first, and subdivides into two tendons, one of which goes to be inserted into the base of the last phalanx of the second toe; the other portion is principally inserted into the middle toe, but also sends off a small tendon to the inner side of the proximal phalanx of the second toe. The second tendon is inserted by distinct portions into the second, third, and last phalanges of the middle toe. The third tendon supplies the outer toe.

The *extensor brevis digitorum*[4] is a small extensor muscle which arises from the insertion of the *tibialis anticus* and sends its tendon to the outer side of that of the great *extensor digitorum*.

Extensor pollicis brevis.[5] This extensor of the small innermost toe arises from the upper and inner side of the tarso-metatarsal bone.

Peroneus medius, Cuv., *Accessorius flexoris digitorum*, Vicq. d'Azyr.[6] This strong penniform muscle arises fleshy from nearly the whole of the outer surface of the fibula, also from the posterior part of the tibia and the interosseous space; the tendon of the biceps perforates its upper part in passing to its insertion. It

[1] xi'. vol. iii. pls. 32, 35, 8. [2] This is ossified in most Birds. [3] Ib. pl. 35, 9. [4] Ib. pl. 35, 10. [5] Ib. pl. 35, 11. [6] Ib. pls. 32, 35, 12.

ends in a strong flat tendon at the lower third of the leg, which tendon runs through a particular sheath at the back part of the tarsal pulley, becomes thickened and expanded as it advances forwards beneath the tarsus, joins the tendon of the *flexor perforatus,* and forms with it the expansion which finally divides into three strong perforating tendons, which bend the last joints of the three long toes.

In the outer, or fourth toe, both the *perforans* and *perforatus* tendons are confined by a double annular ligament; the exterior one being continued from the adjoining toe, the inner and stronger one from the sides of the proximal phalanx of the outer toe. The second and third toes have two perforated tendons; one inserted into the sides of the first, and the other into the sides of the second phalanx.

The chief modification of the skeleton of the hind limb of Birds, in respect of size and proportion, is manifested in its central segment; the ossa innominata being anomalously expanded in order to include, as it were, in their grasp the whole of the very long sacrum required for the support of the horizontal trunk upon a single pair of extremities. The principal modification of the muscles of the leg attached to the ossa innominata might be expected, therefore, to be found in their origins. In the attachment of the fibres of a superficial muscle to the aponeurosis, continued from the outer part of the thigh, over the knee-joint, to the head of the tibia, we recognise the corresponding insertion of the *tensor vaginæ femoris* of Man and Mammals; and no Comparative Anatomist appears to have thought the anomalous developement and extensive origin of this muscle, in Birds, to be any objection to the homology indicated by its insertion, which is the attachment that mainly governs the function of a muscle. Now besides the attachment to the femoral fascia, we find this broad superficial muscle, and especially its middle and posterior fibres, terminating in a strong tendon, implanted into the upper part of the patella, and receiving fibres from the *cruræus* and *vasti* muscles which it immediately covers, and with which it concurs in constituting a *quadriceps extensor* of the leg. Here, therefore, we perceive the normal insertion, the normal function, and the true relative position of the *rectus femoris.* In calling this complex muscle '*abductor magnus,*' it is to be understood as including the homologues of the *tensor vaginæ femoris* and *rectus femoris* in Mammals.

§ 135. *Muscles of the Skin.*—In the *Apteryx* the cutaneous system of muscles presents a distinct and extensive developement connected with the peculiar thickness of the integument, and probably with the burrowing habits of this species, which thereby possesses the power of shaking off the loose earth from its plumage, while busy in the act of excavating its chamber of retreat and nidification.

The whole of the neck is surrounded by a thin stratum of muscular fibres, *constrictor colli*,[1] directed for the most part transversely, and extending from an attachment along the median line of the skin at the back of the neck, to a parallel *raphe* on the median line of the opposite side: this muscle is strongest at its commencement or anterior part, where the fibres take their origin in a broad fasciculus from the outer part of the occipital ridge; these run obliquely downward and forward on each side of the neck, but are continued uninterruptedly with those arising from the dorsal line of the skin above mentioned; the direction of the fibres insensibly changing from the oblique to the transverse. The outer surface of this muscle is attached to the integument by a thin and dense layer of cellular tissue, devoid of fat; the under surface is more loosely connected with the subjacent parts by a more abundant and finer cellular tissue.

Use. To brace the cervical integument, raise the neck feathers, and in combination with the following muscle to shake these parts. This muscle is well developed in the emeu, and acts when the drum-like dilatation of the windpipe is sounded.

The *sterno-cervicalis*[2] arises fleshy from the posterior incurved angular process of the sternum, from the ensiform prolongation and middle line of the outer and posterior surface of the same bone. The fibres pass forward, and, diverging in gently curved lines, ascend upon the sides of the broad base of the neck, and are inserted by a thin but strong fascia into the median line of the dorsal integument. This muscle is a line in thickness at its origin, but becomes thinner as it expands; the anterior part is covered by the posterior fibres of the *constrictor colli*.

Use. To retract the skin of the neck, and brace that portion which covers the base of the neck; when these are the fixed points, it will depress and protract the sternum, and thus aid in inspiration.

Obs. In its position and the general course of the fibres, this muscle is analogous to that which supports and assists in emptying the crop in the common fowl; but the œsophagus presents no partial dilatation in the *Apteryx*.

[1] xi'. vol. iii. pls. 31, 34, *a*. [2] Ib. pl. 31, *b*.

The *sterno-maxillaris*[1] appears at first view to be the anterior continuation of the preceding, but is sufficiently distinct to merit a separate description and name. It arises fleshy from the anterior part of the middle line of the sternum, passes directly forward along the under or anterior part of the neck, expanding as it proceeds, and gradually separates into two thin symmetrical fasciculi, which are insensibly lost in the integument covering the throat and the angle of the jaw. It adheres pretty closely to the central surface of the *constrictor colli*, along which it passes to its insertion. It retracts the fore-part of the skin of the neck, and also the head. Each lateral portion acting alone would incline the head to its own side: the whole muscle in action would bend the neck; but the movements of the head and neck are more adequately and immediately provided for by the appropriate deeper-seated muscles, and the immediate office of the present muscle is obviously connected with the skin. Nevertheless, in so far as this muscle acts upon the head, it produces the same movements as the *sterno-mastoideus* in Mammalia.

The skin covering the dorsal aspect of the lower two-thirds of the neck, besides being acted upon by the *constrictor colli*, is braced down by a thin stratum of oblique and somewhat scattered fibres, *dermo-transversalis*,[2] which take their origins by fasciæ attached to the inferior transverse processes of the sixth to the twelfth cervical vertebræ inclusive; the fibres pass obliquely upward and backward, and are inserted by a thin fascia into the median line of the skin, covering the back of the neck.

The representative of the *platysma myoides*[3] is a thin triangular layer of muscular fibres, taking their origin from the outer side of the ramus of the jaw, and diverging as they descend to spread over the throat, and meeting their fellows at a middle *raphe* of insertion beneath the upper larynx and beginning of the trachea, which they thus serve to compress and support.

The *dermo-spinalis* arises by a thin fascia from the ends of the spinous processes of the three anterior dorsal vertebræ. The fibres slightly converge to be attached to the integument covering the scapular region.

The *dermo-iliacus* arises fleshy from the anterior margin of the ilium. The fibres pass forward and slightly converge to be inserted into the scapular integument.

The *dermo-costalis* is a muscle resembling the preceding in form. It arises fleshy, from the costal appendages of the seventh

[1] xi. vol. iii. pl. 34, *e*. [2] Ib. pl. 34, fig. 1, *d*. [3] Ib. pl. 31, *e*.

and eighth ribs. The fibres pass forward and join those of the preceding muscle, to be inserted into the scapular integument.

The three preceding muscles are broad and thin, but well-defined; they would appear to influence the movements of the rudimentary spur-armed wing through the medium of the integument, as powerfully as do the rudimental representatives of the true muscles of that extremity.

There are also two muscles belonging to the cutaneous series, and inserted directly into the bones of the wing. One of these, the *dermo-ulnaris,* is a small, slender, elongated muscle, which takes its origin from the fascia beneath the *dermo-costalis*; its fibres pass backward, and converge to terminate in a very slender tendon which expands into a fascia, covering the back part of the elbow-joint. It extends the elbow-joint and raises the wing.

The *dermo-humeralis* is also a long and narrow strip, deriving its origin from scattered tendinous threads in the subcutaneous cellular tissue of the abdomen: it passes upward, outward and forward, and is inserted fleshy into the proximal part of the humerus, which it serves to depress.

The cutaneous muscles which spread the plumes of the peacock and raise the hackles of the cock are unusually developed in these birds.

§ 136. *Locomotion of Birds.*—Upon land, the trunk is balanced horizontally on an axis traversing the acetabula perpendicularly to the plane of the medial section. The centre of gravity is brought within the base of support by the advanced position of the thigh-joints in the pelvis, and by the transference of the weight from the femoral heads by the shafts inclining forward to the heads of the tibiæ. The area of the base of support is adjusted to the same end by the anterior extension and divergence of the three longest toes. On this base the stilt-like leg of the crane, rising like a straight slender column to the capital formed by the head of the tibia, is capable, by an outward as well as backward course of the femur to its joint-cup, of sustaining the body singly: the neck of the bird being so bent, and the head so disposed, as to throw the centre of gravity over the vertical line passing through the base of support. Thus sleep the *Grallæ* and allied Palmipeds (Flamingos, Anserines), adjusting by reflex action the superincumbent weight as they may be swayed by the wind on the long and taper pedestal. In standing on two feet, the tibia and metatarse are usually, in Birds, bent at an open angle.

Progression on land is effected in most Birds by alternate advancement of the right and left leg; the body is supported and pushed forward by one as the other leg is raised and swung forward to the step in advance; the centre of gravity oscillating laterally, in a degree corresponding with the breadth of the pelvis and shortness of the legs, and which is such as to cause the 'waddle' of the Duck. The forces acting on the centre of gravity are preserved in equilibrio, during the walk, by movements of the neck and head, conspicuous in poultry; and, in Rails and Coots, also by movements of the tail. Most *Cantores* advance both legs at once, and progress by leaps or hops, the joints being first flexed and the body propelled by their sudden extension. In *Volitores* the legs merely support or suspend the body, and locomotion is wholly performed by the wings. Some birds derive assistance in rapid terrestrial progression by the flapping of the wings, and this is especially the case with the Ostrich, which runs by the alternate advancement of its legs.

The act of *climbing* is performed by means of a peculiar disposition of the toes, the fourth usually being bent back like the first; but sometimes, as in Trogons, the first and second toe are opposed to the third and fourth. The grasp of such 'scansorial' foot may be aided by prehension with the beak, as in the Maccaws and Parrots; or by the prop formed by the stiff tail-feathers, as in the Woodpeckers.

Birds float by the specific levity of their body, arising from the extension of the air-cells and the lightness of the plumage; but to swim requires an expanse of sole, either by marginal membranes of the toes (Water-rails, Coots), or by the extension of webs between and uniting the toes. In such true swimmers the under side of the trunk is boat-shaped, the down is thick and covered by closely imbricated well-oiled feathers, the bulk of the bird being enlarged and its specific gravity diminished by the air intercepted in the plumage. Much of the body is thus sustained above the water by hydrostatic pressure, and muscular action is needed solely for the horizontal movements. The broad oars, acting at the end of a long lever, strike the water backward with great force, the webs being fully expanded; but they collapse, the toes coming together, in the forward movement, and in some of the best swimmers (*Colymbus* e.g.) the metatarsal is compressed to further diminish the resistance in preparing for the next effective stroke. The oar-like action of the hinder legs is still further favoured by their backward position in *Natatores*; and by the metatarse and toes being placed almost on the same perpendicular line with the tibia, an arrangement, however, which is unfavourable for walking. Swans partially expand their wings to the wind while swimming, and thus move along the water by means of sails as well as oars. The act of *diving* is performed by the feet striking the water backward and upward, assisted by the compression of the air-cells: the habitual divers (Penguins, Awks) move in and through the water by the rapid and forcible action of wings, shortened and shaped like paddles, and beating the water as in flight.

Flight, the chief and characteristic mode of locomotion in birds, results principally from the construction and form of the anterior extremities. The form of the body has reference thereto, the trunk being an oval with the large end forward: being also short and inflexible, the muscles act with advantage, and the centre of gravity is more easily changed from above the feet as in the stationary position, to between the wings as during flight. The long and flexible neck compensates for the rigidity of the trunk, and alters the poise according to the required mode of progression, by simply projecting the head forward, or drawing it back. The head of the bird is generally small, and the beak pointed, which is a commodious form for dividing the air. The position of the great pectoral muscles tends to keep the centre of gravity at the inferior part of the trunk. The power which birds enjoy of raising and supporting themselves in the air is aided by the lightness of the body. The large and usually air-filled cavities in the bones diminish their weight without taking away from their strength,—a hollow cylinder being stronger than a solid one of the same weight and length. But the specific levity principally depends on the great air-cells which occupy almost every part of the body. The air which birds inspire distends these cells, and is rarified by the heat of the body. Lastly, the feathers, and especially the quills, from their lightness and elastic firmness, contribute powerfully to the act of flying by the great extent which they give to the wings, the breadth of which is further increased by the expanded integument situated in the bend of the arm and in the axilla.

When a bird commences its flight it springs into the air, either leaping from the ground, or precipitating itself from some elevated point. During this action it raises the humerus, and therewith the entire wing, as yet unfolded; it next spreads the wings horizontally by an extension or abduction of the fore-arm and hand: the greatest extent of surface of the wing being acquired, it is rapidly and forcibly depressed: the resistance of the air thus suddenly struck occasions a reaction on the body of the bird, which is thereby raised in the same manner as in leaping from the ground. The impulse being once given, the bird folds the wings by bending the different joints, and raises them preparatory to another stroke.

Velocity of flight depends upon the rapidity with which the wing-strokes succeed each other; and the ratio of the resistance of the air is not as the velocity simply, but as the square of the velocity. A downward stroke would only tend to raise the bird in the air; to carry it forward the wings require to be moved in an oblique plane, so as to strike backward as well as downward. The turning in flight to the right or to the left is principally effected by an inequality in the vibrations of the wings. To wheel to the right the left wing must be plied with greater frequency or force, and *vice versâ*.

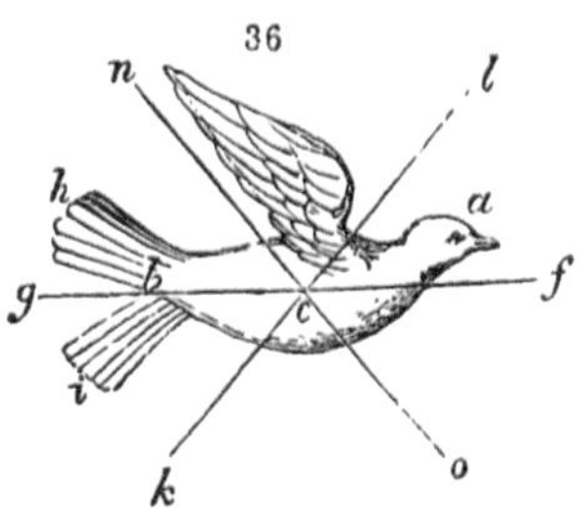

Action of tail in flight. CXXXI.

The outspread tail contributes to sustain the posterior part of the body; and, its plane being horizontal, serves chiefly by its movements to lift or lower the head. If a bird, flying in the direction of its axis, *g*, *f*, fig. 36, brings the tail into the position *b h*, parallel to *o n*, the resistance of the air will depress *b* toward *k*, and, causing the bird to rotate on its centre of gravity *c*, will raise the head from *a* towards *l*. If the tail be moved into the position *b i*, parallel to *l k*, the resistance of the air will raise the point *b* toward *n* and depress the head toward *o*. By partially folding the fan, or bending the tail to one side, it may be made to act like a rudder in the manifold modifications of the course of flight. In Waders and Anserines the tail, represented by the caudal quill feathers, is very short, and the office of the rudder is transferred to the legs, which are extended backward in flight, and counterbalance the long outstretched neck and head.

In descending from a great height birds usually incline the axis of the body obliquely downward, as in fig. 1, the resistance of the air in a vertical direction upward equilibrating the force of gravity acting upon the body vertically downward, so that the motion of the bird becomes uniform without requiring any movement of the wings. 'Another mode of descent is performed with greater celerity by elevating the wings at an angle of nearly 45° above the plane of the horizon, as in fig. 37, by which the resistance of the air, compared with the resistance to the wing when horizontal, is diminished in the ratio of the radius to the cube of the sine of inclination, that is, as *a b* to *d c*; consequently, a bird with its wings elevated at any angle to the horizontal plane

Action of wings in descent. CCIV.

will descend with greater velocity than when they are in the direction of *a b*. Pigeons elevate their wings in this manner until they arrive within a foot or two of the ground, when, to prevent the shock they would otherwise receive, owing to the velocity acquired during their descent, they suddenly turn their axis perpendicular, which had previously been parallel, to the direction of their motion, and by a few rapid strokes of the wing neutralise their momentum, and thus reach the ground with ease and safety.'[1]

The manner of flight varies in different birds: some dart forward by jerks, closing their wings every three or four strokes; the Woodpeckers and Wagtails show this kind of undulatory motion: most birds have an even continuous flight: the Kite and Albatross sometimes buoy themselves in the air without any perceptible motion of the wings. The best flyers often economise their forces by availing themselves of the impetus of a few rapid strokes to scud along with the wings expanded, until the interval of rest requires to be broken by a fresh effort,—a phase of flight beautifully defined by an old observer of nature:—

Mox aëre lapsa quieto
Radit iter liquidum, celeres neque commovet alas.—VIRGIL.

[1] CCIV. p. 428. The principal data requisite for estimating by dynamics the amount of the force employed by birds during flight are briefly stated by Mr. Bishop to be :—'1st, the area of the horizontal section of the body of the bird : 2nd, the area of the wings whilst they are lowered: 3rd, the area of the wings whilst they are raised: 4th, the velocity with which the bird is driven through the air: 5th, the velocity with which the wings are lowered: 6th, the velocity with which the wings are raised: 7th, the respective durations of the elevation and depression of the wings: 8th, the weight of the whole bird: 9th, the weight of an equal volume of air: 10th, the resistance of the air due to the figure and velocity of the bird: 11th, the ratio of the resistance which the air opposes to the wings during their elevation and depression: 12th, the ratio of the resistance of the air to the time of an elevation of the wings and to that of a depression.' Ib. p. 425.

CHAPTER XVI.

NERVOUS SYSTEM OF AVES.

§ 137. *Myelencephalon of Birds.*—The myelon, with its nerves, having led to the developement of protecting arches in the embryo, soon ceases to be coextensive with the neural canal, shrinking from the hind part of it, as the caudal vertebræ begin to be modified, and leaving there but a filamentary trace of its original condition: the neurine accumulates in the sacral region, fig. 38, *s*, as the pelvic members grow, and the central canal there expands, ib. *t*. The myelon becomes more slender in the dorsal region, and again expands near the base of the neck, in connection with the nerves of the wings, ib. *u, v, w*; then, resuming its dorsal dimension, it is continued to the brain, ib. *a*-*e*. The expansions, or at least the pelvic one, present a full transverse ellipse in section, the rest of the chord is cylindrical; it consists of white neurine with grey matter internally originating nerve-roots, and lining a subcentral canal.

The myelon is divisible into ventral and dorsal tracts according to their relative position to the transverse plane of the canal; the ventral tract is actually divided by a longitudinal fissure into symmetrical halves or 'columns,' the fissure extending from the exterior nearly to the canal from which it is separated by a very thin commissural tract uniting the ventral columns. The dorsal ones diverge from each other in the sacrum, forming the cavity above mentioned, called in Ornithotomy 'sinus rhomboidalis,' fig. 38, *s, t*, which is a 'ventricular' dilatation of the myelonal canal. The longitudinal fissure thence continued between the dorsal columns becomes less conspicuous than the ventral fissure, and appears to be obliterated in the neck: but the dorsal columns diverge in the medulla oblongata, as in the sacrum, and again expose the myelonal canal, which is here called 'fourth ventricle,' ib. *d*.

The two expansions of the bird's myelon vary in relative size according to the different developement and powers of the wings and legs: the anterior or alar enlargement is greatest in *Volitores*, especially the Swifts and Humming-Birds: the posterior or

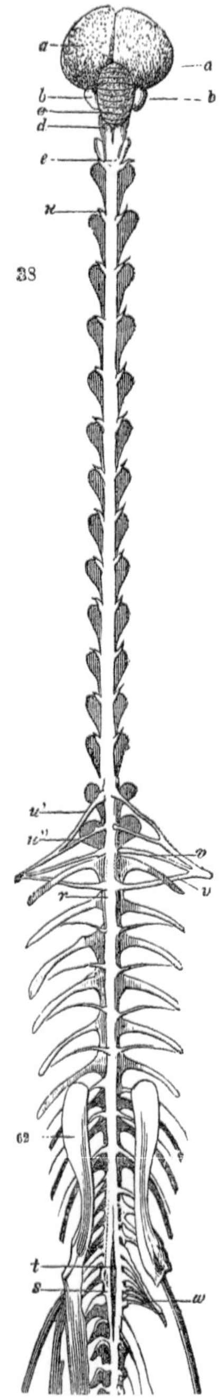

Myelencephalon, or brain and spinal chord, with part of the backbone, of a bird (*Anser*).

pelvic one is greatest in most other birds, and especially in the *Cursores*. The alar enlargement .is due to an accession of white and grey substance, without dilatation of the myelonal canal.

In the brain of a chick at the eighth day of incubation, fig. 39, the 'fourth ventricle' is exposed by divergence of

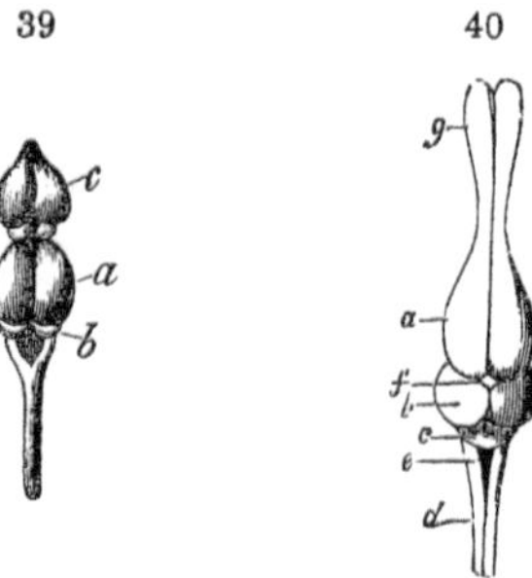

Brain of embryo Bird. Brain of Lizard. CCXVI.

the dorsal myelonal columns which now have the name of 'posterior pyramids:' the plate of neurine developed from them to bridge over the ventricle shows the same incipient state of the cerebellum, ib. *b*, as in the *Batrachia*: it next expands at the middle and represents the condition of the cerebellum in the Lizard, fig. 40: continuing to grow, the cerebellum, fig. 41, *c*, covers, at the sixteenth day of incubation, the fourth ventricle, and has a smooth exterior, as in the Crocodile and Turtle (vol. i. fig. 191). Towards the

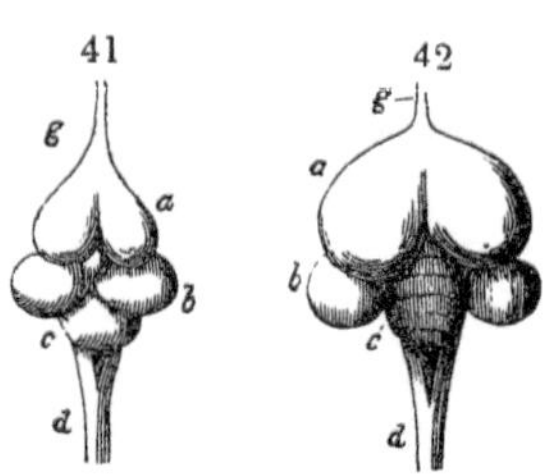

Brain of chick at 16 days. Id. at 20 days. CCXVI.

close of incubation the cerebellum, fig. 42, *c*, presses forward toward the cerebrum, ib. *a*, and seems mechanically to push aside the optic lobes, *b*; the multiplied grey matter of its superficies is disposed in transverse folds: small beginnings of lateral lobes are present in many birds. The white neurine, fig. 45, *q*, continues to accumulate beneath the grey, ib. *d*, and reduces the cavity of the originally vesicular cerebellum to a fissure, ib. *n*, which retains its primitive connection with the fourth ventricle, the floor of which shows the longitudinal groove called 'calamus scriptorius.' The medulla oblongata expands, but its ventral surface, fig. 44, *d*, is not sculptured so as to permit 'anterior pyramids,' 'olivary bodies,' a 'trapezium,' or a 'tuber annulare,' to be defined.

The optic lobes in the embryo, fig. 39, *a*, are smooth vesicles of white neurine, in contact with each other, as in *Reptilia*: they are at first oblong, as in *Batrachia*; next acquire a spheroid figure, as in Lizards, fig. 40, *b*, and then assume their ornithic character by diverging laterally toward the lower plane of the brain, figs. 42, 44, *b*: they maintain their smooth exterior, and their ventricle much reduced in capacity by internal growth of neurine.

The crura cerebri show their first superadditions, forming the optic thalami, in the eight-days embryo, between *a* and *c*, fig. 39, before expanding into the 'hemispheres,' ib. *c*. These progressively increase in size until they acquire the relative dimensions and position shown in fig. 43, *a*.

They are usually of a cordiform shape with the apex directed forward: in the Parrot tribe they present a more elongate, depressed oval figure: they are devoid of convolutions; but a

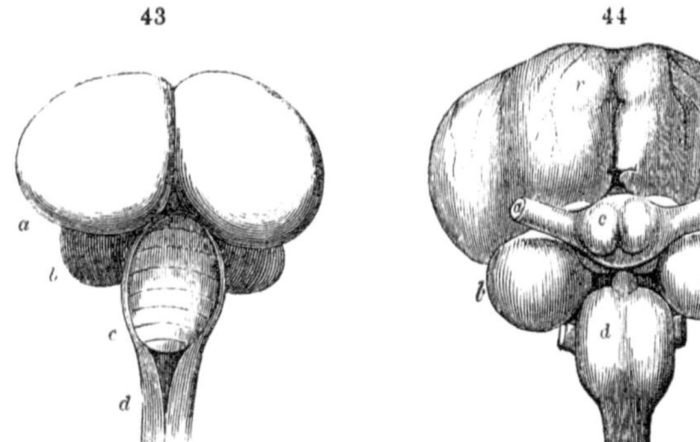

Brain of Sea-gull (*Larus*). Upper view. CCXVI.

Brain of Eagle. Base view. CCII.

shallow longitudinal depression marks off, in some birds, a median from a lateral tract of the upper surface of the hemisphere: in most this surface is uniformly convex. The hemispheres present an undulate surface below; the medial parts being in some birds produced, so as to cause a concavity transversely between them and the lateral borders, as shown in fig. 44. On the lower part of the side of each hemisphere there is a depression which corresponds to the 'fissura magna Sylvii,' and affords the sole indication of a division into lobes. The hemispheres are connected together by means of the round commissure, fig. 45, *k*.

The mesial surfaces of the hemispheres, which are in contact with each other, present striæ which diverge from the commissure. These surfaces are composed of an extremely thin layer of medullary substance, fig. 45, *f*, *g*, forming the internal parietes of the ventricle, and extended outwardly over the corpus striatum, ib. *i*. Like its homologue in *Reptilia* and the mammalian embryo, it does not present the alternate striæ of grey and white matter, which suggested its name in Anthropotomy. This cerebral ganglion is of great relative size in Birds, constituting of itself almost the entire substance of the hemisphere, projecting into the ventricle, ib. *h*, not only from below, but from the anterior and outer sides of the cavity, and being covered by a smooth layer or fold of medullary matter, *f*, which increases in thickness anteriorly. The ventricle does not extend below the corpus striatum to form an 'inferior horn,' or '*cornu ammonis*.' A fold of pia mater enters the bottom of the cerebral ventricle and lies free in the cavity: it is highly vascular, and developes tufts containing plexiform loops of capillaries defended by epithelium, the cells of which are shown at the margin of the villi magnified in fig. 46. The vessel forming the plexus choroides penetrates the ventricle beneath the posterior part of the thin internal wall, and the lateral ventricles communicate together there, and with the third ventricle. They are continued anteriorly to the root of the olfactory nerve, which is itself a continuation of the apex of the hemisphere.

45

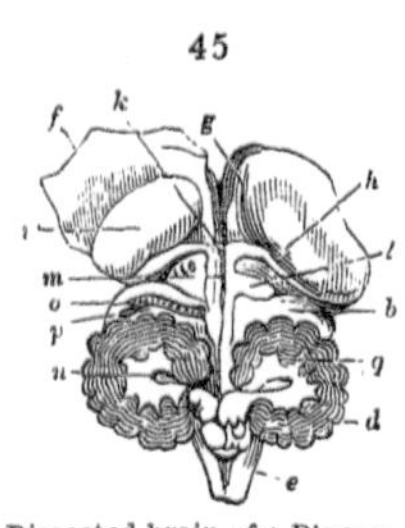

Dissected brain of a Pigeon. XXXIV.

46

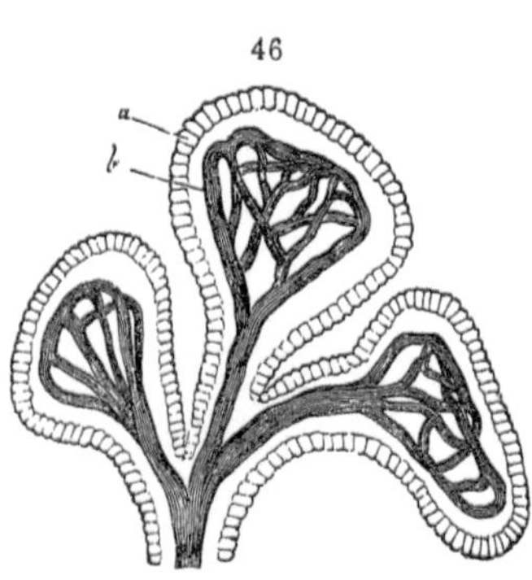

Tuft of the choroid plexus of the brain of a Goose. XXXV.

Just above the orifice of communication there is a smooth flattened projection, rounded externally, which advances into the ventricle from the internal wall; this represents a beginning of the fornix. The round anterior commissure, *k*, is prolonged on either side into the substance of the hemispheres.

The optic thalami, ib. *l*, are of small size, and not united by a soft commissure: between them is the cavity called the third ventricle, ib. *m*; and above and behind they give off the peduncles of the pineal gland. This body does not hang freely suspended by the pedicles, but seems to form a rounded and thickened anterior border of the valvula Vieussenii or lamelliform commissure of the optic lobes. It adheres firmly to the confluence of the great veins situated at the anterior orifice of the aqueduct of Sylvius: it is usually of a conical or pyriform shape. The valve which closes the upper part of the passage from the third to the fourth ventricle, is a thin lamella of great width, in consequenee of the distance at which the optic lobes are separated from one another. Anteriorly the third ventricle communicates with the infundibulum, which terminates in a large hypophysis.

Besides the cavities or ventricles above mentioned, there are also two others situated in the optic lobes, fig. 45, *o*, or bigeminal bodies, each of which, when laid open, is seen to be occupied by a convex body, ib. *p*, projecting from the posterior and internal side of the lobe; these ventricles communicate with the others in the aqueduct of Sylvius.

The brain of the Bird differs from that of the Reptile in the superior size of the cerebrum and cerebellum, together with the folding of the latter, which relates probably to the higher locomotive powers of the Bird: it differs from the brain of the Mammal in the absence or small beginning of the fornix, and of the lateral lobes of the cerebellum: it differs from the brain of every other class in the lateral and inferior position of the optic lobes. In a pigeon weighing eight ounces with and seven ounces without the feathers, or 3360 grains, the myelencephalon weighs 48 grains, the weight of the myelon being 11 grains, and that of the brain 37 grains. The proportion of the weight of the brain to the body is much greater in the Humming-Bird: whilst in the huge *Dinornis*, the brain does not exceed two inches and a half in length, and two inches in width. It thus presents a limited range of size, and much sameness of form and structure in the different orders of the class of Birds.

§ 138. *Nerves of Birds.*—The olfactory or first pair, usually of a simple rounded form, proceed from the small pyriform rhinencephalon, fig. 44, *r*, continued from the apex of the hemisphere, and usually somewhat deflected. The nerve runs along an osseous canal, accompanied by a venous trunk above the orbits, as far as the pituitary membrane of the ethmoturbinals, upon which its filaments are distributed in a radiated manner. In *Apteryx* and *Dinornis*, the rhinencephalon is of large relative size, and sends off the olfactory nerves by many filaments through a 'cribriform plate.'

The optic nerves, fig. 44, *a*, are in general of remarkable size; they arise from the whole of the outer surface of the optic lobes, and from the thalami, the two origins forming by their union the 'radix opticus,' fig. 47, *d*, which expands into the 'chiasma.' Here a partial decussation, ib. *b*, takes place. By removal of the firmly adherent neurilemma, the optic nerve is seen to be com-

47

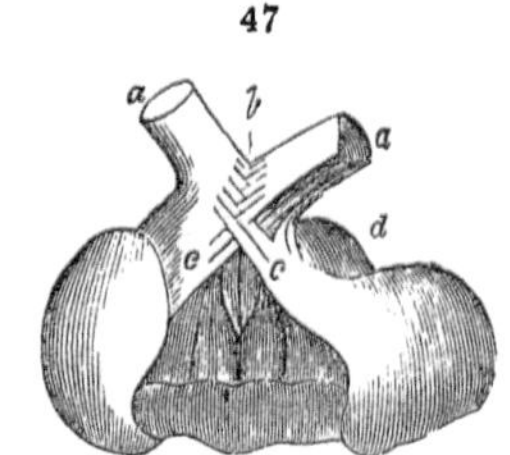

Chiasma of optic nerves. Fowl. xxx'.

48

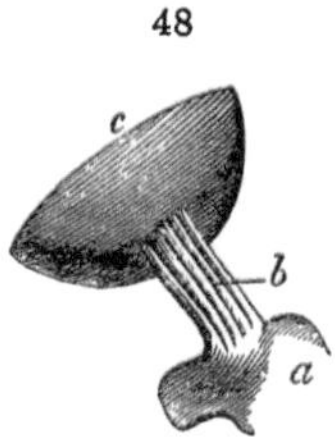

Laminated optic nerve of an Eagle. ccii.

posed of parallel, longitudinal lamellæ, the margins of which are most free on one side, fig. 48, *b*.

The *third*, or oculomotorial nerve, arises behind the hypophysis from the grey matter that lies here between the crura cerebri: it escapes, usually, by a distinct hole, fig. 56, 3, near the foramen opticum, and supplies the superior, inferior, and internal musculi recti, and the obliquus inferior: it also sends off a ciliary branch, which sometimes forms a ganglion before, sometimes after, joining the ramus ciliaris trigemini.

The *fourth* nerve arises from the posterior flattened band, extending over the 'valvula Vieussenii' between the back part of the optic lobes: its course, immediately above the superorbital branch of the fifth pair, is shown at fig. 56, 4*, as far as its termination in the superior oblique muscle, ib. *f*, to which it is, as in other Vertebrates, exclusively distributed.

The *fifth* or trigeminal nerve, fig. 49, 5, 6, 7, has two origins; the 'portio major,' from the fore part of the base of the crus cerebelli, the 'portio minor,' from the prepyramidal tract in advance of the foregoing, which it joins after the reddish ganglionic swelling, fig. 49, 3, has been formed. The two origins are less distinct than in Mammals; but the larger one is more readily traceable, towards the myelon, from not being crossed by a 'trapezium.'

The first or *ophthalmic* division, fig. 49, 5, passes out of the cranium by a canal situated externally to the optic foramen. It is of large size, and describes in its passage through the orbit a curve corresponding to the roof of that cavity; it generally penetrates the substance of the facial bones, fig. 56, 5*, above the nasal fossa, ib. *m*. It divides into three branches; the *first* or superior is the smallest and is lost upon the pituitary membrane; the *second* branch is the largest of the three and the longest; it is received into an osseous canal, and terminates at the extremity of the beak in a great number of divisions; the *third* branch of the ophthalmic nerve is entirely distributed to the skin which covers the circumference of the external nostrils.

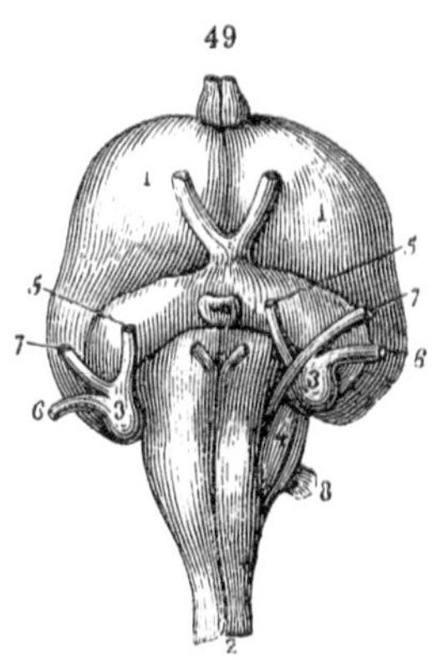

Base of the brain, and origins of cerebral nerves, Goose. CCVIII.

The second division, fig. 49, 6, or *superior maxillary* nerve, passes out of the same foramen as the *inferior* one, ib. 7: it passes forward along the floor of the orbit, and in this part of its course gives off two filaments, of which one joins the ramifications of the ophthalmic nerve, the other ascends, penetrates the substance of the pterygoid muscles and the maxillary bone, to be lost on the lateral parts of the bill. In those Birds, as the *Anatidæ* and other Water-fowl, where the upper mandible is notched on the edge, each denticulation receives four or five nervous filaments, and the nerve is proportionally of large size.

The *inferior maxillary nerve* separates from the superior, and proceeds obliquely downward, dispensing branches to the pterygoid and quadrangular muscles of the jaws; the trunk proceeds outward to the lower jaw, where it divides into two branches, an internal and an external. The internal, which is a continuation of the trunk, penetrates the maxillary canal, and is continued to the anterior end of that mandible. In the *Anatidæ* it gives off nerves to the dentations along the edge of the mandible. The external branch recedes from the internal, perforates the jaw, and is distributed on its external surface beneath the tegumentary or horny substance which sheaths the extremity of the mandible. It supplies no gustatory branch to the tongue, which is an organ of prehension, not of taste, in Birds. The non-ganglionic part of the third division of the fifth is traced on the left side of fig. 49, 7, passing beneath the ganglion. There is no 'otic' ganglion in Birds.

The *facial nerve*, or portio dura, arises immediately anterior to the acoustic from the prepyramidal tracts, enters the petrosal anterior to the acoustic, quits it to pass into the fallopian canal, sends off the 'chorda tympani' to the ramus alveolaris inferior of the trigeminal, and communicates with the sympathetic; it passes out behind the tympanic bone (as in Mammals), gives branches to the digastric and stylohyoid muscles, and combines with the glossopharyngeal, vagus, hypoglossal, and upper cervical nerves, to form the plexus supplying the anterior part of the constrictor colli muscle.

The *auditory nerve*, or portio mollis, is large, soft, and of a reddish colour; it is received into a depression on the petrosal, fig. 56, 7, whence it penetrates by several small foramina to the labyrinth.

The roots of the 'eighth' nerve penetrate the exoccipital by two or three foramina, and unite on their emergence to form the ganglion, from which the glossopharyngeal and the pneumogastric trunks diverge. The glossopharyngeal is large; it communicates more freely with the sympathetic than does the pneumogastric in the neck; it sends off a small internal branch in front of the muscles of the neck; a small posterior twig which unites with the pneumogastric, and a large inferior branch to the anterior part of the neck. The latter is a continuation of the nerve itself; it descends along the œsophagus and divides into two principal branches, of which one passes to the cerato-maxillary muscles, and this branch is remarkably tortuous in the Woodpecker, in order to be accommodated to the extensile motions of the tongue; it supplies the upper larynx, and the surface of the tongue, as far as the tip. The other branch descends along the lateral parietes of the œsophagus, and sends off a twig to join the lingual nerve. The termination of the glossopharyngeal is expended upon the œsophagus.

The pneumogastric, after communicating with the glossopharyngeal, sympathetic and ninth nerves, passes down the neck, along with the jugular vein, and closely connected with the spinal nerves. The right trunk crosses the arch of the aorta, and sends off the recurrent round that vessel, the left trunk reflects its recurrent near the origin of the bronchi; the recurrents supply the lower larynx and part of the trachea, but are chiefly spent upon the œsophagus. The trunks of the two pneumogastrics converge ventrad of the œsophagus and unite above the preventriculus, supplying that part, the gizzard, and ultimately communicating with the splanchnic plexus of the sympathetic. In the Eagle the pneumogastric is recruited by an 'accessorius' nerve arising behind the third cervical.

The *hypoglossal* nerve (9th pair) escapes by one or two precondyloid foramina. It is very slender at its origin; passes to the front of the nervus vagus, partly uniting with, as it crosses over, this nerve, and in that situation it detaches a filament to the hyolaryngeal and long tracheal muscles. The trunk of the hyoglossal next crosses the glossopharyngeal nerve, passing forward to supply the hyoglossal and lingual muscles.

The *spinal nerves* arise by motory (anterior or ventral) and sensory (posterior or dorsal) roots of nearly equal size; but the anterior have more numerous filaments. The ganglion on the posterior root is proportionally large. In the sacral region of the spine, the anterior and posterior roots escape by distinct foramina, and can be separately divided without laying open the bony canal, but they are deeply seated and well protected by the anchylosed processes of the sacrum and the extended iliac bones.

The *cervical* nerves vary with the number of the vertebræ from ten to twenty-three: each nerve divides; the anterior branch supplying the muscles and the skin, the posterior branch the muscles chiefly. Those of the lower cervicals form a plexus, supplying the scapular muscles, and communicate with the lowest cervical nerve going to the brachial plexus. Only the last two or three pairs, fig. 38, *u' u''*, of cervical nerves concur in the formation of this plexus, which is completed by the first pair or two of dorsal nerves *v*. The other *dorsal* nerves, after giving filaments to the intercostals and diaphragmatic muscles, pass to the skin at the sides of the trunk.

The *sacral* nerves have no other peculiarity than their mode of passing out of the spinal canal: they form exclusively the plexus analogous to the lumbar and sacral, fig. 38, *w*. The terminal spinal nerves supply the muscles and skin of the cloaca and tail.

The *brachial plexus*, formed by the two or three last cervical and one or two first dorsal nerves, soon becomes blended into a single fasciculus whence all the nerves of the wing are derived. The internal cutaneous nerve passes from the axilla along the inner and back part of the humerus, bends round the inner (ulnar) side of the elbow joint; it supplies the skin. The next branch distributes filaments to the muscles 22, 24, fig. 35; sends off the 'circumflex' nerve which supplies the latissimus dorsi, deltoid, and shoulder joint; and is then continued as the 'musculo-spiral,' supplying the brachialis internus and biceps, and, as it passes behind the antibrachium, the extensors of the pinion; it also distributes filaments to the skin. The next large branch from the plexus is the 'median nerve,' which sends off the 'external cutaneous' in its course along the biceps, supplying the skin on the outer or radial side of the wing. The 'ulnar' nerve is the next branch, supplying the 'ulnaris internus;' and the continuation of the 'median' gives branches to the muscles on the radius, to those on the pinion, and to the integument.

The nerves of the pelvic limbs are derived from the sacral plexus. The *obturator* nerve, formed by the second and third sacral nerves, passes through the upper part of the foramen ovale, gives off a branch to join the 'saphenus nerve,' and is distributed to the muscles around the hip-joint. The *femoral* nerve passes out of the pelvis in company with an artery, over the front edge of the ilium. It divides into three branches, which are dispersed among the muscles, fig. 35, 40, 42, and integuments on the anterior and inner part of the thigh. One of these filaments represents the 'saphenus,' and descends superficially for a considerable way upon the limb.

The *ischiatic* nerve is derived from five or six of the nerves constituting the sacral plexus, on quitting which, even within the pelvis, it is easily separable into its primary branches. Immediately after it passes through the ischiatic foramen it sends filaments to the muscles on the outer part of the thigh; it then proceeds under the biceps, along the back of the thigh, about the middle of which it becomes divided into the *tibial* and the *peroneal* nerves.

The posterior *tibial* nerve, before it arrives in the ham, separates into several branches, which pass on each side of the blood-vessels, and are chiefly distributed to the muscles, fig. 35, 46, 50, 51, on the back of the leg. Two of these branches, however, are differently disposed of; the one accompanies the posterior tibial artery down the leg, passes over the internal part of the pulley, and is lost in small filaments and anastomoses with a branch of the peroneal nerve on the inner side of the metatarsus; the other branch runs down on the peroneal side of the leg, along the deep-seated flexors of the toes, ib. 52, passes in a sheath formed for it on the outer edge of the moveable pulley of the heel, and proceeds under the flexor tendons along the metatarsal bone, to be distributed to the internal part of the two external toes.

The *peroneal* nerve is directed to the outer part of the leg; it dips above the gastrocnemii muscles, and runs through the same ligamentous pulley that transmits the tendon of the biceps muscle, ib. 41; it then detaches some large filaments to the muscles on the anterior part of the leg, under which it divides into two branches, which proceed close together, in company with the anterior tibial artery, to the fore part of the ankle-joint, at which place they separate; one passes superficially over the outer part of the joint, the other goes first under the transverse ligament which binds down the tendon of the tibialis anticus muscle on the tibia, and then over the inner part of the joint, below which it divides into two branches: the one is distributed to the inner side of the metatarse, and the tibial side of the back toe, *i*, and the next toe; the other turns toward the centre of the metatarsal bone, and penetrates the tendon of the tibialis anticus just at its insertion, and then rejoins the branch of the peroneal nerve it accompanied down the leg. They continue their course together again in the anterior furrow of the metatarsal bone; and at the root of the toes, separate once more, and proceed to the interspaces of the three anterior toes, and each divides into two filaments, which run along the sides of the toes to the claw.

§ 139. *Sympathetic System.*—The superior cervical ganglion is connected with the glossopharyngeal nerve more closely in some birds than in others: it communicates by branches with the portio dura and second division of the fifth, and supplies the lacrymal gland: a second branch accompanies the entocarotid, supplies the harderian gland, and communicates with the first division of the fifth. 'The "cervical portion" of the sympathetic may be compared with that in the Snake in its not having a chord or prolongation accompanying the trunk of the par vagum; it, however, corresponds in some measure also with that of the Turtle, for in the Swan a branch is continued down the neck with each carotid artery, and in its course communicates several times with its fellow.'[1] 'In the Pelican the carotid is a single trunk dividing into two at the upper part of the neck; a branch passes from the superior cervical ganglion with each of these, and becomes united into one near their bifurcation; it gives off branches for the supply of the carotid, and to communicate with the prolongation accompanying the vertebral artery: at the bottom of the neck it dips down in the median line between the anterior cervical muscles, and divides into two branches, each joining the penultimate cervical ganglion.'[2] The sympathetic passes down

[1] LIV. [2] Ib. p. 104.

the cervical vertebræ in a canal with the vertebral artery 'resembling the prolongation in the imperfect canal in the Snake.' (Vol. i. p. 310, fig. 206, s.) 'Also like the chord sent from the first thoracic ganglion and placed at the side of the neck in the Turtle, and that accompanying the vertebral artery in Mammalia.' 'The sympathetic adheres to the anterior trunk of each cervical nerve through a ganglion.' 'Having reached the thorax, the ganglia are connected with those of the dorsal nerves, much as in the Turtle. In the Swan and Pelican a large nerve from the first thoracic ganglion communicates with the pulmonary branches of the par vagum.'[1] The thoracic trunk of the sympathetic is generally double between each ganglion. The anterior ones give off an anterior splanchnic nerve or plexus accompanying the cœliac artery to the gizzard and liver, communicating with the pneumogastric; the posterior splanchnic nerve is intimately combined with the adrenal body, and with the testis or ovarium. Intestinal branches accompany those of the mesenteric arteries; other branches supply the kidneys, and communicate with long branches of the spinal nerves destined for the cloaca and adjoining parts, and thus form a plexus corresponding in some degree with that in Mammalia produced by the junction of the hypogastric plexus with branches of two or three of the sacral nerves. The termination of the sympathetic is formed by a 'ganglion impar' near the end of the caudal vertebræ. The abdominal ganglions in small birds lend themselves favourably to the demonstration of the structure of these centres of the sympathetic system, becoming transparent under pressure, and permitting the nerve-vesicles to be well distinguished from the nerve-chords: the latter only are represented in fig. 50, showing the finer filaments, *c*, that bend round the periphery of the ganglion, as if by resolution of and divergence from the main chords entering at *a* and emerging at *b*, *b*.

50

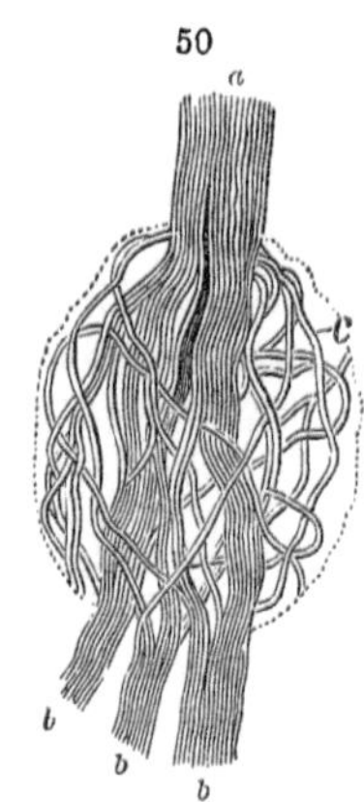

Sympathetic ganglion of a Greenfinch, magnified, as seen under the compressor.

§ 140. *Organ of Touch in Birds.*—The epithelial papillæ[2] sheathed upon vascular ones of the corium[3] on the sole of the toes of most Birds relate to mechanical rather than to sensational

[1] LIV'. p. 104. [2] XX'. vol. iii. p. 239, preps. nos. 1902–1906.
[3] Ib. preps. 1400 (Eagle), 1401 (Ostrich).

ends, giving a closer grasp of the perch or the prey, and a firmer tread of the hard ground. The digital villi are unusually long in the Capercailzie (*Tetrax urogallus*) enabling it to grasp with more security the frosted branches of the Norwegian pine-trees. The integument of the toes is sparingly supplied with nerves in all Birds; but it may be supposed that the more delicately papillose slender flexible digits of the smaller nidificators guide, by sensations analogous to touch, in the complex interweavings of the materials of such beautiful structures as the pensile, domed, and otherwise adaptively perfected nests fabricated by the Tailor-birds, and most *Cantores*.

Actions indicative of tactile exploration have hitherto been observed to be performed by the bill, exclusively, in Birds. Although in most of the class the horny sheath of the bill be hard, sensitive filaments of the 'fifth' nerve (fig. 53, *c*) are traceable to the papillose extensions of the vascular formative surface into such sheath. In the Lamellirostrals the substance of the sheath is softer and its marginal lamellæ are more abundantly supplied by the 'fifth:' from the tactile and selective actions of the bill in those Birds, they are called 'Sifters.' The soft and slightly expanded end of the long and slender bill of some *Grallæ* (Woodcocks, Snipes) is so organised for touch, that it is used as a probe in soft ground to detect the worms, grubs, and slugs that constitute their food.

Peculiar productions of integument, devoid of feathers, such as the 'cere' of Birds of Prey, the 'wattles' of the Cock, and of species of *Philedon, Glaucopis,* and other so-called 'wattle birds,' the cephalic caruncles of the King Vulture and Turkey, &c., have been loosely cited amongst 'organs of touch.'

§ 141. *Organ of Taste.*—The gustatory sense is very imperfectly enjoyed in Birds, which, having no manducatory organs, swallow the food almost as soon as seized. The tongue is organised chiefly to serve as a prehensile instrument, and its principal modifications will be treated of in CHAPTER XVII. It is generally sheathed at the anterior part with horn, and is destitute of papillæ except at its base, fig. 51, *o*, near the aperture of the larynx, *i*; these papillæ are not, however, supplied by a true gustatory nerve, but by filaments of the glossopharyngeal. No branch of the fifth pair goes to the tongue; but the membrane of the palate and fauces is so supplied that the sapid qualities of food may be there appreciated.

The tongue is proportionally largest and most fleshy in the Parrot tribe, and the food is detained in the mouth longer in these than in other Birds. It is triturated and comminuted by the mandibles, and turned about by the tongue, which here seems to exercise a gustatory faculty, since indigestible parts, as the coats of kernels, &c., are rejected. In the Lories the extremity of the tongue is provided with numerous long and delicate papillæ or filaments projecting forwards.

The marginal epithelial papillæ of the tongue of the Toucan, fig. 51, *h*, appear to test, in the way of touch, the ripeness or mellowness of fruit. Similar papillæ at the tip of the tongue of many small birds (Humming-birds, Thrush-tribe, fig. 75, B, Field-fare) exemplify probably the tactile rather than the gustatory faculty.

51

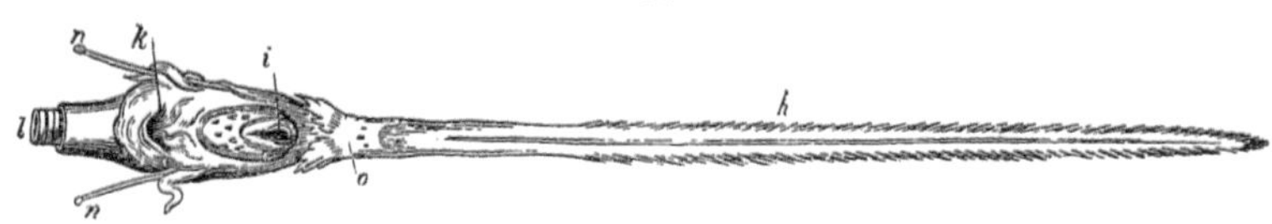

Fauces and tongue of the Toucan (*Ramphastos*). XX'.

§ 142. *Organ of Smell.*—The close affinity subsisting between the cold and warm-blooded Ovipara is manifested in the olfactory organs. The external nostrils are simple perforations, having no moveable cartilages or muscles provided for dilating or contracting their apertures, as in Mammalia. The extent of surface of the pituitary membrane is not increased by any large accessory cavities, but simply by the projections and folds of the turbinals. The olfactory nerve passes out of the skull, as a rule, in Birds, by a single foramen. The Apteryx and Dinornis form the exceptions.

The external nostrils vary remarkably both in shape and position, and serve on that account as zoological characters. They are placed at the sides of the upper mandible in the majority of Birds, but in some species are situated at or above the base of the bill; the latter is the case in the Toucans, fig. 53, *d*; in the *Apteryx australis* they are found at the extremity of the long upper mandible.

In general they are wide and freely open to facilitate the inhalation of air during the rapid motions of the bird, but they are so narrow in the Herons as scarcely to admit the point of a pin; and in some *Pelecanidæ* they are wanting, and the odorous particles get access to the olfactory organ from the palate.

In the *Rasores* the nostrils are partially defended by a scale. In the *Corvidæ* they are protected by a bunch of stiff feathers directed forward. In the Petrels the nostrils are produced in a tubular form, parallel to one another for a short distance along the upper part of the mandible, with the orifices turned forwards, fig. 52, *a*.

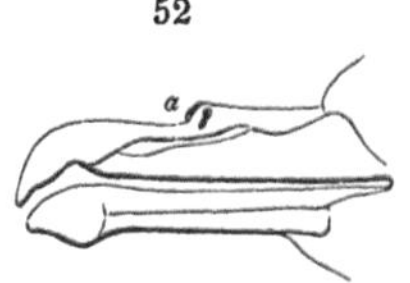

Bill and nostrils of the Petrel.

The septum narium, fig. 53, *e*, *e*, is, in general, complete, and is partly osseous, partly cartilaginous. It is perforated in the Swan just opposite the external nostrils, and in the Toucan, lower down, ib. *b*. The surface of the septum is rugose in this bird, and the pituitary membrane which covers it is highly vascular. The parietes of each of the nasal passages give attachment to three turbinal laminæ. The inferior one is a simple fold adhering to the lower and anterior part of the septum narium; it is partially ossified in some *Rasores*. The middle turbinal is the largest: it is of an infundibular figure, and adheres by its base to the septum and externally to the side-wall of the nose. It is convoluted with two turns and a half in the Anserine Birds, but in many birds it is compressed and forms only one turn and a half. The superior turbinal *t*, 5″,

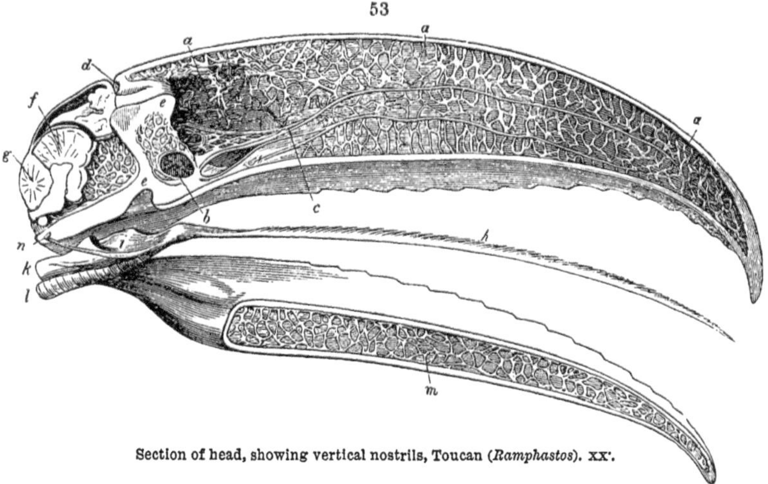

Section of head, showing vertical nostrils, Toucan (*Ramphastos*). xx′.

fig. 56, generally presents the form of a bell; it is more or less ossified at its base, but mostly cartilaginous, and adheres to the upper part of the prefrontal. It is hollow, and divided into two compartments, which are prolonged in a tubular form; the internal one extends toward the orbit, the external terminates behind the middle turbinal in a cul-de-sac. The turbinal supports of the pituitary membrane may be membranous gristly, or bony, and in different proportions. The latter is their texture in the Toucan, in which the olfactory organ is confined to the base of the huge upper mandible, fig. 53, *d, e,* the meatus describing a vertical sigmoid curve. At its commencement it is cylindrical, then dilates forward to receive the outermost turbinal, and bends backward to admit the projection of two ethmoturbinals: after which it descends vertically to the palate, *e.* The pituitary lining of the meatus is not continued or reflected into the contiguous pneumatic structure of the bill, *a, b.*

In most Birds the nasal passages communicate with the palate and pharynx by two distinct but contiguous apertures: in some, e. g., the Cormorant and Gannet, the passages unite and terminate by a single aperture.

The olfactory nerves are distributed to the pituitary membrane of the septum narium, and of the superior and middle, or ethmo-, turbinals; the lower turbinals being supplied by the fifth nerves. The membrane is most vascular and delicate on the ethmoturbinals; and these acquire an unusual size in the Apteryx, where they are attached to the whole outer part of the prefrontals, answering to the 'os planum,' which makes a large convex projection between and below the orbits. This bird appears to be guided by the sense of smell to the worms that form its food, the outer nostrils being at the end of the long probe-shaped bill. The olfactory nerves are proportionally largest in the Apteryx, and are sent off in numerous filaments from the rhinencephalon, by a cribriform plate, to the nose. The extinct *Dinornis* had a similar developement of the organ of smell. In the Vulture the olfactory nerve is single on each side, and continued from an olfactory ganglion, or 'rhinencephalon,' along the upper part of the interorbital space to be distributed upon an upper and middle turbinal, the latter being the largest. In the Turkey the olfactory nerve is one-fifth the size of that in the Vulture, and is ramified on a small middle turbinal, there being no extension of the pituitary membrane over a superior, or ethmo-turbinal.[1] This result of comparative anatomy, and the observed differences in the habits and food of the Vulture and Turkey, point to the greater importance and exercise of the sense of smell in the carrion-eating raptorial bird. But it has been sought to invalidate the inference by certain well-known experiments. Mr. Audubon exposed the skin of a deer, stuffed with hay, and in a few minutes a Vulture flew towards and alighted near it, attacked the seeming carcass in the usual way, and tore open the seams of the skin;

[1] xxxi'. p. 34.

when, finding nothing eatable, the bird flew away. Hence the American ornithologist concludes that the Vulture is led to its game by sight alone. But the truer deduction may be that, having always received impressions from sight, combined with and confirming those, in some cases the first received, from smell, the Vulture was unwilling to disbelieve its own eyes, though the odour was absent. It may often have been led by sight to the carcass of a dying beast, or one dead too soon for any putrefactive emanations to have escaped, and so it mistook the stuffed deer for a recently dead one. In a converse experiment, a dead hog being concealed in a ravine, and covered with briers and cane, 'many Vultures were seen from time to time sailing over the spot where the putrid carcass was hid,' but none of them attempted to expose it: whilst several dogs found their way to it, and devoured the flesh.'[1] The right inference from this experiment is, that the Vultures were attracted by the putrefactive effluvia; but, having always associated sight with smell, and having neither the burrowing power of the dog, nor the habit of hunting exclusively by scent, they were baffled.

§ 143. *Organ of Hearing.*—The general character of this organ resembles that in Reptilia, but, as Hunter well remarks, 'there is a neatness and precision in the structure which is not to be found in the *Tricoilia*.'[2] The whole of the primitive cartilaginous acoustic capsule is ossified and confluent with contiguous elements of cranial vertebræ, and there is a better defined and usually deeper fossa or 'meatus' external to the ear-drum. In most Birds a fold of integument projects from the fore part of the meatus; this is largest in the Owls, but the ear-drum is not protected by one so developed as to form a conspicuous 'conch' or 'auricle.' At most, in some Birds, as the Bustard, fig. 54, *d*, Ostrich and Owls, particular feathers are so developed and arranged around the meatal margin, as to serve the office of an external ear: the auricular feathers being raised and directed so as to catch and concentrate the vibrations of sound that may have excited the bird's attention.

54

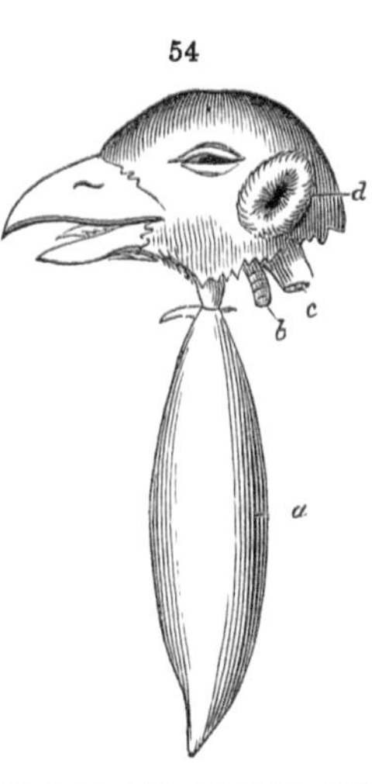

Head and auricle of Bustard, with cervical air-cell. XXXIII'.

[1] XXXII'. vol. ii. p. 34; vol. v. p. 345.

[2] Vol. i. p. 208.

The labyrinth or internal ear consists of the 'vestibule,' three semicircular canals, and beginning of the cochlea. The vestibule is smaller in proportion to the other parts, but is longer than in *Reptilia*. The superior semicircular canal, fig. 55, *h′*, is usually the largest, as in the Owl; but it is relatively smaller in most *Cantores*: the external canal, *i*, inosculates at *m* with the horizontal one *k*, but the chief communications of the canals are through the medium of the vestibule. The ends of the canals where the acoustic nerves enter are expanded into 'ampullæ,' ib. *l*, and the nerves are supported in exquisitely delicate vascular membranes lining the canals, and slightly projecting into the ampullæ.

The *cochlea* is represented by an obtuse osseous conical cavity, fig. 55, *n*, longer than in the Crocodile, very slightly bent, with the concavity directed backward. Its interior is occupied by two small cylinders of fine cartilage, each a little twisted, and united by a thin membrane at their origin and termination. They proceed from the osseous bar, which separates the two foramina, communicating respectively, the one, 'foramen rotundum,' with the vestibule, the other, 'foramen ovale,' with the tympanum. The sulcus, which is left between the cartilages, is dilated near the point, and accommodates the same branch of the auditory nerve, which is sent to the cochlea in Mammals. This nerve spreads in fine filaments upon the united extremity of the cartilaginous cylinders. The cavity is divided by the presence of the cartilages into two 'scalæ,' the anterior of which communicates with the vestibule and is not closed; the posterior scala is shorter, and would communicate with the tympanum by the foramen ovale, were it not closed by a membrane. Besides these parts the cochlea still contains a trace of the cretaceous substance which forms so conspicuous a part of the organisation of the internal ear in Fishes. The Struthious Birds manifest their closer relation to the Reptilia by having the cochlea smaller in proportion to the other parts than in the ears of birds of flight.

55

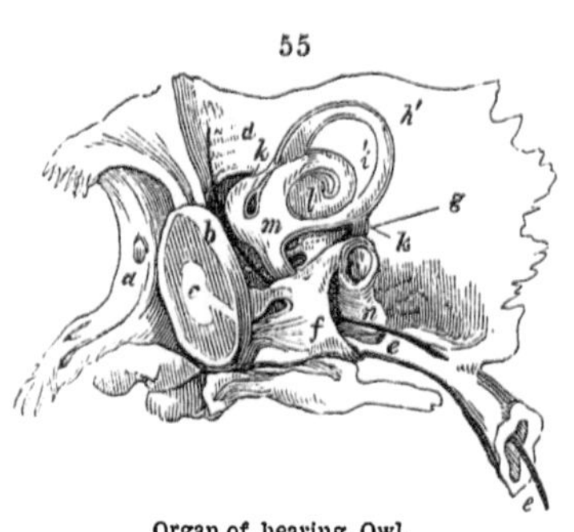

Organ of hearing, Owl.

The cavity of the tympanum has been already described, p. 62: besides the communications with the air-cells of the surrounding bone, it is continued by the 'eustachian' tube, fig. 55, *e*, to the palate: to the membrane closing the 'foramen ovale' is applied the base of the columelliform stapes: the much larger external aperture of the tympanic cavity is closed by the ear-drum, *e*. This is convex outwardly, semitransparent and glistening: the proper 'membrana tympani' is lined by that of the tympanic cavity, which is continued into the eustachian tube; and is covered externally by an epithelial layer, continuous with that of the meatus: the former is more intimately united with the proper membrane. In this may be discerned an outer layer, showing more distinctly a structure of radiating fibres, and one inner, thicker, and less distinctly fibrous layer.

The margin of the ear-drum is set in a groove of bone, afforded by or attached to the tympanic anteriorly, the mastoid above, and the paroccipital behind. One or more points of ossification may be set up in the thick periphery of the drum, which coalesce with the above-named bones. The membrane of the vestibule, passing across the foramen ovale, becomes a little thickened where it adheres to the margin of the disc of the stapes: the connection is such as to admit of a slight movement of the ossicle. From the disk the bone is continued, of a slender form, like a pedicle, to the cartilaginous bifurcation, and this is connected by a larger cartilaginous plate, representing the 'malleus,' to the membrana tympani, at *c*, fig. 55. To the latter cartilage, as to the ossified and coalesced incus and malleus of Marsupials, is attached the chief muscle of the ear-drum, a 'tensor,' fig. 55, *f*: it arises from a depression in the basisphenoid, enters the tympanic cavity above the beginning of the eustachian tube, and by its insertion into and action upon the malleus, tends to push the membrane outward: it is counteracted by two small cords extended to the inner wall of the tympanum: but the muscular character of them is doubtful, and the ear-drum resumes its normal state when the tensor ceases to act. The eustachian tube, fig. 55, *e*, is continued from the lower and back part of the tympanic cavity, grooves the sides of the basisphenoid, as it converges toward its fellow, with which it unites, in most Birds, to terminate by a common aperture behind the posterior or palatal nares.

§ 144. *Organ of Sight in Birds.*—The avian peculiarities of the eye chiefly relate to the extraordinary powers of locomotion in this class, adjusting vision to a rapid change of distance in the objects viewed, and facilitating their distinct perception through a rare medium.

There is no species of Bird in which the eyes are wanting, or rudimentary, as occurs in the other vertebrate classes.

The eyes of Birds are remarkable for their great size, both as compared with the brain and with the entire head, fig. 56, being analogous, in this respect, to the eyes of some of the flying insects. Their *form* is admirably adapted to promote the objects above named. The anterior segment of the eye is more prominent than in any other class of animals, and is in many Birds prolonged into a tubular form, terminated by a very convex cornea, fig. 57, *c*; the Owl furnishes the best example of the disproportion between the anterior and posterior divisions of the globe, the axis of the anterior portion being twice as great as that of the other. This gives room for a greater proportion of aqueous fluid, and by removing the crystalline lens from the retina, causes a greater convergence of the rays of light, by which the nocturnal bird is enabled to discern the objects placed near it, and to see with a weaker light. The anterior division of the eye is least convex in the Swimming Birds. The antero-posterior diameter is to the transverse as 19 to 26 in the Swan, and as 17 to 20 in the Duck.

56

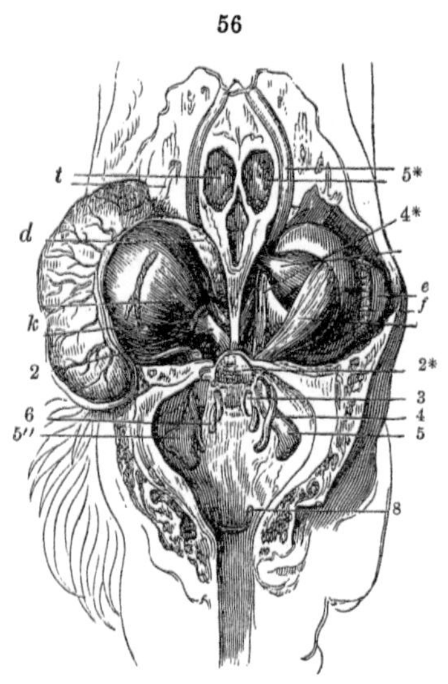

Cerebral nerves, eyes, &c. in situ, of a Goose.

The sclerotic coat, fig. 57, *b*, is divisible into three layers. It is thin, flexible, and somewhat elastic posteriorly, where it presents a bluish shining appearance, but anteriorly its form is maintained by a circle of osseous plates or scales, ib. *a*, fig. 26, 17, interposed between the exterior and middle layers. These plates vary from thirteen to twenty in number, and are situated immediately behind the cornea, with their edges overlapping each other. They are in general thin, and of an oblong quadrate figure, becoming elongated from before backward in proportion as the bird possesses the power of changing the convexity of the cornea. In the Owls they extend from the cornea over the long anterior division of the eye to the posterior hemisphere, which they also contribute to form. The figure of the eye is thus maintained, notwithstanding its want of sphericity.

The bony plates are capable of a degree of motion upon each other, which is, however, restrained within certain limits by the attachments of their anterior and posterior edges to the sclerotic coat; and by their being bound together by a tough ligamentous substance, as it were the continuation of the sclerotic between the edges that overlap each other.

The *cornea*, fig. 57, *c*, possesses the same structure as in Mammalia, but differs with respect to form. When the posterior part of the eye is compressed by the muscles, the humours are urged forward and distend the cornea; which, at that time, becomes more prominent than in Mammalia; and under such circumstances, the eye is in a state for perceiving near objects. When the muscles are relaxed, the contents of the eyeball retire to the posterior part, and the cornea becomes flatter: this is the condition in which we find the eye of a dead bird, but we can have no opportunity of perceiving it during life. It is only practised for the purpose of rendering objects visible that are placed at an extreme distance. From the well-known effects of form upon refracting media, it must be presumed, that the cornea is least convex when a bird which is soaring in the higher regions of the air, and invisible to us, discerns its prey upon the earth; its form will change as the bird descends with unerring flight to the spot, as is customary with many of the rapacious tribe.

On reflecting the sclerotica from the choroid, a grey substance is seen upon the fore part of the latter, like a ring: it consists of fibres showing, like those of the iris, the transverse striæ, and which serve to attach the choroid to the sclerotic plates and contiguous margin of the cornea. These fibres are regarded by the anatomist, who first called attention to their muscular nature in Birds, as helping 'to accommodate the eye to the different distances of objects,' being supposed to act upon the cornea in a manner analogous to that of the muscles of the diaphragm upon its tendinous centre.[1]

57

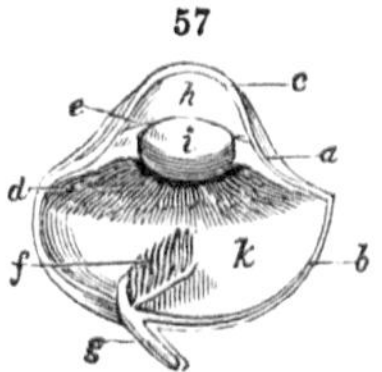

Section of eyeball, *Falco*. XXXIV'.

The *choroid coat* is a loosely cellular and highly vascular membrane, devoid of 'tapetum,' and copiously covered or saturated with a black pigment. Opposite the bony circle the choroid separates into two layers; the external layer is the thinnest, and adheres at first firmly to the sclerotica, after which it is produced freely inwards to form, or be continuous with, the iris.

The *iris*, fig. 57, *e*, is delicate in its texture, which under the lens appears composed of a fine network of interlacing fibres, but it is remarkable for the activity and extent of its movements, which

[1] XXXIV'. p. 170.

seem in some Birds to be voluntary. The contraction and dilatation of the pupil, independently of any change in the quantity of light to which the eye is exposed, is most conspicuous and remarkable in the Parrot tribe, but it has been observed also in the Cassowary and other birds.[1]

The colour of the iris is subject to many varieties, which frequently display great brilliancy, and afford zoologists distinguishing specific characters of Birds; although these cannot always be implicitly relied upon. The breadth of the iris varies in different species, but is greatest in Birds which take their food in the gloom, e. g., Owls and Nightjars, in order that the pupil may be proportionally enlarged to admit as much light as possible to the retina. The ciliary nerves and vessels run in the form of single trunks between the choroid and sclerotica, and terminate anteriorly in several ring-shaped plexuses for the supply of the iris and of the muscular circle of the cornea. The pupil is usually round: in the Goose and Dove it is elongated transversely, and in the Owls is vertically oval.

The inner layer of the choroid is thicker than the external, and is disposed in numerous thickly set plicæ radiating towards the anterior part of the crystalline lens, where they terminate in slightly projecting *ciliary processes*, fig. 57, *d*, the extremities of which adhere firmly to the capsule of the crystalline. These processes are the most numerous, close set, and delicate in the Owl: they are proportionally larger and looser in the Ostrich.

The chief peculiarity in the eye of the Bird is the *marsupium* or *pecten*, ib. *f*, which is a plicated vascular membrane analogous in structure to the choroid, and equally blackened by the pigmentum; situated in the vitreous humour anterior to the retina, and extending from the point where the optic nerve penetrates the eye to a greater or less distance forward, being in many Birds attached to the posterior part of the capsule of the lens. As its posterior point of attachment is not to the choroid but to the termination of the optic nerve, this requires to be first described.

When the optic nerve, ib. *g*, arrives at the sclerotic, it tapers into a long conical extremity, which glides into a sheath of a corresponding figure, excavated in the substance of that membrane, and directed downward and obliquely forward. The central or inner layer of this sheath is split longitudinally, and the plicated substance of the nerve, fig. 48, passes through this fissure. A similar but longer fissure exists in the corresponding part of the choroid: so that the extremity of the optic nerve presents in

[1] VII'. p. 304.

the interior of the eye, instead of a round disc, as in Mammalia, a white narrow streak, from the extremities and sides of which the retina is continued. Branches of the ophthalmic artery, distinct from the vessels of the choroid, and homologous with the *arteria centralis retinæ*, enter the eye between the laminæ of the retina, along the whole extent of the oblique slit above mentioned, and immediately penetrate the folds of the marsupial membrane, upon which they form delicate ramifications. These vessels are shown in fig. 58, representing the excised marsupium unfolded and spread out.

The marsupium is lodged like a wedge in the substance of the vitreous humour, in a vertical plane, directed obliquely forward. In those species in which the marsupium is widest, the angle

58

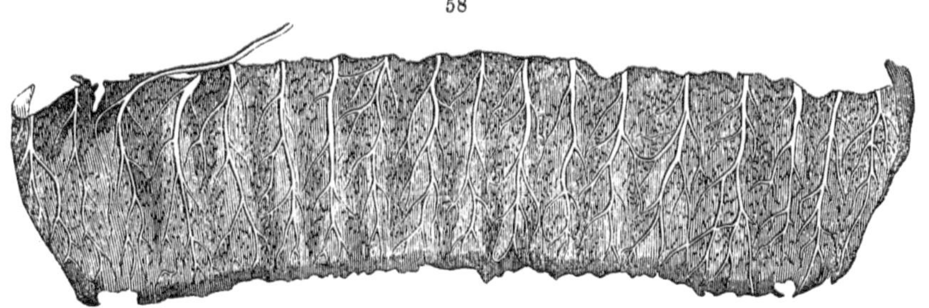

Marsupium of the eye of a Rook, unfolded. XXVII'.

next the cornea reaches the inferior edge of the capsule of the crystalline; but where it is narrow, the whole anterior border is in contact with the same point. This contact is close in some Birds, as the Vulture, Parrot, Turkey, Cassowary, Stork, Goose, and Swan; but in other Birds the marsupium does not extend further than two-thirds of the distance from the back part of the eye, and is attached at its anterior extremity to some of the numerous laminæ of the hyaloid membrane which form the cells for the lodgment of the vitreous humour. In these cases the marsupium can have no influence on the movements of the lens, unless it be endowed with an erectile property, and be so far extended as to push forward the lens. There is no muscular structure in the marsupium; and its changes of form, if such occur in the living bird, must be effected by changes in the condition of the vessels of which it is almost exclusively composed.

The form of the marsupium varies in different Birds; it is broader than it is long in the Stork, Heron, Turkey, and Swan; and of the contrary dimensions in the Owl, Ostrich, and Cassowary. The plicæ of the membrane are perpendicular to the terminal line of the optic nerve; they are of a rounded figure in most species, but in the Ostrich and Cassowary they are compressed, and so far inclined from the plane of the membrane, that their convergence towards its extremity gives it a resemblance to a close-drawn purse.[1] The folds vary in number, being four in the Cassowary, seven in the Great Horned Owl, eight in the Maccaw, from ten to twelve in the Duck and Vulture, fifteen in the Ostrich, sixteen in the Swan and Stork, and still more numerous in the Insessorial Birds, amounting to twenty-eight, according to Soemmerring, in the Fieldfare.

The exact functions of the marsupial membrane are still involved in obscurity. Its position is such that some of the rays of light proceeding from objects laterally situated with respect to the eye must fall upon and be absorbed by it; and Petit accordingly supposed that it contributed to render more distinct the perception of objects placed in front of the eye.

Some physiologists have supposed that this black membrane was extended toward the centre of the eye, where the luminous rays are most powerfully concentrated, in order to absorb the excess of intense light to which Birds are exposed in soaring aloft against the blazing sun. Others have considered it as the gland of the vitreous humour, and that, as this fluid must be rapidly consumed during the frequent and energetic use made of the visual organ by Birds, it therefore might require a superadded vascular structure for its reproduction.

The marsupium may act as an erectile organ, and occupy a variable space in the vitreous humour: when fully injected, therefore, it will tend to push forward the lens, either directly or through the medium of the vitreous humour, which must be displaced in a degree corresponding to the increased size of the marsupium; the contrary effects will ensue when the vascular action is diminished. The nocturnal *Apteryx,* in which the eye is so small, shows also the exception of the absence of the marsupium.

The *retina* is continued from the circumference of the base of the marsupium, and after unfolding its plicæ expands into a smooth layer of medullary matter, which seems to terminate at the periphery of the corpus ciliare. In the Owls not more than half the globe of the eye is lined by the retina; it ceases in fact where the eye loses the spherical form at the base of the anterior cylindrical portion.

[1] The Parisian Academicians, who took their description of this part from the Ostrich, first applied to it the name of *Marsupium* or *Bourse.* XL'.

The humours of the eye no less correspond to the peculiar vision of the Bird, and the rare medium through which it is destined to move, than the shape of the globe and the texture of its coats.

The *aqueous humour* is extremely abundant, owing to the extent of the anterior chamber gained by the convexity of the cornea, and its refractive power must be considerable in the higher regions of the atmosphere. The membrane inclosing it can be more readily demonstrated in Birds than in most Mammals, especially where it adheres to the free edge of the iris. The large size of the ciliary processes may have the same relation to the reproduction of the aqueous, as the marsupium is supposed to have with reference to the vitreous, humour.

The *crystalline lens* is remarkable for its flattened form, especially in the high-soaring Birds of Prey; it is also of a soft texture, and is without the hard nucleus found in Fishes and Reptiles. In the Cormorant and other birds which seek their food in water, the crystalline is of a rounder figure, and this is peculiarly the case in the nearsighted Apteryx and Owls which hunt for prey in obscure light. It is inclosed in a distinct capsule, which adheres very firmly to the depression in the anterior part of the vitreous humour; the capsule is itself lodged between two layers of the membrana hyaloidea, which, as they recede from each other to pass—the one in front and the other behind the lens—leave round its circumference the sacculated canal of Petit.

The vessels of the lens are derived from those of the marsupium, which, as before observed, are ramifications of the homologue of the arteria centralis retinæ: this is not continued as a simple branch from its origin to the marsupium; but, immediately before penetrating the coats of the eye, it breaks into numerous subdivisions, the aggregate of which is greater than the trunk whence they proceed, and these again unite, forming a plexus, *s*, fig. 59, close to the external side of the optic nerve. The artery of the marsupium proceeds from this plexus, and runs along the base of the folds, giving off at right angles a branch to each fold, which in like manner sends off smaller ramuli, fig. 58. The plexus at the origin of the marsupial artery serves as a reservoir for supplying the blood required for the occasional full injection of the marsupium; and a similar but larger plexus, fig. 59, 4, is formed at the origins of the ciliary arteries which supply the erectile tissue of the ciliary processes and iris.

The *vitreous humour* presents few peculiarities worthy of note;

compared with the aqueous humour, it is proportionally less in quantity than in the eyes of Mammals. The outer capsule formed by the hyaloid membrane is stronger, and can be more easily separated from the humour.

The eyeball is moved in Birds by four straight and two oblique muscles. The *Recti muscles* arise from the circumference of the optic foramen, and expand, as they pass forward, to be inserted into the soft middle part of the sclerotic. We have not been able to trace their insertion distinctly to the osseous circle; their aponeurosis cannot be reflected forward from the sclerotica without lacerating that membrane.

59

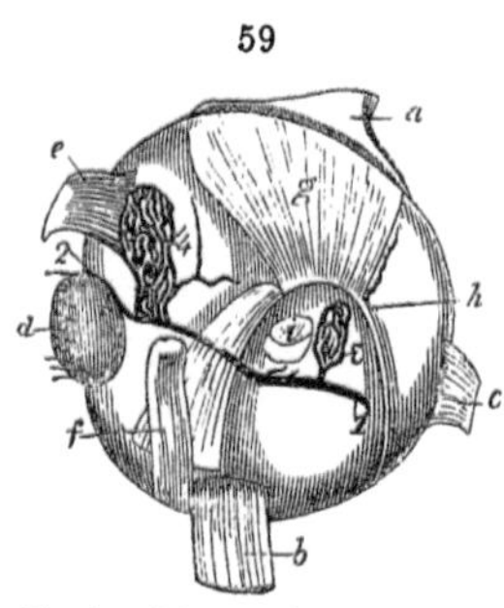

Muscles of the eye, Goose. XLIV'.

The *Obliqui* both arise very near together from the anterior parietes of the orbit, and go to be inserted, the one into the upper, the other into the lower part of the globe of the eye; the superior obliquus does not pass through a pulley, as in Mammalia. All the muscles are proportionally short in this class, but especially so in the Owls, in which the eye, from its large size and close adaptation to the orbit, can enjoy but very little motion. In figs. 56 and 59, *a* is the *rectus superior* or *attollens*; *b* the *rectus inferior* or *deprimens*; *c* the *rectus externus* or *abducens*; *d* the *rectus internus* or *adducens*; *e* the *obliquus superior*; *f* the *obliquus inferior*; *g* the *quadratus*; *h* the *pyramidalis*.

The accessory parts of the eye in Birds are similar to those of the higher Reptiles. There are three eyelids, two of which move vertically, and have a horizontal commissure, while the third, which is deeper-seated, sweeps over the eyeball horizontally, from the inner to the outer side of the globe. The vertical, or upper and lower eyelids, are composed of the common integument, of a layer of conjunctiva, and between these of a ligamentous aponeurosis, which is continued into the orbit, and lines the whole of that cavity. The lower eyelid is the one which generally moves in closing the eye in sleep, and it is further strengthened by means of a smooth oval cartilaginous plate, which is situated between the ligamentous and conjunctive layers.

The *orbicularis* muscle is so disposed as by means of this plate to act more powerfully in raising the lower than in depressing the upper eyelid. In the latter it is continued immediately along the margin: in the lower eyelid the tarsal cartilage intervenes between the muscle and the ciliary margin.

The *levator palpebræ superioris* arises from the roof of the orbit, and is inserted near the external angle of the lid. There is also an express muscle for depressing the lower eyelid, as in the Crocodile. In the Owls and Nightjar (*Caprimulgus*) the eyelids are closed principally by the depression of the upper one. There are but few Birds that possess eyelashes; of these the Ostrich is an example, as also the Hornbills and the Owls, in which they are arranged in a double series; but here they are rather to be considered as feathers with short barbs, than true eyelashes.

The third eyelid, or membrana nictitans, is a thin membrane, transparent in some Birds, in others of a pearly white colour.

Two muscles are especially provided to effect its movements, but are so placed as to cause no obstruction to the admission of light to the eye during their actions. One of these is called the *quadratus nictitantis*, fig. 59, *g*; it arises from the sclerotica at the upper and back part of the globe of the eye, and its fibres slightly converge as they descend towards the optic nerve, above which they terminate in a tendinous sheath, having no fixed insertion. The second muscle, called *pyramidalis nictitantis*, ib. *h*, arises from the lower and nasal side of the eyeball: its fibres converge toward the upper part of the optic nerve, and terminate in a small round tendon which glides through the pulley at the free margin of the *quadratus*; thus, winding over the nerve, it passes down to be inserted into the lower part of the margin of the third eyelid. By the simultaneous action of the two muscles, that nictitating lid is drawn outward and obliquely downward over the fore part of the eyeball. The tendon of the pyramidalis gains the due direction for that action by winding round the optic nerve, and it is restrained from pressing upon the nerve by the counteracting force of the *quadratus*, which thus augments the power of the antagonist muscle, while it obviates any inconvenience from pressure on the optic nerve, which its peculiar disposition in relation to that part would otherwise occasion. The nictitating membrane returns, on the relaxation of its muscles, by virtue of its own elasticity, to the inner corner of the orbit, where it lies folded when not in use.

The lacrymal glands are two, as in Reptiles; but the inner one is the largest, especially subserving the more frequent movements of the nictitating membrane: it is called the 'harderian gland,' fig. 56, *d'*, is situated at the inner or nasal canthus, has a lobulated exterior, and emits its viscid secretion by a short duct which opens beneath the third lid. The 'lacrymal gland,' fig. 59, *d*, lies at the posterior and external part of the eyeball; in the Goose it is of a flattened form, about the size of a pea, and pours its thinner transparent secretion, by a short wide duct, upon the inside of the outer canthus of the eyelids. The naso-lacrymal conduit commences by two apertures at the nasal canthus, and terminates below and a little before the middle turbinal. In the Ostrich there is a glandular prominence at each 'punctum,' analogous to a 'caruncula lacrymalis,' but this structure is not present as a rule in Birds.

Besides the two glands which serve to lubricate and facilitate the movements of the eyeball and eyelids, there exists another gland which from its position in or near the orbit seems to belong to the lacrymal group; but its secretion is exclusively employed upon the pituitary membrane of the nose, and it corresponds rather to the nasal gland of Serpents. In many water and marsh Birds the gland in question is lodged in the superorbital fossa, before described, p. 61; but in most Birds it is situated within the orbit, either beneath the nasal or between it and the maxillary: in the Woodpecker it is found in the subocular air-cell. I have detected it in one or more species of every order of Birds. In the Anserines the gland is large, and seems to complete the upper margin of the orbit, fig. 56, *k*, and is enclosed in a dense fibrous capsule. It is composed of ramified follicles, with cellular walls. In the Albatross and Penguin it sends two or three ducts to the nasal cavity.

CHAPTER XVII.

DIGESTIVE SYSTEM OF BIRDS.

THE digestive function is most potent and rapid in Birds, in order to supply the waste occasioned by their extensive, frequent and energetic motions, and in accordance with the rapidity of their circulation and their high state of irritability.[1]

The parts to be considered with reference to this function are the rostrum or beak, the tongue, the œsophagus, the stomach which is always divided into a glandular and muscular portion, the intestines, and the cloaca: with these are connected the salivary glands, the proventricular follicles, the liver and pancreas.

§ 145. *Beaks of Birds.*—The *beak* consists of an 'upper mandible,' supported by the maxillary and premaxillary bones, and of a 'lower mandible' formed by the lower jaw. In place of teeth these bones are provided with a sheath of horny fibrous material, similar to that of which the claws are composed: this sheath is moulded to the shape of the osseous mandibles, being formed by a vascular substance covering these parts, and its margins are frequently provided with horny processes or laminæ secreted by distinct pulps, analogous in this respect to the whalebone laminæ of the Whale. In a fœtus of a Perroquet nearly ready for hatching, the margins of the bill are beset with white and round tubercles, arranged in a regular order, about seventeen in the upper jaw, the foremost on the mid-line.[2] These tubercles are not, indeed, implanted in the alveolar border, but form part of the sheath of the bill. Under each tubercle, however, there is a gelatinous pulp, like that of a tooth, but resting on the edge of the jaw-bones, and every pulp is supplied by vessels and nerves traversing a canal in the substance of the bone. These tubercles form the first margins of the mandibles, and their remains are indicated by canals in the horny sheath subsequently formed, which contain a softer material, and which commence from small foramina in the margin of the bone.

[1] The Cormorant devours, in captivity, six or eight pounds of fish daily; what may be the amount in its state of wild activity! [2] xxxvi'.

The different degrees of hardness and varieties of form of the beak exercise as much influence upon the nature of Birds as the number and figure of the teeth do upon that of Mammals.

The beak is hardest in those Birds which tear their prey, as Eagles and Falcons; in those which bruise hard seeds and fruits, as Parrots and Grosbeaks; and in those which pierce the barks of trees, as Woodpeckers, in the larger species of which the beak absolutely acquires the density of ivory. The hardness of the covering of the beak gradually diminishes in those Birds which take less solid nourishment, or which swallow their food entire; and it changes at last to a soft skin in those which feed on tender substances, or which have occasion to probe for their food in muddy or sandy soils, or at the bottom of the water, as Ducks, Snipes, Woodcocks, &c.

Cæteris paribus, a short beak must be stronger than a long one, a thick one than a thin one, a solid one than one which is flexible; but the general form produces much variety in the application of the force. A compressed beak with trenchant edges, and a hooked, sharp-pointed end, is the fit instrument for seizing and slaying prey, whether birds, beasts, or fishes; and such 'aduncate' beak is seen in the Frigate-bird, Tropic-bird, Albatross, Petrel, fig. 52, but combined with length in these piscivorous birds. In the *Raptores* the beak is shorter and stronger, and in some genera a tooth-like process on either side of the upper mandible, fig. 60, adds to its destructive power: hence the Falcons, having this armature, are reckoned the more 'noble' or courageous birds of prey.

60

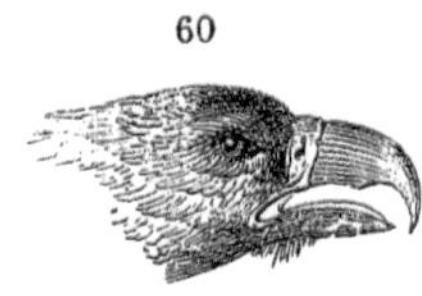

Beak of Falcon.

The Shrike (*Lanius*) and Vanga, which have their bill similarly armed, fig. 61, have the cruel disposition of the Hawk, but take prey proportioned to their small size: and the 'tooth' is confined to the horny sheath, fig. 61, not developed on the bone. As the beak becomes straighter and conical with the margin entire, the bird is less daring in attacks on other kinds, though occasionally predaceous when large and strong (as the Raven and Crow, fig. 62): but most 'conirostrals' are omnivorous, and the rest granivorous, as the 'Hard-billed *Passeres*' of Ray. When the cone is attenuated and lengthened out, fig. 63, it is adapted to extract

61

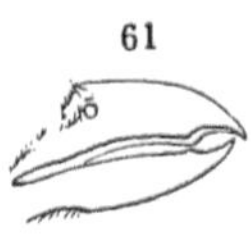

Beak of a Shrike.

62

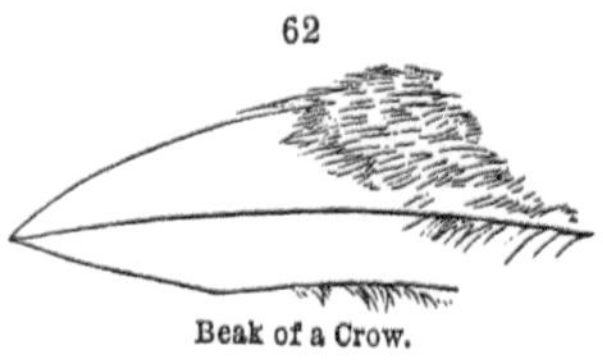

Beak of a Crow.

delicate insects from the recesses of trees and flowers: and the type 'tenuirostrals' (*Trochilidæ*) may suck up, also, the sweet juice of the nectarium.

The *Fissirostrals*, fig. 64, like the Humming-birds, feed on the

63

64

Beak of Humming-bird (*Orthorhyncus*).

Rostrum of the Caprimulgus.

wing, but as their food consists of volant insects, the form of the beak is modified accordingly, and is remarkable for its shortness and the wideness of its gape, fig. 64, especially in the typical families. In these the mode of catching the prey is conformable to their distinguishing characters; they receive it in full flight into the cavity of their mouths, which remain open for that purpose, and where a viscous exudation within, and a strong fence of 'vibrissæ' on the exterior, assist in securing the victim.

A strong, trenchant and pointed, but elongated and straight, bill serves to cut and pierce, and characterises many Waders preying upon reptiles, fishes, and animals that offer some resistance: such a beak is found in the Herons and Bitterns. As it becomes more lengthened and attenuated it is adapted to prey of a lower grade of life, and to get at these it is endowed with a specially sensitive apex. In the Ibis and Curlew such a beak is curved down, fig. 3: in the Jabiru, fig. 65, it is bent up. Some trenchant bills are so compressed as to resemble the blade of a knife; these offer least resistance in the swift pursuit of fishes, and are seen in the Awks, Puffins, and Coulternebs, in which

65

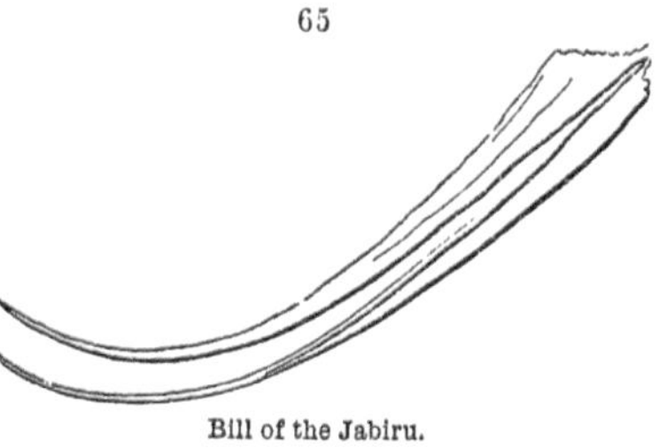

Bill of the Jabiru.

66

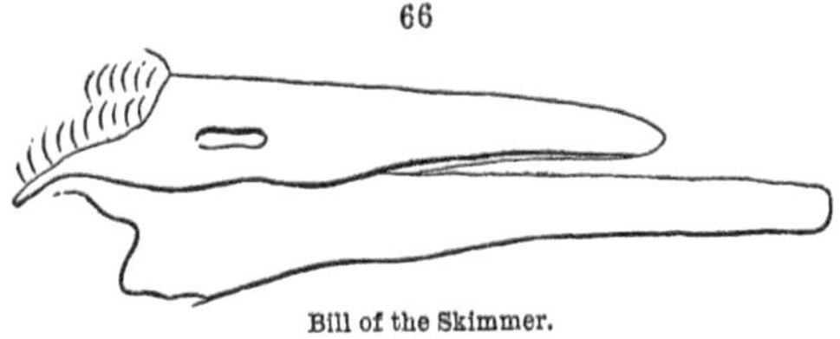

Bill of the Skimmer.

latter the beak may be as deep as it is long. The Skimmer

(*Rhyncops*) has the further peculiarity of an inequality in the length of the two mandibles, the upper one being the shortest, fig. 66, so that this sea-bird gets its food, which consists of floating marine animals, by pushing and tilting them within the action of the upper blade as it swims along.

A sharp-edged beak may be as remarkable for transverse extension and depression, or horizontal flattening: and such a form serves for capturing fishes and reptiles: it is seen in the Boatbills of South America (*Cancroma*), fig. 67, and of Nubia (*Balæniceps*).

67

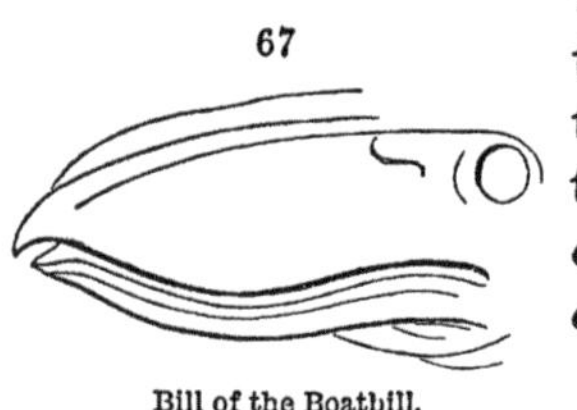

Bill of the Boatbill.

Of the blunt-edged bills we may first notice those which are flattened horizontally. When a bill of this description is long and strong,

68

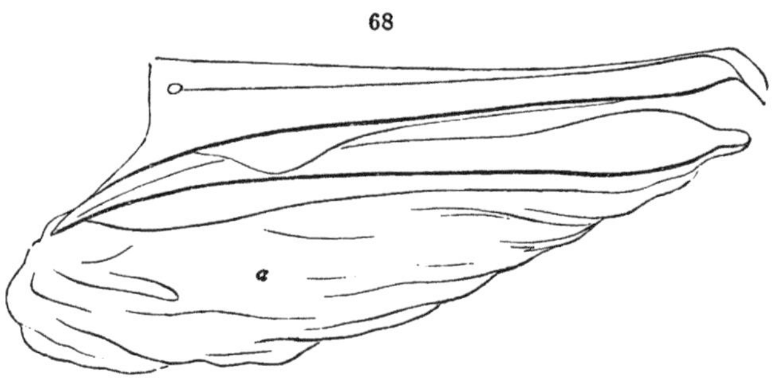

Bill and pouch of the Pelican.

as in the Pelican, fig. 68, it serves to seize large but feebly resisting fishes.

When it is long and weak, as in the Spoonbill, which derives its name from the dilated extremity of the mandibles, it is only available to seize amid sand, mud, or water, very small Crustaceans, Mollusks, &c., fig. 69.

69

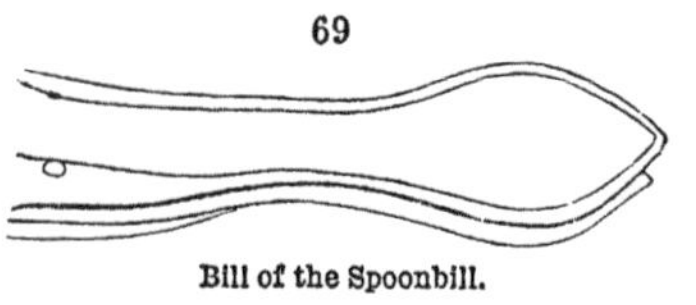

Bill of the Spoonbill.

The more or less flattened bills of Ducks, the more conical ones of Geese and Swans, and that of the Flamingo, of which the extremities of the mandibles are bent downwards abruptly, fig. 70, have all transverse horny laminæ arranged along their edges, which when the bird has seized any object in the water, serve, like the whalebone laminæ of the Whale, to give passage to the superfluous fluid. The aquatic habits of all these birds are in harmony with this structure. But the long-legged palmiped sifts the sand of the sea-shore by raking it up with the bill reversed, as shown in fig. 14. In the Goosanders (*Mergus*, fig. 71), the lateral laminæ are developed into small conical reflected tooth-like processes, which serve to hold fast the fishes on which they feed.

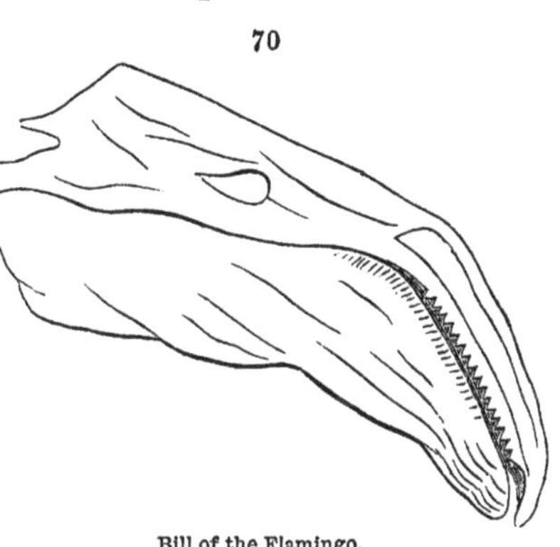

Bill of the Flamingo.

The bills of the Toucans and Hornbills are remarkable for their enormous size, which is sometimes equal to that of the whole bird. The substance of the beak in these cases is extremely light and delicately cellular; yet the osseous portions are adapted to combine, with great bulk, a due degree of strength. The external parietes are extremely thin, especially in the upper beak: they are elastic, and yield in a slight degree to moderate pressure, but present considerable resistance if the force be increased for the purpose of crushing the beak: they gain thickness at the points of the mandibles.

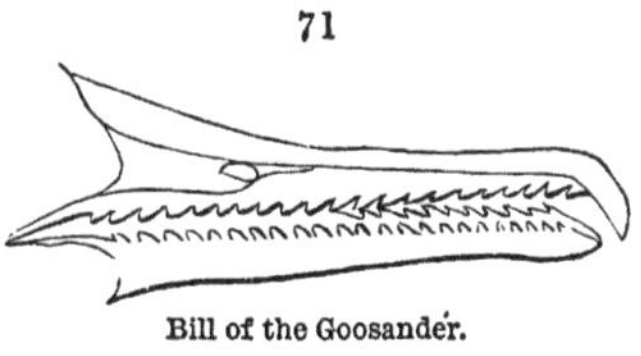

Bill of the Goosander.

On making a longitudinal section of the upper mandible, fig. 53, *a*, its base is seen to include a conical cavity about two inches in length and one inch in diameter, with the apex directed forward. The walls of this cone consist of an osseous network, intercepting irregular angular spaces, varying in diameter from half a line to two lines. From the parietes of the cone a network of bony fibres is continued to the outer parietes of the mandible, the fibres which immediately support the latter being almost invariably at right angles to the part in which they are inserted. The whole of the mandible anterior to the cone is occupied with a similar network, the meshes of which are largest in the centre of the beak, in consequence of the union which takes place between different small fibres as they pass from the circumference inwards. The principle of the cylinder is introduced into this structure: the smallest of the supporting pillars are hollow. The structure is the same in the lower mandible, ib. *m*, but the fibres composing the network are in general stronger than those of the upper mandible.

The air is admitted to the interior of the upper mandible from a cavity, ib. *b*, situated anterior to the orbit, which communicates at its posterior part with the air-cell continued into the orbit, and at its anterior part with the maxillary cavity. The nasal cavity is closed at every part except at its external and internal apertures by the pituitary membrane, and has no communication with the interior of the mandible.[1]

The horny sheath of the mandibles in the Hornbills and Toucans is so thin that it often becomes irregularly notched at the edge from use. The Hornbills have, besides, upon their enormous beak, horn-like prominences of the same structure and of different forms, the use of which is not known.

The Trogons, Touracos, Buccos, &c., exhibit forms of the bill which are intermediate to that of the large but feeble bill of the Toucans, and the short, but hard, strong, and broad bill of the Parrot-tribe, which is also hooked, so as to assist in climbing, like a third foot, fig. 30.

The short, conical, and vaulted beak of the *Rasores*, fig. 72, serves to pick up with due rapidity the vegetable seeds and grains which constitute their food, as well as small insects, as ants, &c., with which the young are frequently nourished. The tooth-billed pigeon of the Samoan Isles has the lower mandible deeply cleft into three points near the top, and the upper mandible hooked, the better for seizing fruit and denuding palm-nuts and other strongly coated kinds.

72

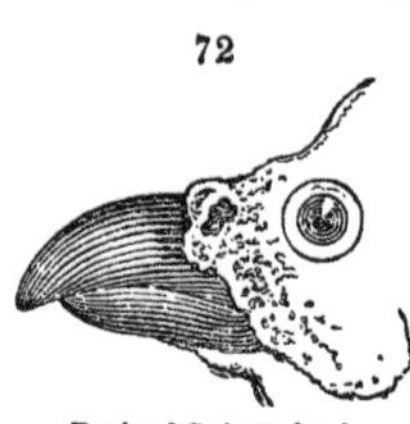

Beak of Guinea-fowl.

The bills of the small Insessorial or Passerine birds present every gradation of the conical form, from the broad-based cone of the Hawfinch to the almost filamentous cone of the Humming-bird, fig. 63, and each of these forms influences the habits of the species in the same manner as in the larger birds. The short and strong-billed Insessores live on seeds and grains; those with a long and slender bill on insects or vegetable juices. If the slender bill be short, flat, and the gape very wide, as in Swallows, the bird takes the insects while on the wing; if the bill be elongated and endowed with sufficient strength, as in the Hoopoes, it serves to penetrate the soil and pick out

[1] xx'.

worms, &c. One kind of Humming-bird, feeding on spiders, has the end of the bill finely toothed.

Of all bills, the most extraordinary is that of the Cross-bill, in which the extremities of the mandibles curve towards opposite sides and cross each other at a considerable angle—a disposition which at first sight seems directly opposed to the natural intention of a bill. With this singular disposition, the Cross-bill, however, possesses the power of bringing the points of the mandibles into contact with each other; and can pick up the smallest seeds, and shell or husk larger kinds like other birds. But the disposition and power of the muscles is such that the bill gains by its very apparent defect the requisite power for breaking up the pine-cones and wrenching out the seeds that constitute its usual food.

§ 146. *Tongues of Birds.*—The tongue, as has been already observed, can hardly be considered as an organ of taste in Birds, since, like the mandibles, it is generally sheathed with horn. It is principally adapted to fulfil the offices of a prehensile organ in association with the beak, and it presents almost as many varieties

73

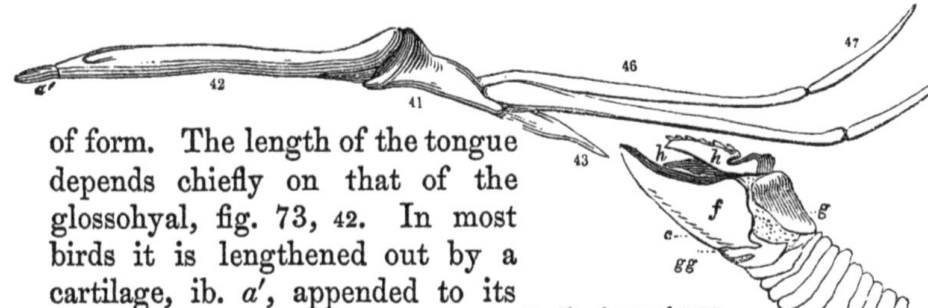

Hyoid and upper larynx, Swan.

of form. The length of the tongue depends chiefly on that of the glossohyal, fig. 73, 42. In most birds it is lengthened out by a cartilage, ib. *a'*, appended to its extremity. This is remarkable in the Swan and other *Lamellirostres*. The ceratohyals are obsolete. The basihyal, 41, contracts as it recedes to support the urohyal, 43, and the hypo- 46, and cerato- 47, branchials are modified to form the posterior cornua or 'thyro-hyals,' which are of moderate length. The tongue supported by the glossohyal is broad, and furnished with a series of retroverted spines, fig. 75, D. In the Humming-bird the horny sheath of the glossohyal is divided at its extremity into a pencil of fine hairs. In the Toucan's tongue, fig. 51, the sheath gives off from the lateral margins stiff bristle-like processes which project forward: this structure is continued to the apex, and the tongue so provided becomes an instrument for testing the softness and ripeness of fruit, and the fitness of other objects for food, thereby acting as a kind of antenna or feeler. A similar but less developed structure is found in the tongue of the frugivorous Touraco.

In the Woodpeckers the apex of the horny sheath, fig. 74, 77, *a*, gives off at the sides short pointed processes directed backward, converting it into a barbed instrument for holding fast the insects which its sharp point has transfixed, after the strong beak has dislodged them from their hiding places. The cornua (thyrohyals, ib. 46, 47) wind round the back of the head, and converge as they pass forward to be inserted in a canal generally on the right side of the upper mandible, ib. *e*.

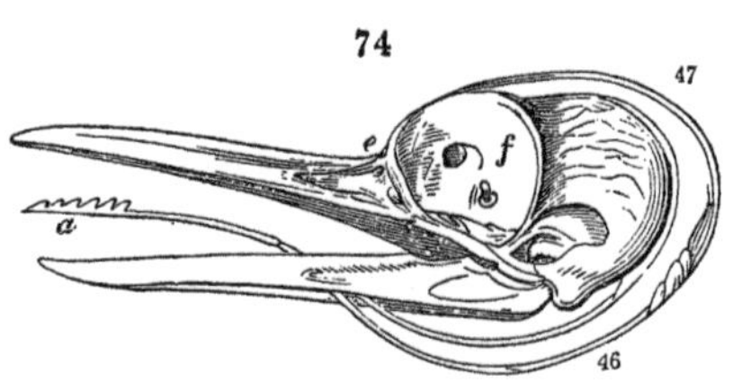

Cranium and tongue of a Woodpecker.

The tongue of the Flamingo is almost cylindrical, slightly flattened above, and obliquely truncate anteriorly, so as to correspond with the form of the inferior mandible. The pointed

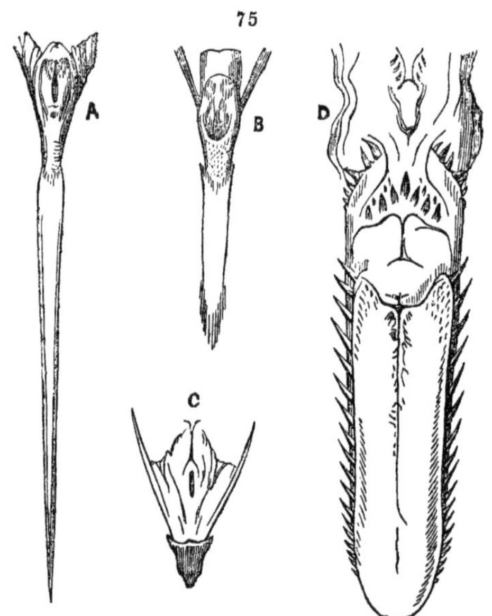

Tongue A. Snipe. B. Fieldfare. C. Kingfisher. D. Goose. CCXL'.

extremity of the truncated part is supported beneath by a small horny plate. Along the middle of the upper surface there is a moderately deep and wide longitudinal furrow; on either side of which there are from twenty to twenty-five recurved spines, from one to three lines in length. These spines are arranged in an irregular alternate series: the outer ones being the smallest, which may almost be considered as a distinct row. At the posterior part of the tongue there are two groups of smaller recumbent spines directed towards the glottis. The substance of the tongue is not muscular, but is chiefly composed of an abundant elastic cellular substance, permeated by an oily fat.[1] Of like nature is the tongue in Anserines: but the retroverted spines are marginal, fig. 75, D. The tongue of the great Penguin is beset with horny spines like a hedgehog's skin.

In the *Raptores* the tongue is of a moderate length, broad, and somewhat thick, and has a slight division at the tip. In the Vultures its sides can be voluntarily approximated so as to form a canal, and its margins are provided with retroverted spines. In the Raven it is bifid at the apex: it is more deeply cleft in the 'Nutcracker.'

In the Struthious Birds, in many of the Waders, and in the *Pelecanidæ*, the tongue is remarkably short, as it is likewise in the Kingfisher, fig. 75, C. In the Snipe it is as remarkable for its length and slenderness, ib. A. In the Fieldfare (*Turdus pilaris*) the sheath is resolved into fine filaments at the apex of the tongue, ib. B.

In the Parrots the tongue is thick and fleshy, is terminally tufted in Lories, serves admirably to keep steady the nut or seed upon which the strength of the mandibles is exerted, and is applied to the kernel so extracted, as if to ascertain its sapid qualities.

The following are the muscles of the tongue in Birds.

76

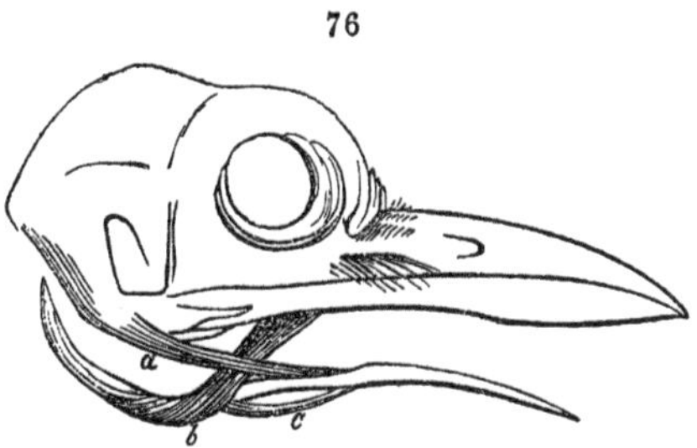

Muscles of the tongue of the Fieldfare (*Turdus pilaris*).

1st. The *Mylo-hyoideus*: this is a thin layer of fibres attached to the lower and inner border of the lower jaw, and running transversely to a mesial tendon which separates them, and extends to the urohyal. It raises the tongue towards the palate.

2nd. The *Stylo-hyoideus*, fig. 76, *a*, arises from the upper and back part of the lower jaw, and is inserted into the thyrohyal at

[1] XXI.

its junction with the basihyal. In some birds it divides into three or more portions: the *posterior* descends obliquely forward, and is inserted into the tendinous commissure of the mylohyoideus: the *middle* portion is inserted into the urohyal: the *anterior* fasciculus is inserted into the side of the basihyal above the transverse hyoglossus. The actions of these different portions vary according to their insertion; the first and second depress the apex of the tongue by raising the urohyal, the third raises the tongue and draws it to one side when it acts singly.

3rd. The *Genio-hyoideus,* fig. 76, *b*: this arises by two fleshy bands from the lower and internal edge of the lower jaw; these unite, pass backward, and surround the cornua (thyrohyals); and as they draw them forward protrude the tongue from the beak.

4th. The *Cerato-hyoideus*: this passes from the thyrohyal to the urohyal, and is therefore subservient to the lateral movements of the tongue.

5th. The *Sterno-hyoidei*: these are replaced by a slip of muscle which extends from the anterior surface of the upper larynx to be attached to the base of the glossohyal.

6th. A small and short muscle, which is single or azygos; it passes from the basihyal to the under part of the glossohyal; it depresses the tip of the tongue and elevates its base.

7th. A short muscle, fig. 75, *c*, which arises from the junction of the basihyal with the urohyal, and is inserted into the thyrohyal.[1]

All these muscles are remarkably large in the Woodpecker, in which there is a singular pair of muscles that may be termed *Cerato-tracheales* (fig. 77, *h*). They arise from the trachea about eight lines from the upper larynx, twist four times spirally round the trachea, and then pass forward to be inserted into the base of the thyrohyals. This is the principal retractor of the singular tongue in this species.

§ 147. *Salivary Glands.*—The salivary organs, being in general developed in a degree corresponding to the extent of the changes which the food undergoes in the mouth, and the length of time during which it is there detained, are by no means so conspicuous a part of the digestive system in Birds as in Mammals. Glands which pour out their secretion upon the food prior to deglutition are, however, met with in every bird, but vary in number, position, and complexity of structure.

In some species, as the Crow, they are of the simplest structure, consisting of a series of unbranched, cone-shaped follicles or

[1] Dr. Salter proposes the name of 'Cerato-glossal' for this muscle. CCXL'. p. 1140.

tubules, opening separately upon the mucous membrane of the mouth, along the sides of which cavity they are situated. They pour out a viscid mucus, and are the only traces of a salivary system met with in this bird.

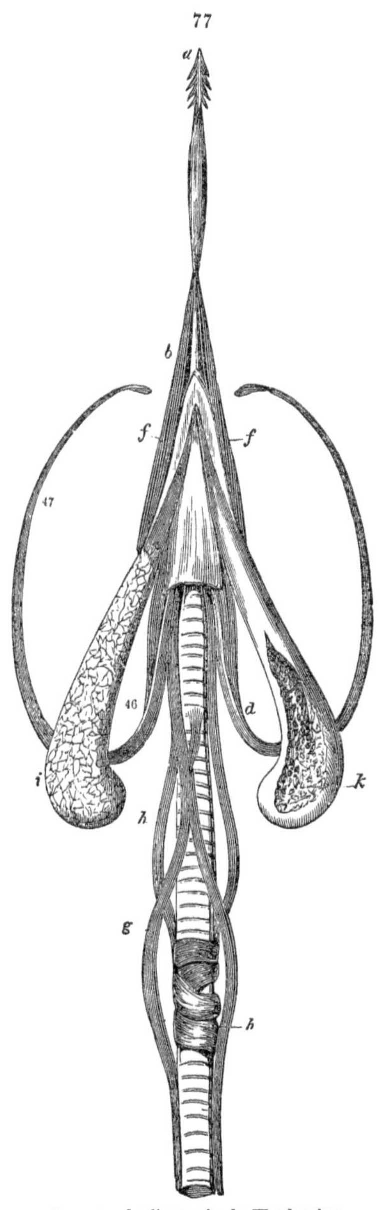

Tongue and salivary glands, Woodpecker.

In many other birds, and especially in the Scratching, Wading, and Swimming Orders, glands of the conglomerate structure are found beneath the lower jaw, answering to the submaxillary glands of quadrupeds.

In the Goose they occupy the whole of the anterior part of the space included by the rami of the lower jaw, being of an elongated form, flattened and closely united together at the middle line. On either side of this line the mucous membrane of the mouth presents internally a series of pores, each of which is the terminal orifice of a distinct gland or aggregate of ramified ducts.

A third and higher form of salivary gland, in which the secretion of the conglomerate mass is conveyed into the mouth by a single duct, is found in the Woodpeckers and some species of the Rapacious Order. In the latter birds these glands are termed, from their situation, *anterior palatine*: in the *Picæ* they correspond to the parotid and sublingual of Quadrupeds.

The sublingual glands of the Woodpecker, fig. 77, *i*, *k*, are of extraordinary size, extending from the angle to the symphysis of the lower jaw. The single ducts of each gland unite just before their termination, which is a simple orifice at the apex of the mouth.

Besides the preceding, which may be considered as the true salivary glands, there are numerous accessory follicles in different parts of the oral apparatus of Birds. In the Waterhen (*Gallinula chloropus*) there is a series of cœcal glandular tubes along each side of the tongue: similar elongated follicles are situated along the margin of the lower jaw, resembling in their parallel pectinated disposition the branchiæ of Fishes. In the Goose the corresponding follicles are longer and wider, and are situated near the sides of the tongue. In the Raven these mucous follicles are narrower but longer. The glandular structures supplying the mouth in Birds may be summed up under the following heads: 'folliculi linguales,' 'glandulæ sublinguales,' 'glandulæ submaxillares' (*Pici, Raptores, Rasores, Aptenodytes*), 'glandulæ anguli oris' (Swan, *Cantores,* Diurnal Raptores); 'folliculi preglottidei;' 'folliculi post-nasales,' i.e., opening behind the posterior nostrils; 'amygdalæ,' or close-set groups of follicles, in two rows, opening behind the eustachian outlet.

§ 148. *Alimentary Canal.*—The food, after being imbued with the secretion of the preceding glands, is poised upon the tongue and swallowed, partly by means of the pressure of the tongue against the palate, partly by a sudden upward jerk of the head. The posterior apertures of the nostrils being generally in the form of narrow fissures are undefended by a soft palate or uvula; and the laryngeal aperture, which is of a similar form, is in like manner unprovided with an epiglottis, but is defended by the retroverted papillæ at the base of the tongue. In many Birds, indeed, as the Albatross and Coot, there is a small cartilage in the usual place of an epiglottis, but insufficient to cover more than a very small part of the laryngeal aperture.[1] The surface of the mouth is rarely smooth above, commonly provided with retroverted papillæ: similar mechanical helps to the right course of the food occur at or near the fauces, in addition to those already noted on the tongue. The width of the mouth in *Caprimulgus,* and the length and depth due to the mandibular pouch in the Pelican, are remarkable. The extensibility of the membrane between the rami of the lower jaw admits of its formation into a bag, fig. 68, *a,* which is calculated to contain ten quarts of water, and serves as a receptacle for fishes, making in that state a conspicuous appendage to the huge bill;

[1] For these structures in Birds, see XXXVII'. p. 613.

when empty it can be contracted so as to be hardly visible. By means of this mechanism a quantity of food can be transported to the young; and, as in disgorging the bleeding fishes the parent presses the bottom of the sac against her breast, this action has probably given rise to the fable of her wounding herself to nourish the young with her own blood.

The Swift presents an analogous dilatation of the faucial membrane at the base of the lower jaw and upper part of the throat: it is most developed at the period of rearing the young, when it is generally found distended with insects in the old birds that are shot while on the wing. A similar structure obtains in the Rook, and probably in other Insectivorous Birds. It is notable in the Nutcracker (*Caryocatactes*); which, descending from its favourite snowy altitudes, may be seen to return with a swelling like an enormous goitre as big as the head, formed by the gular pouch, crammed with nuts.[1]

The œsophagus, *H*, fig. 94, *a*, fig. 78, like the neck, is usually very long in Birds: as it passes down, it generally inclines toward the right side; it is partially covered by the trachea, *G*, fig. 94, and connected to the surrounding parts by a loose cellular tissue. It is wide and dilatable, corresponding to the imperfection of the oral instruments as comminutors of the food. In the rapacious, and especially in the piscivorous Birds, it is of great capacity, enabling the latter to swallow the fishes entire, and serving also in many Waders and Swimmers as a temporary repository of food.

When the Cormorant has by accident swallowed a large fish, which sticks in the gullet, it has the power of inflating that part to its utmost, and while in that state the head and neck are shaken violently, in order to promote its passage. In the Gannet the œsophagus is extremely capacious, and, as the skin which covers it is equally dilatable, five or six herrings may be contained therein. In both these species it forms one continued canal with the stomach. In the Flamingo, on the contrary, the diameter of the gullet does not exceed half an inch, being suited to the

[1] When writing the article Aves for the 'Cyclopedia of Anatomy,' in 1835, I had not dissected a male Bustard, and introduced the old figure from 'Edwards's Nat. Hist. vol. ii. tab. 73 (1747),' fig. 54, with the current story of the sub-gular water-pouch, which Edwards derived from the anatomist Douglas. In 1848 I had the desired opportunity and made the preparation, No. 772, Q, described in the 'Physiological Catalogue of the Hunterian Collection.' The supposed gular pouch is a large cervical air-cell, fig. 54, *a*, capable of inflation and singularly swelling out the neck in the amorous male Bustard. See XXXVIII'.

smallness of the objects which constitute the food of this species.

Besides deglutition, the œsophagus is frequently concerned in regurgitation; and in the Birds in which this phenomenon occurs, the muscular coat of the gullet, like that in Ruminants, is well developed. The *Raptores*, for example, habitually regurgitate the bones, feathers, and other indigestible parts of their prey, which, in the language of Falconry, are called 'castings.' I have observed a Toucan to regurgitate partially digested food, and after submitting it to a rude kind of mastication by its enormous beak, again to swallow it.

The œsophagus possesses an external cellular covering, a muscular coat, an internal vascular tunic, and a cuticular lining. The muscular coat consists of two layers of fibres; in the external stratum they are transverse, fig. 81, *a*, in the internal longitudinal, ib. *b*. The mucous coat is generally disposed in longitudinal folds, rarely connected by transverse folds; still more rarely villous, as in the Ostrich.[1]

In those Birds which are omnivorous, as the Toucans and Hornbills, in the frugivorous and insectivorous Birds, and in most of the *Grallatores*, which find their food in tolerable abundance and take it in small quantities without any considerable intermission, it passes at once to the stomach to be there successively digested, and the gullet presents no partial dilatations to serve as a temporary reservoir or macerating receptacle. But in the larger Raptorial Birds, as the Eagles and Vultures, which gorge themselves at uncertain intervals from the carcases of bulky prey, the œsophagus does not preserve a uniform width, but undergoes a lateral dilatation anterior to the furculum at the lower part of the neck. This pouch is termed the *ingluvies* or crop, fig. 78, *b*.

78

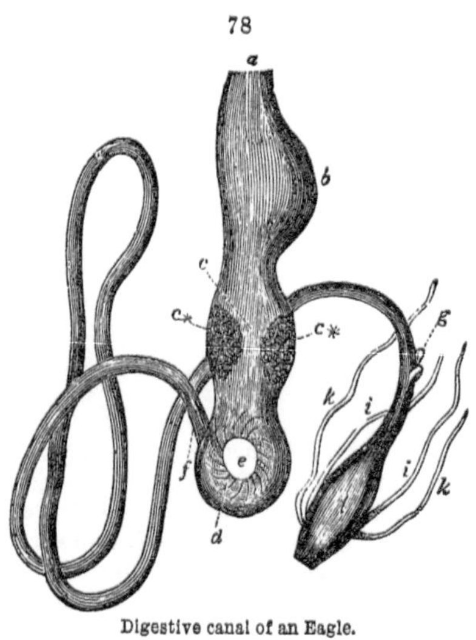

Digestive canal of an Eagle.

In those Birds, again, the food of which is exclusively of the vegetable kind, as grains and seeds, and of

[1] xx'. vol. i. p. 125, prep. no. 458.

which consequently a great quantity must be taken to produce the adequate supply of nutriment, and where the cavity of the gizzard is very much diminished by the enormous thickness of its muscular coat, the crop is more developed, and takes a more important share in the digestive process. Instead of a gradual lateral dilatation of the gullet, it assumes the form of a globular or oval receptacle appended to that tube, and rests upon the elastic fascia which connects the clavicles or two branches of the furculum together.

79

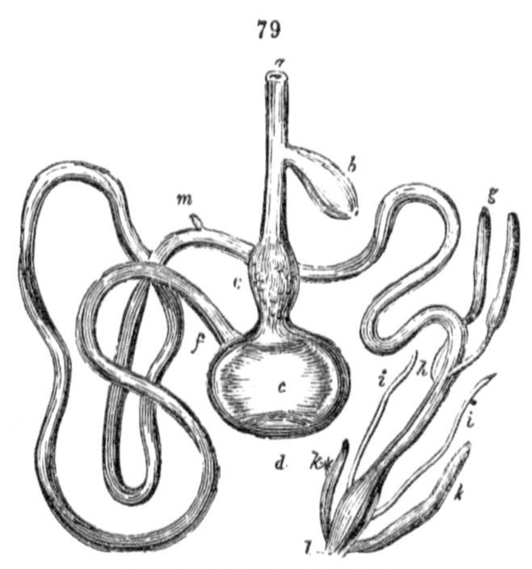

Digestive canal, Common Fowl.

In the Common Fowl the crop is of large size and single, fig. 79, *b*, but in the Pigeon it is double, consisting of two lateral oval cavities, fig. 80, *b*, *c*.

The dilatation of the œsophagus to form the crop is more gradual in the Ducks than in the Gallinaceous Birds. The crop is wanting in the Swans and Geese; but is present in that modified Anserine, the Flamingo.

80

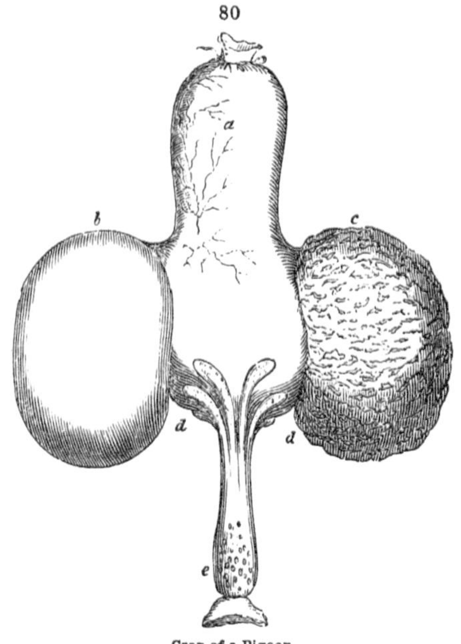

Crop of a Pigeon.

The disposition of the muscular fibres of the crop is the same as in the œsophagus, but the muciparous follicles of the lining membrane are larger and more numerous. This difference is most conspicuous in the ingluvies of the granivorous Birds, where it is not merely a temporary reservoir, but in which the food is mixed with the abundant secretion of the glands, and becomes softened and macerated, and prepared for the triturating action of the gizzard and the solvent power of the gastric secretion.

The change which the food undergoes in the crop is well known to bird-fanciers. If a Pigeon be allowed to swallow a great quantity of peas, they will swell to such an extent as almost to suffocate it.

The time during which the food remains in the crop depends upon its nature. In a common Fowl animal food will be detained about eight hours, while half the quantity of vegetable substances will remain from sixteen to twenty hours. Hunter made many interesting observations on the crop of Pigeons, which takes on a secreting function during the breeding season, for the purpose of supplying the young pigeons in the callow state with a diet suitable to their tender condition.[1] An abundant secretion of a milky fluid of an ash-grey colour, which coagulates with acids and forms curd, is poured out into the crop and mixed with the macerating grains. This phenomenon is the nearest approach in the class of Birds to the characteristic mammary function of a higher class; and the analogy of the 'pigeon's milk' to the lacteal secretion of the Mammalia has not escaped popular notice. In fig. 80, one side of the crop, *b*, shows the ordinary structure of the parts, the other, *c*, the state of the cavity during the period of rearing the young. The secretion consists of proteine with oil, but contains no sugar of milk nor fluid caseine.

The canal continued from the ingluvies to the stomach is called the lower œsophagus; at its commencement it is narrower and more vascular than that part which precedes the crop, but gradually dilates into the first or glandular division of the stomach, which is termed the 'proventriculus' (*ventriculus succenturiatus, bulbus glandulosus, echinus, infundibulum*), figs. 78, 79, 80, *c*.

The proventriculus of the Bird, like the spiral valve of the Shark, is an alimentary surface packed into the smallest space: in the latter the membrane is chylific, in the former chymific or digestive: every follicle is, in fact, a portion of the peptic secreting surface, with its gastric tubuli at right angles thereto; the surface being moulded to form either a simple or compound cavity.

In birds with a wide œsophagus, fig. 78, *a*, the commencement of the proventriculus is not indicated by any change in the direction or diameter of the tube, but only by its greater vascularity, by the difference in the structure of the lining mem-

[1] XCIV*. p. 124.

brane, and by the stratum of glands which open upon its inner surface, and which are its essential characteristic, fig. 81, *c*. Hence it is by some comparative anatomists regarded as a part of the œsophagus.

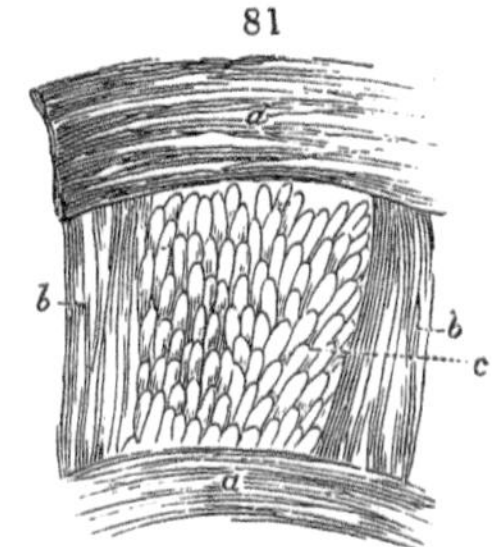

Part of the proventriculus of a Swan dissected.

The proventriculus varies, however, in form and magnitude in different Birds. In the *Rasores* it is larger than the œsophagus, but much smaller than the gizzard. In *Euphones*[1] it forms almost the entire stomach, the gizzard being minute: in Alcedo opposite proportions prevail. In the *Psittacidæ* and *Ardeidæ* (Parrot and Stork tribe) it is larger than the gizzard, and of a different form. In the Ostrich the proventriculus is four or five times larger than the triturating division of the stomach, being continued down below the liver, and then bent up upon itself towards the right side before it terminates in the gizzard, which is placed on the right and anterior part of this dilatation.

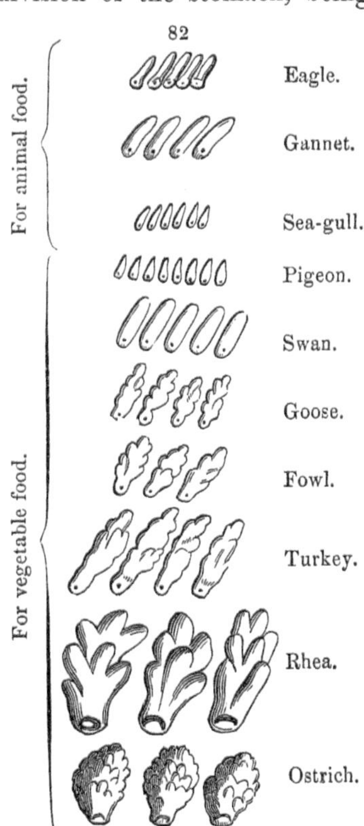

In the majority of Birds the gastric follicles are simple, having no internal cells, dilated fundus, or contracted neck; but from their external blind extremity proceed with an uniform diameter to their internal orifice. This form obtains in the zoophagous and omnivorous Birds. In the Dove-tribe the follicles are of a conical shape; in the Swan they are tubuliform; in the Goose and Turkey they present internal loculi; in the Ostrich and Rhea these loculi are so developed that each gland forms a racemose group of follicles, terminating by a common aperture in the proventriculus.

The subjoined figures show the different forms of the solvent or proventricular glands in different Birds.

[1] xxxix'.

The gastric glands are variously arranged.

Among the *Raptores,* we find them in the Golden Eagle disposed in the form of a broad compact belt; in the Sparrowhawk this belt is slightly divided into four distinct portions.

In the *Insessores* the glands are generally arranged in a continuous zone around the proventriculus; but in some of the *Syndactyli,* as the Hornbill, the circle is composed of the blending together of two large oval groups.

Among the *Scansores* the Parrots have the gastric glands disposed in a continuous circle, which is at some distance from the small gizzard. In the Woodpeckers the glands are arranged in a triangular form, with the apex towards the gizzard. In the Toucan they are dispersed over the whole proventriculus, but are more closely aggregated near the gizzard; the lining membrane of the cavity is reticulate, and the orifices of the glands are in the interspaces of the meshes.

Among the *Rasores* the Pigeon shows its affinity to the Passerine Birds in having the gastric glands of a simple structure, and arranged in a zonular form: they are chiefly remarkable for their large cavity and wide orifice. In the Common Fowl and Turkey the glands are more complex, and form a complete circle.

In the *Cursores* the arrangement of the glands is different in almost every genus. In the Ostrich they are of an extremely complicated structure, and are extended in unusual numbers over an oval space on the left side of the proventriculus, which reaches from the top to the bottom of the cavity, and is about four inches broad. The Rhea has the solvent glands aggregated into a single circular patch, which occupies the posterior side of the proventricular cavity. In the Emeu the gastric glands are scattered over the whole inner surface of the proventriculus, and are of large size; they terminate towards the gizzard in two oblique lines. In the Cassowary the glands are dispersed over the proventriculus with a similar degree of uniformity; but they are smaller, and their lower boundary is transverse. In the Apteryx the glands occupy its whole circumference, opening in the meshes of a reticulate surface.[1]

Among the *Grallatores,* the Marabou (*Ciconia argala*) has the nearest affinity to the Rhea in the structure and disposition of the gastric glands; they are each composed of an aggregate of five or six follicles, terminating in the proventriculus by a com-

[1] XI.

mon aperture; and they are disposed in two compact oval masses, one on the anterior, the other on the posterior surface of the cavity. In the Heron (*Ardea cinerea*) the solvent glands are of more simple structure, and are more dispersed over the proventriculus; but still they are most numerous on the anterior and posterior surfaces. In the Flamingo the gastric glands are short and simple follicles, arranged in two large oval groups, which blend together at their edges.

The *Natatores* present considerable differences among themselves in the disposition of the solvent glands. In the Cormorant (*Phalacrocorax carbo*) they are arranged in two circular spots, the one anterior, and the other posterior; while in the closely allied genus *Sula*, or Gannet, they form a complete belt of great width, and consequently are extremely numerous. In this respect the Gannet, or Solan Goose, shows a nearer affinity to the Pelican.

In the Sea-Gulls the gastric glands form a continuous zone; and in the Little Awk (*Alca alle*) they are spread over a great proportional extent of surface, and the form of the digestive organs is peculiar. The proventriculus is continued from the œsophagus, with very gradual enlargement, below the liver, and is then bent up to the right side, and terminates in the gizzard. The solvent glands are situated at the anterior or upper part of the cavity everywhere surrounding it, but lower down they lie principally upon the posterior surface, and where it is bent upward toward the right side they are entirely wanting. In the graminivorous lamellirostral Water-birds, as the Swan, Goose, &c., the gastric glands have a simple elongated exterior form, but have an irregular or cellular internal surface: they are closely arranged so as to form a complete zone.

In general the muscular or pyloric division of the stomach called 'gizzard' (*gigerium, ventriculus bulbosus*), immediately succeeds the glandular or cardiac division; but in some Birds, as the Awk and Parrots, there is an intervening portion without glands.

The *gizzard* is situated below or sacrad of the liver, on the left side of the abdomen, generally resting on the mass of intestines. In the Owl the gizzard adheres to the membrane covering the internal surface of the abdominal muscles; but in most Birds it has a more dorsal position.

In all Birds the gizzard forms a more or less lengthened sac, having at its upper part two apertures; one of these is of large size, communicating with the proventriculus, figs. 83, 84, *a*, the second is in close proximity with, and to the right side of the preceding, leading to the duodenum, ib. *o*; below these apertures the cavity extends to form a cul-de-sac, *c*. At the middle of the anterior and posterior parts of the cul-de-sac there is a tendon, figs. 78, 79, *e*, from which the muscular fibres radiate.

83

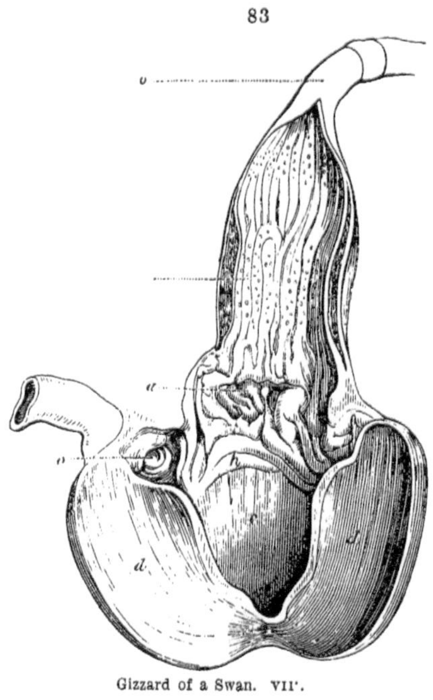

Gizzard of a Swan. VII'.

The differences in the structure of the gizzard resolve themselves into the greater or less extent of the tendons, and the greater or less thickness of the muscular coat, and of the lining membrane.

In the *Raptores* the gizzard, fig. 78, *d*, assumes the form of a mere membranous cavity, in accordance with the animal and easily digestible nature of their food. The muscular coat is thin; the fibres principally radiate from small tendons, ib. *e*, and there are some longitudinal fibres beneath the radiating or external layer.

In the *Rasores* and lamellirostral *Natatores* it exhibits the structure to which the term gizzard can be more appropriately applied, figs. 83, 84. The muscular fibres are distinguished by their unparalleled density of texture and deep colour, and are arranged in four masses; two are of a hemispherical form, and their closely-packed fibres run transversely to be connected to very strong anterior and posterior tendons, fig. 84, *e*; they constitute the sides of the gizzard, and are termed the digastric muscles or 'musculi laterales,' fig. 83, *d*: between these, at the end of the gizzard, are the two smaller and thinner muscles called 'musculi intermedii,' fig. 84, *f*. There are likewise irregular bands placed about the circumference of the gizzard.

Fig. 83 shows the relative thickness of the musculi laterales in the gizzard of a Swan, and fig. 84 that of the musculi intermedii and tendon.

The internal coat of the gizzard, fig. 84, *c*, *h*, is extremely hard and thick, and being of a horny nature, it is liable to be increased by pressure and friction, and as it is most subject to these influences at the parts of the gizzard opposite the musculi laterales two callous buttons are there formed, ib. *g*, *g*. It is here that the fibrous structure of the lining membrane can be most plainly

84

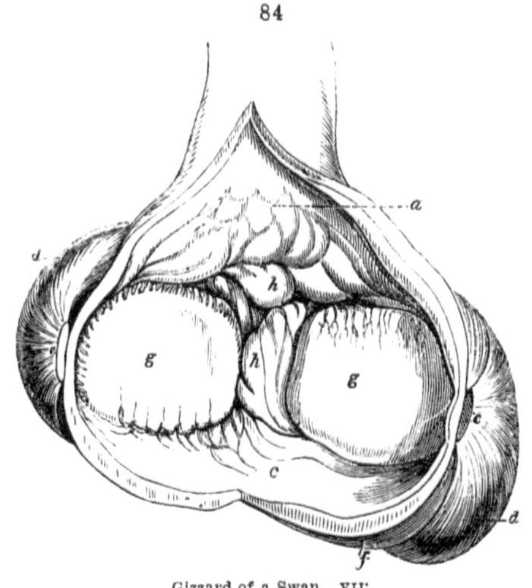

Gizzard of a Swan. VII'.

seen: and it is worthy of observation that the fibres are not perpendicular to the plane of the muscles, but oblique, and in opposite directions, on the two sides. Elsewhere the cuticular lining is disposed in ridges and prominences, figs. 84, 85, *h*, which vary in different birds, but are pretty constant in the same species. In a Petrel (*Procellaria glacialis*), the lining membrane is disposed in a pavement of small square tubercles, like the gastric teeth of some Mollusks.

The cavity of the gizzard is so encroached upon by the grinding apparatus, that it is necessarily very small, the two horny callosities having their internal flat surfaces opposed to one another, like 'millstones.' A crop is as essential an appendage to this structure as is the 'hopper' to the mill; it receives the food as it is swallowed, and supplies it to the gizzard in small successive quantities as it is wanted.[1]

Between the stomach of the carnivorous Eagle and that of the graminivorous Swan there are numerous intermediate structures,

[1] Thus we find in Parrots, where the gizzard is remarkably small, that a crop is present. A like receptacle exists also in the Flamingo, in which the gizzard is small but strong.

but it is necessary to observe that the animal or vegetable nature of the food cannot always be divined from the different degrees of strength in the gizzard. Hardcoated coleopterous insects, for example, require thicker parietes for their due comminution than pulpy succulent fruits.

In the subgenus *Euphones*, among the Tanagers, the muscular or pyloric division of the stomach is remarkably small and not separated from the duodenum by a narrow pylorus.

The parietes of the gizzard, like those of other muscular cavities, become thickened when stimulated to contract on their contents with greater force than usual. In the Hunterian collection this fact is well illustrated by preparations of the gizzard of the Sea-Gull in the natural state, and that of another Sea-Gull which had been brought to feed on barley. The digastric muscles in the latter are more than double the thickness of those in the Sea-Gull which had lived on fish.[1]

The immediate agents in triturating the food are hard foreign bodies, as sand, gravel, or pebbles.

Pigeons carry gravel to their young. Gallinaceous Birds grow lean if deprived of pebbles; and no wonder, since experiment[2] shows that unless the grains of corn are bruised, and deprived of their vitality, the gastric juice will not act upon or dissolve them. The observations and experiments of Hunter have completely established the truth of Redi's opinion, that the pebbles perform the vicarious office of grinding teeth.

Hunter inferred from the form of hair-balls occasionally found in the stomach of Cuckoos,[3] that the action of the great lateral muscles of the gizzard was rotatory. Harvey appears to have first investigated, by means of the ear, as it were in anticipation of the art of auscultation, the actions which are going on in the interior of an animal body, in reference to the motions of the gizzard. He observes (*De Generatione Animalium*, in *Opera Omnia*, 4to, p. 208), 'Falconibus, aquilis, aliisque avibus ex

[1] xx'. vol. i. p. 149, prep. 522, D, and 523.

[2] Grains of barley, inclosed in strong perforated tubes, pass through the alimentary canal unchanged. Dead meat, similarly introduced into the gizzard, is dissolved.

[3] The hairs of caterpillars devoured by this bird are sometimes pressed or stuck into the horny lining of the gizzard, instead of being collected into a loose ball. They are then neatly pressed down in a regular spiral direction, like the nap of a hat, and have often been mistaken for the natural structure of the gizzard. One of these specimens, exhibited as such to the Zoological Society, was sent to me for examination, when, upon placing some of the supposed gastric hairs under the microscope, they exhibited the peculiar complex structure of the hairs of the larva of the Tiger-moth (*Arctia caja*), and the broken surface of the extremity which was stuck into the cuticular lining was plainly discernible. See *Proceedings of Zool. Soc.* 1834, p. 9.

preda viventibus, si aurem prope admoveris dum ventriculus jejunus est, manifestos intus strepitus lapillorum illuc ingestorum, invicemque collisorum, percipias.' And Hunter observes (*Animal Œconomy*, 4to, p. 198), 'The extent of motion in grindstones need not be the tenth of an inch, if their motion is alternate and in contrary directions. But although the motion of the gizzard is hardly visible, yet we may be made very sensible of its action by putting the ear to the sides of a fowl while it is grinding its food, when the stones can be heard moving upon one another.' Tiedemann believed that the muscles of the gizzard were in some degree voluntary, having observed that when he placed his hand opposite the gizzard, its motions suddenly stopped.

The pyloric orifice of the gizzard is guarded by a valve in many Birds, especially in those which swallow the largest stones. This valve in the Ostrich is formed by a rising of the cuticle divided into six or seven ridges, which close the pylorus like a grating, and allow only stones of small size to pass through. In the *Touraco* the pylorus projects into the duodenum in a tubular form. There is a double valve at the pyloric orifice in the Gannet, and a single large valvular ridge at the same part in the Gigantic Crane. In this species and some other Waders, as the Heron and Bittern; also in the Pelican, and, according to Cuvier, in the Penguin and Grebe, there is a small but distinct cavity interposed between the gizzard and intestine. The analogous structure has been described in the Crocodile (vol. i. p. 442, fig. 298, *g*).

The *intestines* reach from the stomach to the cloaca; in relative length they are much shorter than in the Mammalia. In the Toucan, for example, the whole intestinal canal scarcely equals twice the length of the body, including the bill. The canal is divided into small and large intestines, sometimes by an internal valve, sometimes by the insertion of a single cœcum, but most generally by those of two cœca, which are always opposite to one another. In a few instances there is no such distinction. The small intestines and cœca are longest in the vegetable feeders. The large intestine is, with one or two exceptions, very short and straight in all Birds.

The course of the small intestine varies somewhat in the different orders of Birds; it is always characterised by the elongated fold or loop made by the duodenum, fig. 85, *f*, *f*, which fold receives the pancreas, ib. *q*, *q*, in its concavity.

In the *Raptores* the intestines are generally disposed as follows:—

The duodenum forms a long and broad fold, the lower part of which is commonly bent or doubled upon itself; the intestine then passes backward on the right side of the abdomen, crosses to the left, and is disposed in deep folds upon the edge of a scolloped

85

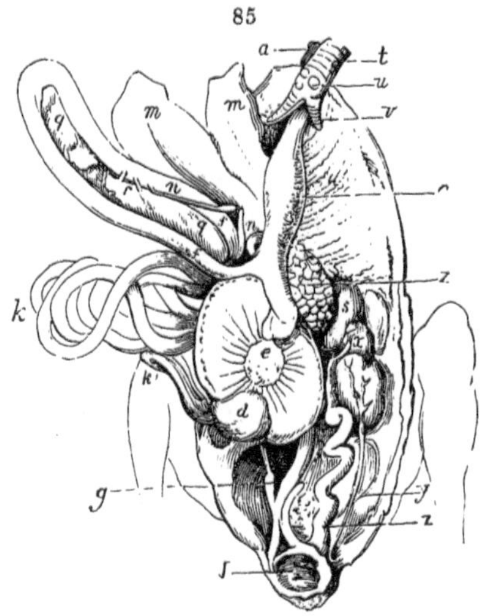

Abdominal viscera of a Pigeon. XXXIV'.

mesentery; towards its termination the ileum passes up behind the stomach and adheres to it, having here but a narrow mesentery; then passing down the posterior part of the abdomen the ileum makes another loose fold and ends in the rectum, which is continued straight to the cloaca.[1] In the Owls, the last fold of the ileum is nearly as long as the duodenal fold, and the cœca adhere to each side of the fold.

In the *Diurnal Raptores* the intestinal canal is only twice the length of the body, except in the fish-eating Osprey, in which the intestines are very narrow, and are to the length of the Bird itself as eight to one.

In the *Cantores* the scolloped folds of the small intestine are narrower and longer than in the *Raptores*, and the ileum generally adheres to the duodenal mesentery and pancreas instead of to the stomach, prior to passing down to form its last fold and to terminate in the rectum. In the Raven the small intestines are disposed at their commencement in concentric folds.

Among the *Scansores* the Cuckoo presents the following dis-

[1] In fig. 78, the intestines are not represented according to their natural arrangement.

position of the intestinal canal: after the usual long and narrow duodenal fold, the ileum[1] makes a fold which is widened at the end; it then forms a close fold upon itself, at the termination of which the rectum commences. In the Maccaw the course of the small intestine is somewhat peculiar: after forming the duodenal fold, it is disposed in three distinct packets of folds: the intestine, after forming the first two, passes alternately from one to the other, describing shorter folds upon each; it then forms the third distinct fold, which is a long one, at the termination of which the ileum adheres closely to the right side of the gizzard, and then passes backward and dilates into the rectum.

In the *Rasores* the Dove-tribe have the small intestines disposed in three principal folds; the first is the duodenal fold, fig. 85, *f*, *f*; the second is a long and narrow fold, coiled and doubled upon itself, with the turns closely connected together, ib. *k*; the third is also a long fold, which is bent or twisted, ib. *k'*. In the Common Fowl the duodenum is disposed in a long simple loop; the ileum passes toward the left, and is disposed in loose folds on the right and lower edge of the mesentery; the ileum before its termination passes up behind the preceding folds, and is accompanied as far as the root of the mesentery by the two cœca, which there open into the commencement of the large intestine.

The Ostrich presents the most complicated course of the intestinal canal in the whole class of Birds. The duodenal fold is about a foot in length, and the returning part makes a bend upon itself before it reaches the pylorus; the intestine then turns down again behind the duodenal folds and gradually acquires a wider mesentery. The ileum after a few folds ascends toward the left side, accompanied by the two long cœca, and becomes again connected with the posterior part of the duodenal mesentery; beyond which the cœca enter the intestine behind the root of the mesentery, and the large intestine commences. This part differs from the rectum in other Birds in its great extent, being nearly double the length of the small intestines, and being disposed in folds upon a wide mesentery. It terminates by an oblique valvular aperture in a large urinary receptacle. In the Bustard the rectum is a foot in length, which is the nearest approach to the Ostrich which the rest of the class make in this respect.

The small intestines in the *Grallatores* are characterised by their small diameter and long and narrow folds; these are sometimes extended parallel to one another, as in the Crane and Coot;

[1] There is seldom any part of the small intestine empty so as to merit the name of *jejunum*.

or folded concentrically in a mass, as in the Curlew and Flamingo. In the latter species the duodenal fold is four inches in length; then the small intestines are disposed in twenty-one elliptical spiral convolutions, eleven descending towards the rectum and ten returning towards the gizzard in the interspaces of the former.

Many of the *Natatores* present a concentric disposition of the folds of the small intestines similar to the Flamingo.

The arrangement of the muscular fibres of the intestine is the same as in the œsophagus, the external layer being transverse, the internal longitudinal.

The villi of the lining membrane manifest an analogy with the covering of the outer skin, being generally much elongated, so as to present a downy appearance when viewed under water. There are, however, great varieties in the shape and length of the villi. In the Emeu they consist of small lamellæ of the lining membrane folded like the frill of a shirt. In the Ostrich the lamellæ are thin, long, and numerous. In the Flamingo they are short and arranged in parallel longitudinal zig-zag lines.

In many Birds a diverticulum is observed in the small intestine, which indicates the place of attachment of the pedicle of the yolk-bag in the embryo, fig. 79, *m*. We have found this process half an inch in length in the Gallinule, and situated seventeen inches from the pylorus: in a Bay Ibis (*Ibis falcinella*) the vitelline cœcum was an inch in length: in a young female Apteryx it dilated into a sac, about an inch in diameter, with a yellowish stratum of the remains of the yolk.[1]

The Birds in which the *cœca coli* have been found wanting are comparatively few, though such examples occur in all the orders. These exceptions are most frequent among the *Scansores*, in which the cœca are absent in the Wrynecks, the Toucans, the Touracos, the Parrot-tribe, and according to Cuvier in the Woodpeckers.[2] In the *Insessores* the cœca are deficient in the Hornbill and the Lark. Among the *Grallatores*, we have found them wanting in a Spoonbill. In the *Natatores* they are absent in the Cormorant. The Herons, Bitterns, and, occasionally, the Grebes afford the rare examples of a single cœcum, which is also remarkably short.

In the *Raptores* the diurnal and nocturnal tribes differ remarkably in the length of the cœca. They are less than half

[1] xr'.

[2] In the Popinjay (*Picus viridis*, Linn.) we have found two small cœca, so closely adhering to the intestine as easily to be overlooked.

an inch in length in the Eagles and Vultures, but are occasionally wanting in the latter. Cuvier states that the cœca are deficient in the greater part of the Diurnal Raptores, but we have observed them in the *Haliætus Albicilla, Aquila Chrysætos, Astur palumbarius,* and *Buteo nisus.* They seldom exceed the length above mentioned, fig. 78, *g*, and in the Secretary Vulture they form mere tubercles. In the Barn Owl the cœca severally measure nearly two inches in length, and are dilated at their blind extremities; they are proportionally developed in the larger *Strigidæ.*

In the *Cantores* they are invariably very short where present. Among the *Scansorial* genera which possess the cœca, these parts are found to vary in length, measuring in the Cuckoo and Wattle-bird (*Glaucopis*), each half an inch; while in the *Scythrops*, or New Holland Toucan, the cœca are each two inches long, and moderately wide.

In the *Rasores* the cœca present considerable varieties. In the Pigeons, fig. 85, *g*, they are as short as in the Insessorial order, and are sometimes wanting altogether as in the Crown-Pigeon. In the Guan (*Penelope cristata*) each cœcum is about three inches in length: while in the Grouse each cœcum measures a yard long, being thus upwards of three times the length of the entire body. The internal surface of these extraordinary appendages to the alimentary canal is further increased in the Grouse by being disposed in eight longitudinal folds, which extend from their blind extremities to within five inches of their termination in the rectum. We have always found the cœca in this species filled with a homogeneous pultaceous matter without any trace of the heather buds, the remains of which are abundant in the fæcal matter contained in the ordinary tract of the intestines.

In the Peacock the cœca measure each about one foot in length; in the Partridge about four inches; in the Common Fowl and other *Phasianidæ* the cœca are each about one-third the length of the body; they commence by a narrow pedicle, which extends about half their length, and then they begin to dilate into reservoirs for the chyme, fig. 79, *g*.

In the *Cursores* the cœca again present very different degrees of developement. In the Apteryx the cœca are each five inches in length. In the Emeu they are narrow and short. In the Cassowary they are wholly deficient; while in the Ostrich they are wide, upwards of two feet each in length, and their secreting and absorbing parietes are further increased by being produced into a spiral valve, analogous to that which exists in the long cœcum of the Hare and Rabbit.

In the *Grallatores* the two cœca are generally short where present; they attain their greatest developement in this order in the Demoiselle, where the length of each cœcum is five inches; and they are also large in the Flamingo, where they each measure nearly four inches, and are dilated at their extremities, presenting with the gizzard, crop, lamellated beak, and webbed feet, the nearest approach to the *Anatidæ* of the following order.

In the *Natatores,* the cœca, where they are present, vary in length according to the nature of the food, being very short in the fish-eating Penguin, Pelican, Gull, &c., and long in the Duck, Goose, and other vegetable-feeding *Lamellirostres.* In the Crested Grebe (*Podiceps cristatus*), each cœcum measures 3-16ths of an inch in length. In the Canada Goose the cœca are each nine inches in length, and in the Whitefronted Goose the same parts measure severally thirteen inches. They have the same length in the Black Swan. In the Wild Swan the cœca measure each ten inches in length, while in the tame species they are each fifteen inches long.

As digestion may be supposed to go on less actively in the somnolent, night-flying Owls, than in the high-soaring Diurnal Birds of Prey, an additional complexity of the alimentary canal for the purpose of retaining the chyme somewhat longer in its passage, might be expected; and the enlarged cœca of the Nocturnal Raptores afford the requisite adjustment in this case. For, although the nature of the food is the same in the Owl[1] as in the Hawk, yet the differences of habit of life call for corresponding differences in the mechanism for its assimilation.

In the *Rasorial Order,* where the nature of the food differs so widely from that of the Birds of Prey, the principal modification of the digestive apparatus obtains in the more complex structure of the crop, proventriculus, and above all the gizzard; but with respect to the cœca, as great differences obtain in their developement as in the *Raptores.* Now these differences are explicable on the same principle as has just been applied towards the elucidation of the differences in the size of the cœca in the *Raptores.* Where the difference in the locomotive powers is so great in the Dove-tribe and the common Fowl; where the circulating and

[1] The indigestible parts of the prey of the Owl do not pass into the intestine, but are regularly *cast* or regurgitated from the stomach; the length of the cœca cannot, therefore, be accounted for on Macartney's supposition of their being receivers of those parts.

respiratory systems must be so actively exercised to enable the Pigeon to take its daily flights and in some species their annual migrations—a less complicated intestinal canal may naturally be supposed with such increased energy in the animal and vital functions to do the business of digestion, than in the more sluggish and terrestrial vegetable feeders; and accordingly we find that the requisite complexity of the intestinal canal is obtained by an increased developement of the cœcal processes in the *Gallinæ*, while in the *Columbidæ* the cœca remain as little developed as in the *Insessores*, which they resemble in powers of flight. If we regard the cœca as excretive organs, their differences in the above orders may be in like manner explained by their relations to the locomotive and respiratory functions.

In the *Cursores* the developement of cœca seems to have reference to the quantity of food, and the ease with which it may be obtained, according to the geographical position of the species. In the Cassowary, which is a native of fertile and tropical islands, New Guinea, North of Australia, New Britain, &c., vegetable food of a more easily digestible nature may be selected, and it need not be detained long, where a fresh supply can be so readily procured. But in the Ostrich, which dwells amidst arid sands and barren deserts, every contrivance has been adopted in the structure of the digestive apparatus to extract the whole of the nutritious matter of the food which is swallowed.

In the *Grallatores*, where no material differences of locomotive powers or means of obtaining food exist, the cœca present in their developement a direct relation to the nature of the food, and are most developed in the *Gruidæ*. The same holds good in the *Natatores*.

Why the increased extent of intestinal surface in the above different cases should be chiefly obtained by the elongation of the cœca, will appear from the following considerations. In consequence of the stones and other foreign bodies which birds swallow, it is necessary that there should be a free passage for these through the intestinal canal, which is therefore generally short and of pretty uniform diameter. In the omnivorous Birds of the tropics, as the Hornbills, Toucans, Touracos, and Parrots, which dwell among ever-bearing fruit-trees, the rapid passage of the food is not inconsistent with the extraction of a due supply of nourishment, but is compensated by the unfailing abundance of the supply. But where a greater quantity of the chyle is to be extracted from the food, and where, from the nature of the latter, a greater proportion of foreign substances is required for its trituration,—while the advantages of a short intestinal tract are obtained, the chyme is at the same time prevented from being prematurely expelled by the superaddition of the two cœcal bags which communicate with the intestines by orifices that are too small to admit pebbles or undigested seeds, but which allow the chyme to pass in. Here, therefore, it is detained, and chylification assisted by the secretion of the cœcal parietes, and the due proportion of nutriment extracted.

The large intestine is seldom more than a tenth part of the length of the body, and, except in the Ostrich and Bustard, is continued straight from the cœca to the cloaca; it may therefore be termed the rectum rather than the colon. It is usually wider than the small intestine, and its villi are coarser, shorter, and less numerous. The rectum, fig. 86, *a*, terminates by a valvular circular orifice, *b*, in a more or less dilated cavity, which is the remains of the allantois, and now forms a rudimental urinary bladder, *c*, *d*. The ureters, *h*, *h*, and efferent parts of the generative apparatus, *f*, *g*, open into a transverse groove at the lower part of the urinary dilatation; and beyond this is the external cavity which lodges, as in Reptiles, Marsupials and Monotremes, the anal glands and the exciting organs of generation. The anal follicles in Birds are lodged in a conical glandular cavity, *k*, which communicates with the posterior part of the outer compartment of the cloaca, and has obtained from its discoverer the name of *Bursa Fabricii*.

86

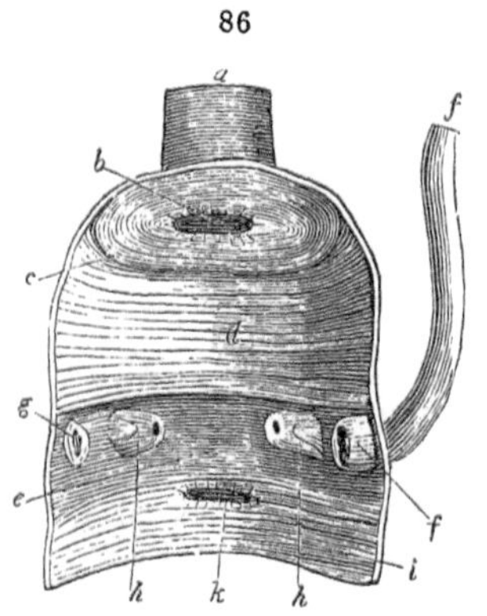

Cloaca of the Condor. XVIII'.

§ 149. *Liver of Birds.*—On the hypothesis of chylification, or on that of the formation of blood-discs, or on that of the production of grape-sugar in relation to the raising of animal heat, being essential functions of the liver, it might have been expected, since digestion is so much more active and the blood so much more abundant and rich, and the temperature so much higher in Birds than in Reptiles, that the liver would be proportionally larger; but it is not so. 'Carefully ascertained upon delicate balances,' the proportionate weight of the liver to the body is the same in a Vulture as in a Tortoise, in an Owl as in a Bull-frog, in a Curlew as in a Corn-snake, in a Turkey as in an Alligator,

&c.'[1] My own observations show the liver to be relatively largest in the less active aquatic and land birds, smallest in the birds that fly best and breathe most: compared in the limits of the class the liver seems to be developed inversely as the lungs and their appendages, and so far as it is associated with the lungs in eliminating waste elements from the blood, to have less to do in that way, as the breathing organs perform most.

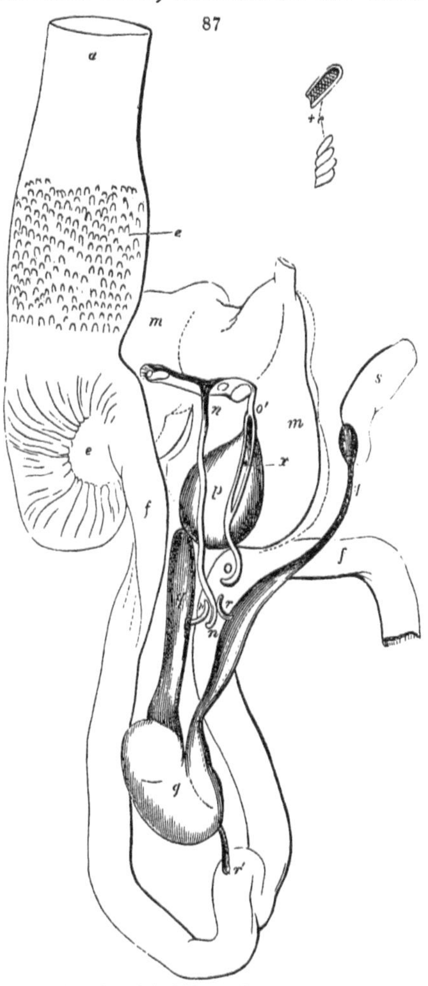

Posterior view of the biliary and pancreatic duct in the Hornbill. VII'.

The liver, figs. 85, 87, *m, m*, is situated a little above the middle of the thoracic abdominal cavity, with its convex surface towards the abdominal parietes, and its concavity turned towards the subjacent viscera: the right lobe covers the beginning of the duodenal loop, pancreas, and part of the small intestines; the left lobe covers the proventriculus and part of the gizzard; and the apex of the heart is received between the upper ends of these principal lobes. The liver is, as it were, moulded upon all these parts, and presents corresponding depressions where it comes in contact with them.

It is generally divided into two nearly equal lobes (*Raptores*,

[1] Number of times the weight of the liver in that of the body:—
Cathartes atratus, 47; *Chelonia caretta*, 47:
Syrnium nebulosum, 56; *Rana Catesbiana*, 55:
Tantalus loculator, 64; *Coluber guttatus*, 64:
Meleagris gallopavo, 70; *Alligator lucius*, 73:
and see other examples, in CCXLV. p. 113.

Stork), which are often separated for a short extent, and connected together by a narrow isthmus of the glandular substance In some Birds, however, as in the Pigeon, Cormorant, Swan, and Goose, there is a third, smaller lobe, situated at the back of the liver between the lateral lobes, which from its situation appears analogous to the 'lobulus Spigelii' of Mammalia. In the Common Fowl the left lobe is occasionally cleft from below so deeply as to form two lobes on that side. In most species the right lobe exceeds the left in size; this is remarkably so in the Bustard, in which the right lobe extends into the pelvis.[1] In the Eagle, however, the left lobe is sometimes the largest. Each lobe is invested by a double membranous tunic, one embracing it closely, the other surrounding it loosely, like the pericardium of the heart. They are formed by laminæ of the peritoneum, and by the air-cells. The two adherent layers are continued from the base of the liver, one over the anterior, the other over the posterior surface, closely adhering to the proper capsule: the loose layers are formed by the hepatic air-cell, surrounding each lateral lobe, the thin border of which is usually free.

The principal ligament of the liver is formed by a large and strong duplicature of the peritoneum, which divides the abdomen longitudinally like the thoracic mediastinum in Mammalia. It is reflected from the linea alba and middle line of the sternum upon the pericardium, and passes deeply into the interspace of the lobes of the liver; it is attached to these lobes through their whole extent, and connects them below to the gizzard on one side, and to the duodenal fold on the other: the lateral and posterior parts of the liver are attached to the contiguous air-cells; and the whole viscus is thus kept steady in its situation during the rapid and violent movements of the bird. The ligament first described is analogous to the falciform ligament of Mammalia; and, although there is no free margin enclosing a round ligament, yet the remains of the umbilical vein may be traced within the duplicature of the membranes forming the septum. As the muscular septum between the thorax and abdomen is wanting, there is consequently no coronary ligament; but the numerous membranous processes which pass from the liver to the surrounding parts amply compensate for its absence.

The liver is of a lighter colour in Birds of flight than in the heavier Waterfowl, where it is of a deep livid brown. Each

[1] The French Academicians (xl'. 2de partie, pp. 99–109) saw this in some of their Bustards: but in the male dissected by me the hepatic lobes were equal, and both were long.

lobe has its hepatic artery and vena portæ. The hepatic arteries are proportionally small, but the portal veins are of great size, being formed not only by the veins of the intestinal canal, pancreas, and spleen, but also by the inferior emulgent and sacral veins. The blood, which has circulated in the liver, is returned to the inferior cava by two venæ hepaticæ. There are occasionally some smaller hepatic veins in addition to the two principal ones. The coats of the portal and hepatic veins appear to be equally attached to the substance of the liver. A duct arises by two roots from each lobe, and the biliary secretion is carried out of the liver by these two and sometimes by three ducts; one duct always terminates directly in the intestine, and is an 'hepatic duct,' fig. 87, *n*, *n*; the other enters the gall-bladder, and is an 'cyst-hepatic duct,' ib. *o′*; the cystic bile is conveyed to the duodenum by a 'cystic duct,' ib. *o*. Where, as in a few instances, the gall-bladder does not exist, both hepatic ducts terminate separately in the duodenum, fig. 85, *n*, *n*; but in no case is there a single ductus communis choledochus as in Mammalia.

The *gall-bladder*, fig. 87, *p*, is situated near the mesial edge of the concave or under side of the right lobe, and is commonly lodged in a shallow depression of the liver; but sometimes, as in the Eagle, Bustard, and Cormorant, only a very small part of the bag is attached to the liver. It has no visible muscular tunic: its inner surface is delicately reticulated.

The gall-bladder is present in all the *Raptores*, *Insessores*, and *Natatores*. It is wanting in a great proportion of the *Scansores*, as in the genus *Rhamphastos* and in almost all the *Psittacidæ* and *Cuculidæ*. Among the *Rasores* the gall-bladder is constantly deficient in the *Columbidæ* or Dove-tribe alone, in which the cœca are shorter than in any other vegetable feeder. The gall-bladder is occasionally absent, according to the French Academicians,[1] in the Guinea-fowl; and they also found it wanting in two out of six Demoiselles (*Anthropoides Virgo*). The gall-bladder is small and sometimes absent in the Bittern: I found it absent in one out of three Kivis (*Apteryx*): it is always wanting in the Ostrich, but is present in the Emeu and Cassowary.

The bile, as before observed, passes directly into the gall-bladder, and not by regurgitation from a ductus choledochus; the cyst-hepatic duct arises from the right lobe, and is continued in some birds along that side of the bag which is in contact with the liver, where it penetrates the coats of the cyst and terminates about one-third from the lower or posterior end. In the

[1] XL.

Hornbill[1] I found it passing over the upper end of the bladder to the anterior or free surface, and the cystic duct continued from the point where the cyst-hepatic duct opened into the bladder; so that the cystic duct had a communication both with the reservoir and the cyst-hepatic duct; being somewhat analogous to the ductus communis choledochus; (fig. 87, where *x* represents the orifice by which the bile passes both in and out of the gall-bladder).

In the Goose the cyst-hepatic duct terminates by a very small orifice, surrounded by a smooth projection of the inner membrane, which, aided by the obliquity of the duct, acts as a valve and prevents any regurgitation towards the liver. The cystic duct here passes abruptly from the posterior extremity of the gall-bladder, which is not prolonged into a neck. The duct makes a turn round the end of the bag, and is so closely applied to it, as to require a careful examination to determine the true place of its commencement.

The hepatic duct, fig. 87, *n*, arises by two branches from the large lateral lobes of the liver, which unite in the fissure or 'gates' of the gland. Two hepatic ducts have been found in the Curassow; but these and the cystic duct terminate separately in the duodenum. Of the two hepatic ducts in Pigeons, one, the right and larger, enters the beginning of the duodenum, the other near its termination. The place of termination of the cystic and hepatic duct is generally, as shown in fig. 87, pretty close together at the end of the fold of the duodenum; but in the Ostrich one of the hepatic ducts, which is very large and short, terminates in the commencement of the duodenum about an inch from the pylorus; while the other enters with the pancreatic duct at the termination of the duodenum. Both the cystic and hepatic ducts undergo a slight thickening in their coats just before their termination. The passage of the bile-ducts in Birds through the coats of the intestine is oblique, and they terminate upon a valvular prominence of the lining membrane of the gut.

§ 150. *Pancreas of Birds.*—The non-mastication of food in the mouth is associated with a low condition of the salivary glands; a large pancreas is concomitant with gastric mastication, in Birds. This organ, figs. 85, 87, *q*, *q*, consists of two (*Picus, Certhia, Upupa, Caprimulgus, Grus, Colymbus*), and sometimes of three (*Oriolus*), distinct portions; but these are so closely applied together at some point of their surface as to appear like one continuous gland, fig. 88. It is of a narrow, elongated, trihedral form, lodged in the inter-

[1] xxv'.

space of the duodenal fold, and generally bent upon itself like the duodenum, as in the Hornbill, fig. 87, *q*. It is there supported by the gastro-hepatic and gastro-colic omenta.

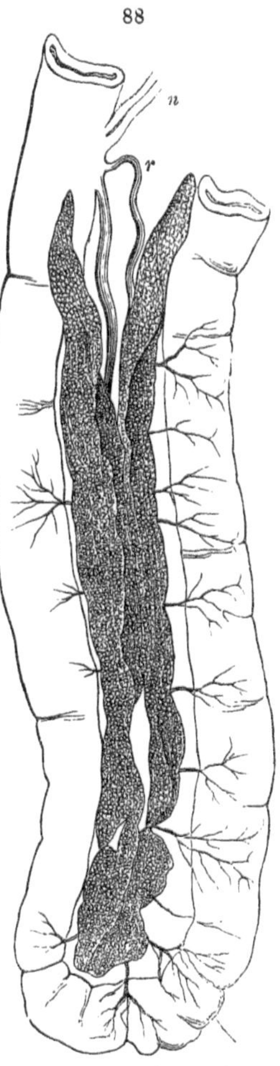

Pancreas and duodenum of the Goose. CCXXXI.

The structure of the pancreas is conglomerate, like that of the salivary glands, but the ultimate follicles are differently disposed. In the salivary glands these are irregularly branched, while those of the pancreas in Birds diverge in the same plane from digitated and pinnatifid groups. The substance is firmer than in Reptiles, of a pinkish, yellowish, or brownish colour.

The ducts, figs. 85, 87, 88, *r*, *r*, formed by the reiterated union of the efferent branches from the component follicles of the pancreas, are in general two in number, which terminate separately in close proximity to the hepatic and cystic ducts, *n*; but occasionally there are three pancreatic ducts, as in the Fowl, Pigeon, Raven, and Hornbill; in which case the third duct commonly terminates at a distance from the other two: in the Hornbill it proceeds from an enlarged lobe of the pancreas at the end of the duodenal fold, and enters that part, at *r′*, fig. 87. As a rule, the pancreatic secretion is the first poured into the gut, the cystic bile is the last.

CHAPTER XVIII.

ABSORBENT SYSTEM OF BIRDS.

§ 151. The absorbents of Birds, as of Reptiles, differ from those of Mammals in having fewer valves, which are also less perfect, being so loose as frequently to permit for a certain extent a retrograde passage of the injected fluid. The lacteals, lymphatics, and thoracic ducts have very thin parietes, so as easily to be ruptured: they are composed of two tunics, of which the internal is the weakest.

The lymph resembles that of Mammals, but the chyle differs essentially in its transparency and want of colour. The lacteals have, however, been observed to contain an opaque white fluid in a Woodpecker that had been killed after swallowing a quantity of ants.

With respect to the disposition of the absorbents, they do not form in Birds two strata, as in Mammals; at least those only have been observed which correspond to the deep-seated absorbents which accompany the large vessels.

The lymphatic glands or ganglions are few in Birds; the most constant and conspicuous are those at the anterior part of the chest or the root of the neck. Small ones have been seen in the axilla and groin of sea-birds (*Aptenodytes*). In other parts of the body the absorbent glands are replaced by plexuses of lymphatic vessels surrounding the principal bloodvessels.

The absorbents of Birds terminate principally by two thoracic ducts, one on either side, which enter the right and left jugular veins by several orifices; the plexuses of the posterior part of the body communicate with the contiguous sacral and renal veins.

The lymphatics of the foot unite to form the vessels which are found running along the sides of each toe, fig. 89, 1. In the *Palmipedes* there are anastomosing branches which pass from the lateral vessel of one toe to that of the adjoining toe, forming arches in the uniting web, 2. These branches form a small plexus, 3, at the anterior part of the digitometatarsal joint, from which three or four lymphatics are continued. The anterior and internal branches, 4, accompany the bloodvessels, and form a network around them; the posterior and external branches, 5, receive the lymphatics of the sole of the foot, then ascend along the metatarse, and form at its proximal articulation a close network, 6, from which the vessels climb the tibia, forming a plexus, 7, around it as far as the middle of the leg; from this proceed two branches, of which the smaller passes along the anterior part of the depression between the tibia and fibula as far as the knee-joint, where it joins the other branch which accompanies the bloodvessels. The trunk formed by the union of the two preceding branches accompanies the femoral vessels, forming plexuses in its course, 8, which receive tributary absorbents from the surrounding muscles, and a large branch, 9, corresponding to the deep-seated femoral vessels.

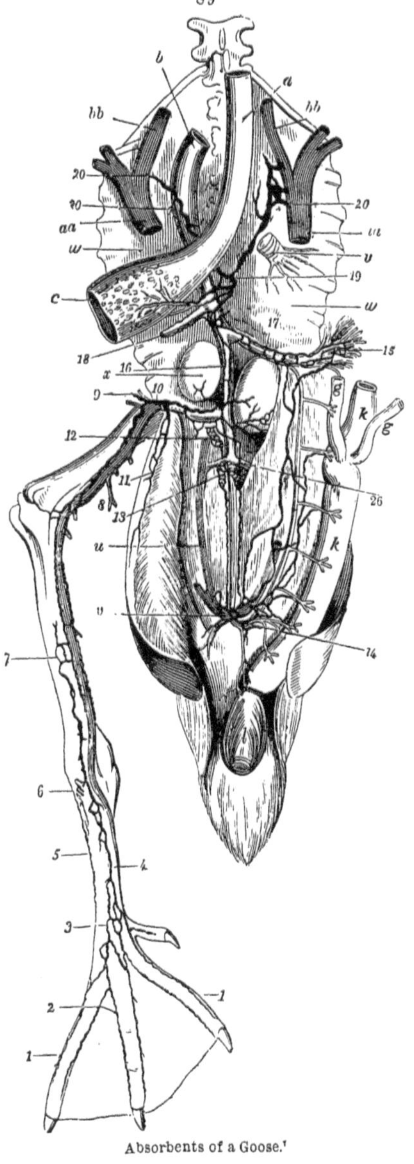

Absorbents of a Goose.[1]

The iliac trunk, 10, accompanies the great femoral vein into the abdomen, which it enters anterior to the origin of the pubis; it there receives branches from the lateral parts of the pelvis, 11, and afterwards separates into two divisions.

The posterior division receives some lymphatics from the anterior lobes of the kidneys, and those of the ovary or testi-

[1] From Lauth's Monograph, Annales des Sciences Nat. t. iii. pls. 23 and 25.

cles; it communicates anteriorly with a branch from the absorbents which surround the great mesenteric artery, and posteriorly with large vesicular plexuses or receptacles, 12, 13, surrounding the aorta and its branches, and which receives the lymphatics from the renal plexus, and those accompanying the arteria sacra media, 14.

The sacral or pelvic plexiform vesicles of the lymph are two in number, situated in the posterior region of the body, in the angle between the tail and the thigh. Each vesicle is little more than half an inch long and a quarter of an inch broad, and is shaped somewhat like a kidney-bean in the Goose.[1] They have muscular coats with striated fibre, distinctly recognisable in those of the Ostrich, where these 'lymph-hearts' are attached to the contiguous bone. In the Cassowary, Stork, and Goose, they lie free. The pulsations correspond with the motions of respiration.

The anterior division of the femoral lymphatic trunk, 16, accompanies the aorta, upon which it forms a plexus with the branch of the opposite side, and with the intestinal absorbents, 15. These vessels commence from a plexiform continuous network situated between the mucous and muscular coats of the intestine; they are larger here than when they quit the intestine to pass upon the mesentery. They accompany the branches of the superior mesenteric artery, there being many absorbents for one artery, which by their anastomoses form plexuses surrounding the blood-vessels. Before reaching the aorta, the lacteals communicate with the inferior or posterior division of the femoral trunk, and with the absorbents of the ovary or testicles, after which they pass upon the aorta, 16, 17, where they receive the lymphatics of the pancreas and duodenum, and terminate by uniting around the cœliac axis, 18, with the lymphatics of the liver, the proventriculus, *c*, the gizzard, and the spleen, forming a considerable plexus, from which, according to Lauth,[2] it is by no means rare to see branches passing to terminate in the surrounding veins.

The aortic plexus, 19, which answers to the 'receptaculum chyli' of Mammals, gives origin to two thoracic ducts, 20, 20, of varying calibre, but often, as in the Goose, exceeding a line in diameter. They are situated at their origin behind the œsophagus, *a*, and in front of the aorta, *b*; they advance forward, diverging slightly from each other, pass over the lungs, *w*, *w*, from which they receive some lymphatics, and terminate respectively, after being joined by the lymphatics of the wing, in the jugular

[1] XLII'. tab. ix. fig. 3. [2] XLI'. p. 381.

vein of the same side. The left thoracic duct, before entering the vein, receives the trunk of the lymphatics of the left side of the neck; the right thoracic duct receives only a branch of those of the same side.

The lymphatics of the wing follow the course of the brachial artery, forming a plexus around it, especially at the elbow-joint. Their principal trunk, to which all the collateral branches are united about the upper third of the humerus, is here of large size, but its diameter soon begins to be diminished, and it is very small at the head of the humerus. When it reaches the parietes of the chest, it receives two or three large lymphatics from the pectoral muscles, and a branch which accompanies the brachial plexus. Soon after a small lymphatic gland is sometimes formed on the trunk, which lastly unites with the thoracic duct of its own side.

The lymphatics of the head accompany the branches of the jugular vein, and are readily discerned upon those which are situated between the rami of the lower jaw. They form, by uniting with the cervical absorbents, two lateral branches on each side, which accompany the corresponding jugular vein, being situated, one in front, the other behind that vessel. These lymphatics communicate together, at the anterior and posterior parts of the neck, by transverse or oblique branches. They receive in their progress absorbents from the muscles, and from the peculiar glands which are seen beneath the skin of the neck. The internal branch on the left side receives also a considerable absorbent from the œsophagus. At the lower part of the neck both branches receive a notable branch which accompanies the carotid arteries, and a little further on they form on each side a lymphatic gland situated on the jugular vein. On the right side the trunk of the cervical lymphatics terminates in the jugular vein, after having furnished a communicating branch with the thoracic canal of that side; on the left side it terminates at once in the corresponding thoracic duct.

CHAPTER XIX.

CIRCULATING SYSTEM OF BIRDS.

§ 152. *Blood of Birds.*—The blood is hot and of a deep red colour. The blood-discs are more abundant than in the cold-blooded Vertebrates, save, perhaps, in some Ophidia: they are nucleated, elliptic, and flattened in form; averaging in size, in long diameter $\frac{1}{2100}$th, in short diameter $\frac{1}{3806}$th, of an inch; with the following observed extremes: — Humming-bird, long diameter $\frac{1}{2666}$, short diameter $\frac{1}{4000}$; Ostrich, long diameter $\frac{1}{1649}$, short diameter $\frac{1}{3000}$. Milne-Edwards notes decimally the following range of size in different species of the class:—'Long diameter, maximum, 1·59; minimum, 1·105: short diameter, maximum, 1·110; minimum, 1·158.'[1] (Metrical system.)

The blood-discs are largest in the embryo, losing size as the respiration gains in activity and extent in the progress of the individual to maturity. The smaller size of the blood-discs of Birds as compared with those of cold-blooded Ovipara exemplifies the same inverse ratio of their size to the amount of respiration. The proportion of organic matters contained in the water of the blood is greater than in the *Hæmatocrya*, as will be seen by comparing the subjoined Tables with those in vol. i. pp. 463, 464:—

AVES	Water	Clot	Albumen and Salts
Anas Boschas (Duck)	765	150	85
Ardea cinerea (Heron)	808	133	59
Columba livia (Pigeon)	797	156	47
Gallus domesticus (Fowl)	780	157	63
Corvus Corax (Raven)	797	146	56[2]

	MOIST BLOOD-DISCS			PLASMA		
	Total weight	Water	Solid matter	Total weight	Water	Solid matters
Ardea nycticorax (Night Heron) . . .	315·84	236·88	78·96	684·16	639·01	48·15
Syrnium nebulosum (Barred Owl)	427·36	320·52	106·84	572·64	519·14	·50
Cathartes atratus (Black Vulture) . . .	626·88	470·16	156·72	373·12	329·01	44·11[3]

[1] CCXXXIX. p. 86. [2] CCLXV. p. 64. [3] Ib. p. 27.

The fibrine in the blood of Birds is soft and very lacerable. The serum is usually of a light yellow colour, but is golden in the *Cathartes atratus*. Dr. Jones notices the strong musky odour of the blood of this Vulture, like that of the living bird.[1]

§ 153. *Heart of Birds.*—This organ consists of two ventricles and two auricles; the septum of both being complete.

The form of the heart is always that of a cone, sometimes wide and short, as in the Ostrich and Crane; sometimes more elongated, as in the Emeu, fig. 90, and Vulture; or still more acute, as in the Curlew and Common Fowl.

90

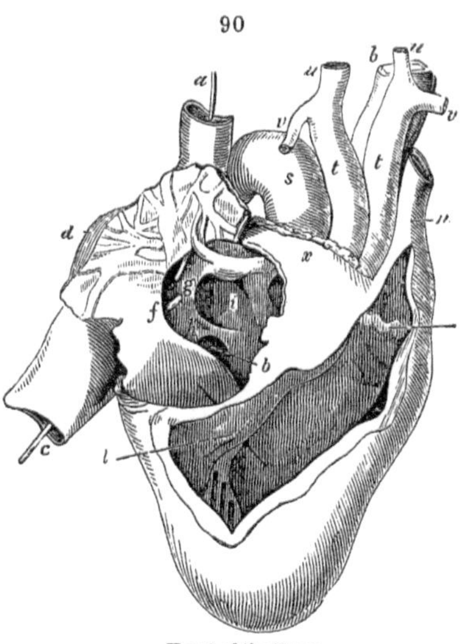

Heart of the Emeu.

Its situation is more anterior and mesial than in Mammalia, and its axis is always parallel with the axis of the trunk. It is not contained with the lungs in an especial cavity, but its apex is lodged between the lobes of the liver; the diaphragm, as a rule, not being so far developed as to separate the chest from the abdomen.

As the lungs are confined to the dorsal part of the chest, the whole of the anterior surface of the *pericardium* is exposed when the sternum of the bird is removed. The pericardium is thin, but of a firm texture, and adheres by its external surface to the surrounding air-cells. It is of considerable size, and commonly prolonged for some way between the lobes of the liver.

The *auricles* of the heart in Birds have not externally such free appendices as in Mammals. The right auricle is much larger than the left; it is more distinctly divided internally into a sinus, ib. *d*, and auricle proper, *b*, *i*, *x*, than in Mammals, and these parts are separated by a more complete valvular structure, but less definitely developed than in the Crocodile.

Three veins terminate in the sinus, there being in Birds always two precavals, as in Reptiles. The right precaval, ib. *a*, which returns the blood from the right wing and right side of the neck, terminates in the upper and anterior part of the sinus; the left

[1] CCXLV'. p. 16.

precaval, ib. *b*, winds round the posterior part of the left auricle to open into the lower part of the sinus; just before its termination it receives the coronary vein, so that this does not open separately into the auricle as in most Mammals. The postcaval vein, ib. *c*, terminates in the sinus just above the orifice of the left precaval, and a semilunar valvular fold, ib. *h*, analogous to that of the coronary vein in man, is extended forward between these orifices so as to separate them, and afford a protection to the mouth of the left precaval, in addition to that which it derives in common with the other veins from the larger valves at the mouth of the sinus.

The disposition of the valves between the sinus and auricle seems more especially destined to prevent regurgitation into the sinus, when the pulmonary circulation may be impeded. A strong oblique semilunar muscular fold, ib. *g*, commences in the Emeu by a band of muscular fibres running along the upper part of the auricle, and expanding into a valvular form extends along the posterior and left side of the sinus, terminating at the lower part of the fossa ovalis, ib. *i*. A second semilunar muscular valve, ib. *f*, of equal size, extends parallel with the preceding along the anterior border of the orifice of the sinus, its lower extremity being fixed to the smooth floor of the auricle, its upper extremity being continued into a strong muscular column running parallel to the one first mentioned across the upper and anterior part of the auricle, and giving off from its sides the greater part of the *musculi pectinati*. From this structure it results that the more powerfully the musculi pectinati act in overcoming the obstacle to the passage of the blood from the auricle to the ventricle, the closer will the valves be drawn together, and the stronger will be the resistance made to them by the regurgitation of the blood from the auricle into the sinus. The valves *f* and *g* are homologous with the pair dividing the auricle, *o*, from the sinus, *s*, in the Crocodile's heart (vol. i. fig. 339). The parietes of the auricle in the interspaces of the muscular fasciculi are thin and transparent, consisting in many parts only of the lining membrane of the cavity and the reflected layer of the pericardium blended together. The *fossa ovalis*, fig. 90, *i*, is a deep depression situated behind the posterior semilunar valve, which, we may observe, bears nearly the same relation to the fossa as the annulus ovalis in the human heart. The membranous septum closing the foramen ovale is complete and strong, but thin and semitransparent. The *appendix auriculæ*, ib. *x*, is the most muscular part of the cavity;

it does not project freely in front of the great vessels arising from the ventricles, but is tightly tied down to them by the reflected layer of the pericardium. The auriculo-ventricular orifice is an oblique slit, fig. 92, *k*; a bristle is passed through it in fig. 90. The manner in which regurgitation by this orifice is prevented is one of the chief peculiarities in the heart of Birds. The right ventricle, fig. 91, *k*, is a narrow triangular cavity, applied as it were to the right and anterior side of the left ventricle, but not extending to the apex of the heart. The parietes are of pretty uniform thickness, except at the septum ventriculorum, and are weaker in comparison to those of the left ventricle than in Mammals. Short fleshy columns extend from the septum to the free wall of the ventricle at the angle of union of these two parts, leaving deep cells between them; a strong column, fig. 92, *m*, also extends from the right side of the base of the pulmonary artery to the upper extremity of the auriculo-ventricular valve; but these are the only 'columnæ carneæ' in the right ventricle; there being none of a pyramidal form projecting into the cavity, nor any 'chordæ tendineæ.' The principal valve which guards the auricular aperture is a strong muscular fold, figs. 90, 91, 92, *l*, nearly as thick as the wall of the ventricle itself, extending from the fleshy column above mentioned obliquely downward and backward to the angle formed between the septum and the free wall of the ventricle at the lower and posterior part of the cavity. The convex edge of this muscular valve is turned toward the convex projection made by the septum, and must be forcibly applied to this part during the systole of the ventricles; so that, while all reflux into the auricle is prevented, additional impulse is given to the flow of blood through the pulmonary artery; the muscular parietes of the ventricle being thus complete at every part except at the orifice of the artery. This valve is strongest in the Diving Birds, weakest in

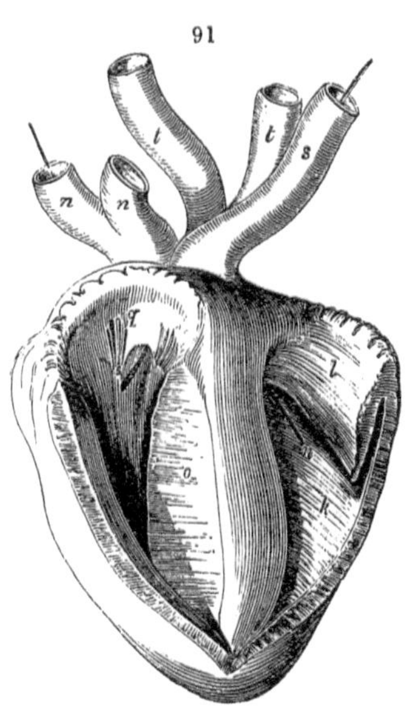

Ventricles of the heart of a Swan.

the Struthious, and especially in the Apteryx, in which it is partially membranous and has its margin tied by a few tendons to a fleshy process from the fixed ventricular wall.

The small muscular column, fig. 92, *m*, at the upper part of the auricular orifice is analogous in its position to the single valve which guards the corresponding orifice in Reptiles; the Crocodiles alone present a second muscular valve (vol. i. p. 510, fig. 339, *r*) homologous with the larger valve in Birds.

The right ventricle is remarkable for the smoothness and evenness of its inner surface. The pulmonary artery is provided at its origin with three semilunar valves, fig. 92, *n*. It divides, as usual, into two branches, fig. 168, one for each lung; the right branch passes under the arch of the aorta.

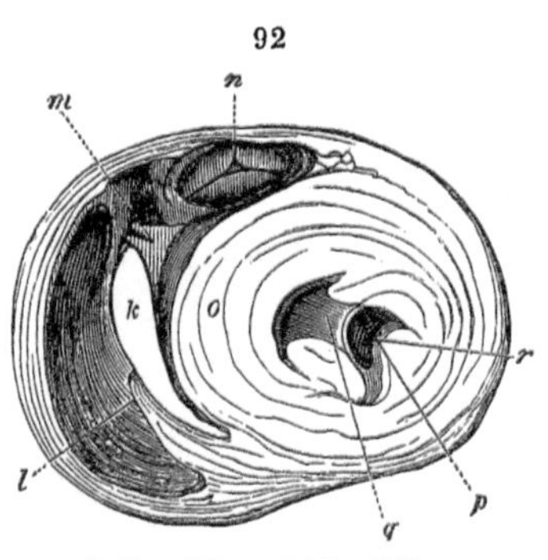

Section of the ventricles, Pelican.

The aerated blood is returned from the lungs by two veins which open into the back part of the left auricle; a strong semilunar ridge, which is hardly sufficiently produced to be called a valve, divides the cavity of the auricle in which the veins terminate from the muscular part or appendix. The fleshy columns are very numerous and complicated in this part of the auricle, which is closely tied down to the ventricle by the serous layer of the pericardium and dense cellular tissue.

The left ventricle, figs. 91, 92, *o*, is an elongated conical cavity, the parietes of which are three times as thick as those of the right ventricle, and exhibit strong fleshy columns extending from the apex towards the base; two of the largest of these columns present in the Emeu a short convex eminence towards the auriculo-ventricular orifice, fig. 92, *r*, and give off short thick tendons to the margin and ventricular surface of two membranous folds, figs. 91, 92, *p*, *q*, which correspond to the 'mitral valve' in Mammals. Of these valves, the one next the aorta, *q*, corresponds to the single valve which guards the auricular opening in the heart of Reptiles, and is most developed in Birds; the opposite valve is of much less size. In many Birds the chordæ tendineæ pass from the valves at once to the parietes of the ventricle, and are not attached to columnæ carneæ. The surface of the ventricle formed by the septum is smooth from the orifice of the aorta down to the apex of the heart. The aorta is provided with three semilunar valves, not with two only as in Reptiles. The extremities of these valves are connected to small, firm, and sometimes ossified styles imbedded in the fibrous coat of the vessels.[1]

The arrangement of the muscular fibres of the ventricle in Birds is such that the right ventricle appears to be formed by a partial secession of the outer from the inner layers of the parietes of the left ventricle at the anterior and right side of that cavity. See the transverse section, fig. 92.

§ 154. *Arteries of Birds.*—The arterial system of Birds mainly differs from that of Mammals in the following points:—The division of the aorta into three principal branches, almost immediately at its origin: the course of the arch of the aorta over the right instead of the left bronchus to become the descending aorta: the basilar artery being formed by the entocarotids, not by the vertebral: the great length of the common carotid, which is a single median trunk in some birds: and the origin of the arteries of the posterior extremities, which do not come off from a single branch, or 'external iliac,' but from two arteries which are detached successively from the aorta at a great distance from each other, and pass from the pelvis by two separate apertures. In these differences the closer affinity of Birds to Reptiles is shown.

The aortic trunk, fig. 93, 1, is so short that it is only brought into view after the reflections of the pericardium and the adjoining vessels are detached by dissection. The first branch is to the left side and, after it is sent off, the trunk affects to turn over the auricle before it gives the branch of the right side; these two branches pass in a curved manner from the heart towards the axilla, and may be regarded as two *arteriæ innominatæ*, fig. 90, *t*, *t*. After these branches are parted with, the arterial trunk, ib. *s*, fig. 93, 2, is continued over the right bronchus, and, on reaching the back part of the heart, becomes the 'descending aorta.'

The *arteria innominata*, fig. 93, 3, first sends off the common trunk of the carotid and vertebral arteries, ib. 4, which before its division gives off one or two small branches; one of these runs down upon the lungs in company with the par vagum, and appears to supply branches to the aponeurosis of the lungs, and the air-cells at the upper part of the thorax; the other branch, after supplying the lymphatic gland of the neck with several small arteries, ascends upon the side of the œsophagus, to which, and the inferior larynx, the divisions of the trachea, and to the parts

[1] Home Clift, in VII'. p. 331.

and integuments of the side of the neck, its branches are distributed, anastomosing with the superior œsophageal and tracheal arteries. Sometimes, in the Duck, the *supra-scapular* artery, which is usually divided from the vertebral, is a branch of the common trunk, as it is also in the Apteryx, where it supplies the muscles at the back of the base of the neck.

Birds, as a rule, are peculiar in sleeping with their long neck much bent or twisted, and this position might be expected to exercise some effect on the vessels subject thereto. Accordingly we find that the *carotid* arteries, ib. 4, fig. 90, *u*, *u*, are frequently of unequal size; in the Dabchick the left is the largest, fig. 93, 2; in an Emeu I found it the smallest. One or other carotid may be obliterated, according, perhaps, as the bird habitually sleeps with the head under the right or left wing. In the Apteryx I found that the left carotid alone passed to the usual place in the neck, and divided at the third cervical vertebra to supply the head in the usual way. In the Flamingo the right carotid was single and bifurcated at the upper part of the neck. In the Common Fowl, each carotid, after parting from the vertebral artery, ib. 6, proceeds to the middle of the neck and soon disappears; becoming covered by the muscles of the anterior part of the neck, and entering the canal formed by the hypapophyses, fig. 25, *h*, within which it lies hidden, and in close contact with its fellow of the other side, to very near the head. In the Bittern the two carotids are situated one behind the other, and adhere so intimately together in this situation that they seem like a single trunk.

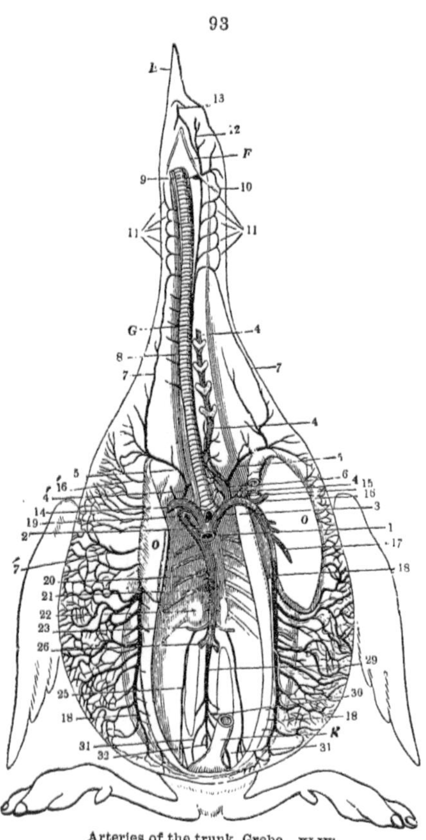

Arteries of the trunk, Grebe. XLIV'.

In the following details of the vascular system, I adopt, with little modification, the words of MACARTNEY, by whom it was first and best described in his Article 'BIRDS,' in 'Rees's Cyclopædia.' The carotid artery emerges from between the muscles of the neck, at about the third or fourth vertebra from the head (9); and after giving off the *arteriæ cutaneæ colli laterales*, fig. 93, 10, 11, to the lateral muscles and integuments of the neck, it runs along the outer edge of the rectus major anticus to behind the angle of the jaw, where it divides into its several branches.

An artery (*arteria occipitalis*) first goes off posteriorly, which passes a little forward under the thyrohyal, and after sending some blood to the muscles of the neck, makes a turn backward, enters the foramen in the transverse process of the second vertebra, and terminates by anastomosing with the vertebral artery.[1]

The next branch is the entocarotid; it passes behind the muscles of the jaw close to the basioccipital, sends a branch upward, which penetrates the tympanum; and another through the articulation of the jaw, to unite with the ophthalmic, and contribute to the plexus at the back of the orbit (*rete ophthalmicum* of Barkow). The entocarotid then enters an osseous canal, which runs along the side of the basisphenoid between the tables of the bone; and at the lower and back part of the orbit, the artery receives a remarkable anastomosing branch of the internal maxillary, which almost equals in size the carotid itself, and these two vessels produce by their union one which passes almost directly to the cranium, entering at the deep 'sella turcica.' It forms within the skull an anastomosis similar to the circle of Willis; but the branch which occupies the place of the *basilar artery* is very small, and appears to be furnished entirely from the anastomosis of the carotids, and designed only to supply the medulla oblongata. The branches of the entocarotid are spread in an arborescent form upon the surfaces of the brain; some on the outside and others on the ventricles and the fissure between the two hemispheres: the tufted termination of the vessels of the choroid plexus is shown in fig. 46. The orbital plexus formed by the carotid sends off the *inferior palpebral, ethmoidal, lacrymal*, and *ophthalmic* arteries. The ophthalmic artery forms two remarkable plexuses at the posterior part of the globe of the eye; the first, fig. 59, *ε*, is situated close beside the inner side of the optic nerve, is formed by an artery answering to the *arteria centralis retinæ*,

[1] Barkow, XLIV., has established the accuracy of this observation of Macartney's (XLIII.), having found this singular anastomosis of the occipital with the vertebral artery in all the birds which he injected.

and supplies the marsupial membrane; the second plexus, ib. 4, is situated more exteriorly, and gives off the ciliary arteries.

After the carotid has sent off the entocarotid, it passes for a little way downward and forward behind the angle of the jaw, and divides at once into different branches, corresponding to those of the ectocarotid in Mammals; the first of which might be called the œsophageal or laryngeal artery. This vessel sends a branch to the muscles upon the thyrohyal, and then turns downward and divides into two branches, one to the trachea, fig. 93, G 1, and the other to the œsophagus, upon the side of which parts they descend to near the thorax, forming a series of arches, ib. 11, 11, and ultimately inosculate with the tracheal and œsophageal branches of the common trunk of the carotid and vertebral arteries.

The external *maxillary artery*, ib. 12, dips in between the pterygoid and digastric muscles; it then passes behind the tympanic, and gives twigs upward to the muscles of the jaws, and to the plexus at the back of the orbit: upon emerging from behind the tympanic, it lies under the zygomatic arch, and sends an artery upward, which is distributed to the temporal and masseter muscles, and proceeding under the triangular tendon that comes from the inferior margin of the orbit to the lower jaw, it divides into two principal branches; one of these passes along the side of the upper jaw, gives a branch upward to the fore part of the orbit which unites with the ophthalmic artery, and is lost at the top of the head. This branch is very large in birds with combs, as, in conjunction with the ophthalmic, it furnishes numerous vessels to these vascular parts. The artery then goes on and supplies branches to the sides of the head before the orbits, and to the integuments and substance of the upper mandible, inosculating with the palatine branches of the internal maxillary artery. The second portion of the external maxillary proceeds to the lower jaw, to which, and the lower part of the masseter muscle, it is distributed. The external maxillary supplies the place of the *temporal*, *labial*, *angular*, *nasal*, and *mental* arteries of mammals.

The *laryngeal* or *posterior palatine* artery is a little branch of the ectocarotid, which is sent off posteriorly opposite to the external maxillary artery. Its branches are exhausted upon the back part of the fauces, the muscles for moving the upper jaw, and posterior nares.

The *lingual* or submaxillary artery, ib. 13, passes under the muscles which connect the hyoid to the lower jaw, and close upon the back of the membrane of the lower part of the mouth, it sends a branch to the œsophagus and trachea, supplies the muscles of the hyoid, ib. F, the tongue, E, the lower surface of the mouth, and furnishes the artery which enters the substance of the lower jaw.

'Just at the origin of the submaxillary artery there is another little branch of the carotid, which is lost upon the muscles of the hyoid arch.

'The *internal maxillary* artery is, as usual, the continuation of the trunk of the ectocarotid; it runs forward between the pterygoid muscle and the lining of the mouth, upon the side of the long muscle for moving the upper jaw, and divides into two principal branches; one of them proceeds under the tendon of the long muscle to get upon the palate, where it forms two branches, of which one runs along the external side of the palate, between the membrane and the bone of the mandible to the extremity of the bill, where it becomes united to the same branch of the opposite side, as also to the middle artery of the palate. The other branch lies also superficially under the membrane which lines the mouth. It passes onward to meet its corresponding vessel of the opposite side, with which it becomes actually incorporated, and by their union a single artery is generated, which runs along the middle line of the palate to the end of the mandible, where it unites with the lateral branches, as already mentioned. At the junction of the vessel of each side to form the middle palatine artery, two branches go off, which are lost upon the lining of the mouth, and the interior of the organ of smell.

'The other branch of the internal maxillary artery is reflected upward toward the orbit, below which it divides and unites again, forming a triangle, through which the vein passes: at this place it produces a remarkable plexus of vessels, like the rete mirabile of the carotid artery of quadrupeds, which is increased by branches from the ophthalmic and the palatine arteries, and from which the back part of the organ of smell receives its supply of blood.

'The internal maxillary artery then runs directly backward below the orbit, passes between the radiated or fan-shaped muscle which moves the upper jaw and the pterygoid; and turning inward round the basis of the cranium, becomes incorporated with the *internal carotid* artery just as it enters the bony canal which conducts it to the brain.[1]

'The *vertebral artery*, fig. 93, 6, soon after it parts from the

[1] Barkow describes the internal maxillary artery as wanting in birds, and its place being supplied by branches of both the external and internal carotids and the facial artery, all of which sometimes unite to form the maxillary plexus of vessels, which is very conspicuous in the Goose and Duck. XLIV'.

carotid, sends off a branch backward, which passes over the neck of the scapula and is lost among the muscles on the posterior part of the shoulder, inosculating with the articular and other arteries about the joint: this branch might be called the *supra-scapular*, ib. 5. In the *Duck* we have observed it, before it makes the turn over the scapula, to send an artery upward along the muscles of the neck. The trunk of the vertebral artery proceeds obliquely upward, and having entered the foramen in the transverse process of the penultimate cervical vertebra, gives off a large branch downward, which is distributed between the vertebræ, and to the spinal canal, in the manner of the intercostal arteries, with which it anastomoses upon arriving in the thorax. The remainder of the vertebral artery is continued upward in the canal formed by the pleur- and di-apophyses of the cervical vertebræ, diminishing gradually in consequence of the branches it sends off between each vertebra to the myelon and the muscles of the neck. Near the head the artery is found considerably reduced, and within the uppermost foramen in the transverse processes terminates entirely by inosculation with the reflected occipital branch of the carotid, as before noticed.

'After the common trunk of the carotid and vertebral is detached from the arteria innominata, this vessel may assume the name of the *subclavian*, fig. 93, 14. While passing under the coracoid it sends off some important branches: the first might be called the *pectoral artery*; it proceeds upward upon the internal surface of the pectoralis minor muscle, which it supplies, and then dividing into two branches, one passes over the anterior edge of the coracoid, and under the pectoralis medius, between which and the sternum it runs, detaching its branches to the muscle; the other sends first along the under side of the coracoid a branch which is again subdivided and distributed to the outside of the shoulder-joint and to the deltoid muscle, in which it inosculates with the articular artery. The vessel then passes between the coracoid and the furculum, and on a ligament which connects the head of the coracoid to that of the scapula, and disperses its branches upon the upper part of the shoulder-joint, forming anastomoses with the neighbouring arteries.

'The next branch of the subclavian is the *humeral artery*, ib. 15; it arises from the upper side of the vessel, and makes a slight curve to reach its situation on the inside of the arm in order to disperse its branches in the manner hereafter described.

'The *internal mammary* artery, fig. 94, 24, is given off just as the subclavian leaves the chest. It divides into three branches;

one ramifies upon the inner surface of the sternum, another upon the sternal ribs and the intercostal muscles, and the third runs along the anterior extremities of the vertebral ribs, supplying the intercostal muscles, &c.

'The chief peculiarity of the arteries of the superior extremities in birds consists in the great magnitude of the vessels which supply the pectoral muscles; these, instead of being inconsiderable branches of the axillary artery, are the continuations of the trunk of the subclavian, of which the humeral is only a branch.

'The *great pectoral* or *thoracic artery* passes out of the chest over the first rib and close to the sternum, and immediately divides into two branches. One of them, fig. 93, 16, ramifies in the superior part of the pectoralis major, and the other, ib. 17, is exhausted in the lower part of the muscle, and sends off a branch analogous to the long thoracic artery of Mammalia.' Fig. 93 shows the distribution of these arteries to the skin after perforating the *pectoralis* muscle.

'The *humeral artery*, while within the axilla, gives a small branch backward to the muscles under the scapula, and upon reaching the inside of the arm produces an artery that soon divides into the articular and the profunda humeri. The *articular* artery passes round the head of the humerus, underneath the extensors; its branches penetrate the deltoid muscle, and anastomose with the other small arteries around the joint.

'The *profunda humeri*, as usual, turns under the extensor muscles to reach the back of the bone, at which place, in birds, it separates into two branches, of which one descends upon the inside, and the other upon the outside of the articulation of the humerus with the radius and ulna, and there inosculate with the recurrent branches of the arteries of the fore-arm.

'After the humeral artery has sent off the profunda, it descends along the inner edge of the biceps muscle, detaching some branches to the neighbouring parts; upon arriving at the fold of the wing, it divides into two branches; one of these is analogous to the *ulnar* artery, and the other from its position deserves to be called rather the *interosseous* than the radial artery.

'At the place where the humeral produces the two arteries of the fore-arm, a small branch is sent off, which is lost upon the fore-part of the joint, and in anastomoses with the recurrent of the ulna and profunda humeri.

'The *ulnar artery* is the principal division of the humeral; it proceeds superficially over the muscles which are analogous to the pronator, sends a large recurrent branch under the flexor ulnaris to the back of the joint, upon which it ramifies and forms anastomoses with the profunda humeri. The artery then proceeds along the inner edge of the ulnar muscles, to which it distributes branches. It is afterwards seen passing over the carpal bone of the ulnar side, and under the annular ligament, at which place it sends off some branches which spread upon the joint and inosculate with the similar ones of the interosseous artery. Very soon after the ulnar artery gets upon the metacarpus, it dips in between the bones, and reappears upon the opposite side, lying under the roots of the quills, to each of which it sends an artery; it preserves this situation to the end of the metacarpal bones, where it passes between the style analogous to the little finger and the principal or fore-finger, and pursues its course along the edge of the latter, to the extremity of the wing, supplying each of the true quills with an artery, and sending at each joint of the finger a cross branch to communicate with the anastomosing branches on the opposite side.

'The *interosseous* artery detaches first a branch of some size to the membrane which is spread in the fold of the wing, upon which it forms several ramifications, fig. 94, *o*. After this the artery dips down behind the flexor muscles to get into the space between the ulna and radius. It here gives a branch backward to communicate with the others about the joint, and proceeds in the interosseous space as far as the carpal joint, during which course they become much diminished from giving off several branches which are distributed to the integuments and the quills placed upon the outside of the ulna. The remainder of the interosseous artery is expended in small branches upon the back of the carpal joint, the bastard quills, and along the radial edge of the metacarpal and bones of the fore-finger, where it forms communications with the cross branches of the ulnar artery already mentioned.

'From this description it will be perceived that no artery exists in birds strictly analogous to the radial; that there are no palmar arches; and that the size of the interosseous artery, and the course of the ulnar, along the outside of the metacarpus, are peculiarities which arise from the necessity of affording a large supply of blood to the quills during their growth.

'The *descending aorta*, fig. 93, 19, makes a curve round the right auricle and right bronchus, in order to get upon the posterior surface of the heart, after which its course is close along the spine, in which situation it is bound down by cellular substance, and the strong membrane or aponeurosis, which covers the lungs on their anterior part. The first branches which this vessel appears to send off are *bronchial arteries*; they arise from the fore part of the aorta, just when it arrives upon the spine; and having entered the lungs, their ramifications accompany those of the pulmonary arteries. They appear also to send branches to the spine and the spaces between the ribs.

'The *intercostal arteries* do not take their origin from the aorta in numerous and regular branches as in Mammals, but consist originally of but few vessels, which are multiplied by anastomoses with each other, and with the arteries which come out of the spinal canal. An arterial plexus is thus formed round the heads of the ribs, from which a vessel is sent to each of the intercostal spaces. Many of these branches, besides supplying the intercostal muscles and ribs, are continued into the muscles upon the outside of the body and the integuments. The anastomosis of the intercostal arteries round the ribs is very similar to the plexus, which is produced by the great sympathetic nerve in the same situation.

'The aorta produces no branch which deserves the name of the *phrenic artery*, as birds do not possess that muscular septum of the body to which the artery of this name is distributed in other animals.

'The *cœliac artery*, fig. 93, 20, is a very large single trunk, and arises from the fore part of the aorta, even higher than the zone of the gastric glands. It descends obliquely for a short way, and then gives off a branch which soon divides into two or three others that are spread upon the lower part of the œsophagus, and the side of the zone of the gastric glands, uniting with the other arteries of the œsophagus above, and extending downwards upon the posterior side of the ventricle, and anastomosing with the anterior gastric artery. The trunk of the cœliac now divides into two very large branches, which from their distribution we have chosen to call the posterior and the anterior gastric arteries.

'The posterior *gastric artery*, almost as soon as it is formed, detaches the *splenic* artery; and very soon after it furnishes from the posterior side of the vessel the right *hepatic artery*. This branch proceeds to the right lobe of the liver, which it enters on the side of the hepatic duct; after having divided into two or three minute arteries on its way to the liver, it supplies the hepatic duct with a branch which accompanies the duct to the intestine, and is there lost. The posterior gastric artery then runs down upon the back of the gizzard, and opposite to the origin of the first intestine it sends off an artery, which proceeds directly to one of the cœca (in the Fowl), upon which and the side of the next intestine it is expended, inosculating at the end of the cœcum with branches of the mesenteric artery, which are distributed to the adjoining portion of the small intestine. The posterior gastric then furnishes a large vessel which runs upon the gizzard, and divides into two chief branches, which penetrate the substance of the digastric muscle, in which they are lost.

'The next branch of the posterior gastric artery is the *pancreatic.* It runs between the two pancreatic glands, dispensing branches to each and to the duodenum. After this the trunk of the posterior gastric divides into two branches, which furnish twigs to the muscular parietes of the ventricle, and run along the margins of the upper and lower portions of the digastric muscle, supplying them with numerous twigs, and anastomosing with the ramifications of the other gastric arteries.

'The *anterior gastric artery* descends to the angle formed by the bulbus glandulosus and the gizzard, and there sends off a small branch which spreads upon the zone of the gastric glands, and inosculates with the first ramifications of the cœliac, and immediately afterwards it detaches a large artery, which runs round the superior margin of the digastric muscle, which it furnishes with many twigs, and communicates freely with the corresponding branch of the posterior gastric artery.

'Three *small hepatic arteries* take their origin from this branch of the anterior gastric, just as it passes over the highest part of the margin of the gizzard; these vessels enter the fissure in the left lobe of the liver. The anterior gastric artery now proceeds along the fore part of the gizzard, sending one or two branches into the muscular substance, and near the tendon it terminates in two large vessels, one of which is distributed upon the left side of the digastric muscle, and the other passes a little over the tendon, and then divides into two arteries, which produce several branches that disappear in the substance of the gizzard, and between the digastric muscles and the parietes of the ventricle, anastomosing with the vessels of the posterior side.

'The *superior mesenteric artery*, fig. 93, 21, takes its origin from the fore part of the aorta, a little below the cœliac, and proceeds for some way without detaching any branches; after which it experiences the same kind of division and subdivision that takes place in Mammalia; and the numerous arteries which are thus ultimately produced are spent upon the small intestines. One of the first and largest branches of the superior mesenteric, however, is allotted to supply one of the cœca, and to establish a communication with the inferior mesenteric and gastric arteries. This branch, soon after it leaves the trunk of the superior mesenteric, divides into two. One descends upon the rectum, where it meets with the inferior mesenteric artery, with which it produces a very remarkable anastomosis, similar to the mesenteric arch in the human subject; this united artery supplies the rectum and origin of the cœca. The second portion of this branch of the superior mesenteric runs in the space between the last part of the small intestine and the cœcum of one side, sending numerous branches to each, and at the end of the cœcum communicates in a palpable manner with another branch of the superior mesenteric artery, which runs upon the adjoining part of the small intestine.

'A branch, *arteria spermatica*, fig. 93, 22, arises from the anterior part of the aorta, just below the lungs; it is designed for the nutrition of the organs of generation, and except in the season for propagation, it is so small as to be discovered with difficulty; but when the testicles become enlarged, it is considerably increased in size in the male bird, and much more so in the female, when the ovary and oviduct are developed for producing eggs. It nearly equals the superior mesenteric artery during the period of laying, in which state we shall describe it. It is a single artery, like the cœliac and mesenteric, proceeds at a right angle from the aorta, and soon sends off a branch, which goes into the kidney of the left side, to which it gives some twigs, and afterwards emerging from the kidney, it runs in the membrane of the oviduct, upon which it is distributed. After this branch is detached, the artery projects a little farther forward into the cavity, and divides into two branches; one of these goes to the ovary, in which it ramifies, and furnishes an artery of some size to each of the cysts containing the ova. The other is distributed in numerous branches to the membrane and superior parts of the oviduct, and inosculates with the other arteries of the oviduct. It deserves to be remarked, that this and all the other arteries which are furnished to the oviduct have a tortuous or undulating course, in the same manner as the vessels of the uterus of the human subject.

'There are no regular *emulgent* arteries in birds; the kidneys deriving their blood from various sources, which will be pointed out as they occur.

'The inferior extremity is supplied with two arteries, which have a separate origin from the aorta. One corresponds to the femoral, and the other to the ischiadic artery.

'The *femoral artery*, figs. 93, 94, 23, is a small trunk, which takes its origin from the side of the aorta, opposite to the notch in the bones of the pelvis immediately under the last rib. This notch is formed into a round hole in the recent subject by a ligament which is extended from it to the rib; and it is through this hole that the femoral artery makes its exit from the pelvis; just before it passes out upon the thigh, it sends off a long branch, 25, which runs backward the whole length of the margin of the pelvis, dispensing arteries to the abdominal muscles on one side, and the obturator internus on the other. This branch also appears to supply one to the oviduct. The femoral artery, immediately after leaving the pelvis, separates into two branches; one goes upward and outward, ramifying amongst the muscles in that situation; the other turns downward, and is distributed to the flexors of the limb and round the joint, and sends an artery to the edge of the vastus internus, which can be traced as far as the knee. The kidneys appear to derive some irregular inconsiderable branches from the femoral artery while it is within the pelvis.

'The *ischiadic artery*, figs. 89, 93, 26, is the principal trunk of the lower extremities, exceeding very much in size the femoral. When it is produced by the aorta, it appears to be the continuation of that trunk; the remaining part of the aorta becomes so much and so suddenly diminished, and seems, as it were, to proceed as a branch from the back part of the vessel.

'The ischiadic artery, while in the pelvis, is concealed by the kidneys, in which situation it gives a branch from its lower side, which divides into three others that are distributed to the substance of the kidneys; one of these on the left side is continued out of the kidney to be lost upon the oviduct. The artery leaves the pelvis by the ischiadic foramen in company with the great nerve; while within the foramen it gives a branch obliquely downward under the biceps to the muscles lying in the pelvis; and as it passes over the adductor it sends off another along the lower edge of that muscle, which is chiefly lost in the semi-membranosus. It then detaches several small branches to the muscles on the outer and fore part of the thigh, some of which anastomose round the joint with the branches of the femoral artery. Just as the ischiadic arrives in the ham, it furnishes a very large branch downward, which divides into two; one goes under the gastrocnemius, to which and the deep-seated flexors its branches are distributed as far as the heel: the other is analogous to the *peroneal artery*; it goes to the outside of the leg, supplies the peroneal muscles posteriorly, and passes along the outer edge of the flexors of the toes to the heel, above which, and behind the flexor tendon, it divides, running on each side of the heel, and forming several articular arteries around the joint, and communicating with the other branch, and with the anterior tibial, and the metatarsal branch of the plantar artery.

'The *articular arteries* go off next from the artery in the ham; the two principal ones are deep-seated. One proceeds under the vastus internus to the external part of the joint; the other is large, and situated upon the inside. It forms two vessels: one is the true articular artery, and spreads upon the ligaments of the joint; the other is distributed in the substance of the flexor of the heel, which is placed upon the inside and fore part of the leg, and comes out upon the edge of this muscle to be lost in the integuments.

'The *posterior tibial artery*, fig. 94, 28, is extremely small; it only supplies muscular branches to the internal head of the gastrocnemius, and some of the flexors of the toes; it is lost on the inside of the heel in anastomoses with the peroneal artery, and other small superficial branches.

'The trunk of the artery of the leg now gets upon the posterior surface of the tibia, and sends off, through the deficiency left between the tibia and fibula at the superior part, a branch which is distributed to all the muscles upon the fore part of the leg. The artery then creeps along the back of the bones for some way, and passing between them above, where the fibula is anchylosed with the tibia, it reappears on the anterior part of the leg in the situation of the *anterior tibial artery*; at this place it detaches some very small branches, which frequently divide and unite again, to produce a most singular reticulation or plexus of vessels, which closely adheres to the trunk of the artery, and is continued with it as far as the articulation of the tibia with the metatarsal bone, where it disappears without seeming to answer any useful design. This plexus resembles in appearance exactly the division of the arteries of the extremities, which has been described by Mr. Carlisle in the tardigrade quadrupeds, but differs from it in this circumstance, that the trunk of the artery is preserved behind it, without suffering any material diminution of its size.

'The anterior tibial artery furnishes no branch of any importance during the time it is proceeding along the fore part of the leg. It passes under the strong ligament which binds down the tendons of the anterior muscles of the leg, and over the fore part of the joint on the inside of the tendon of the tibialis anticus, at which places it distributes some branches which inosculate with the other arteries round the joint; it then pursues its course in the groove along the anterior surface of the metatarsal bone, and covered by the tendon of the flexor digitorum. On coming near the foot it sends off an artery, which divides, behind the joint of the internal toe, into two branches; one goes between the internal and middle toes, ramifies upon both their joints, and unites with the artery in the sole of the foot; the other is distributed between the internal toe, and the pollex or toe which occupies the place of the great toe; the main artery now passes to the sole of the foot through a hole in the metatarsal bone, left for the purpose, when the original parts of this bone were united by ossification. In this situation the artery might receive the name of the *plantar*. It has scarcely passed through the bone, when it divides into six branches; three of these are distributed to the tendons and ligaments, &c., on the outside of the foot and the back of the metatarsus, anastomosing with the descending branches of the peroneal artery; the fourth branch supplies the pollex, and also sends a branch from the metatarsus. The remaining branches are designed for the three principal toes; one dips in between the internal and middle toe, unites with the anterior branch of the metatarsal artery, and is distributed to the sides of these toes as far as their extremity. The other divides, between the external and middle toe, into two branches, which run upon the opposite side of each of these toes to the end.

'When the feet are webbed, the digital arteries send off numerous branches, which, ramifying in the membrane between the toes, establish a communication with each other. The present description has been taken from birds which possess three principal toes, and the back toe or pollex; but no material difference can be expected in those with a greater number of toes.

'After the trunk of the aorta has detached the ischiadic arteries, it is continued along the spine as the *arteria sacra media*, fig. 93, 29, sending off small branches answering to *lumbar* arteries, one of which ascends upon the rectum, supplies the place of the *inferior mesenteric*, ib. 30, and unites with the superior mesenteric as already mentioned. The aorta separates above the coccygeal vertebræ into three branches; two of these (the *hypogastric arteries*, ib. 31, proceed laterally, and are distributed to the neighbouring parts, and to the kidneys and oviduct; the third branch (the *coccygeal artery*, ib. 32) descends to the very point of the tail, upon the muscles and quills of which its branches are exhausted.

'The arterial system of Birds, besides the distinguishing characters above mentioned, differs from that of Mammals chiefly in the frequent anastomoses, which exist more especially amongst the arteries of the head and the viscera. Similar communications occur between the veins.'[1]

Besides the remarkable arterial plexuses mentioned in the general description, as the orbital, the temporal, the spermatic plexuses, &c., that which Barkow[2] has described under the name of the plexus of the organ of incubation (*Brütorgane*) deserves special notice. It is represented at 17, 18, fig. 93, and is composed of branches coming from the posterior thoracic, abdominal, cutaneous, and ischiadic arteries, which ramify beneath the integument of the abdomen, and form, by their unions, a rich network of vessels which becomes truly extraordinary in the time of hatching. At this period many birds pluck off the feathers from the seat of incubation, probably thereto impelled by the great degree of heat caused by the influx of blood into the incubating plexus.

94

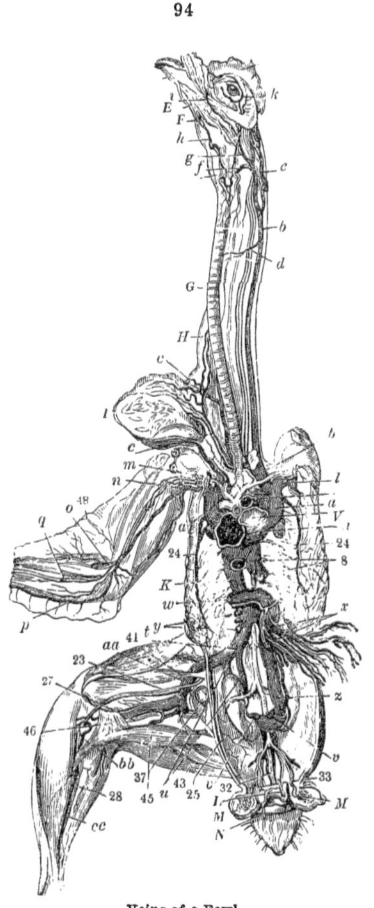

Veins of a Fowl.

§ 155. *Veins of Birds.*—The venous blood is returned to the heart by means of three trunks; two of these are precavals, fig. 90, *a*, *b*, and one postcaval, ib. *c*. Each precaval, fig. 94, *a*, is composed of the jugular and vertebral, and the veins of the wing.

'The *vertebral vein* is lodged in the same canal with the vertebral artery; it anastomoses between the vertebræ with the veins of the myelonal membranes. It also freely communicates at the base of the cranium with the jugular vein, and receives blood from the muscles of the neck.

'The *jugular vein*, fig. 94, *b*, is a single trunk in birds, and

[1] XLIII'. [2] XLIV'.

does not admit of the distinction into external and internal; it proceeds superficially along the side of the neck in company with the par vagum. The vein of the right side exceeds the other in size; it is often twice as large. The jugular vein receives several lateral branches from the muscles and integuments of the neck, ib. *d*, the œsophagus, &c. (the veins from the crop joining the jugular are shown at *c*): one of these near the head is much longer than the rest, ib. *e*; it lies deep amongst the muscles, and appears to communicate with the vertebral vein. There is a branch of the jugular which goes to the superior larynx amongst the muscles of the tongue and of the hyoid, and another for the muscles within the jaws and the integuments in the back of the mouth; these might be called the *lingual, thyroid,* and *submaxillary* veins, ib. *g, h, i.*

'The jugular veins form a remarkable communication with each other immediately below the cranium, by means of a cross branch, generally of an equal size with the trunks themselves. From each side of the arch thus formed there issues a large vessel, which is made up of the veins of the external part of the head; one of these passes round the tympanic, and apparently penetrates the joint of that bone with the lower jaw; it appears in several branches upon the side of the cheek, and contributes to form a plexus of veins below the posterior part of the orbit, ib. *k*, similar to the arterial plexus already described in that situation. The principal branch of the veins of the head passes obliquely round the pterygoid bone, and below the orbit divides into several large vessels, one of which belongs to the back part of the palate; another ascends on the orbit, and unites with the ophthalmic vein; and a third is distributed to the interior of the organ of smell, the palate, and the external parts of the upper and lower jaws. These branches produce plexuses along the base of the orbit and the external edge of the palate, which correspond to those of the arteries before described.'[1]

The sinuses of the brain are irregular in form, and consist of flattened canals. The principal ones, besides those upon the cerebellum, are the superior longitudinal, and one which runs along the lower edge of each hemisphere of the cerebrum; there appears to be also one upon the side of the cerebellum, corresponding to the lateral sinus. All these sinuses communicate with each other on the back of the cerebellum, and seem to discharge their contents principally into some veins which lie in the

[1] XLIII'.

myelonal sheath, and these appear to dispose of their blood gradually, as they descend in the neck, by means of lateral communication with the vertebral veins. The superior longitudinal sinus is continued at its anterior part under the frontal and nasal bones, and anastomoses with the ophthalmic and nasal veins. There are other small sinuses in the several duplicatures of the dura mater.

The *veins of the wings*, which are derived from the parts within the chest, the muscles about the scapula, and the pectoral muscles, accompany the arteries of the same parts so regularly that their course does not require description.

The *axillary vein*, fig. 94, *l*, lies considerably lower in the axilla than the artery, but still continues to receive corresponding branches (*m* indicates the great pectoral vein). The trunk of the vein descends in the course of the humeral artery, but more superficially; in this situation it may be called the *humeral* vein, ib. *n*. Branches of this vein accompany the articular and profunda arteries, and at the middle of the humerus a large branch of the vein enters the bone; there are also two very small branches which lie in close contact with the humeral artery, which they accompany nearly its whole length.

The principal vein of the wing divides into two, opposite to the joint of the humerus with the fore-arm. One of these branches, ib. *o*, belongs to the sides of the radius; it receives blood from the muscles and skin on the upper part of the fore-arm, but its chief vessels lie between the integuments of the fold of the wing. The other branch of the humeral vein, ib. *p*, crosses the fore-arm, just below the articulation, in company with the nerve, and running along the inferior edge of the ulna, receives a branch from between the basis of each quill, is continued along the ligament which sustains the rest of the quills to the extremity of the wing, receiving many veins of the joints from the opposite side of the fingers. Besides these large superficial veins of the fore-arm, there appears to be one, and sometimes two, small accompanying veins to the ulnar and interosseous arteries, ib. *q*.

The *inferior vena cava*, ib. K, before it enters the auricle, A, receives as usual the *hepatic veins*, ib. *s*; these are numerous, and open into the cava as it passes behind the liver, or more frequently within the substance of that viscus in the back part.

The trunk of the *vena cava* is very short in the abdomen; it separates into two great branches analogous to the primary *iliac veins*, ib. *t*, opposite to the adrenals; these turn to each side, and experience a very singular distribution. On coming near the edge of the pelvis, each of these two veins forms two branches; one of which collects the blood of the lower extremity, as hereafter described; the other passes straight downward imbedded in the substance of the kidney, and admits the several emulgent veins, which are very large, and are seen to pass for some way obliquely in the kidney before their termination. Sometimes the *emulgent veins* are double, as in the figure, ib. *u*. The limb-vein sends off a descending branch into the renal tissue which, when arrived at the lower end of the kidney, divides into three branches; one receives the blood of the muscles of the tail and parts adjacent; another accompanies the ureter to the side of the rectum, and is distributed about the anus and parts of generation, answering to the *hæmorrhoidal* veins; the third, ib. *v, v*, passes inward to the middle line between the kidneys, and there unites with the corresponding branch of the opposite side. These are the branches which have been supposed to carry venous blood *into* the kidneys, for the purpose of supplying material for the urinary secretion. The vessel which is in this manner produced, ib. *z*, receives all the blood of the rectum from the anus to the origin of the cœca, anastomosing below with the branches of the hæmorrhoidal veins; and at the upper part of the rectum, it becomes continuous with the trunk of the veins of the small intestines, ib. *x*, forming the most remarkable anastomosis in the body, both on account of its consequences and the size of the vessels by which it is effected. By means of this communication, the blood of the viscera and the external parts of the body flows almost indifferently into the vena cava and vena portæ, *w*; for the anastomosing vessels are sufficiently large to admit the ready passage of a considerable column of blood in proportion to the whole mass which circulates in the body of the bird; for instance, in the Goose the communicating veins of the pelvis are equal in size to a goose-quill, and in the Ostrich and Cassowary they are as thick as a finger. Besides their anastomoses the principal visceral veins are remarkable for their large size in the Diving Birds.

'The anastomosis of the pelvic veins, in being the means of conveying common venous blood into the liver, goes to prove that the blood of the vena portæ does not require any peculiar preparation by circulation in the spleen or other viscera to fit it for the secretion of bile.

'The *vena portæ*, ib. *w*, belongs almost exclusively to the right or principal lobe of the liver. It is formed by three branches. The *splenic vein* is the smallest, and is added to the vena portæ, just as it penetrates the liver on the side of the hepatic duct. The next is made of two branches; of which one returns the blood of the posterior gastric artery, and therefore may be called the *posterior gastric vein*; and the other is furnished by the pancreas and duodenum, and is the *pancreatic* vein. The third and largest branch of the vena portæ is the *mesenteric vein*, ib. *x*, which not only collects the blood from all the small intestines, but likewise receives the *inferior mesenteric*, ib. *z*, or vein of the rectum, which forms the communication that has been described with the pelvic veins.

'The *veins of the left lobe of the liver* are furnished in the Goose by those which accompany the anterior gastric artery, and some branches from the head of the duodenum.

'The *anterior gastric veins* produce two small trunks, which enter at the two extremities of the fissure, in the concave surface of the left lobe of the liver, as it lies upon the edge of the gizzard; the veins from the head of the duodenum furnish a small vessel which passes backward to penetrate the posterior part of the fissure in the left lobe.

'In the *Cock* the veins that the left lobe of the liver derives from the anterior gastric, are more numerous than in the *Goose*.

'The veins of the zone of gastric glands, and of the lower portion of the œsophagus, do not contribute to the secretory vessels of the liver, but proceed to the superior part of that viscus, to terminate in the vena cava, as does also the umbilical vein.

'The vein which returns the blood of the inferior extremities is divided in the pelvis into two branches, which correspond with the femoral and ischiadic arteries; the one passes through the ischiadic foramen, and the other through the hole upon the anterior margin of the pelvis; but the proportion they bear to each other in magnitude is the very reverse of what occurs in the arteries; for the anterior vein is the principal one, whilst the other is not a very considerable vessel, and receives its supply of blood from the muscles at the posterior part of the joint.

'The *femoral vein*, ib. *a a*, immediately without the pelvis, gives branches on both sides, which receive the blood of the extensor and adductor muscles at their superior part: the trunk passes obliquely under the accessory muscle of the flexor digitorum, and over the os femoris, where it lies superficially; it then winds under the adductor muscles, and gets into the ham, *b b*, where it receives many muscular branches, and comes into company with the artery and nerve. It here divides into the *tibial*, *c c*, and *peroneal* veins. The first is joined by some branches from the surface of the joint answering to the articular arteries; it also receives the *anterior* tibial vein which accompanies the artery of the same name. The tibial vein proceeds down the leg along with the artery on the inside of the deep-seated flexors of the heel: it turns over the fore part of the articulation of the tibia with the metatarsal bone, in order to get upon the inner side of the metatarsus; above the origin of the pollex, it receives a communicating branch from the peroneal vein, and immediately after two branches from the toes: one of them comes from the inside of the internal toe; the other arises from the inside of the external and middle toes, unites at the root of the toes in the sole of the foot, and is joined by a branch from the pollex, before its termination in the internal vein of the metatarsus.

'The *peroneal vein* derives its principal branches along with those of the peroneal artery, from the muscles on the outside of the leg. The trunk of the vein comes out from the peroneal muscles, and passes superficially over the joint of the heel, and along the outside of the metatarsus; near the pollex, or great toe, it sends a branch round the back of the leg, to communicate with the tibial vein; after which it is continued upon the outside of the external toe to the extremity, receiving anastomosing branches from the tibial vein.

'Where the veins run superficially upon the upper and lower extremities, they seem to supply the place of the branches of the *cephalic*, *basilic*, and the two *saphenæ*; but the analogy is lost upon the upper arm and thigh, these branches forming deep-seated trunks; this constitutes the greatest peculiarity,[1] as compared with Man and many Mammals, in the distribution of the veins in the extremities of Birds.'

[1] XLIII'.

CHAPTER XX.

RESPIRATORY SYSTEM OF BIRDS.

§ 156. *Lungs of Birds.* Notwithstanding the extent and activity of the respiratory function in Birds, the organs subservient thereto manifest more of the Reptilian than of the Mammalian type of formation.

The lungs are confined, as in the Tortoise, to the back part of the thoracic-abdominal cavity, being firmly attached to the ribs and their interspaces; and, as in the Serpent, they communicate with large membranous cells which extend into the abdomen and serve as reservoirs of air. In the Apteryx alone they do not penetrate the diaphragm.

In those aquatic Birds which are deprived of the power of flight, as the Penguin, the air-receptacles are confined to the abdomen; but in the rest of the class they extend along the sides of the neck, and, escaping at the chest and pelvis, accompany the muscles of the extremities. They also penetrate the medullary cavities and diploë of the bones, extending in different species through different proportions of the osseous system, until in *Volitores*, even in the Hornbill, every bone of the skeleton is permeated by air. There is, indeed, no class of Animals so thoroughly penetrated by the medium in which they live and move as that of Birds.

95

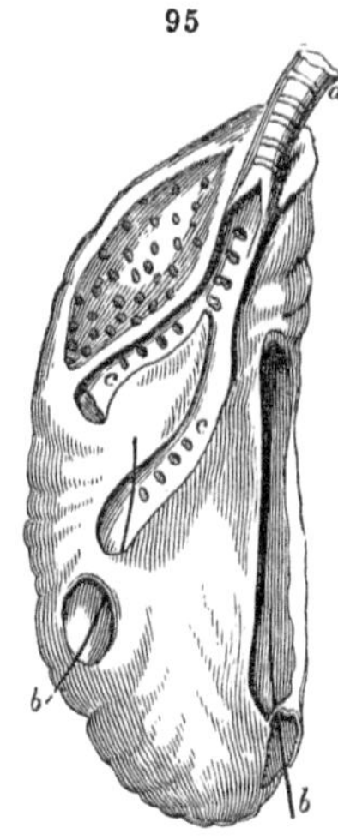

Right lung of a Goose.

The *lungs* are two in number, of a lengthened, flattened, oval shape, fig. 95, extending along each side of the spine from the second dorsal vertebra to the kidneys, and laterally to the junction of the vertebral with the sternal ribs. They are not suspended freely, nor divided into lobes, as in Mammals; but are confined to the back part of the chest by cellular membrane, and the pleura is reflected over the sternal surface only, to which the strong aponeurosis of the diaphragmatic muscles is attached. They are consequently smooth and even on that surface, but posteriorly are accurately moulded to the inequalities of the ribs and intercostal spaces: the bosses varying in number from four to seven (*Apteryx*) or eight (Emeu).

The lungs in general are of a bright red colour, and of a loose spongy texture. The bronchi, fig. 85, *v*, fig. 95, *a*, penetrate their mesial and anterior surfaces about one-fourth from the upper extremities, become membranous, dilate, give off branches, which diverge as they run along the anterior surface; and the trunk divides into the two which open at *b*, *b*, into the thoracic-abdominal air-receptacles. These orifices are oblique, and are partially covered by a slight projection of membrane. Some cartilaginous traces are found through their entire extent.

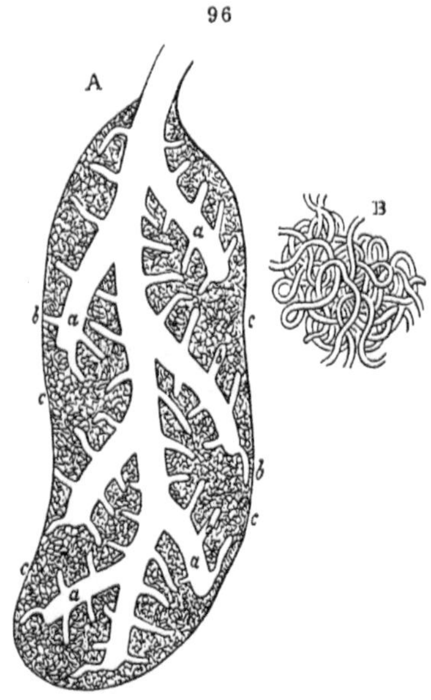

A. Lobule of the lung of a bird represented in ideal longitudinal section. CCLXVIII.

The pulmonary artery divides, almost immediately after its origin, into two branches, one to each lung; the ramifications of each artery form plexuses, fig. 96, B, which chiefly compose the pulmonary tissue: the pulmonary veins leave each lung by a single trunk, and the two trunks unite into one before terminating in the left auricle.

The superficial primary branches of the bronchi, fig. 95, *c*, *c*, send off deeper-seated secondary ones, fig. 96, *a*, *a*, which maintain a uniform diameter to their cæcal terminations: the tertiary bronchi, ib. *b*, *b*, distributed penniformly, also maintain a regular diameter, and open upon a dense labyrinth of blood-vessels, ib. B. The mucous ciliated lining of the bronchi ceases with them; and the capillaries of the pulmonary tissue are covered only by a hyaline epithelium, so as to appear naked.[1] The ultimate pulmonary capillaries do not form a network lining definitely bounded air-cells, but each capillary crosses

[1] CCLXVIII. p. 277.

an air-space of its own; they interlace in every direction, forming a cubic mass of capillaries permeated everywhere by air. In fig. 97, *a* is the cavity of a bronchial tube, *b* its lining membrane supporting blood-vessels with large areolæ; *c*, *c*, perforations in the membrane at the orifices of the lobular passages, *d*, *d*: *e*, *e*, are interlobular spaces containing the terminal branches of the pulmonary vessels supplying the capillary plexus, *f*, *f*, to the meshes of which the air gets access by the lobular passages.

97

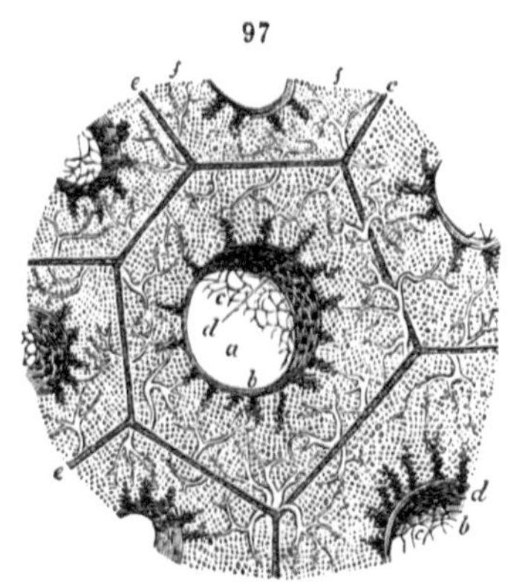

Section through a bronchial tube, lung of Bird, magn. XLV'.

§ 157. *Air-cells of Birds.*—The thoracic-abdominal cavity is subdivided and intersected by a number of membranes; the greater part of the cells thus formed are filled with air. The texture of their parietes possesses considerable firmness in the larger birds, as the Ostrich and Cassowary.

The innermost layer of the air-receptacles can be separated from the outer layer, and is a continuation of the lining membrane of the bronchial tube; the outer layer is a serous membrane, and appears to form the cells by a series of reflections of what may be regarded as the pleura or peritoneum.

These large membranous receptacles, into which the extremities of the bronchial bifurcation and also some of the preceding branches open, are disposed with sufficient general regularity to admit of a definite description and nomenclature.

The first or *interclavicular* air-cell, fig. 98, *a*, extends from the anterior part of each lung, forward to the interspace of the furculum, anterior to which it dilates in the Gannet and many other birds into a large globular receptacle. In the Vultures it is divided into two lateral receptacles, between which the large crop is situated. A thin fan-shaped muscle is extended from the anterior edge of the furculum, over the interclavicular air-cell in these and some other birds.

The *anterior thoracic* cell, ib. *b*, contains the lower larynx and bronchi, and the great vessels with their primary branches to the head and wings. It is traversed by numerous membranous septa, which connect the different vessels together, and maintain them in their situations. The air passes into the posterior part of this receptacle by two openings at the anterior part of the lungs. The deep-seated air-cells of the neck are continued from it anteriorly.

The *lateral thoracic* cells, ib. *d*, are continued on each side from a foramen on the inner edge of the lung, situated just opposite the base of the heart; they are covered by the anterior thoracic air-cell, and from them the air passes into the *axillary* and *subscapular* cells, into those of the wing, and into the humerus, ib. *e*. They also communicate with the *cellula cordis posterior*, ib. *c*, behind the heart and bronchi, which cell is often subdivided into several small ones.

98

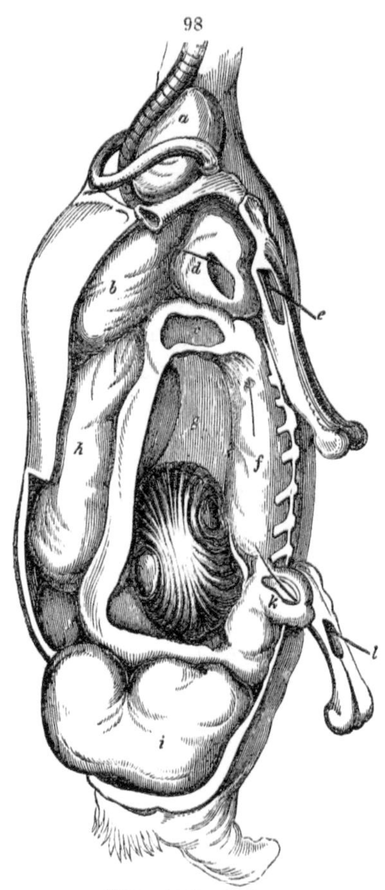

Air-receptacles of a Swan.

The *cellulæ hepaticæ* are of much larger size; they are two in number, of a pyramidal figure, with their bases applied to the lateral thoracic cells, and their apices reaching to the pelvis: they cover the lower portions of the lungs and the lobes of the liver; they receive air from several foramina situated near and at the external edge of the lungs.

The *cellulæ abdominales* commence beneath the cellulæ hepaticæ at the inferior extremity of the lungs, where the longest branches of the bronchiæ open freely into them. (A bristle is passed through one of these openings in the figure.) They are distinguished into *right* (*h*) and *left* (*f*): the former is generally the largest receptacle in the body; it extends from the last ribs to the anus, and covers the greater part of the small intestines, the suprarenal gland, and kidney of the same side. The left abdominal cell, *f*, contains the intestines of its own side, and is attached to the gizzard. In some large Birds, as the Gannet, it is separated from the right receptacle by a mediastinal membrane, *g*, which is continued on from the gizzard to the anus.

Both the abdominal receptacles transmit air to the *pelvic* cells, *i*, *k*, of their respective sides, and to several small and extremely delicate cells between and behind the coils of intestine. One of these is continued round the fold of the duodenum and pancreas to the gizzard, and has been termed the *duodenal cell.*

From the inguinal cell are continued the intermuscular *glutæal* and *femoral* cells, which surround the head of the femur, and communicate with that bone by an aperture, *l*, situated immediately anterior to the great trochanter, except in those Birds in which the femur retains its medulla.

The *cervical* air-cells are continued from the large clavicular cell, and form in the Argala and Bustard, fig. 54, *a*, a singular appendage or pouch, contained in a loose fold of integument, which the bird can inflate at pleasure.

In the Pelican and Gannet extensive air-cells are situated beneath almost the whole of the integument of the body, which is united to the subjacent muscles only here and there by the septa of the cells and the vessels and nerves which are supported by the septa in their passage to the skin. The large pectoral muscles and those of the thigh present a singular appearance, being, as it were, cleanly dissected on every side, having the air-cavities above and beneath them. The axillary vessels and nerves are also seen passing bare and unsupported by any surrounding substance through these cavities. Numerous strips of *panniculus carnosus* pass from various parts of the surface of the muscles to be firmly attached to the skin; a beautiful fan-shaped muscle is spread over the interclavicular or furcular air-cell. The use of these muscles appears to be to produce a rapid collapse of the superficial air-cells, and an expulsion of the air, when the bird is about to descend, in order to increase its specific gravity, and enable it to dart with rapidity upon a living prey.

The air-receptacles of the thoracic-abdominal cavity present varieties in their relative sizes and modes of attachment in different birds. In the *Raptores* they are principally attached posteriorly to the ribs, the diaphragmatic aponeurosis covering the lungs, and to the kidneys; while in the *Grallatores* they have anterior attachments to the intestines in many places.

The singular extension of the respiratory into the osseous system was discovered almost simultaneously by Hunter and Camper, and ably investigated by them through the whole class of Birds. The air-cells and lungs can be inflated from the bones, and Hunter injected the medullary cavities of the bones from the trachea. If the femur into which the air is admitted be broken, the bird is unable to raise itself in flight. If the trachea be tied, and an opening be made into the humerus, the bird will respire by that opening for a short period, and may be killed by inhaling noxious gases through it. If an air-bone of a living bird, similarly perforated, be held in water, bubbles will rise from it, and a motion of the contained air will be exhibited, synchronous with the motions of inspiration and expiration.

The proportion in which the skeleton is permeated by air varies in different Birds. In the *Alca impennis*, the Penguins (*Aptenodytes*) and the *Apteryx*, air is not admitted into any of the bones. The condition of the osseous system, therefore, which all birds present at the early periods of existence, is here retained through life.

In the large Struthious Birds, which are remarkable for the rapidity of their course, the thigh-bones and bones of the pelvis, the vertebral column, ribs, sternum and scapular arch, the cranium and lower jaw, have all air admitted into their cavities or cancellous structure. In the Ostrich the humeri and other bones of the wings, the tibiæ and distal bones of the legs, retain their marrow. Most Birds of Flight have air admitted to the humerus: the Woodcock and Snipe are exceptions. The Pigeon tribe, with the exception of the Crown Pigeon, have no air in the femur, which retains its marrow. In the Owls also the femur is filled with marrow; but in the Diurnal Birds of Prey, as in almost all other Birds of Flight, the femur is filled with air. In the Pelican and Gannet the air enters all the bones with the exception of the phalanges of the toes. In the Hornbill even these are permeated by air.

Hunter has given the following characters as distinguishing the bones which receive air. They may be known—'first, by their less specific gravity; secondly, by their retaining little or no oil, and, consequently, being more easily cleaned, and when cleaned, appearing much whiter than common bones: thirdly, by having no marrow, or even any bloody pulpy substance in their cells; fourthly, by not being in general so hard and firm as other bones; and, fifthly, by the passage that allows the air to enter the bones.'[1] The openings by which the air penetrates the bones, may be readily distinguished in the recent bone, since they are not filled up by blood-vessels or nerves, but have their external edges rounded off.

In the dorsal vertebræ the air-orifices are small, numerous, and irregular; situated along the sides of the bodies, and the roots of

[1] XCIV. p. 178.

the spinous processes, the air passes into them directly from the lungs. In the two or three lower cervical vertebræ the air-holes are in the same situation, but receive the air from the lower cervical or clavicular air-cells: in the remainder of these vertebræ the air-holes are situated within the canal lodging the vertebral artery, and communicate with the lateral air-cells of the neck.

The air-holes of the vertebral ribs are situated at the internal surface of their vertebral extremities, and appear, like those of the contiguous vertebræ, to have an immediate communication with the lungs. The sternal ribs have also internal cavities which receive air from the lateral thoracic cells by means of orifices placed at their sternal extremities.

The orifices by which air is admitted to the sternum are numerous, but are principally situated along the mesial line of the internal surface, opposite the origin of the keel, forming a reticulation at that part; the largest foramen is near the anterior part of the bone; some smaller ones occur at the costal margins. All these orifices communicate with the thoracic air-receptacles.

The scapula is perforated by several holes at the articular extremity, which admit air into its cancellous structure from the axillary cell. The coracoid has small air-holes at both extremities; the largest is situated on its inner surface, where it is connected with the clavicle or furculum. The furculum receives air principally by a small hole in the inner side of each of its scapular extremities, which communicates with the clavicular air-cell.

The air-hole of the humerus is of large size, and situated at the anconal or back part below the head of the bone, in the hollow of the ulnar or inner tuberosity. It communicates with the axillary air-cell, and transmits the air to the cavity of the bone by several cribriform foramina.

The air-holes of the pelvic bones are situated irregularly on the inner surface upon which the kidneys rest, and must therefore receive air from continuations of the abdominal receptacles around the kidneys.

A depression at the anterior part of the base of the great trochanter receives air from the glutæal cell, and transmits it by several small foramina into the interior of the femur. In the Ostrich, the air-holes are situated at the posterior part of the bone at both its extremities.

The cavities of the long bones into which air is thus admitted are proportionally larger than in the corresponding bones of Mammalia, and are characterised by small transverse osseous columns which cross in different directions from side to side, and are more numerous near the extremities of the bone; they abut against and strengthen, like cross-beams, the parietes of the bone. The membrane lining these cavities, is not very vascular.

The lower jaw receives its air by an orifice situated upon each ramus behind the tympanic articulation, from an air-cell which surrounds the joint. The bones of the cranium and upper jaw receive air admitted to the tympanic cavity by the Eustachian tube, not from the nasal passages. With these, however, the subocular air-cell communicates; and in the Coot, Water-hen, Goose, and other water-birds, entozoa (*Monostoma mutabile*, e. g.) gain access to that air-cell.

The extension from the lungs of continuous air-receptacles throughout the body is subservient to the function of respiration, not only by a change in the blood of the pulmonary circulation effected by the air of the receptacles on its repassage through the bronchial tubes; but also, and more especially, by the change which the blood undergoes in the capillaries of the systemic circulation, which are in contact with the air-receptacles. The free outlet to the air by the bronchial tubes does not, therefore, afford an argument against the use of the air-cells as subsidiary respiratory organs, but rather supports that opinion, since the inlet of atmospheric oxygenated air to be diffused over the body must be equally free.

A second use may be ascribed to the air-cells as aiding mechanically the actions of respiration in Birds. During the act of inspiration the sternum is depressed, the angle between the vertebral and sternal ribs made less acute, and the thoracic cavity proportionally enlarged; the air then rushes into the lungs and into the thoracic receptacles, while those of the abdomen become flaccid: when the sternum is raised or approximated towards the spine, part of the air is expelled from the lungs and thoracic cells by the trachea, and part driven into the abdominal receptacles, which are thus alternately enlarged and diminished with those of the thorax. Hence the lungs, notwithstanding their fixed condition, are subject to due compression through the medium of the contiguous air-receptacles, and are affected equally and regularly by every motion of the sternum and ribs.

A third use, and perhaps the one which is most closely related to the peculiar exigencies of the bird, is that of rendering the whole body specifically lighter; this must necessarily follow from the desiccation of the marrow and other fluids in those spaces which are occupied by the air-cells, and by the rarefaction of the contained air from the heat of the body.

Agreeably to this view of the function of the air-cells, it is found that the quantity of air admitted into the system is in proportion to the rapidity and continuance of the bird's flight; and, where it is limited, the air is distributed to those members which are most employed in locomotion; thus the air is admitted into the wing-bones of the Owl, but not into the femur; while in the Ostrich the air penetrates the femur, but not the humerus or other bones of the wing.

A fourth use of the air-receptacles relates to the mechanical assistance which they afford to the muscles of the wings. This was suggested by observing that an inflation of the air-cells in a Gigantic Crane (*Ciconia Argala*) was followed by an extension of the wings, as the air found its way along the brachial and antibrachial cells. In large birds, therefore, which, like the Argala, hover with a sailing motion for a long-continued period in the upper regions of the air, the muscular exertion of keeping the wings outstretched will be lessened by the tendency of the distended air-cells to maintain that condition. It is not meant to advance this as other than a secondary and probably partial service of the air-cells. In the same light may be regarded the use assigned to them by Hunter, of contributing to sustain the song of Birds, and to impart to it tone and strength. It is no argument against this function that the air-cells exist in birds which are not provided with the mechanism necessary to produce tuneful notes; since it was not pretended that this was the exclusive and only office of the air-cells.

§ 158. *Air-passages in Birds.*—The air-passages in Birds commence by a simple *superior larynx*, from which a long *trachea* extends to the anterior aperture of the thorax, where it divides into the two *bronchi*, one to each lung. At the place of its division there exists, in most birds, a complicated mechanism of bones and cartilages moved by appropriate muscles, and constituting the true organ of voice: this part is termed the *inferior larynx*.

The tendency to ossification, which is exemplified in the bony condition of the sternal ribs and tendons of the muscles, is again manifested in the framework of the larynx and the rings of the trachea, which, instead of being cartilaginous, as in Reptiles and Mammals, are in most birds of a bony texture.

The *superior larynx*, figs. 73, *e*—*h*, 99, and 100, is situated behind the root of the tongue, and rests upon the urohyal, fig. 73, 43, to which it is attached by dense cellular texture.

It is composed of several bony and cartilaginous pieces, varying in number from four to ten. The largest of these pieces constitutes the anterior part of the larynx. It is of an oval or triangular form, according as its superior termination is more or less pointed, and answers to the thyroid cartilage, fig. 73, *f*. The cricoid cartilage is represented by three osseous pieces, which are situated at the posterior and inferior part of the upper larynx; the middle one, fig. 73, *g*, is of an oblong form, and varies in size, being larger than the lateral ones in the *Anatidæ*, but smaller in the *Cantores*. The lateral pieces are connected at one extremity with the thyroid piece, and at the other to the middle oblong piece above described, which completes the circle of the laryngeal framework posteriorly: the first two incomplete tracheal rings, ib. *g*, *g*, may represent the anterior part of the cricoid. The arytenoid bones, ib. *h*, rest upon the middle oblong portion of the cricoid, and extend forward, being connected at their outer edge by means of elastic cellular substance to the thyroid, and attached by their anterior extremities

99

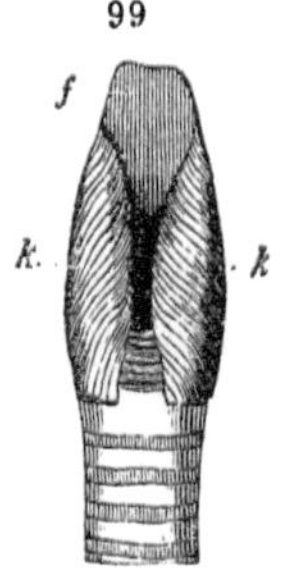

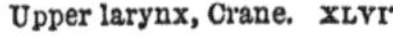
Upper larynx, Crane. XLVI'.

100

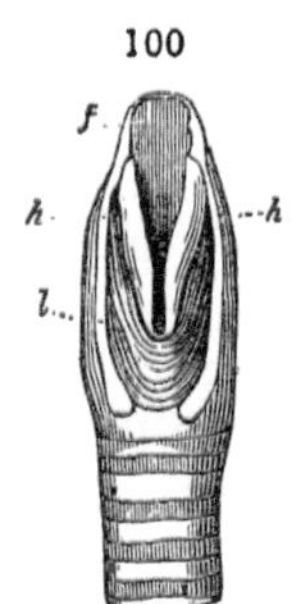

Upper larynx, Crane. XLVI'.

to the urohyal by means of two small ligaments: they form, by their inner margins, the *rima glottidis* or laryngeal fissure.

This fissure, fig. 51, *i*, being thus bounded by inflexible rigid substances, is only susceptible of having its lateral diameter varied according to the degrees of separation or approximation to which the arytenoid bones are subject. These different states are produced by appropriate muscles, one pair of which may be regarded as *Thyreo-arytenoidei*, and the other may be termed *Constrictores glottidis*. The former, fig. 99, *k*, *k*, arise from the sides and posterior surface of the thyroid, and are inserted into the whole length of the inner edge of the arytenoids, which they draw outward, and consequently open the laryngeal fissure. The *Constrictores glottidis* in the Gigantic Crane arise from the middle of the internal or posterior surface of the thyroid, and are inserted into the arytenoids: they close the laryngeal opening with such accuracy as to supersede the necessity of an epiglottis. From the simplicity of the structure just described, from the situation of the superior larynx with relation to the rictus or gape of the bill, and from the absence of lips by which this might be partially or entirely closed, it is plain that it cannot be considered as influencing the voice, otherwise than by dividing or articulating the notes after they are formed by the lower larynx. The superior larynx presents, indeed, but few varieties in the different species of Birds; and these relate chiefly to certain tubercles in its anterior, which vary in number, and do not exist at all in some species, as the Singing Birds; being chiefly present in those birds which have a rough unmusical voice. In the Pelican, the Gigantic Crane, and most of the *Rasores*, a process extends backward into the cavity of the upper larynx from the middle of the posterior surface of the thyroid cartilage, and seems destined to give additional protection to the air-passage.

The *trachea*, figs. 93, 94, G, is proportionally longer, in consequence of the length of the neck in Birds, than in any other class of animals, its length being further increased in many species by convolutions varying in extent and complexity. A species of Sloth (*Bradypus tridactylus*) among Mammals, and a species of Crocodile (*Crocodilus acutus*) among Reptiles, present an analogous folding of the trachea.

The trachea is composed in Birds of a series of bony, and sometimes, as in the Ostrich, of cartilaginous rings, included between two membranes. In those cases in which they are of a bony structure, the ossification is observed to commence at the anterior part of each ring, and gradually to extend on both sides to the opposite part.

The tracheal rings, whether bony or cartilaginous, are, with the exception of the two uppermost, always complete, and not, as in most quadrupeds, where the windpipe bears a different relation to the organ of voice, deficient posteriorly. They differ in shape, being sometimes more or less compressed. They are generally of uniform breadth, but in some species are alternately narrower at certain parts of their circumference and broader at others, and in these cases the rings are generally closely approximated together, and, as it were, locked into one another. This structure is most common in the *Grallatores*, where the rings are broadest alternately on the right and left sides.

With respect to the diameter of the tracheal rings, this may sometimes be pretty uniform throughout, and the trachea will consequently be cylindrical, as in the *Cantores*, the *Grallatores* which have a shrill voice, the females of the *Natatores*, and most *Raptores* and *Rasores*: or the rings may gradually decrease in diameter, forming a conical trachea, as in the Turkey, the Heron, the Buzzard, the Eagle, the Cormorant, and the Gannet; or they may become wider by degrees to the middle of the trachea, and afterward contract again to the inferior larynx; or lastly, they may experience sudden dilatations for a short extent of the trachea; the Golden-eye (*Anas clangula*), the Velvet-duck (*Anas fusca*), and the Merganser (*Mergus serrator*), present a single enlargement of this kind, in which the bony rings are entire, and of the same texture as in the rest of the tube. In the Golden-eye the trachea is four times larger at the dilatation than at any other part. In the Goosander (*Mergus merganser*), the trachea presents two sudden dilatations of a similar structure to that above described. The trachea of the Emeu (*Dromaius ater*) is also remarkable for a sudden dilatation, but in this instance the cartilaginous rings do not preserve their integrity at the dilated part, but are wanting posteriorly, where the tube is completed by the expanded membranes only.

With regard to the windings of the windpipe, in an Australian Snipe (*Rhynchæa australis*), the convolutions, which are short, are external to the chest, between the skin and the fore part of the pectoral muscles. In the same position lie the long double coils of the windpipe in the Semipalmate Goose (*Anas semipalmata*), and the long single fold in *Ortalida Parraqua*. In the Crested Pintado (*Numida cristata*), the apex of the furculum forms a bony cup which receives a loop of the trachea. In the crestless Guan (*Penelope Mirail*), the Demoiselle (*Grus virgo*), and Stanley Crane (*Grus Stanleyanus*), the trachea forms a curve sinking into the upper and fore part of the sternum. In the common Crane (*Grus cinerea*), and Serass Crane (*Grus Antigone*), the keel of the sternum is more deeply hollowed for the lodgment of more extensive coils of the trachea. In the male wild Swan (*Cygnus ferus*), the windpipe describes a double vertical coil within the long and deep keel of the sternum: in Bewick's Swan (*Cygnus Bewickii*), the distal part of the coil lies horizontally within the body of the sternum: the entry and exit of the intrasternal coils are shown in fig. 101.

§ 159. *Lower Larynx in Birds.*—The main or essential organ of voice is situated at the bifurcation of the trachea, ib. *a*, into the bronchi, ib. *b*, *b*; and herein may be discerned an analogous relation to convenient stowage, which the position of the masticatory apparatus shows: for even the muscles of the organs of voice and the bony drum of the larynx, &c., are brought beneath the centre of gravity, at the base of the neck, not accumulated at its anterior extremity. In general the rings of the bronchi are incomplete. In the King-Vulture the entire rings are continued a little way along the bronchial divisions of the trachea, without any modifications, external or internal, indicative of a laryngeal structure. The same may be seen in the Ostrich, where the bronchi are provided with entire slender rings rapidly diminishing in size as they approach the lungs: but the terminal rings of the trachea are thickened and protrude outward, forming a cavity on each side, the lining substance of which projects into the area of the tube above the commencement of each bronchus.[1]

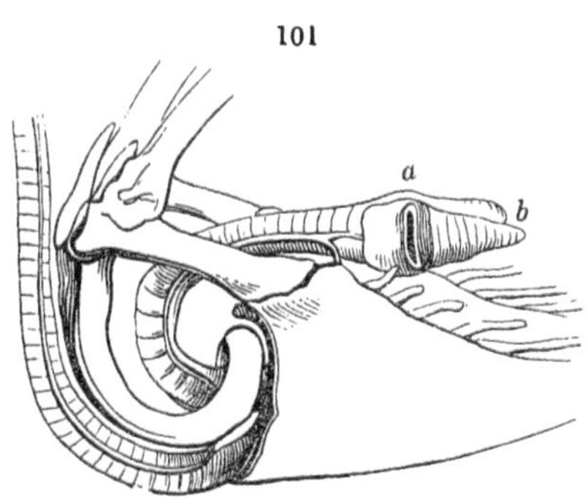

Tracheal coils and lower larynx, Bewick's Swan. XLVIII'.

In most Birds the bronchi, figs. 85, *v*, 101, *b*, are straight, compressed, and easily lacerable tubes, strengthened by half-rings on the outer side, the inner side being formed by a membrane ('membrana tympaniformis'). Usually the bronchi rapidly contract as they approach the lungs.

The muscles of the trachea are the 'sterno-tracheales,' fig. 104, *d*, a long pair, arising from the costal processes of the sternum and converging to ascend along the sides of the windpipe. To these are sometimes added a second pair from the furculum, called 'cleido-tracheales.' In *Cursores* and most *Rasores* the sterno-tracheales alone are present. In most *Raptores* and *Grallatores*, a muscle, broncho-trachealis, situated on each side of the lower part of the trachea, descends to be inserted into the first or second bronchial half-ring: in *Alcedo* and *Caprimulgus* it descends to the third half-ring; in some of the Owls its insertion is still lower, and the degree of tension of the tympaniform membrane will be proportionally varied. In *Colopterus cristatus* an azygos muscle occupies the anterior interspace of the broncho-tracheales.[2]

In other Vocal Birds there is a double glottis, usually produced by a bony bar, 'pessulus,' 'os transversale,' fig. 102, *i*, which traverses the lower end of the trachea from before backward: it supports a thin membrane which ascends into the tracheal area and, terminating there by a free concave margin, is called the 'membrana semilunaris,' ib. *h*. This is most developed in Singing

[1] XX. vol. ii. p. 103 (1834). [2] XLIX'.

Birds, and, being vibratile, forms an important part of their trilling vocal apparatus. The air passes on each side the membrana semilunaris and its sustaining bone to and from the bronchi and lungs. The walls of the lower larynx are formed by modified rings and half-rings of the end of the trachea and beginning of the bronchi.

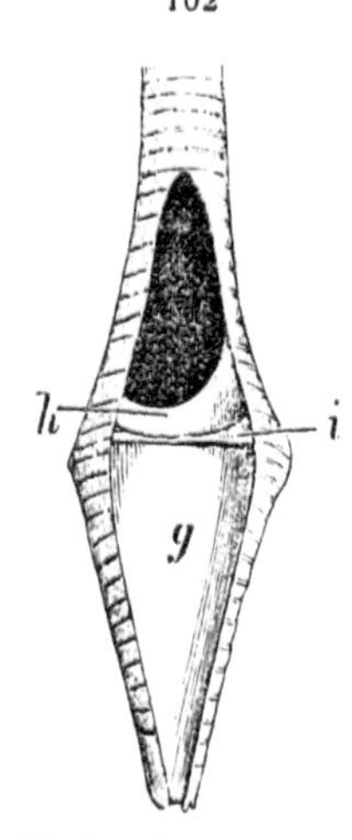

Side view of cavity of lower larynx, Raven. xxx.

The last ring of the trachea, fig. 103, *t*, usually expands as it descends, with its fore and hind parts produced, and the lower lateral borders concave: the extremities of the pessulus, fig. 102, *i*, abut against the produced angles, and expand to be there connected, also, with the fore and hind terminations of the first half-ring of the bronchus, fig. 103, *a*, strengthening and clamping together the upper parts of the vocal framework. The second bronchial half-ring, ib. *b*, is flattened and curved with the convexity outward, like the first, but is more moveable. The third half-ring, ib. *c*, is less curved and further separated from the second, to the extremities of which its own are connected by ligament, and, for the intervening extent, by membrane; its inner surface supports the fibrous chord or fold which forms the outer lip of the glottis of that side; it is susceptible of a rotatory movement on its axis, and is an important agent in the modulation of the voice. All the above parts, *t*, *a*, *b*, *c*, fig. 103, are bony. The bronchial half-rings and their connecting ligaments and membranes form the outer convex wall of the tube: the inner wall is a flat membrane, stretched like a drum-head, between the extremities of the half-rings, and attached above to the cross-bar, and through it to the semilunar membrane. The outer part of the lower tracheal and bronchial rings, being cut away in fig. 102, exposes the central surface of the 'membrana tympaniformis,' *g*, with its upper connexions with the cross-bone *i*, and the 'membrana semilunaris,' *h*. Part of the peripheral surface of the tympaniform membrane is seen in the front view of the lower larynx and bronchi, fig. 104, A, *g*. A small appendage to the inner margin of the half-ring, fig.

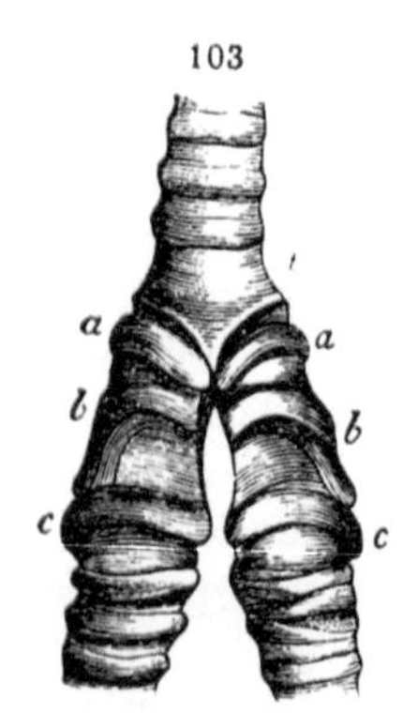

Lower larynx, Raven. xxx.

103, *b*, making a prominence where the external vocal fold is continued over it, in the starling, thrush, nightingale, &c., has been called 'arytenoid cartilage,' from its analogy to that of the upper larynx of Mammals. The proper muscles of the lower larynx, as seen in the Raven, are shown in fig. 104, in front view A, and side view B.

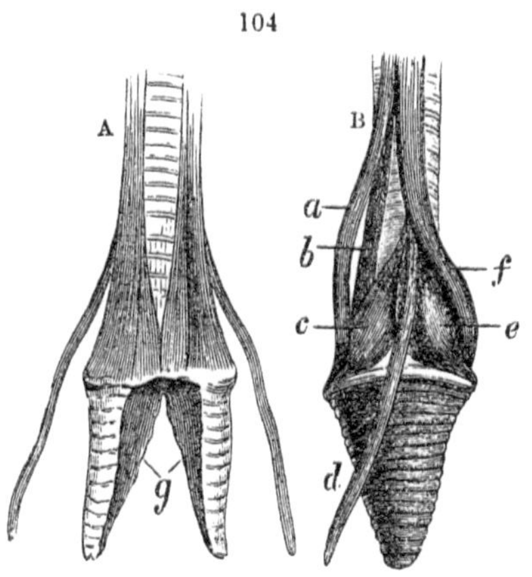

A front, B side, view of lower larynx, Raven. XXX'.

The muscle answering to the 'tracheo-lateralis' in *Volitores* expands toward the lower end of the trachea and divides into two fasciculi which diverge, the one, *f*, to the fore, the other, *a*, to the back part of the bronchus, to be inserted into the corresponding extremities of the third half-ring, fig. 104, *c*. The fasciculus, fig. 104, B, *f*, is the 'broncho-trachealis anticus:' the fasciculus, *a*, is the broncho-trachealis posticus. Beneath this is a shorter muscle, ib. *b*, the broncho-trachealis brevis, which is inserted into the posterior end of the second bronchial half-ring.

The remaining two muscles are enlarged divisions or differentiated fasciculi of the common laryngeal muscle (Kehlkopfmuskel, Müller[1]) of *Volitores*: the 'bronchialis posticus,' ib. *c*, arising from the lower and lateral border of the last tracheal ring, swells into a 'venter,' and contracts as it passes backward to be inserted into the hinder end of the second half-ring. The 'bronchialis anticus,' ib. *e*, is partly covered by the 'broncho-trachealis anticus,' and is thick and ventricose: it arises from the last tracheal ring and passes forward to its insertion into the fore ends of the first and second half-rings and into the supplemental (arytenoid) cartilage.

All the foregoing muscles tend to tighten the whole or parts of the tympaniform membrane which is below their points of insertion, and to relax the part above the insertion. They lengthen the part of the bronchus below or beyond their insertion and shorten the part above, by approximating to the trachea the half-rings they are attached to. The chief antagonistic power is the

[1] XLIX'.

elasticity of the membranes so put on the stretch: but there is a direct 'relaxor' of the tympaniform membrane in the 'sterno-trachealis,' ib. *d*, which, passing from the side of the trachea to the sternum, shortens the whole bronchus as it draws down the wind-pipe. This is the most constant of all the muscles affecting the lower larynx. It is reckoned by Savart as the sixth pair of vocal muscles, but not by Cuvier, since it is not directly attached to any part of the lower larynx, and exists in Birds, as, e. g., the Vulture and Ostrich, in which that larynx is not developed.

The manifold ways and degrees in which the several parts of the complex vocal organ in *Cantores* may be affected, each of the principal bony half-rings, as one or other end may be pulled, being made to perform a slight rotatory motion, are incalculable: but their effects are delightfully appreciable by the rapt listener to the singularly varied kind and quality of notes trilled forth in the stillness of gloom by the Nightingale.

In many of the *Volitores* there is a single pair of 'broncho-tracheales,' and a single pair of short ventricose 'bronchiales.' In *Thamnophilus* each sterno-trachealis bifurcates to send a small strip to the lower larynx, and the rest to the side of the trachea, as usual. In *Furnaria* the sterno-trachealis is inserted into the upper end of a long appendage to the upper bronchial half-ring.

The Parrot tribe have a single glottis bounded by a lateral pair of vibratile membranes; each membrane, connecting together, and occupying the interspace between, the last tracheal and first bronchial half-rings. These have each one margin concave, with the concavity turned towards each other, and are moveably joined together at their fore and hind extremities. These half-rings expand, and stand out from the end of the trachea. A narrow muscle, 'tensor longus glottidis, fig. 105, *a*, passes from the side of the trachea to the upper (tracheal) half-ring; and, by raising it, makes tense the elliptical elastic membrane: a broader 'tensor brevis glottidis,' ib. *b*, passes from the lower rings of the trachea to the same half-ring, diverging to its extremities: a third narrow muscle passes from the tracheal to the bronchial half-rings, ib. *c*, and, by approximating them, relaxes the membrane occupying the elliptical interspace. These membranes, projecting on each side into or below the termination of the air-tube, leave a narrow chink between them, through which the air passes to and from the lungs; and when, in forcible expiration, the membranes are put into a sufficient

105

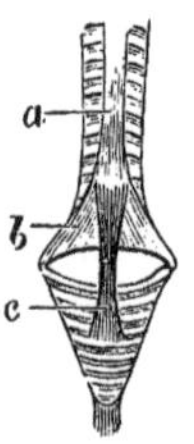

Lower larynx, Parrot. xxx'.

state of tension, they vibrate, and the vocal air is driven along the trachea through the upper larynx, where some modification of sound may be made. The tongue of the Parrot is more fleshy

106

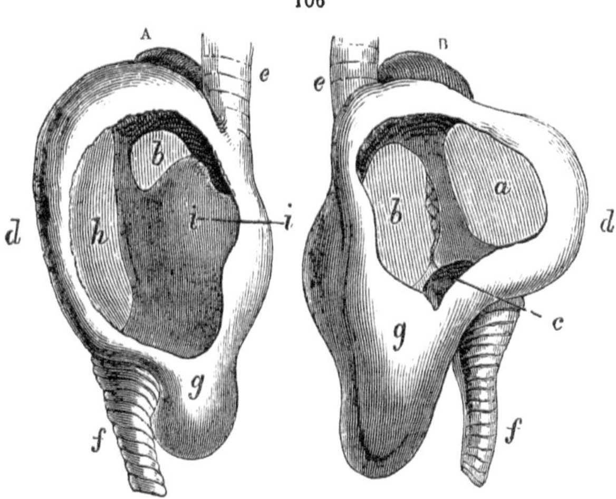

Lower larynx: A, right side; B, left side. *Mergus serrator.* CCCXX.

than in most Birds. These structures, concomitant with the single glottis and pair of vocal folds in the lower or true larynx, relate to the faculty, so remarkable in these singular birds, of imitating human speech.

107

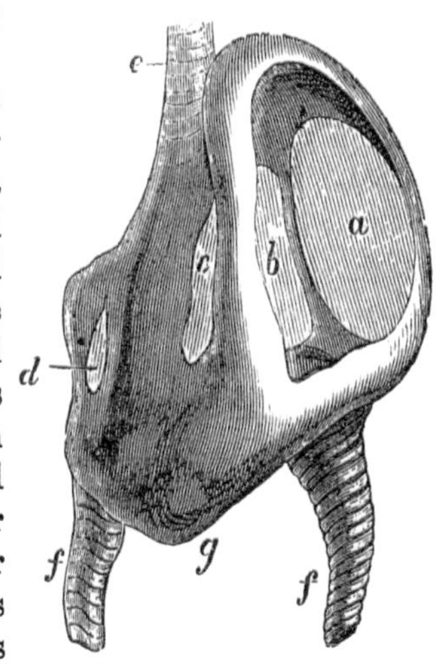

Lower larynx. *Mergus Merganser.* CCCXX.

In the males of the Mergansers and of most Ducks a certain number of the terminal rings of the trachea are welded together and expanded into an irregular bony case, divided into two unequal cavities. In the *Mergus serrator*, fig. 106, the broad 'pessulus,' *i*, leaves a passage at its upper part, *b*, by which the air from the right bronchus, *f*, can pass to and from the trachea, *e*: part of the outer wall of the right laryngeal chamber is formed by membrane, *h*: this chamber is extended by the osseous cavity, *g*. A similar but somewhat more complex lower larynx exists in the male *Anas clangula*. These modifications relate to the power rather than to the variety of the voice.

CHAPTER XXI.

URINARY SYSTEM OF BIRDS.

§ 160. *Kidneys of Birds.*—The urinary excretion is early provided for in the bird: about the third day of incubation a series of short parallel cæcal tubes, fig. 108, *c*, are developed in the blastema beneath the vertebral column, and pass transversely to a longitudinal canal, ib. *d*, which conveys the excretion to the cloaca. These are the primordial or transitory kidneys. Behind them subsequently appear the secondary or persistent kidneys, ib. *a*, together with the genital glands, ib. *e*, and adrenals, ib. *f*. The proper ducts of the kidneys, or ureters, ib. *b*, soon follow the appearance of the true renal tissue, and as this proceeds in its developement, the primordial glands disappear, or yield up their duct and a remnant of their tissue to form the epididymis and vas deferens in the male.

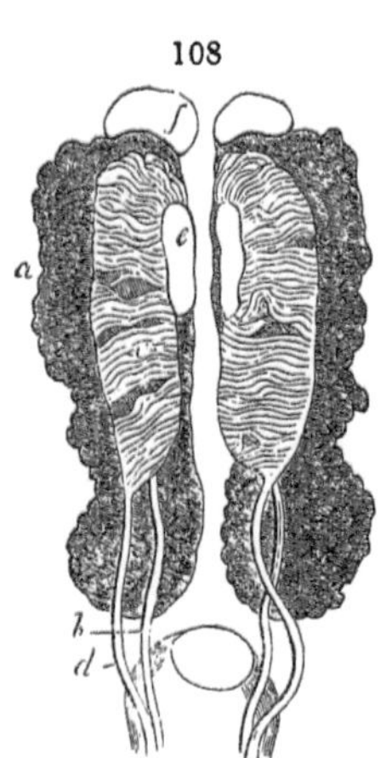

Kidneys, Wolffian bodies, and testes of an embryo Bird, magnified. LXXIV.

In the mature bird the urinary system consists of the kidneys, ureters, and a more or less incomplete urinary receptacle.

As in Reptiles the kidney is distinguished from that of the Mammal by the homogeneity of its substance, which is not divided into a cortical and medullary part, and by the tubuli uriniferi extending to the surface of the gland there to form by reiterated unions the ureter, and not terminating in a cavity or pelvis in the interior of the kidney, from which the ureter commences.

The *kidneys*, fig. 85, *x*, are two in number, of an elongated form, commencing immediately below the lungs, and extending along the sides of the spine as far as the termination of the rectum; in which course they are impacted in, and as it were moulded to, the cavities and depressions of the pelvis. From this fixed condition it results that they are generally symmetrical in position, not placed one higher than the other, as in the Mammalia. The posterior surface of the kidney presents inequalities corresponding to the risings and depressions of the pelvis; the anterior surface is smoothly convex or flattened; but rising into a series of prominences which correspond, not to the eminences, but to the cavities of the bones on which they rest; their inner or mesial side is generally pretty regular and straight, but the external edge is more or less notched. They are relatively larger than in most Mammals; resembling in this respect the kidneys of Whales and of the cold-blooded Ovipara, where there is no perspiration from the skin.

The kidneys vary in size in different birds, being, for example, smaller in most of the *Grallatores*, as the Bustard and Heron, where the pelvis is short, than in the Rasorial Order, in which it is of great extent. Where they are short they are in general more prominent, and this is so remarkable in some Birds, as the Owls, that in them they resemble somewhat in their superficial position the kidneys of Mammals.

As might be expected from their relations to the pelvis, the kidneys in Birds present as many varieties of outward configuration as there are differences in the part of the skeleton to which they are moulded. In some Aquatic Birds, as the Grebe (*Podiceps*) and the Coot (*Fulica*), the kidneys are more or less blended together at their lower extremities, as in most Fishes: in *Colymbus* the extent of the union is greater; in *Platalea* they have been observed to be joined by a middle band. In the rest of the class they are distinct from one another.

The principal lobes are in general three in number: the anterior or highest one is, in some cases, the largest; while in others, as the Pelican, the contrary obtains, the lowest division being most developed in this bird. In the *Tern* each kidney is divided by fissures into seven or eight square-shaped lobes: in the *Eagle* they each present four divisions; but in these cases there are not distinct ureters to each lobe as in the subdivided kidneys of Mammals. In the Emeu (*Dromaius ater*) the kidney presents only two lobes; the superior or anterior one is the broadest and most prominent, being of a rounded figure, and constituting one-third of the whole; the lower division is flattened, and gradually tapers to a point. In one specimen I found the left kidney half an inch longer than the right. In the small *Cantores* the exposed superficies of the kidney is rarely lobular.

Each kidney is invested by its proper capsule, a thin membrane, which also extends into the substance of the gland, between its divisions: a layer of peritoneum is reflected over their anterior surfaces.

The texture of the kidneys is much more frail than in Mammalia, readily yielding under the pressure of the finger, to which they give a granular sensation as their substance is torn asunder.

In colour they resemble the human spleen. Besides being divided into lobes, the surface of the kidneys may be observed to be composed of innumerable small lobules, separated by continuous gyrations like the convolutions of the cerebral substance.

The *tubuli uriniferi* originate from every part of the internal substance of the lobules, extending to the gyrations, uniting in the pinnatifid form, and coursing to the margins of the lobules, all the inflexions of which they follow. The pinnatifid ramification of the uriniferous tubules is sometimes 'opposite,' sometimes 'alternate,' sometimes the branches are simple, sometimes dichotomously divided; but these ramuli appear scarcely smaller than the branches from which they spring, and never intercommunicate.[1] The uriniferous ducts from the convoluted lobules unite dichotomously, and ultimately escape by a single duct—the ureter.

The arteries and veins of the kidneys have already been described. Where the entire stream of the venous blood is not sent to the lungs, but part is diverted to the arterial system, then also a portion of the venous blood circulates through the kidneys before it reaches the heart; but in Birds, where not only the whole venous current is sent to the lungs, but with peculiar energy and frequency, such vicarious office of removing effete particles directly from the venous blood is not required. A certain retention of the oviparous type in the apparent entry of veins into the lower ends of the kidneys is shown, but a reniportal vein does not exist: the connection of the lower veins coming from the kidneys with the iliaco-mesenteric is of such a kind that the renal venous blood may flow to the portal system of the liver when that system and digestion are at work; or it may flow by the upper emulgent veins to the inferior cava and so to the lungs, when respiration is unusually active.

The *ureter*, figs. 85, *y*, 108, *b*, is continued down along the anterior surface of the kidney toward the mesial side; here and there imbedded in its substance, forming a series of dilatations corresponding to the principal lobes or enlargements of the gland, and receiving the branches of the tubuli uriniferi as it passes along. Below the kidney the ureters pass behind the rectum, becoming connected to, and after a short distance involved in, its coats; they ultimately terminate upon valvular eminences in a depression at the lower part of the urinary sac, ib. *d*; the terminal papillæ of the ureters are situated with the orifices of the genital

[1] CXXII. p. 92.

ducts, in the same segment of the cloaca, which is therefore termed the urogenital cavity, fig. 109, *e*.

The space intervening between the urogenital cavity and the valvular termination of the rectum, ib. *c*, forms a cavity more or less developed in different Birds, but always distinct in the smoothness of its lining membrane from the rectum, which has a more vascular and villous internal tunic. The Birds in which this rudimental urinary bladder presents the largest capacity are the Owls, many of the Aquatic Birds, as the Pelican, Willock, Grebe, Swan, &c.; some of the Wading Order, as the Bittern and Bustard, but more especially the Ostrich, among the *Cursores*, in which the urinary receptacle is represented as laid open at *d*, fig. 109.

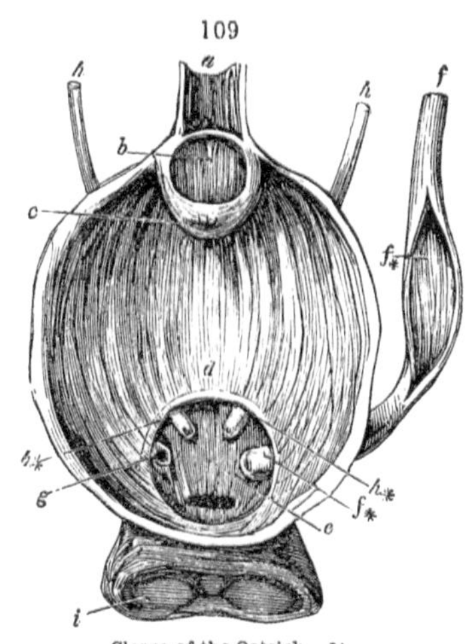

Cloaca of the Ostrich. L'.

§ 161. *Adrenals of Birds.*—The adrenals, *d*, *d*, figs. 117, 127, are small bodies, usually of a bright yellow colour, situated on the mesial or inner side of the superior extremities of the kidneys; closely attached to the coats of the contiguous large veins and in contact with the testes in the male; and the left one adhering to the ovary in the female. They vary in shape, being sometimes of a round, flattened, oval, or irregularly triangular figure. They are proportionally smaller than in Mammals, being in the Goose each about the size of a pea. They are sometimes confluent.

They present, like the kidneys, a homogeneous texture throughout, and do not exhibit the alternate strata of different-coloured substances as in Mammalia. In the Gigantic Crane we found the texture of the suprarenal glands to be coarsely fibrous; in the Hornbill they were granular, similar to the kidney; in the Pelican they were of a granular but more pulpy texture.

There is no cavity in the suprarenal glands. The veins which return the blood from them are of proportionally large size, as in all the parenchymatous bodies without excretory ducts. The suprarenal glands have been found to present a slight enlargement corresponding with the increased developement of the sexual organs. Their relative size and position to the testes in the male embryo are shown at *f*, fig. 108.

§ 162. *Spleen of Birds.*—The spleen, figs. 85, 87, *s, s*, is comparatively of small size in Birds; it is generally of a round or oval figure, but sometimes presents an elongated and vermiform shape, as in the Sea-Gull, or is broad and flat as in the Cormorant. It is situated beneath the liver, on the right side of the proventriculus. It is, however, somewhat loosely connected to the surrounding parts, so that its position has been differently described by different authors. A process of the pancreas commonly passes into close contact, and is connected with the spleen by a continuation of vessels, as in the Hornbill, fig. 87, *q, s*. The texture of the spleen is closer in Birds than in Mammals; but a minute examination proves that the blood of the splenic artery is ultimately deposited in cells, from which the splenic veins arise. These veins in the Swan and some other Lamellirostres form a network on the exterior surface of the spleen, as in the Chelonian Reptiles.

In many Birds, as e.g. Vultures, Falcons, the Starling, Magpie, Heron, Bustard, and in most Aquatic Birds, two small bodies are found, one on each side of the trachea, very near the lower larynx and frequently attached to the jugular veins. They may be homologues of the 'thyroid gland.' In addition to these there are two similar bodies, in the Gannet, attached to the upper part of the commencement of each bronchus.

§ 163. *Peculiar Secretions.*—The unctuous fluid with which Birds lubricate their feathers is secreted by a gland situated above the coccyx or uropygium. This gland consists of two lateral moieties conjoined. As might be expected, it is largest in the birds which frequent the water. In the Swan it is an inch and a half in length, and has a central cavity, which serves as a receptacle for the accumulated secretion. Each lateral portion is of a pyriform shape, and they are conjoined at the apices, which are directed backward, and are perforated by numerous orifices, encircled in some birds by a crown of feathers. The longitudinal central cavities present numerous angular openings, in which there are still smaller orifices of the secerning follicles. These consist of close-set almost parallel straight tubules, extending to the superficies of the gland, without ramifying or intercommunicating, and preserving an equable diameter to their blind extremities. The tubules are longest at the thickest part of the gland, and become shorter and shorter towards the apex.

The follicles to which is due the peculiar odour of certain birds, as e. g. the Hoopoe, Muscovy Duck, Black Vulture, &c., are probably somewhat diffused on parts of the integument.

CHAPTER XXII.

TEGUMENTARY SYSTEM OF BIRDS.

§ 164. *Composition of the Tegument.*—This is composed, as in Mammalia and Reptilia, of the corium or derm, the epiderm and its appendages, and an intermediate layer of unhardened epiderm with colouring matter, called 'rete mucosum.'

The *corium*, or true skin, is very thin and lacerable, but vascular. In some Birds it adheres to the subcutaneous muscles by cellular tissue, which is frequently the seat of accumulation of dense yellow fat. In the Penguin the layer of subcutaneous cellular tissue adheres to the corium, but is separated from the muscles, and has a smooth internal surface: long vessels, like threads, connect this layer to the muscles. The skin is moved by muscles which at the same time raise and ruffle the plumage which it supports. In most Birds the skin is more or less separated from the muscles of the trunk by the interposed air-cells; as in the Batrachians it is by the lymph receptacles. It adheres, however, to a larger proportion of the osseous system than in other classes; as, e.g., to the upper and lower jaws, the feet and part of the tibiæ, the pinion bones. The corium has extensions beyond the covering of the body, to form the webs for swimming and the broader folds at the axillæ and bend of the arm for flight: it developes the papillæ beneath the toes, the vascular comb and wattles of the Cock, the caruncle and pendent ornaments of the Turkey, &c.

The *rete mucosum* rarely contains any colouring matter where the feathers grow; at this part the skin is of a pale greyish colour, or pink, from the colour of the blood which circulates in it. But in the naked parts of the integument, as the cire, the lore, the comb, the wattles, the naked parts of the head and neck in some Birds, and the tarsi and toes, the rete mucosum frequently glows with the richest crimson, orange, purple, green, black, and a variety of other tints, of which the *planches coloriées* and the different zoological monographs of geographical groups and families of Birds afford numerous examples.[1]

[1] Amongst these merit highest mention the works of our countryman GOULD on the Birds of Australia, Europe, Asia, Great Britain, &c.; and his magnificent monographs on the Humming-Birds, Trogons, and Toucans.

The *epiderm* is in some places continued as a simple layer over the corium, following its wrinkles and folds, as around the naked necks of some Vultures. It is moulded upon the bony mandibles to form the beak, and in some Birds adheres to osseous protuberances on the cranium, where it forms a species of horn; and it is remarkable that these instances occur chiefly in those orders of Birds, the *Cursores* and *Rasores*, which are most analogous to the Ruminantia among quadrupeds: the Cassowary and Helmeted Curassow are examples. The Hornbills are, however, instances in the Volitorial, and the Kamichi in the Grallatorial Order. The cuticle is sometimes developed into spines or spurs, as upon the wing of the Snake Vulture, Cassowary, *Palamedea*; and upon the leg of many Gallinaceous Birds. The claws which sheath the unguial phalanges of the feet assume various forms adapted to the habits and manner of life of the different orders. A remarkable artificial form is given to the claw of the middle toe in certain Birds; the inner edge being produced and divided into small parallel processes like the close-set teeth of a comb, fig. 110. These teeth are not reflected or recurved, as they might be expected to be, if they had been intended to serve as holders of a slippery prey, but are either placed at right angles to the claw or are inclined towards its point. The Common Barn-Owl (*Strix flammea*), the Night-jar genus (*Caprimulgus*), the Heron and Bittern kind (*Ardeidæ*, Vig.), afford examples of this structure; and as each species of bird appears to be infested by its peculiar louse (*Nirmus*), the solution of the final intention of so singular a contrivance, which is limited to so few species, and these of such different habits, may yet be afforded by the entomologist.

110

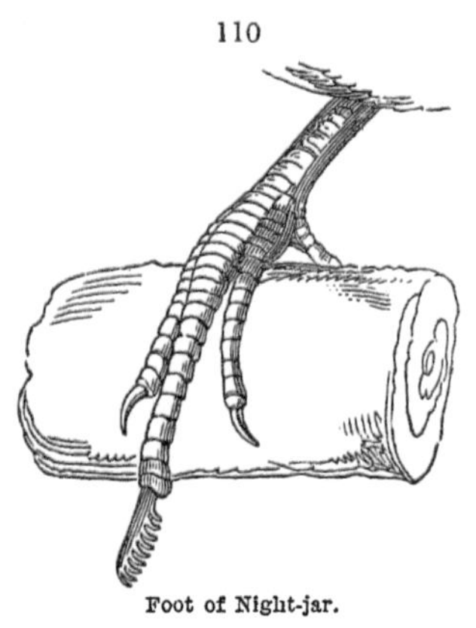

Foot of Night-jar.

With respect to the scales which defend the naked parts of the legs of birds, they do not differ from those of Reptiles. Their form and disposition, as has been already observed, have afforded distinctive characters to the zoologist. In most of the *Raptores*, the *Psittacidæ*, the *Rasores*, the *Grallatores*, and the *Natatores*, the scales are polygonal, small, and disposed in a reticulate form; the birds so characterised formed the *Retipedes* of Scopoli. In the rest of the class the tarsi are covered anteriorly with unequal semi-annular scales, ending on each side in a longitudinal furrow, and these birds he termed the *Scutipedes*. In one section of the *Tyranni*, Cuv., the scuta surround the tarsi as complete rings. Where the carneous parts of the muscles are continued low down upon the legs, as in the Owls, a covering of feathers is co-extended to preserve their temperature.

§ 165. *Appendages of the Tegument.*—The Vertebrate classes have each their characteristic external covering: the cold-blooded Ovipara are naked, or their external surface is defended only by hard scales or plates (*squamæ* and *scuta*); but the warm-blooded classes require to be invested by an integument better adapted to maintain the high degree of temperature peculiar to them: hence quadrupeds are clothed with fur and hair, and birds with down and feathers.

111

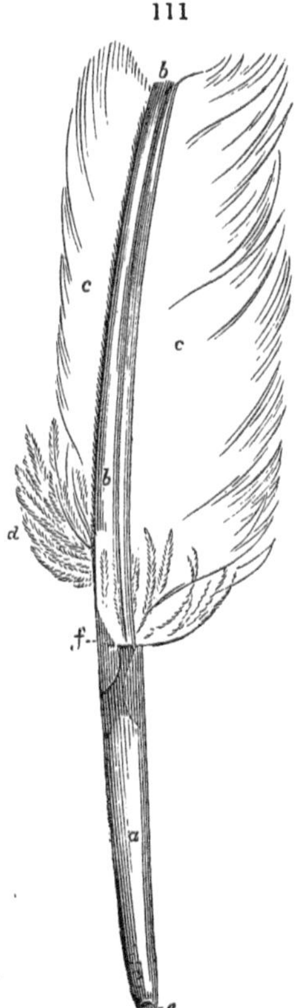

Feathers are the most complicated of all the modifications of the epidermic system, and are quite peculiar to the class of Birds. They are proverbially light; and, as the eloquent Paley well observes, 'every feather is a mechanical wonder;' 'their disposition, all inclined backward, the down about the stem, the overlapping of their tips, their different configuration in different parts, not to mention the variety of their colours, constitute a vestment for the body so beautiful, and so appropriate to the life which the animal is to lead, as that, I think, we should have had no conception of anything equally perfect, if we had never seen it, or can now imagine anything more so.'[1]

Notwithstanding the varieties of size, consistence, and colour, all feathers are composed of a *quill* or *barrel*, fig. 111, *a*, a *shaft*, *b*, *b*, and a *vane* or *beard*, *c*, *c*; the vane consists of *barbs*, fig. 112, *e*, and *barbules*, *f f*.

The quill (*calamus*), by which the feather is attached to the skin, is larger and shorter than the shaft, is nearly cylindrical in form and semi-transparent; it possesses

[1] LXVI'. p. 234.

in an eminent degree the opposite qualities of strength and lightness. It terminates below in a more or less obtuse extremity, which is pierced by an orifice termed the *lower umbilicus*, fig. 111, *e*; a second orifice, leading into the interior of the quill, is situated at the opposite end, at the point at which the two lateral series of barbs meet and unite; this is termed the *upper umbilicus*, ib. *f*. The cavity of the quill contains a series of conical capsules fitted one upon the other, and united together by a central pedicle.

The shaft (*scapus*) is more or less quadrilateral, and gradually diminishes in size from the upper umbilicus to its distal extremity. It is always slightly bent, and the concave side is divided into two surfaces by a middle longitudinal line continued from the upper umbilicus; this is the *internal surface*, fig. 112, *c*. The opposite, or *external surface*, ib. *b*, is smooth, and slightly rounded; both sides are covered with a horny material similar to that of which the quill is formed, and they inclose a peculiar white soft elastic substance, called the *pith*, ib. *a*.

112

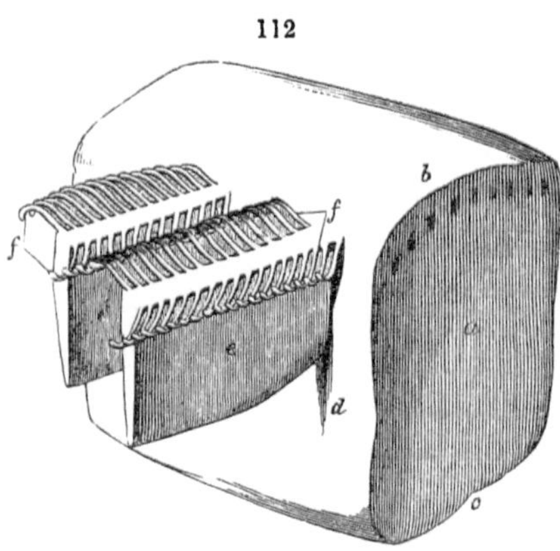

Diagrammatic section of the shaft and vane.

The barbs (*rami*) are attached to the sides of the shaft near the external surface, and consist of laminæ, varying as to thickness, breadth, and length. They are arranged with their flat sides toward each other, and their margins in the direction of the external and internal sides of the feather; consequently they present a considerable resistance to being bent out of the vane's plane, although readily yielding to any force acting upon themselves in the line of the stem: (*e*, *e*, fig. 112, are the bases of two barbs of a feather magnified). The barbules (*radii, hamuli*), ib. *f*, *f*, are given off from either side of the thicker margin of the barbs, and are sometimes similarly barbed themselves, as may be seen in the barbules of the great feathers of the Peacock's tail. In these feathers and in the plumes of the Ostrich, the barbules are long and loose; but more commonly they are short and close-set, and by their form and disposition constitute the mechanism by which the barbs are united together. The barbules arising from the upper side of the barb, or that next the extremity of the feather, are curved downward or toward the internal surface of the shaft; those which arise from the under side of the barb are curved in the contrary direction: so that the two adjoining series of hooked barbules lock into one another in a manner which has been compared to the fastening of a latch of a door into the catch of the door-post. There is much complicated variety in the interlocking mechanism here generally explained.

Besides the parts which constitute the perfect feather, there is an appendage attached to the upper umbilicus, called the accessory plume (*hyporachis*). It is usually a small downy tuft, but varies both in different species, and even in the feathers of different parts of the body of the same bird. In the quill-feathers of the wings and tail, it retains the state of a small tuft of down; but in the body-feathers of Hawks, Grouse, Ducks, Gulls, &c., it is to be found of all sizes, sometimes equal to that of the feather from which it is produced.

In the Ostrich and Apteryx the feathers have no accessory plume; in the Rhea it is represented by a tuft of down; in the Emeu it rivals in size and structure the original feather; and in the Cassowary, besides the double feather, there is a second accessory plume, so that the quill supports three distinct shafts and vanes.

The feathers vary in form in different parts of the bird according to their functions, and afford zoological characters for the distinction of species; they have, therefore, received in Ornithology distinct names. The ordinary imbricated feathers which cover the body are called 'clothing feathers:' the larger ones for special uses, 'quill-feathers.' Those which surround or cover the external opening of the ear are termed the 'auriculars.' Those which lie above the scapula and humerus are called the 'scapulars.' The small feathers which lie in several rows upon the bones of the antibrachium are called the 'lesser coverts' (*tectrices primæ*). Those which line the under or inner side of the wings are the 'under coverts.' The feathers which lie immediately over the quill-feathers are the 'greater coverts' (*tectrices secundæ*). The quill-feathers supported by the wings are the '*remiges*,' or 'rowing feathers.' The largest of these remiges, which arise from the bones of the hand, are termed the 'primaries' (*primores*). Those which rise from the ulna, towards its distal end, are the 'secondaries' (*secundariæ*). Those which are attached to its proximal extremity are the 'tertiaries' (*tertiariæ*). These in some Birds, as the Woodcock and Snipe, are so long as to give them the appearance, when flying, of having four wings. The quill-feathers which grow from the phalanx representing the index, form what is termed the bastard wing (*alula spuria*). Those forming what is called the 'tail' of the bird, and supported by the coccyx, are the '*rectrices*,' or steering quills. The overlying feathers are the 'tail-coverts' (*calypteria*); these bear the ornamental 'eyes' and are so developed in the Peacock as to form what is called the 'tail' or 'train' of that gorgeous bird.

In considering the structures which determine the powers of flight in different Birds, it is necessary to take into account the texture, forms, and proportions of the wing-feathers, as well as the developement of the bones and muscles which support and move them; as much depends upon the mechanical advantages resulting from the shape of the expanded wing. When the primary quill-feathers gradually increase in length as they are situated nearer the extremity of the pinion, they give rise to the acuminated form of wing, as in the Swifts and Humming-Birds, in which the first primary is the longest; and in the true Falcons, in which the second primary is the longest. In the Hawks the wing is of a less advantageous form, in consequence of the fourth primary being the longest. When the primaries gradually decrease in length towards the end of the pinion, they give rise to a short rounded form of wing, such as characterises the Gallinaceous Order; in which, although the pectoral muscles are immensely developed in order to counteract the disadvantage resulting from the disposition of the primaries, yet they are only able, in consequence of the form of the wing, to carry the bird rapidly forward for a comparatively short distance, and that with an exertion and vibratory noise well known to every sportsman.

The texture of the quill-feathers has also a material effect on the powers of flight. In the Falcons each primary quill-feather is elongated, narrow, and gradually tapers to a point; the webs are entire, and the barbs closely and firmly connected together.[1] In the Owls the plumage is loose and soft, filaments from the barbules extend upon the outer surface of the vane, and one edge of the primaries is serrated; so that, while they are debarred from so swift a flight as the Hawk, they are enabled, by the same mechanism, to wing their way without noise, and steal unheard upon their prey.

§ 166. *Developement of Feathers.*—The first covering of the bird

[1] Of so much consequence are the quill-feathers to the Falcons, that when any of them are broken the flight is injured and the falconers find it necessary to repair them; for this purpose they are always provided with perfect pinion and tail feathers regularly numbered.

is a partial and temporary one, consisting of fasciculi of long filaments of down, which on their first appearance are enveloped in a thin sheath, but this soon crumbles away after being exposed to the atmosphere. The down-fasciculi, which diverge each from a small quill, are succeeded by the feathers, which they guide as it were through the skin; and after the first plumage, at each succeeding moult, the old feathers serve as the 'gubernacula' to those which are to follow. It is to be observed that feathers do not grow equally from every part of a surface of a bird; they are not developed, for example, at those parts which are subject to friction from the movements of the wings and legs. They first appear in clumps upon the parts of the skin which are least affected thereby, as, e. g., upon the head, along the spine, upon the exterior surface of the extremities, at the sides of the projecting sternum and of the abdomen.

113

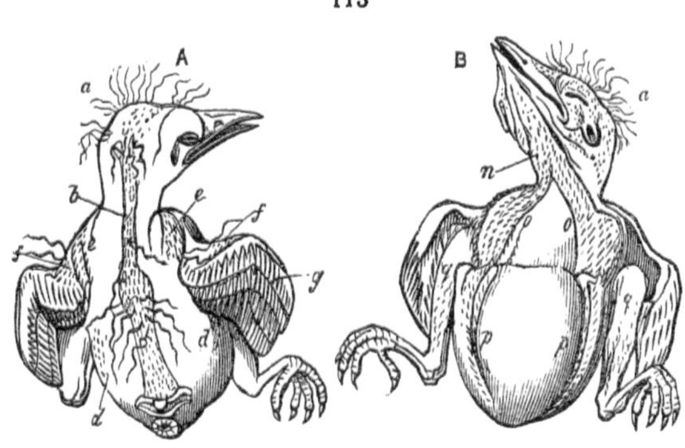

Young Blackbirds, showing primary down and growing feather-clumps. XX.

In fig. 113, Hunter[1] designates them as follows: *a*, 'cranial clump' (*pteryla capitis*, Nitzsch); *b*, 'posterior cervical' and 'dorsal clumps' (*pt. spinalis*, N.); *d*, 'lumbar clumps' (*pt. femorales seu lumbales*, N.); *e*, 'brachial clumps' (*pt. humerales*, N.); *f*, 'antibrachial,' and *g*, 'carpal clumps' (*pt. alarum*, N.); *q*, 'femoral clumps' (*pt. crurales*, N.); *n*, the 'anterior cervical,' and *o*, 'pectoral clumps' (*pt. colli laterales*, N.); *p*, 'abdominal clumps' (*pt. gastræi*, N.), &c. Nitzsch[2] illustrates the affinities of Birds by the characters of the 'pterylæ,' exhaustively followed out in LIV'.

The matrix, or organ by which the perfect feather is produced, has the form of an elongated cylindrical cone, and consists of a

[1] XX. vol. iii. p. 311. [2] LIV'.

capsule, a bulb, and intermediate membranes which mould the secretion of the bulb into its appropriate form. The matrix is at first an extremely minute cone, attached by a filamentary process to a follicle or papilla of the skin; but it is not a developement of that part, being of a different structure and adhering to it by a small part only of its circumference. The matrix progressively increases in length; its base sinking deeply into the corium, and acquiring a more extended connection by enlarged vessels and nerves, while its apex protrudes to a greater or less extent from the surface of the integument, when the capsule drops off to give passage to the feather which it incloses, and the formation of which has, in the meanwhile, been gradually proceeding from the apex downward. The capsule of the matrix, *a, a,* fig. 114, is composed of several layers, the outermost of which is of the nature of epiderm; the inner ones are more compact and pulpy. The sides of the capsule which correspond to the outer and inner sides of the growing feather within are indicated by a white longitudinal line.

114

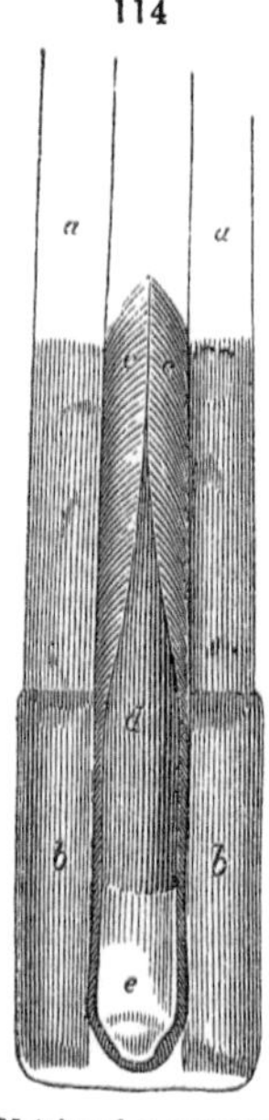

Matrix of a growing feather, with the capsule laid open. LI'.

The axis of the capsule is occupied by a medulla or bulb, ib. *e*, also of a cylindrical form, and of a soft fibrous texture, adhering by its base to the parts beneath, and there receiving numerous bloodvessels and a nerve.

Between the medulla and the capsule there are two parallel membranes, one internal, ib. *d*; the other external, ib. *b*; from the latter membrane a number of close-set parallel laminæ extend obliquely from one of the white longitudinal lines above mentioned to the other on the opposite side of the cylinder. The two membranes seem to be united together by the oblique septa. In the long and narrow spaces between these septa, the matter of the vane, ib. *c*, is deposited and formed into barbs and barbules. The deposition of the material of the barbs commences at the apex of the bulb, and the stem is next formed in the following manner.

The external longitudinal line from which the oblique laminæ are continued, receives and moulds on the inner surface of the external capsule the horny covering of the back of the feather, or that longitudinal band to the two sides of which the barbs are attached; and on the opposite surface of the internal membrane are formed the pith or substance of the shaft, and the horny pellicle which incloses it on the inner surface. The internal longitudinal line has no other use than to establish a solution of continuity between the extremities of the barbs of one side and those of the other, which meet at that part, and thus curve round and completely inclose the formative bulb. In fig. 115, the capsule of the matrix of a growing feather, *c*, has been laid open, and the nascent barbs, *d*, *d*, which surrounded the bulb, have been unfolded, exposing that part at *a*, *b*. A portion of the barbs and stem have been completed and protruded, and the bulb is beginning to undergo a process of absorption at that part, which will hereafter be described. The shaft and barbs at the apex of the cylinder are the first parts which acquire consistence, and the molecules composing the remainder are less compactly aggregated as they are situated nearer the base of the matrix. As the gelatinous medulla increases at the base, the first-formed shaft and barbs are protruded through the extremity of the capsule, the bulb continuing to furnish the secretion which is moulded between the two striated membranes until the entire feather is completed. If the striated membrane inclosing the bulb be attempted to be reflected from below upward, it will be found to be connected with a series of membranous cones, *a*, *b*, *c*, *d*, *e*, fig. 116, ranged one upon the other throughout the whole length of the bulb, and connected together by a tube running through its centre. In this figure the pulpy matter which occupied the interspaces of the cones has been removed to show their central connecting tube.

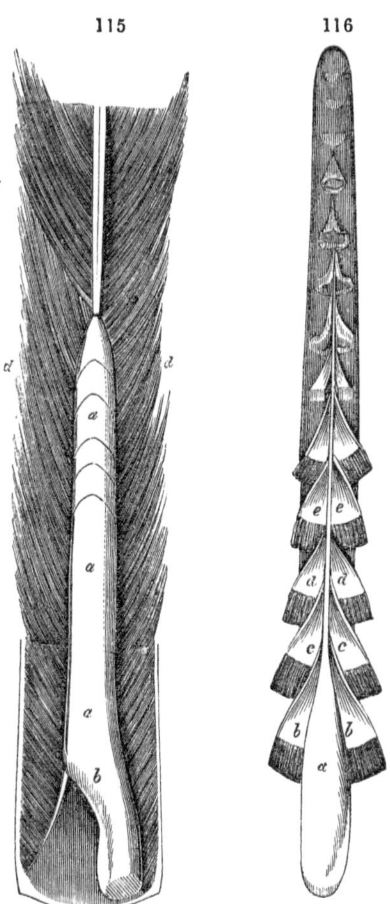

Growing feather. LT. Structure of the bulb. LT.

As the developement of the feather advances, the pulpy matter disappears from the summit of the medulla, and only the membranous funnel-shaped caps remain, which are protruded from the theca and the centre of the new-formed barbs, and fall off as these expand. The theca which incloses the whole is of a firm texture where the new-moulded barbs are yet pulpy and tender, but it becomes thinner as these acquire consistency, and, lastly, dries and crumbles away after it has been exposed to the action of the atmosphere. The bulb itself, when examined in a half-formed quill-feather, is composed of two parts, corresponding to the external and internal aspects of the feather. The internal part represents a semi-cylinder or case, inclosing the external part, which is of a conical form; the latter extends from the base of the bulb, and gradually diminishes to a point where the shaft is completed and the barbs begin to expand. Its office is to deposit the pith within the shaft, and it is absorbed in proportion as this is effected. The internal part or case also commences at the base of the bulb, and adheres closely to the cone, with which, indeed, its substance is continuous; it increases in thickness as the cone diminishes, its margins are beautifully scolloped or crenate, and the crenations are lodged in the interspaces of the oblique laminæ or moulds, and deposit in them the material of the vane. The horny sides of the shaft are lodged and formed in the grooves between the external and internal parts of the bulb, and correspond in degree of formation to the depths of those grooves; and being progressively brought into contact from above downwards, the shaft is thus completed, leaving the longitudinal line at the internal side. When all the grooves (wherein are formed the barbs, and the portion of the shaft which carries them) are filled by the horny matter, and the barbed part of the feather is finished, this horny matter lastly expands uniformly around the medulla, and forms the quill of the feather.

When the quill of the feather has acquired the due consistence, the internal medulla becomes dried up, and is resolved, as before, into membranous cones arranged one upon the other; but these latter never pass out, for the quill, which is now hardened and closed by the shaft at the extremity opposite to the lower umbilicus, will not permit their egress; they remain, therefore, inclosed, and constitute the light dry pith which is found in the interior of the quill. The last remains of the bulb are seen in the ligament which passes from the pith through the lower opening of the quill and attaches it to the skin.

There is a close analogy between the formation of a feather and that of a tooth; but a tooth may take years to be perfected, and there are but two series produced in one part of the jaw, and only one in the other, in any warm-blooded animal. Feathers, on the other hand, are developed in the course of some days; they attain a length of from one to two feet or more in many Birds, and they are almost all renewed every year,—in some species even twice a year. It may be conceived, then, how much vital energy the organisation of Birds must exercise, and how many dangers must accompany so critical a period as that of the moult.

The plumage is commonly changed several times before it attains that state which is regarded as characteristic of the adult bird. The time required for this varies from one to five years, and several birds rear a progeny before they acquire the plumage of maturity.

When the male bird assumes a vestment differing in colour from the female, the young birds of both sexes resemble the latter in their first plumage (Blackbird); but when the adult male and female are of the same colour, the young have then a plumage peculiar to themselves (Swan). When adult birds assume a plumage during the breeding season decidedly different in colour from that which they bear in winter, the young birds have a plumage intermediate in the general tone of its colour compared with the two periodical states of the parent birds, and bearing also indications of the colours to be afterwards attained at either period (Ruff). When both males and females are alike in colour, but species of the genus differ widely in colour, as e. g. the Black and White Swans, the young of such species are alike and of an intermediate hue.

Changes in the appearance of the plumage of birds may be produced:—

By the feather itself becoming altered in colour;

By the bird's obtaining a certain number of new feathers without shedding any of the old ones;

By the wearing off of the lengthened lighter-coloured tips of the barbs of the feathers on the body, by which the brighter tints of the plumage underneath are exposed;

By an entire or partial moulting, at which old feathers are thrown off and new ones produced in their places.

The first three of these changes are observed in adult birds at the approach of the breeding season; the fourth change is partial in spring and entire in autumn.

CHAPTER XXIII.

GENERATIVE SYSTEM OF BIRDS.

§ 167. *Male Organs and Semination.*—The few varieties of structure which the generative organs present in the Class of Birds, are principally met with in those of the male.

The organs in this sex exhibit all the essential characteristics of the oviparous type of structure. The testes are situated high up in the abdomen, whence they never descend into an external scrotum. The intromittent organ is either double, as in Serpents, when, however, each penis is extremely small; or it is single, but in this case, to whatever extent it may be developed, it is simply grooved along the upper surface or dorsum for the passage of the fecundating fluid. As there is no true urethral canal, so neither are the glands of Cowper or the prostatic glands present.

The *testes*, figs. 89, *x*, 117, *a*, *a*, are two in number; in form more or less oval, situated near the upper extremities of the kidneys. They vary remarkably in colour in different birds; I have seen them white in the Peregrine Falcon and Dove; pale yellow in the Horn-Owl and Gallinule; of a brighter yellow in the Magpie, Bay Ibis, Ruff, and Oyster-catcher; of a black colour in the Chough, Partridge, Heron, Seagull, but whitish toward the lower end in the last two. They are invested with a strong and dense 'albuginean' tunic, and are fastened or suspended by a fold of peritoneum. The contorted seminiferous tubules are very slender, and are separated into packets by delicate and membranous septa, continued from the inner surface of the tunica albuginea. The arteries spread in an arborescent form beneath that capsule. The vas deferens, fig. 117, *c*, *c*, is continued from the posterior or 'dorsal' and internal or 'mesial' part of the gland.

The periodical variations of size which the testicles undergo are very remarkable in the Class of Birds; and the limited period during which their function is in activity is compensated by the frequency and energy with which it is exercised.

The proportional size which the testes acquire at the breeding season is immense, as may be seen in the subjoined figures of the testes of the House-Sparrow, which commences with the glands as they appear in January, when they are no bigger than pins' heads, and ends with their full developement in April.

It rarely happens that both testes are developed in exactly the same degree: the left is commonly the largest; but sometimes

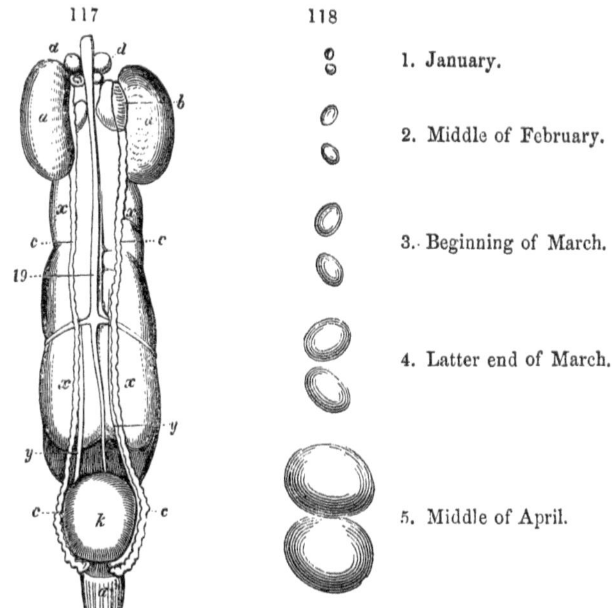

Urinary and male organs of a Cock. Testes of the House-Sparrow. XCIV.

the right exceeds the left; and I have seen an example, in a Rook, where it alone had taken on the action of sexual increase, and had acquired a bulk compensating for the want of developement in the left testis.

In most Birds, the only appearance of an epididymis, fig. 117, *b*, is a remnant of the primordial kidney, fig. 108, *c*. This part frequently presents a colour strikingly different from that of the testes: thus it has been observed in the Bustard and Curassow to be black; in the Cassowary, yellow; and in the Demoiselle (*Anthropoides Virgo*) to be of a green colour. In the Ostrich the epididymis is folded upon itself at the side of the testis.

The vas deferens, fig. 117, *c*, commonly passes down to the cloaca by the side of the ureters without undergoing any remarkable convolution; but in the Common Cock it is bent upon itself in short transverse folds from side to side almost from its commencement; the folds gradually but slightly increase as they approach the cloaca, both in extent and in the diameter of the tube composing them; and they are so closely compacted, and inclosed by a covering of peritoneum, as to present in a longitudinal section the appearance of a series of cells, which are capable of retaining, as in a vesicula seminalis, a quantity of the seminal secretion. In the Sparrow there is a dilatation at the end of each vas deferens, which opens, as in the Common Cock, on a papilla, situated in the urogenital division of the cloaca anterior to the insertion of the ureter.

119

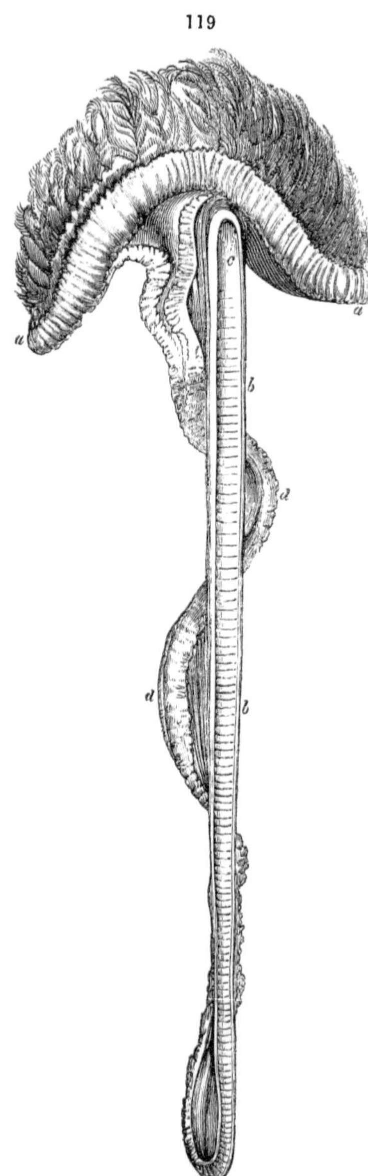

Penis of a Drake. XXVII.

The base of each papilla is surrounded by a remarkable plexus of arteries and veins, M, M, fig. 94, which serve as an erectile organ during the venereal orgasm, when the fossa of the turgid papilla is everted, and the semen brought into contact with the similarly everted orifice of the oviduct in the female, along which the spermatozoa pass by undulatory movements of their ciliary appendage or 'tail.'

In some Natatores which copulate in water there is provision for a more efficient coitus than by simple contact of everted cloacæ, and in the *Anatidæ* a long single penis is developed, fig. 119. It is essentially a saccular production of a highly vascular part of the lining membrane of the cloaca, continued from the fore-part of that cavity, ib. *a*, *a*; and in the passive state is coiled up like a screw by the elasticity of associated ligamentous structure, *b*, *b*. The vascular membrane gives off many small pointed processes, which, in the Gander, are arranged in transverse rows on either side the urethral groove, *d*, and near the extremity of the penis are inclined backward. The elastic band, *b*, *b*, has been cut open lengthwise in the figure given by Home:[1] it is surrounded by cavernous tissue, and terminates in the blind end of the sac which can be everted. A groove, ib. *d*, *d*, commencing widely at the base, follows the spiral turns of the sac to its termination: the sperm-ducts open upon papillæ at the base of this groove. This form of penis has a muscle by which it can be everted, protruded, and raised.

The base of the penis in the Ostrich is attached to the fore wall of the cloaca, the conical body is bent in a recess, out of which it can be drawn and into which it can be returned by muscles. It consists of two solid fibrous bodies, the fissure between which is covered by cavernous erectile tissue, bounding the seminal groove; but it has no evertible sacciform part: there is a third elastic substance internal to the cavernous substance which produces the twisted form.

The Drake's penis is formed after the type of that of Lizards and Serpents. The Ostrich's penis is like that of the Tortoise and Crocodile.[2]

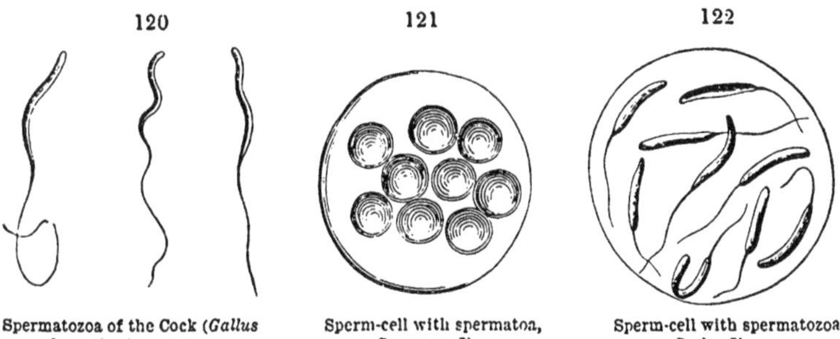

120 — Spermatozoa of the Cock (*Gallus domesticus*). CCCVI.

121 — Sperm-cell with spermatoa, Sparrow. Ib.

122 — Sperm-cell with spermatozoa, Cock. Ib.

The spermatozoa of Birds, like those of Lizards, have a long cylindrical body; generally straight or wavy, obtuse anteriorly, and tapering behind into a filamentary tail of varying length according to the species, fig. 120; but in the *Cantores* the body is twisted spirally in three to five or more turns, pointed anteriorly and terminating in a usually long filamentary tail, fig. 123. The sperm-cell contains many spermatoa, fig. 121, and in these the spermatozoa are developed and usually excluded

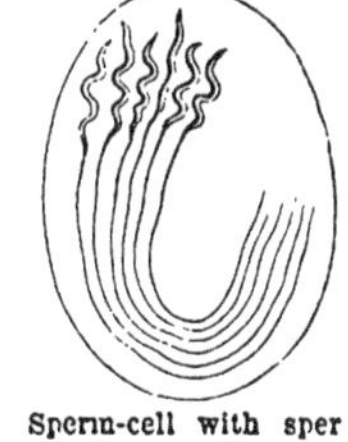

123 — Sperm-cell with spermatozoa, Sparrow. CCCVI.

[1] XXVII'. and Phil. Trans. 1802.

[2] LXVII'.

within the common sperm-cell, fig. 122: here they are agglutinated together, either in irregular groups; or, as in the *Cantores*, in a regular bundle, with the spiral bodies at one end and the tails extending, parallel, to the other, fig. 123. In both cases the spermatozoa are set free by rupture and solution of the sperm-cell: in the *Cantores* they are then found fasciculate in the 'tubuli testis,' whilst in other birds they are irregularly dispersed.

124

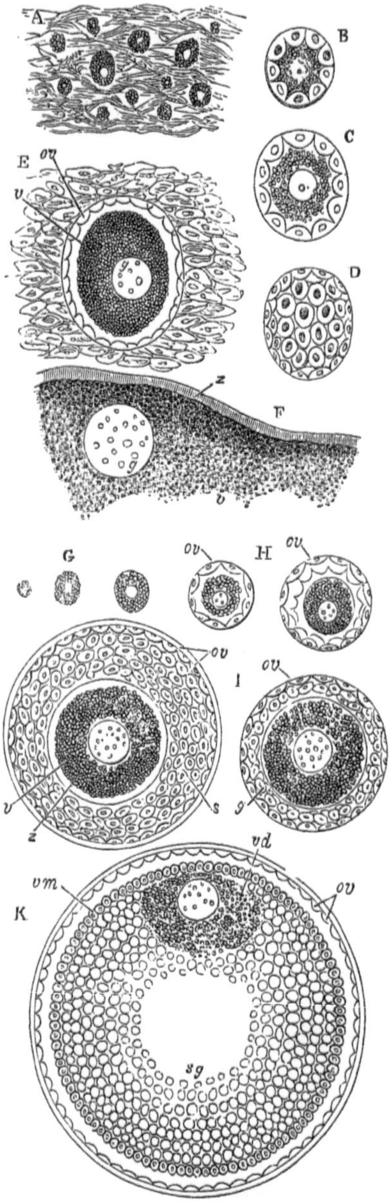

Earliest stages of the formation of the ovarian egg in the Bird. CCCVIII.

§ 168. *Female Organs and Ovulation of Birds.*—The ovarium of the Bird consists essentially of the germ-cells, with the stroma or blastema modified by their presence, and the vitelline matter superadded to the germ-cell. The formative processes are most clearly traceable in the smaller singing-birds. In fig. 124, A, the small clusters of granules indicate the beginning of the ova in the ovarian stroma: in larger clusters a clear point appears, which in the largest assumes the character of a germ-cell surrounded with opaque minute granules. The almost contemporaneous formation of the 'ovisac' (Barry) soon manifests itself by its lining of epithelial cells, ib. B, at which period the germ-cell manifests, by its macula, the ordinary characters of the 'germinal vesicle.' This is shown, in focus, at C; the epithelium of the ovisac is shown in focus at D: in E, *ov* is the ovisac with its epithelial lining, *v* the granular yolk surrounding, *g* the germinal vesicle or developed 'germ-cell.' F is part of an ovule of $\frac{1}{40}$ of an inch in diameter, highly magnified: *v*, minutely granular or primitive yolk-substance; *g*, germinal vesicle; *z*, 'thick consolidated membranous layer which formed a vesicular covering for the primitive ovule, and which corresponds to the zona pellucida of the mammiferous ovum.'[1]

In G and H, Prof. Allen Thomson gives diagrammatic figures of the earliest stages of formation of the ovarian ovum in a Blackbird: figs. I and K 'are intended to illustrate, diagrammatically, the view, that after the disappearance of the zona, and the formation of the larger granular yolk-cells, the outer layer of the cells of this substance forms the permanent vitelline membrane of the bird's egg; *vd*, 'remains of minutely granular yolk, forming the vitelline disc round the germinal vesicle; *sg*, large corpuscles of the yolk; *vm*, outer layer of the cells of the same, on which the vitelline membrane is afterwards formed.'[2]

The germinal vesicle, with the firmer primitive vitelline granules ('germ-yolk,' K, *vd*), moves from the centre to the periphery of the ovum, which then begins to expand by the addition of the softer 'food-yolk,' ib. *sg*: this seems to be due to cells thrown off by, and to fluid exuding from, the inner surface of the ovisac, *ov*, the cells greatly and rapidly increasing in number and acquiring the characteristic yellow or orange colour of the yolk in birds.

At the earlier stages of the developement of the ova the ovarium appears as a flattened solid, granular body, attached by a fold of peritoneum, or of air-cell, to the bodies of the middle dorsal vertebræ, fig. 125, *a*.

At first, the right and left ovaria are similar in size, fig. 127, *c*: but the symmetry is soon disturbed by concentration of developement in the left ovarium (fig. 125, *a*), the right one, *a'*, remaining stationary and ultimately, in most birds, disappearing.

The enlargement of the ovarian ovum is now due to the accumulation of the yellow or 'food' yolk, with concomitant distension of the membrana vitelli and of the ovarian capsule, or 'calyx,' fig. 126, *a*, *d*, which maintains its connection with the rest of the ovarium by a contracted base or pedicle.

The calyx consists of two membranes, united together by lax tissue and blood-vessels: these ramify as in fig. 126, *c*, converging toward a white transverse line or band across the most prominent

[1] CCCVIII'. p. 76. [2] Ib.

part of the calyx, where the vessels become suddenly so minute, as to seem to be wanting: fig. 128, *c*. This part, called the 'stigma,' begins to appear when the ova have attained, in the Common Fowl, the diameter of an inch: it increases in breadth, and the membranes there become thinned, as the ovum acquires its

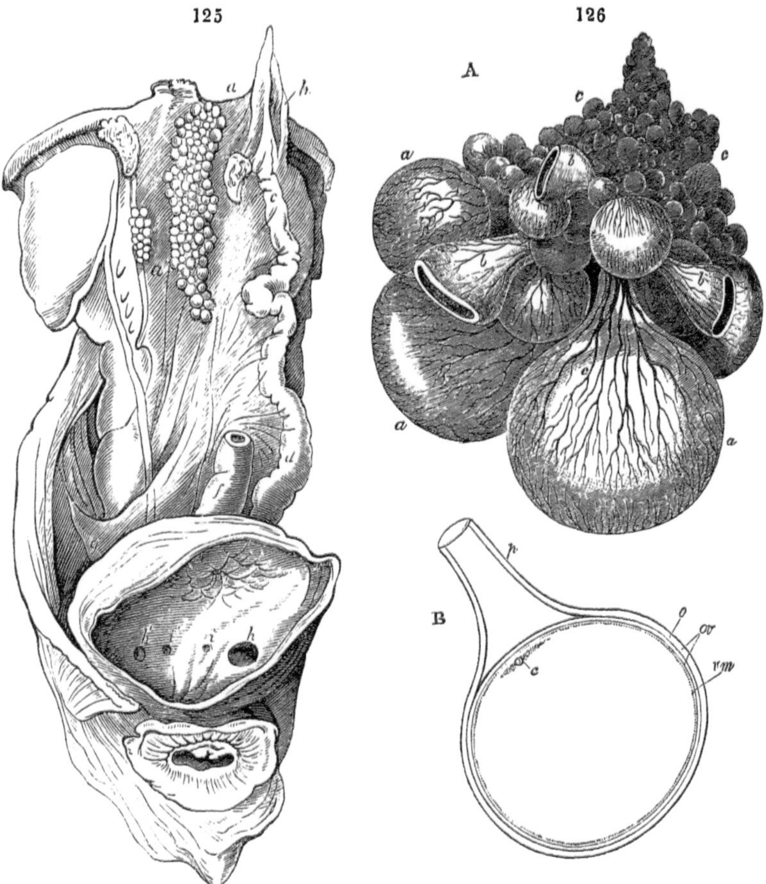

Female organs, Fowl, at non-breeding season.

Relation of the ova to the ovary in Birds. CCCVIII.

full size; when they readily yield and are rent by the compressing force of the infundibular opening of the oviduct, fig. 128, *e*, whereupon the ovum slips out of the calyx into the efferent passage.

The empty calyx collapses, as at *b*, *b*, fig. 126, and *d*, fig. 128, rapidly shrinks, and is ultimately absorbed.

In birds that have few young at a brood, as the *Apteryx*,[1]

[1] XI. vol. iii. p. 310, pl. 36.

Eagle, Dove, &c., the number of enlarged ovarian ova or 'yolks' is correspondingly small; but in the more prolific species, as the Common Fowl, fig. 126, A, they are more numerous. The number of young produced may be, by this means, in some degree inferred, if the female of a rare species happen to be killed during the breeding season.

In the diagrammatic section of a full-sized ovarian ovum, B, fig. 126, *o* is the outstretched ovarian capsule and stroma forming the 'calyx,' *p* its peduncular connection with the rest of the ovarium; *c* is the common position of the germ-cell and discoid germ-yolk; *ov*, the two layers of the ovisac into which the blood-vessels penetrate; *vm*, the vitelline membrane. This membrane is sufficiently strong and ductile to permit the ovarian ovum being compressed into an elliptical form to facilitate its passage through the contracted part of the oviduct. Certain changes now occur in the ovarian ovum, and much addition is made to it; but, before entering upon these, the canal through which it passes and in which the egg is completed must be described.

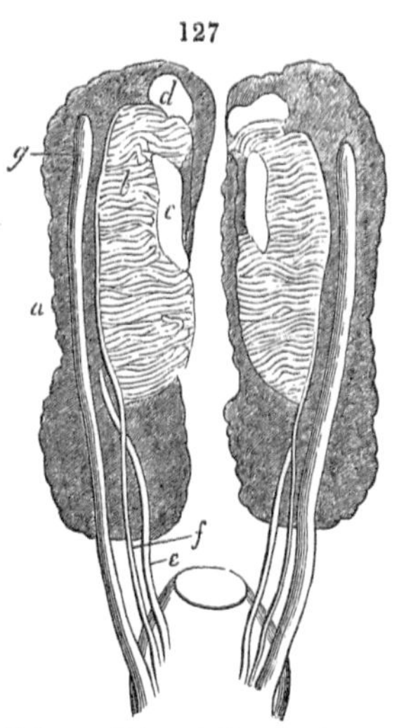

Kidneys, Wolffian bodies, ovaries, and oviducts of a fœtal bird, at a period when both oviducts are still of nearly equal size. Magnified. LXXIV.

In the female embryo the basis or stroma of the ovarium, fig. 127, *c*, appears in a similar relation to the primordial kidneys (ib. *b*), as the testis in the male. At the period when the permanent kidneys, ib. *a*, have sent the ureters, ib. *e*, to the cloaca, the oviducts, *g*, have been developed as prolongations from that part, and, to a certain point of developement, they are of equal size and length. Subsequently the left oviduct alone proceeds to grow; the right is stationary, or shrivels: occasionally it may be discerned as a rudiment in the mature bird, but usually all trace of it has disappeared. The left oviduct expands above or at its free end into the infundibular orifice, fig. 125, *b*, where its parietes are very thin; as it descends, these increase in thickness, and the efferent tube gradually acquires the texture and form of an intestine. Like this, it is attached to and supported by a duplicature of peritoneum called the *mesometrium*, but which also includes muscular fibres, to be presently described.

The oviduct in the quiescent state is generally straight, but at the period of sexual excitement it is augmented in length as well as capacity, and describes three principal convolutions before reaching the cloaca, fig. 128, *n*. The lining membrane presents a different character in different parts of the oviduct; at the infundibulum, ib. *e*, the surface is longitudinally rugous: lower down the lining membrane begins to be disposed in oblique ridges, ib. *g*, beset with follicular glands: at the more contracted part, or 'isthmus,' they become longitudinal and subside: in the terminal dilatation, ib. *k*, the lining membrane is beset with large flattened villi, containing the follicles concerned in the secretion of the shell. The whole oviduct is lined by vibratile epithelium. The shell-forming part has been termed the 'uterus,' but the ovum is never developed in it. The rest of the canal, *l*, which, by the same loose analogy, is termed 'vagina,' opens into the urogenital segment of the cloaca, anterior to the orifice of the left ureter, and its termination, figs. 86, 109, *f*, is provided with a sphincter.

128

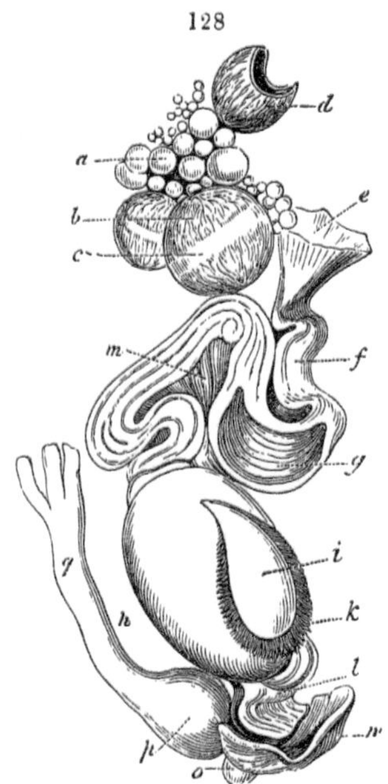

Female organs, Fowl, at breeding season. XXXIV.

The mesometry, fig. 128, *m*, differs most from the mesentery when the female organs are in full sexual action. It presents at that period a muscular structure, but the fibres are not striated. It is divided into two parts, one superior, the other inferior. The inferior mesometry has its point of attachment at the lower part of the uterine portion of the oviduct, and forms a somewhat dense and cruciform plexus of muscular fibres radiating from that part. The transverse fasciculi are spread out on either side and around the uterus. The lower fasciculus surrounds the vagina more laxly, and contributes to the expulsion of the ovum. The upper fasciculus spreads out like a fan upon the oviduct from its insertion into the uterine portion to the commencement of the infundibulum.

The superior mesometry commences by a firm elastic ligament, which is attached to the root of the penultimate rib of the left side, whence the muscular fibres are continued to the upper part of the oviduct, upon which they form a delicate muscular tunic, whose fibres embrace the oviduct for the most part in the transverse or circular direction, except at the infundibular aperture, where they affect the longitudinal direction, which enables them to dilate that orifice. Longitudinal muscular fibres begin again to be distinctly seen in the uterine portion of the oviduct, whence they are continued along the so-called vagina. An internal stratum of circular fibres is also situated immediately behind the calcifying membrane of the 'uterus.' In the vagina the circular fibres are concentrated at its termination to form the sphincter above mentioned.

The 'clitoris' of the Ostrich is continued from the anterior margin of the preputial cavity of the cloaca, and is grooved like the penis of the male: it is furnished with corresponding muscles. A smaller clitoris exists in those birds of which the males have a well-developed intromittent organ.

§ 169. *Fecundation in Birds, and Structure of the laid Egg.*—*In coitu* spermatozoa enter the cloaca and penetrate the oviduct, ascending to the ovarium. The germinal vesicle, on the reception of the ovum by the oviduct, is no longer visible, as such. A discoid aggregate of cells constitutes an opaque white circular spot on the part of the periphery of the yolk to which the germ-cell and germ-yolk had passed, and this was known to the older embryologists as the 'cicatricula.' It consists of a central clearer and of a peripheral denser portion, fig. 129, B: beneath the clear centre is a group of minute opaque granules called 'nucleus cicatriculæ.' In the diagrammatic figure A, *a* is the vitelline membrane; *d* the clear tract leading from the 'nucleus,' *c*, to the centre of the yolk,—the trace of the excentric course of the germ-cell: *b*, *b*, are the minute granules forming the denser part of the cicatricula; *e*, *e*, are the larger yolk corpuscles. The 'nucleus cicatriculæ,' *c*, is the 'germ-mass,' the result of the same series of spontaneous divisions of the impregnated germ-cell, as affected the entire yolk in the Batrachian (vol. i. fig. 452); to which the ovum of the Bird offers the opposite condition in the preponderance of the 'food-

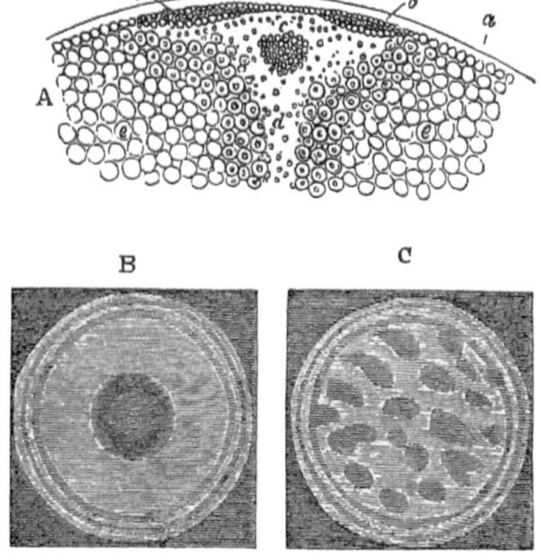

Structure of the cicatricula in a laid Fowl's egg. CCCVIII

yolk' over the 'germ-yolk.' In fig. 129, B is an enlarged view of the cicatricula as seen from above on the surface of the yolk in an impregnated egg: the dark central space is the 'transparent area' surrounded by the 'opaque area,' and by one or two delicate 'halones.' C is the cicatricula of an unfecundated laid egg: instead of the central transparent area a number of rather irregular transparent spots are seen.

The yolk forms an ellipsoid mass, somewhat flattened on the cicatricular surface, and consists of the external coloured part, fig. 130, B, *d*, in concentric layers indicative of successive deposit, and of a central lighter-coloured part, ib. *c*, about one-fourth of

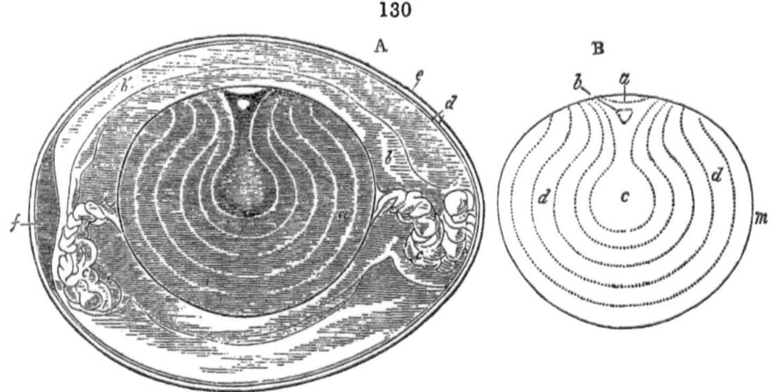

Fowl's egg and structure of the yolk as exhibited by a section. CCVIII.

the diameter of the whole. The margins of yolk-layers, interrupted by the 'cicatricula' and its canal, may form the 'halones.' The yolk-layers, *d, d*, usually show some diversity of tint.

The ripe ovarian ovum, having passed into the oviduct, is propelled by the peristaltic action of that tube in a rotatory course to the 'uterus.' The contact of the membrana vitelli stimulates the exudation of the product of the lining membrane in a denser state than usual, which forms a kind of accessory tunic, and is continued, thread-like, from near each pole of the ellipsoid, usually a little toward that half which is opposite the one supporting the cicatricula: these filaments, fig. 130, A, *c*, are the 'chalazæ,' and the layer of dense albumen from which they are continued is called the 'membrana chalazifera.' During the passage of the egg and its acquisition of successive deposits of the ordinary albuminous secretion, the chalazæ become twisted in opposite directions, fig. 131, B, and ultimately the one next the small end of the egg contracts some adhesion to the membrane lining the shell there. In fig. 131, A shows the ovum from the upper part of the oviduct, with the coating of dense albumen continued into the chalazæ; B, the outstretched chalazæ from opposite sides of the yolk, showing the opposite turns of the spiral; C, an egg from above the middle of the oviduct, with the first layers of soft albumen deposited upon the chalaziferous membrane and chalazæ.

The albumen is rapidly added in the more glandular and vascular part of the oviduct, by the ridges of follicles which correspond in direction with the spiral course of the egg; and, when it has arrived at the narrower part of the oviduct called the isthmus, denser layers of albumen are again excreted, forming the 'membrana putaminis,' fig. 130, A, *d*. So inclosed, and having acquired its ovate form with the small end toward the cloaca, the egg passes into the 'uterine' or shell-forming dilatation, fig. 128, *k*.

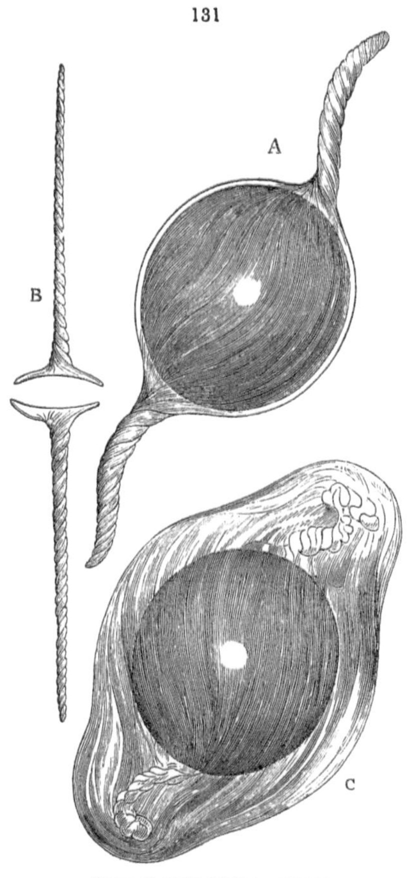

Stages of egg in oviduct. CCCVIII.

Artificial coagulation of the albumen or 'white' of an egg enables one to demonstrate its disposition in spirally deposited layers. It is at the latter stage of the egg's formation that the spiral structure of the chalazæ becomes apparent. The time of the passage of the egg from the infundibulum to the uterus, in the Common Fowl, is from four to six hours. Here it may remain from twelve to twenty hours. On entering the 'uterus,' a thickish white fluid exudes from the inner surface of the cavity and condenses on the 'membrana putaminis,' forming thereon a cellular matrix in which soon appear particles of calcareous matter, which from the shape they assume in the interstices of the matrix appear to be crystalline.

Allen Thomson has given the annexed illustration, fig. 132, of the structure of the lining membrane of the shell and of the proper shell-membrane. A shows the 'lining membrane of the shell; *a*, thick matter or felty portion; *b*, thin shred of the torn margin, showing the peculiar fibrous tissue of which the various layers are composed; B, outermost layer of the same, which is incorporated with the shell; some of the angular corpuscles of the shell lying upon the fibrous substance and firmly united with it. C, small portion of the calcareous shell, which has been steeped in dilute hydrochloric acid, showing the remains of opaque calcareous substance in the centre: here and there clear oval cells seen, as at *a*, *a*.'[1]

132

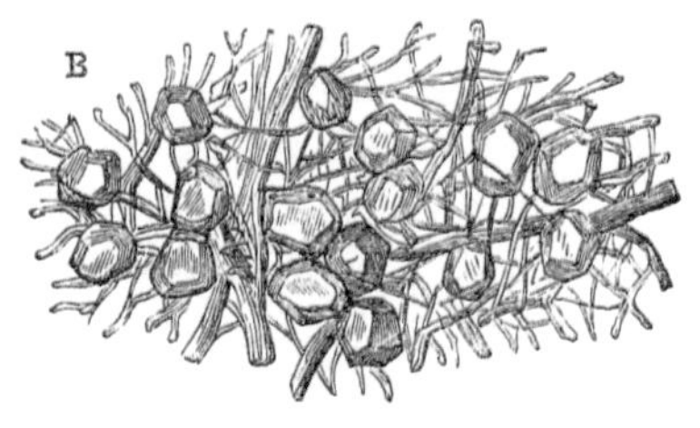

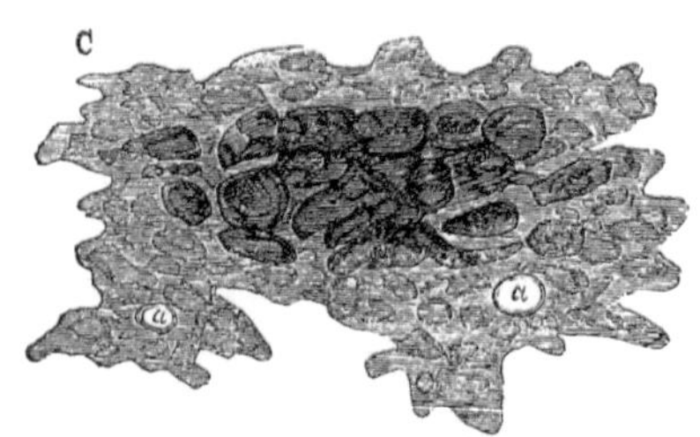

Structure of the shell and shell-membrane in the Fowl's egg. CCCVIII.

The colour of the egg-shell depends on pigmental matter secreted by particular follicles of the villous membrane of the 'uterus;' and either incorporated uniformly with the outermost layer of the shell, as in the Thrush: or deposited in cells more or less dispersed or aggregated in patches. The shell consists in great part of carbonate of lime, with a little carbonate of magnesia and phosphate of lime and magnesia.

The appearance to the unaided eye of pores on the surface of the shell is due to the impressions of the villi of the formative membrane: the permeability of the shell by the atmosphere depends on a more minutely porous texture. The first effect of this permeability is penetration of air between the layers of the lining membrane as the contents of the egg condense by cold and eva-

[1] CCCVIII. p. 63.

poration after it is laid. The air accumulates between layers of the 'membrana putaminis' at the great end of the egg, fig. 130, A, *f*; and in increased quantity as the other contents become condensed into the tissues of the chick, when it is averred to contain rather more oxygen than in ordinary atmospheric air. Such is the complex structure of the egg of a bird prior to its becoming subject to the influence of incubation.

It differs from the egg of the cold-blooded, non-incubating Ovipara, in the presence of the chalazæ and of the air-chamber, in the firmer and more complex structure of the shell, and in the greater proportion of albumen: in all which differences may be discerned a prospective adaptation to the business of hatching.

The cicatricula, or germ, is on the uppermost part of the floating yolk, the thinner part of which, occupying the nuclear tract, fig. 130, B, *c*, makes that half of it the lightest. Pressure of the upfloated germ against the shell-wall is moderated by the weight of the denser albumen forming the chalazæ, ib. A, *c*; and their usual attachments, a little below the axis of the yolk, help also to make the cicatricular half the lightest and uppermost. Under ordinary circumstances rotation of the egg takes place on its long axis, and, if a fresh egg be so turned round, 'the cicatricula will keep its position upwards for one turn or a little more, and then, by the twisting of the chalazæ, the yolk is carried completely round, and balances itself again with the cicatricula uppermost in its new position.'[1] The main function of the chalazæ is to keep the yolk more steady in the albumen, and to moderate the effects of any violent movement or rotation of the egg. The domed form of the hard shell enables it to bear the superincumbent weight of the brooding mother. How these modifications of the oviparous egg in anticipatory relation to the needs and conditions of incubation can be brought about by 'selective' or other operations of an unintelligent nature is not conceivable by me.

Birds differ in the number of eggs which they lay at one breeding season, in the relative size, in the shape, colour, surface, and thickness of the shell of the laid eggs. The Frigate Bird, Albatross, Penguin, Fulmar, Petrel, Awks, and some other sea-birds that brood on bare rocks, severally lay and hatch but one egg at a season: the Skua Gulls (*Lestris*) have two eggs; the Common Gulls (*Larus*) three eggs; the *Lamellirostres* and most *Gallinæ* hatch many eggs at a brood.

The Cuckoo has the smallest egg in proportion to its size, the

[1] CCCVIII. p. 65.

Apteryx the largest: in this species it weighs 14½ oz.; the entire bird 60 oz.; so that the egg is nearly equal to one-fourth of the parent. The hugest known egg of a bird is that of the extinct *Æpyornis* of Madagascar. The following are comparative admeasurements of this egg and that of an Ostrich:—

	Æpyornis.		Struthio.	
	In.	Lines.	In.	Lines.
Length of major axis	12	3	5	10
„ „ minor axis	9	4	5	0
Greater circumference	34	2	17	10
Smaller circumference	29	2	16	6

The contents of the egg of the Æpyornis are computed to equal those of 6 Ostriches' eggs, and 148 hen's eggs.

The eggs of most Owls, of some Penguins (*Spheniscus*), of the King-fishers (*Alcedo, Halcyon*), of the Plantain-eaters (*Musophaga*), and Bee-eaters (*Merops*), are those that have, or nearly approach to, the spherical shape: those that furthest depart from it, or have the longest shape, are the eggs of the Megapode and Albatross. The oval, or ovate, is the common form in birds (fig. 130, and vol. i. p. 599, fig. 420, C): the eggs with the narrowest small end, ib. D, are those of the Plovers, Snipes, Sand-pipers and allied Waders, which usually lay four eggs, packed in the smallest compass by the meeting of the small end of each in the centre. The egg of the Chinese Jacana (*Parra sinensis*) is like a top in shape. The eggs of Grebes, Cormorants, Pelicans, are elliptic. The shell of the Emeu's and Ostrich's egg has a rough exterior: that of the Gangas (*Pterocles*) has a glossy smoothness. The shell of the egg of the Ostrich, Emeu, and Cassowary is relatively thicker than that of the Apteryx, Mound-bird, and Dinornis.

§ 170. *Accessory Generative Structures and External Sexual Characters.*—The exception to the rule of incubation is given by the Megapodial birds of the Australasian Islands. A huge mound of decaying vegetable matter is raised: the eggs are deposited vertically in a circle at a certain depth, near the summit, and the chick is developed with the aid of the heat of fermentation. The large size of the egg relates to affording a supply of material sufficing for an unusually advanced state of developement of the chick at exclusion; whereby it has strength to force its way to the surface of the hatching-mound, with wings and feathers sufficiently developed to enable it to take a short flight to the nearest branch of an overshadowing tree.[1]

A steady continuous temperature of about 100° Fahr. is the

[1] LVI'. and LVIII'.

requisite condition of successful incubation: the heat of the sun alternating with the cold of night would hatch no bird's egg. The Ostrich deposits about fifteen eggs in a hollow of the sand: the male bird incubates, and the young are excluded in from fifty to sixty days. The following are the periods of incubation in some birds of the different orders of the class: the female sitting where not otherwise stated:—

Species		No. of days
American Ostrich	(*Rhea americana*) male	35
Mooruk	(*Casuarius Bennettii*) male	48
Emeu	(*Dromaius Novæ Hollandiæ*) male	54
Puffin	(*Fratercula arctica*)	30
Guillemot	(*Uria troile*)	30
Hooded Merganser	(*Mergus cucullatus*)	31
Sheldrake	(*Tadorna vulpanser*)	30
Muddy Wildrake	(*Casarca rutila*)	30
Summer Duck	(*Aix sponsa*)	30
Mandarin Duck	(*Aix galericulata*)	30
Sandwich Island Goose	(*Bernicla sandvicensis*)	31
Cereopsis Goose	(*Cereopsis Novæ Hollandiæ*)	35
Black Swan	(*Cygnus atratus*)	35
White Stork	(*Ciconia alba*)	31
Heron	(*Ardea cinerea*)	28
Dotterel	(*Charadrius morinellus*)	20
Capercailzie	(*Tetrao urogallus*)	28
Californian Quail	(*Callipepla californica*)	21
Purple Kaleege	(*Gallophasis Horsfieldii*)	24
Impeyan Pheasant	(*Lophophorus Impeyanus*)	28
Crown Pigeon	(*Goura coronata*)	28
Ringdove	(*Columba palumba*)	16
Cuckoo	(*Cuculus canorus*) by Hedge-Sparrow or other Passerines	14
Belted Kingfisher	(*Alcedo alcyon*)	16
Martin	(*Hirundo urbica*)	13
Skylark	(*Alauda arvensis*)	15
Chaffinch	(*Fringilla cœlebs*)	13
Wren	(*Troglodytes vulgaris*)	10
Bullfinch	(*Pyrrhula vulgaris*)	15
Starling	(*Sturnus vulgaris*)	16
Raven	(*Corvus corax*)	20
Golden Eagle	(*Aquila chrysaetos*)	30

Most birds nidify, i.e. prepare a receptacle for the eggs, to aggregate them in a space that may be covered by the incubating body (sand-hole of Ostrich), or superadd materials to keep in the warmth. The most complex 'nests' are made by birds of the singing order: and of these the pendent nests of the Weaver-Birds (*Ploceidæ*) are, perhaps, the most perfect and remarkable examples of nidification. Not only does the female construct her nest for incubation; but the male makes his, in the form of a bec-hive, open at the bottom, which is crossed by a perch of strong woven material, upon which he sits, sheltered from the tropical sun or storm by the dome above, which is suspended to a branch near that to which is attached the nest of the female, whom he solaces during her confinement with his song.

Certain conirostral *Cantores* still practise in the undisturbed wilds of Australia the formation of marriage-bowers distinct from the later-formed nesting-place.[1] The Satin Bower-Bird (*Ptilonorhynchus holosericeus*), and the Pink-necked Bower-Bird (*Chlamydera maculata*), are remarkable for their construction on the ground of avenues, over-arched by long twigs or grass-stems, the entry and exit of which are adorned by pearly shells, bright-coloured feathers, bleached bones, and other decorative materials, which are brought in profusion by the male, and variously arranged to attract, as it would seem, the female by the show of a handsome establishment. For receiving and incubating her eggs the female builds a nest, like that of the Magpie, in the concealment of a tree.[2]

Most birds, on reaching maturity, show external sexual characters. In Diurnal *Raptores* the female is larger than the male; in *Gallinaceæ* and most other polygamous birds, she is less. In this suborder the male is most conspicuous by the richness and beauty of his colours; and a difference in this respect is the most common sexual character in birds, with the frequent addition of a peculiar size and shape of certain feathers, especially at the breeding season, when, e.g., the male of *Machetes pugnax* becomes the 'Ruff,' the female the 'Reeve.' There is a sexual difference in the length of the beak in the Hook-billed Parrots (*Nestor*), in the *Apteryx*, and in the singular genus of Humming-Bird (*Androdon*, Gould), in which the end of the longer bill of the male is dentated. The comb and wattles of the Cock exemplify sexual characters of certain cutaneous appendages: his spur and that of other *Gallinæ* and *Phasianidæ*, including *Meleagris*, is a weapon of combat, analogous to the horns of Mammalian Herbivores.

Swifts, Swallows, Doves, Crows, King-fishers, Parrots, and the majority of the Waders are examples of birds in which the sexes are alike.

[1] LVII. and LVIII.

[2] It is possible that the old propensity of the Magpie, Jackdaw, and some others of our Conirostrals, to which the Australian Bower-Birds are allied, to pilfer glittering objects, may be the remnant of a similar instinct which the increase of human population has scared out of them: the conditions of cultivation reducing the birds to the constructions which are essential to the continuance of the species.

CHAPTER XXIV.

DEVELOPEMENT OF BIRDS.

§ 171. The heat-force being converted into movements of the parts of the germ thereto subjected, the expansion of the pellucid area, fig. 133, *a*, is the first sign of such change: in this area appears the embryonal trace, in the form of the parallel lines called 'plicæ primitivæ,' which diverge to form the cephalic dilatations. Concurrently with the appearance of the myelencephalous columns, ib. *p*, *p*, the blood-lakes expand in the surrounding halones, and tracts, ib. *h*, *h*, along which pass colourless blood-particles, extend from below the cephalic expansion, *b*, to the peripheral sinuses: as the proto-vertebræ, ib. *v*, *v*, begin to appear at the sides of the myelon, the red colour is acquired by the blood, and the heart is made more manifest, by its movements, as the 'punctum saliens,' ib. *c*. A distinct membrane, 'serous layer,' ib. *s*, *s*, is formed upon the germ and blastoderm:[1] the cephalic end of the embryo rises from the surface of the blastoderm, and then curving down, sinks into it, forming for itself a kind of hood of the serous layer: it is reflected at *b*, to show the fossa, *f*. This hood gradually extends from the margin of the fossa over the body, and, meeting a similar fold formed by the projecting and incurved tail, closes over the germ on the upper side, 'making a circumscribed cavity which is the amnios,'[2] fig. 134, *a*. The progress of differentiation of layers of the blastoderm has gone on beneath: in fig. 183, the 'serous layer' *b* is

133

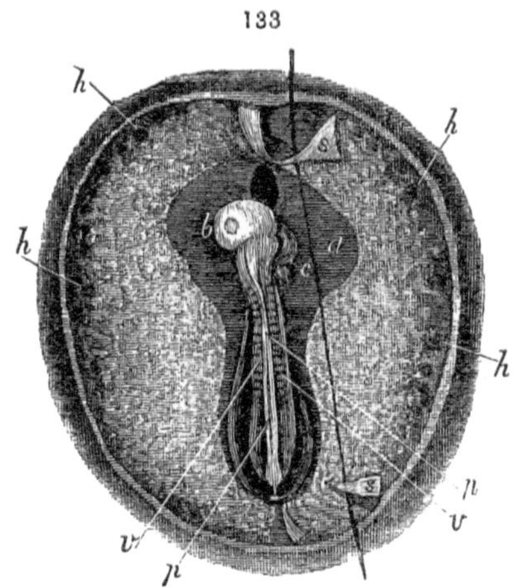

Embryo of Chick thirty-fifth hour. LV*.

[1] xx. vol. v. p. xx. Ib. pl. lxix. fig. 7.   [2] Ib. p. xx.

partially reflected from the 'vascular' and 'mucous layers.'[1] The mucous layer is concerned in the formation of the intestinal canal; and beyond this part, which is at first an open groove, the mucous layer expands over the yolk, which it ultimately incloses, the margins of the 'vitellicle' so formed, fig. 134, *c*, contracting and uniting at the side opposite to the embryo at a sort of 'cicatrix, to which the last part of the slime adheres.'[2] The vitellicle is richly vascular, and the surface next the yolk is augmented by rugæ, the yolk in contact with which becomes more liquid, and loses its coagulability.

At about the fortieth hour in the Common Fowl the limbs begin to bud forth, and a vesicle to protrude from near the anal end of the intestine which, rapidly expanding, fig. 134, *b*, spreads over the embryo, acquiring a close adhesion to the amnios, ib. *a*, but remaining distinct from the vitellicle, ib. *e*, *c*, over which it spreads, finally inclosing the albumen, and interposing itself between the latter and the lining membrane of the shell. Bloodvessels called 'umbilical,' fig. 134, *i*, are coextended with this bag, which Hunter 'called "allantois," from its containing urine.'[3] But that it 'answers other important purposes, must appear evident from its extent being far beyond what would answer that purpose. I conceive that the side of the bag which surrounds and is in contact with the albumen, acts as the chorion or placenta, for it must be by this surface that the albumen is absorbed and the chick supported. The external part of the bag, which comes in contact with the shell, I conceive to act as lungs, for it is the only part that comes in contact with the air: and on opening an egg with the chick pretty far advanced I find that the blood in the veins is scarlet, while it is of the Modena colour in the arteries of the bag.'[4] Schwann's experiments show that the developement of the chick may go on without oxygen to the fifteenth hour, and that the life of the germ is not destroyed till between the twenty-fourth and thirtieth hour, but that the presence of oxygen is essential to further developement.[5] As the embryo

134

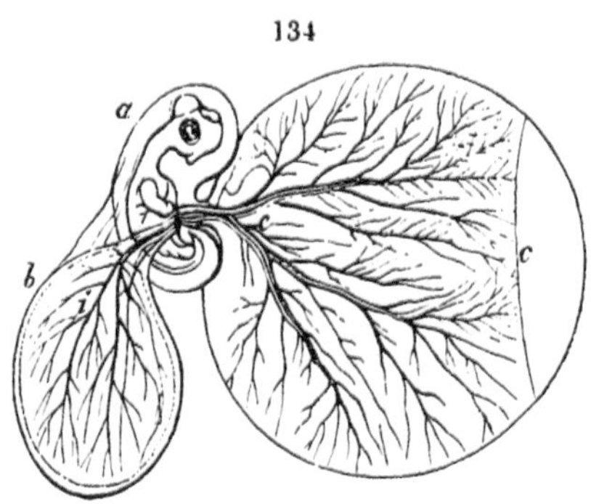

Membranes of the Chick third day. LV'.

[1] The mucous layer, *f*, is shown reflected from the vascular area, *g*.
[2] XX. vol. v. p. xxi. [3] XX. p. xxiv. [4] Ib. [5] LXIX'.

grows it turns upon its left side, exhibiting a profile view; it then indents the yolk, and finally almost divides it into two portions.

The formation of the digestive tube and glands closely follows the course described in Vol. I. pp. 604, 605, 606.

The embryo of the bird is that which best admits the observation of the commencement of the developement of the organ of hearing by a superficial depression of the cephalic blastema, fig. 135, *f*, to meet the process from the epencephalon, ib. *e*, which forms the acoustic nerve. The lining of the depression becomes, on the closure of the slit, the proper tunic of the labyrinth.[1]

The vesicle of the labyrinth, *f*, swells into four dilatations, of which three are 'ampullar,' and the fourth 'cochlear:' the ampullar dilatations extend into very slender canals, at first almost in the same plane, by which they are brought into mutual communication: as the canals expand and elongate, they assume their characteristic relative positions as external, superior, posterior: the hinder end of the external canal being extended beneath the posterior canal. The cochlear dilatation curves as it elongates: an inner layer becomes distinct from the common membrane, and forms the acoustic lamina.

135

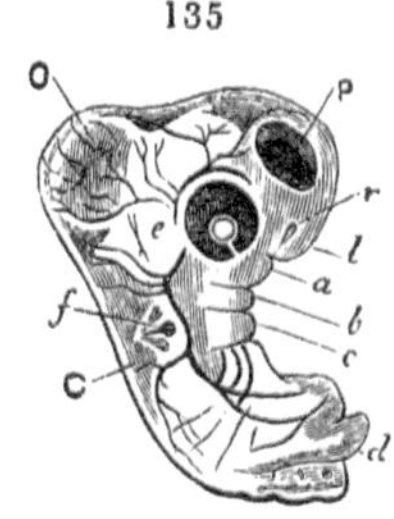

Fore-part of embryo Chick second day.

As in the developement of the eye, the production of the nerve-process from the cerebral centre is the first step, the infolding of the superficial blastema to meet the nerve is the next: the so-called 'cutaneous follicle' becomes a circumscribed sac or vesicle, in which the changes and developements next proceed, converting the vesicle into 'acoustic labyrinth' or 'eyeball.' In each case neural elements of two vertebræ become modified to lodge and protect the sense-organs, forming respectively the recesses called 'otocrane' and 'orbit,' the one between the occipital and parietal vertebræ, the other between the frontal and nasal vertebræ. The part of the outer blastemal layer of the head which sinks to meet the process from the mesencephalic dilatation, rapidly changes its follicular into a vesicular state: the vesicle elongates, bending round the cell-mass in which the crystalline lens is formed (as in the Fish, Vol. I. fig. 423), and by the meeting of the two ends, the 'choroid fissure,' at the lower part of the eyeball, figs. 134, 135, results. The mesencephalic process, or 'optic nerve,' expands at

[1] In xx. pl. lxx. fig. 3, embryo of the Goose at the thirtieth hour of incubation, the open state of the acoustic sac is erroneously described as 'meatus:' but the sac becomes closed, and the tympanum and its passage are later developements.

the back part of the circular sac, and, in the course of its mutation into eyeball, lines its posterior part with the layer called retina, interrupted only by the cicatrix of the inferior and rapidly blended ends of the primitive eye-sac. The transparent layer covering the fore-part of that sac and the inclosed lens is metamorphosed into cornea. Other layers of the sac are differentiated into choroid, ciliary processes, iris; and a fold of the vascular layer protrudes through the choroid fissure as a persistent structure in birds, in which the 'pecten' significantly marks a curious step in the developement of the eye in all Vertebrates. Of the appendages of the eye the membrana nictitans, fig. 137, *a*, is the first to appear, the lower lid and then the upper lid follow. It is a mistake to speak of the labyrinth or eyeball as being formed by the integument, or beginning as 'cutaneous follicles,' for the structures of the skin are not differentiated when they first appear; a layer of cellular or primitive blastemal tissue represents the integument, and a greater number of cells is aggregated at the points which tend inward to meet the productions from the nervous centres. After the essential organs of sense are established, then is the skin developed and modified more or less for their protection, forming the outer ear and the eyelids: but both passages are closed by transparent membranes, as 'ear-drum' and 'cornea.' Only in the case of the olfactory organ does the primitive depression, fig. 135, *r*, retain its outlet, and in the bird and other air-breathers, it also communicates with the air-passage: having the tegument superadded and modified, in most, as external nostril and nose.

136

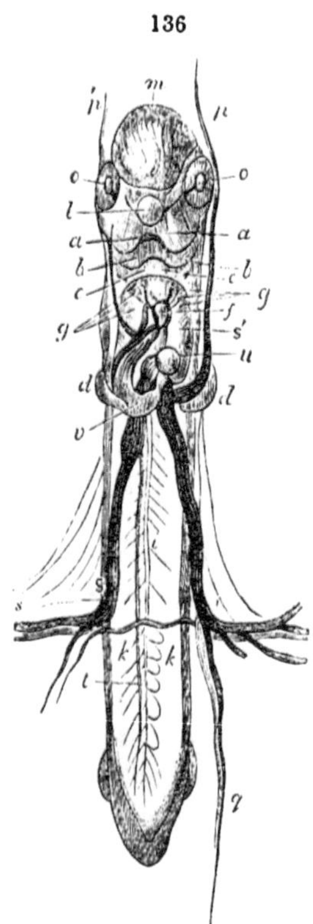

Primitive blood-vessels of embryo Bird second day. LV*.

As in the Lizard and Snake (Vol. I. fig. 444), so in the Bird, the four vertebral segments constituting the head are shown by the embryological characters and course of formation of the 'maxillary' arch, figs. 135, 136, *a*, the 'mandibular arch,' ib. *b*, the 'hyoidean arch,' ib. *c*, and the scapular arch, ib. *d*. The transitory branchial arteries, fig. 136, from the aortic bulb traverse the tissue between the hyoidean and scapular arches.

The channels which return the blood from the vitellicle are the 'transverse' and 'longitudinal vitelline' veins: the first are so called because these trunks pass to the embryo at right angles to its axis; they are the largest returning channels: the longitudinal veins run parallel with the axis of the embryo and are of smaller size. The right anterior longitudinal vein, fig. 136, *p'*, becomes the right precaval and receives the remains of the right transverse vitelline vein, ib. *s*, as the right vena azygos. The left anterior longitudinal vitelline vein, ib. *p*, is also persistent as the left precaval, and enters in the mature bird, as in the embryo, the posterior or lower (sacral) part of the auricle. The left transverse vitelline vein, ib. *r*, is also subsequently reduced, by receiving only the vertebral veins of that side, *t*, to the condition of a so-called 'azygos vein.' The main trunk of the postcaval is the result of the returning channels from the abdominal viscera and the hind-limbs, at a later stage of developement. There is but one principal posterior longitudinal vitelline vein, ib. *q*, which anastomoses with the left transverse vein as it enters the embryo: the homotype of the right side appears as one of the ordinary small tributaries of the right transverse vein.

The auricle which by the dilatation of the left side, ib. *u*, appears to be double, receives the venous blood at its right division. The left one, subsequently receiving the veins from the lungs, is ultimately separated from the left precaval and right auricle to which that vein is conducted and restricted.

The ventricular part of the heart, ib. *v*, at the second day of incubation, is in the form of a bent tube, curving from behind downward, forward, to the right and upward, continued insensibly into the part representing the 'aortic bulb,' ib. *f*, in which the septum first appears, ultimately dividing the ventricle into two.

At this stage the piers of the maxillary arch, ib. *a*, appear as buds from beneath the eyeballs; the naso-premaxillary process, ib. *l*, is above their interspace; the piers of the mandibular arch, ib. *b*, *b*, and those of the hyoidean arch, ib. *c*, *c*, follow in close succession. The blastemal base of the scapular arch, ib. *d*, *d*, slightly projects at the sides of the 'fovea cardiaca:' the piers, now separate, ultimately meet in front of the heart, and accompany it in its retrograde course. The mesencephalon, ib. *m*, is the largest segment of the brain, in connection with the eyeballs, *o*, *o*.

When the heart has assumed its form, as such, distinct from the great trunks rising from it, the arteries from the base of the ventricle appear, during the fœtal circulation of the chick, to be two: that to the right bifurcates, one division supplying the head and wings, the other winds over the right bronchus: that to the left also bifurcates: its left division arches over the left bronchus and anastomoses with the right arch a little below and behind the apex of the heart: its right division arches over the back of the heart, bending rather to the right, and anastomoses with the right aortic arch, just above the other 'ductus arteriosus.' Each of these divisions of the left primary arterial trunk sends off a branch to its corresponding lung, and as the lung expands, and especially begins to act as such, toward the close of incubation, the blood is diverted into the pulmonary vessels, and the channels below them shrink and disappear. The left primary artery is retained as the trunk of the pulmonaries, and, through the changes in the interior of the ventricle, this arises exclusively from the ventricle answering to the 'right' in Mammals, whilst the retained aorta rises from the 'left' ventricle. It arches, however, over the right bronchus. There is no left aorta in birds distinct (as in fig. 335, A, Vol. I. p. 509) from the trunk (ib. *p*) which gives off the artery to the left lung: only one arterial trunk arises from the right ventricle instead of two.

The air-cells begin at the lower point of the lungs, like a small hydatid, and extend further and further into the abdomen, before the kidneys: they are at first full of a fluid; as they extend, they are, as it were, squeezed among the intestines and at last fill with them the whole abdomen. Soon after other air-cells are forming. The lungs are, at first, free, as in Reptiles, but afterwards begin to be attached to the ribs and spine. In the female embryo we first 'observe two oviducts, one on each side' (as in fig. 127, *g*); but 'before hatching the right seems to decay.'[1] 'There are two kinds of down on the chick, one long, which comes first, about two or three days before hatching; a second, or fine, down forms at the roots of the other.' 'The little horny knob at the end of the beak, fig. 137, *b*, with which it breaks the shell when arrived at full time, is also gradually forming into a more regular and determinate point, the progress of which is seen from the first figure to the

137

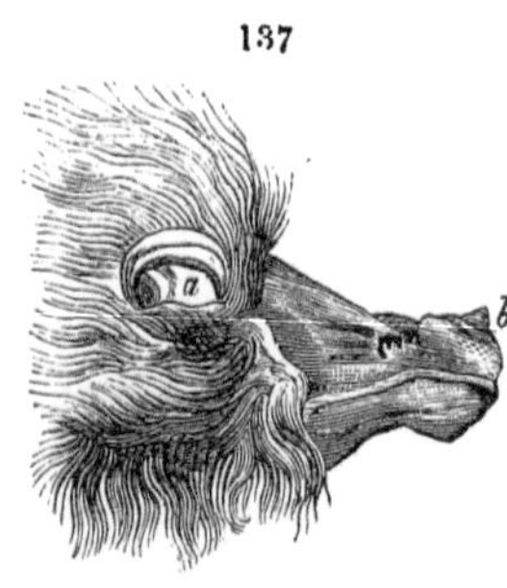

Head of Gosling. LV'.

[1] xx. vol. v. p. xxvi.

sixth.'[1] As the contents of the egg become condensed by embryonal developement, the air-cavity, fig. 130, *f*, expands. 'The chick some time before birth has a kind of mixed action of life, for it breathes, and we can hear it pip and chirp in the egg; and we find the adult circulation through and out of the heart is formed before birth: yet it is receiving its nourishment from the remaining slime.'[2]

The 'slime' or albumen is reduced to the small mass adhering to the cicatrix of the vitellicle, and with this and its yolk little decreased in bulk, it is taken into the abdomen, where it serves to nourish the chick in the first feeble days of its free life: the pedicle of the vitellicle communicates with a loop of the small intestine. The allantois is left, lining the shell: the urachus is obliterated. The anus has a dorsal position near the hind end of the trunk in the nestlings, fig. 113, A.

The degree of developement under which the young bird quits the egg differs in different groups of the class. It is naked or covered with down only, and is dependent on the parents for shelter and support, in the orders RAPTORES, SCANSORES, VOLITORES, CANTORES, in the Rasorial suborder GEMITORES (Doves, p. 9); in the Grallatorial *Cultrirostres* (Herons, &c. p. 9); in the Natatorial *Longipennatæ* (Gulls, &c. p. 9), and *Totipalmatæ* (Pelicans, &c. p. 9). The young bird is excluded well clothed and able to run or swim about and provide food for itself in the suborder GALLINACEÆ, in the CURSORES, in the Natatorial *Brevipennatæ* (Penguins, Awks, &c.) and *Lamellirostres* (Duck, Goose, &c.), and in all the GRALLATORES save the Cultrirostral group or part of it. Of these 'precocious' birds (*Præcoces*) most are polygamous, and the females hatch many young; whereas in the 'nursing' groups (*Altrices*) the species are monogamous, and have few young.[3]

[1] Three of the figures here referred to by Hunter are engraved in xx. vol. v. pl. lxxvi. figs. 16, 17, 18. [2] Ib. p. xxvii. [3] VII'. p. 265.

CHAPTER XXV.

CHARACTERS AND PRIMARY GROUPS OF THE CLASS MAMMALIA.

§ 172. *Class Characters.* — Mammals are outwardly distinguished by a covering of hair, entire or partial, and (with two exceptions) by teats, fig. 138, *a*, *b*, whence the name of the class.[1] All possess mammary glands and suckle the young: the embryo or fœtus is developed in the womb. The male has a penis, and impregnation is preceded by intromission. The lungs, fig. 139, *lg*, minutely cellular throughout, are suspended freely in a thoracic cavity separated by a musculo-tendinous partition or 'diaphragm,' ib. *d*, from the abdomen.

138

New-born fœtus and teats, Kangaroo (*Macropus major*).

Mammals, like Birds, have a heart, ib. *h*, composed of two ventricles and two auricles, and have warm blood: they breathe quickly; but inspiration is performed chiefly

139

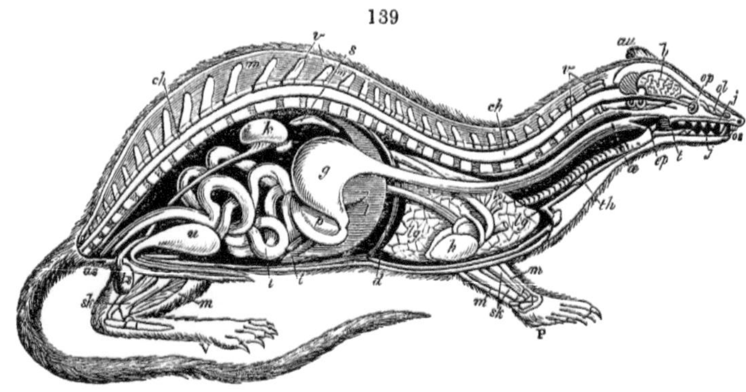

Ideal section of a Mammalian animal.

by the agency of the diaphragm; and the inspired air acts only on the capillaries of the pulmonary circulation.

[1] *Mamma*, a teat. The Monotremes have mammary glands without teats. The fœtal Cetacea show tufts of hair on the muzzle.

The blood-discs are smaller than in Reptiles, and, save in the Camel-tribe, are circular. The right auriculo-ventricular valve is membranous, at least never entirely fleshy; and the aorta bends over the left bronchial tube. The abdominal aorta terminates by dividing beyond the kidneys into the iliac arteries, from which spring both the femoral and ischiadic branches: if continued beyond, it is as a caudal or sacro-median artery.

The kidneys, ib. *k*, are relatively smaller and present a more compact figure than in the other Vertebrate classes; their parenchyme is divided into a cortical and medullary portion, and the secreting tubuli terminate in a dilatation of the excretory duct, called the pelvis. They derive their secretion exclusively from the arterial system. Their veins, commencing by minute capillaries in the renal parenchyme, terminate generally by a single trunk on each side in the abdominal vena cava: they never anastomose with the intestinal veins.

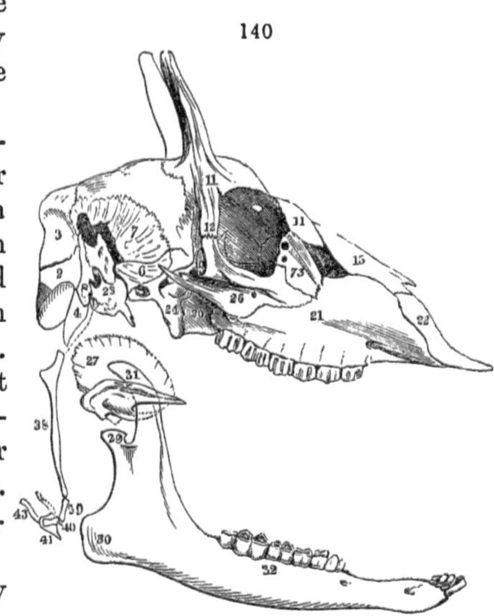

Skull of Mammal, Deer.

The liver, ib. *l*, is generally divided into a greater number of lobes than in Birds. The portal system is formed by veins derived exclusively from the spleen and chylopoietic viscera. The cystic duct, when it exists, always joins the hepatic, and does not enter the duodenum separately. The pancreatic duct is commonly single.

The mouth is closed by soft flexible muscular lips: the upper jaw is composed of palatine, fig. 140, 20, maxillary, 21, and premaxillary, 22, bones, and is fixed; the lower jaw consists of two rami, formed each by one bony piece, ib. 30–32, and articulated by a convex or flat condyle, ib. 29, to the squamosal, ib. 27, not to the tympanic, ib. 28.

The jaws of Mammals, with few exceptions, are provided with teeth. These are limited to the premaxillary, maxillary, and mandibular bones, and are there arranged in a single row; they are lodged in sockets, not anchylosed with the substance of the jaw. Only in the present class are teeth implanted by two or more fangs: when they are of limited growth, and usually molars: ever-growing teeth require the base to be kept open for the persistent pulp. Some Mammals are 'monophyodont,'[1] or have but one set of teeth: the majority are 'diphyodont,'[2] or have two sets: none have more. The tongue is large, fleshy, with the apex more or less free. The posterior nares are protected by a soft palate, and the larynx by an epiglottis, fig. 139, *ep*: the rings of the trachea are generally cartilaginous and incomplete behind: there is no inferior larynx. The œsophagus, ib. *œ*, is continued without partial dilatations to the stomach, ib. *g*, which varies in its structure according to the nature of the food, or the quantity of nutriment to be extracted therefrom.

141

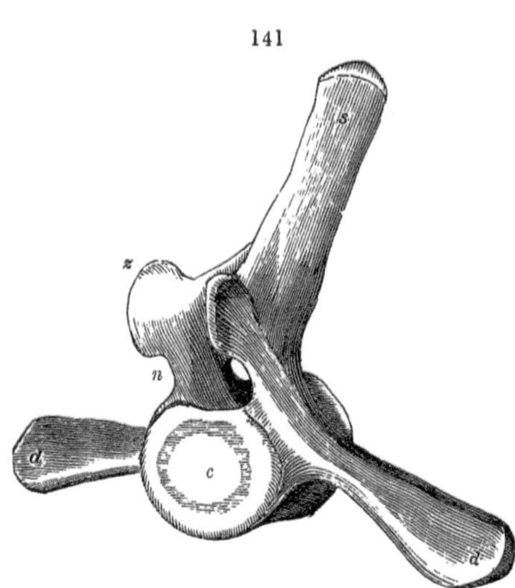

Mammalian vertebra, lumbar of Whale.

The vertebral bodies, fig. 141, *c*, are ossified from three centres, and present for a longer or shorter period of life a discoid epiphysis at each extremity. They are articulated by concentric ligaments with interposed glairy fluid, fig. 199, forming what are called the intervertebral substances; the articulating surfaces are generally flattened, but, in the neck of certain Ungulates, they are concave behind and convex in front. The cervical vertebræ, in all Mammals save two, are seven in number, neither more nor less. The atlas is articulated by concave zygapophyses to two convex condyles, which are developed from the neurapophyses (exoccipitals) of the last cranial vertebra.

The scapula is generally an expanded plate of bone; the coracoid, with two (monotrematous) exceptions, appears as a small process of the scapula. The sternum is usually narrow, and consists of a simple longitudinal series of bones: the sternal ribs are generally cartilaginous. The centrums of two cranial vertebræ (basisphenoid and presphenoid) preserve their distinctness to a late period of growth, in the species where they ultimately coalesce.

[1] μόνος, once; φύω, I generate; ὀδούς, tooth.

[2] δίς, twice; φύω, and ὀδούς.

The brain has a cerebellum with large lateral lobes, fig. 142, *d*, and the grey superficies much folded; the commissural fibres form, as they cross the under surface of the epencephalon, a defined tract or prominence called 'pons Varolii,' fig. 143, *p*. The optic

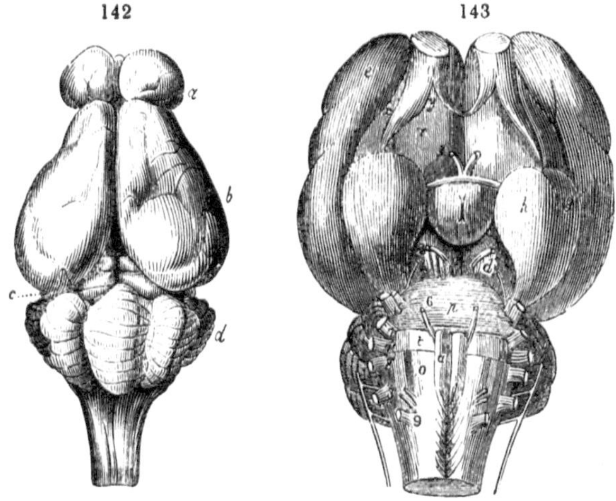

Upper surface of brain, Dasyure. LXX'. Under surface of brain, Porcupine.

lobes, fig. 142, *c*, are medial in position and divided by a transverse furrow. The cerebral lobes, ib. *b*, are not only united by a round

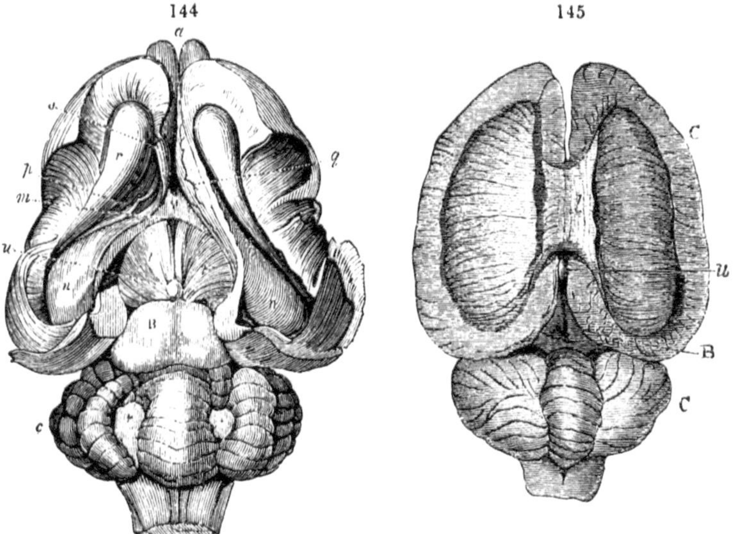

Lyencephalous brain, Wombat (*Phascolomys fusca*). LXX'. Corpus callosum, brain of Beaver. LXX'.

commissure, but by a 'lyra' and hippocampal commissure, fig. 144, *m*; from which is developed, in the majority of the class, a 'corpus callosum' or great commissure, fig. 145, *l*. The rhinencephalon, fig. 142, *a*, is in contact with the prosencephalon, *b*, and sends off numerous olfactory nerves which perforate a 'cribriform' plate of the prefrontal.

§ 173. *Mammalian Subclasses.*—The primary subdivisions of the present class are characterised by conditions of the brain.[1]

When the hemispheres are connected by the 'round commissure' and 'hippocampal commissure' only, fig. 144, *m*, this cerebral condition is associated with the absence of a vascular chorion or placenta, and with prematurely born young, compared with the rest of the class (fig. 138 shows the natural size of the new-born Kangaroo of the largest species).

The cerebral hemispheres are usually without folds, and leave, as in fig. 142, the cerebellum, *d*, olfactory lobes, *a*, and part of the optic lobes, *c*, exposed. The subclass so characterised is called LYENCEPHALA.[2] Mammals of this low type existed as far back, in time, as the oolitic and triassic periods, and are the oldest known.[3]

146

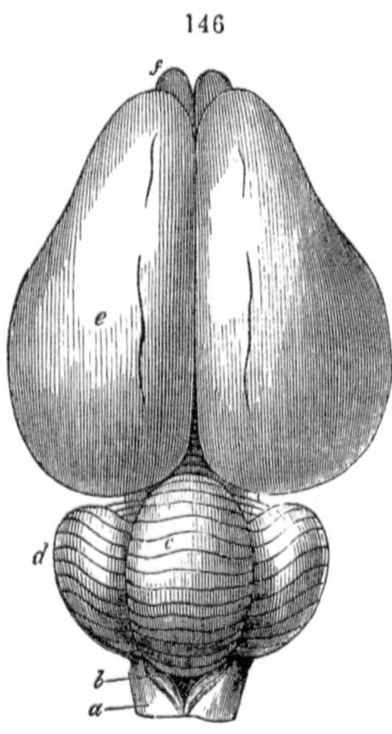

Upper surface of brain, Agouti.

The next stage of complexity in the Mammalian brain is where the 'corpus callosum,' fig. 145, *l*, is present; but connects hemispheres as little advanced in relative bulk or outward character as in the preceding subclass; the surface being smooth, fig. 146, *e*, or with folds, in the largest members of the group, not more numerous or complex than in the larger Lyencephala. The hemispheres, ib. *e*, leave the cerebellum, *c*, *d*, and part of the olfactory lobes, *f*, exposed. The subclass so characterised is called LISSENCEPHALA.[4] In the species with this condition of brain the testes remain in the abdomen, or are protruded into a temporary scrotum only at the breeding period, to be again retracted: in most there is a common external urogenital aperture: there are two precaval veins. The squamosal in most, and the tympanic in many, retain their primitive condition as distinct bones. The orbits have not an entire

[1] LXVIII'. and LXIV'.

[2] λύω, I loose; ἐγκέφαλος, brain. LXX'.

[3] XVII'. p. 338.

[4] λισσός, smooth; ἐγκέφαλος, brain.

rim of bone. Besides these general characters of affinity to Birds and Reptiles, there are other striking indications of the same low position in particular orders or genera of the subclass. Such, e.g., are the cloaca, convoluted trachea, supernumerary cervical vertebræ and their floating ribs, in the Three-toed Sloth; the irritability of the muscular fibre, and persistence of contractile power in the Sloths and some other Bruta; the long, slender, beak-like edentulous jaws and gizzard of the Anteaters; the imbricated scales of the equally edentulous Pangolins, which have both gizzard and gastric glands like the proventricular ones in Birds; the dermal bony armour of the Armadillos like that of loricated Saurians; the quills of the Porcupine and Hedgehog; the proventriculus of the Dormouse and Beaver; the prevalence of disproportionate developement of the hind-limbs in the *Rodentia*; coupled, in the Jerboa, with confluence of the three chief metatarsals into one bone, as in Birds; the keeled sternum and wings

147

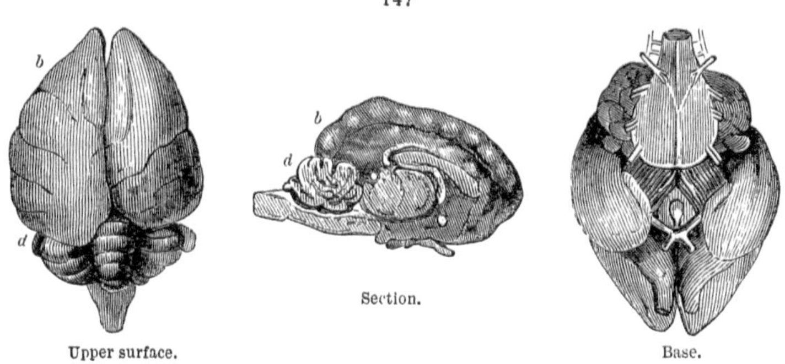

Brain of Lemur (*Stenops tardigradus*). LXIX'.

of the Bats; the aptitude of the *Cheiroptera*, *Insectivora*, and certain *Rodentia* to fall, like Reptiles, into a state of torpidity, associated with a corresponding faculty of the heart to circulate carbonised or black blood:—these, and the like indications of co-affinity with the *Lyencephala* to the Oviparous air-breathing Vertebrata, concur with the cerebral character in demonstrating the low position of the *Lissencephala* in the Mammalian class.

The third leading modification of the Mammalian brain is such an increase in the relative size of the cerebrum, fig. 147, *b*, that it extends over half or more of the cerebellum, and of the olfactory lobes. The surface of the hemispheres may be smooth, or with few and simple folds, in the smallest species; but, as a rule, it is disposed in many gyri or convolutions, fig. 148, whence the name GYRENCEPHALA,[1] proposed for this third subclass of Mammalia.

In this subclass, there are no such marks of affinity to the Oviparous Vertebrates as have been instanced in the preceding. The testes are concealed, in adaptation to aquatic life, in Cetacea; but, in the rest of the subclass, with the exception of the Elephant, they pass out of the abdomen, and the Gyrencephalous quadrupeds, as a general rule, have a scrotum. The vulva is externally distinct from the anus. With the exception, again, of the Elephant, the blood from the head and anterior limbs is returned to the right auricle by a single precaval trunk. The Mammalian modification of the Vertebrate type attains its highest physical perfections in the *Gyrencephala*, as manifested by the bulk of some, by the destructive mastery of others, by the address and agility of a third order. And, through the superior psychological faculties—an adaptive intelligence predominating over blind instinct—which are associated with the higher developement of the brain, the *Gyrencephala* supply those species which have ever formed the most cherished companions and servitors, and the most valuable sources of wealth and power, to Mankind.

148

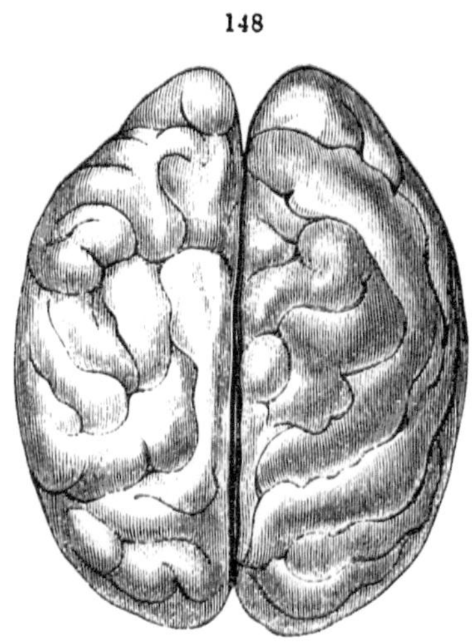

Upper surface of the brain of the Orang-œtan. (After Sandifort.)

149

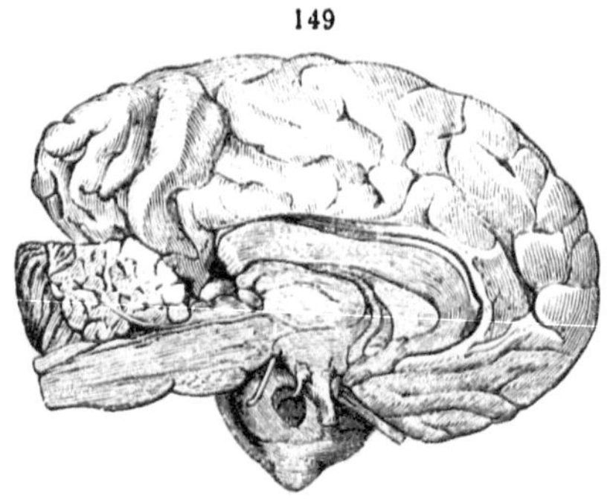

Vertical section of the brain of the Orang-œtan. (After W. Vrolik.) Half nat. size.

In Man the brain presents an ascensive step in developement, higher and more strongly marked than that by which the preceding subclass was distinguished from the one below it. Although in the highest *Gyrencephala* the cerebrum, figs. 148,

[1] γυρόω, I wind about; ἐγκέφαλος, brain. LXIV'.

149, *b*, may extend over the cerebellum, *d*, in Man not only do the cerebral hemispheres, fig. 149, *b*, overlap the olfactory lobes and cerebellum, *d*, but they extend in advance of the one, and further back than the other. Their posterior developement is so marked, that anatomists have assigned to that part the character of a third lobe; it is peculiar, with its proportionally developed posterior ventricular horn and 'hippocampus minor,' to the genus *Homo*.[1] Concomitantly with the correspondingly developed anterior lobes of the cerebrum, the ventricle is, in like manner, produced into a

[1] Kuhl in *Ateles Belzebuth*,[a] Tiedemann in the Macacque[b] and Orang,[c] Vrolik in the Chimpanzee,[d] and myself in the Gorilla,[e] have severally shown all the homologous parts of the human cerebral organ to exist, under modified forms and low grades of developement, in *Quadrumana*.

Kuhl rightly characterises the homologue of the posterior cornu, which he found in a platyrrhine monkey, 'Anfang des hintern, dritten Horns des Seitenventricels' (*op. cit.* p. 70)—'the beginning of the posterior or third horn of the lateral ventricle.' Tiedemann, with equal accuracy, defines the answerable part in the catarrhine quadrumana, as, 'Scrobiculus parvus loco cornu posterioris' (*op. cit.* p. 14). In regard to the posterior cornu in the brain of the Orang he is silent as to any 'hippocampus minor.' It exists, however, in the condition described by Vrolik, in that Ape and in the Chimpanzee, as 'une éminence que nous croyons avoir le droit de nommer indice de pes hippocampi minor' (*Versl. en Mededeel. der Kon. Akad.* 1862, p. xiii.) These 'beginnings' and 'indications' of structures which reach their full developement in Man in no way affect the value of the latter as zoological characters. In propounding them as such to the Linnæan Society in 1857, I forbore to encumber my memoir with reference to facts known to all who possessed the elements of Comparative Anatomy. Tiedemann's definition was the accepted one:—'Pedes hippocampi minores vel ungues, vel calcaria avis, quæ a posteriore corporis callosi tanquam processus duo medullares proficiscuntur, inque fundo cornu posterioris plicas graciles et retroflexas formant, in cerebro Simiarum desunt; nec in cerebro aliorum a me examinatorum mammalium occurrunt; Homini ergo proprii sunt.' (Ib. p. 51.) In like manner Cuvier had characterised the species of his order *Quadrumana* as having, 'Pouce libre et opposable au lieu du grand orteil.' And he rightly affirms: 'L'homme est le seul animal vraiment *bimane* et *bipède*.' (*Règne Animal*, i. p. 70.) To adduce beginnings of structures in one group which reach their full developement in another, as invalidating their zoological application in such higher group, is puerile; to reproduce the facts of such incipient and indicatory structures as new discoveries is ridiculous; to represent the statement of the zoological character of a higher group as a denial of the existence of homologous parts in a lower one is disgraceful. Mr. Flower was not the first to see in the hippocampal commissure the beginning of the corpus callosum: the homologues of 'cornu posterius' and of 'hippocampus minor' were known in the Orang before Prof. Rolleston: and the homologies of the bones of the hind foot in mammals had been determined before Prof. Huxley propounded them to show that the hind thumb of the Ape was a great toe, and that Man was not the only animal who possessed two hands and two feet.

[a] Beiträge zur Zoologie und vergleichenden Anatomie, 4to, 1820, zweite Abtheilung, p. 70, tab. vii. [b] Icones cerebri Simiarum, fol. 1821, p. 14, fig. iii. 2. [c] Treviranus, Zeitschrift für Physiologie, Bd. ii. s. 25, Taf. iv. [d] Nieuwe Verhandlungen der erste Klasse vom het Köningl. Nederlandsche Institut. Amsterdam, 1849. [e] Fullerian Lectures, Royal Institution (March 18, 1861), reported, with copies of diagrams, in 'Athenæum,' March 23rd, 1861, p. 395.

hornlike form, in advance of the 'corpus striatum.' The superficial grey matter of the cerebrum, through the number and depth of the convolutions, attains its maximum of extent in Man.

Peculiar mental powers are associated with this highest form of brain, and their consequences strikingly illustrate the value of the cerebral character; according to my estimate of which, I am led to regard the genus *Homo* as not merely a representative of a distinct order, but of a distinct subclass of the Mammalia, for which I have proposed the name of '*ARCHENCEPHALA*.'[1]

150

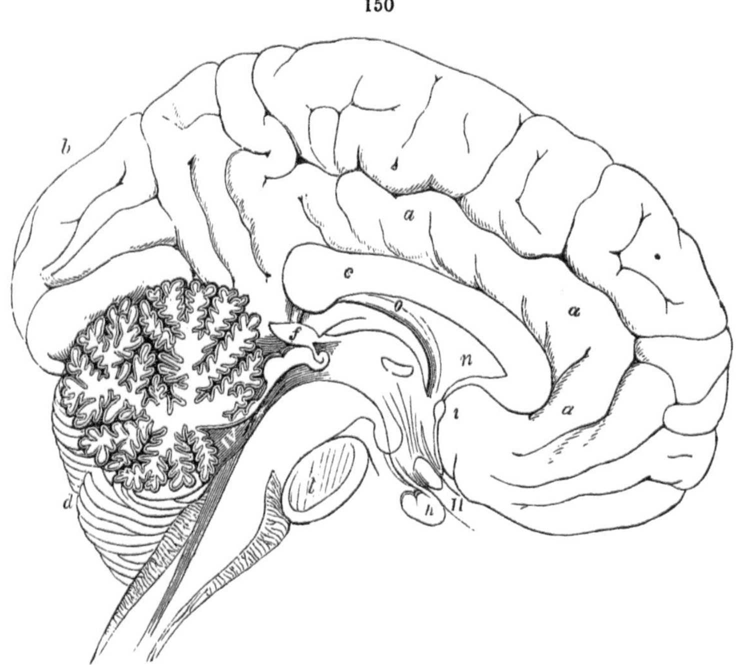

Vertical section of the adult human brain. (After Arnold.) Half natural size.

With this preliminary definition of the organic characters, which guide to a conception of the natural primary groups of the class *Mammalia*, I next proceed to define those of secondary importance, or the subdivisions of the foregoing subclasses.

§ 174. *Characters of Orders.*—In the Lyencephalous Mammalia some have the optic lobes less definitely divided into 'corpora quadrigemina' than others. Those with the more simple optic lobes are 'edentulous' or without calcified teeth, are devoid of external ears, scrotum, nipples, and oviducal fimbriæ; they are

[1] ἄρχω, I overrule; ἐγκέφαλος, brain.

true 'testiconda,' and are ovoviviparous, fig. 151: they have a coracoid bone extending from the scapula to the sternum, and also an epicoracoid and episternum, as in Lizards; they are unguiculate and pentadactyle,[1] with a supplementary tarsal bone supporting a perforated spur in the male. The order so characterised is called 'MONOTREMATA,' in reference to the single excretory and generative outlet, which, however, is not peculiar to them among Mammalia. The Monotremes are insectivorous, and are limited to Australia and Tasmania; where they are represented by the Platypus or Duck-Mole (*Ornithorhynchus*), and by the Spiny Anteater (*Echidna*).

151

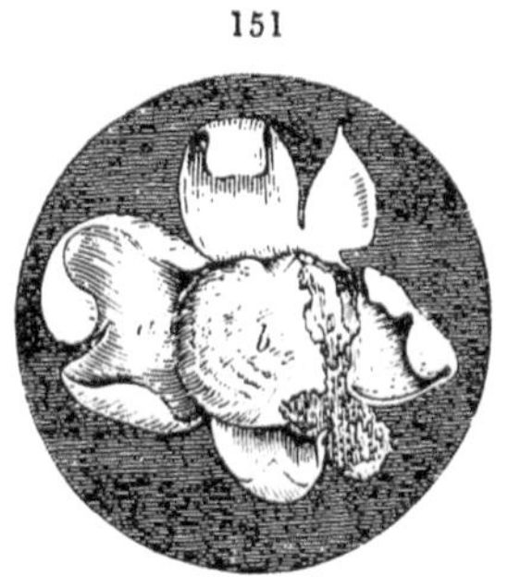

Uterine ovum, dissected, Ornithorhynchus. LXXVII'.

The MARSUPIALIA are Mammals distinguished by a peculiar pouch or duplicature of the abdominal integument, which in the males is everted, forming a pendulous bag containing the testes, and in the females is inverted, forming a hidden pouch containing the nipples and usually sheltering the young for a certain period after their birth: they have the marsupial bones, fig. 152, *m*, in common with the Monotremes; a much-varied dentition, especially as regards the number of incisors, but usually including four true molars; and never more than three premolars: the angle of the lower jaw is more or less inverted.

152

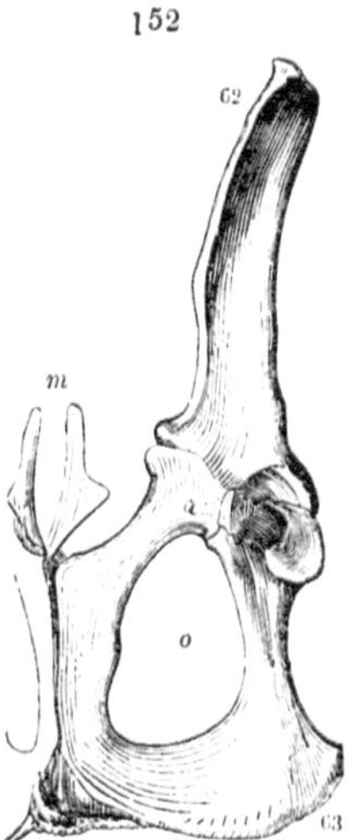

Pelvis and marsupial bones, Kangaroo. LXXV'.

With the exception of one genus, *Didelphys*, which is American, all the known existing Marsupials are Melanesian, i. e. belong to Australia, Tasmania, New Guinea, and some adjacent isles. The grazing and browsing Kangaroos are rarely seen abroad in full daylight, save in dark rainy weather. Most of the Marsupialia are nocturnal. Zoological wanderers in Australia, viewing its plains and scanning its scrubs by broad daylight, are struck by the seeming absence of mammalian life; but during the brief twilight and dawn, or by the light of the moon, numerous forms are seen to emerge from their hiding-places and illustrate the variety of marsupial life with which many parts of the continent abound. We may associate with their low position in the mammalian scale the prevalent habit amongst the Marsupialia of limiting the exercise of the faculties of active life to the period when they are shielded by the obscurity of night.

The Lissencephala or smooth-brained Placentals form a group, equivalent to the Lyencephala or Implacentals, and include the following orders, *Rodentia*, *Insectivora*, *Cheiroptera*, and *Bruta*. The RODENTIA are characterised by two large and long curved incisors in each jaw, fig. 153, *i*, separated by a wide interval from the molars; and these teeth are so constructed, and the jaw is so articulated, as to serve in the reduction of the food to small particles by

153

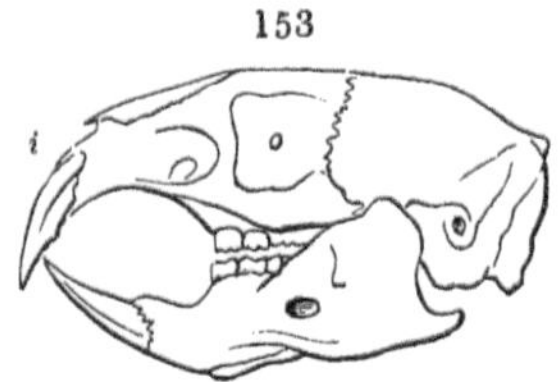

Skull of a Rodent (*Jerboa*).

154

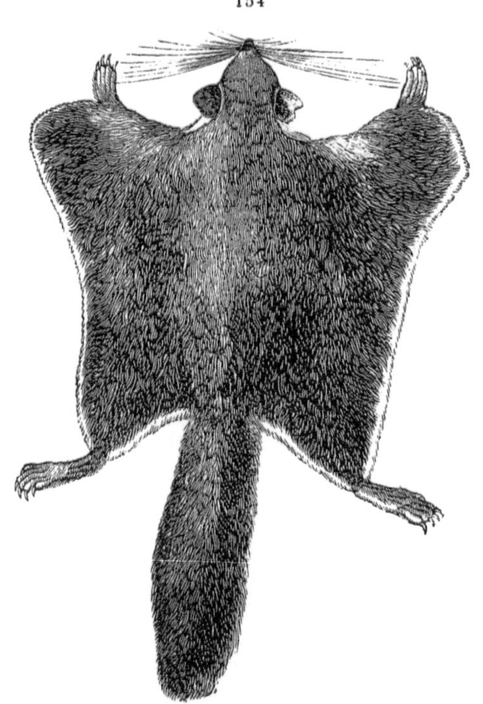

Pteromys Volucella.

acts of rapid and continued gnawing, whence the name of the order. The orbits, ib. *o*, are not separated from the temporal fossæ. The testes pass periodically from the abdomen into a temporary scrotum, and are associated with prostatic and vesicular glands. The placenta is commonly discoid, but is sometimes a circular mass (Cavy), or flattened and divided into three or more lobes (Lepus). The Beaver and Capybara are now the giants of the order, which chiefly consists of small, numerous, prolific and diversified unguiculate genera, subsisting wholly or in part on vegetable food.[1] Certain squirrels achieve short flights by means of expansions of skin between the fore and hind limbs, fig. 154. Some Lemmings perform remarkable migrations, the impulse to which, unchecked by dangers or any surmountable obstacles, seems to be mechanical. Many Rodents build very artificial nests, and a few manifest their constructive instinct in association. In these inferior psychical manifestations we are reminded of Birds. Many Rodents hibernate like Reptiles. They are distributed over all continents. They have not been found in older deposits than eocene tertiary.

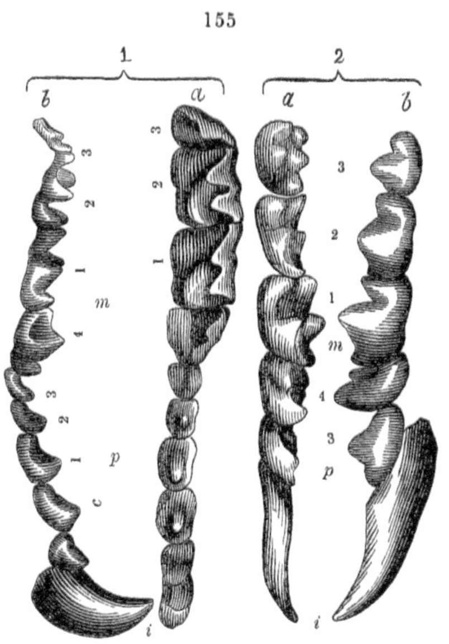

Dentition of a Shrew (*Sorex*). Magnified.

The transition from the Marsupials to the Rodents is made by the Wombats; and the transition from the Marsupials is made, by an equally easy step, through the smaller Opossums to the INSECTIVORA. This term is given to the order of small smooth-brained Mammals, the molar teeth of which are bristled with cusps, fig. 150, *m*, *p*, and are associated with canines and incisors: they are unguiculate, plantigrade, and pentadactyle, and they have complete clavicles. The testes pass periodically from the abdomen into a temporary scrotum, and are associated with large prostatic and vesicular glands: like most other *Lissencephala*, the Insectivora have a discoid or cup-shaped placenta. Their place and office in South America and Australia are fulfilled by Marsupialia; but true Insectivora exist in all the other continents.

[1] CLI.

The order CHEIROPTERA, with the exception of the modification of their digits for supporting the wide webs that serve as wings, fig. 156, repeat the chief characters of the Insectivora; but a few of the larger species are frugivorous and have corresponding modifications of the teeth and stomach. The mammæ are pectoral in position, and the penis is pendulous, in all Cheiroptera.

The most remarkable examples of periodically torpid Mammals are to be found in the terrestrial and volant Insectivora. The frugivorous Bats differ much in dentition from the true Cheiroptera, and would seem to conduct, through the Colugos or Flying Lemurs, directly to the Quadrumanous order. The

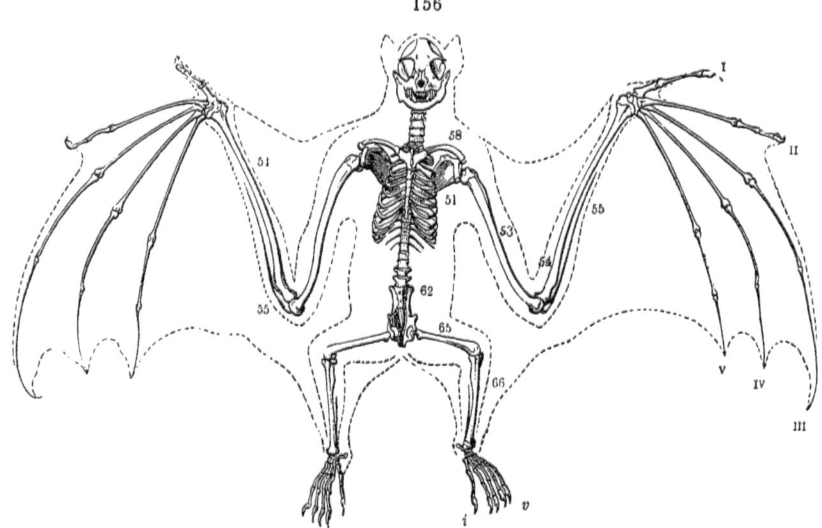

Skeleton of a Bat (*Pteropus*).

Cheiroptera are cosmopolitan. They have not been found in older deposits than eocene tertiary.

The order BRUTA (*Edentata* of Cuvier) includes two genera which are devoid of teeth, figs. 157 and 158; the rest possess those organs, which, however, have no true enamel, are never displaced by a second series, and are very rarely implanted in the premaxillary bones. All the species have very long and strong claws. The ischium as well as the ilium unites with the sacrum; the orbit is not divided from the temporal fossa. Besides the illustration of affinity to the oviparous Vertebrata which the Three-toed Sloths afford by the supernumerary cervical vertebræ supporting false ribs and by the convolution of the windpipe in the thorax, it may be remarked that the unusual number—three and twenty pairs—of ribs, forming a very long dorsal, with a short lumbar, region of the spine in the Two-toed Sloth, recalls a lacertine structure. The same tendency to an inferior type is shown by the abdominal

157

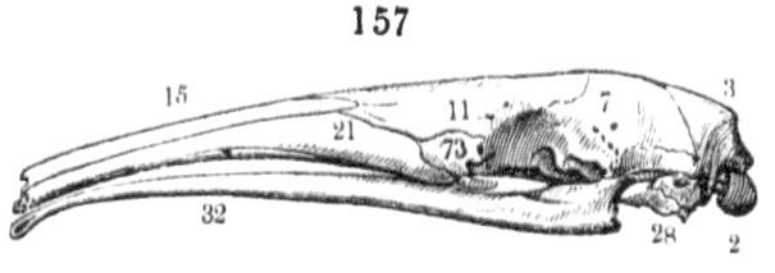

Skull of Anteater (*Myrmecophaga*).

testes, the single cloacal outlet, the low cerebral developement in all *Bruta*, by the bony scutes of the Armadillos and the horny scales of the Pangolins, fig. 158; by the absence of medullary canals in the long bones in the Sloths, and by the great tenacity of life and long-enduring irritability of the muscular fibre, in both the Sloths and Anteaters.

The order Bruta is but scantily represented at the present period. One genus, *Manis* or Pangolin, is common to Asia and Africa; the *Orycteropus* is peculiar to South Africa; the rest of the order, consisting of the genera *Myrmecophaga*, or true Anteaters, *Dasypus* or Armadillos, and *Bradypus* or Sloths, are confined to South America. The earliest known fossil of this order is of miocene age.[1]

158

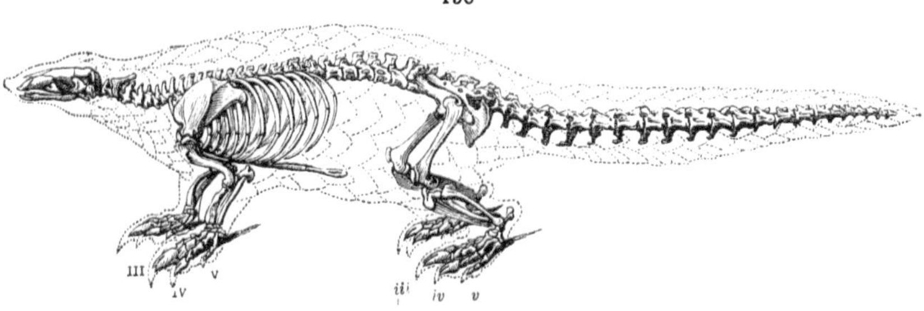

Skeleton of Scaly Anteater (*Manis*). LXXIII'.

In proceeding to consider the subdivisions of the Gyrencephala, we seem at first to descend in the scale in meeting with a group of animals in that subclass, having the shape and life of Fishes; but a high grade of mammalian organisation is masked beneath this form. The Gyrencephala are primarily subdivided, accord-

[1] CLI. t. v. pt. 1, p. 193.

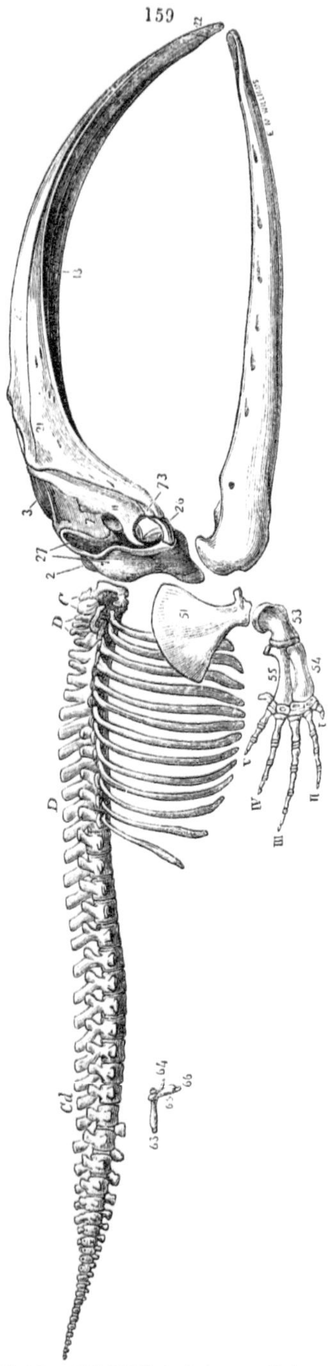

Skeleton of Right Whale (*Balæna mysticetus*). LXV'.

ing to modifications of the locomotive organs, into three series; viz. *Mutilata*, *Ungulata*, and *Unguiculata*, the maimed, the hoofed, and the clawed series; and these are of higher value than the ordinal divisions of the Lissencephala; just as those orders are of higher value than the representative families of the Marsupials.

The *Mutilata*, or maimed Mammals with folded brain, are so called because their hind-limbs seem, as it were, to have been amputated, fig. 159, 66; they possess only the pectoral pair of limbs, and these in the form of fins, ib. 54: the hind end of the trunk expands into a broad, horizontally flattened, tegumentary caudal fin. They have large brains with many and deep convolutions, are naked, and have neither neck, scrotum, nor external ears. Like the wingless group among Birds, the present includes species allied to, or derived from, different types.

The first order, called CETACEA, in this division are either edentulous, fig. 159, or monophyodont: the latter have teeth of one kind and usually of conical shape: the pectoral digits, ib. III., may have more than three phalanges. They are testiconda and have no 'vesiculæ seminales.' The mammæ are pudendal; the placenta is diffused; the external nostrils —single or double—are on the top of the head, and called spiracles or 'blow-holes.' They, for the most part, range the ocean;

though with certain geographical limits as respects species. They feed on fishes or marine animals. Some undoubted Cetacean fossils are of eocene age: and there are indications of the order in the upper oolitic period.[1]

160

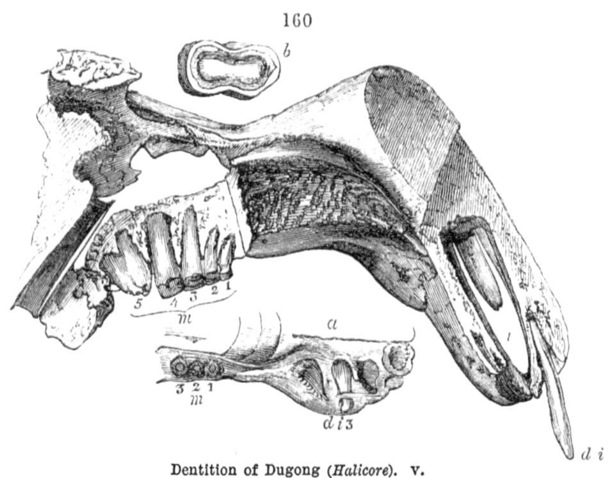

Dentition of Dugong (*Halicore*). v.

The second Mutilate order, called SIRENIA, have teeth of different kinds, incisors, *i*, fig. 160, which are preceded by milk-teeth, *d i*, and molars, *m*, with flattened or ridged crowns, adapted for vegetable food. No digit has phalanges in excess of the mammalian number, three. The nostrils are two, situated at the upper part of the snout; the lips are beset with stiff bristles; the mammæ are pectoral; the testes are abdominal, but are associated with vesiculæ seminales. The Sirenia exist near coasts or ascend large rivers; browsing on fuci, water plants, or the grass of the shore. The oldest known Sirenian is of miocene age. There is much in the organisation of this order that indicates its affinity to members of the succeeding division.

161

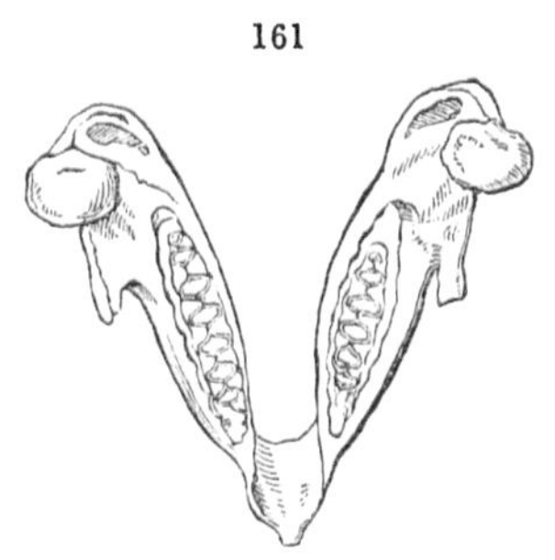

Molars of lower jaw, African Elephant.

In the *Ungulata* the four limbs are present, but that portion of the toe which touches the ground is incased in a hoof, figs. 162 and 163, which blunts its sensibility and deprives the foot of prehensile power. With the limbs restricted to support and

[1] XVIII', pp. XV. 520.

locomotion, the Ungulata have no clavicles; the fore-leg is prone: the molar teeth are massive, with inflected folds of enamel: they feed on vegetables.

A remarkable order, most of the members of which have passed away, is characterised by two incisors in the form of long tusks; in one genus (*Dinotherium*) projecting from the under jaw, in another genus (*Elephas*) from the upper jaw, fig. 162, *i*, and in some of the species of a third genus (*Mastodon*) from both jaws. There are no canines; the molars are few, large, and

162

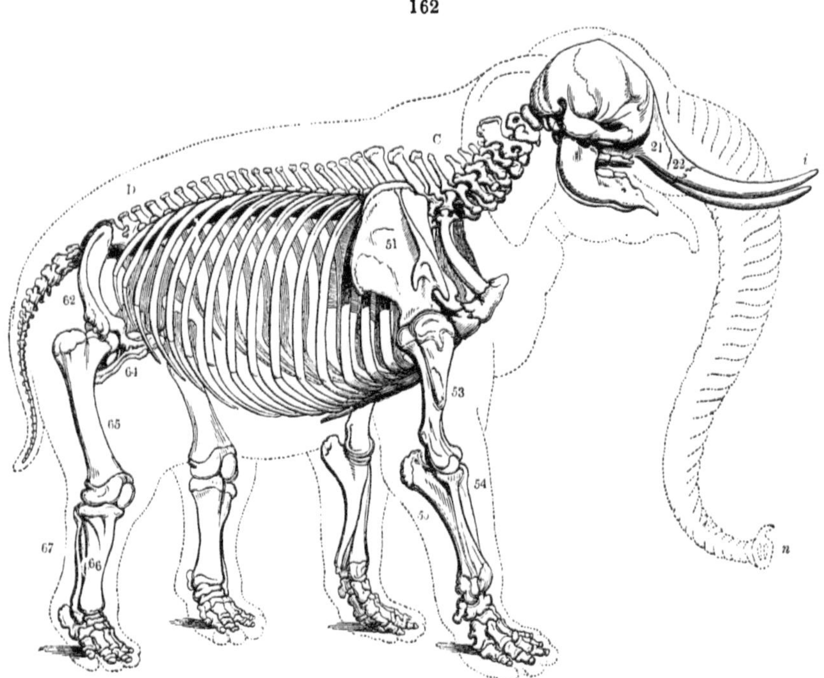

Skeleton of Elephant (*Elephas Indicus*)

transversely ridged, fig. 161, the ridges sometimes few, sometimes mammillate, often numerous and with every intermediate gradation. The nose is prolonged into a cylindrical trunk, flexible in all directions, highly sensitive, and terminated by a prehensile appendage like a finger, fig. 162, *n*: on this organ is founded the name PROBOSCIDIA given to the order. The feet are pentadactyle, but the digits are outwardly indicated only by divisions of the hoof; the testes are abdominal; the placenta is annular; [1] the mammæ are pectoral.

[1] Besides the annular placenta there is a subcircular villous patch at each pole of

The present order rests with the *Ungulata* mainly upon its hoofs: the dentition and some other particulars of the organisation of the Elephant, indicate an affinity to the Rodentia: the abdominal testes, the two precavals and exposed cerebellum, are characters of the inferior subclasses: but the cerebrum, concomitantly with the bulk of the mammal, is large and well convoluted, and the psychical qualities correspond. The earliest known evidences of Proboscidian Ungulates are from miocene strata.

The typical Ungulate quadrupeds are divided, according to the odd or even number of the toes, into PERISSODACTYLA and ARTIODACTYLA.[1] In the former the hoofs may be one (Horse) or three (*Rhinoceros*, fig. 163): in the latter the hoofs may be two (Giraffe), or four (*Hippopotamus*), or two functional and two rudimental (most Ruminants, fig. 164).

163

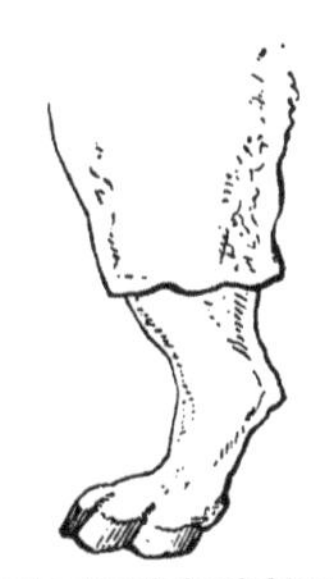

Perissodactyle hoofed limb
Hind leg, Rhinoceros.

In the Perissodactyle Ungulata—odd-toed in regard to the hind-foot in all, and with the fore-foot unsymmetrically tetradactyle in the Tapir—the dorso-lumbar vertebræ, fig. 165, C, D, differ in number in different species, but are never fewer than twenty-two; the femur has a third trochanter, ib. 65; and the medullary artery penetrates the back-part of its shaft. The fore-part of the astragalus is divided into two very unequal facets. The os magnum and the digitus medius which it supports are large, in some disproportionately so, and the digit is symmetrical: the same applies to the ectocuneiform and the digit which it supports in the hind-foot. If the species be horned, the horn is single: or, if there be two, they are placed on the median line of the head, one behind the other, each being thus an odd horn. The nasals expand posteriorly.[2] There is a well-developed post-tympanic process which is separated by the true mastoid from the paroccipital in the Horse, but unites with the lower part of the paroccipital in the Tapir, and seems to take the place of the mastoid in the Rhinoceros and Hyrax. The hinder half, or a larger proportion

164

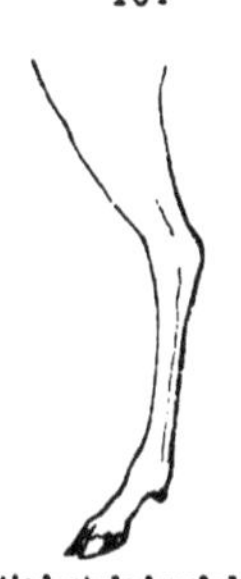

Artiodactyle hoofed limb:
Hind leg, Antelope.

the chorionic bag, by which it derived additional attachment to the uterus, in the Elephant. LXIII'. p. 347, pl. xvi.

[1] From περισσοδάκτυλος, qui digitos habet impares numero; and ἄρτιος, par, δάκτυλος, digitus.

[2] LXXI. p. 398.

of the palatines enters into the formation of the posterior nares, the oblique aperture of which commences in advance either of the last molar, or, as in most, of the penultimate one. The pterygoid process has a broad and thick base, and is perforated lengthwise by the ectocarotid. The crown of from one to three of the hinder premolars is as complex as those of the molars :[1] that of the last lower milk-molar is commonly bilobed. To these osteological and

165

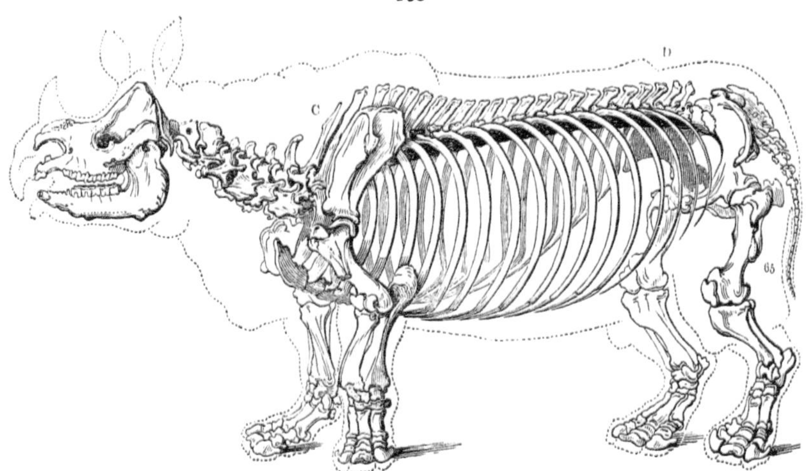

Perissodactyle skeleton (*Rhinoceros Indicus*). LXXIII'.

dental characters may be added some important modifications of internal structure, as, e.g., the simple form of the stomach and the capacious and sacculated cæcum, which equally evince the mutual affinities of the Perissodactyle hoofed quadrupeds, and their claims to be regarded as a natural group of the *Ungulata*. The placenta is replaced by a diffused vascular villosity of the chorion in all the recent genera of this order, excepting the little *Hyrax*, in which there is a localised annular placenta, with decidua, as in the Elephant. But the diffused placenta occurs in some genera of the next group, showing the inapplicability of that character to exact classification. The oldest known Perissodactyles are from the lowest tertiary strata. Many extinct genera, e.g. *Coryphodon*, *Pliolophus*, *Lophiodon*, *Tapirotherium*, *Palæotherium*, *Ancitherium*, *Hipparion*, *Acerotherium*, *Elasmotherium*, &c., have been discovered, which once linked together the now broken series of

[1] Some early tertiary extinct forms (*Pliolophus*, *Coryphodon*, *Lophiodon*) offered exceptions to this rule.

Perissodactyles, represented by the existing genera *Rhinoceros*, *Hyrax*, *Tapirus*, and *Equus*.

In the even-toed or 'artiodactyle' Ungulates, the dorso-lumbar vertebræ are the same in number, as a general rule, in all the species, being nineteen, fig. 166, *d*, *l*. The vertebral formulæ of the Artiodactyle skeletons show that the difference in the number of the so-called dorsal and lumbar vertebræ does not affect the

166

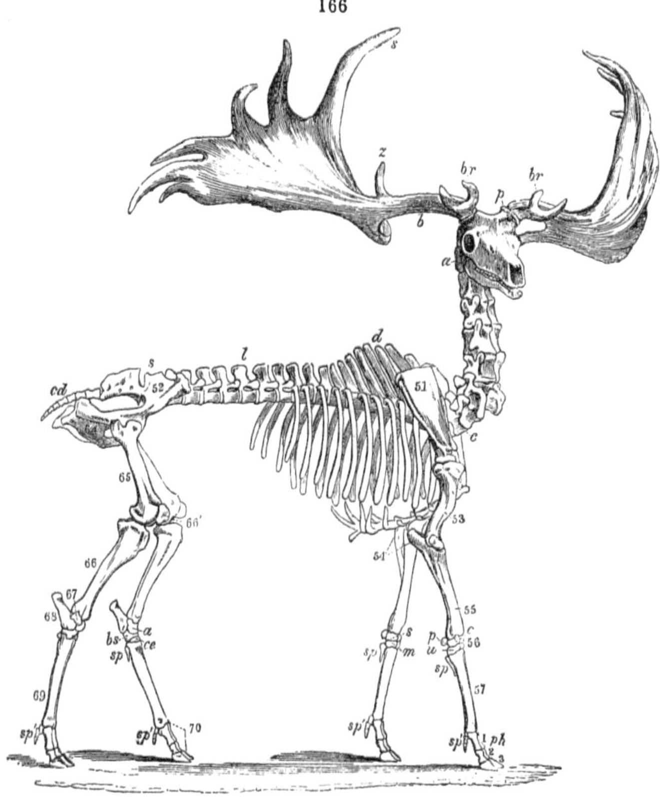

Artiodactyle skeleton (*Cervus Megaceros*). XVIII'.

number of the entire dorso-lumbar series: thus, the Indian Wild Boar has *d*. 13, *l*. 6=19; the Domestic Hog and the Peccari have *d*. 14, *l*. 5=19; the Hippopotamus has *d*. 15, *l*. 4=19; the Gnu and Aurochs have *d*. 14, *l*. 5=19; the Ox and most of the true Ruminants have *d*. 13, *l*. 6=19; the aberrant Ruminants have *d*. 12, *l*. 7=19. The natural character and affinities of the Artiodactyle group are further illustrated by the absence of the third trochanter in the femur, ib. 65, and by the place of perforation of the medullary artery at the fore and upper part of the shaft, as in the Hippopotamus, the Hog, and most of the Ruminants. The fore part of the astragalus is divided into two equal or sub-equal facets: the os magnum does not exceed, or is less than the unciforme, in the carpus; and the ectocuneiform is less, or not larger, than the cuboid, in the tarsus. The digit answering to the third in the pentadactyle foot is unsymmetrical, and forms, with that answering to the fourth, a symmetrical pair. If the species be horned, the horns form one pair or two pairs; they are never developed singly, of symmetrical form, from the median line. The post-tympanic does not project downward distinctly from the mastoid, nor supersede it, in any Artiodactyle; and the paroccipital always exceeds both those processes in length. The bony palate extends further back than in the Perissodactyles;[1] the hinder aperture of the nasal passages is more vertical and commences posterior to the last molar tooth. The base of the pterygoid process is not perforated by the ectocarotid artery. The crowns of the premolars are smaller and less complex than those of the true molars, usually representing half of such crown. The last milk-molar is trilobed.

To these osteological and dental characters may be added some modifications of internal structure, as, e.g., the complex form of the stomach in the Hippopotamus, Peccari, and Ruminants; the comparatively small and simple cæcum and the spirally folded colon in all Artiodactyles. The placenta is diffused in the Camel-tribe, Chevrotains,[2] and Non-ruminants; is cotyledonal in the true Ruminants. The oldest known Artiodactyles were non-ruminants, and from eocene beds. Many of the extinct genera, e. g. *Chœropotamus*, *Anthracotherium*, *Hyopotamus*, *Entelodon*, *Dichodon*, *Merycopotamus*, *Xiphodon*, *Dichobune*, *Anoplotherium*, *Microtherium*, &c., linked together the now broken series of Artiodactyles, represented by the existing genera, *Hippopotamus*, *Sus*, *Dicotyles*, *Camelus*, *Auchenia*, *Moschus*, *Camelopardalis*, *Cervus*, *Antelope*, *Ovis*, and *Bos*.

A well-marked, and at the present day very extensive subordinate group of the Artiodactyles, is called *Ruminantia*, in reference to the second mastication to which the food is subject after having been swallowed; the act of rumination requiring a peculiarly complicated form of stomach. The Ruminants have the 'cloven foot,' i.e. two hoofed digits on each foot forming a symmetrical pair, as by the cleavage of a single hoof; in most species two small supplementary hoofed toes are added, fig. 166, *sp.*

[1] LXXI'. p. 399. [2] CCXXXVI'. vol. ii. p. 135; and LXXII'.

The metacarpals of the two functional toes coalesce to form a single 'cannon-bone,' fig. 166, 57, as do the corresponding metatarsals, ib. 69. The Camel-tribe have the upper incisors reduced to a single pair; in the rest of the Ruminants the upper incisors are replaced by a callous pad, figs. 167, 168. The lower canines, fig. 168, *c*, are contiguous, and, save in the Camel-tribe, similar, to the six lower incisors, forming part of the same terminal series of eight teeth,

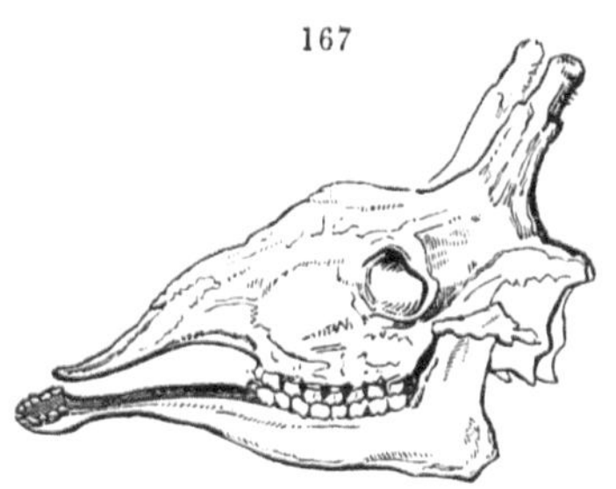

Ruminant skull, Giraffe.

between which and the molar series there is a wide interval. The true molars have their grinding surface marked by two double crescents, the convexity of which is turned inward in the upper, fig. 168, *, and outward in the under jaw, fig. 169, *.

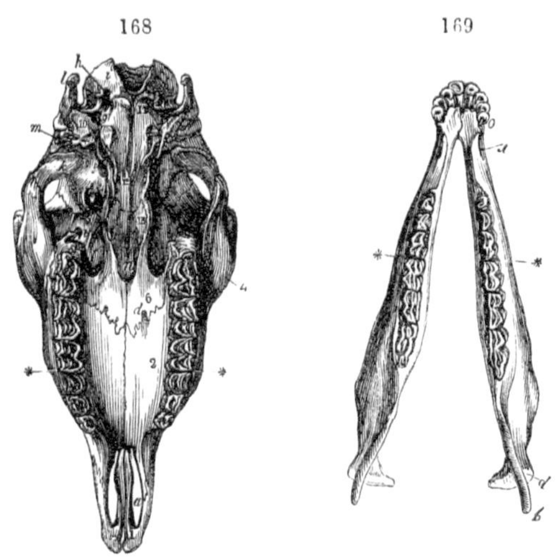

Ruminant dentition, Sheep.

Many fossil Artiodactyles, with similar molars (*Dichodon*, *Microtherium*, &c.), appear to have differed from the existing Ruminants chiefly by retaining structures which in them are transitory and embryonic, as, e.g., upper incisors and canines, first premolars, and separate metacarpal and metatarsal bones; these are among the lost links that once connected more intimately the Ruminants with the Hog and Hippopotamus.

The third division of the *Gyrencephala* enjoy a higher degree of the sense of touch through the greater number and mobility of the digits, and the smaller extent to which they are covered by horny matter. This substance forms a single plate, in the shape of a claw or nail, which is applied chiefly to one of the surfaces of the extremity of the digit, leaving the other, usually the lower, surface possessed of its tactile faculty, fig. 170; whence the name *Unguiculata*, which, in the present classification, is restricted to this group. All the species are 'diphyodont,' and the teeth have a simple investment of enamel.

170

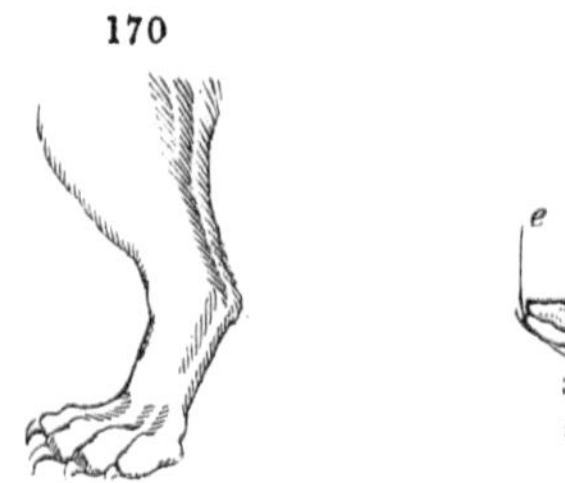

Unguiculate limb, Lion.

171

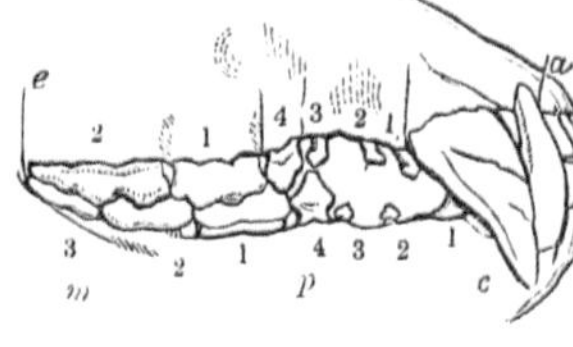

Carnivorous dentition, Bear.

The first order, CARNIVORA, includes the beasts of prey, properly so called. With the excèption of a few Seals, the incisors, fig. 171, *i*, are $\frac{3-3}{3-3}$ in number; the canines, ib. *c*, $\frac{1-1}{1-1}$, always longer than the other teeth, and usually exhibiting a full and perfect developement as lethal weapons; the molars, ib. *p*, *m*, graduate from a trenchant to a tuberculate form, in proportion as

172 173

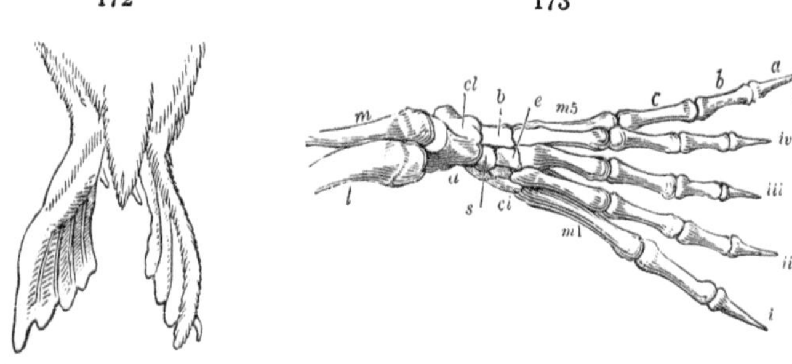

Pinnigrade foot: Hind limbs, Seal. Bones of do.

the diet deviates from one strictly of flesh to one of a more miscellaneous kind. The clavicle is rudimental or absent; the innermost digit is often stunted or absent; there are no vesiculæ seminales; the teats are abdominal; the placenta is zonular. The Carnivora are divided, according to modifications of the limbs, into 'pinnigrades,' 'plantigrades,' and 'digitigrades.' In the

Pinnigrades (Walrus, Seal-tribe) both fore and hind feet are short, and expanded into broad, webbed paddles for swimming, fig. 173, the hinder ones being fettered by continuation of integument to the tail, fig. 172. In the Plantigrades (Bear-tribe) the whole or nearly the whole of the hind foot, fig. 174, forms a sole, and rests

174

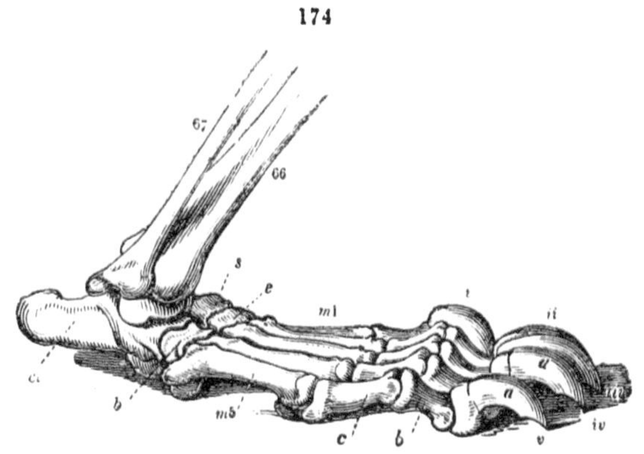

Plantigrade foot, hind limb, Bear.

on the ground. In the Digitigrades (Cat-tribe, Dog-tribe, &c.) only the toes touch the ground, the heel, *cl*, being much raised, fig. 175.

175

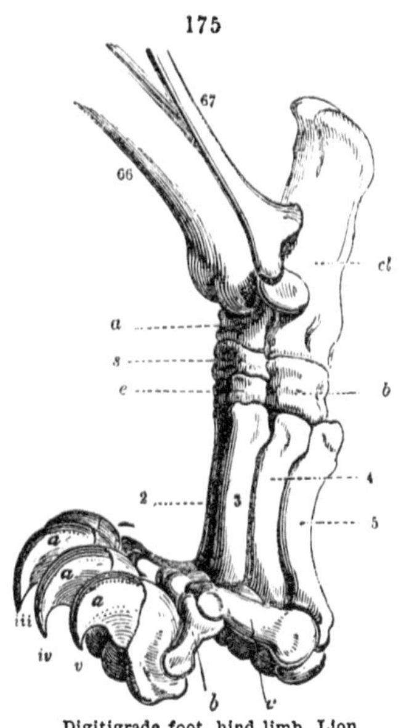

Digitigrade foot, hind limb, Lion.

It has been usual to place the Plantigrades at the head of the Carnivora, apparently because the higher order, Quadrumana, is plantigrade; but the affinities of the Bear, as evidenced by internal structure, e. g. the renal and genital organs, are closer to the Seal-tribe; the broader and flatter pentadactyle foot of the plantigrade is nearer in form to the flipper of the Seal than is the more perfect digitigrade, retractile-clawed, long and narrow hind foot of the feline quadruped, which is the highest and most typical of the Carnivora. The oldest known species of the order are of eocene tertiary date.

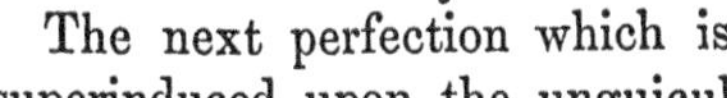

The next perfection which is superinduced upon the unguiculate limb is such a modification in the size, shape, position, and direction of the innermost digit, that it can be opposed, as a thumb, to the other digits, thus constituting what is termed a 'hand.' Those Unguiculates which have both fore and hind limbs so modified, or at least the hind limbs, figs. 176, 180, form the order QUADRUMANA. The incisors are commonly $\frac{2-2}{2-2}$, and the molars $\frac{3-3}{3-3}$, broad and tuberculate; they have perfect clavicles, an os penis, pectoral

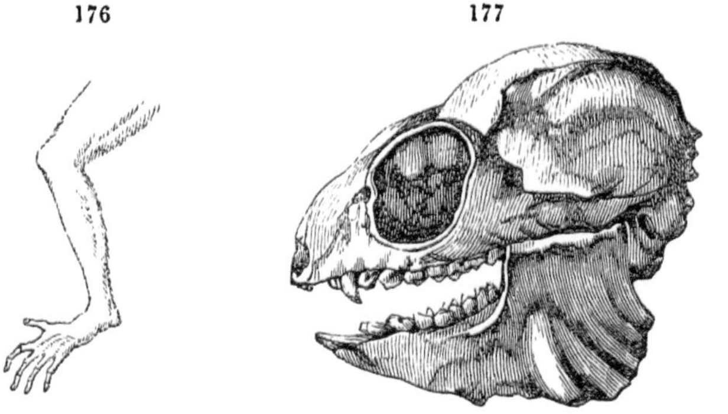
176 Pelvic limb, Ape.

177 Dentition of Woolly Lemur.

mammæ, vesicular and prostatic glands, a simple or slightly bifid uterus, and a discoid, sometimes double, placenta. The Quadrumana have a well-marked threefold geographical as well as structural division. The Strepsirhines are those with curved or twisted terminal nostrils, with much-modified incisors, commonly $\frac{3-3}{3-3}$; premolars $\frac{3-3}{3-3}$, *Lichanotus*, fig. 177, or $\frac{2-2}{2-2}$ in number, and

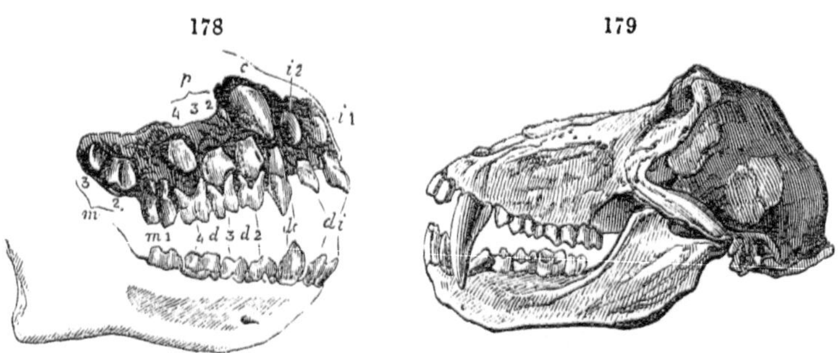

178 Platyrhine dentition (*Cebus*).

179 Catarhine dentition (*Papio*).

molars with sharp tubercles; the second digit of the hind limb has a claw. This group includes the Galagos, Pottos, Loris, Aye-Ayes, Indris, and the true Lemurs; the three latter genera being restricted to Madagascar, whence the group diverges in one direction to the continent of Africa, in the other to the Indian Archipelago. The Platyrhines are Quadrumana with the nostrils simple, subterminal, and wide apart; premolars $\frac{3-3}{3-3}$ in number, the molars with blunt tubercles, fig. 178; the thumbs of the fore-hands not opposable, or wanting; the tail in most prehensile; they are peculiar to South America. The Catarhines have the nostrils oblique and approximated below, and opening above and behind the muzzle: the premolars are $\frac{2-2}{2-2}$ in number, fig. 179; the thumb of the forehand is opposable. They are restricted to the Old World, and, save a single species on the rock of Gibraltar, to Africa and Asia. The highest organised family of Catarhines is tailless, and offers in the Gorilla, or, as some contend, the Siamang, fig. 180, the nearest approach to the human type.

180

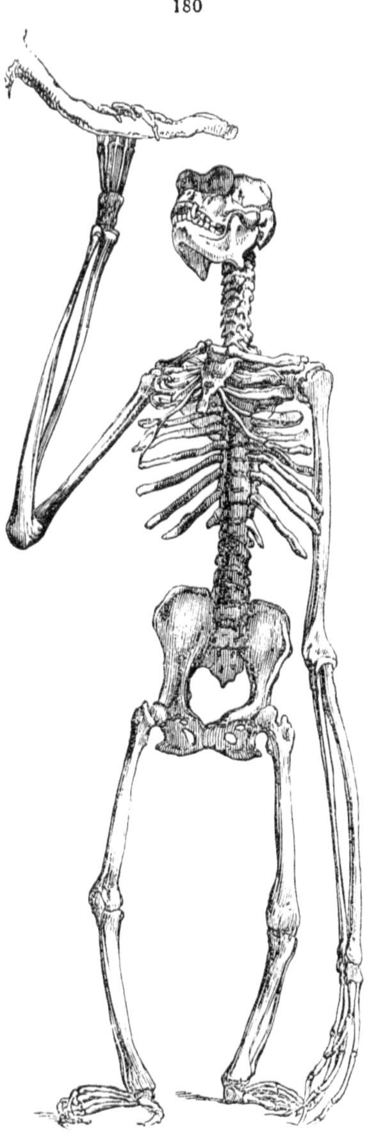

Quadrumanous skeleton, Siamang (*Hylobates syndactylus*).

In all the tailless Apes the pelvic limbs are short, and, like the longer pectoral ones, are organised for grasping. The pelvis is long and narrow; the spine shows one curve, and articulates with the hinder part of the skull. There is a sexual distinction in the teeth, the canines being long and laniariform in the males. All Quadrumana are clothed with hair. The oldest known species of the Quadrumanous order are of miocene date.

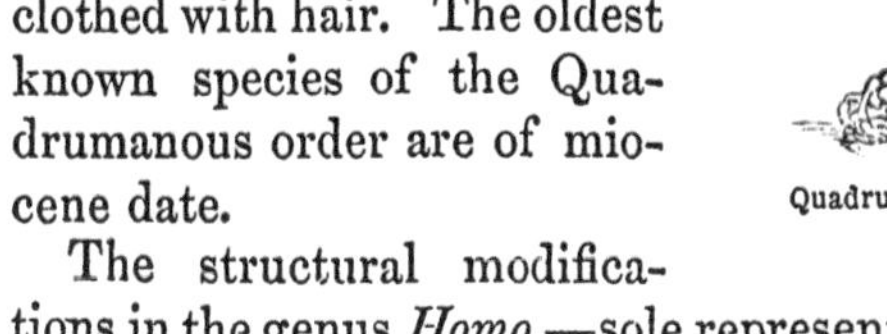

The structural modifications in the genus *Homo*,—sole representative of the *Archencephala*,

—more especially those of the pelvic limbs, by which the erect stature and bipedal gait are maintained,—are such as to claim for MAN, on merely external zoological characters, ordinal distinction, at least. The consequences of the liberation of one pair of limbs from all service in station and progression, due to the extreme modification of the other pair for the exclusive discharge of those functions, are greater, and involve a superior number and quality of powers, than those resulting from the change of an ungulate into an unguiculate condition of limb: and they demand, therefore, an equivalent value in a zoological system. But, as I have elsewhere argued, Man's psychological powers, in association with his extraordinarily developed brain, entitle the group which he represents to rank with the primary divisions of the class *Mammalia* founded on cerebral characters. In this subclass Man forms but one genus, *Homo*, and that genus but one order, called BIMANA, on account of the opposable thumb being restricted to the upper pair of limbs. In every Ape the pelvic limb is terminated by a 'hand,' fig. 176; in every Man by a 'foot,' fig. 181.[1] In *Bimana* the testes are scrotal; their serous sac does not communicate with the abdomen; they are associated with vesicular and prostatic glands. The penis is pendulous, without bone, and the prepuce has a frænum. The mammæ are pectoral. The placenta is a single, subcircular, cellulo-vascular, discoid body.

181

Pelvic limb, Man.

Man is naked, and is the sole terrestrial Mammal in that predicament: of the partial growths of hair, the chief protects the head, and is distinctive of sex.

[1] The fact of the homologous bones being determinable in the pelvic limb, as in other parts of the skeleton, of Mammals, does not make the grasping organ of the Ape, fig. 176 the less a 'hand,' nor does it prove the lacerating organ of the Lion, fig. 175, to be no 'paw,' nor the swimming organ of the Seal, fig. 172, to be no 'fin.' Prof. Huxley, however, by pointing out those homologies between Man and the Ape, under colour of a new element in the question, probably persuaded the 'working men' for whom, as 'Government Professor' in the School of Science, he selected such subject of instruction, that it was an important argument in favour of their Ape-origin. So speciously indeed was this old elementary fact in zootomy set forth, that the propounder succeeded in deceiving some non-anatomical authors into a belief that he had really made a discovery. See CRAWFURD, 'Antiquity of Man,' 8vo. 1863: 'Prof. Huxley has very satisfactorily shown that the designation of "quadrumane," or four-handed, is incorrectly applied to the family of monkeys. Their feet are real feet, although prehensile ones; but the upper limbs are true hands,' &c., p. 18; also LYELL, 'Antiquity of Man,' 8vo. 1863, p. 476 et seq.: whom I would refer to CUVIER, 'Leçons d'Anatomie Comparée,' 8vo. 1805, tom. i. p. 376, 'Des os du coude-pied.'

The dentition of the genus *Homo* is reduced to thirty-two teeth by the suppression of the outer incisor and the first two premolars of the typical series on each side of both jaws, the dental formula being:—

$$i.\ \frac{2-2}{2-2},\ c.\ \frac{1-1}{1-1},\ p.\ \frac{2-2}{2-2},\ m.\ \frac{3-3}{3-3} = 32.$$

The teeth are of equal length, show no sexual distinctions, and there is no break in the series; they are subservient in Man not only to alimentation, but to beauty and to speech, fig. 182.

The human foot is broad, plantigrade, with the sole, not inverted as in *Quadrumana*, but applied flat to the ground; the

182

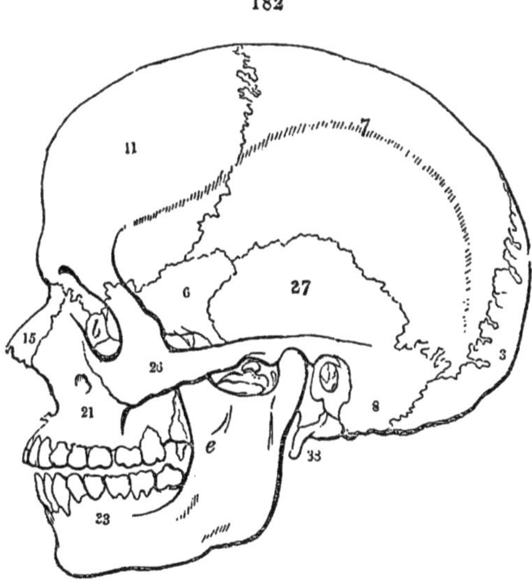

Bimanous dentition (*Homo*).

leg, fig. 183, 66, bears vertically on the foot; the heel, 68, is expanded beneath; the toes are short, but with the innermost, *i*, longer and much larger than the rest, forming a 'hallux' or great toe, which is placed on the same line with, and cannot be opposed to, the other toes: the pelvis, 62, 63, is short, broad, and wide, keeping well apart the thighs; and the neck of the femur is long, and forms an open angle with the shaft, 65, increasing the basis of support for the trunk. The whole vertebral column, with its slight alternate curves, and the well-poised, short, but capacious subglobular skull, are in like harmony with the requirements of the erect position. The widely-separated shoulders, with broad scapulæ and complete clavicles, 58, give a favourable position to the upper limbs, now liberated from the service of locomotion, with complex joints for rotatory as well as flexile movements, and terminated by a hand of matchless perfection of structure the fit instrument for executing the behests of a rational intelligence and a free will. Hereby, though naked, Man can clothe himself, and rival all native vestments in warmth and beauty; though defenceless, Man can arm himself with every variety of weapon, and become the most terribly destructive of animals. Thus he fulfils his destiny as the master of this earth, and of the lower Creation.

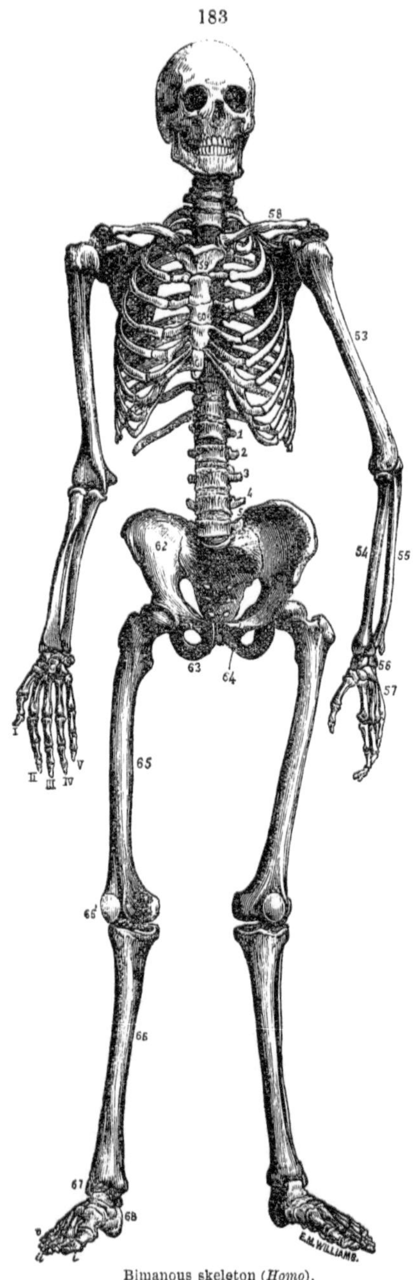

Bimanous skeleton (*Homo*).

The system of Cuvier being still in use in some estimable works, and the one according to which groups of Mammals are most commonly referred to in physiological and paleontological propositions, an outline thereof, as applied to that class, is here appended, with a similar outline of the classification adopted in the present work.

TABLE OF THE SUBCLASSES AND ORDERS OF THE MAMMALIA, ACCORDING TO CUVIER.[1]

CLASS	SUBCLASS		ORDER	GENUS OR FAMILY	EXAMPLE
MAMMALIA	UNGUICULATA	With three kinds of teeth	BIMANA	*Homo*	Man
			QUADRUMANA	*Catarrhina*	Ape
				Platyrrhina	Marmoset
				Strepsirrhina	Lemur
			CARNARIA [2]	*Cheiroptera*	Bat
				Insectivora	Hedgehog
					Shrew
					Mole
				Carnivora	Bear
					Dog
					Seal
			MARSUPIALIA	*Didelphys*	Opossum
				Phalangista	Phalanger
				Macropus	Kangaroo
				Phascolomys	Wombat
		Without canines	RODENTIA	*Claviculata*	Rat
				Non-claviculata	Hare
		Without incisors	EDENTATA	*Bradypus*	Sloth
				Dasypus	Armadillo
				Myrmecophaga	Anteater
				Monotremata	Echidna
					Ornithorhynchus
	UNGULATA		PACHYDERMATA	*Proboscidia*	Elephant
				Ordinaria	Hog
					Tapir
				Solidungula	Horse
			RUMINANTIA		Sheep
	MUTILATA ('point du tout d'extrémités postérieures')		CETACEA	*Herbivora*	Dugong
				Ordinaria	Whale

[1] xxiv. vol. i. p. 65.

[2] Written '*Carnassiers*' by Cuvier, ib.

TABLE OF THE SUBCLASSES AND ORDERS OF THE MAMMALIA, ACCORDING TO THE CEREBRAL SYSTEM.

CLASS	SUBCLASS		ORDER	GENUS OR FAMILY	EXAMPLE
MAMMALIA	ARCHENCEPHALA		BIMANA	*Homo*	Man
	GYRENCEPHALA	Unguiculata	QUADRUMANA	*Catarrhina*	Ape
				Platyrrhina	Marmoset
				Strepsirhina	Lemur
			CARNIVORA	*Digitigrada*	Dog
				Plantigrada	Bear
				Pinnigrada	Seal
		Ungulata	ARTIODACTYLA	*Omnivora*	Hog
				Ruminantia	Sheep
			PERISSODACTYLA	*Solidungula*	Horse
				Multungula	Tapir
			PROBOSCIDIA	*Elephas*	Elephant
				Dinotherium	Dinothere
		Mutilata	SIRENIA	*Manatus*	Sea-cow
				Halicore	Dugong
			CETACEA	*Delphinidæ*	Porpoise
				Balænidæ	Whale
	LISSENCEPHALA		BRUTA	*Bradypodidæ*	Sloth
				Dasypodidæ	Armadillo
				Edentula	Anteater
			CHEIROPTERA	*Frugivora*	Roussette
				Insectivora	Bat
			INSECTIVORA	*Talpidæ*	Mole
				Erinaceidæ	Hedgehog
				Soricidæ	Shrew
			RODENTIA	*Non-claviculata*	Hare
				Claviculata	Rat
	LYENCEPHALA		MARSUPIALIA	*Rhizophaga*	Wombat
				Poephaga	Kangaroo
				Carpophaga	Phalanger
				Entomophaga	Opossum
			MONOTREMATA	*Echidna*	Echidna
				Ornithorhynchus	Duck-mole

CHAPTER XXVI.

OSSEOUS SYSTEM OF MAMMALIA.

§ 175. *General Characters of the Skeleton.*—The osseous tissue and the bone-cells characteristic of it in the higher members of the Mammalian class are shown in Vol. I. p. 23, figs. 14, 15. In the *Lyencephala* (*Ornithorhynchus, Echidna,* Kangaroo, Rat, Beaver, Sloth, Hedgehog, Mole),[1] the Haversian canals resemble those of Birds in their smaller relative size, as do the bone-cells in the number and peculiar branchings of their canaliculi, compared with higher *Mammalia*; in these the radiated disposition of the canaliculi, concomitantly with the shorter and wider form of the cells, becomes more marked, as shown in fig. 14, Vol. I. In the larger Cetacea the bone-cells have a larger size and less regular shape, and send off long branching canaliculi.[2] The osseous tissue in Mammals is less dense and compact than in Birds: the long bones have medullary cavities, as a rule, relatively larger than in *Reptilia*, smaller and with thicker walls than the homologous pneumatic cavities in Birds. In the *Cetacea* and the Sloths, recent and extinct, the long, like the other, bones are solid, the central tissue being cancellous: in the *Sirenia* the bone of the thick ribs is dense and compact throughout: the hardest bone in the present class is that which is accordingly termed 'petrosal,' especially in the Whale-tribe, in which its specific gravity reaches 2·433, that of ivory being 1·744.

The proportion of the Mammalian skeleton which is pneumatic is noticed in Vol. I. p. 25. The vertebral bodies and the limb-bones have the articular surfaces, in the growing state, supported on distinct plates, called 'epiphyses,' which usually coalesce with the rest of the bone, at maturity. Examples of the exoskeleton are seen in the Armadillos and their huge extinct congeners the Glyptodons: small detached bony nodules were also developed in parts of the thick tegument of the Megatherioids.[3] The lacrymal is properly a mucous scale-bone. The bone of the heart

[1] ccxciii. pls. xi. and xii. [2] Ib. p. 151. [3] Burmeister, *MS.*

in large Ruminants is referable to the 'splanchnoskeleton;' the ossified tendons in some small Musk-Deer to the 'scleroskeleton.'

184

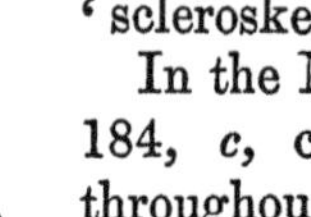

Mammalian type vertebra, from the thorax.

In the Mammalian class the centrum, figs. 141, 184, *c*, coalesces with the neural arch, ib. *n*, throughout the vertebræ of the trunk. In the seven anterior vertebræ, fig. 185, 1–7, the pleurapophyses are short and commonly coalesce with the centrum and diapophysis, circumscribing the lateral foramina for the 'vertebral' arteries. In the Monotremes they retain, as in Reptiles, their individuality, fig. 186, *a*. In *Cetacea* the interspace between the cervical par- and di-apophyses is not always closed by bone. Occasionally the pleurapophyses of the seventh, fig. 185, A, *b*, and, more rarely, also of the sixth, *a*, vertebræ, manifest their rib-like nature by increase of length, and freedom of articulation, even in Man; but these segments are not completed by the hæmapophyses and hæmal spine. This resumption of type takes place in the eighth vertebra, ib. *c*, *d*; and the dorsal series of vertebræ here begins, as a rule, in Mammals. The most marked exception occurs in the Ai (*Bradypus tridactylus*); and if the vertebræ, fig. 185, B, 8, 9, supporting the pleurapophyses, *a*, *b*, be regarded as homologous with the first two dorsals in other Mammals, the exception is so far saved: but the presence of short pleurapophyses in all the cervical vertebræ and their occasional developement in the last two, as in fig. 185, A, support the recognition of the tenth vertebra in the Three-toed Sloth, ib. B, 10, *c*, *d*, as the first dorsal. The pleurapophyses of the dorsal ver-

185

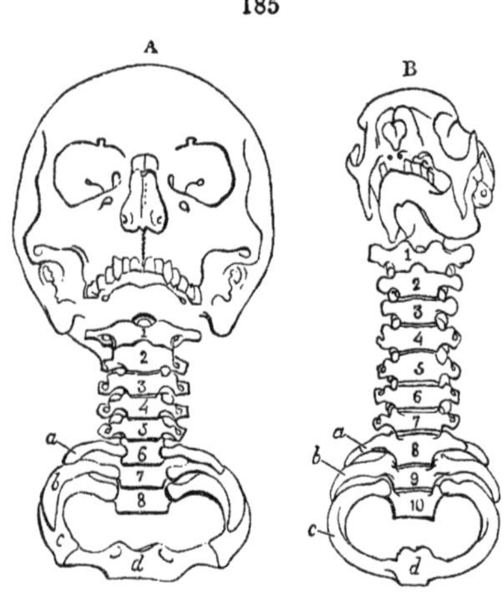

Cervical vertebræ, A Man, B Sloth.

186

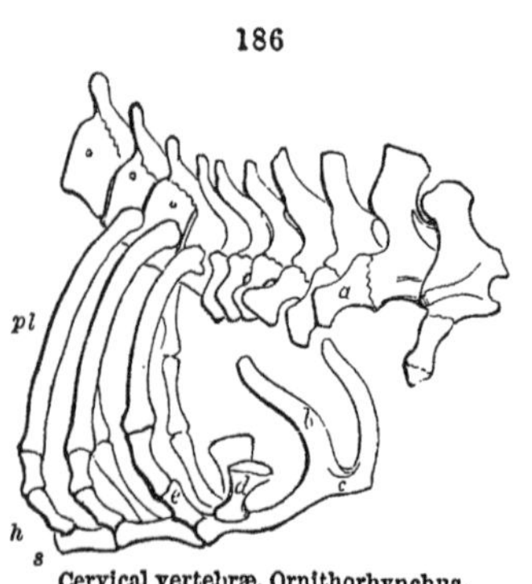

Cervical vertebræ, Ornithorhynchus.

tebræ, figs. 184, and 186, *pl*, are subject to slight displacement, and their articulations, like those of the neurapophyses in the Bird's sacrum, extend over the interspace between their own and a contiguous centrum. The hæmapophyses, ib. *h*, are rarely ossified: the exceptions occur in the lowest subclasses (Duck-Mole, Armadillo, Sloth): in the Monotremes a portion of cartilage intervenes between the pleur- and hæmapophyses. Some of the posterior hæmapophyses have no hæmal spine, but terminate freely, fig. 187, 7, or in connection with each other. The segments typically completed, as in fig. 184, are called 'vertebræ with true ribs,' those not so completed 'vertebræ with false ribs,' in Anthropotomy.

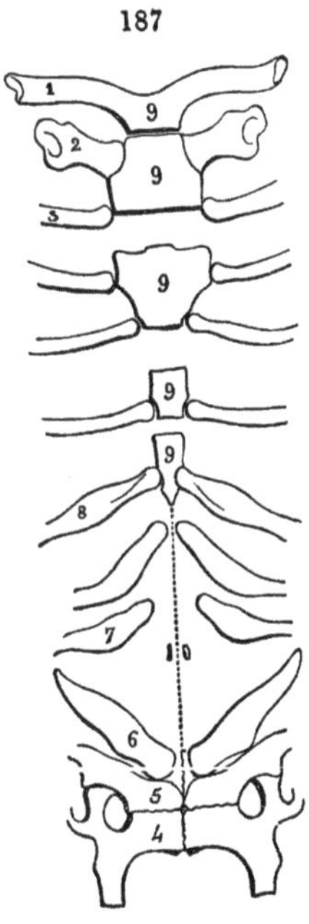

Hæmapophyses and spines of trunk, Mammal.

The hæmal spine of each thoracic segment is separately developed. They commonly remain distinct, fig. 187, 9, 9, forming a chain of ossicles, answering in number to those anterior dorsal segments which they complete: they coalesce with each other in some Mammals, and form collectively the 'sternum.' Only in Monotremes is there an episternum, figs. 186, *b*, *c*, 187, 9, *i*, or hæmal spine of a cervical segment, to which the clavicles articulate. As the dorsal vertebræ recede in position the pleurapophyses become shorter, return to their proper segment, and usually become appended to its diapophysis. When it becomes confluent therewith, or replaces that process, the 'dorsal' series ends and the 'lumbar' one, figs. 166, *l*, 183, 1–5, begins. These vertebræ are commonly more numerous in Mammals than in Reptiles. Their hæmapophyses—the abdominal ribs of Reptiles—are represented by the 'intersectiones tendineæ musculi recti,' &c., the lowest pair are partially ossified as 'marsupial bones,' fig. 187, 6, in *Lyencephala*. In the *Mutilata* the 'sacrum' is defined by the reappearance of the ossified hæmapophyses, fig. 159, 63, of a segment at the end of the trunk. In *Cetacea* it is suspended beneath its segment, as in Fishes, and may support some rudiment of a pelvic or ventral fin, ib. 65, 66.

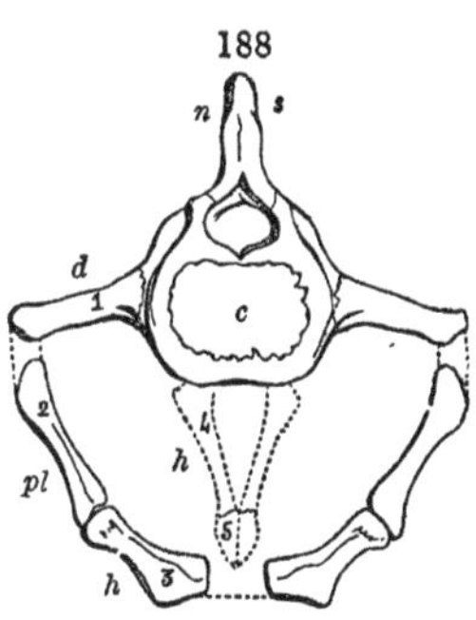

Type of pelvic and caudal segment, Mammal.

In *Sirenia*, the hæmapophyses, fig. 188, *h*, 3, are connected by pleurapophyses, ib. *pl*, 2, completing the hæmal arch with the diapophyses *d*, 1; and the Batrachian condition of pelvis (Vol. I. p. 48, p. 163, fig. 101, D) is resumed. In the rest of the class two or more segments following the lumbar series become, like those in the head, the seat of modifications by anchylosis of the centrums, flattening and broadening with an expanse of the neural canal: and with these modifications is associated great developement of the hæmal arches of two of those segments, fig. 187, 4 and 5, which, therefore, have got special names, as 'ischium,' ib. 4, and 'pubis,' ib. 5. These are, however, connected with their respective segments by a concomitant expanse of the single pleurapophysial element, fig. 188, *pl*, which, so modified, has the name of 'ilium,' and in some *Lyencephala* (Sloths, Megatherioids, Armadillos) resembles that bone in Birds, by the number of sacral segments with which it articulates or coalesces.

The caudal segments in Mammals are characterised by the abrupt cessation of the pleurapophysial developement forming the ilium, by the retention of the riblet, or beginning of the pleurapophysis, anchylosed, as a diapophysis, fig. 188, *d*, and by the approximation of the hæmapophyses, ib. *h*, 3, to the under surface of the centrum, *c*, as at *h*, 4, the divergent bases articulating therewith, and the apices converging to unite with, or develope, a hæmal spine, ib. 5. The wider pelvic hæmal canal encompassed terminal parts of the generative and intestinal canals; the narrower caudal one has only to defend the main blood-vessels of the tail. The terminal caudal vertebræ are progressively reduced in size and complexity, and vary greatly in number: anchylosis is an exception (*Dasypus*, e.g.) in this region.

§ 176. *General Characters of the Skull.*—Pursuing the survey of the Mammalian modifications of the Vertebrate archetype as they appear in the segments of the skeleton forming the skull, with the light of the stage of developement manifested in an immature Mammal when a certain growth has proceeded from the several points of ossification established in the primordial membranous and cartilaginous basis, we find that the neural arch of the occipital vertebra, fig. 189, NI, 1, 2, and 3, agrees with that of the Bird and Crocodile in the connation of the diapophysis, 4, with the neurapophysis, 2; but the process, called 'paroccipital,' now descends from the lower part of the arch, and, in many Mammals, is of great length. An articular condyle is developed from each neurapophysis, 2, which articulates with the concave anterior zygapophysis of the atlas, and is the homotype of the posterior zygapophysis in the trunk-vertebræ. The centrum, 1, is reduced to a compressed plate, and its hinder articular surface is not more developed than is the front one of the centrum of the atlas, with which it is connected by ligament. The expanse of

189

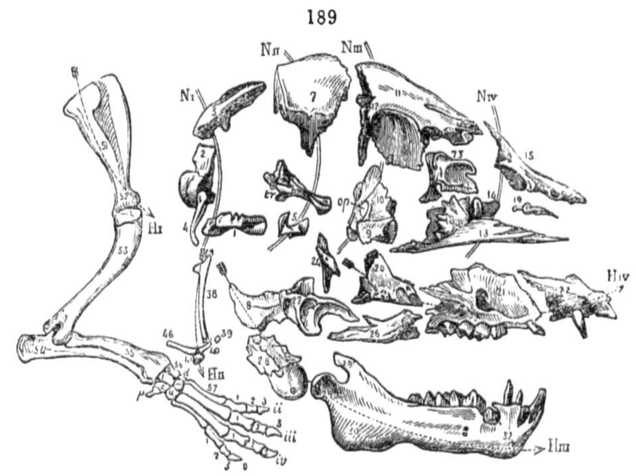

Side view of cranial vertebræ and appendages, Hog.

the occipital spine, 3, has been governed by the superior developement of the cerebellum in the Mammalian class.

The hæmal arch of the occipital vertebra is here represented, like those of the cervical vertebræ, by the pleurapophysial elements only; but these are developed into broad triangular plates with outstanding processes: that called 'spine,' 51, is exogenous; but that called 'coracoid' is developed from an independent osseous centre, which is a rudiment of the hæmapophysis, coalesces with the pleurapophysis, and, in the present class, only attains its normal proportions, completing the arch at figs. 186, *d*, 187, 2, with the hæmal spine, ib. 9, in the Monotremes. The diverging appendage (fore-limb, 53–57) of this arch, though retaining the general features of its primitive radiated form, has been the seat of great developement and much modification and adjustment of its different subdivisions in relation to the locomotive office it is now called upon to perform.

With the exception of this excess of developement of the appendage, the defective developement and displacement of the hæmal arch, and the coalescence of the diapophyses in the neural arch, there are few points of resemblance which are not sufficiently salient between the segment represented by the bones, NI, 1, 2, and 3, in the Mammal, and that so marked in the Fish, Vol. I.

fig. 81. And, if the interpretation of the more normal or archetypal condition of this segment in the lower Vertebrate animal, fig. 101, A, Vol. I., be accepted, so also must be the explanation here given of the nature of the modifications of the special homologues of the constituents of the occipital segment by which that archetype is masked in the Mammal. A single nerve supplies the appendage, 53–57, in *Protopterus*; subsequent developement of that appendage in higher forms presses more nerves from other centres into its service; these do not originate the complex conditions calling for them. And if the simple limb, fig. 101, A, 53–57, be the special homologue of the complex one, fig. 189, 53–57, neither the number of nerves, of vessels, or of terminal rays can affect the conclusions deducible from fig. 101, as to its general nature in relation to the Vertebrate archetype.

In the second segment of the skull, NII, the centrum, 5, is long distinct from both 1 and 9; and the hæmal arch (hyoid bone) retains its natural connection with the rest of the segment, and by means of a more complete developement of the pleurapophyses, 38, than in any of the inferior air-breathing Vertebrates. In the Hog, as in other Mammals, may be separated, without artificial division of any compound bone, the entire parietal segment, but with it is brought away the petrified capsule of the acoustic organ and the anchylosed distal piece, 27, of the maxillary appendage, which more or less conceals the typical character of the neural arch of the parietal vertebra in every Mammal: least so, however, in the Monotremes and Ruminants. The neurapophyses, 6, of the parietal vertebræ have coalesced with the centrum, 5, but retain much of the proportions they present in the cold-blooded classes; for the mesencephalic segment of the brain is, in fact, but little more developed in the Mammal: they are notched in the present example, but are perforated in the Sheep, by the larger divisions of the trigeminal, and they send down an exogenous process, which articulates and sometimes coalesces with the appendage, 24, of the palato-maxillary arch, and with the pleurapophysis, 20, of the same arch. The neural spine, 7, always developed from a pair of centres in Mammals, often vastly expanded, and sometimes complicated with a third, intercalary or interparietal osseous piece, in subserviency to the large size of the prosencephalon, is occasionally uplifted and removed from the neurapophyses by the interposed squamous expansion of the bone, 27; but this, which reminds one of the occasional separation of the neural arch from the centrum of the atlas in Fishes, is a rare modification in the Mammalian class. The diapophysis, 8, always commences as an autogenous element by a distinct centre of ossification; in most Mammals it speedily coalesces with the petrosal, but not in the Babyroussa,[1] e.g.: it usually coalesces with the squamosal, 27, as in the Hog; but retains its distinctness in the Echidna; its apophysial character is usually well-marked, and it is known as the 'mastoid process' in Anthropotomy. In most Mammals the pleurapophysis, 38, retains its primitive independency and rib-like form, with usually the 'head' and 'tubercle;' but by reason of its arrested growth it has been called 'styloid' bone or process. Sometimes it is separated from the short hæmapophysis, 40, by a long ligamentous tract, sometimes is immediately articulated with it, or by an intervening piece. The hæmal spine, 41, is usually small, and always single. The rudiments of hypobranchial elements, 46, are retained as diverging appendages of the parieto-hæmal arch in all Mammals, and have received the special names of 'posterior cornua,' or 'thyrohyals,' from their subservient relationship to the larynx.

In the frontal segment, NIII, the centrum, 9, and neurapophyses, 10, very early coalesce. Two separate osseous centres mark out the body, and each neurapophysis has its distinct centre, the optic foramina, *op*, being first surrounded by the course of the ossification from these points. The superior developement of the neurapophysial plates, 10, as compared with those of the parietal vertebra, 6, in most Mammals, harmonises with the greater developement of the prosencephalon; but the chief bulk of this segment is protected by the expanded spines of the frontal, 11, and parietal, 7, vertebræ, and the intercalated squamosal, 27. This appendicular piece not only fulfils some of the functions of the proper cranial neurapophyses, but, likewise, the normal office of the frontal pleurapophysis, 28, in the support, viz., of the distal elements of the hæmal arch, 29–32, which now articulate directly with 27, in place of 28, as in all oviparous Vetebrates. The true pleurapophysis of the frontal vertebra, 28, is almost restricted in the Mammalian class to functions in subserviency to the organ of hearing; is sometimes, as in the Hog, swollen into a large bulla ossea, like the parapophyses and pleurapophyses of the cervical vertebræ of *Cobitis*; is sometimes produced into a long auditory tube, and sometimes reduced to the ring supporting the tympanic membrane. Yet, under all these changes, since its special homology is demonstrable with 28 in the Bird, fig. 26, Turtle, fig. 91, Vol. I., and Crocodile, fig. 92, Vol. I., as well as with the teleologically compound bone, 28, *a*, *b*, *c*, *d*, in the Fish, fig. 81, Vol. I., so likewise must its general

[1] XLIV. p. 556, no. 3338, in which the petrosal is instructively distinct from all the surrounding vertebral elements composing the otocrane.

homology be equally recognised. The frontal hæmapophysis, fig. 189, 29, and the corresponding half of the hæmal spine, ib. 32, are connate on each side in all Mammals. The arch, as in other air-breathing Vertebrates, has no diverging appendage.

The nasal segment, NIV, is chiefly complicated by the confluence of parts of the enormously developed olfactory capsules, 18, and its typical character is further masked by the compression and mutual coalescence of the neurapophyses, 14. The centrum is usually much elongated, as at 13, and soon coalesces with both neurapophyses, 14, and with the nasal capsules, 18. The neural spine, 15, is bifid. The pleurapophysis, 20, or proximal element of the hæmal arch of the nasal vertebra has its real character and import almost concealed by the excessive developement of the second element of the arch, 21, which resumes in Mammals all those extensive collateral connections which it presented in the Crocodile; and to which are sometimes added attachments to the expanded spine of the frontal vertebra, as well as to that of its own segment. The pleurapophysis, however, besides its normal attachment to its centrum, 13, sends up a process to the orbit, in order to effect a junction with its neurapophysis. The hæmal spine, 22, is developed in two moieties, which never coalesce together, although, in the higher Apes, and at a very early period in Man, each half coalesces with the hæmapophysis, and repeats the simple homogeneous character of the corresponding elements of the succeeding (mandibular) arch.

The appendicular element, 24, which diverges from the pleurapophysis, 20, contributes to fix and strengthen the palato-maxillary arch by attaching it to the descending process of the parietal centrum, 6: with which, in most Mammals, it ultimately coalesces. The other elements of the diverging member of the arch correspond in number and in the point of their divergence with those in Birds, Chelonians, and Crocodiles. They are two in number, succeeding each other, and both become seats of that expansive developement which is followed by the multiplication of the points of connection; thus the proximal piece, 26, 'malar bone,' is connected in the Hog not only with the hæmapophysis, 21, from which it diverges, but likewise with the muco-dermal bone, called 'lacrymal,' 73. The distal piece of the appendage, 27, expands as it diverges, and fixes the naso-hæmal arch not only to the frontal pleurapophysis, 28, and parietal parapophysis, 8, but also to the frontal, parietal, and, sometimes, occipital neurapophyses and spines: it also affords, in the Hog, as in other Mammals, an articular surface to the frontal hæmapophysis, 29.

The steps by which the bony capsule of the otic organ is finally differentiated and individualised in Mammals are instructive examples of that character of advance in organisation. The ex- and par-occipitals which contribute a partial bony support to the back part of the gristly capsule in Fishes and Reptiles, and coalesce with that fully ossified capsule in Birds, remain distinct from the petrosal in all Mammals. The alisphenoid, which contributes a partial bony support to the fore part of the gristly otic capsule in Hæmatocrya, and coalesces with the same part of the bony capsule in Birds, has likewise permanently liberated itself therefrom in Mammals. The mastoid, which contributes a bony support to the outer part of the otic capsule in cold-blooded Vertebrates, and is extensively confluent with the same part of the ossified capsule in Birds, retains such confluence in some Mammals, but instructively manifests its primitive independency in others. In the Cetacea, where the mastoid and paroccipital are distinct from the petrosal, this capsule coalesces with the tympanic, which, having lost its mandibular function, is fixed and contracts anchylosis with the petrosal. The Babyroussa exemplifies the essential individuality of the acoustic capsule, the petrosal not only being ossified from its own centre, but remaining distinct from every bone of the otocrane.[1]

190

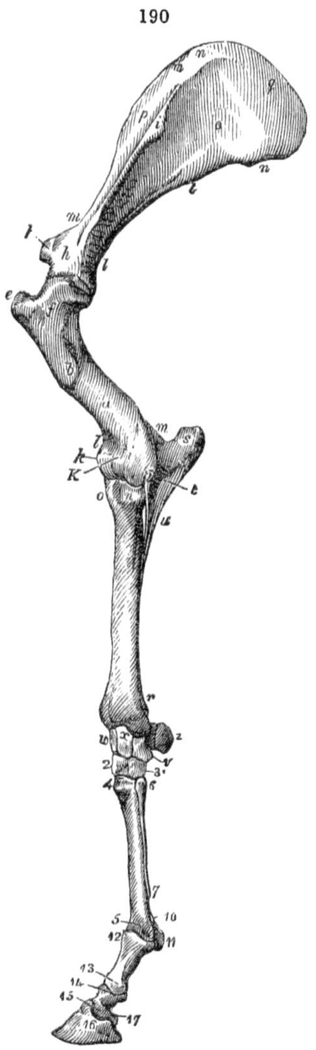

Bones of fore-limb, Horse.

§ 177. *General Characters of the Limbs.*—The diverging appendage of the occipital vertebra is never absent in Mammals, and

[1] XLIV. p. 556, nos. 3388, 3345.

offers its most simple condition in the Horse, fig. 190. It is essentially, as in *Protopterus*, a jointed ray, but every part is adaptively modified for special ends and reciprocal adjustment and interplay; in the monodactyle Mammal it is, in fact, the result of simplification from a more complex ancestral condition of limb, in reference to the application of that limb to a more vigorous kind of locomotion. Viewing the framework of such limb, here, in merely its archetypal relations, we remark that the supporting arch is incomplete, as in most Mammals; the pleurapophysis, *a*, *h*, is expanded into a 'scapula,' with its coalesced hæmapophysis as a 'coracoid' process, ib. *k*. The first segment of the appendage is modified as 'humerus,' *a*; the second segment as 'radius,' *o*, with which has coalesced the process *s*, *u*, developed in most Mammals as 'ulna.' In the blastema between the second and third ray have been formed a cluster of ossicles called 'carpal,' *w*, *z*, 2, 3; the third segment, 4, 5, is a metacarpal, and with it are connected two

191

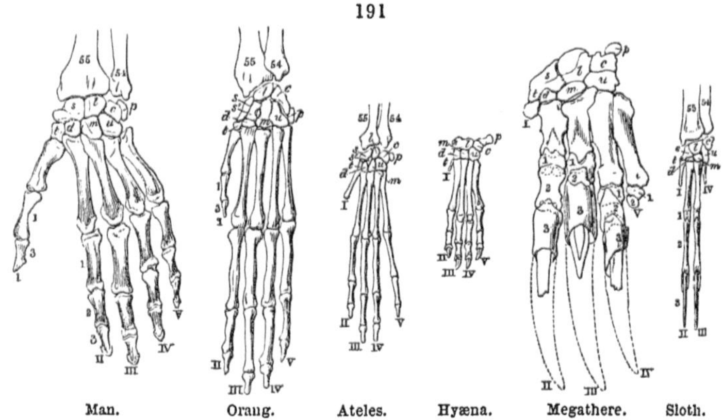

Man. Orang. Ateles. Hyæna. Megathere. Sloth.

styliform appendages, 6, 7; the abortive remnants of other metacarpals. Next follow the three terminal shorter segments of the limb-ray called 'phalanges,' 13, 14, 15; the whole forming the 'digit' which answers to the middle finger, III, in the pentadactyle foot of beast and in the hand of man. Gradational steps to this perfect condition of 'hand' are selected from the Mammals with claws, in fig. 191. In the Unau or Two-toed Sloth (*Bradypus didactylus*), the digits which are functionally developed answer to the second, II, and third, III, in Man; the fourth, IV, and first, I, are represented by styliform beginnings of their metacarpals. The carpal ossicles include one, *s*, *t*, answering to the separate scaphoid, *s*, and trapezium, *t*, in Man, a 'lunare,' *l*, and 'cuneiform,' *c*, a 'trapezoides,' *d*, supporting the metacarpal of the second digit; a 'magnum,' *m*, supporting that of the middle digit, and an 'unciforme,' *u*, limited to the rudiment of IV.

In the huge extinct congener of the Sloths (*Megatherium*), the fourth, IV, as well as the third and second digits are developed, as in the *Bradypus tridactylus*, for the support of claws. There is a metacarpal of the fifth digit, supporting stunted rudiments of the first, 1, and second, 2, phalanges; the first digit is still represented by a like rudiment of its metacarpal, I. The carpal ossicles include, as in Sloths, a 'scapho-trapezium,' *s*, *t*, with a well-marked 'pisiforme,' *p*, and a larger 'unciforme,' *u*. In the *Hyæna* the fifth digit, V, is functionally developed: the first, I, retains the rudimental state. The scaphoid and lunare, *s*, *l*, have here coalesced: the trapezium, *t*, is distinct, but very small: the unciforme supports, as usual, the metacarpals of the fourth, IV, and fifth, V, digits. In the Spider-Monkey (*Ateles*), the metacarpal representative of the first digit, I, is longer: the scaphoid, *s*, is distinct, and the 'intermedium,' *s*′, is a dismemberment thereof, answering to *e*, fig. 173, Vol. I.

In the Orang the carpus also has the dismembered scaphoid, *s*, or 'intermedium,' *s*′. The inner digit, I, is short and feeble, but with the usual mammalian number of two phalanges. In the hand of Man, this digit, which is the last to be completed in that class, attains its highest functional developement: it is articulated in such a way and at such an angle as to be opposable to any of the joints of any of the other digits. Of these the third, III, which is the most constant in the class, is the longest. The carpus consists of eight bones in two rows; the first including the undivided 'scaphoides,' *s*, 'lunare,' *l*, 'cuneiforme,' *c*, 'pisiforme,' *p*; the second including 'trapezium,' *t*, 'trapezoides,' *d*, 'magnum,' *m*, 'unciforme,' *u*. These names, suggested by the shapes and proportions of the carpal bones in the human skeleton, become arbitrary signs of their homologues in lower animals.

192

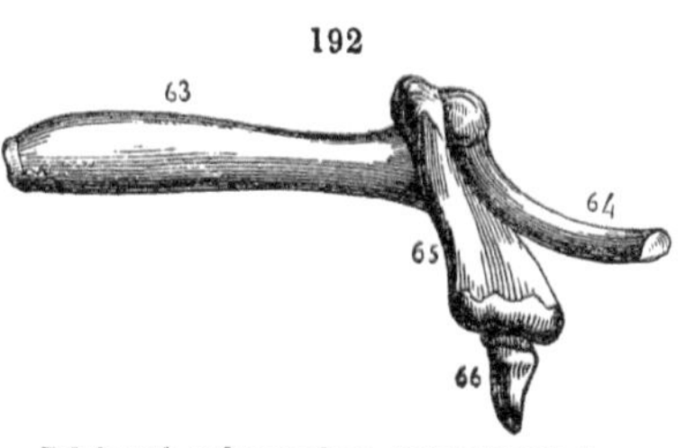

Pelvic arch and appendage, *Balæna Mysticetus*. LXV′.

The appendage of the pelvic arch may be wholly wanting, as in *Sirenia* and most *Cetacea*, or represented by a two-jointed ray, as in the Right Whale, fig. 192, and fig. 159, 65, 66; articulated to two elements, 63 and 64, of the pelvic arch, which, as in Fishes, are loosely suspended in the flesh.

The successive gradational steps by which the pentadactyle condition of the limb or appendage is attained are selected from the series of hoofed Mammals in fig. 193.

The pelvic limb, fig. 195, shows the same monodactyle simplicity as the pectoral one, in the Horse. The ossicles developed in the connective substance between the second and third principal segments of the long-jointed ray, are the 'astragalus,' *a*, 'calcaneum,' *cl*, 'naviculare,' *s*, 'mesocuneiforme,' *cm*, 'ectocuneiforme,' *ce*, 'cuboides,' *b*. The metacarpal supporting the three joints or 'phalanges' of the digit articulates chiefly with the ectocuneiform, *e*, which accords in size. The largely developed digit, or continuation of the main limb-ray, fig. 193, answers to the third, III, of the pentadactyle foot. At its base are rudiments of the metatarsals of the second, II, and fourth, IV, digits. In the Ox the naviculare, ib. *s*, is connate with the 'cuboides,' *b*: and, as this supports one-half of the single metatarsal, such half is held to be the developed homologue of the rudimental fourth metatarsal in the Horse: whilst the half supported by the 'ectocuneiforme,' *ce*, in the Ox, is held to answer to the metatarsal of the developed digit, III, in the Horse.

193

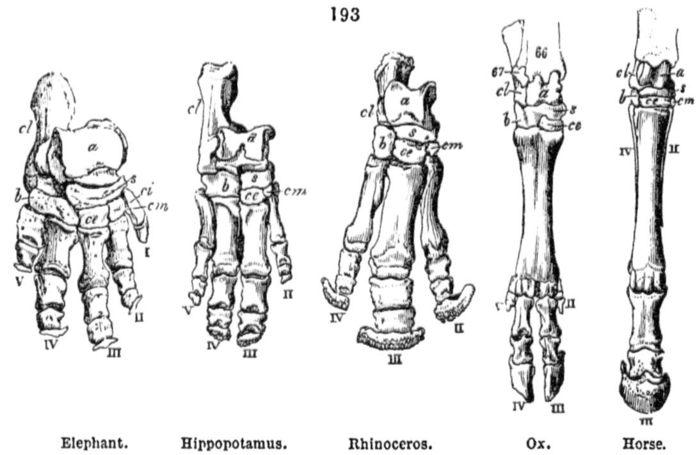

Elephant. Hippopotamus. Rhinoceros. Ox. Horse.

Embryology here lends partial proof to this view: the so-called 'cannon-bone' being developed from a single centre and epiphyses in the Horse, and from a pair of shafts or centres and epiphyses in the Ox: it accordingly supports a pair of toes, which answer to the third, III, and fourth, IV, in the pentadactyle foot. The Camel and Giraffe have not rudiments of any other toes: in the Ox such rudiments of the distal parts of the second, II, and fifth, V, digits are appended to the coalesced metatarsals of the functional pair of toes, III and IV.

In the miocene fossil Horse (*Hipparion*, fig. 194) a similar pair of 'spurious' hoofs, *ii*, *iv*, dangled behind the main toe, *iii*, completing the digits, *ii* and *iv*, indicated by the 'splint-bones' or proximal parts of the metatarsals in the modern Horse, fig. 193, II, IV, but stunted in growth. In the eocene *Palæotherium* these digits were nearly equal in size to the middle one. The Rhinoceros at the present day preserves these proportions of the toes, II, III, IV, but with shorter and more massive proportions of the whole foot. Accordingly, in fig. 193, it will be seen that the 'mesocuneiforme,' *cm*, and 'cuboides,' *b*, have a larger proportional size than in the Horse; but the structure of the tarsus is essentially the same: the cuboid, *b*, articulates directly with the calcaneum, *cl*; the naviculare, *s*, intervenes between the two cuneiform bones and the astragalus, *a*. The affinity of these 'perissodactyles' is obvious, and the closer links of affiliation are supplied by the extinct forms above cited. In like manner we find the affinity of the Ox and Hippopotamus illustrated in the structure of the hind-foot, the Hog holding a similar intermediate step in the developement of the toes, IV and V. In the tarsus the cuboid, *b*, and naviculare, *s*, show the same near equality of size, but they are distinct bones in the Hippopotamus as in all Artiodactyles except the restricted or horned Ruminants: a mesocuneiforme, *cm*, now supports the metatarsal of the toe, II, that of the fifth, V, articulates with the cuboid. In the Elephant the innermost digit, I, is present—the last to appear in the ungulate as in the unguiculate series, and the tarsal group shows the completeness which it manifests in Man. The human anatomist will recognise the astragalus, *a*, calcaneum, *cl*, naviculare, *s*, extended transversely and presenting articular facets to the three 'cuneiform' bones, 'internal,' 'middle,' and 'external,' which for convenience of definition I have called 'entocuneiform,' *ci*, 'mesocuneiform,' *cm*, 'ectocuneiform,' *ce*; the 'cuboides,' *b*, supports as usual the metatarsals of the fourth and fifth toes. The toe, I, has a short metatarsal and some bony representative of a phalanx imbedded in the innermost part of the hoof: the other toes have the normal complement of phalanges, which, in Mammalia, do not exceed (save in *Cetacea*) three in number, nor two in the innermost digit, I, in both pectoral and pelvic limbs.

194

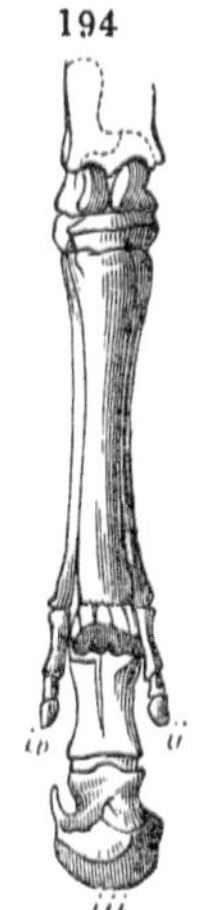

Foot of extinct Horse (*Hipparion*).

The 'serial homology' of the parts of the respective arches of these limbs is illustrated in Vol. I. p. 188. In the limbs themselves, or appendages of the arches, the femur, fig. 195, *a*, answers to, or is the homotype of, the humerus, fig. 190, *a*; the tibia, fig. 195, *u*, is the homotype of the radius, fig. 190, *o*; the fibula, fig. 195, 1, 2, of the ulna, *s*, *u*: the tarsus repeats the carpus, the metatarsus the metacarpus, and the three phalanges, as respectively named 'proximal,' 'middle,' and 'distal' or 'ungual.' In the tarsus it will be seen that the cuboid, in the 'Elephant,' fig. 193, *b*, supports the two outer metatarsals, as does the unciforme the two outer metacarpals, in the Orang and Man, fig. 191, *u*: the ectocuneiform in the tarsus, *ce*, and the 'magnum' in the carpus, *m*, respectively support the middle digit, III: the mesocuneiforme, *mc*, holds the same 'serial' relation to the trapezoides, *d*, and the 'entocuneiform,' *ec*, to the 'trapezium,' *t*. The bone of the carpus, fig. 191, *s*, in Man articulates with the three innermost carpals of the second row; and, in the Orang, but in a divided state, *s* and *s′*, leaves a larger share of the wrist-joint with the radius to the bone *l*, and in the same degree tends to repeat in the carpus the position and connections of the bone *s* in the tarsus: so I infer that the carpal scaphoid and tarsal naviculare are homotypes: the carpal lunare, fig. 191, *l*, answers to the tarsal astragalus, and the carpal cuneiforme and pisiforme to the tarsal calcaneum, in which bone the lever-process forming the 'heel' more immediately repeats the pisiforme, which also in many quadrupeds, fig. 191, *p*, Hyæna, makes a 'heel-like' projection in the carpus.

195

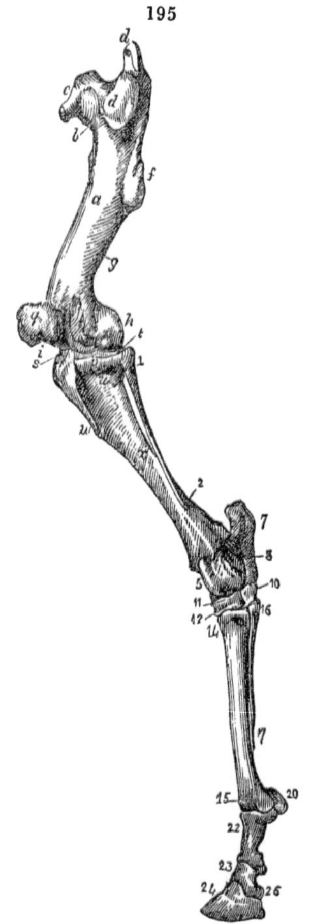

Bones of hind-limb, Horse.

§ 178. *Special Homologies.*—As that which is engendered by a Mammal is mammalian from its beginning, each step of its building up has the finishing of the Mammal for its end, and shows it the more as it nears the goal. The developemental phenomena of the head neither supersede nor can supply the better evidences of homology afforded by relative position and connections any more than do those of the foot. The cannon-bone of the ox is developed from three terminal and two middle centres of ossification; but embryology does not show which of these signify bones distinct in other Mammals: it is neither here nor elsewhere the criterion of homology. In the foregoing account of the Mammalian modifications of the Vertebrate skeleton, the general and serial homologies are given, as determined in my work on the Vertebrate Archetype. But as a few of the special homologies of cranial bones are still unaccepted by fellow-labourers in this field of anatomy, I offer the following remarks in excuse for the retention of my opinions on such moot-points.

To rightly determine the cranial bones in Mammals, as in Birds, we must pass to their investigation from the previously determined bones in the skull of an inferior Vertebrate. Thus, placing the skull of a young Ostrich or Apteryx, showing the sutures, by the side of that of the low, bird-like Monotreme (*Echidna*, fig. 197), we find that the transversely extended, medially notched occipital condyle, in the Bird, fig. 27, has become bisected or divided into two in the Mammal, fig. 202; each moiety being developed wholly (*Echidna*) or in great part (some *Cetacea*) from the exoccipital, 2. The basioccipital either wants, or developes only the lower end of, the divisions of the occipital condyle. The exoccipital, in most Mammals, sends off a 'paroccipital' process, 4, as in Birds. The basi-, ex-(2), and super-(3)occipitals coalesce into one bone, but rarely are fused, as in Birds, with the sense-capsule and segment in advance. The basisphenoid, fig. 202, 5, differs from that of the Bird, figs. 27, 32, in not being coossified with the presphenoid, 9: laterally it contributes to form part of the otocrane and tympanum, in advance of which it articulates with the alisphenoid, figs. 196, 197, 6. In the *Echidna* a bone, fig. 197, 8, coossified with or anchylosed to the outside of the petrosal, expands beyond it to articulate with the ex-(2) and super-(3)occipitals, with the parietals, 7, and the alisphenoids, 6. This bone, in many other Mammals, developes a 'mastoid' process, as in Birds: it is developed, as in them, in and from the lateral cranio-cartilage enveloping the otic capsule: it is plainly the homologue of 8 in the Bird, fig. 196. Between 8 and 6 in *Echidna* there is a vacuity in the bony skull. The parietal, 7, is relatively larger, the frontal, 11, is smaller, than in the Bird. The nasal, 15, is simply elongate, in *Echidna* as in *Rhea*: it does not bifurcate anteriorly by sending down a maxillary prong or process as in the Ostrich, fig. 196, 15, and most Birds: but it is longer and articulated, or is united, throughout its length, with its fellow in the Mammal. The premaxillary, 22, is correlatively shorter in the Mammal, not medially confluent nor sending off a nasal process from the symphysis, as in the Bird. The maxillary, 21, is larger, and the nasal process, of which the beginning is shown in *Struthio*, is a broad and high plate in *Echidna* and most other Mammals. The hind part of the maxillary unites with a malar,

196

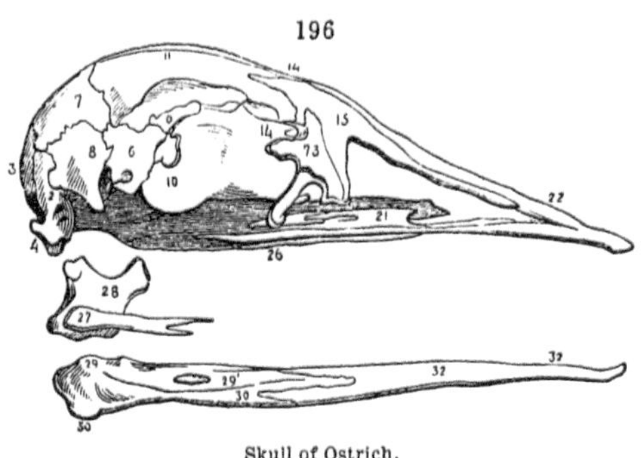

Skull of Ostrich.

fig. 197, 26, styliform in *Echidna* and some *Bruta*, as it is in Birds, fig. 196, 26. The bone, 27, articulates with 26, but expands in *Echidna*, as in Chelonians, as it extends backward, and applies

197

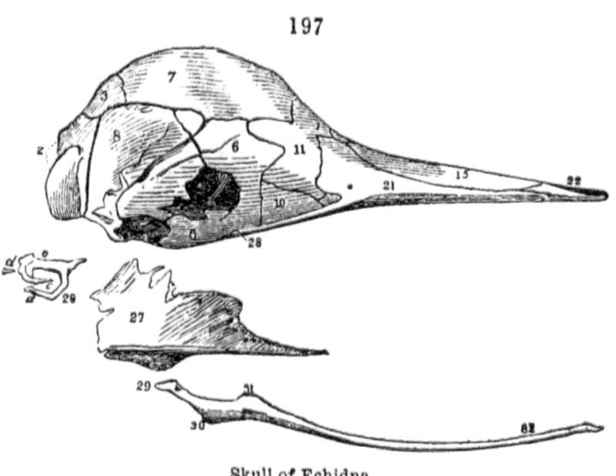

Skull of Echidna.

itself, in most Mammals, to close the gap in the side-wall of the cranium left between 8 and 6, before articulating with the tympanic, 28: it also developes the articular surface for the mandible, 29–32. This is one of the marked modifications of the squamosal in the Mammalian class. The retroduced part or appendage of the upper jaw again affords the joint to the lower jaw, as in the Plagiostomous Fishes: but the common pedicle, 28, is reduced in the Mammal mainly to the support of the ear-drum, the accessory function with which it is charged wholly or in part in all air-breathing Vertebrates.

Remove 27 and 28 from the cranium of the Bird and Monotreme, as in figs. 196 and 197, and the homology of the remaining cranial bones, especially of 2, 3, 8, 6, is unmistakable. The mastoid, 8, in both Bird and Monotreme, is developed from cartilage; articulates posteriorly with 2, 3, superiorly with 7, anteriorly with 6; coossifies internally with the petrosal, and gives attachment inferiorly to the bone, 28, which supports wholly or in part the 'membrana tympani.' The squamosal, 27, is a backward prolongation of the bar, 26, attaching the upper jaw to the tympanic; it is developed in the embryonal scaffolding external to the proper cranial cartilage; it articulates posteriorly with the tympanic, 28. It forms no part of the outer wall of the cranium in Birds, and is equally excluded from that cavity in *Cetacea*, fig. 198, in most Ruminants, fig. 140, and in many Rodents: the supplementary function of completing such cranial wall is peculiarly mammalian, and does not supersede the share taken in such lateral wall by 8 and 6, in all Vertebrates. Moreover, 27 constitutes the hinder and major part of the zygomatic arch in both Birds and Mammals, as in most Reptiles; with such homologically unimportant modifications of shape as are exemplified in the Turtle and Crocodile (Vol. I. figs. 91 and 95, 27), and in the figures 26, 140, 196, 197, 27, of the present Volume.

198

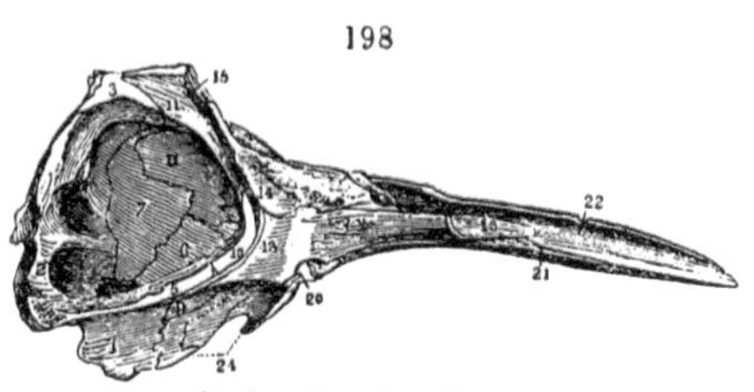

Section of cranium, Porpoise.

The bony pedicle which suspends the mandible to the side-processes of the cranium, is that which is marked 28 in the Fish (Vol. I. figs. 81, 84), the Serpent (figs. 96, 97), the Tortoise (figs. 91 and 92), and the Crocodile (figs. 93, 95). As those side-processes are homotypes of the transverse processes (par- di-apophyses) of the trunk-vertebræ, so 28 bears the same serial relation to the 'pleurapophyses,' the mandibular rami completing 'hæmapophysially' the inferior or hæmal arch of the cranial segment. This vertebral character is shown in the developement of the Vertebrate skull: the simple rib-like cartilage formed in the second (counting backward) of the embryonal, 'visceral,' or hæmal arches, manifests always its upper or 'pleurapophysial' and its lower or 'hæmapophysial' portions: and these are more equal in length in Birds and Hæmatocrya than in Mammals; for the embryo of a highly modified and advanced class early shows the characters of its class, which become deceptive when exclusively used as a light to general vertebral homologies. The true guide to the homology of 28 is its articular connections to one or more cranial diapophyses: in Fishes and Crocodiles, e.g., to the post-frontal and mastoid, in Lizards and Snakes to the mastoid, in Birds to the mastoid and paroccipital, in Mammals to the mastoid. The connection with the squamosal is later and supplementary in the Vertebrate series.

The tympanic pedicle undergoes various and extreme modifications in relation to the functions, as various, allotted to the second hæmal arch (counting backward) in the head. In Fishes, much of the mechanical part of the respiratory functions is performed by the 'tympano-mandibular' arch: hence the length, subdivision, and resultant elasticity of the suspensory piers or pedicles. In air-breathing *Hæmatocrya* the branchial duty ceases; but a special organ of sense, claiming more direct relation with the air, presses the tympanic pedicle into a service unknown to it in the water-breathers. In *Chelonia*, fig. 91 (Vol. I.), the tympanic, 28, is developed to form a frame for the ear-drum, and it contributes more or less of that frame in Crocodiles, Lizards, and Birds: it has least concern with the tympanum in Serpents; and, as these are exclusively air-breathers, 28 is restricted to its function of suspending the mandible, and retains most of its simple rib-like form as it descends from the lengthened diapophysis, 8, fig. 97 (Vol. I.) to the dentigerous hæmapophysis, 31. The proximal articular end of the tympanic may have a double condyle, as in some Fishes and Birds, a single condyle, as in Lizards and Serpents, or a sutural margin for fixed junction, as in *Chelonia*, fig. 91, 28, and *Crocodilia*, fig. 95, 28. Such is its mode of articulation in all Mammals, in which class it manifests its extreme simplicity of function and reduction of size. To the ear-drum, which it sustains, is articulated, in Birds, a columelliform 'stapes,' by the intermedium of a cartilage; and in Monotremes and Marsupials, fig. 197, *d*, by the intermedium of a bone, *c*. This ossicle in higher Mammals is divided into 'incus' and 'malleus,' which, like the columelliform 'stapes' in Birds and Reptiles, is developed, as in fig. 444, B, *e* (Vol. I.), in connection with, but not like the tympanic (*d*) and mandible in and from, the periphery of the primary 'visceral' or hæmal cartilaginous arch, called, from its discoverer, 'Meckel's cartilage.'

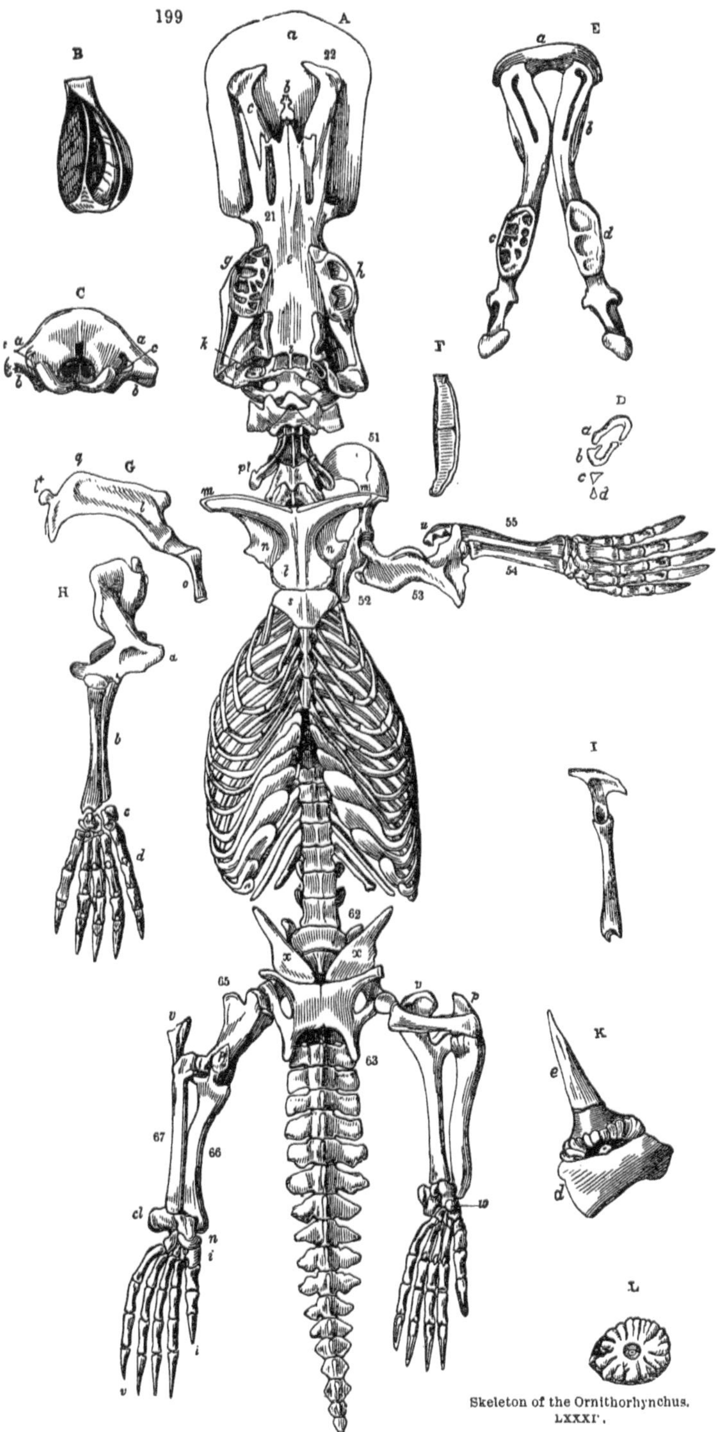

Skeleton of the Ornithorhynchus.
LXXXIV.

Having premised so much in reference to the Mammalian skeleton generally or typically, its main modifications as exemplified in the several orders of the class, will next be noticed.

§ 179.—*Skeleton of Monotremata.*—A. *Vertebral Column.*—The principal osteological characters of this order are:—The extension of the 'coracoid,' fig. 199, 52, *o*, as in Birds and Lizards, from the scapula, 51, to the sternum, *s*, and anchylosing at full growth with the scapula, as at G, fig. 199; the epicoracoid, ib. *n*, as in Lizards; the marsupial bones, ib. *x*, *x*; the supplementary tarsal bone, ib. *d*, supporting the perforated spur, *e*, in the male; the long persistence of distinct pleurapophyses, *pl*, in the vertebra dentata.

Both the genera have twenty-six 'true vertebræ,' of which seven are cervical; but the Ornithorhynchus has seventeen and the Echidna sixteen dorsals, the lumbar vertebræ being three in the latter, and reduced to the lacertian number two in the Ornithorhynchus.

200

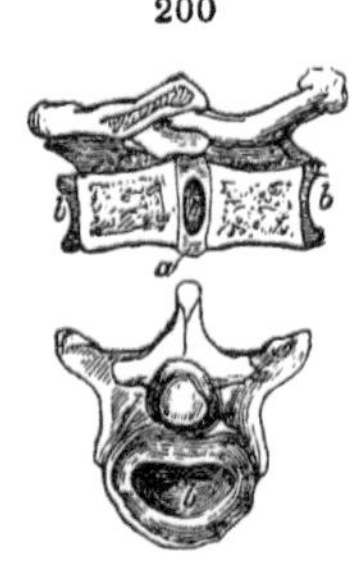

Lumbar vertebræ, intervertebral cavities, Echidna.

The intervertebral substance is dense and fibrous at its periphery, fig. 200, *a*, but the fluid central part, *b*, fills a more definite cavity in the Echidna than in higher Mammals.[1]

In the dorsal vertebræ the nerves perforate the neurapophyses; but escape, as usual, at their intervals in the cervical and lumbar regions. The dorso-lumbar neural spines are short and subequal, fig. 201. The ribs of the first six dorsals have ossified sternal portions which articulate with the sternum; in the succeeding vertebræ to the fifteenth the sternal portions are cartilaginous, expanded, and overlap each other, fig. 199; the last two pairs of ribs terminate freely. Most of the vertebral ribs articulate over the interspace of their own and the antecedent centrum; a small tubercle defines the neck of the rib, save in the last four; but, save in the first and second, does not articulate with the diapophysis. The first dorsal pleurapophysis is broad, the others are cylindrical and slender; cartilage is interposed between the bony pleur- and hæm-apophyses of the anterior dorsal vertebræ, as in the Crocodile. The sternum consists of four bones in *Ornithorhynchus*, and of five in *Echidna*. The first, fig. 199, *s*, is an unusually expanded 'manubrium,' receives the hæmapophyses of the first and second ribs, and supports a large T-shaped

[1] LXXIX'. p. 375.

episternum, ib. *t*. The sacrum consists of two vertebræ in *Ornithorhynchus*, and of three in *Echidna*.

There are thirteen caudal vertebræ in the Echidna, fig. 201. The first is the largest, with broad transverse processes, the rest progressively diminishing, and reduced, in the six last, to the

201

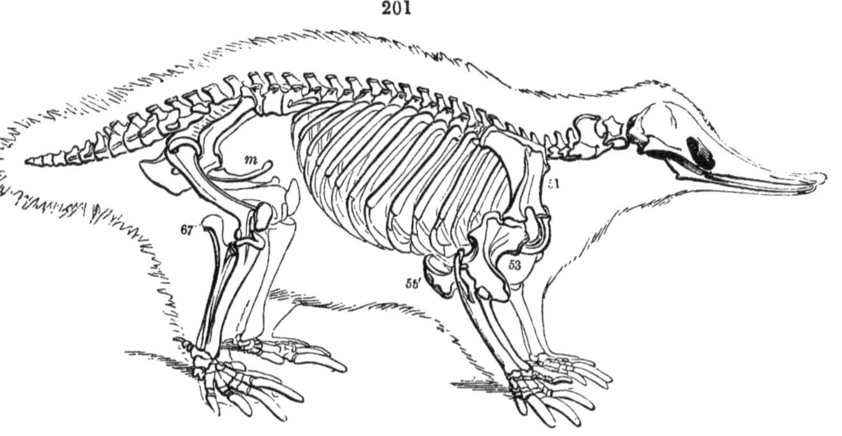

Skeleton of Echidna. LXXIII.

central element. The Ornithorhynchus, fig. 199, has twenty-one caudal vertebræ, of which all but the last two have transverse processes, and the first eleven have also spinous and articular processes. The pleurapophysial parts of the transverse processes are distinguishable in half-grown animals. The transverse processes are broad and depressed; they gradually increase in length to the tenth caudal, then as gradually diminish to the twentieth; their extremities are expanded, and, from the fifth backward, are thickened and tuberculate. The spinous processes progressively diminish in height from the first caudal. Hypapophyses are developed from the bodies of the third to the nineteenth caudal vertebra inclusive; but there are no hæmapophyses articulated to the vertebral interspaces, as in many Marsupials. In the Echidna hypapophyses are absent; but rudiments of hæmapophyses are connected with the interspaces of one or two of the middle vertebræ of the tail. The caudal vertebræ in the Ornithorhynchus are of nearly the same length to the two last; they progressively diminish in vertical diameter as they recede from the trunk, and are chiefly remarkable for their breadth and flatness; resembling in this respect the caudal vertebræ of the Beaver and of the Cetacea; the horizontally extended tail having a similar relation to the frequent need which an aquatic animal with hot blood and a quick respiration of air has to ascend rapidly to the surface of the water.

The cervical vertebræ, fig. 186, have short and broad centrums confluent with neurapophyses; the former developing a par- the latter a di-apophysis: the pleurapophysis, *pl*, is short and broad, and circumscribes the 'vertebrarterial' canal by junction with both the transverse processes: which joints in the last five cervicals are obliterated earlier in *Ornithorhynchus* than in *Echidna*. In the latter not any of the cervicals have zygapophyses save the atlas. The true centrum of this vertebra supports a great part of its neural arch, and long continues distinct from that of the axis: it has a long 'odontoid' process. The lower part of the ring of the atlas sends off in *Ornithorhynchus* a pair of long divergent hypapophyses.

B. *Skull.*—The skull in both genera of Monotremata is long and low, but characterised by a relatively larger cranium in proportion to the face than in most Marsupials. The parietes of the expanded cerebral cavity are rounded, and their outer surface is smooth. These characters are most conspicuous in the Echidna, in which the jaws are slender, elongated, and gradually diminish forward to an obtuse point, so that the whole skull resembles the half of a pear split lengthwise. The facial angle of the Echidna is 36°, that of the Ornithorhynchus 20°, being almost the lowest in the mammiferous class. The cranial bones and their constituent pieces continue longer distinct in the Echidna than in the Ornithorhynchus, in which they ultimately coalesce to a degree resembling that in Birds.

202

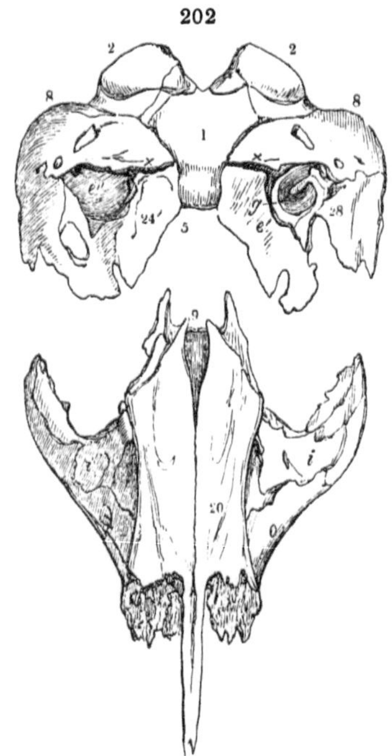

Base view of partially disarticulated cranial segments, Echidna. LXXIX'.

In the Echidna the basioccipital, fig. 202, 1, is flat and hexagonal, with the hind-border notched to complete below the large vertical 'foramen magnum,' and contributing to the lower part of each condyle, 2, 2: these are large and formed chiefly by the exoccipitals, fig. 203, 2, 2, which are separated above the foramen magnum by a notch, closed by membrane, in the recent state. The superoccipital, ib. 3, is a transversely oblong quadrilateral plate, articulated by 'harmonia,' not only with the exoccipitals, but with the large mastoids, 8, and anteriorly with the parietals, fig. 197, 7. The basisphenoid, fig. 202, 5, supports laterally a pair of alisphenoids, fig. 197, 6, which are notched posteriorly by the trigeminal nerves, and expand as they rise to articulate with the parietals, 7, the mastoid, 8, and anteriorly with the orbitosphenoid and frontal, 11. The mastoid, 8, is chiefly conspicuous by its great size, in the Echidna, and the share which it takes, conjointly with the petrosal, in the formation of the lateral, lower and posterior parts of the cranial cavity: in this character it retains much of its ornithic condition, fig. 196, 8. The small vacuity, left in the Monotreme, between the mastoid and alisphenoid, is closed by the application thereto of the posteriorly expanded squamosal, fig. 197, 27. The presphenoid, fig. 202, 9, is connate with orbitosphenoids, fig. 197, 10, pierced by the small optic nerves: the frontals, 11, expand as they rise, but without developing superorbital ridges, and meet at a toothless suture along the middle of the narrow forehead. The vomer and prefrontals are chiefly remarkable for their connection with enormous and obscuring turbinals, supporting an olfactory organ of vast extent. The anterior part of the frontals is largely overlapped by the bases of the nasal bones, which encroach upon the interorbital space. These, fig. 197, 15, receive the upper edge of the maxillary into a groove at their outer margin, and articulate anteriorly with the premaxillaries, ib. 22, which meet above the nasal canal in front of the nasal bones for an extent of about three lines, and thus exclusively form the boundary of the single, oval, and terminal external nostril. The lower or palatal process of the premaxillary extends backward in the form of a long and slender pointed process which is wedged into a fissure of the maxillary. The incisive fissure is narrow and extends from the premaxillary symphysis some way between the palatine plates of the maxillaries. The palatines, fig. 202, 20, are long and entire where they form the hinder half of the roof of the mouth, diverging

203

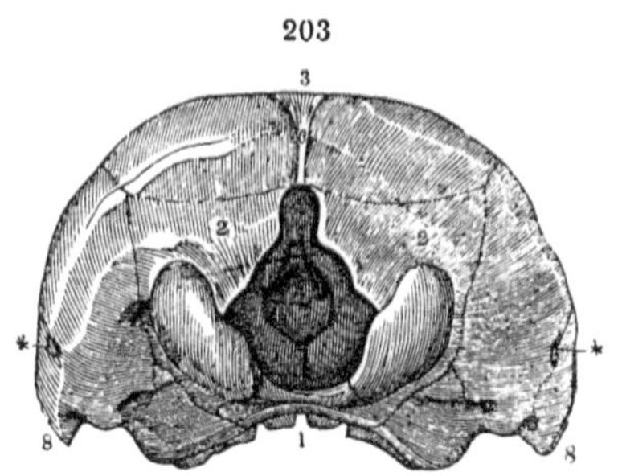

Occiput of Echidna. LXXIX'.

posteriorly to form the narrow median nasal opening. The roof is continued by the pterygoids, ib. 24 and 16′, which articulate, as in many Birds, with the tympanic, *e*, 28, and the basisphenoid, 5. Another mark of ornithic affinity is the confluence of the malar and squamosal, fig. 197, 27: unless the slender process of the maxillary, ib. 26, may represent the malar. The tympanic cavity is excavated in the petromastoid and partly closed by the slender tympanic, fig. 202, 28, *e*, which sends forward a short homologue of the orbital process of that of the bird: about three-fourths of the ear-drum are attached to the tympanic, and one-fourth to the mastoid: the plane of the drum is nearly horizontal and looks downward. The 'stapes' is columelliform, fig. 197, *d*: one crus of the incus anchyloses with the reduced tympanic at *o*; the other is confluent with the malleus, *c*.

The lower jaw consists in the Echidna, fig. 197, 29–32, of two long and slender styliform rami without a symphysial joint, but loosely connected together at their anterior extremities. An angular process, 30, divides the horizontal from the ascending

204

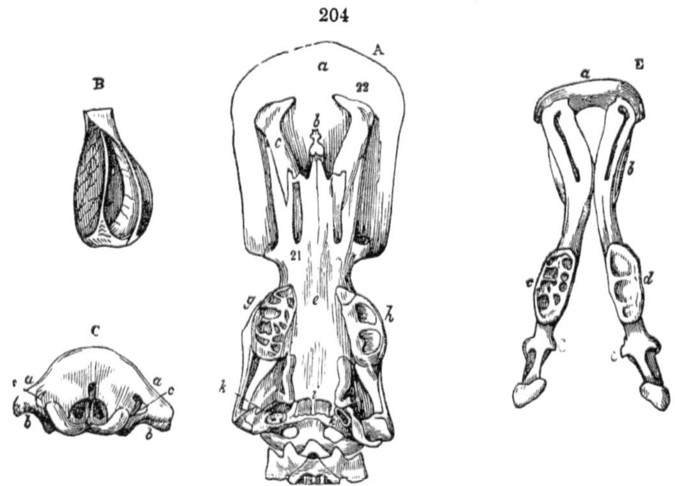

Skull of Ornithorhynchus, half natural size. LXXXI'.

ramus, which rises at an open angle and terminates in a small oblong convex condyle, 29. A short obtuse coronoid process, 31, extends from the upper part of the horizontal ramus as far in advance of the angle as the condyle is behind it. The rest of the ramus is rounded like a rib, and diminishes to the anterior extremity. The dental canal commences below the coronoid process and divides in its progress, one branch terminating near the middle of the smooth alveolar border, the other close to the end of the ramus. In no mammiferous animal does the lower jaw bear so small a proportion to the skull or to the rest of the skeleton as in the Echidna.

In the Ornithorhynchus the lower jaw, fig. 204, E, is much more developed. Each ramus commences posteriorly by a large convex condyle. The ascending ramus is nearly horizontal, flattened below, and continued upward in the form of a low vertical compressed plate, on each side of which there is a deep fossa. The ascending is continued by a gentle curve into the horizontal ramus, and the angle of the jaw is very feebly indicated. The horizontal ramus suddenly expands and sends off above in the same transverse line two short obtuse processes, both of which might be termed 'coronoid;' this structure is peculiar to the Ornithorhynchus. The innermost process, *c*, although the largest, is the superadded structure, as it affords insertion to the internal pterygoid. The socket, *d*, *e*, for the horny grinder, is shallow; its floor is perforated by several large foramina. The dental canal divides; one branch opens by a wide elliptical foramen on the outside of the ramus immediately anterior to the alveolus, the other terminates at the lower part of the end of the ramus. The rami of the jaw converge and are united at a symphysis of more than half an inch in length; there they become expanded and flattened, then again disunite, and are continued forward as two spatulate processes, *b*, which diverge from each other to their broad rounded terminations, and are situated just behind the inflected extremities of the similarly separated premaxillaries, ib. A, and fig. 205, 22. On the outer sides of the upper surface of the broad symphysis are the long and narrow sockets of the two anterior trenchant horny teeth. The Monotremes differ from the Marsupials in the absence of the inflected process developed from the angle of the lower jaw.

205

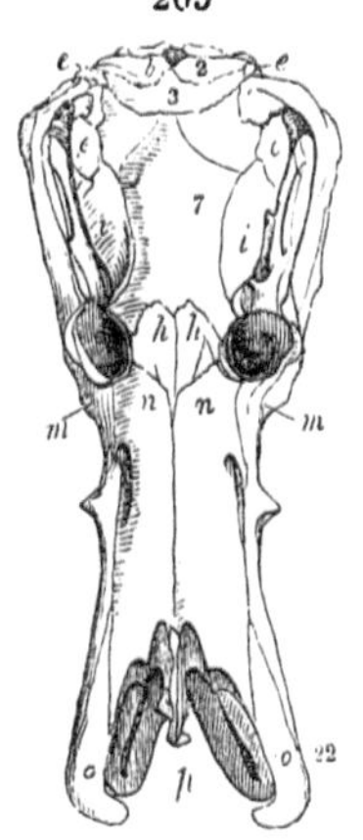

Skull of young Ornithorhynchus, nat. size.

The exoccipitals, fig. 205, 2, *b*, and superoccipital, ib. 3, are separate in the skull of the young Ornithorhynchus here figured of the natural size. The mastoid, ib. *e*, *e*, contributes to part of the occipital surface, and advances anteriorly to the small cranial expansion of the squamosal at *f*. This expansion does not exceed in size the glenoid process, which it meets at a right angle, and from the union of which the zygomatic process is continued forward to join the malar process of the maxillary: there is no distinct malar bone. The parietal, ib. 7, is long and large, undivided by a sagittal suture, from the place of which a bony falx is developed internally, fig. 204, B. The frontals are small, and in fig. 205, *h*, *h*, retain the frontal suture. The nasals, ib. *n*, *n*, are long and large: they contribute to the rim of the orbit, and form the posterior half of the large bony nostril, *p*. The maxillary, ib. *m*, after sending off a process which curves over the antorbital foramen, extends forward, diverging from the nasal to form the angular fissure which receives the premaxillary, *o*, 22. Each of these bends inward at the anterior extremity, but is separated by a wide space. There is a small prenasal ossicle at *p*, fig. 205, and *b*, fig. 204, A. The vomer forms a bony vertical septum dividing the nasal cavity from the presphenoid forward. The palatine plate of the maxillary, fig. 204, A, 21, is pierced by large oblique canals for the transmission of palatine branches of the trigeminal nerve. The bony palate is continued backward entire between the large shallow alveoli, *g*, *h*, of the upper horny molars to the posterior nostrils, *i*, which resemble those of the Crocodile in their backward position. The sutures defining the palatines and pterygoids are soon effaced.

The Ornithorhynchus differs from the Echidna in the large vacuities behind and in front of the tympanic cavity, the one representing the combined jugular and precondyloid foramen, the other the foramen ovale. The notch above the foramen magnum, fig. 204, C, is better defined; as is also the orbit.

There is a small lacrymal foramen at the anterior and inner part of the orbit in both the genera of Monotremes; a little lower down is the commencement of the antorbital canal. This canal branches in the Echidna, and terminates on the outer side of the maxillary bone by a succession of small foramina; but in the Ornithorhynchus, where it transmits a much larger sensitive nerve, it divides into three canals, of which one emerges beneath the uncinated process of the maxillary above mentioned; a second descends and opens upon the palate; and the third passes forward into the substance of the facial fork, and terminates by a large foramen at the outside of the premaxillary bone.

On the exterior of the cranium the ridges indicating the extent of the temporal muscles are clearly developed in the Ornithorhynchus, and correspond with the stronger zygomata and the more complete apparatus for mastication in this Monotreme.

Four linear impressions upon the upper surface of the skull diverge from the middle of the lambdoidal ridge, and terminate at the temporal ridges.

The interior of the skull offers many unusual modifications. The sella turcica is elongated and narrow in both Monotremes; it is bounded by two very distinct lateral walls in the Echidna. The posterior clinoid processes are chiefly remarkable for their height in the Ornithorhynchus. The semicircular canals stand out in high relief in this species, as in Birds. In the Echidna the olfactory capsule encroaches upon the anterior part of the cranial cavity in the form of a large convex protuberance, and a very extensive cribriform plate is developed. In the Ornithorhynchus the olfactory tract is comparatively small in the form of a depression, and the nerve escapes by a single foramen in the prefrontal: this is likewise an interesting mark of affinity to the Bird and Reptile. But the most remarkable feature in the interior of the skull of the Ornithorhynchus is the bony falx, fig. 204, B. This is not present in the Echidna. The tentorium is membranous in both Monotremes.

C. *Bones of the Limbs.*—The *scapulæ*, fig. 199, G and 51, are compressed curved plates, vertical in position, like the other pleurapophyses: they have coalesced with their hæmapophyses, the coracoids, 52 and G, *o*, which articulate below to the expanded hæmal spine, called 'episternum,' *t*, and also with the succeeding spine, called 'manubrium,' *s*, or first bone of the true sternum. A dismemberment of the coracoid, *n*, extends its attachments also to the elongated T-shaped episternum.

The whole scapula is broader, thicker, and less curved in the Echidna, fig. 201, 51, than in the Ornithorhynchus. In both Monotremes, the posterior margin or costa is concave, most so in the Ornithorhynchus, and in both it is turned toward the trunk, so that the subscapular surface looks obliquely forward and inward. The articular surface is divided into two facets: the one, internal and flat, articulates with the coracoid; the other, external, is slightly concave, and contributes, with a similar but narrower concave surface of the coracoid, to form the glenoid cavity for the humerus.

The *coracoid*, fig. 199, G, *o*, and 52, early coalesces with the scapula in the Ornithorhynchus; it maintains its independent condition to a later period in the Echidna. In both it is a strong, subcompressed, subelongate bone, expanded at both ends: one of these is articulated and anchylosed with the scapula, as above described; the other is joined to the anterior and external facet of the manubrium sterni. The posterior margin of the coracoid is concave and free; the anterior margin is straight and articulated with a narrower 'epicoracoid' in the Echidna than in the Ornithorhynchus.

The *clavicles*, fig. 199, *m*, *m*, are two curved styles, extending from the acromion along the transverse bar of the episternum, *t*. The *humerus*, ib. 53, is remarkable for its shortness and breadth, especially of its two extremities. There is a small sesamoid ossicle, above the internal tuberosity, answering to the 'os humero-capsulare' in the shoulder-joint of Birds (p. 67). The proximal expansion terminates by a broad thick convex border, the middle part of which is developed into the articular head, which is so adapted to the glenoid cavity, that the bone is maintained in a horizontal position, and the distal expansion is nearly vertical. The deltoid and pectoral crests are strongly developed; both condyles are remarkably produced, especially the internal one, which is perforated, fig. 199, H, *a*. The distal articular surface scarcely occupies a fourth part of that broad termination of the humerus: it presents, in the Echidna, fig. 201, 53, a convex tubercle, which is broadest in front for the articulation of the radius, narrow behind for that of the ulna. The articular surfaces of both antibrachial bones are concave: so that the elbow-joint admits freely of flexion and extension, abduction and adduction, but is restricted in the movement of rotation.

The *radius*, fig. 199, 54, and *ulna*, ib. 55, are in contact and pretty firmly connected together through nearly their whole extent; the interosseous space being reduced to a slight fissure. The *ulna* is chiefly remarkable for the olecranon, fig. 199, I, *u*, which is bent forward upon the humerus, and transversely expanded at its extremity, especially in the Ornithorhynchus, in which the lower or inner angle of the expanded extremity is considerably produced. The shaft of the ulna is compressed, and increases in breadth, in the Echidna, as it approaches the broad carpus. In the Ornithorhynchus it is bent like the italic *ſ*, is more cylindrical, and more suddenly expanded at the distal end. The *radius* offers little worthy of notice, except that in the Ornithorhynchus it is flattened next the ulna, and so applied to that bone as to prevent altogether a rotation of the hand upon the ulna. In the Echidna the distal articular surface of the ulna, fig. 207, *n*, presents two convex trochleæ separated by a median concavity; that of the radius, ib. *r*, offers a reverse condition; here two concavities are divided by a median convex ridge: all the four facets at the carpal joint of the antibrachium are in the same transverse line. The two radial concavities receive the two articular convexities of the broad scapholunar bone, fig. 206, *a*; the two convex trochleæ of the ulna play upon two concavities, one-half of each of which is contributed by the cuneiform, ib. *b*, and pisiform, *c*. This complicated joint limits the movement of the hand upon the fore-arm to flexion and extension.

Notwithstanding the confluence of the scaphoid with the lunar bone in the *carpus* of the Echidna, as in that of the Marsupials and Carnivora, it includes eight ossicles, a small sesamoid bone, fig. 206, *x*, being developed in the tendon of the flexor carpi radialis, and articulated with the scapholunar bone, *a*, and radius. The distal series of the carpus includes the four nor-

206

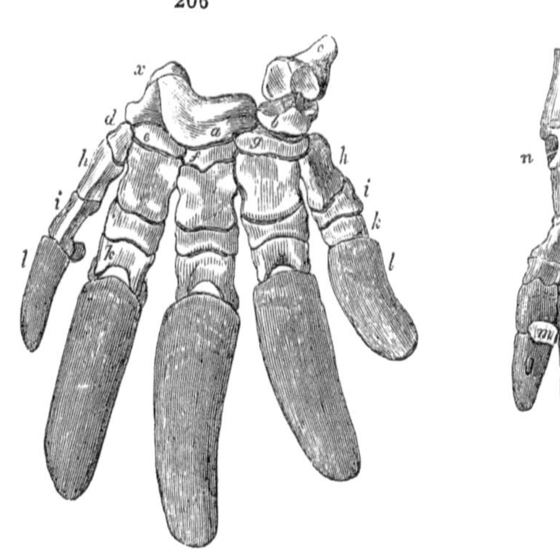

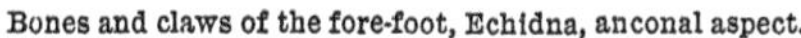
Bones and claws of the fore-foot, Echidna, anconal aspect.

207

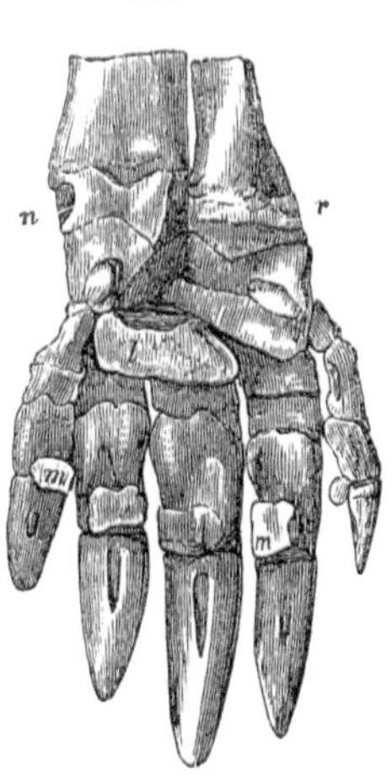

Bones of the fore-foot, Echidna, palmar aspect.

mal bones, the trapezium, ib. *d*, supporting the innermost digit or pollex, the trapezoides, *e*, the index, the os magnum, *f*, which is almost the smallest, sustaining the medius, and the unciforme, *g*, the two outer digits: in the Ornithorhynchus the os magnum contributes a greater share to the articulation with the ring-finger.

In the Echidna all the bones of the fore-extremity are relatively larger and stronger than in the Ornithorhynchus, but this difference is especially remarkable in the metacarpals and two first rows of phalanges, fig. 206, *h*, *i*, *k*, which are singularly short, broad, and thick. The palm is strengthened by two large sesamoids developed in the flexor tendons; these are sometimes confluent, fig. 207, *l*. The number of phalanges in both Monotremes is the same as in other Mammals, viz. two to the thumb and three to each of the fingers. This is not the case in any Saurian, and the retention of the Mammalian type at the peripheral segment of the limb, with the singular deviation from it at the central supporting arch, is not one of the least remarkable points in the osteology of the Monotremes.

There is a sesamoid bone at the palmar aspect of each of the distal articulations of the phalanges in the Echidna, fig. 207, *m*, and at all the digital articulations in the Ornithorhynchus, fig. 199, H, *d*. The ungual phalanges are long, depressed, nearly straight, of great strength in the Echidna, in which each of them is perforated at the palmar aspect, fig. 207.

The pelvis of the Monotremes bears a resemblance to that of Reptiles in the length of time during which the three components of each os innominatum remain distinct, especially in the Echidna; and in the great developement of the ilio-pectineal spine, which equals in size that of the Tortoise, in the Ornithorhynchus: the pelvis of the Echidna resembles that of Birds in the perforation of the acetabulum, fig. 208, *g*; but the pelvis in both Monotremes chiefly resembles that of the higher Lyencephala in the presence of the marsupial bones, ib. *e*, fig. 199, *x*.

208

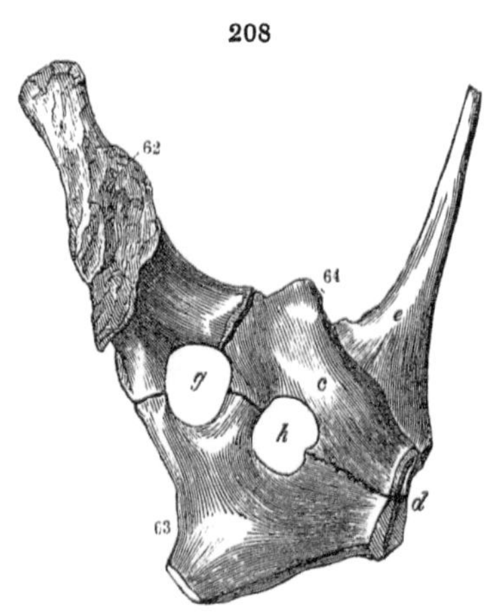

Internal view of pelvis, Echidna.

The *ilium*, figs. 199, 208, 62, is a short, strong, trihedral bone, with the upper extremity expanded and everted in the Ornithorhynchus: the *ischium*, ib. 63, has its tuberosity prolonged backward in an obtusely-pointed form: the *pubis* in the same animal, besides having the spinous process directed forward, gives off a second smaller process, which projects outward; this process is present, but less developed, in the Echidna, fig. 208, 64. The pubis and ischium contribute an equal share to the formation of the foramen obturatorium, ib. *h*, and to the symphysis, *d*, which closes the pelvis below.

The *marsupial bones*, fig. 199, *x*, *x*, 208, *e*, are relatively larger and stronger in the Monotremes than in the ordinary Marsupialia, the Koala excepted; their base extends along the anterior margin of the pubis from the symphysis outward to that of the spinous process: they are relatively longer in the Echidna than in the Ornithorhynchus; they always remain moveably articulated with the brim of the pelvis.

The *femur*, fig. 199, 65, is short, broad, and flattened; its head rises, like that of the humerus, from the middle of a broad expanded proximal end, having on each side a strong process, the outer one representing the great, the inner one the small, trochanter. In the Echidna a projecting ridge extends from the great or outer trochanter beyond the middle of the bone; the whole of the inner part of the shaft is bounded by a trenchant edge; both outer and inner margins of the bone are trenchant in the Ornithorhynchus. The distal end of the femur is expanded transversely, but compressed from before backward. The rotular trochlea is flat transversely, convex vertically, in the Echidna; it is hardly definable when the cartilage is separated from the bone; but the *patella* itself is well developed, and ossified in both Monotremes, fig. 199, *p*.

The *tibia*, ib. 66, is straight in the Echidna, but bent, with the convexity next the fibula, in the Ornithorhynchus; its cristæ are slightly marked.

The *fibula* is slightly bent in the Echidna, fig. 201, 67, but is straight in the Ornithorhynchus, fig. 199, 67; in both Monotremes it is longer than the tibia by the extent of a process, ib. *v*, which rises upward beyond the proximal articulation of the fibula, and strongly expresses the homotypal relation of this bone with the ulna: this process reaches halfway up the back of the femur in the Ornithorhynchus, and, like the olecranon, is greatly expanded at its termination.

209

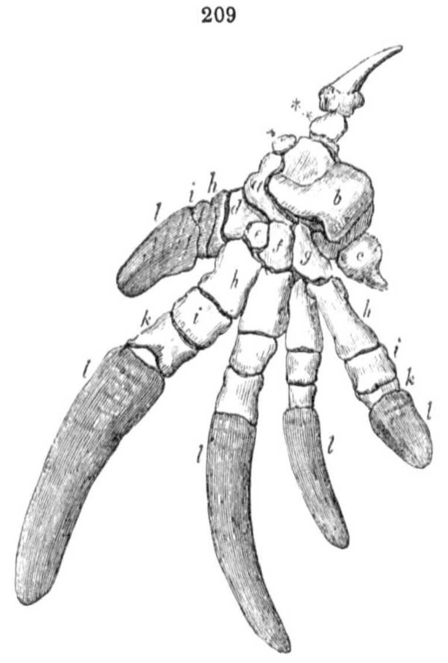

Bones and claws of the hind-foot, rotular aspect, Echidna.

The *tarsus*, figs. 209, 210, consists of a naviculare, *a*, astragalus, *b*, a calcaneum, *c*, three cuneiform bones, *d*, *e*, *f*, and a cuboid, *g*, in the Echidna; but the cuboid in the Ornithorhynchus is divided into two bones, as in some Reptiles, one for the fourth and the other for the fifth metatarsals. In both Monotremes there is a sesamoid bone, fig. 209, *, placed at the interspace between the astragalus and the naviculare; a second supernumerary bone, ib. **, is articulated to the posterior part of the astragalus, and supports the perforated spur which characterises the male sex, fig. 199, K, *d*, *e*. The calcaneum of the Ornithorhynchus, *cl*, terminates by sending outward a short obtuse tuberosity; in the Echidna this part is more slender, and is singularly directed inward and forward nearly in a line with the digits, fig. 210, *c*.

The astragalus in the Ornithorhynchus presents a double trochlea above for the tibia and fibula, and a depression on its inner side, which receives the incurved malleolus of the tibia, almost as in the Sloths. The toes have the same number of bones as in other Mammals; their size and form are more alike in the two Monotrematous genera than those of the fingers: the ungual phalanges, like the claws they support, are more curved than those on the fore-foot, but like them they are perforated on their inner and concave side, fig. 210.

210

Bones of hind-foot, plantar aspect, *Echidna setosa*.

§ 180. *Skeleton of Marsupialia.*—A. *Vertebral Column.*—The number of 'true' vertebræ is the same in all the *Marsupialia*, viz. 26; that of the dorsal and lumbar series varying according to the number of long and free ribs, e. g. *d* 12 *l* 7 in Petaurists, *d* 15 *l* 4 in Wombats, *d* 13 *l* 6 in other genera; the cervicals are seven in all.

In the Koala the length of the spine of the first dorsal hardly exceeds that of the last cervical, but in all other Marsupials the difference is considerable, the first dorsal spine being much longer; those of the remaining dorsal vertebræ progressively diminish in length and increase in breadth and thickness. They slope toward the centre of motion, which is shown by the verticality of its spine: this, in the *Perameles*, is at the eleventh dorsal vertebra, in Potoroos and Kangaroos, fig. 211, at the ninth-twelfth, in the Petaurists at the thirteenth vertebra. In the Phalangers, Koala, and Wombat, the flexibility of the spine is much diminished, and the centre of motion is not defined by the convergence of the spinous process toward a single vertebra, but they all incline slightly backward, fig. 212.

The metapophyses which begin to increase in length in the three posterior dorsal vertebræ, attain a great size in the lumbar vertebræ, and are locked into the interspace between the anapophyses and post-zygapophyses. The diapophyses of the lumbar vertebræ progressively increase in length as the vertebræ approach the sacrum; they are most developed in the Wombat, where they are directed obliquely forward. In the Kangaroos, Potoroos, and Perameles, they are curved forward and obliquely downward. The length of these and of the metapophyses is relatively least in the Petaturists, Phalangers, and Opossums.

In the Wombat the metapophysis rises suddenly from the outside of the prezygapophysis of the twelfth dorsal, increases in length to

211

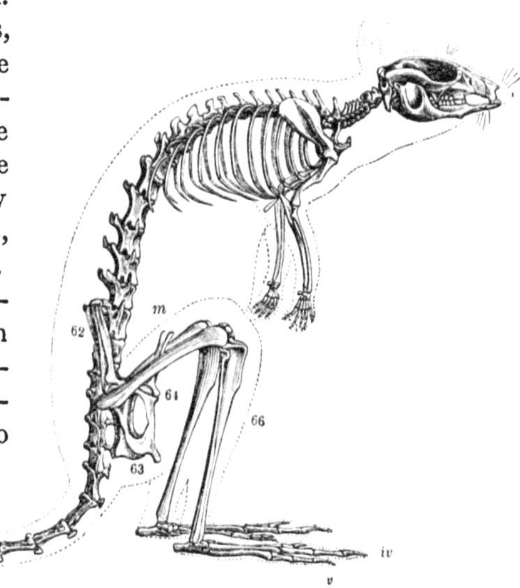

Kangaroo.

the second lumbar, diminishes by degrees to the second sacral, and is rudimental in the following sacral and caudal vertebræ. A rudiment of the anapophysis is first discernible on the eleventh dorsal: the process gradually increases to the last dorsal, diminishes in the lumbar, and disappears in the last of that series. The diapophysis, moreover, is not suppressed on the last dorsal vertebra. The serial homology of the transverse processes of the lumbar vertebræ is here manifested in the most unequivocal way; both metapophyses and anapophyses coexist with diapophyses in the last four dorsal and the first three lumbar vertebræ. Whether, therefore, the metapophysis or the anapophysis be the part called 'tubercle' by some Anthropotomists, neither of them is, in the lumbar vertebræ, the process named 'transverse' in the thoracic vertebræ: that process, to which the name 'diapophysis' is restricted in the present work, is continued distinctly into the lumbar region, and is there lengthened out by a superadded 'pleurapophysis,' which is ossified from a distinct centre in the Wombat.

The free or thoracic ribs consist of bony pleurapophyses and gristly hæmapophyses, acquiring bone-earth only in aged Marsupials: in the Wombat the six anterior pairs articulate directly with the sternum, in the nine following the hæmapophyses are attached to one another. The pressure which the trunk of the Wombat must occasionally have to resist in its burrowing work may be the condition of the unusual number of bony arches of the trunk. In the Kangaroo seven anterior pairs of ribs join the sternum; several of the posterior pairs terminate freely, fig. 211.

212

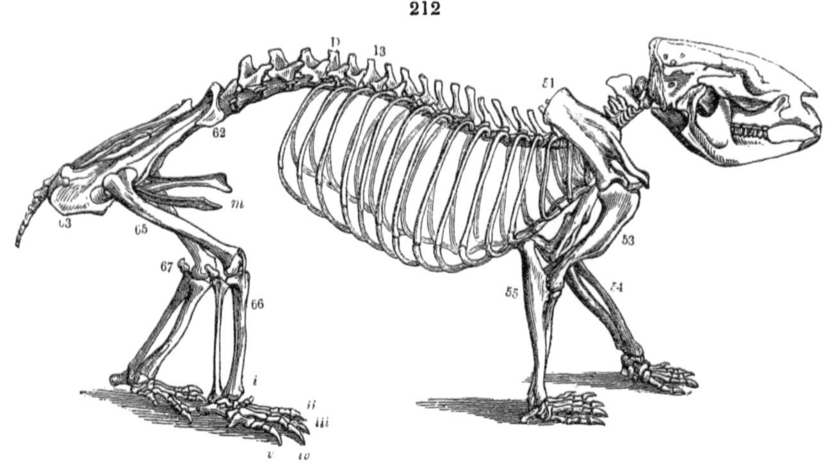

Wombat.

The sternum consists of a longitudinal series of four bones in the Wombat, of five in the Petaurist, and of six in most other Marsupials. The first, or 'manubrium,' is the largest; in many a longitudinal crest is developed from the middle of its outer surface, which in Wombats and Phalangers is produced, and gives attachment to the clavicles. The first pair of ribs abut upon the anterior angles of the triangular manubrium of the Wombats, but in Dasyures, Opossums, Petaurists, and Kangaroos, the manubrium is compressed and elongated, and the clavicles join the produced anterior end. The hæmapophyses articulate at the interspaces of the succeeding sternebers, the last of which supports a broad flat 'xiphoid' cartilage.

The number of vertebræ succeeding the lumbar which are anchylosed together in the sacral region of the spine, amounts in the Wombat to seven, fig. 213; but if we regard those vertebræ only as sacral which join the ossa innominata, then there are but four. In the Phalangers there are generally two such sacral vertebræ, but in the *Phalangista Cookii* the last lumbar assumes the character of the sacral vertebræ both by anchylosis and partial junction with the ossa innominata.

In the Kangaroos and Potoroos the impetus of the powerful hinder extremities is transferred to two anchylosed vertebræ. In the Perameles there is only a single sacral vertebra, the spine of which is shorter and thicker than those of the lumbar vertebræ, and is turned in the contrary direction, viz. backward.

213

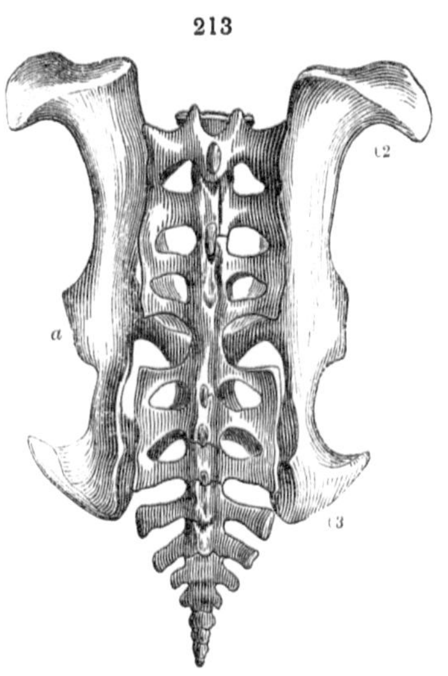

Pelvis of the Wombat.

In the Myrmecobius there are four sacral vertebræ by anchylosis, two of which join the ilia. In Mauge's Dasyure, two sacral vertebræ are anchylosed, but it is to the expanded transverse processes of the anterior one only that the ossa innominata are joined. The same kind of union exists in the Viverrine Dasyure, but three vertebræ are anchylosed together in this species. In the Phalangers and Petaurists there are two sacral vertebræ. In *Petaurus macrurus* three are anchylosed together, though only two join the ilium. In the Wombat, fig. 213, the transverse processes of the numerous anchylosed vertebræ are remarkable for their length and flatness: those of the first four are directed outward and are confluent at their extremities; the remaining ones are turned in a slight degree backward, coalesce, and very nearly reach the tuberosities of the ischia: behind these they gradually diminish in size and disappear in the three last caudal vertebræ. The transition from the sacral to the caudal vertebræ is here very obscure. If we limit the sacral to the three or four which join the ilium, then there remain twelve vertebræ for the tail. The spinal canal is complete in all but the last three, which consist only of the body. There are no hæmal spines, and as only the six posterior vertebræ, which progressively diminish in length, extend beyond the posterior aperture of the pelvis, the tail is scarcely visible in the living animal. In the Koala, fig. 227, the tail is also very short. In the Chæropus it seems to be wanting. In one species of Perameles I find eighteen caudal vertebræ; in another twenty-three. In two species of Potoroo there are twenty-four caudal vertebræ, but the relative length of the tail differs in these by one third, in consequence of the different length of the bodies of the vertebræ. In *Hypsiprymnus ursinus* there are more than twenty-six caudal vertebræ. In the Great Kangaroo there are twenty-two caudal vertebræ. In Bennett's Kangaroo there are twenty-four caudal vertebræ, which are remarkable for their size and strength. In the *Phalangista vulpina,* there are twenty-one caudal vertebræ. In the *Petaurus macrurus* I find twenty-eight caudal vertebræ, while in the *Pet. sciureus* there are but twenty; the bodies of the middle caudal vertebræ in both these species are remarkably long and slender. The Myrmecobius has twenty-three caudal vertebræ: in *Didelphys cancrivora* there are thirty-one; in the Virginian Opossum there are twenty-two caudal vertebræ. In the latter species the spinal canal is continued along the first six; beyond these the neural spines cease to be developed, and the body gives off, above, only the two anterior and two posterior zygapophyses which are rudimental, and no longer subservient to the mutual articulation of the vertebræ. The transverse processes are single on the first five caudal vertebræ, and are nearly the breadth of the body, but diminish in length from the second caudal, in which vertebræ they are generally the longest. In the other vertebræ a short obtuse process is developed at both extremities of the body on either side, so that the dilated articular surfaces of the posterior caudal vertebræ present a quadrate figure.

In most of the Marsupials which have a long tail, this appendage is subject to pressure on some part of the under surface. In the Kangaroo, fig. 211, this must obviously take place to a considerable degree when the tail is used as a fifth extremity, to aid in supporting or propelling the body. In the Potoroos and Bandicoots the tail also transmits to the ground part of the superincumbent pressure of the body by its under surface, when the animal is erect, but it is not used as a crutch in locomotion as in the Kangaroos. In the Phalangers and Opossums the tail is prehensile, and the vessels situated at the under surface are liable to compression when the animal hangs suspended by the tail. To protect these vessels, therefore, as well as to afford additional attachment to the muscles which execute the various movements for which the tail is adapted in the above-mentioned Marsupials,

V-shaped bones, or hæmal arches, are developed, of various forms and sizes, and are placed beneath the articulations of the vertebræ, a situation which is analogous to that of the neural arches in the sacral region of the spine in Birds, and in the dorsal region of the spine in the Chelonian Reptiles. The two crura of the hæmal arch embrace and defend the bloodvessels, and the spinous process continued from their point of union presents a variety of forms in different genera. In Cook's Phalanger the hæmapophyses commence between the second and third caudal vertebræ, increase in length to the fourth, and then progressively diminish to the end of the tail; the penultimate and antepenultimate presenting a permanent separation of the lateral moieties, and an absence of the spine, fig. 214. In the Great Kangaroo the spine of the first hæmal arch only is simple and elongated, the extremities of the others are expanded, and in some jut out into four obtuse processes, two at the sides, and two at the anterior and posterior surfaces.

214

Terminal caudal vertebræ, Phalanger.

The cervical vertebræ, seven in number in all Marsupials, show usually to the last the circumscription of the vertebrarterial foramen by confluence of a short pleurapophysis, fig. 216, *pl*, with di- and met-apophyses: but I have seen the pleurapophyses still unanchylosed in a full-grown Perameles. In Dasyures, Opossums, Phalangers, and Perameles, the seventh cervical has the diapophysis only: in the Kangaroos both atlas and dentata may have the transverse process merely grooved by the vertebral arteries: in the Koala and Wombat the atlas presents only the perforation on each side of the superior arch. In the Perameles and some other Marsupials, the neurapophyses of the atlas, fig. 216, *n*, are distinct from the hypapophysis, fig. 215, *h*, as well as from their proper centrum, the odontoid, fig. 216, *c a*. In the Koala and Wombat the hypapophysis remains cartilaginous, and the lower part of the vertebral ring is completed, in the skeleton, by dried gristly substance, fig. 216. In the Petaurists, Kangaroos, and Potoroos, the atlas is completed below by an extension of ossification from the neurapophyses into the cartilaginous hypapophysis, simulating the body, and the ring of the vertebra is for a long time interrupted by a longitudinal fissure in the middle line, the

215

Atlas of *Perameles lagotis*.

breadth of which diminishes with age. In all the Marsupials the spine of the dentata is well developed both in the vertical and longitudinal directions, but most so in the Virginian and Crab-eating Opossums, fig. 217, where it increases in thickness posteriorly; in these species also the third, fourth, and fifth cervical vertebræ have their spines remarkably long and thick, but progressively diminishing from the third, fig. 218, which equals in height and thickness, but not in longitudinal extent, the spine of the dentata. These spines are four-sided, and being closely impacted together, one behind another, must add greatly to the strength, while they diminish the mobility, of this part of the spine. The structure of the transverse processes of the cervical vertebræ, fig. 218, *d*, is also adapted to the strengthening and fixation of this part of the vertebral column: they are expanded nearly in the axis of the spine, but so that the posterior part of one transverse process overlaps the anterior part of the succeeding. This structure is exhibited in a slighter degree in the cervical vertebræ of the Dasyures, Phalangers, and Great Kangaroo. In the Petaurists, Potoroos, Wombat, and Koala, the direction and simpler form of the transverse processes allow of greater freedom of lateral motion. In the Koala and Wombat a short obtuse process is given off from the under part of the transverse process of the sixth cervical vertebræ. In the Potoroos, Kangaroos, Petaurists, Phalangers, Opossums, and Dasyures, this process is remarkably expanded in the direction of the axis of the spine. In the Bandicoots corresponding processes are observed, progressively increasing in size, on the fourth, fifth, and sixth cervical vertebræ.

216

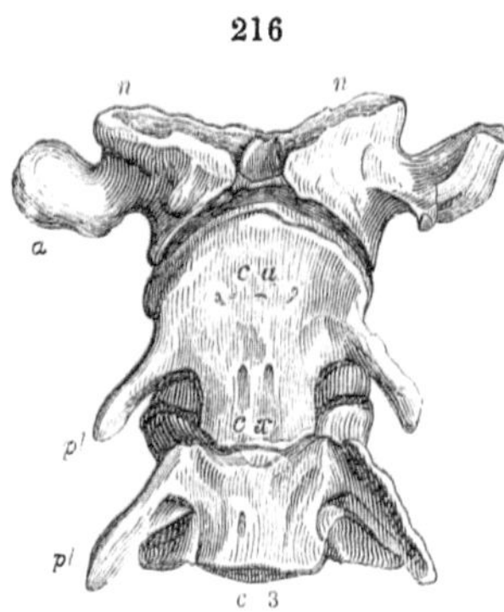

Atlas, axis, and third cervical vertebra, Koala.

217

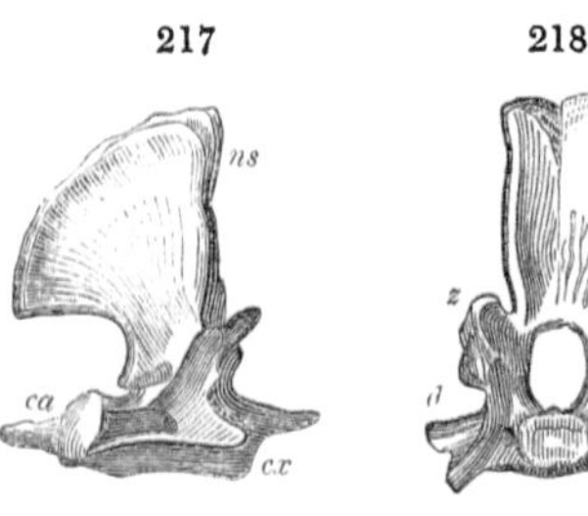

Vertebra dentata, *Didelphys Virginiana.*

218

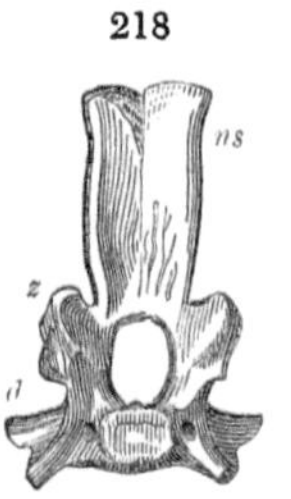

Third cervical vertebra, *Didelphys Virginiana.*

B. *Skull.*—The form of the skull varies much in different Marsupials, but it may be said, in general terms, to resemble an elongated cone, being terminated by a vertical plane surface behind, and in most of the species converging toward a point anteriorly: it is also generally more depressed or flattened than in the placental Mammalia. The skull is also remarkable in all the Marsupial genera for the small proportion which is devoted to the protection of the brain, and for the great expansion of the nasal cavity immediately anterior to the cranial cavity.

In the stronger carnivorous Marsupials the exterior of the cranium is characterised by bony ridges and muscular impressions, but in the smaller herbivorous and insectivorous species, as the Petaurists, Potoroos, and *Myrmecobius*, the cranium presents a smooth convex surface as in Birds, corresponding with the smooth unconvoluted surface of the simple brain contained within, fig. 219.

The breadth of the skull in relation to its length is greatest in the Wombat,[1] Ursine Dasyure,[2] and Petaurists, in which it equals three-fourths of the length, and is least in the *Perameles lagotis*, in which it is less than one-half.

The occipital region, which is generally plane, and vertical in position, forms a right angle with the upper surface of the skull, from which it is separated by an occipital or lambdoidal crista. This crista is least developed in the Myrmecobius, Petaurists, and Kangaroos, and most so in the Thylacine and larger Opossums, in which, as also in the Koala, the crest curves slightly backward, and thus changes the occipital plane into a concavity for the firm implantation of the strong muscles from the neck and back.

219

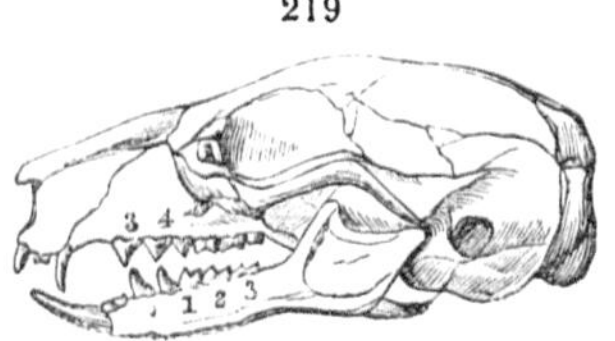

Petaurus pigmæus, magnified.

The upper surface of the skull presents great diversity of character, which relates to the different developement of the temporal muscles, and the varieties of dentition in the different genera.

The extinct Nototherium offers the singular exception of an expansion of the facial part of the skull, vertically and transversely, from the orbits to the terminal muzzle.[3]

In the Wombat the coronal surface offers an almost flattened tract bounded by two slightly elevated temporal ridges, which are upwards of an inch apart posteriorly, and slightly diverge as they extend forward to the anterior part of the orbit. In the skull of the Virginian Opossum the sides of the cranium meet above at an acute angle, and send upward from the line of their

[1] As 15 to 20. [2] As 10 to 14.
[3] LXXXIII'. p. 170, pl. vii.

union a remarkably elevated sagittal crest, which, in mature skulls, is proportionally more developed than in any of the placental Carnivora, not even excepting the strong-jawed Hyæna. The Thylacine and Dasyures, especially the Ursine Dasyure, exhibit the sagittal crest in a somewhat less degree of developement. It is again smaller, but yet well marked, in the Koala and Perameles. The temporal ridges meet at the lambdoidal suture in the larger *Phalangistæ* and in the *Hypsiprymni*, but the size of the muscle in these does not require the developement of a bony crest. In the Kangaroo, the temporal ridges, which are very slightly raised, are separated by an interspace of the third of an inch. They are separated for a proportionally greater extent in the Petaurists, especially *Petaurus flaviventer*; and in the smooth and convex upper surface of the skull of *Petaurus sciureus, Pet. pigmæus, Myrmecobius,* the impressions of the feeble temporal muscles almost cease to be discernible.

The zygomatic arches are, however, complete in these as in all the other genera; they are usually, indeed, strongly developed; but their variations do not indicate the nature of the food so clearly, or correspond with the differences of animal and vegetable diet in the same degree, as in the placental Mammalia. And this is not surprising when we recollect that no Marsupial animal is devoid of incisors in the upper jaw, like the ordinary Ruminants of the placental series: accordingly the more complete dental system with which the herbivorous Kangaroos, Potoroos, Phalangers, &c., are provided, and which appears to be in relation to the scantier pasturage and the dry and rigid character of the herbage or foliage on which they browse, requires a stronger apparatus of bone and muscle for the action of the jaws, and especially for the working of the terminal teeth. There are, however, well-marked differences in this part of the Marsupial skull; and the weakest zygomatic arches are those of the Insectivorous *Perameles* and *Acrobates,* in which structure we may discern a correspondence with the Edentate Anteaters of the placental series. Still the difference in the developement of the zygomata is greatly in favour of the Marsupial Insectivora.

The *Hypsiprymni* come next in the order of developement of the zygomatic arches; which again are proportionally much stronger in the true Kangaroos. The length of the zygomata in relation to the entire skull is greatest in the Koala and Wombat. In the former animal they are remarkable for their depth and straight and parallel course, as well as for their longitudinal extent, fig. 221. In the Wombat, fig. 220, they have a considerable curve outward, so as greatly to diminish the resemblance which otherwise exists in the form of the skull between this Marsupial and the Herbivorous Rodentia of the placental series, as, e. g., the Viscaccia.

In the carnivorous Marsupials the outward sweep of the zygomatic arch, which is greatest in the Thylacine and Ursine Dasyure, is also accompanied by a slight curve upward, but this curvature is chiefly expressed by the concavity of the lower margin of the zygoma, and is by no means so well marked as in the placental Carnivora. It is remarkable that this upward curvature is greater in the slender zygomata of the Perameles than in the stronger zygomata of the Dasyures and Opossums. In the Koala and Phalangers there is also a slight tendency to the upward curvature; in the Wombat the outwardly expanded arch is horizontal. In the Kangaroo the lower margin of the zygoma describes a slightly undulating curve, the middle part of which is convex downward.

In many of the Marsupials, as the Kangaroo, the Koala, some of the Phalangers, Petaurists, and Opossums, the superior margin of the zygoma begins immediately to rise above the posterior origin of the arch. In the Wombat an external ridge of bone commences at the middle of the lower margin of the zygoma, and gradually extends outward as it advances forward, and being joined by the upper margin of the zygoma, forms the lower boundary of the orbit, and ultimately curves downward in front of the antorbital foramen, below which it bifurcates and is lost. This ridge results, as it were, from the flattening of the anterior part of the zygoma, which thus forms a smooth and slightly concave horizontal platform for the eye to rest upon.

The same structure obtains, but in a slighter degree, in the Koala. In the Kangaroo the anterior and inferior part of the zygoma is extended downward in the form of a conical process, which reaches below the level of the grinding-teeth; it is developed from the maxillary. A much shorter and more obtuse process is observable in the corresponding situation in the Phalangers and Opossums.

The relative length of the facial part of the skull anterior to the zygomatic arches varies remarkably in the different Marsupial genera. In the Wombat it is as six to nineteen; in the Koala as five to fourteen; in the *Petaurus sciureus* and *Petaurus Bennettii* it forms about one-fourth of the entire skull; in the Phalangers about one-third; in the carnivorous Dasyures and Opossums more than one-third; in the Thylacine nearly one-half; in *Perameles*,

Macropus, and *Hypsiprymnus murinus*, the length of the skull anterior to the orbit is equal to the remaining posterior part; but in a species from Van Dieman's Land (*Hypsiprymnus myosurus*, Ogilb.), the facial part of the skull anterior to the orbit exceeds that of the remainder, and the arboreal *Hypsiprymni* from New Guinea present a still greater length of muzzle. In most Marsupials the skull gradually converges toward the anterior extremity; the convergence is more sudden in the *Petaurists*, especially *Pet. Bennettii*; but in the *Perameles lagotis* the skull is remarkable for the sudden narrowing of the face anterior to the orbits, and the prolongation of the attenuated snout, preserving the same diameter for upwards of an inch before it finally tapers to the extremity of the nose. In the Koala the corresponding part of the skull is as remarkable for its shortness as it is in the *Per. lagotis* for its length, but it is bounded laterally by parallel lines through its whole extent. In nearly all the Marsupials two long parapophyses project downward from the inferior angles of the occipital region. These processes are longest in the Kangaroos and Koala; in the Wombat they coexist with the true mastoids, which are of larger size, fig. 220, 8. In the Opossums and Dasyures the paroccipitals are short and obtuse; in *Acrobates* they cease to exist, but they are present in the larger *Petaurists.*

The elements of the occipital neural arch remain longer distinct in Marsupials than in most other Mammals. In the skull of a half-grown Thylacine the basioccipital has coalesced with the exoccipitals, which almost meet above the foramen magnum. The lateral sinus impresses the fore part of each exoccipital, and then sinks into a canal which communicates or opens into the precondyloid canal: from this another canal extends forward through the side of the basioccipital. The superoccipital has coalesced with the parietals and interparietal. The basisphenoid has coalesced with the alisphenoids and the presphenoid, but not with the pterygoids: it has no 'sella' nor clinoid processes: it is perforated by the entocarotid at its back and outer angle: the canals converge forward and slightly upward, and terminate above the middle of the basisphenoid. The alisphenoids have the foramen ovale near their posterior borders: the foramen rotundum is a longer canal. The posterior angles of the alisphenoid expand into tympanic bullæ: pterapophyses are sent off in advance which join both pterygoids and palatines. The parietals have coalesced with each other, with the frontal, with the interparietal, and the superoccipital. The orbitosphenoids are very small; their coalesced bases arch backward over the optic nerves and presphenoidal prolongation of the basisphenoid, as in the bird, and their under part is grooved (not perforated) by the optic nerves, which escape by the fissura lacera anterior.

The nasal portion of the coalesced frontals is more expanded than the cerebral one: the frontal sinuses extend to the coronal suture and raise the outer far above the vitreous table: in this table the frontal and coronal sutures remain, but they are obliterated in the outer table. The vomer is carinate below. The nasals are distinct from each other and from the frontals: they are grooved externally for the premaxillaries. The petromastoid, tympanic, and temporal bones continue permanently separate, though confluent ossification proceeds to blend the occipital, parietal, and frontal into one bone. The petrosal is small, its tentorial ridge or angle is sharp, and its cerebellar fossa very deep, though small: a branch of the lateral sinus perforates the petromastoid and the adjoining part of the temporal to open behind the root of the zygoma: the mastoid part is compressed and abuts against the outer side of the base of the paroccipital. The tympanic is a simple scoop-shaped bone, or half-cylinder, cut obliquely. The palatine process of the premaxillary is very deeply notched, and is excavated behind the outer incisor.

In the skull of the mature Wombat, fig. 220, the exoccipitals were still unanchylosed; the left is figured separate at 3.

In the skull of a *Perameles nasuta* the exoccipitals are separated by an interspace, so that a fissure is continued from the upper part of the foramen magnum to the superoccipital element. The same structure may be observed in the Great Kangaroo, and it is very remarkable in the young skulls of this species. In the Wombat the corresponding fissure is very wide, and the lower margin of the superoccipital is notched, so that the shape of the foramen magnum somewhat resembles that of the trefoil leaf. In the Opossum, the exoccipitals meet above and complete the foramen magnum. The petrosal and mastoid are commonly confluent. So loose is the connection of the tympanic, that without due care it is liable to be lost in preparing the skulls of the Marsupials. In the Kangaroo and Wombat, it forms a complete bony tube, fig. 220, 28, and in the Potoroo the bony circle is incomplete at the upper part; in the Perameles and Dasyures the tympanic bone forms a semicircle, the posterior part being deficient, and the tympanic membrane being there attached to the mastoid, as in Birds. In the Dasyures, Petaurists, Perameles, Potoroos, and Koala, fig. 221, 6, there is a large *bulla ossea* for increasing the extent of the auditory cavity, formed by the expansion of the base of the sphenoid. In *Acrobates* and *Perameles lagotis*, there is also an external dilatation of the petrosal, fig. 222, 16, which thus forms a second and smaller bulla on each side, behind the large bulla ossea formed by the alisphenoid, *b*. In other Marsupials the petrous bone is of small size, generally limited to the office of protecting the parts of the internal ear, and sometimes, as in the Koala, is barely visible at the exterior of the base of the skull. The mastoid portion appears in the occipital region of the skull of the Koala, fig. 221, 8, between the exoccipital and squamous portion of the temporal. In the Wombat the mastoid sends outward the strong compressed process, fig. 220, 8, which terminates the lateral boundaries of the occipital plane of the cranium; but this process is entirely due to the exoccipitals in the Koala, fig. 221, 4, and other Marsupials.

220

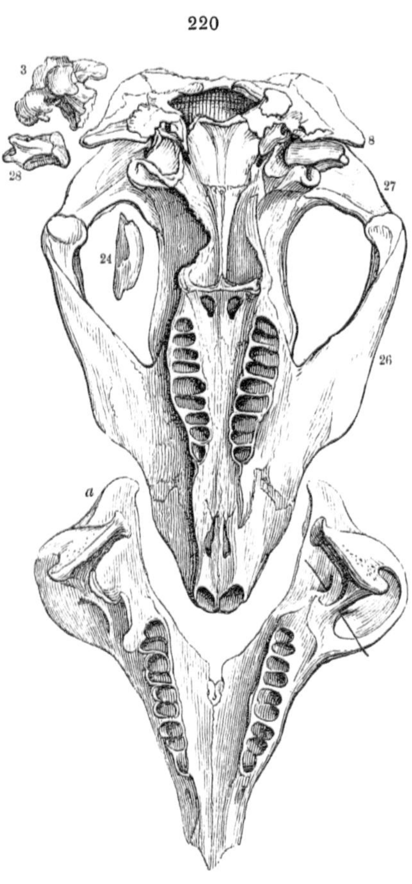

Phascolomys.

The auditory chamber of the ear is augmented in the Phalangers, the Koala, the Kangaroos, and Potoroo, by a continuation of air-cells into the base or origin of the zygomatic process; but the extent of the bony air-chambers communicating with the tympanum is proportionally greatest in the Flying Opossums, where, besides the sphenoid bulla, the mastoid and the whole of the zygomatic process of the squamosal are expanded to form air-cells with very thin and smooth walls, fig. 219.

The direction of the bony canal of the organ of hearing corresponds with the habits of the species. The meatus is directed outward and a little forward in the carnivorous Dasyures; outward and a little backward in the Perameles and Phalangers; outward, backward, and upward in the Kangaroos, and directly outward in the Petaurists and Wombat; but the differences of direction are but slightly marked in these timid vegetarians.

The squamosal generally reaches half-way from the root of the zygoma to the sagittal ridge or suture; it is most developed in the Wombat, in which its superior margin describes a remarkably straight line. The zygomatic process is generally compressed and much extended in the vertical direction in the Opossum, Dasyure, Phalanger, Koala, fig. 221, 27, and Kangaroo. In the Wombat it curves outward from the side of the head in the form of a compressed and almost horizontal plate, fig. 220, 27; it is then suddenly twisted into the vertical position, to be received into the notch of the malar portion of the arch, ib. 26. In *Macropus* the back part of the zygoma is perforated by a pneumatic foramen which receives air from the tympanum.

The cavity corresponding to the sphenoidal bulla ossea in other Marsupials is in this species excavated in the lower part of the squamosal at the inner side of the articular surface for the lower jaw. This articular surface, situated at the base of the zygomatic process, presents in the marsupial as in the placental Mammalia various forms, each manifesting a physiological relation to the structure of the teeth and adapted to the required movement of the jaws in the various genera. In the herbivorous Kangaroo the glenoid cavity forms a broad and slightly convex surface, as in the Ruminants, affording freedom of rotation to the lower jaw in every direction. In the Phalangers and Potoroos the articular surface is quite plane. In the Perameles it is slightly convex from side to side, and concave from behind forward. In the Wombat it is formed by a narrow convex ridge considerably extended, and slightly concave, in the transverse direction. This ridge is not bounded by any descending process posteriorly, so that the jaw is left free for the movements of protraction and retraction. In the Koala the glenoid cavity is a transversely oblong depression with a slight convex rising at the bottom, indicating rotatory movements of the jaw. In the carnivorous Dasyures it forms a concavity still more elongated transversely, less deep than in the placental Carnivora, but adapted, as in them, to a ginglymoid motion of the lower jaw. In all the genera, save in the Wombat, retraction of the lower jaw is opposed by a descending process of the temporal bone immediately anterior to the meatus auditorius and tympanic bone.

The glenoid cavity presents a characteristic structure in most of the Marsupialia in not being exclusively formed by the squamosal. With the exception of the Petaurists, the malar bone forms the outer part of the articular surface for the lower jaw, and in the *Thylacinus*, *Dasyurus Maugei*, *Dasyurus ursinus*, *Perameles*, *Hypsiprymnus*, and *Macropus*, the alisphenoid forms the inner boundary of the same surface. In the Kangaroo, Dasyures, Koala, and Wombat, the alisphenoid articulates with the parietal, but by a very small portion in the two latter species: in the Perameles and Potoroos the alisphenoid does not reach the parietal.

221

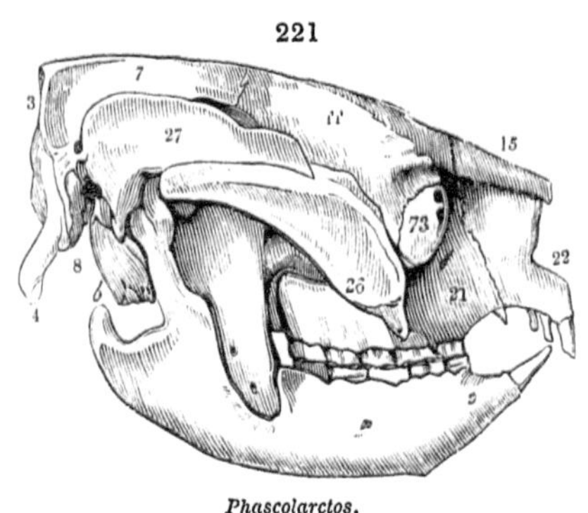

Phascolarctos.

In the *parietals*, fig. 221, 7, the sagittal suture is obliterated in those species in which a bony crista is developed in the corresponding place. They present a singularly flattened form in the Wombat, in an aged skull of which, and in a similar one in the Kangaroo, I observe a like obliteration of the suture. In the Kangaroo, Potoroo, Petaurus, Phalanger, and Myrmecobius, there is a triangular interparietal bone. The corresponding bone I find in three pieces in the skull of a Wombat.

The *frontals*, ib. 11, are chiefly remarkable for their anterior expansion and the great share which they take in the formation of the nasal cavity. In the Thylacine the part of the cranium occupied by the frontal sinuses exceeds in breadth the cerebral cavity, from which it is divided by a constriction. The coronal suture presents in most of the Marsupials an irregular angular course, forming a notch in the frontals on each side which receives a corresponding triangular process of each parietal bone. A process corresponding to the posterior frontal augments the bony boundary of the orbit in the Thylacine, the Ursine Dasyure, and in a slighter degree in the Virginian Opossum; it is relatively most developed in the skull of the *Myrmecobius fasciatus*, where the orbit is large; but the bony boundary of the orbit is not complete in any Marsupial. In the Myrmecobius there is a deep notch at the middle of the superorbital ridge. A corresponding but shallower notch is present in the skull of *Petaurus sciureus*. I have found the frontal suture obliterated in old specimens of the Thylacine, the Virginian Opossum, Cook's Phalanger, the taguanoid, and yellow-bellied Petaurists; but the frontal suture exists in *Petaurus sciureus*, *Acrobates*, and other Marsupials. The interorbital space is concave in the Phalangers and in the *Petaurus taguanoides*, but is quite flat in the other Petaurists.

The *lacrymals* vary in their relative size in different Marsupials. In the Koala, fig. 221, 73, they extend upon the face about a line beyond the anterior boundary of the orbit, and at this part they present a groove with one large and two or three small perforations. In the Wombat their extent upon the face is slightly increased; it is proportionally greater in the Kangaroos, Potoroos, Phalangers, and Petaurists, in which this part of the lacrymal bone presents two perforations close to the orbit. In the Thylacine, besides the two external holes there is a large perforation within the orbital margin. This carnivorous Marsupial, as compared with the Wolf, presents a greater extent of the facial portion of the lacrymal bone, and thus indicates its inferior type. In the *Myrmecobius* the lacrymal bone exhibits its greatest relative developement. The extraorbital lacrymal foramen is a good marsupial character: it is present in the *Thylacoleo,* where it is single, as in *Dasyurus ursinus.*

The *malars*, figs. 220 and 221, 26, are very strong and of great extent in almost all the Marsupialia. They are least developed in *Acrobates*, fig. 219, *Myrmecobius*, and *Perameles lagotis*. In the latter, fig. 222, the malar bone presents a singular form, being bifurcate at both extremities: the *processus zygomaticus maxillæ superioris* is wedged into the cleft of the anterior fork; the corresponding process of the squamosal fills up the posterior notch. The anterior bifurcation of the malar bone is not present in the Marsupials generally: the external malo-maxillary suture forms an oblique and almost straight line in the Wombat, Phalanger, Opossum, Dasyures, and Kangaroo. Owing to the inferior developement of the zygomatic process of the superior maxillary in the Wombat, the malar bone is not suspended in the zygomatic arch as in the Rodentia. It is also of relatively much larger size and of a prismatic form, arising from the developement of the oblique external ridge above described. In the Kangaroos, Potoroos, Great Petaurus, and Phalangers, it is traversed externally by a ridge showing the attachment of the masseter, of which muscle the extent of origin is augmented by the descending zygomatic process of the maxillary; this is most developed in the gigantic fossil Notothere and Diprotodon.

The *nasal bones* vary in their form and relative size in the different genera; they are longest and narrowest in the Perameles, shortest and broadest in the Koala, fig. 221, 15. Their most characteristic structure is the expansion of their upper and posterior extremity, which is well marked in the Wombat, Myrmecobius, Petaurists, Phalangers, Opossums, and Dasyures. In the Potoroos the anterior extremities of the nasal bones converge to a point which projects beyond the premaxillaries. In the *Perameles lagotis* the bony case of the nasal passage is further increased by the presence of two small *rostral bones*, resulting, as in the Hog, from ossification of the nasal cartilage.

The *premaxillaries* always contain teeth, and the ratio of the developement of these bones corresponds with the bulk of the dental apparatus which they support. They are consequently largest in the Wombat, where they extend far upon the side of the face and are articulated to a considerable proportion of the nasals, but do not, as in Rodentia, reach the frontal or divide the maxillary bone from the nasal. They present a somewhat lower degree of developement in the Koala, fig. 221, 22, but both in this species and in the Wombat they bulge outward, and thus remarkably increase the transverse diameter of the osseous cavity of the nose. In *Hypsiprymnus* and *Macropus* the incisive palatal foramina are chiefly in the premaxillaries, but a small proportion of their bony circumference is due to the anterior extremity of the palatal process of the maxillary: the same structure obtains in the Wombat, Koala, and Opossums. In the Dasyures and Phalangers a greater proportion of the posterior boundary of the incisive foramina is formed by the maxillaries; in the Petaurists they are entirely surrounded by the maxillary bones, while in the Perameles they are, on the contrary, entirely included in the maxillaries. They always present the form of two longitudinal fissures, fig. 222, *a*.

The *maxillary*, fig. 221, 21, in the Koala and Wombat sends upward a long, narrow, irregular nasal process, which joins the frontal and nasal bones, separating them from the premaxillaries. The antorbital foramen does not present any marked variety of size, which is generally moderate. It is much closer to the orbit in the carnivorous Marsupialia than in the corresponding placental quadrupeds. It is relatively largest in the Ursine Dasyure. It presents the form of a vertical oblique fissure in the Wombat. I have observed it double in the Kangaroo. In this and some other herbivorous Marsupials the malar process of the maxillary sends down a process for increasing the power and size of the masseter muscle.

In *Phalangista Cookii*, in *Petaurus flaviventer*, and *Petaurus sciureus*, in *Macropus major*, and some other Great Kangaroos, the bony palate is of great extent and presents a smooth surface, concave in every direction toward the mouth; it is pierced by the two posterior palatine foramina at the anterior external angles of the palatine bones, either within or close to the transverse palato-maxillary sutures. Behind these foramina, in the Kangaroo, there are a few small irregular perforations. The bony palate is similarly entire in the *Hypsiprymnus ursinus*. In *Macropus Bennettii* there are four orifices at the posterior part of the bony palate. The two anterior ones are situated upon the palato-maxillary suture, and are of an ovate form with the small end forward. The two posterior foramina are of a less regular form and smaller size. In the Brush Kangaroo (*Macropus Brunii*, Cuv.) the posterior palatal foramina present the form of two large fissures placed obliquely and converging posteriorly. They encroach upon the posterior borders of the maxillary plate. Anterior to these vacancies there are two smaller foramina, and posterior to them are one or two similar foramina.

In the Potoroos, Wombat, and Koala, the posterior palatal openings are large and oval, and situated entirely in the palatal bones. In *Hyps. setosus* they extend as far forward as the interspace between the first and second true molars; in *Hyps. murinus* they reach to that between the second and third true molars: posterior and external to these large vacuities there are two small perforations. In the Phalangers, with the exception of *Ph. Cookii*, the palatal openings are proportionally larger; they extend into the palatal process of the maxillaries, and the thin bridge of bone which divides the openings in the Potoroo, &c., is wanting; the two perforations at the posterior external angles of the palatine bones are also present. In the Virginian Opossum the bony palate presents eight distinct perforations, besides the incisive foramina; the palatal processes of the palatine bone extend as far forward in the median line as the third molars: a long and narrow fissure extends for an equal distance (three lines) into the palatal processes both of the palatines and maxillaries: behind these fissures and nearer the median line are two smaller oblong fissures; external and a little posterior to these are two similar fissures, situated in the palato-maxillary suture; lastly, there are two round perforations close to the posterior margin of the bony palate.

In the Ursine Dasyure a large transversely oblong aperture is situated at the posterior part of the palatal processes of the maxillary bones, and encroaches a little upon the palatines. In Mauge's Dasyure there are two large ovate apertures crossing the palato-maxillary sutures separated from each other by a broad plate of bone; posterior to these are two apertures of similar size and form, which, being situated nearer the mesial line, are divided by a narrower osseous bridge; each posterior external angle of the bony palate is also perforated by an oval aperture. In the Viverrine Dasyure the two vacancies which cross the palato-maxillary suture are in the form of longitudinal fissures, corresponding to the fourth and fifth grinders; the posterior margin of the bony palate has four small apertures on the same transverse line.

222

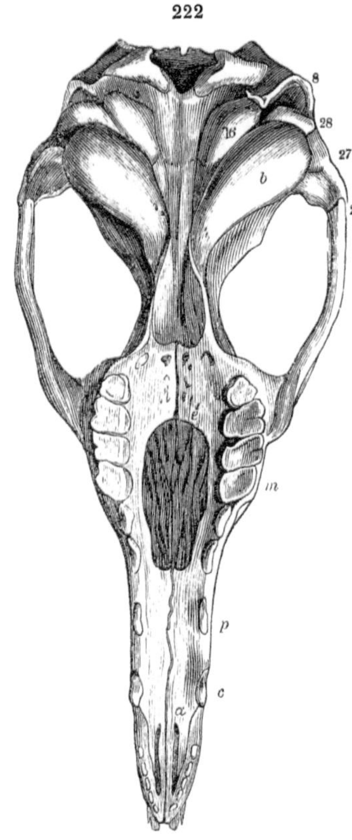

Perameles lagotis.

Since the defective condition of this part of the cranium is one of the characteristics of the skull of the Bird, it might be expected that some approximation would be made to that structure in the animals which form the transition between the Placental and Oviparous Vertebrates. We have already noticed the large vacuities which occur in the bony palate of nearly all the Marsupials; but this imperfectly ossified condition is most remarkable in the *Perameles lagotis,* in which, fig. 222, the bony roof of the mouth is perforated by a wide oval space extending from the second premolars to the penultimate molars, exposing to view the vomer and the convolutions of the inferior spongy bones in the nasal cavity. The pterygoids, fig. 220, 24, long maintain their individuality; and repeat the connections they present in Birds.

The parietes of the cranial cavity are remarkable for their thickness in some of the Marsupial genera. In the Wombat the two tables of the parietal bones are separated posteriorly for the extent of more than half an inch, the interspace being filled with a coarse cellular diploë; the frontal bones are about two and a half lines thick. In the Ursine Dasyure the cranial bones have a similar texture and relative thickness. In the Koala the texture of the cranial bones is denser, and their thickness varies from two lines to half a line. In the Kangaroo the thickness varies considerably in different parts of the skull, but the parietes are generally so thin as to be diaphanous, which is the case with the smaller Marsupials, as the Potoroos and Petaurists. The union of the body of the second with that of the third cranial vertebra takes place in the marsupial as in the placental Mammalia at the sella turcica, which is overarched by the backward extension of the orbitosphenoids. The optic foramina and the fissuræ laceræ anteriores are all blended together, so that a wide opening leads outward from each side of the sella. Immediately posterior and external to this opening are the foramina rotunda, from each of which in the Kangaroo a remarkable groove leads to the fossa Gasseriana at the commencement of the foramen ovale; the same groove is indicated in a slight degree in the Dasyures and Phalangers, but is almost obsolete in the Wombat and Koala. The entocarotid canals pierce the basisphenoid, as in Birds, and terminate in the cranial cavity very close together behind the sella turcica, which is not bounded by a posterior clinoid process. The sphenoidal bulla, which forms the chief part of the tympanic cavity in the *Perameles lagotis*, forms a large convex protuberance on each side of the floor of the cranial cavity in that species. The petrosal in the Kangaroo, Koala, and Phalangers, is impressed above the meatus auditorius by a deep, smooth, round pit, which lodges the lateral appendage of the cerebellum, as in Birds. The corresponding pit is shallower in the Dasyuri, and is almost obsolete in the Wombat. The middle and posterior fissuræ laceræ have the usual relative position, but the latter are small. The condyles are each perforated anteriorly by two foramina in most of the Marsupials, the Thylacinus forming the exception and showing only one. The foramen magnum is of great size in relation to the capacity of the cranium; the aspect of its plane is backward and slightly downward. A venous canal leads from the lateral sinus between the upper part of the petrosal and the squamosal, and perforates the latter behind or above the root of the zygoma.

In the Kangaroo and Phalanger a thin ridge of bone extends for the distance of one or two lines into the periphery of the tentorial process of the dura mater, and two sharp spines are sent down into it from the upper part of the cranium in the *Phalan-*

gista vulpina. The tentorium is supported by a thick ridge of bone in the Thylacine; but it is not completely ossified in any of the Marsupials: in some species, indeed, as the Dasyures, the Koala, and the Wombat, the bony crista above described does not exist. There is no ossification of the falciform ligament as in the Ornithorhynchus.

The rhinencephalic division of the cranial cavity is well defined from the prosencephalic one. It is relatively smallest in the Koala. In all Marsupials it is bounded anteriorly by the cribriform plate, which is converted into an osseous reticulation by the number and size of the olfactory apertures. The cavity of the nose, from its great size and the complication of the turbinal bones, forms an important part of the skull. It is divided by a complete bony septum to within one-fourth of the anterior aperture; the anterior margin of the septum is slightly concave in the Koala; describes a slight convex line in the Wombat, Kangaroo, and Phalanger, and a sigmoid flexure in the Dasyure. A longitudinal ridge projects downward from the inside of each of the nasal bones, and is continued posteriorly into the superior turbinal; this bone extends into the dilated space anterior to the cranial cavity, which corresponds with the frontal sinuses. The convolutions of the middle turbinal are extended chiefly in the axis of the skull; the processes of the anterior turbinal are arranged obliquely from below upward and forward. The nasal cavity communicates freely with large maxillary sinuses, and finally terminates by wide apertures behind the bony palate. In the skull the nasal cavity communicates with the mouth, as before mentioned, by means of the various large vacuities in the palatal processes.

In the carnivorous Marsupials, as the Thylacine, the lower maxillary bone resembles in general form that of the corresponding species in the placental series, as the Dog: a similar transverse condyle is placed low down near the angle of the jaw, on a level with the series of molar teeth; a broad and strong coronoid process rises high above the condyle, and is slightly curved backward; there is the same well-marked depression on the exterior of the ascending ramus for the firm implantation of the temporal muscle, and the lower boundary of this depression is formed by a strong ridge extended downward and forward from the outside of the condyle. But in the Dog and other placental *Carnivora* (some Seals excepted), a process, representing the angle of the jaw, extends directly backward from the middle of the above ridge, which process gives precision and force to the articulation of the jaw, and increases the power by which the masseter acts upon the jaw. Now, although the same curved ridge of bone bounds the lower part of the external muscular depression of the ascending ramus in all the Marsupials, it does not in any of them send backward, or in any other direction, a process corresponding to that just described in the Dog. The angle of the jaw itself, in the Marsupials, is as if it were bent inward in the form of a process, encroaching in various shapes and various degrees of developement in the different genera upon the interspace of the rami of the lower jaw. On looking directly upon the lower margin of the jaw, we see, therefore, in place of the margin of a vertical plate of bone, a more or less flattened triangular surface extended between the external ridge, and the internal process or inflected angle. In the Opossums the internal angular process is triangular and trihedral, directed inward, with the point slightly curved upward, and more produced in the small than in the large species. In the Dasyures it has a similar form, but the apex is extended into an obtuse process. In the Thylacine the base of the inverted angle is proportionally more extended, and a similar structure is presented by the fossil Phascolothere. In the Perameles the angle of the jaw forms a still longer process; it is of a flattened form extended obliquely inward and backward and slightly curved upward. It presents a triangular, slightly incurved, and pointed form in the Petaurists, in which it is longest and weakest in the pigmy species (*Acrobates*, Desm.). It is shorter and stronger in the *Myrmecobius*, fig. 223, *a*. In the Potoroos and Phalangers the process is broad with the apex slightly developed; it is bent inward and bounds the lower part of a wide and deep depression in the inside of the ascending ramus. In the Great Kangaroo the internal margin of this process is turned upward, so as to augment the depth of the internal depression above mentioned. The internal angular process arrives at its maximum of developement in the Wombat, fig. 220, *a*, and the breadth of the base of the ascending ramus very nearly equals the height of the same part. In the Koala the size of the process in question is also considerable, but it is compressed, and directed backward with the obtuse apex only bending inward, so that the characteristic flattening of the base of the ascending ramus is least marked in this species. There is no depression on the inner side of the ramus of the jaw in the Koala, but its smooth surface is simply

223

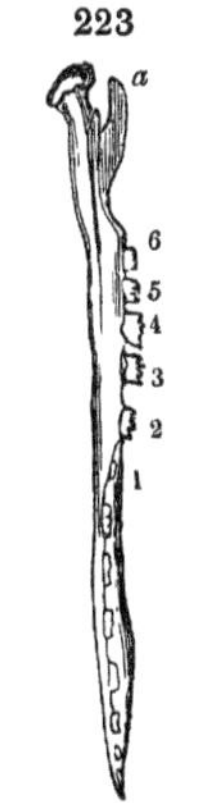

Lower jaw, Myrmecobius.

pierced near its middle by the dental nerve and artery. The surface of the external muscular depression bounded below by a broad angular ridge, as above described, is entire in the Dasyures, Opossums, Bandicoots, Petaurists, and Phalangers; but in the Wombat the outer surface of the ascending ramus is directly perforated by a round aperture immediately posterior to the commencement of the dental canal:[1] the corresponding aperture is of larger size in the Kangaroo. But in the Potoroos both the external and internal depressions of the ascending ramus lead to wide canals, or continuations of the wide depressions which pass forward into the substance of the horizontal ramus, and soon uniting into one passage, leave a vacant space in the intervening bony septum.

In the Thylacine, Ursine Dasyure, and the allied fossil carnivores called Phascolothere, Thylacoleo, and Plagiaulax,[2] the condyle of the lower jaw is placed low down, on a level with the molar series: it is raised a little above that level in the smaller Dasyures and Opossums, and ascends in proportion to the vegetable diet of the species.

In all those Marsupials which have few or very small incisors the horizontal rami of the jaw converge toward a point at the symphysis. The angle of convergence is most open in the Wombat, in which the symphysis is longest. The suture becomes obliterated in aged individuals. In other Marsupials, the rami of the lower jaw are less firmly united at the symphysis; they permit independent movements of the right and left incisors in the Kangaroos: and in the Opossum, both the rami of the lower jaw and all the bones of the face are remarkable for the loose nature of their connections.

C. *Bones of the Limbs.*—The *scapula* varies in form in the different Marsupials. In the Petaurists it is a scalene triangle, with the glenoid cavity at the convergence of the two longest sides. In the Wombat, fig. 212, 51, it presents an oblong quadrate figure, the neck being produced from the lower half of the anterior margin, and the outer surface being traversed diagonally by the spine, which in this species gradually rises to a full inch above the plane of the scapula, and terminates in a long narrow compressed acromion arching over the neck to reach the clavicle.

In the Koala (fig. 224), the superior costa does not run parallel

[1] A bristle is represented passing through this aperture in fig. 220.

[2] XVII'. pp. 341, 353, figs. 113, 119, 173.

with the inferior, *a, d,* but recedes from it as it advances forward, and then passes down, forming an obtuse angle, *c,* and with a gentle concave curvature, to the neck of the scapula; a small process extends from the middle of this curvature. In the Potoroo, the upper costa is at first parallel with the lower, but this parallel part is much shorter; the remainder describes a sigmoid flexure as it approaches the neck of the scapula. In the Great Kangaroo, the Perameles, Phalangers, Opossums, and Dasyures, the whole upper costa of the scapula describes a sigmoid curve, the convex posterior position of which varies as to its degree and extent. The subscapular surface is remarkable in the Perameles for its flatness, but presents a shallow groove near the inferior costa. In most other Marsupials it is more or less convex or undulating.

224

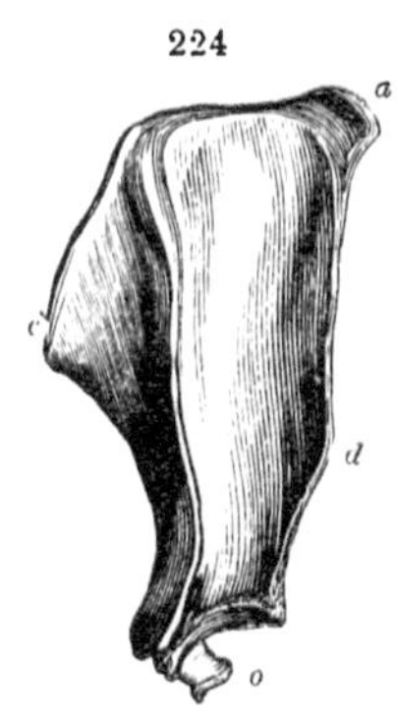

Scapula of Koala.

In the Kangaroos, fig. 211, the supraspinal fossa is of less extent than the space below the spine, and the spine is inclined upward. In the Perameles and Dasyures the proportions of the supra- and infra-spinal surfaces are reversed, and the whole spine is bent downward over the infraspinal surface. In the Potoroos and Phalangers the acromion is bent downward so as to present a flattened surface to the observer; in the Potoroos and Opossums this appearance is produced by a true expansion of the acromion. In the Perameles the coracoid process is merely represented by a slight production of the superior part of the glenoid cavity. In the Kangaroo and Potoroo it forms a protuberance on the upper part of the head of the scapula. In the other Marsupials it assumes the character of a distinct process from the same part, and attains its greatest developement in the Wombat and Koala, in the latter of which it is forcibly curved downward and inward, fig. 224, *o.*

The *clavicles* are present in all the Marsupials, with the exception of the genus *Perameles* and probably also the *Chœropus.* In the claviculate Marsupials they are relatively strongest and longest in the burrowing Wombat, weakest and shortest in the Great Kangaroo. In the latter they are simply curved with the convexity forward, and measure only two inches in length. In the Wombat they are upwards of three inches in length, and have a double curvature; they are expanded and obliquely truncate at the sternal extremity, where the articular surface presents a remarkably deep notch; they become compressed as they approach the acromion, to which they are attached by an extended narrow articular surface.

In the Koala the clavicles are also very strong, but more compressed than in the Wombat, bent outward in their whole extent, and the convex margin formed, not by a continuous curve, but by three almost straight lines, with intervening angles; progressively diminishing in extent to the outermost line which forms the articular surface with the acromion. In the Myrmecobius the clavicles are subcompressed and more curved at the acromial than at the sternal end. In most of the other Marsupials the clavicle is a simple compressed elongated bone, with one general outward curvature.

The *humerus* in most Dasyures resembles that of the Dog-tribe in the imperforate condition of the inner condyle, but differs in the more marked developement of the muscular ridges, especially of that which extends upward from the outer condyle for the origin of the great supinator muscle. This ridge is terminated abruptly by the smooth tract for the passage of the musculo-spiral nerve.

In all the other genera of Marsupials that I have examined the internal condyle of the humerus is perforated. But in some species of *Petaurus*, as *Petaurus sciureus*, the foramen is represented by a deep notch; and in the *Phalangista Cookii*, both foramen and notch are wanting.[1] The ridge above the external condyle is much developed in the *Petaurus macrurus* and *sciureus*, and notched at its upper part, but this notch does not exist in *Pet. taguanoides*. I find similar differences in the developement of the supinator, or outer ridge, in the genus *Perameles*; in the *Per. lagotis* it is bounded above by a groove; in *Per. Gunnii* it is less developed and less defined. In the Kangaroos, Potoroos, Wombat, and Koala (fig. 225), the outer condyloid ridge extends in the form of a hooked process above the groove of the radial nerve. In all these, and especially in the Wombat, the deltoid process of the humerus, fig. 212, 53, is strongly developed; it is continued from the external tuberosity down the upper half of the humerus; except in the Petaurists, where, from the greater relative length of the humerus, it is limited to the upper third. The interspace of the condyles is occasionally perforated, as in the *Perameles lagotis* and Wombat. The articular surfaces at both extremities of the humerus have the usual form; but it may

[1] In the other species of *Phalangista*, and in the *Petaurus taguanoides* and *macrurus*, the internal condyle of the humerus is perforated. In a Thylacine I found it perforated; and in one Ursine Dasyure in the left humerus, but not in the right.

be observed in some Marsupials, as the Koala, that at the distal articulation the external convexity for the radius has a greater relative extent than usual, and the ulnar concavity is less deep.

225

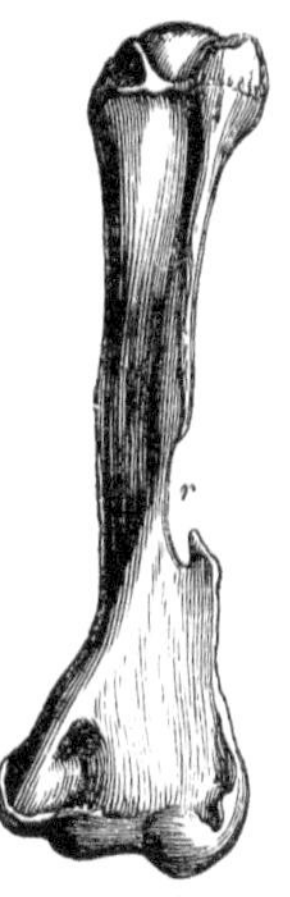

Humerus of the Koala.

The bones of the fore-arm are always distinct and well developed, and their adaptation to pronation and supination is complete. The prehensile faculty and unguiculate structure of the anterior extremities appear to have been indispensable to animals where various manipulations were required in the economy of the marsupial pouch. When, therefore, such an animal is destined like the ruminant to range the wilderness in quest of pasturage, the requisite powers of the anterior members are retained and secured to it, as has been already observed, by an enormous developement of the hinder extremities, to which the function of locomotion is restricted.

We find, therefore, that the bones of the fore-arm of the Kangaroo differ little from those of the burrowing Wombat, the climbing Koala, or the carnivorous Dasyure, save in relative size. They present the greatest proportional strength in the Wombat, and the greatest proportional length and slenderness in the Petaurists or Flying Opossums, in which the radius and ulna are in close contact through a great portion of their extent, and thus lend a firmer support to the outstretched dermal parachute. They are also long and slender in the Koala. In general the radius and ulna run nearly parallel, and the interosseous space is very trifling. It is widest in the Potoroos. The olecranon is well developed in all the Marsupials. In the Virginian Opossum and Petaurists we find it more bent forward upon the rest of the ulna, than in the other Marsupials. In the Wombat, where the acromion is the strongest, and rises an inch and a half above the articular cavity of the ulna, it is extended in the axis of the bone. The distal end of the radius in this animal is articulated to a bone representing the os scaphoides and os lunare.

The ulna, which in the same animal converges toward a point at its distal end, has that point received in a depression formed by the cuneiform and pisiform bones; these are bound together by strong ligaments, and the pisiform then extends downward and backward for two-thirds of an inch. The second row of the carpus consists of five bones. The trapezium supports the inner digit, and has a small sesamoid bone articulated to its radial surface. The trapezoides is articulated to the index digit, and is wedged between the scapholunar bone and os magnum; this forms an oblique articular surface for the middle digit; but the largest of the second series of carpal bones is the cuneiform, which sends downward an obtuse rounded process, and receives the articular surface of the fifth, and the outer half of that of the fourth digit, the remainder of which abuts against the oblique proximal extremity of the middle metatarsal bone. The five metacarpal bones are all thick and short, but chiefly so the outermost. The innermost digit, or pollex, has two phalanges, the remainder three; the ungual phalanx of all the digits is conical, curved, convex above, expanded at the base, and simple at the opposite extremity. In the Perameles the ungual phalanx of the three middle digits of the hand, and of the two outer digits of the foot, are split at the extremity by a longitudinal fissure commencing at the upper part of the base. This structure, which characterises the ungual phalanges in the placental Anteaters, has not been hitherto met with in other Marsupial genera.[1] The terminal phalanges of the Koala are large, much compressed and curved; the concave articular surface is not situated, as in the Cats, on the lower part of the proximal end, but, as in the Sloths, at the upper. The claws which they support are long.

In the Great Kangaroo the first row of the carpus is composed, as in the Wombat, of three bones, but the apex of the ulna rotates in a cavity formed exclusively by the cuneiforme. There are four bones in the second row; of which the unciform is by far the largest, and supports a part of the middle, as well as the two outer digits. In the Potoroos I find but three bones in the distal series of the carpus, the trapezoides being wanting, and its place in one species being occupied by the proximal end of the second metacarpal bone, which articulates with the os magnum. In the Perameles there are four bones in the second carpal row, although the hand is less perfect in this than in any other Marsupial genus, *Chæropus* excepted, the three middle toes only being fully developed. In the Petaurus the carpus is chiefly remarkable for the length of the pisiforme.

It would be tedious to dwell on the minor differences observable in the bony structure of the hand in other Marsupials. I shall therefore only observe that though the inner digit is not situated

[1] It would be instructive to examine the skeleton of the rare *Chæropus*, with reference to this structure.

like a thumb, all the fingers enjoy lateral motion, and that those at the outer can be opposed to those at the inner side so as to grasp an object and perform, in a secondary degree, the function of a hand. In the Koala the two inner digits are more decidedly opposed to the three outer ones than in any other climbing Marsupial. But some of the Phalangers, as the *Ph. Cookii* and *Ph. gliriformis* of Bell, present in a slighter degree the same disposition of the fingers, by which two out of the five have the opposable properties of a thumb. I have observed a similar disposition of the digits in the act of climbing in the Dormouse, and it probably is not uncommon in other placental Mammalia of similar habits and which have long, slender, and freely moveable fingers. As a permanent disposition of the digits, the opposition of three to two is most conspicuous in the prehensile extremities of the Chameleon.

The pelvis, figs. 152, 226, 227, in the mature Marsupials is composed of the os sacrum, the two ossa innominata, and the characteristic supplemental bones, attached to the pubis, called by Tyson the *ossa marsupialia* or *Janitores Marsupii, m.*

We seek in vain for any relationship between the size of the pelvis and that of the new-born young, the minuteness of which is so characteristic of the present tribe of animals. The diameters both of the area and apertures of the pelvic canal are always considerable, but more especially so in those Marsupialia which have the hinder extremities disproportionately large; as also in the Wombat, where the pelvis is remarkable for its width. The pelvis is relatively smallest in the Petaurists; but even here the diameter of the outlet is at least six times that of the head of the new-born young. The anterior bony arches formed by the ossa pubis and the ischia are always complete, and the interspace between these arches is divided, as in other Mammalia, into the two obturator foramina by an osseous bridge continued from the pubis to the ischium on each side of the symphysis.

In the Kangaroos, Potoroos, Phalangers, and Opossums, the ilia offer an elongated prismatic form. They are straight in the Opossum, but gently curved outward in the other Marsupial genera. In the Dasyures there is a longitudinal groove widening upward in place of the angle at the middle of the exterior surface of the ilium. The ilia in the Petaurists are simply compressed, with an almost trenchant anterior margin. They are broader and flatter in the Perameles, and their plane is turned outward. But the most remarkable form of the ilia is seen in the Wombat, in which they are considerably bent outward at their anterior extremity, fig. 226, 62. In the Kangaroos and Potoroos the eye is arrested by a strong process given off from near the middle of the ilio-pubic ridge, and this process may be observed less developed in the other Marsupialia. The tuberosity of the ischia inclines outward in a very slight degree in the Dasyures, Opossums, Phalangers, Petaurists, and Perameles, in a greater degree in the Kangaroos and Potoroos, and gives off a distinct and strong obtuse process in the Wombat, fig. 226, 63, which not only extends outward but is curved forward. In the Potoroos the symphysis of the ischia, or the lower part of what is commonly called the symphysis pubis, is produced anteriorly. The length of this symphysis, and the straight line formed by the lower margin of the ischia, is a characteristic structure of the pelvis in most of the Marsupials.

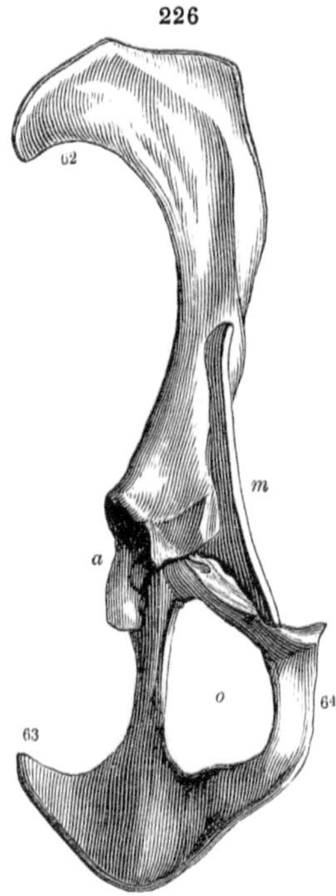

Right os innominatum and marsupial bone, Wombat.

The marsupial bones, figs. 226, 227, *m*, are elongated, flattened, and more or less curved, expanded at the proximal extremity, which sometimes, as in the Wombat, is articulated to the pubis by two points: they are relatively straightest and most slender in the Perameles; shortest in the Myrmecobius, where they do not exceed half an inch in length; longest, flattest, broadest, and most curved in the Koala, where they nearly equal the iliac bones in size. They are always so long that the cremaster muscle winds round them in its passage to the testicle or mammary gland, and the uses of these bones will be described in treating of that muscle. Homologically they are the last pair of lumbar hæmapophyses advanced, as in many Reptiles, from the sclerous to the osseous states: teleologically they belong to the category of the trochlear ossicles, commonly called sesamoid, and are developed in the tendon of the external oblique which forms the mesial pillar of the abdominal ring, as the patella is developed in the tendon of the *rectus femoris*. I cannot, however, participate in the opinion of Laurent and De Blainville,[1] that the marsupial bones are superadded to the abdominal muscles to aid in an unusually energetic

[1] 'Bulletin des Sciences Médicales' of Férussac, 1827, No. 77, p. 112, and the 'Annales d'Anat. et de Physiologie,' 1839, p. 240.

compression required to expel the uterine fœtus. It is not in the females of those animals which give birth to the smallest young that we should expect to find auxiliary parts for increasing the power of the muscles engaged in parturition. The bones in question are, moreover, equally developed in both sexes: and they are so situated and attached that they add to the power of the muscles which wind round them, and not of those implanted in them. They are not, however, merely subservient to add force to the action of the 'cremasteres,' but give origin to a great proportion of the so-called 'pyramidales.'[1]

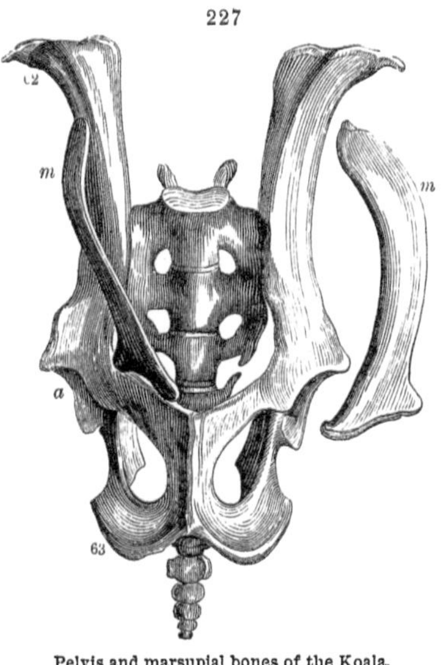

Pelvis and marsupial bones of the Koala.

The osteogenesis of the marsupial pelvis derives some extrinsic interest from the not yet forgotten speculations which have been broached regarding the homologies of the marsupial bones. These have been conjectured to exist in many of the placental Mammalia, with a certain latitude of altered place and form, disguised, e.g., as the bone of the penis in the Carnivora, or appearing as the supplemental ossicles of the acetabulum, which exist in the young of many of the Rodentia. In the os innominatum of the immature Potoroo the curved prismatic ilium contributes to form, by the outer part of its base, the upper or anterior third of the acetabulum; the rest of the circumference of this cavity is completed by the ischium and pubis, excepting a small part of the under or mesial margin, which is formed by a distinct ossicle or epiphysis of the ilium (*a*, fig. 152), answering to that described by Geoffroy St. Hilaire as the rudimental marsupial bone in the Rabbit. Now here there is a coexisting marsupial bone: but besides the five separate bones just mentioned, there is a sixth distinct triangular ossicle, which is wedged into the posterior interspace of the ischiopubic symphysis. The circumference of the acetabulum is always interrupted by a deep notch opposite the obturator foramen, which is traversed by a ligamentous bridge,

[1] See the abstract of a paper on the anatomy of the *Dasyurus*, Proc. Zool. Soc. January, 1835.

and gives passage to the vessels of the Harderian gland lodged in the wide and deep acetabular fossa.

The *femur* is a straight, or nearly straight, long, cylindrical bone, having a hemispherical head supported on a very short neck, especially in the Petaurists, and situated here almost in the axis of the shaft, above and between the two trochanters, which are nearly of equal size. In the Kangaroos and Potoroos the head of the thigh-bone is turned more inward, and the outer or greater trochanter rises above it. In other Marsupials the great trochanter is less developed. In most of the species a strong ridge is continued downward to within a short distance from the trochanter, and this ridge is so produced at the lower part in the Wombat as almost to merit the name of a third trochanter. In the Wombat and Koala there is no depression for a ligamentum teres. The shaft of the bone presents no *linea aspera.*

The canal for the nutrient artery commences at the upper third and posterior part of the bone in the Koala, and extends downward, contrariwise to that in most other marsupial and placental Mammalia.

At the distal extremity of the femur the external condyle is the largest, the internal rather the longest. The intermediate anterior groove for the patella is well marked in the Perameles, where the patella is fully developed, but is broad and very shallow in the Phalangers and Dasyures, where the tendon of the rectus muscle is merely thickened or offers only a few irregular specks of ossification; and the corresponding surface in the Petaurists, Wombat, and Koala is almost plane from side to side; in these Marsupials and in the Myrmecobius the patella is wanting. I find a distinct but small bony patella in the *Macropus Bennettii.* There is a sesamoid bone above and behind the external condyle of the femur in the Myrmecobius and some other Marsupials.

In the knee-joint, besides the two crucial ligaments continued from the posterior angles or cresses of the semilunar cartilages—one to the outer side of the inner condyle, the other to the interspace of the condyles—there is a strong ligament which passes from the anterior part of the tibial protuberance backward to the inner side of the fibular condyle, and a second continued from the same point along the outer margin of the outer semilunar cartilage to the head of the tibia.

The *tibia*, fig. 228, 66, presents the usual disposition of the articular surface for the condyles of the femur, but in some genera, as the Wombat and Koala, the outer articular surface is continuous with that of the head of the fibula. In the Kangaroos and Potoroos the anterior part of the head is much produced, and in the young animal its ossification commences by a centre distinct from the ordinary proximal epiphysis of the bone. A strong ridge is continued down from this protuberance for about one-sixth the length of the tibia. In the Koala a strong tuberosity projects from the anterior part of the tibia at the junction of the upper with the middle third. In this species and in the Wombat, as also in the Opossums, Dasyures, Phalangers, and Petaurists, the shaft of the tibia is somewhat compressed and twisted; but in the Kangaroos, Potoroos, and Perameles the tibia is prismatic above and sub-cylindrical below. The internal malleolus is very slightly produced in any Marsupial, but most so in the Wombat.

The *fibula*, ib. 67, is complete, and forms the external malleolus in all the Marsupials. In one species of *Hypsiprymnus* and in one species of *Perameles* (*P. lagotis*) it is firmly united to the lower part of the tibia, though the line of separation be manifest externally. In a second species of each of the above genera it is in close contact with the corresponding part of the tibia, but can be easily separated from that bone. In the Great Kangaroo the fibula is also a distinct bone throughout, but it is remarkably thinned and concave at its lower half, so as to be adapted to the convexity of the tibia, with which it is in close attachment. In each of these genera, therefore, in which locomotion is principally performed by the hinder extremities, we perceive that their osseous structure is so modified as to insure a due degree of fixity and strength; while in the other Marsupial genera, as *Phascolarctos*, *Phascolomys*, *Phalangista*, *Petaurus*, *Didelphys*, and *Dasyurus*, the tibia and fibula are so loosely connected together and with the tarsus, that the foot enjoys a movement of rotation analogous to the pronation and supination of the hand. This property is especially advantageous in the Petaurists, Phalangers, Opossums, and Koala, because in these the inner toe is so placed and organised as to perform the office of an opposable thumb, whence these Marsupials have been termed *Pedimana*, or foot-handed (fig. 228).

It is to this prehensile power that the modifications of the fibula chiefly relate. In the Wombat, fig. 212, 67, Koala, Petaurists, and Phalangers, it expands to nearly an equal size with the tibia, 66, at the distal extremity, and takes a large share in the formation of the tarsal joint; but the articular surface is slightly convex, while that of the tibia is slightly concave. The proximal extremity of the fibula is also much enlarged, but compressed and obliquely truncated, and giving off two tuberosities from its exterior surface; to the superior of these a large sesamoid bone, fig. 228, 67′, is articulated; a similar sesamoid 'fabella' is attached to the upper end of the fibula in a *Dasyurus macrurus* and *Petaurus taguanoides*. M. Temminck figures it in the *Didelphys ursina* and *Didelphys Philander*. This sesamoid and the expanded process to which it is attached form the homotype of the olecranon, fig. 212, 55; and the correspondence of the fibula with the ulna is very remarkably maintained in the *Pet. taguanoides*, in which the proximal articular surface of the fibula is divided into two facets, one playing upon the outer condyle of the femur, the other concave, vertical, and receiving an adapted convexity on the outer side of the head of the tibia, which rotates thereupon like the radius in the lesser sigmoid cavity of the ulna.

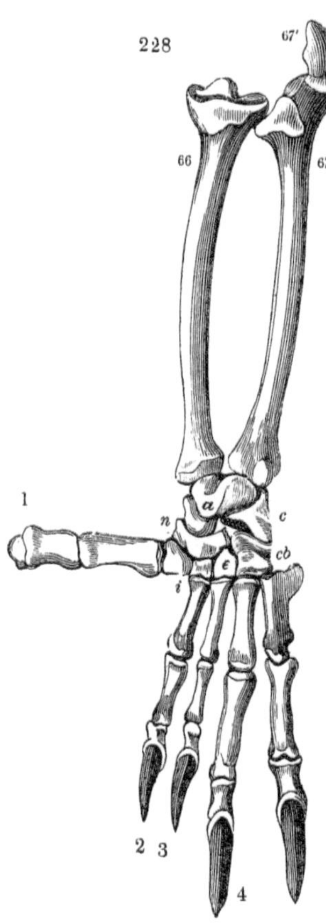

Bones of the leg and foot, *Phalangista*.

In the scansorial and gradatorial Marsupials the bones of the hinder and fore extremities are of nearly equal length, but in the saltatory species the disproportion in the developement of the bones of the hind leg is very great, especially in the Kangaroos and Potoroos, fig. 211. However, in those singular species of *Hypsiprymnus* which inhabit New Guinea and take refuge in trees, the organisation of the Kangaroo is modified and adapted so as to make climbing a possible and easy action. The fore and hind legs are here more equally developed, and the claws on the two larger toes of the hind feet are curved instead of straight. In a skeleton of one of these scansorial Potoroos, the *Hypsiprymnus ursinus*, in the Museum at Leyden, in which the humerus is three inches and a half long, the femur does not quite equal five inches in length: the ulna is nearly four inches, the fibula nearly five inches in length. The fibula is also less firmly connected with the tibia than in the Great Kangaroo.

The following is the structure of the tarsus in the Wombat and Phalanger, fig. 228. The *astragalus*, *a*, is connected as usual with the tibia, fibula, calcaneum, *c*, and naviculare, *n*. The upper articular surface for the tibia is as usual concavo-convex, the internal surface for the inner malleolus flattened and at right angles with the preceding, but the outer articular surface presents a triangular flattened form, and instead of being bent down parallel with the inner articular surface, slopes away at a very open angle from the upper surface, and receives the articular surface of the fibula, 67, so as to sustain its vertical pressure. A small proportion of the outer part of the inferior surface of the astragalus rests upon the calcaneum: a greater part of the superincumbent pressure is transmitted by a transversely extended convex anterior surface to the naviculare, *n*, and cuneiform bones, *i*, *e*. This form of the astragalus is also characteristic of the Koala, Petaurists, Dasyures, and the Pedimanous Marsupials generally. In the Kangaroos, Potoroos, and Perameles which have the *pedes saltatorii*, the fibular articular surface of the astragalus is bent down as usual at nearly right angles with the upper tibial surface.

The calcaneum in the Wombat presents a ridge on the outer surface which serves to sustain the pressure of the external malleolus, which is not articulated to the side of the astragalus. The internal surface which joins the astragalus is continuous with the anterior slightly concave surface which articulates with the cuboides. The posterior part of the bone is compressed, it projects backward for nearly an inch, and is slightly bent downward and inward. This part is relatively shorter in the Koala, Phalangers, Opossums, and Petaurists, but it is as strongly developed in the *Dasyuri* as in the Wombat. The anterior part of the calcaneum of the Phalangers is shown at *c*, fig. 228.

In the *Dasyurus macrurus* a small sesamoid bone is wedged in between the astragalus, tibia, and fibula at the back part of the ankle-joint. In the *Petaurus taguanoides* there is a supplemental tarsal bone wedged in between the naviculare and cuboides on the plantar surface.

The homotypy of the carpal and tarsal bones is very clearly illustrated in the Phalanger. The *lunare* and *scaphoid* of the hand correspond with the *astragalus* and *naviculare* of the foot, transferring the pressure of the *focile majus* upon the three innermost bones of the second series. The long, backward-projecting *pisiform* bone of the wrist closely resembles the posterior process of the *os calcis*; the articular portion or body of the *os calcis* corresponds with the *cuneiforme* of the carpus; the large carpal *unciform* represents the tarsal *cuboides*, and performs the same function, supporting the two outer digits; the three *cuneiform* bones of the foot are obvious homotypes of the *trape-*

zium, *trapezoides*, and *os magnum*. The entocuneiform bone is the largest of the three in the Wombat, although it supports the smallest of the toes. It is of course more developed in the Pedimanous Marsupials, where it supports a large and opposable thumb.

In the Wombat the metatarsals progressively increase in length and breadth from the innermost to the fourth; the fifth or outermost metatarsal is somewhat shorter but twice as thick, and it sends off a strong obtuse process from the outside of its proximal end. A corresponding process exists in the Phalangers, fig. 228. The innermost metatarsal of the Wombat, fig. 212, *i*, supports only a single phalanx; the rest are succeeded by three phalanges each, progressively increasing in thickness to the outermost; the ungual phalanges are elongated, gently curved downward, and gradually diminish to a point.

In the Myrmecobius the tibial or innermost toe is represented by a short rudimental metatarsal bone concealed under the skin. In the Dasyures the innermost toe has two phalanges, but it is the most slender and does not exceed in length the metatarsal bone of the second toe. In the Petaurists it is rather shorter than the other digits, but is the strongest, and in *Petaurus taguanoides* the terminal phalanx is flattened and expanded; the toes are set wide apart in this genus. In the Opossums and Phalangers the innermost metatarsal bone is directed inward apart from the rest, and together with the first phalanx is broad and flat. The second phalanx in the Opossums supports a claw, but in the Phalangers is short, transverse, unarmed, singularly expanded in *Ph. Cookii*, but almost obsolete in *Ph. ursina* (fig. 228, 1). In all the preceding genera there are two small sesamoid bones on the under side of the joints of the toes, both in the fore and hind feet.

The commencement of a degeneration of the foot which is peculiar to Marsupial animals may be discerned in the Petaurists, in the slender condition of the second and third toes, as compared with the fourth and fifth. In the Phalangers this diminution of size of the second and third toes, counting from the hallux, is more marked. They are, also, both of the same length and have no individual motion, being united together in the same sheath of integument as far as the ungual phalanges, whence the name of *Phalangista* applied to this genus (fig. 228, 2 and 3).

In the saltatorial genera of Marsupials the degradation of the corresponding toes is extreme; but though reduced to almost filamentary slenderness they retain the usual number of phalanges, and the terminal one of each is armed with a claw. These claws being the only parts of the rudimental digits which project freely beyond the integument, they look like little appendages at the inner side of the foot for the purpose of scratching the skin and dressing the fur, to which offices they are exclusively designed. The removal of the innermost toe, corresponding with our great toe and the hallux of the *Pedimana*, commences in the Perameles. In one species I find the metatarsal bone of this toe supports only a single rudimental phalanx which reaches to the end of the next metatarsal bone, and the internal cuneiform bone is elongated. In another species the internal toe is as long as the abortive second and third toes, and has two phalanges, the last of which is divided by the longitudinal fissure characteristic of the ungual phalanges in this genus. In the *Perameles lagotis* the innermost toe is represented by a rudimentary metatarsal bone, about one-third the length of the adjoining metatarsal.

In the Poephagous Marsupials no rudiment of the innermost toe exists. The power of the foot is concentrated in all these genera on the fourth and fifth or two outer toes, but especially the fourth, which, in the Great Kangaroo, is upwards of a foot in length, including the metatarsal bone and the claw. This formidable weapon resembles an elongated hoof, but is three-sided and sharp-pointed like a bayonet, and with it the Kangaroo stabs and rips open the abdomen of its assailant: with the anterior extremities it will hold a powerful dog firmly during the attack, and firmly supporting itself behind upon its powerful tail, deliver its thrusts with the whole force of the hinder extremities. The cuboid bone which supports the two outer metatarsals is proportionally developed. The internal cuneiform bone is present, though the toe which is usually articulated to it is wanting. It is also the largest of the three, and assists in supporting the second metatarsal; posteriorly it is joined with the navicular and external cuneiform bones, the small middle cuneiform occupying the space between the external and internal wedge-bones and the proximal extremities of the two abortive metatarsals. The great or fourth metatarsal is straight and somewhat flattened; the external one is compressed and slightly bent outward; the toe which this supports is armed with a claw similar to the large one, but the ungual phalanx does not reach to the end of the second phalanx of the fourth toe, and the whole digit is proportionally weaker. In the climbing Potoroos (*Hypsiprymnus ursinus* and *Hypsiprymnus dorcocephalus*), the two outer toes are proportionally shorter than in the leaping species, and are terminated by curved claws by which they gain a better hold on the branches and inequalities of trees.

§ 181. *Skeleton of Rodentia.*—A. *Vertebral Column.*—The Rodentia have seven cervical, and, as a rule, nineteen dorso-lumbar vertebræ. The agile Hares with flexile trunk have long loins, viz. D 12, L 7, fig. 229. The Jerboas, fig. 232, that bear the

229

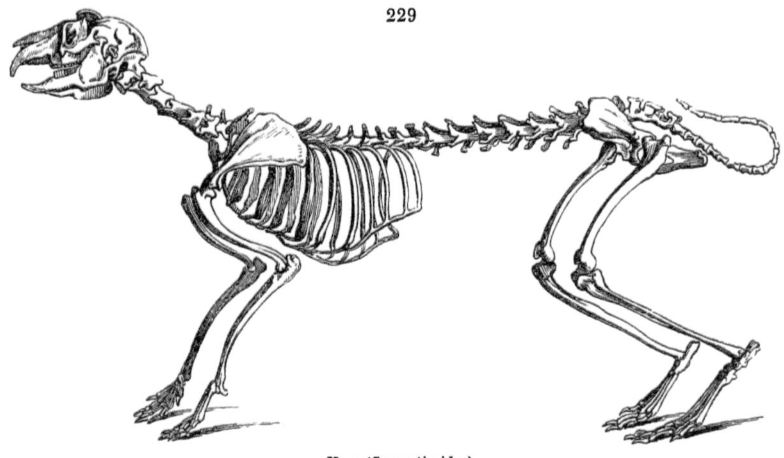

Hare (*Lepus timidus*).

trunk aloft, like the Kangaroos, have also twelve dorsal and seven lumbar vertebræ: the burrowing Porcupines and swimming Beavers, fig. 230, have their trunk braced by a greater

230

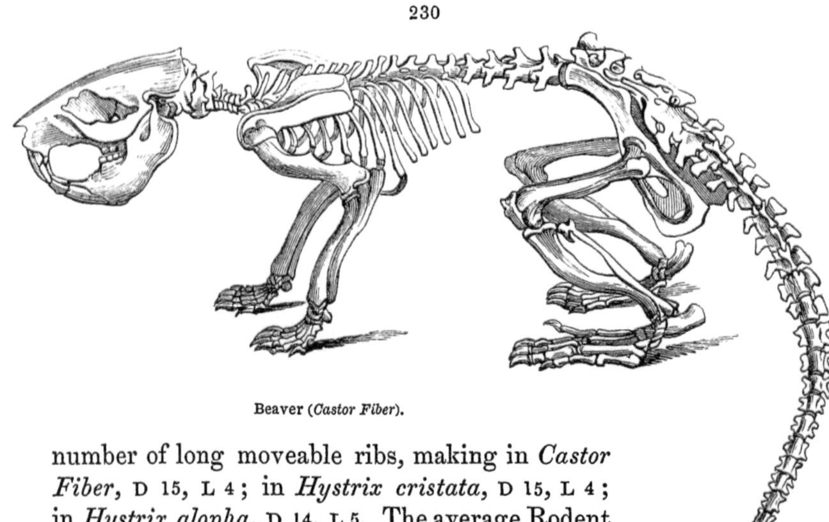

Beaver (*Castor Fiber*).

number of long moveable ribs, making in *Castor Fiber*, D 15, L 4; in *Hystrix cristata*, D 15, L 4; in *Hystrix alopha*, D 14, L 5. The average Rodent number is D 13, L 6. Exceptions are few, and fewer than at first sight, if well looked to; thus, a skeleton of *Dasyprocta Acuchy*, showing D 13, L 7, has the supplemental lumbar vertebra with sacral characters and connection on the left side: Cuvier assigns to the Dormice (Loirs and Lerots) D 13, L 7: the burrowing Cape Mole-Rats have twenty or twenty-one dorso-lumbars: in these I have found 13–7, 14–6, and 14–7, and the latter is the number of dorsal and lumbar vertebræ respectively: the Australian Water-Rat (*Hydromys chrysogaster*) has D 14, L 7: the best-marked exception is that of the Capromys, which has D 16, L 7=23. In some Rodents only one, in most but two, vertebræ join the ilia: three and four are common numbers of anchylosed sacrals. In the seemingly tailless Cavies and Pacas the caudal vertebræ may be but seven, eight, or ten in number: in the Black Rat and *Hapalotis albipes* I have counted as many as thirty. The Great Jerboa has twenty-nine caudals, which also have the proportions and perfections of those in the Kangaroo.

The met- and an-apophyses commence by a common tubercle at the fore part of the dorsal series; the anapophysis, fig. 231, D, *a*, begins to be distinct at the back part of the series, and the metapophysis, ib. *m*, to project from above the anterior zygapophysis, *z*: both processes are usually well developed in the posterior dorsal and lumbar vertebræ, ib. L: the diapophysis, *d*, subsides in the posterior dorsals and is lengthened in the lumbars, L, by a coalesced riblet (*pleurapophysis*), ib. *d*. In

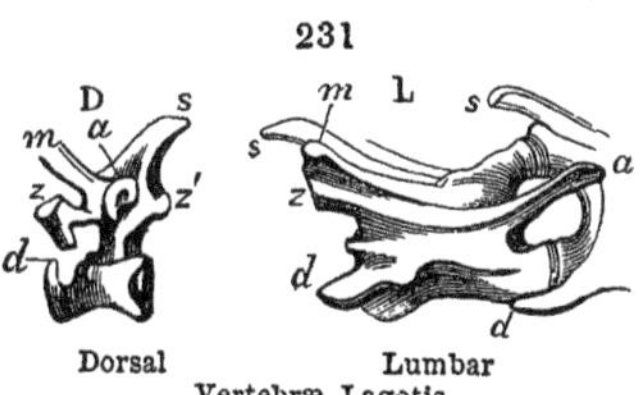

Vertebræ, Lagotis.

the Great Jerboa (*Helamys*) the diapophysis is unusually long and strong in the first dorsal: the anapophysis first projects from the back part of the eighth dorsal, and the metapophysis from the fore part of the ninth: both processes are long in the first five lumbars. The neural spines progressively increase in length to the last lumbar, and are strongly inclined forward toward that of the eleventh dorsal, fig. 232, D: the antecedent spines incline backward to the same vertebra, the spine of which is vertical, and indicates the centre of motion of the trunk. This arrangement of the neural spines is well marked in all the agile flexible-bodied Rodents. In the Hare, fig. 229, the neural spine of the ninth vertebra, D, indicates the centre. The anapophyses begin on the eighth, the metapophyses on the ninth, dorsal: these increase and are continued throughout the lumbar region, where they are very long. The anapophysis assumes the form of a ridge in the last dorsal and lumbar vertebræ. The lumbar di-pleur-apophyses, ib. *d*, are long and incline forward and downward. Long hypapophyses, ib. *h*, are also developed.

The thoracic ribs consist of bony pleur- and gristly hæmapophyses: of these the seven anterior pairs, as a rule, directly join the sternum, which then consists of six bones or 'sternebers.' In the Beaver, Porcupine, Coypu, and a few others, there are eight pairs of 'true ribs:' in an Acuchy with this number I found nine sternal bones, the foremost representing an 'episternal' articulated to the 'manubrium.' The first rib is the shortest, unusually so in *Hydromys*, and has often a partial articulation with the last cervical vertebra. The neural spine of the second dorsal is commonly the longest.

232

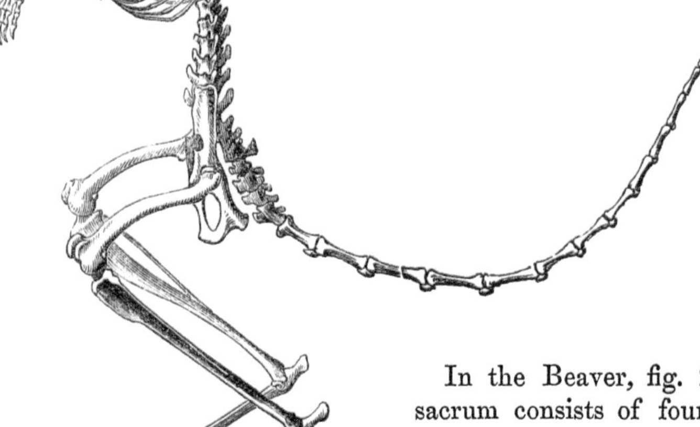

Jerboa (*Dipus Sagitta*).

In the Beaver, fig. 230, the sacrum consists of four anchylosed vertebræ: the articular surface for the ilium is almost confined to the transverse process of the first of these vertebræ: those of the last are the longest. The sacral nerves directly perforate the neurapophyses of the last two vertebræ, anterior to the vacuity left between the bases of the transverse processes. The neural arches of the first six caudal vertebræ are similarly perforated. Their transverse processes are long, horizontally flattened, and terminally expanded; and the vertebræ, after these processes subside, are remarkable for their large size, and a certain degree of correspondence of shape with the broad, flat, scaly tegumentary tail which they support. In most Rodents with long tails, hæmapophyses are developed beneath the intervertebral spaces, as in the Jerboas, fig. 232, *h*. In one member of the Porcupine family (*Cercolabes*), and in one species of *Capromys* (*C. prehensilis*), the tail has a prehensile extremity.

The seventh cervical vertebra has an imperforate transverse process in some Rodents, a perforate one in others: in the Hare I have observed this difference in different individuals. The pleurapophyses early anchylose to form the vertebrarterial foramen in the sixth–second cervicals. The neural spine is usually longest in the second and seventh; it is obsolete in the intermediate cervicals in many Rodents. In the Hare the transverse processes of the atlas are perforated longitudinally by the vertebral arteries, which then perforate the neural arch. The hypapophysis, or so-called body, is ossified, and a small tubercle extends backward from its under part. In the atlas of the Chinchilla the transverse process is pierced both horizontally and obliquely, and the vertebral artery also perforates the neural arch.

B. *Skull.*—As in the Marsupialia, the confluence of the elements of the epencephalic arch is late, and that of the tympanic is restricted to the petrosal and mastoid. The squamosal maintains its individuality, and also much of its long slender proportions, and the malar is suspended in the middle of the zygomatic arch,

233

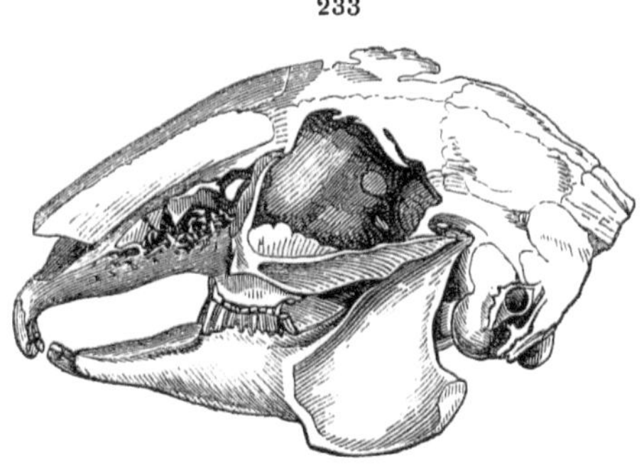

Skull of the Hare.

as in Birds: other characteristics of the Rodent skull will be exemplified in the following species.

In the Hare (*Lepus timidus*, fig. 233) the superoccipital is surmounted by a square platform of bone—originally a distinct interparietal—the posterior angles of which project backward in the form of two tubercles, from between which a vertical crest descends to the foramen magnum. The paroccipitals arch downward and outward in close connection with the descending process of the large subquadrate mastoid, 8, which anchyloses with the petrosal and tympanic. The long bony 'meatus auditorius' ascends obliquely backward—the direction in which this timid Rodent is most concerned in ascertaining the sounds that may warn it of an approaching enemy. The tympanic cavities intercommunicate by a sinus traversing the basisphenoid. The outer part of the alisphenoid is perforated by the ectocarotid artery. The entocarotid pierces the tympanic bulla. The petromastoid is articulated in a peculiar manner to the squamosal, which, after expanding beyond its zygomatic part to be applied to the parietal and alisphenoid, resumes the form of a narrow thin plate of bone, applied to a shallow depression upon the mastoid, and thus clamping it, as it were, to its place. The frontal sends outward a large aliform curved plate above each orbit, the extremities of which form postorbital and antorbital processes, the notches which divide the anterior from the posterior part of the frontal being unusually deep. The common outlet of the optic nerves extends forward, so as to occasion a small vacuity at the back part of the interorbital septum. Each orbit presents a wide vacuity at its fore part, which leads into the lateral nasal cavity, bounded externally by the singularly reticulate nasal plate of the maxillary, 21. The zygomatic arch, which is slightly curved downward but scarcely at all outward, developes a small prominence both from its front and hind extremity below the points of suspension. The articular surface for the lower jaw is broad and concave transversely, narrow and convex longitudinally. The bases of the sockets of the superior molars form a strong prominence in the orbit below the anterior vacuity. The nasal bones, 15, are remarkable both for their length and breadth: they extend further back than the long slender nasal processes of the premaxillaries, 22. The bony palate is extensively encroached upon by the prepalatal apertures, which blend together to form a narrow heart-shaped vacuity with the apex directed forward, largely exposing the vomer and the nasal cavities. The palatal processes of the maxillaries and palatines form a bridge, or platform, extending across opposite the three anterior molar teeth. The nasal processes of the palatines are of unusual height. The angle of the lower jaw forms a broad compressed plate, with the lower border rounded and thickened, so as to project a little beyond both the outer and inner surface of the ascending plate: the outer ridge is continued forward to the horizontal ramus, bounding the large masseteric fossa. The petrotympanics form 'bullæ osseæ.' The pterygoids develope both external and internal plates: the outer plate is widely perforated at its base; the inner plate terminates in a hamular process.

The common foramen opticum, the wide palatal vacuities, the transversely extended glenoid cavity, and the inflected mandibular angle, indicate affinity to the Marsupials.

In a skull, seven inches long, of a Capybara, fig. 234, with the entire series of permanent teeth in place, the sutures between the elements of the occipital bone still remain. The compressed paroccipitals, 4, are of enormous length. The basioccipital contributes to each condyle its lower extremity. The exoccipitals almost meet above the foramen magnum, the plane of which is nearly vertical. The basisphenoid is perforated by a median vertical canal, and is notched laterally by the entocarotids. The squamosals are distinct, and essentially like those in the Hare, sending backward the long compressed lamina which clamps the

234

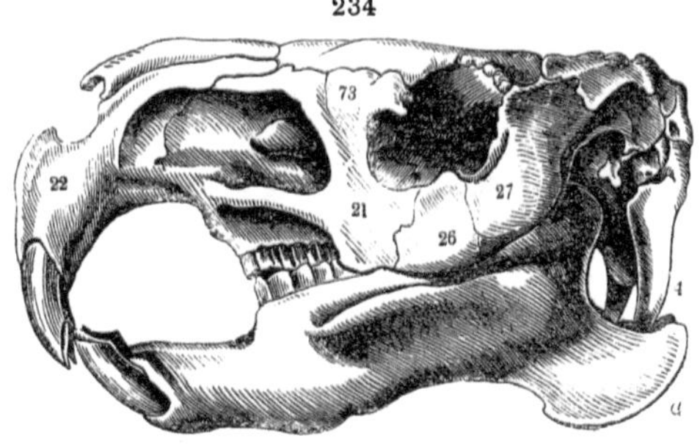

Skull of the Capybara.

tympanic and mastoid to the side of the cranium. A venous sinus issues from beneath this process of the squamosal. The longitudinal groove forming the articular cavity for the lower jaw is angular, and completed externally by the malar bone, 26. The meatus auditorius is unusually contracted, is cleft below, and bounded there by two small tuberosities. The temporal and orbital fossæ are blended together, as in all Rodents. The lacrymal bone is of unusual size, and extends forward upon the side of the face between the frontal, 11, and maxillary, 2.. The antorbital vacuity is immense. The nasal bones, 15, are long, large, and of nearly equal breadth throughout. The nasal processes of the premaxillaries, 22, are coextensive with them. The sagittal suture is obliterated, as well as a great part of the frontal suture. There is no trace of interparietal bone. A single foramen incisivum is situated anterior to the two large normal prepalatine apertures; the postpalatine foramina are in the centre of the bony palate, between the palatines and maxillaries. The palatines are large. The cribriform plate and its median ridge or 'crista galli' project backward into the large rhinencephalic fossa. Pterygoid sinuses are formed anteriorly by the proper pterygoids, and posteriorly by the ecto- and ento-pterygoid plates of the sphenoid. The ectopterygoid plate is perforated by an 'interpterygoid' canal, above which is a smaller 'ectocarotid' canal. The lower jaw shows a strong ridge or platform outside the molar alveoli. The coronoid and condyloid processes rise very little above the grinding surface of the molars. The chief process of the lower jaw is the angle, *a*, which is broad, compressed, and produced far backward, where it terminates obtusely. The upper surface of the skull is flat, and its contour deviates little from a straight line, slightly descending toward the occiput

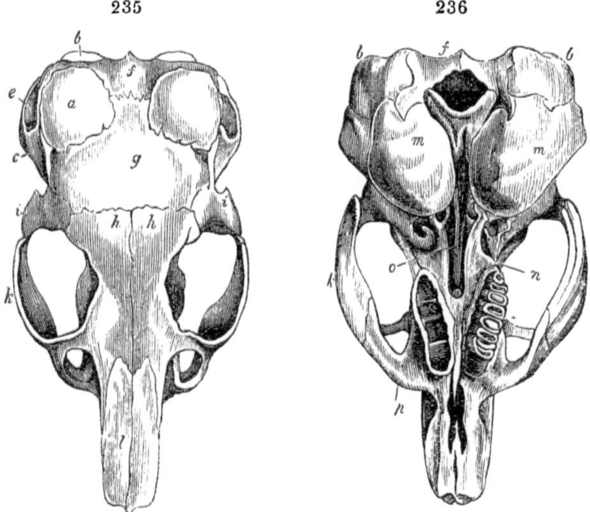

Skulls of the Chinchilla.

and the end of the nasals. The zygomatic arch is compressed but deep, especially below the fore part of the orbit. The acoustic bullæ are comparatively small.

In *Chinchilla lanigera*, figs. 235, 236, the mastoid portion, *b*, of the large tympanic bulla, *a*, *b*, *m*, rises to the upper surface of the cranium, as at *a*, but it is girt by a process of the superoccipital, *f*, which extends outward to articulate with the extremity of the slender process of the squamosal, *e*. The vacuity which intervenes between the alisphenoid, parietal, and tympanic, and which, in other Mammals, is closed by the more expanded squamosal, is here, through the retention by that bone of its primitive form as a diverging slender ray, left uncovered. The meatus is long, wide, infundibuliform, with the outlet obliquely truncate and directed upward and a little backward: the petrosal bulla, *m*, continued from its lower extremity, seems to describe a semicircular curve downward and backward, circumscribing the large foramen, which directly pierces the bulla beneath the meatus. The paroccipital is slender; its point does not extend below the level of the tympanic bulla. The articular groove for the lower jaw is deep, and is completed externally by the malar, *k*. An almost circular piece seems to be cut out of the zygoma, above the junction of the malar with the squamosal. The facial part of the lacrymal extends half-way across the antorbital root of the zygoma, where the zygomatic part of the maxillary articulates by suture with the nasal process of the same bone, circumscribing a large antorbital vacuity. The nasal processes of the premaxillaries slightly expand at their extremities, which extend beyond the corresponding ends of the nasals, *l*. A strong and long oblique ridge traverses the inner side of the ramus of the lower jaw. The outer side is irregularly swollen by the bases of the sockets of the curved molars, but has not the distinct ridge which characterises that part in the Cavies.

237

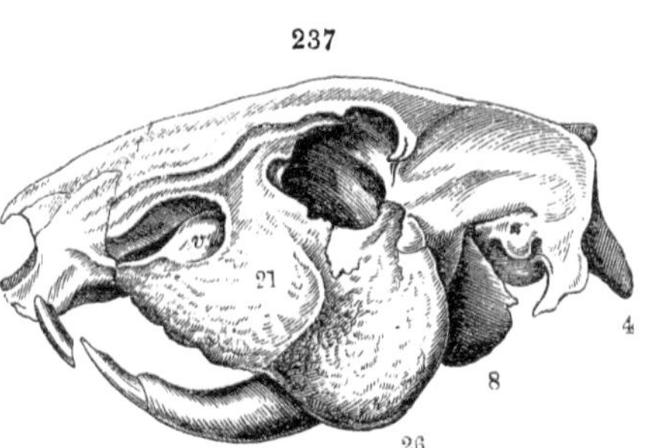

Skull of the Cœlogenys.

In the skull of an adult Paca (*Cœlogenys*, fig. 237), with the mature dentition, the sutures between the elements of the occipital, as likewise the sagittal suture, are obliterated. There is no trace of interparietal bone. The basioccipital, basisphenoid, and presphenoid have coalesced to form a continuous bony floor for the cranial cavity. The third division of the fifth notches the alisphenoid posteriorly, the foramen ovale being an irregular fissure between the ali- and basi-sphenoids and the petrosal. The petrotympanic is free from the squamosal, and rather loosely suspended beneath the overarching posterior lamella of the squamosal, which bends down external to the mastoid and paroccipital, 4. The malar, 26, is a slightly curved plate, twice as deep as it is long, and forms the posterior third part of the zygomatic expansion, the rest being formed by the maxillary, 21, which is unusually and enormously developed. The squamosal forms only the base of the zygoma; it is grooved below for the mandibular joint, to which the malar contributes the outer part. The nasal processes of the premaxillary do not extend so far back as the nasals: the large antorbital vacuity, *v*, is reduced by the maxillary zygomatic plate to a crescentic form. The zygomatic expansion of the maxillary, 21, is deeply excavated on the inner side; it forms, in the recent animal, a large bony capsule on each side of the mouth, communicating therewith and lined by the buccal membrane. A vertical sinus terminating below in two small foramina, communicating with the orbit, divides the rhinencephalic from the prosencephalic fossa. A branch of the lateral sinus leads from above the petrosal to between the squamosal and tympanic externally. The olfactory cavity extends backward beneath the rhinencephalic one, but not above it. The ectopterygoid process joins the proper pterygoid, and, with the entopterygoid plate, completes a wide interpterygoid canal. The base of the ectopterygoid is perforated by an ectocarotid foramen.

The squamosal is excluded from the cranial cavity by a fissure which widens as it descends between the squamosal and petrotympanic: a venous sinus occupies this fissure. A horizontal septum divides an upper from a lower compartment of the anterior half of the tympanic bulla. The sella turcica is shallow, and not defined by clinoid processes; the chiasmal platform is subquadrate, and leads to a fossa, perforated by the two large and approximated elliptical optic foramina; a deep and narrow groove extends from the optic fossa to the rhinencephalic compartment, where it divides to terminate at the orbito-ethmoidal foramina. The foramen rotundum and foramen lacerum anterius combine to form a large subquadrate vacuity. The cerebellar fossa on the upper part of the petrosal is very deep. The meatus internus is extremely shallow, and almost immediately divides into the cochlear and vestibular canals.

238

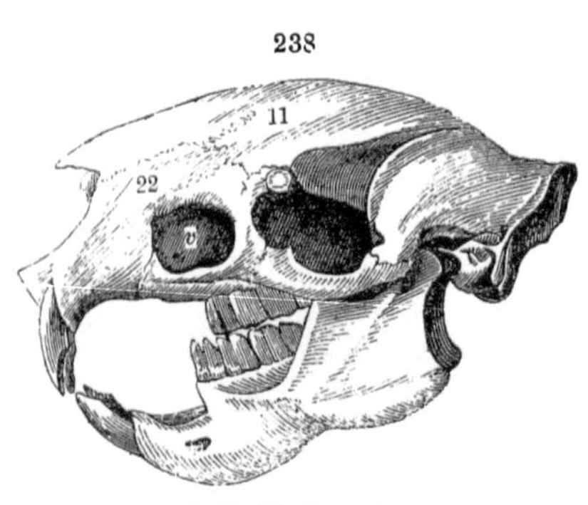

Skull of the Porcupine.

In the Porcupine (*Hystrix cristata*), fig. 238, the occipital region is nearly flat; the paroccipitals descend only to the level of the occipital condyles. The mastoid forms but a rough ridge. The auditory bullæ are moderately developed; the external meatus is short, directed outward and a little forward, and is notched behind. A fissure, which widens at both ends, divides the tym panic from the clamping process of the squamosal: this articulates behind by a suture with the mastoid. The parietals, fig. 239, 7, are broad, but short, and pinched in, as it were, by the temporal fossæ, which almost meet at the line of the sagittal suture, which is obliterated. The frontals, ib. 11, are more than double the size of the parietals, and are greatly swollen by the enormous sinuses. The most remarkable feature of the Porcupine's cranium is the magnitude of the nasal bones, 15, especially their great posterior expanse, which terminates behind on the same vertical parallel as the middle of the zygomatic arch. This character is contrasted in fig. 239 with the small size of the nasals, 15, in the Manatee and Capuchin Monkey. The thick anterior pier by which the zygomatic arch is suspended is formed by the maxillary and lacrymal. The slender

239

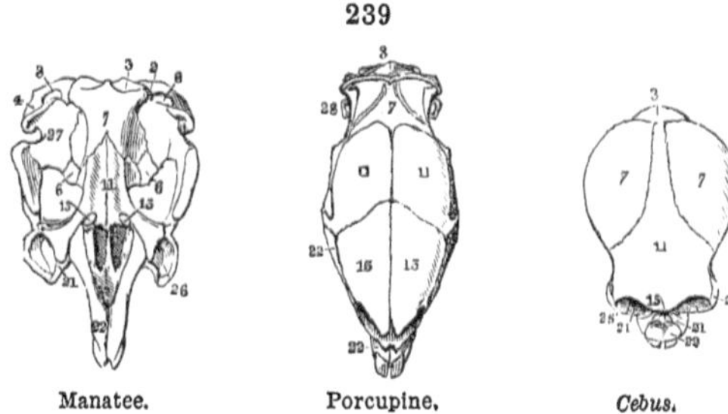

Manatee. Porcupine. *Cebus.*

horizontal process of the maxillary, which bounds the lower part of the antorbital vacuity, fig. 238, *v*, appears like a second zygoma. The premaxillaries progressively contract as they pass backward and join the frontals, nearly an inch in advance of the hinder border of the nasals. The bony palate terminates by a thick rounded border between the last molar teeth. The pterygoids send backward and upward a hamular process, which joins the tympanic bulla. The cerebellar depression upon the petrosal is very shallow: the fore part of the petrosal presents a large protuberance. The rhinencephalic fossa is relatively of large size, and is defined by a well-marked ridge from the rest of the cranial cavity. Two vascular canals are continued into its lower part from above the optic foramina, instead of an open groove, as in the Agouti. The coalesced prefrontals are compressed. The vomer is deeply cleft posteriorly, and has coalesced with the ethmoturbinals, and its anterior part articulates with the median ascending process of the premaxillary arching over the wide vacuities which lead from the nasal passages to the prepalatine apertures, as in most Rodents.

The cranial air-cells continued from the nasal and tympanic cavities reach the occiput. The tympanum is divided by a horizontal partition into an upper and lower chamber, intercommunicating posteriorly above the membrana tympani, which is situated in the lower division, where the meatus auditorius externus terminates in a narrow oblique slit. The extraordinary extent of the air-sinuses surrounding the fore part of the cranial cavity and developed in the orbitosphenoids, alisphenoids, squamosals, and frontals, with the radiating bony septa of those sinuses, are peculiarities of the Porcupine.

In an almost full-grown Beaver, fig. 240, the elements of the occipital bone are still unanchylosed; the lower third of each condyle is formed by the basioccipital, the under surface of which presents a large and deep excavation. The upper part of the foramen magnum is completed by the broad superoccipital. The mastoid is larger than in the Porcupines, and articulates anteriorly with both the parietal and squamosal: it is anchylosed to the petrosal. There is a perforation in the suture between the superoccipital and mastoid. The interparietal is large, and wholly upon the upper surface of the cranium. The squamosal is perforated behind and below the root of the zygoma. The frontals are small and almost flat above. The nasal bones extend further back than the premaxillaries, in the European Beaver, beyond the transverse line which extends between the antorbital tuberosities. The anterior root of the zygoma formed by the maxillary is a simple plate which appears to be imperforate, the orifice of the slender antorbital canal being concealed by a vertical ridge of the maxillary, which inclines forward over the maxillo-premaxillary suture.

240

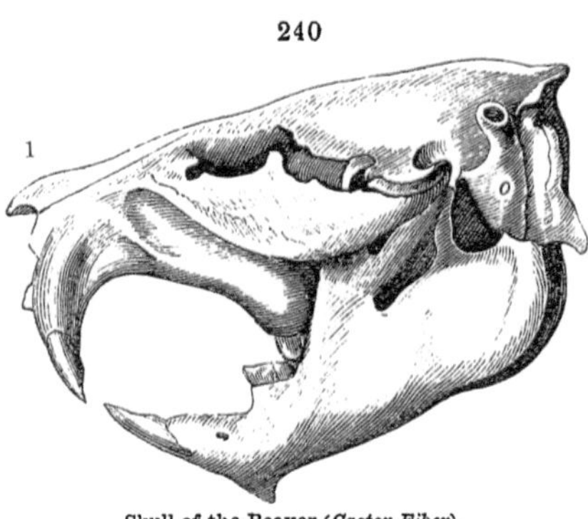

Skull of the Beaver (*Castor Fiber*).

The epencephalic compartment is lower and broader than in the Porcupine. The cerebellar fossa of the petrosal is larger and deeper. The upper compartment of the tympanum is much less. The length and direction of the auditory meatus is shown, fig. 240, *o*: it changes its form into a transverse fissure, as it approaches the membrana tympani, the plane of which is almost parallel with that of the meatus itself. There are no nasal air-sinuses in the cranial bones of this aquatic Rodent, and their texture is denser than in most of the order. The sella turcica is extremely shallow, and without clinoid processes: the middle of the basioccipital is reduced by the excavation on its under surface to extreme thinness. A small vacuity in the basisphenoid communicates with the cranial cavity close to the 'fissura lacera anterior.' The presphenoid is perforated transversely. The rhinencephalic fossa is well marked. The anterior end of the vomer articulates with both the maxillary and premaxillary bones, as in the Rat.

In the skull of the Ondatra or Musk Vole (*Fiber zibeticus*, fig. 241), the basioccipital is not excavated, as in the Beaver, but there is the same perforation between the mastoid and superoccipital, and a large vacuity in the posterior process of the squamosal communicating directly with the cranial cavity. The squamosal is unusually expanded above the zygomatic process, and articulates largely with both frontal and parietal. The zygomatic process of the maxillary reaches almost to that of the squamosal, and supports a great part of the malar bone. The antorbital foramen, *v*, is larger than in the Beaver, but is bounded externally, as in it, by a nearly vertical ridge of the maxillary. The interorbital septum is perforated behind, beneath the orbitosphenoid. There is no distinct lacrymal bone; but the turbinal bones appear at the fore part of the orbit between the two processes of the maxillary which join the frontal, and above the aperture communicating with the nasal cavity. The anterior part of the maxillary, in front of the antorbital foramen, is swollen, and forms a curved canal commencing by an oblique aperture superiorly, and descending outward and backward round the socket of the superior incisor to terminate in the nasal meatus: this part may probably protect the lacrymal sac and duct. The interparietal is a transversely quadrate bone. The sagittal suture is retained, and the upper surface of the parietal is smooth, and nearly flat: the temporal ridges meet and develope a crest upon the narrow frontals, obliterating the frontal suture. The back part of each ramus of the lower jaw is trident-shaped from the

241

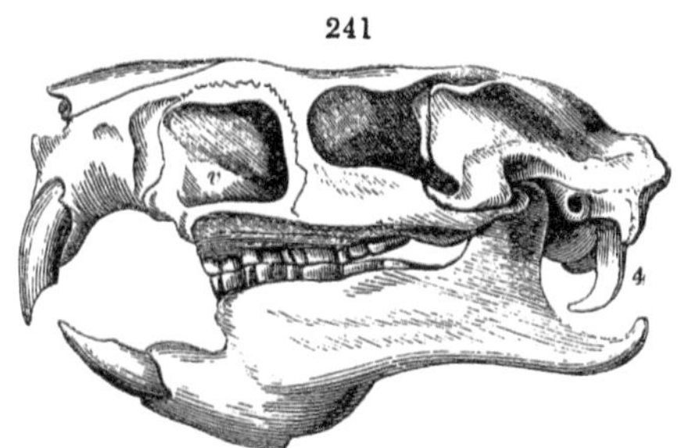

Skull of the Ondatra (*Fiber zibeticus*).

almost equal developement of the coronoid and angular processes, on each side the base of the narrow process supporting the condyle.

In the Great Mole-Rat (*Orycteropus capensis*), the occipital region of the skull is very broad and low. The compressed paroccipitals project downward and backward. The auditory bulla is pyriform, its apex articulating with the pterygoids. The temporal fossæ meet along a well-developed crista extending from the interorbital region to the strong transverse superoccipital crest. The squamosal forms a horizontal plate, with a curved border extending from the root of the zygoma to above the 'meatus externus,' which is directed upward and forward. The zygomatic arches are strongly curved outward. The premaxillaries extend further backward than the nasals: these are very long and narrow. In the Blind Mole-Rat (*Spalax typhlus*), the orbit, fig. 242, *o*, is not defined: the great antorbital vacuity, *v*, might be mistaken for it.

242

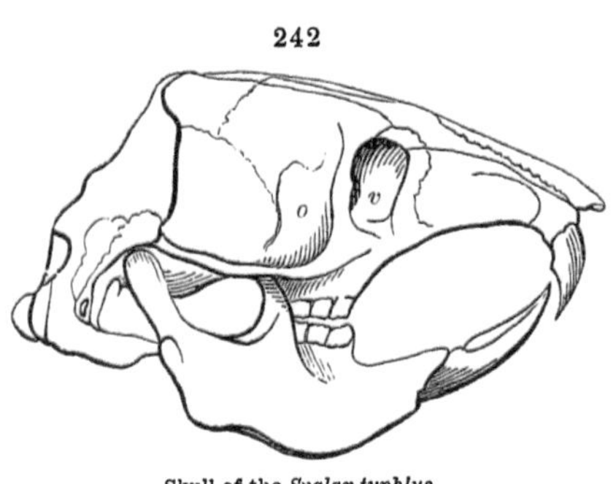

Skull of the *Spalax typhlus*.

In the skull of the Cape Jerboa (*Helamys capensis*), the occipital region, owing to the enormous developement of the acoustic bullæ, appears as a broad shallow depression between them at the back part of the skull. The paroccipitals are small, slender, subelongate, and project downward, distinct from the bullæ. The broader mastoid processes are applied to the outer side of the petrosal portion of the bullæ: the swollen bases of the mastoids form a tract upon the upper surface of the cranium larger than the interparietal bone, on each side of which they are situated. The slender posterior clamping processes of the squamosals impress the outer sides of the bullæ which they support, above the 'meatus externus:' this canal is directed upward and a little outward. The parietals are pushed by the squamosals entirely to the upper region of the cranium: the sagittal suture remains, as well as the frontal one. The temporal

243

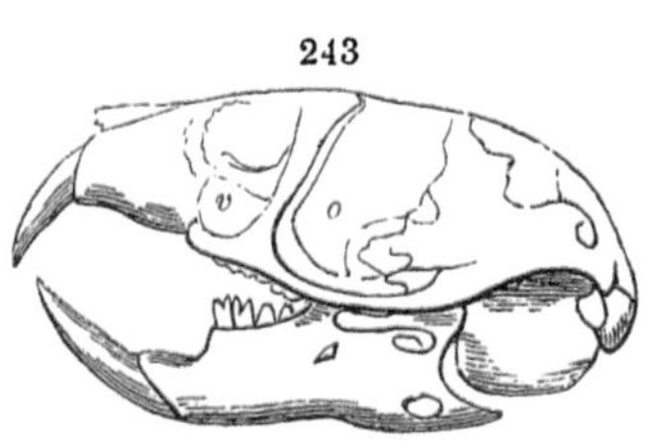

Skull of the Jerboa (*Dipus Sagitta*).

muscles seem to have been unusually small in this Rodent: their fossæ impress only the small squamosals. The coronoid process of the lower jaw is obsolete. The movements of the jaws appear to have been chiefly committed to the masseteric and pterygoid muscles. The zygomatic arch, which extends from the squamosal to the premaxillary, is very broad below the orbit, and is traversed externally by a ridge indicating the powerful origin of the masseter. The antorbital vacuity and the maxillary depression, bounded externally by the two roots of the zygoma, are larger than the orbits: the front root of the zygoma is formed by a combination of the frontal, lacrymal, maxillary, and malar bones. The slender extremities of the premaxillaries terminate on nearly the same transverse line with the back part of the broad nasals. These are bent down anteriorly, so as to form the sides of the external nostrils. The deep sockets of the rootless teeth form protuberances at their bases, where the osseous case becomes absorbed, converting the socket into a canal open at both ends, the persistent matrix of the tooth being attached to the periosteum, and protected by the contiguous soft parts. In all the Rodents with the wide antorbital vacuities, the fore part of the masseter takes its origin from the facial bones anterior thereto, and traverses the vacuity in its oblique course beneath the fore part of the zygoma, to expand and blend with the normal part of the masseter.

The lower jaw is modified for the lodgement of the pair of long, curved, scalpriform incisors, the sockets of which may extend to the middle (Hare) or even to the hind part (Beaver, Porcupine) of the ramus: in the latter case the prominent inner wall curves beneath the molar alveoli and forms, as in figs. 238, 241, 242, *c*, the lower part of the horizontal ramus. The condyle, crowning this, rises usually high above the grinders; it is lowest in the Capybara and some Cavies: in all Rodents the condyle is convex transversely and extended longitudinally. The chief work of the teeth being by horizontal movements to and fro, all that part of the ascending ramus serving for the implantation of the masseter is expanded, while that for the temporal muscle is reduced, so that the 'coronoid' process is very small, and may be a mere tubercle (*Lagomys*), while the angle of the jaw usually forms the whole base of the ascending ramus, projecting below its fore part, angularly in the Hare, fig. 233, *a*; and behind its back part, extensively in Cavies, fig. 234, *a*, and Voles, fig. 241, *a*. In most of these it is long and pointed; but is obtuse and compressed in *Dolichotis*: it is subquadrate in Squirrels,

Rats, Marmots. In many Rodents the angle is extended outward and subsides, advancing, as a ridge upon the outer side of the horizontal ramus, as in fig. 242: in *Ctenomys* the breadth of the mandible exceeds the length. Most Cavies show, also, the external ridge noted in the Capybara's jaw, below the molar series. The upper jaw is similarly modified in relation to the masseter, e.g., in those Rodents which have the fore part of the muscle passing through the wide antorbital vacuity, *v*, to its peripheral ridges.

C. *Bones of the Limbs.*—In this extensive and ubiquitous order, which includes three-fourths of the known species of Mammals, some have limbs giving power in running, some in swimming, some in burrowing, some in leaping, some in climbing, and a few show modifications in relation to parachute-like expansions of integument for a kind of flight.

In the Hare, fig. 229, the scapula is long and narrow, traversed externally by a spine extending into an acromion at an unusual distance beyond the glenoid cavity, and there developing a retroverted process; the coracoid is compressed and introverted. The clavicular ossicles are freely suspended, allowing full swing to the fore-limb. The humerus, long, slender, and sigmoid, has a large intercondyloid vacuity. The radius and ulna are in close contact; the latter is grooved for the reception of the radius. Their ginglymoid joint with the humerus restricts the movements to one plane. The carpus has the 'os intermedium,' fig. 191, *s'*. There are five digits, the innermost very short, though with the normal number of phalanges. The fore limbs are relatively shorter and stronger in the burrowing Rabbits; the ungual phalanges are less compressed, and afford a closer attachment of the broader claws by being cleft on the upper surface. In all *Leporidæ* the ilia are long and subprismatic where they articulate with the sacrum, the joint being limited to the first vertebra, fig. 245, *a*, *b*. They extend in advance of this on each side the last lumbar, ib. *d*, expanding into a crista, *c*, which is rough and slightly everted: the ilia form with the lumbar series an angle of 165°, fig. 229. The ischia have a process, fig. 244, *e*, above the terminal tuberosities: the pubic bones are long and slender, meeting at a long symphysis produced into a ridge, *f*: there is a 'pectineal' process, *d*, near the acetabular end of the pubis. The iliopubic angle is about 120°.

The femur has a third trochanter near the base of the great one. The medullary artery pierces the inner side of the proximal third of the bone, and the canal extends downward. The fibula is anchylosed along its distal half to the tibia: its proximal end projects beyond the tibia, and a 'fabella' is wedged between it and the outer condyle of the femur; there is a similar sesamoid behind this condyle, and a third behind the inner condyle. The patella is ossified.

The tarsus shows the naviculare, astragalus, calcaneum with a long lever: the meso- and ecto-cuneiform bones, and the cuboid. There is a supplemental ossicle beneath the astragalus. The naviculare has a large process. The inner digit is wanting, and the base of the metatarsal of the second is extended backward, like an entocuneiform, to join the naviculare.

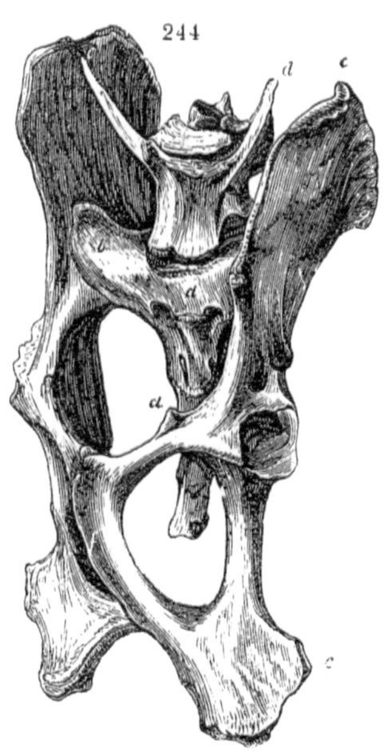

Pelvis of the Hare, anterior view.

In the Hare-like Cavies of South America (*Dasyprocta*) the clavicles are represented, as in the Hares, by slender ossicles: the supra- and infra-spinal fossæ of the scapula are of equal depth: the humerus is perforated between the condyles: the radius and ulna have become anchylosed, reducing the interosseous space to a narrow chink near their proximal ends in the Acouchy: in an Agouti I found this confluence not complete. The fore foot is pentadactyle. The first row of carpals is formed by the scapholunar, the cuneiform, and a large pisiform. There is an 'intermedium' between the os magnum and trapezoides. The pollex is shorter in the Agouti than in the Acouchy. The fifth finger is much reduced in size, but has the normal number of phalanges. The ungual phalanges are notched at their apex. The femur gives a feeble indication of the third trochanter at the middle of its outer side. The tibia and fibula are distinct; a fabella is attached to each femoral condyle. The foot has but three digits. The long entocuneiform bone has coalesced with the inner side of the metacarpal of the second toe—here the innermost. The supplementary ossicle crossing the articulation between the astragalus and scaphoides is present. There is a distinct sesamoid beneath the joint of the cuboid with the external metatarsal (*iv*): both the naviculare and cuboid send strong processes to the plantar side of the tarsus. There are trochlear sesamoids beneath the metatarso-phalangial joints: the ungual phalanges are notched.

In the prolific Guinea-Pig, the pelvis, fig. 245, is long and laterally compressed, the passage being much narrower than the diameter of the head of the mature fœtus. Prior to parturition the symphysial ligaments become soft and extensile, and the innominata, gliding on the sacro-iliac joints, diverge at the symphysis to the extent shown in fig. 246 during parturition. After this process the symphysis quickly returns to its former or normal

245

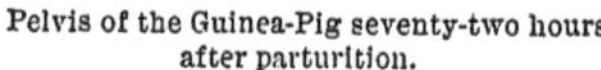

Pelvis of the Guinea-Pig seventy-two hours after parturition.

246

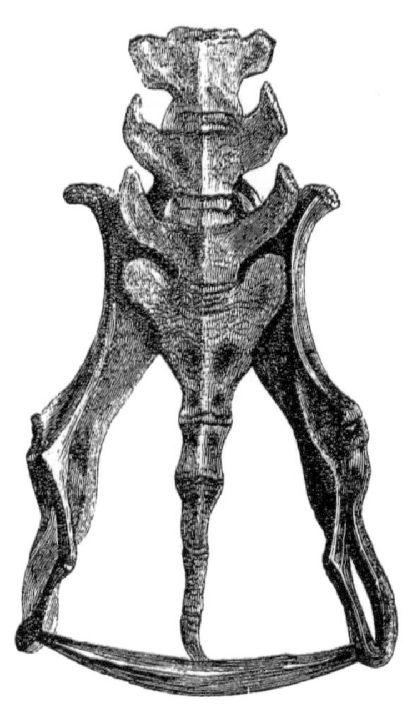

Pelvis of the Guinea-Pig at the time of parturition.

state, and in a few days presents only a little thickness and mobility. The young of the Guinea-Pig are far advanced at birth, some of the deciduous teeth are shed *in utero*, and they run about and begin to eat soon after they see the light.

In the Water-Hog, or Capybara, there is no complete clavicle. The acromion is long and slender, and bifid at its extremity, with the longer division directed downward. The humerus is widely perforated between the condyles, but not above the inner condyle: both this bone and the ulna are solid. The scaphoides and lunare are connate. The pollex is wanting in the fore feet, and both the hallux and the fifth toe are wanting on the hind feet. The ungual phalanges are short, obtuse, and broad.

The Beaver, fig. 230, is a member of that great division of the Rodentia in which the clavicles are complete: the acromion scapulæ bends toward and joins that bone. In the humerus the deltoid ridge has a tuberosity: both the intercondyloid space and the internal condyle are imperforate: a coarse cancellous structure occupies the middle of the shaft. The radius and ulna are distinct. The femur shows the slender neck and lofty trochanter common to most Rodents; it has a third trochanter, and has no medullary cavity. The rotular surface is distinct from that of the condyles. A section of the tibia and fibula also shows the absence of that cavity, and the complete confluence of the compact walls of the two bones at the lower third of the fibula. The projecting part of the calcaneum is depressed. The toes are longer and stronger than the fingers, they support a broad foot which is webbed, and the second toe has a double oblique nail or broad claw.

In our Water-Vole (*Arvicola amphibia*) the acromion of the scapula is long and bent downward; its inferior process is feebly developed. The deltoid process of the humerus is prominent and well-defined, compressed, and bent downward. There is a minute perforation between the condyles, but none above the inner one. The bones of the fore-arm are in contact and closely united, except at the narrow space near their proximal ends. The pollex is represented by its metacarpal bone. The femur has a third trochanter, with two patellæ in front of, and two fabellæ behind, the condyles. I have found, also, a small ossification at the anterior end of each semilunar cartilage. The fibula is anchylosed to the tibia at both its extremities. The entocuneiform is long, and applied to the inner side of the base of the second metatarsal, but it supports a short metatarsal with the first and ungual phalanx of its proper digit, the hallux.

Rodents burrow chiefly for concealment, rarely for food: the Rabbit needs but a slight modification of the limbs, as compared with the surface-dwelling Hare, to excavate, in loose soil, its retreat. Perhaps the 'Mole-Rats' of the Cape are the best burrowers of the order. In *Bathyergus* the upper border of the scapula describes an open angle; its outer surface is nearly equally bisected by the spine, which rises to an unusual height, and sends off a remarkably long subtrihedral acromion, the extremity of which appears as a thick epiphysis bent toward the long and strong clavicle with which it articulates. A well-marked deltoid process stands out from the middle of the shaft of the humerus, which is imperforate at its distal end. The olecranon is unusually thick and expanded. The femur shows a rudiment of a third trochanter. The fibula is anchylosed to the tibia. A remarkable accessory ossicle, articulated to the tarsal os naviculare, projects inward like an accessory or sixth digit of the hind foot. As in other burrowing animals, the lumbar and pelvic regions are narrow.

In the Marmot (*Arctomys*) the clavicles are complete and strong. The acromion is long and bifurcate, the anterior division curves to the clavicle. The humerus shows a thick, but not prominent, deltoid ridge: it is perforate between the condyles and above the inner condyle. The antibrachial bones admit of rotation. In the femur there is a rudiment of the third trochanter: the tibia is not confluent with the fibula. In all the Rat-tribe the clavicles are entire: the distal part of the fibula coalesces with the tibia. In the Black Rat (*Mus rattus*) the deltoid ridge is angular, and commences near the upper end of the humerus, which is imperforate at the lower extremity. A strong ridge represents the third trochanter of the femur. There is a fabella behind each condyle. In an Australian Rat (*Hapalotis albipes*) the humerus is perforated between the condyles. The radius and ulna are moveably united. There is a third trochanter in the femur. In the *Hydromys* the deltoid ridge projects from the fore part of the proximal half of the humerus, and is prominent below. The humerus is imperforate. The ulna sends a process to abut against the radius across the middle of the interosseous space. The fore foot is pentadactyle, but the pollex does not exceed the length of the metacarpus of the index. The femur has a third trochanter and a fabella behind each condyle. The hallux extends to the second phalanx of the next toe. The strength of the hinder half of the skeleton, with the size of the hind extremities, contrasts with the slenderness of the fore part in most Rats, and especially in this large Australian aquatic kind.

Among the leaping Rodents the following noteworthy characters of the limb-bones are seen in the Great Jerboa of the Cape (*Helamys*). The lower costa of the scapula forms an acute angle with the base, and the infraspinal fossa is much broader than the supraspinal one, the spine of the scapula curving toward the upper angle. The acromion is moderately long and slender, the tuberosity answering to the lower division of that in the *Caviadæ*. The clavicles are strong, and curved backward at their outer half. The humerus is perforated at the inner condyle, but not between the condyles. The bones of the fore-arm have a long and wide interosseous space, and allow of free pronation and supination. The hand is pentadactyle, and the whole anterior extremity much shorter than the posterior one. The iliac bones extend upward considerably above their junction with the anterior sacral vertebræ, and curve outward. The tuberosities of the ischia are unusually developed. The obturator vacuities are very extensive, the size of the pelvis according with that of the hinder extremities. The great trochanter is of unusual length, is expanded and slightly bent at its extremity. The fossa upon the neck of the femur is unusually deep; there is no third trochanter. The medullary artery enters on the inner side of the base of the small trochanter. The slender fibula coalesces with the lower third of the tibia, but both its extremities are free, and the lower one is detached, as in the Chevrotain, from the rest of the bone. The calcaneum, astragalus, and cuboid are all remarkable for their length: the scaphoid sends a long and thick process downward and forward to beneath the middle cuneiform and the base of the inner metatarsal. There are four distinct metatarsals and four toes. An oblong ossicle, attached to the inner side of the base of the inner metatarsal, may be a rudiment of the metatarsal of the hallux.

In the Smaller Jumping Mouse (*Dipus Sagitta*), fig. 232, may be noticed the large size of the ischium, as compared with the ilium, and the coalescence of the metatarsals of the three middle toes into one bone, *m*, as in Birds. Both hallux and little toe are wanting. The lower half of the slender fibula is anchylosed to the tibia.

The climbing Squirrels (*Sciuridæ*) have four digits on the fore foot, and five digits on the hind foot, conversely to the *Helamys*. All possess complete collar-bones. In *Sciurus maximus* the acromion is bent almost at right angles with the spine of the scapula, and it terminates in three prominences: the coracoid is unusually long. The humerus is perforate above the inner condyle, but not between the condyles. In the femur the small trochanter is unusually prominent: there is also a trochanterian ridge below the base of the great trochanter. In the Grey Squirrel (*Sc. cinereus*) the scapula is remarkable for the number and strength of the intermuscular cristæ: of these, that which is commonly called the 'spine' is the largest, its breadth being equal to that of the infraspinal fossa: this fossa is bounded by a second ridge, formed anteriorly by the outwardly bent lower costa, but being distinct from the costa at its posterior third. The two principal masses of the 'subscapularis' muscle are divided by a longitudinal crest, like the spine, rising from the inner surface of the scapula. Both the acromion and coracoid are well developed.

The humerus is perforated above the inner condyle: this is a tuberosity which appears to be supported by four converging columnar ridges or processes. The deltoid and supinator ridges are well marked. The shaft of the ulna is much compressed: its distal portion coalesces with that of the radius. In the carpus an accessory 'intermedium' is wedged between the scaphoid and trapezium. The bones of the pollex support the tubercle that outwardly represents that digit; the Squirrel's favourite nut is mainly held between the 'thumb tubercles' when operated on by the chisel-teeth. In the pelvis the epicotyloid tubercle is strongly developed: the ilia articulate with the first sacral vertebræ exclusively, but the ischia abut against the long transverse processes of the first caudal: beyond this vertebra the ischia develope on each side two tuberosities, one at the usual place, the other and stronger one near the lower end of the symphysis. The femur shows an almost equal developement of the three trochanters. The medullary artery enters on the inner side of the shaft, just below the small trochanter.

The tibia and fibula coalesce distally. There is an inter-articular ossicle in the knee-joint; a patella; and two fabellæ, one behind each femoral condyle. In the tarsus an accessory ossicle is wedged between the calcaneum, astragalus, naviculare, and entocuneiforme.

247

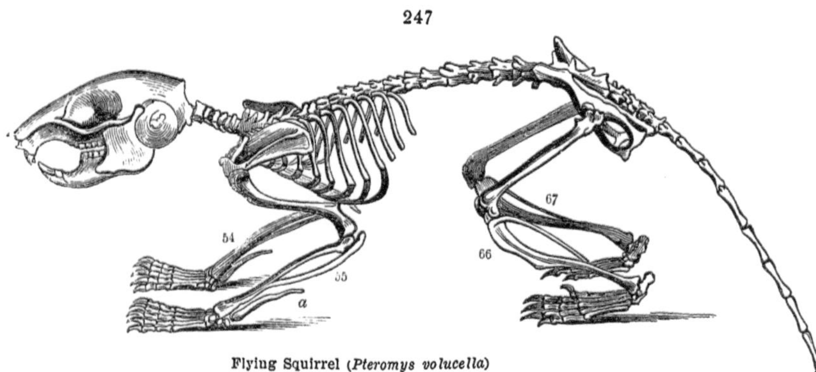

Flying Squirrel (*Pteromys volucella*)

The little Flying Squirrel (*Pteromys volucella*), fig. 247, is chiefly remarkable for the long and strong accessory cartilage, *a*, projecting from the ulnar side of the carpus, which aids in supporting the lateral fold of integument serving as a parachute to support this light and delicate species of Rodent, fig. 154, in its long flight-like leaps from bough to bough. Increased stiffness and resistance are imparted to the bones of the arm by the anchylosis of the radius, 54, and ulna, 55, at their distal halves. The tibia, 66, and fibula, 67, are similarly united.

There are few generalisations deducible from the limb-bones of *Rodentia*. The absence of clavicles accords, in the main, with natural groups; but *Lagomys* is an exception among *Leporidæ* and *Chinchilla* among *Hystricidæ*, Wth.: most non-claviculate Rodents have the tibia and fibula distinct.

Among the bones of the splanchnoskeleton may be noted the 'os penis,' which is present in most members of the Rodent order.

§ 182. *Skeleton of Insectivora.*—The present like the preceding Lissencephalous order has species organised, not only, as in Hedgehogs, for ordinary terrestrial progression, but also for leaping and swimming, and in a more especial degree for burrowing and flying.

A. *Vertebral Column.*—In the Hedgehog the vertebral formula is:—7 cervical, 15 dorsal, 6 lumbar, 3 sacral (or L 5, S 4), and 14 caudal. The transverse processes of the last cervical are not perforated. All the processes are small throughout the vertebral column, and offer no impediment to the free inflection of the spine required in the defensive array of the prickly integument. The sacrum is narrow and articulates by three vertebræ with the ilia: these form an angle of 130° with the spinal column: the ilio-pubic angle is about 150°: the symphysis is short and the pelvic outlet large. The neural canal is widest in the cervical region, contracts towards the middle of the back, and expands a little in the loins. Seven pairs of ribs directly join, by hæmapophyses often ossified, the sternum, which consists of four bones. The cancellous structure of the vertebræ is light and open.

The Tenrecs (*Centetes*) have 19 dorsals with 5 lumbar vertebræ, and the neural spines are longer on the anterior ones and the contiguous cervical vertebræ, in relation to the larger skull and more powerful jaws of these tropical Hedgehogs.

In the leaping *Macroscelides*, with *d* 13, *l* 7, the neural spines of the hinder dorsal and lumbar vertebræ are longer, and, with those anterior to them, indicate 'a centre of motion' of the trunk. The caudal vertebræ are more numerous and have hæmapophyses in part of the series. This part of the skeleton is also well developed in the climbing Tupaias.

In the burrowing Mole, fig. 248, the first sternal bone, or manubrium, is of unusual length, being much produced forward, and its under surface downward in the shape of a deep keel for extending the origin of the pectoral muscles. Seven pairs of ribs directly join the sternum, which consists of four bones, in

248

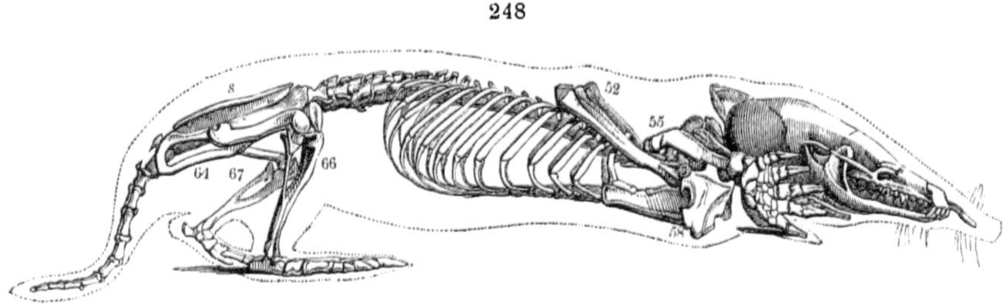

Mole (*Talpa europæa*).

addition to the manubrium and an ossified ensiform appendage. The neural spines, which are almost obsolete in the first eight dorsals, rapidly gain length in the rest, and are antroverted in the last two dorsal vertebræ. The diapophyses, being developed in the posterior dorsals, determine the nature of the longer homologous processes in the lumbar vertebræ. In these the neural spines are low, but of considerable antero-posterior extent: the diapophyses are bent forward in the last four vertebræ: a small, detached, wedge-shaped hypapophysis, fig. 249, C, *a*, is fixed into the lower interspace of the bodies of these vertebræ.

249

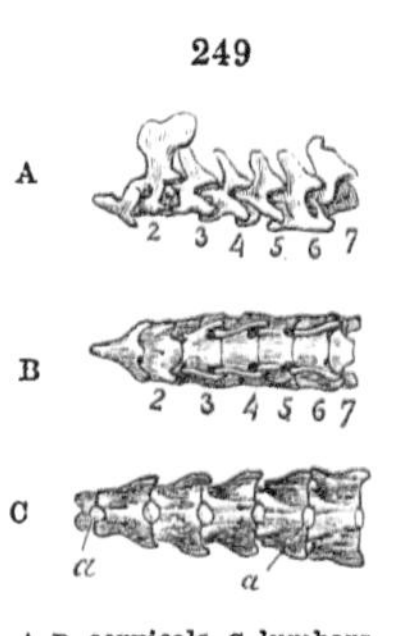

A, B, cervicals, C, lumbars, of Mole. LXXXVI'.

The ossa innominata have coalesced with the sacrum, fig. 248, *s*, but not with each other, the pubic arch, 64, remaining open. The bodies of the sacral vertebræ are blended together and are carinate below: their neural spines have coalesced to form a high ridge. The acetabula look almost directly outward. In the cervical series the odontoid process shows a sharp hypapophysis: the neural spine of the dentata, fig. 249, A, 2, is large and extended back over the third vertebra: the neural arches of this and the succeeding vertebræ form, above the zygapophyses, thin simple arches, without spines; the transverse processes of the fourth, fifth, and sixth cervicals are produced forward and backward, and overlap each other, ib. 4—6: in the seventh those processes are reduced to tubercular diapophyses which are not perforated: the bodies of the vertebræ are depressed and quadrate, ib. B.

Among the volant Insectivora the vertebral formula, in *Vespertilio murinus*, gives—7 cervical, 12 dorsal, 7 lumbar, 3 sacral, and 12 caudal. The chief characteristics of the trunk-skeleton in Bats are:—the gradual diminution of size of the spinal column from the cervical to the sacral regions; the absence of neural spines, a keeled sternum, and a feeble slender pelvis. In a frugivorous Bat (*Pteropus fuscus*, fig. 156) I find the following vertebral formula:—7 cervical, 14 dorsal, 4 lumbar, and 6 sacral. The keel of the large manubrium sterni is produced into a process at each angle: the three succeeding sternal bones are carinate: seven pairs of ribs directly join the sternum. The narrow subcylindrical ilia, fig. 250, *a*, coalesce with the sacral vertebræ, and are parallel with the spinal column: the pubis, ib. *b*, is continued in a line with the ilium to the symphysis, ib. *c*, which is but slightly closed in the male, and remains open in most female Bats. There is a 'pectineal' process, ib. *h*, in *Pteropus*. The ischium, ib. *d*, joins the last sacral vertebra and defines a sacrosciatic foramen. The acetabula look backward (dorsad) as well as outward.

250

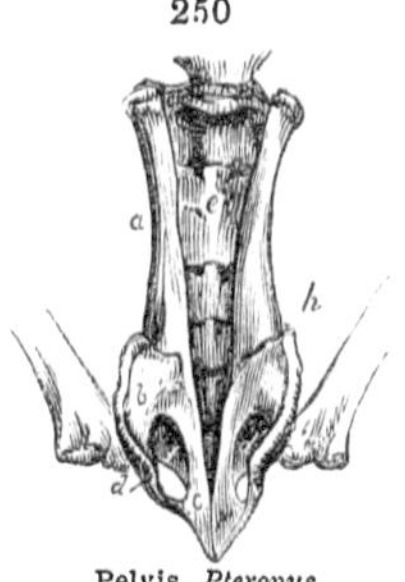

Pelvis, *Pteropus*.

251

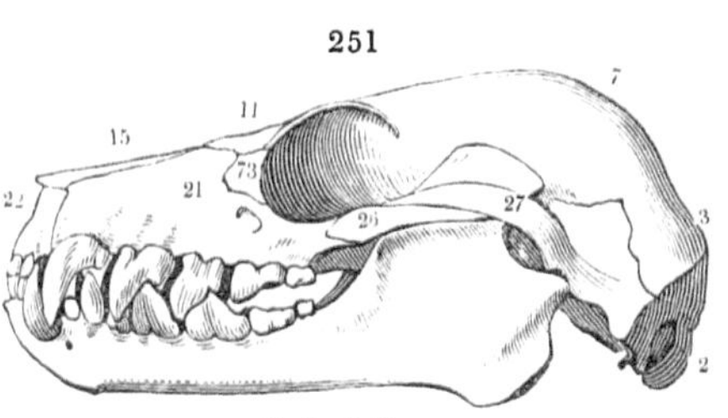

Skull of *Pteropus*.

B. *The Skull.*—This, in *Pteropus* and *Galeopithecus*, fig. 253, manifests the lissencephalous affinity by the squamosal being perforated by a venous canal behind the root of the zygoma, by the suspension of the malar, 26, in the zygoma, by the distinct petrotympanic, 28, by the vertical occiput, small cranial cavity, and blended orbital and temporal fossæ. The orbit is partly defined behind by long and slender processes of the frontal, ib. 11, which is perforated by a superciliary foramen. The parietals, ib. 7, usually coalesce at the sagittal suture, but rarely develope a crest. The occipital condyles, ib. 2, are terminal, and the plane of the foramen is vertical. The basioccipital, 1, is a subquadrate plate. The lacrymal, ib. 73, is in great part facial. The palatines, 20, are entire. In *Pteropus* the palatal processes of the maxillary, 21, show a long fissure: those of the premaxillaries, ib. 22, are divided by a triangular 'foramen incisivum.' The mandible has a broad and high coronoid process: the angle is rounded. In *Galeopithecus* the coronoid is small.

252

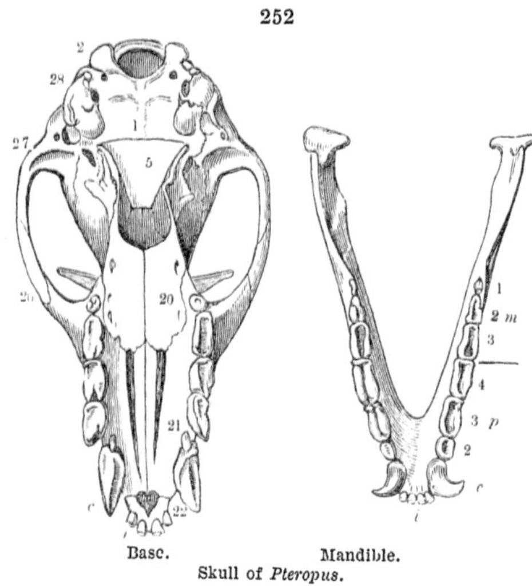

Skull of *Pteropus*.

In the cranial cavity the rhinencephalic fossa is large and well defined. The petrosal shows a deep cerebellar fossa, overarched by the vertical semicircular canal. The 'sella' has no clinoid processes.

In most insectivorous Bats the occipital condyles are subterminal, the superoccipital, fig. 254, 3, sloping backward, and contributing to the crista continued forward by the interparietal and parietal bones. The occipital foramen is very large. The mastoid, 8, is large and distinct, giving attachment to the tympanic, 28. The basisphenoid is broad and flat as in *Pteropus*, fig. 252, 5. The frontal has no postorbital process. The zygomatic parts of the squamosal, 27, and malar, 26, are slender. The premaxillaries are very small: in some Bats they are wanting (*Rhinolophus*), or are represented by separate moieties attached to the fore-part of the maxillaries. The mandible

253

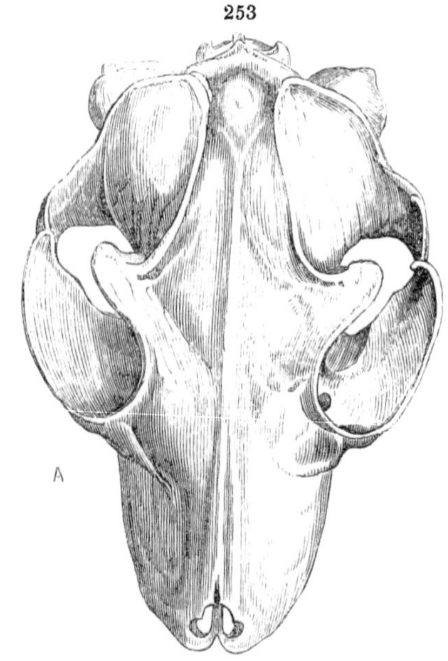

Skull of Galeopithecus.

has the angle usually produced, fig. 254, A. The malleus and incus are united; the crura of the stapes are long and slender.

The skull of the Mole (*Talpa*, fig. 255) is subdepressed, pyriform, large behind, tapering to the fore-part, which is prolonged by the prenasal ossicle, *n*. The outer surface of the cranium is smooth and devoid of crests: it is remarkable for the extension of the superoccipital upon its upper part, and for the expanded mastoids. The very slender zygomata show no distinct malar bones. The petrosal is largely and deeply excavated by the cerebellar fossa. The rhinencephalic fossa is large and well defined. The basioccipital and basisphenoid are thick and of a fine spongy texture. The orbit is no way defined from the temporal fossa: the antorbital foramen is large. In the Cape Mole (*Chrysochloris*) the cranium resembles that of the bird in its thin smooth convex walls, its great transverse and vertical diameters, its allocation at the back of the skull, and by the transverse crest extending, as in some sea-birds, from one mastoid, over the vertex, to the other. The base of the zygomatic process of the squamosal is deeply excavated anteriorly, and the zygomata converge, straight, to the maxilla. Some Shrews (*Amphisorex*) have no zygomata.

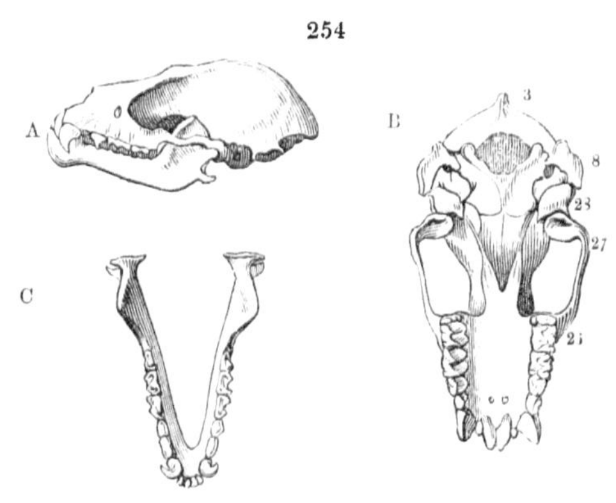

Skull of Bat (*Phyllostoma*).

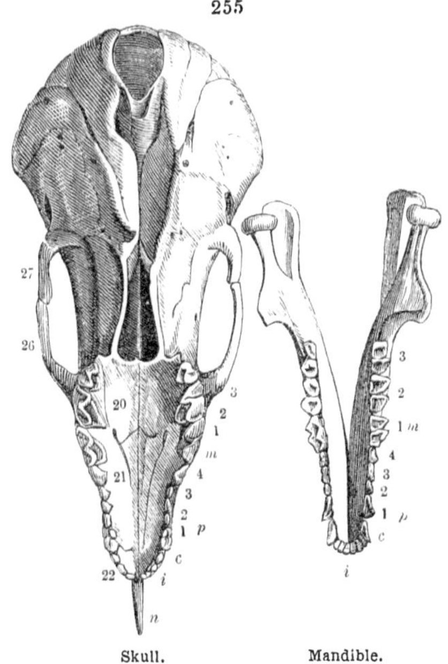

Skull. Mandible.
Mole: magnified.

In the Hedgehog the squamosal is traversed by a vertical venous canal. The malar is applied, like a splint, along the outer and under side of the junction of the zygomatic with the maxillary. The cranial cavity shows the rhinencephalic compartment to be nearly equal in size with the epencephalic one: the petrosal is impressed by a cerebellar fossa. The nasal passage terminates behind in a hemispheric excavation of the basisphenoid; and this bone expands outwardly to form the floor of the tympanic cavity. There are two oblong vacuities in the palatal bones.

In the *Gymnura* the bony palate is entire. The premaxillaries join the anterior half of the nasals. The lacrymal perforation is in a small fossa at the fore-part of the orbit, which is not defined from the temporal fossa. The zygomatic process of the squamosal is long and slender, joining that of the maxillary. The basisphenoid expands to form the floor of the tympanic cavity. The superoccipital and parietal crests are well developed. The pterygoid is pierced lengthwise by an ectocarotid canal.

C. *Bones of the Limbs.*—All the Insectivora have perfect clavicles. The scapula of the Hedgehog is almost as long as the humerus; the acromion is bilobed: the coracoid produced and thick. The humerus is perforated between the condyles. The antibrachials are distinct, but closely connected together: the ulna being the larger and more compressed. The carpus consists of a scapholunar bone, a cuneiforme and large pisiforme, a trapezium, trapezoides, magnum and unciforme. A sesamoid is attached to the outside of the base of the metacarpal of the digitus minimus. The fibula coalesces at its distal end with the tibia. The ectocuneiform and cuboides are elongated. The foot is pentadactyle and plantigrade.

In *Amphisorex* the radius and ulna are closely united, and the fibula appears as a slender process ascending from the middle of the tibia. In the proboscidian Shrew (*Rhynchocyon*) the pollex is wanting, and the fifth digit has but two phalanges; but the index, medius, and annularis present the normal number of phalanges supported on long metacarpals. Besides the usual eight carpals, there is an 'intermedium' between the scaphoid, trapezium, trapezoides, and magnum. In the hind-foot there is a rudiment of the metatarsal of the first toe, and the fifth has the usual number of phalanges.[1]

The Moles exhibit the extremes of modification of the fore-limb in relation to the power of making progress in earth. In *Talpa*

[1] LXXXIV'.

europæa the scapula, fig. 256, 52, combines ornithic proportions with unusual strength; its length exceeds its extreme breadth by six times: it is trihedral, save at the middle, which is cylindrical: the spine is co-elongate, and developes an acromion ligamentously connected, for freedom of movement, with the clavicle. This bone, ib. 58, is cubical—an unique form in Vertebrata. The humerus, fig. 256, 53, fig. 257, is a subquadrate, lamelliform bone, with a proximal articulation for the clavicle, 58, as well as for the scapula, *b*, *c*. The inner tuberosity swells out with the deltoid, *d*, and pectoral, *p*, ridges into an enormous convex crest, divided by a short and deep emargination from the inner epicondylar process, *i*, the base of which is perforated by the median nerve. The outer tuberosity, *t*, is produced and unciform; and a long and deep emargination divides it from the retroverted production of the radial condyle, *o*. This condyle offers a convexity to the head of the radius, fig. 257, 54. The olecranon, fig. 257, 55, expands transversely at its extremity, and the back part of the ulna is produced into a strong ridge of bone. The

256

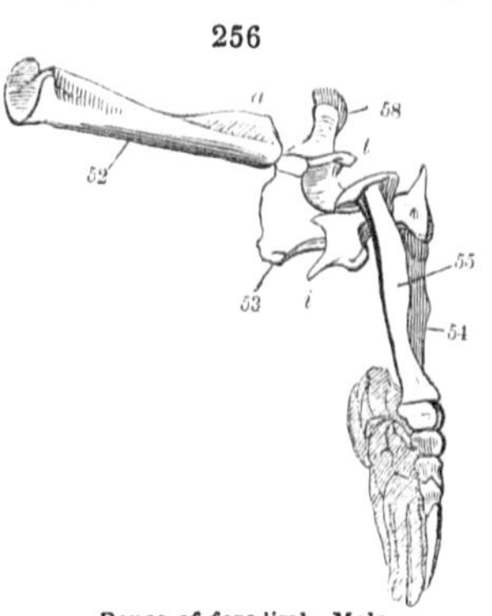

Bones of fore-limb, Mole.

257

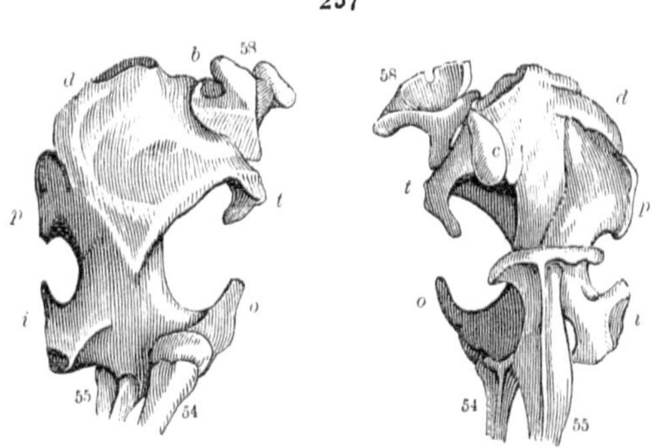

Humerus and its connections, Mole: magnified.

carpus, fig. 258, consists of the usual eight bones in two rows, viz. 'scaphoid' *s*, lunare *l*, cuneiforme *c*, pisiforme *p*; trapezium *t*, trapezoides, magnum *m*, unciforme *u*: with an 'intermedium,' *o*, and a second sabre-shaped accessory ossicle at the radial side of the carpus. The ungual phalanges are bifid.

In the Cape Mole (*Chrysochloris*) the clavicle is long: the humerus is short and arcuate, with a single proximal articulation. The radius and ulna coalesce. The carpus consists of a scaphoid, usually confluent with a trapezoid, fig. 259, *s*, *t*; a lunare, of small size, articulated to both radius, 54, and ulna, and presenting the opposite and larger surface to the magnum, *m*; a still

258

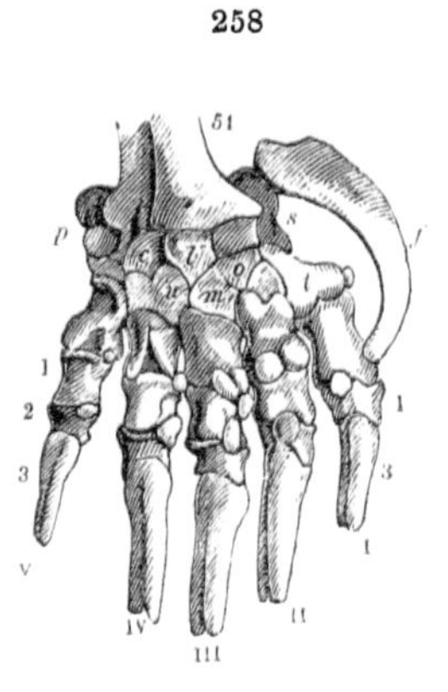

Bones of the fore-foot, Mole: magn.

259

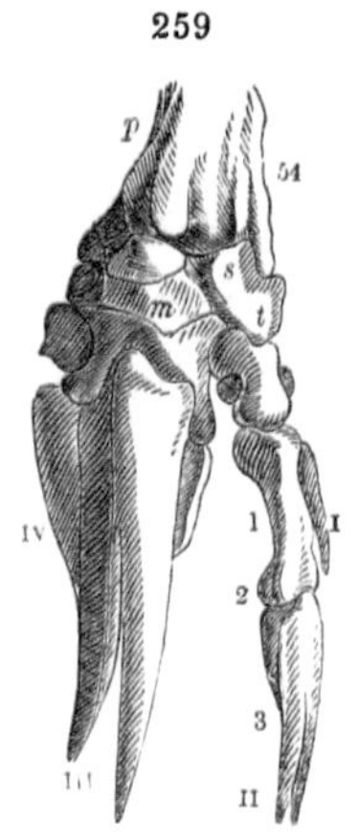

Bones of the fore-foot, *Chrysochloris*: magn.

smaller cuneiforme; and a pisiforme, *p*, in the form of a long subcylindrical bone extending from the carpus to the humerus, and simulating a third antibrachial bone. On the outer side of the magnum, *m*, is a small unciforme supporting a rudiment of a fourth metacarpal. The functional digits are but two: the pollex is represented by a metacarpal and two short phalanges, the second being styliform, I: the index, II, consists of a short metacarpal, a phalanx representing the first and second confluent, 1, 2, and a larger ungual phalanx, 3, cleft at the end. The medius, III, is of monstrous proportions: its metacarpal is broader than long, to which articulates an enormous ungual phalanx, III, bifurcate through the depth of the terminal cleft. A metacarpal representative of the fourth digit, IV, is firmly articulated with, and strengthens the base of, the third digit.

The volant Insectivora are as remarkable for the length and slenderness of the arm- and finger-bones, as the fossorial species for the opposite proportions. The Common Bat (*Vespertilio murinus*) has long and strong bent clavicles: broad scapulæ: elongated humeri: still more elongated and slender radius and metacarpals and phalanges of the four fingers, which are without claws, the thumb being short and provided with a claw: the pelvis is small, slender, and open at the pubis: the fibula is absent, like the ulna in the fore-arm: and a long and slender styliform appendage to the heel helps to sustain the caudo-femoral membrane.

In the frugivorous Bat (*Pteropus*, fig. 156) the clavicles, 58, are long, arched, and very powerful. The humerus, 53, is long, slender, gently sigmoid. The ulna, 55, is slender, and terminates in a point at the lower third of the radius, 54: the olecranon is a detached sesamoid ossicle. The index, II, has a claw as well as the pollex, I: the ungual phalanx is wanting in the other three digits, in which the second phalanx is long, slender, and terminates in a point. The femur, 65, is straight, half the length of the humerus: the tibia, 66, is more slender, rather longer than the femur: the fibula is in the form of a slender style ascending from the outer malleolus and terminating above in a point. The inner digit of the foot, *i*, is a little separated from the other four, *v*, which are of equal length, and unguiculate for suspending the body.

In the Colugo (*Galeopithecus*) the ulna terminates in a point at the lower fourth of the radius: all the five digits of the hand, like those of the foot, have claws supported on deep compressed ungual phalanges.

Amongst the most remarkable bones of the scleroskeleton is the ossification of the raphé between the lateral masses of the muscles of the nape, forming a styliform bone coextensive with the cervical vertebræ, in the Mole. The patella in the triceps extensor cruris, and the fabellæ in the tendinous origins of the gastrocnemii, are present in most Insectivora. The os penis is also found in this order.

§ 183. *Skeleton of Bruta.*—A. *Vertebral Column.*—In the loricate or Armadillo family this is remarkable for the prevalence of anchylosis in unusual parts, e. g. the cervical region, and throughout the dorso-lumbar regions in the great extinct Glyptodonts, which have their cuirass in one piece.

In the Nine-banded Armadillo (*Dasypus Peba*, fig. 260), the vertebral formula is:—7 cervical, 10 dorsal, 5 lumbar, 8 sacral, and 16 caudal. The spine of the dentata is compressed, lofty, and developed backward beyond those of the third and fourth cervicals, with which it has partially coalesced: a corresponding coalescence has taken place between the bodies of these vertebræ, which are unusually broad and flat below. The diapophysial part of the transverse processes of the last cervical abuts against the tubercle of the first broad dorsal rib: the pleurapophysial part of the same transverse process is broad and short, and extends down-

260

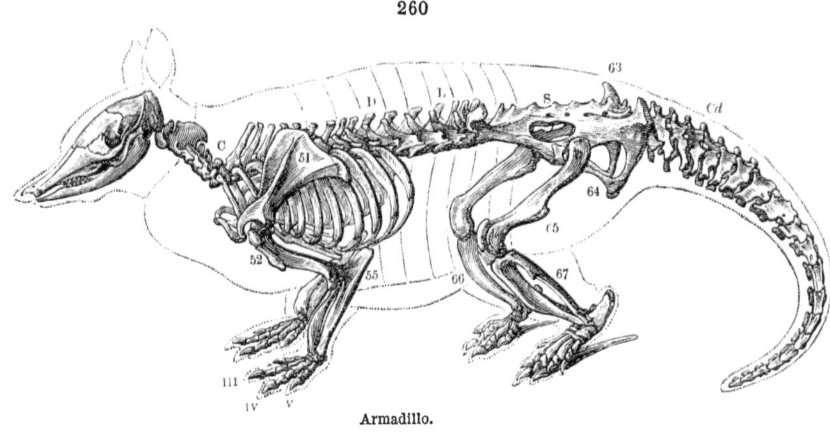

Armadillo.

ward in front of the same rib. The last three cervicals, ib. C, have no spinous processes; that of the first dorsal rises to a considerable height, and those of the remaining dorsals, D, and lumbar vertebræ, L, attain the same horizontal line. The metapophysis is first fully developed upon the seventh dorsal, and progressively elongates to the last lumbar, fig. 261, *m*: it presents an articular surface at the under and fore part of its base to be articulated with the anapophysis of the antecedent vertebra. The anapophyses increase in thickness rather than in length in the succeeding vertebræ, and upon the last dorsal present an articular surface at their under part for connection with a parapophysis, ib. *p*. These accessory joints coexist with the ordinary articulations between the anterior and posterior zygapophyses, and there are consequently twelve joints between each pair of vertebræ, in addition to the ligamentous one between the bodies of the vertebræ, ib. *c*. This mechanism is designed to give great strength and fixedness to the vertebræ of the trunk in relation to the support of the bony carapace, ib. *b*, *b*, and to

261

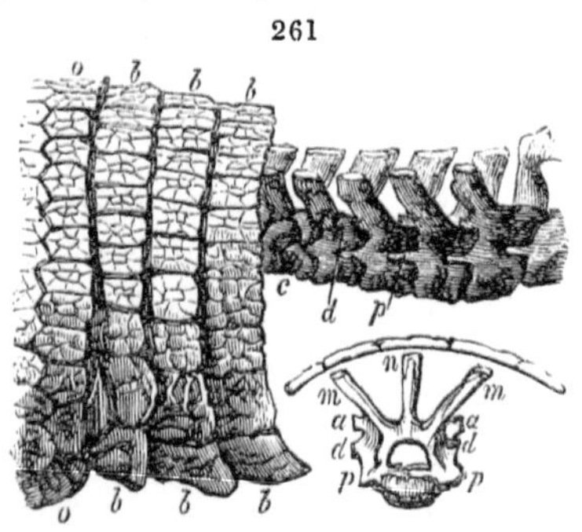

Vertebral architecture for support of carapace, Armadillo.

the affording a firm fulcrum or centre to the powerful muscular forces exercised by the limbs in the act of burrowing. The elongated metapophyses have a more direct relation to the support of the carapace, the spinous processes representing the 'king-posts,' ib. *n*, and the metapophyses the 'tie-beams,' ib. *m*, in the architecture of a roof. The sacral vertebræ progressively increase in breadth after the second, to form an extensive juncture with the ischial bones. The tuberosities of the ischia, fig. 260, 63, and similar tuberosities at the fore-part of the ilia, fig. 277, 62, bend outward and upward, to afford four strong additional supports to the bony carapace: the long diapophyses of the first caudal vertebra abut against those of the last sacral vertebra and the tuberosities of the ischia. The metapophyses reappear upon the second caudal vertebra, and continue to the antepenultimate one, where they are reduced to ridges upon the anterior zygapophyses. Hæmal arches, with short terminally expanded and flattened spines, are present beneath the intervals of many tail-vertebræ. In *Glyptodon* the caudals coalesce.

The posterior dorsal ribs are deeply excavated upon their external surface; five pairs directly join the sternum, which consists of six bones, a very small one being interposed between the fourth and the long one supporting the ensiform cartilage.

In the Cape Anteater (*Orycteropus*), the vertebral formula is:—7 cervical, 13 dorsal, 8 lumbar, 6 sacral, and 25 caudal: anchylosis is limited to the sacral region: the cervical transverse processes overlap each other; the costal part of the sixth is a broad plate. The dorsal and lumbar neural spines are much longer than those of the cervical, and are subequal: they slightly converge to that of the twelfth dorsal, which is vertical, indicating a greater extent of inflection of the trunk than in the South American Anteater; increased freedom of motion is likewise favoured by the less complex character and mode of union of the vertebræ. An accessory tubercle is developed upon the diapophysis of the seven anterior dorsal vertebræ, which divides near the eighth into metapophysis and anapophysis. These progressively increase and diverge from one another in the succeeding dorsals, and in the first lumbar vertebra the metapophysis projects upward, outward, and forward upon the outside of the anterior zygapophysis; whilst the anapophysis extends backward from the back part of the diapophysis, which it equals in length. The anapophysis decreases in size in the following lumbar vertebræ and disappears in the last; the metapophysis also decreases in size, but is continued throughout the lumbar series and along part of the sacral. The sacral spines coalesce, leaving intervening foramina, fig. 262, *a*: the transverse processes of the three anterior sacrals join the ilia, *c*, *g*; those of the three posterior ones coalesce to form a broad depressed plate, with the posterior angles produced, ib. *b*, but not joining the ischia. These have a long and broad tuber ischii, ib. *k*, *l*. The pubis, *d*, is long and slender: the pectineal spine, ib. *h*, is long; the symphysis pubis is short; the lumbo-iliac angle is 140°. Metapophyses are developed from the outside of the anterior zygapophyses, as far as these extend along the caudal series, viz. to the eighth vertebra; beyond these the metapophyses are developed, independently of the zygapophyses, to near the termination of the tail. The hæmal arches commence below the interspace between the second and third caudals, and are continued as far as that between the sixteenth and seventeenth. The neural arch disappears upon the sixteenth caudal vertebra.

262

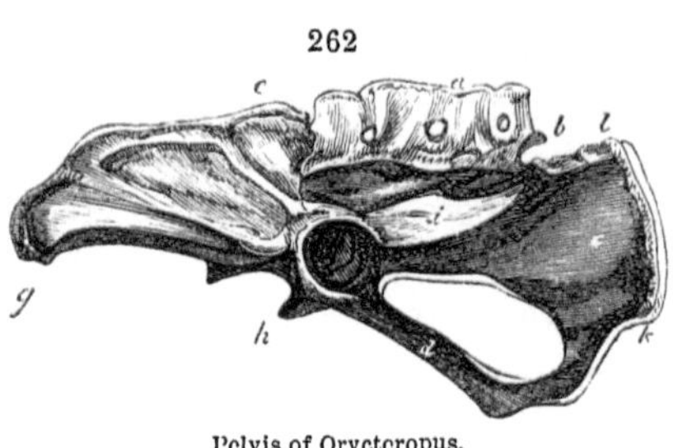

Pelvis of Orycteropus.

In the Scaly Anteater (*Manis pentadactyla*, fig. 158), the transverse process of the seventh cervical is perforated: its spine is longer than that of the others. There are 13 dorsal vertebræ. The nine anterior pairs of ribs directly articulate with the sternum, which consists of ten bones. The tenth is of unusual length, and supports a still longer and much-expanded xiphoid cartilage. The metapophyses commence as tubercles on the first dorsal vertebra, and rapidly increase in size in the succeeding vertebræ. The anapophyses are not developed in this genus. There are 4 lumbar, 4 sacral, and 26 caudal vertebræ. The metapophyses continue to be developed from the sacral series. The transverse processes of the last sacral suddenly expand both in length and breadth, and articulate with the tuberosities of the ischia. Well-developed hæmal arches are articulated to the inferior interspaces of the caudal vertebræ as far as the penultimate one. The anterior zygapophyses cease upon the fourteenth vertebra, but the metapophyses are continued as far as the penultimate caudal. The neural arch gradually subsides and disappears upon the twentieth caudal vertebra, which consists of centrum, diapophyses, metapophyses, and the hæmal arch.

In the skeleton of the Great Anteater (*Myrmecophaga jubata*, fig. 263), the vertebral formula is:—7 cervical, 15 dorsal, 3 lumbar, 5 sacral, and 35 caudal. The atlas is pierced in two places obliquely at the fore-part of the neural arch on each side. The axis has a transverse perforation on each side the neural arch anterior to the transverse process, which is imperforate. The transverse processes of the three succeeding cervicals are imperforate, the vertebral artery entering the neural canal behind, and perforating obliquely the base of the neurapophysis, anteriorly. In the sixth cervical, the canal for the vertebral artery runs through the base of the transverse process. These processes are much extended antero-posteriorly in all the cervicals and overlap

263

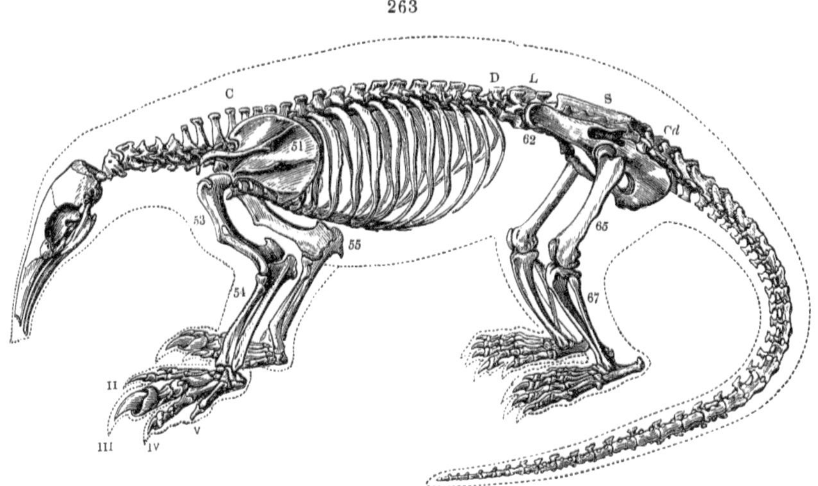

Myrmecophaga jubata.

each other: their di- and pleur-apophysial portions are very distinct in the fifth and sixth cervicals. The spine of the seventh is longer than the rest and truncate above; it is much exceeded in antero-posterior diameter by the spine of the first dorsal. A metapophysial tubercle is developed from the outer side of the prozygapophysis in all the five posterior cervicals. It is placed more outwardly in the first and second dorsals, and gets upon the top of the diapophyses in the succeeding dorsals. In the eleventh dorsal the metapophysis begins to resume its former position, and developes an articular surface from its under part, which joins the upper articulating surface of the anapophysis of the preceding vertebra. In the thirteenth dorsal, the metapophysis is half-way between the diapophysis and anterior zygapophysis, and repeats the same articulation with the anapophysis. In the last two dorsal vertebræ, the base of the metapophysis developes a second articular surface from its inner side, which joins a new or accessory articular surface on the outside of the posterior zygapophysis of the antecedent vertebra. This tenon-and-mortice articulation of the metapophysis with the zygapophysis on the inner side and with the anapophysis on the outer side, is repeated throughout the whole lumbar series. The anapophysis begins to be developed from the anterior dorsal vertebra, and even there presents an articular surface at its under part to join a corresponding surface on a parapophysis developed from the fore and outer part of the neural arch of the succeeding vertebra. In the tenth dorsal a second articular surface is established in the upper part of the anapophyses for the inferior metapophysial one of the succeeding vertebra; here, therefore, the anapophysis begins to be morticed between the parapophysial and metapophysial articular surfaces, which surfaces continue to the antepenultimate lumbar vertebra, from which, forward, to the eleventh dorsal, there are sixteen joints between each pair of vertebræ. But this complication goes further; for, in the penultimate lumbar vertebra, a third articular surface is developed from the under and outer part of the anapophysis, which joins an articular surface on the upper and fore part of the diapophysis of the last lumbar: and this vertebra is united in a similarly complex manner with the first sacral vertebra, which would make eighteen synovial joints, in addition to those at the ends of the centrum, but that those between the normal articular processes, or zygapophyses, are now suppressed. The true serial homology of the processes as 'parapophyses,' developed from the fore part of the base of the neural arch to articulate with the under part of the anapophyses, is well illustrated by the vertebræ of the Great Anteater, as in the Megatherium, in which the true diapophyses are better developed than in the Armadillos.

The spines of the sacrals, ib. *s*, blend into a bony ridge; the transverse processes of the last three join the ischia. Hæmal arches articulate with the intervals of most of the caudal vertebræ.

In the little Two-toed Anteater the dorsal pleurapophyses show a chelonian expansion; but overlap, or join by squamous instead of dentate sutures.

In the Ai (*Bradypus tridactylus*, fig. 263), the vertebral formula is:—C 9, D 16, L 3, S 6, C*d* 11. The neural arch of the atlas is perforated by the vertebral artery anteriorly, and by the cervical nerve posteriorly. The spines of the cervical vertebræ are moderately and equably developed. The pleurapophysial part of the transverse process of the eighth cervical, fig. 265, *a*, *pl*, is more extended antero-posteriorly than in the preceding cervicals, and long remains free. The pleurapophysis of the ninth cervical, ib. *b*, *pl*, retains its freedom, and is more extended in the direction of its length, but is very short as compared with the homologous part, *pl*, of the following vertebra. The slender neck and head of this little rib, joining the fore part of its centrum, occasions the perforated character, as in the antecedent cervical vertebra. A short metapophysis is developed from the fore part of the diapophysis of the penultimate dorsal vertebra, increases in size in the

264

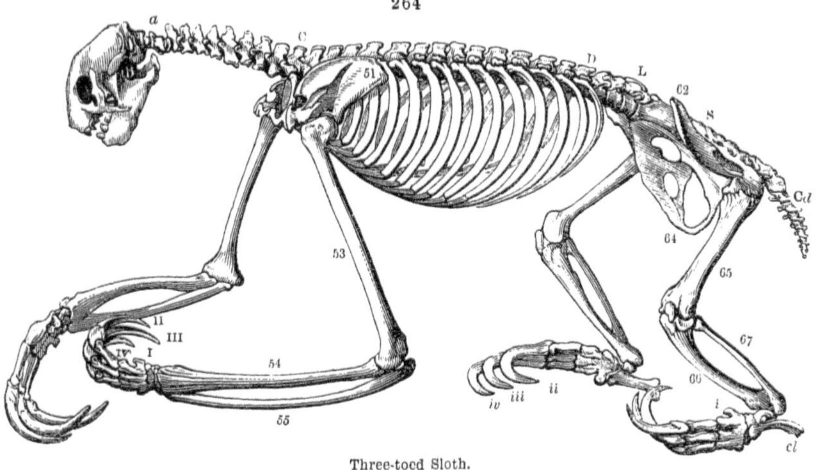

Three-toed Sloth.

last dorsal, and ascends upon the base of the prozygapophysis of the third lumbar vertebra. The anapophysis is also developed from the last dorsal and from the three lumbar vertebræ; it is short, with an articular surface applied to the outer side of the prozygapophysis of the succeeding vertebra. The spinous processes gradually subside in the posterior dorsals, fig. 264, D, and become obsolete in the lumbar vertebræ, L. The first pair of dorsal ribs, *pl*, *h*, is anchylosed to the manubrium, *s*: nine pairs directly articulate with the sternum, which consists of eight bones; these are compressed, and progressively increase in depth; the hinder ones are divided into a larger posterior and a smaller anterior part, between which are four articular facets on each side for the bifurcated extremities of two of the ossified cartilages. There is a pair of hypapophyses on the fifth, sixth, and seventh caudal vertebræ. The pelvis consists of five or six sacral vertebræ coalesced with each other, also (1–4) with the ilia, and (5–6) with the ischia, leaving wide ischiadic foramina, fig. 266, *d*. The sacral spines are obsolete, and the centrums much depressed. The ilia are short and broad, forming an anterior concavity. The ischial tuberosities are small, and the part joining the pubis to circumscribe the large obturator foramina is slender. The pubis is also slender, and forms a very short symphysis, *c*. The pelvic outlet is wide.

265

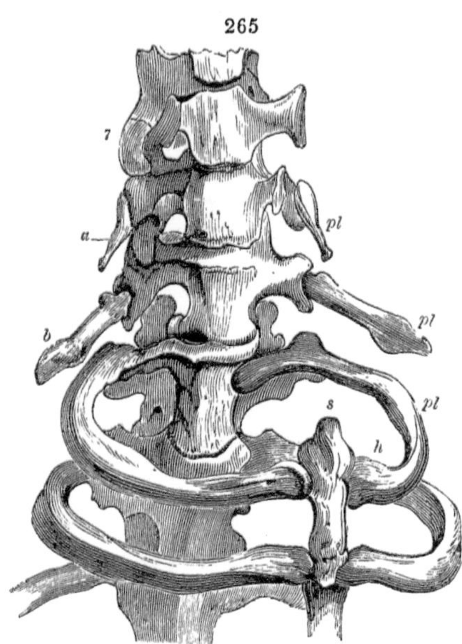

Cervical and dorsal vertebræ, Ai.

In the Two-toed Sloths (*Cholœpus*) the vertebral formula is C 7, D 23, L 3, S 8, C*d* 4, or D 24, L 2; or D 23, L 4, S 7, the number being essentially the same. The second and third cervicals sometimes coalesce. *Cholœpus Hoffmanni* has only six cervicals.[1] Twelve pairs of ribs join the sternum, which consists of eleven bones. Not any of the great extinct Ground Sloths have more than seven cervical vertebræ; but in the number of dorsal vertebræ, as in every other bradypodal character, they manifest their true affinities. The Mylodon, fig. 267, offers a singular contrast with the Mole, fig. 248, in the proportions of the sacrum: this, by anchylosis with the three lumbar and last dorsal, *dls*,

266

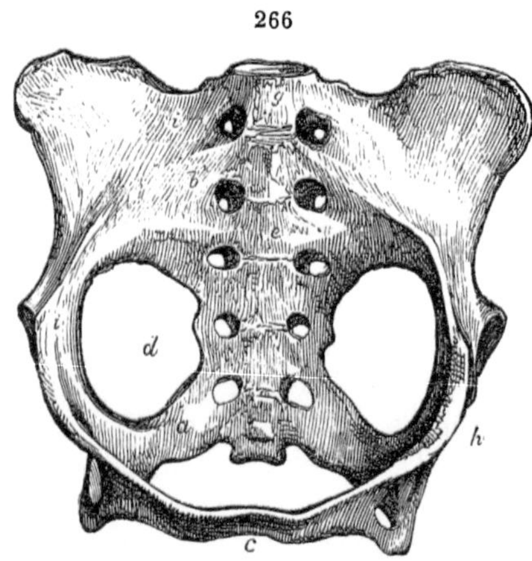

Pelvis of the Ai, anterior view.

[1] LXXV'.

includes eleven vertebræ, and forms one strong and continuous bony mass along the whole lumbar region. Its total length is two feet four inches, and it gradually increases in breadth to the sacro-iliac union, which is formed by the first, second, and third true sacral vertebræ, and there presents its greatest breadth. It then contracts slightly, and, at the sixth and last, expands again to join the ischia, fig. 268, *c*, with which and

267

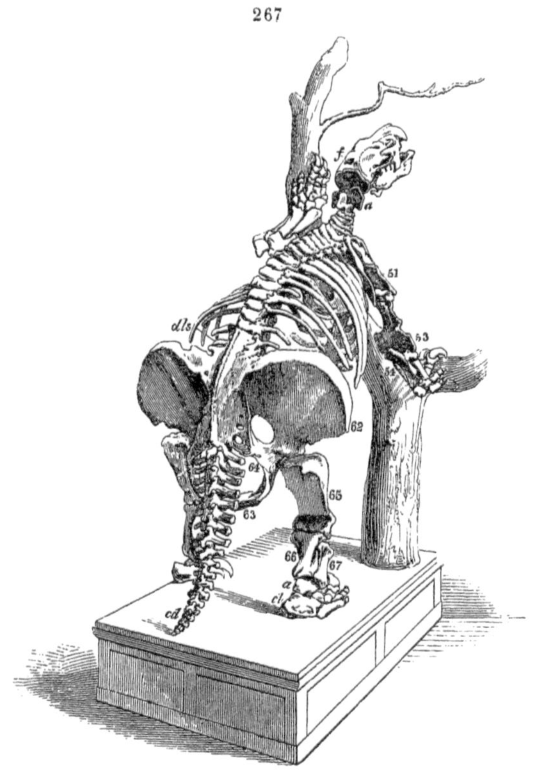

Mylodon robustus. XC'.

the ilia it coalesces. Its anterior surface is curved both laterally and vertically. The spinal canal is very wide, and the foramina passing from it mark the primary segments. Their neural spines form a curved crest, ib. *g*. There are twenty-one *caudal* vertebræ, fig. 267, *cd*. The iliac crest is arched, thickened, and rough; the alæ notably expanded. The ischia, after effecting the sacral confluence, at *c*, arch outward, of slender form, and expand, at *h*, fig. 268, to join the still more slender pubis, *b*, and complete the wide obturator hole, *o*. The tuberosity, *k*, is not well marked.

The vertebral column shows neither in *Mylodon* nor *Megatherium* any apophysial developements related, as in the Armadillos, to the support of a bony carapace.

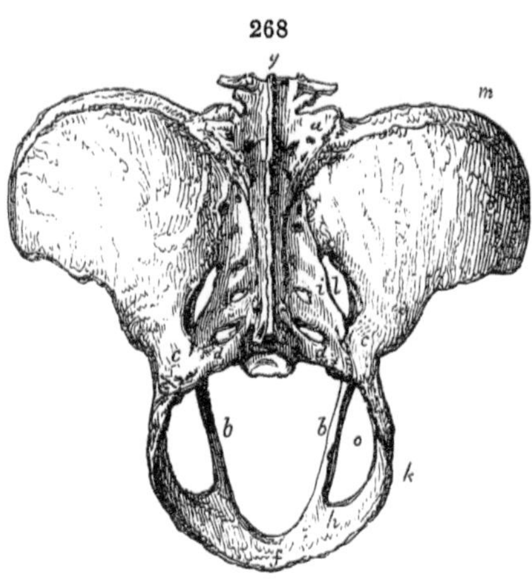

Pelvis of the *Mylodon robustus*, posterior view. xc'.

The Megatherium has C 7, D 16, L 3, S 5, *Cd* 18; fig. 279. The neural spines elongate in the last two cervical, and increase in both length and breadth in the anterior dorsal vertebræ. The dorsal hæmapophyses are bony as in the Mylodon, Sloths, Anteaters, and Armadillos. Nine pairs of ribs directly articulate with the sternum, which consists of eight bones. Most of the dorsal vertebræ are peculiar in having a third, medial, zygapophysis between the ordinary anterior and posterior pairs: they likewise present three surfaces to the pleurapophysis, one on the centrum for the 'head of the rib,' one on the neurapophysis for the 'neck,' and one on the diapophysis for the 'tubercle.' The hæmapophyses present analogous complex joints with the sternebers, having a pair of condyles, synovially articulated with two pairs of cavities on contiguous sternebers, each such sterneber presenting ten articulations, one for the antecedent, another for the succeeding sterneber, and two pairs of hæmapophysial cavities on each side. The posterior dorsal and lumbar vertebræ present, besides the articular surfaces for the centrums, and those of the zygapophyses, also a pair of metapophysial and a pair of anapophysial articulations.[1] In the confluence of the anterior dorsal pleur- and hæm-apophyses the Megathere[2] resembles the Sloth. The caudal vertebræ supporting, as in *Mylodon*, a powerful column, serving as a prop to the massive hind part of the trunk when the fore part is raised, as in fig. 267, have long and strong di- and hæmapophyses: the latter separate in the first caudal, but coalesced at their distal ends in the succeeding caudals to near the end of the tail.[3]

[1] xcı'. Pl. iii. figs. 4 and 5. [2] Ib. Pl. ix. [3] Ib. Pls. ii. and viii.

The large pelvis, the union of the ischia with the sacrum, and the speedy osseous confluence of the several pelvic elements, are common characteristics of the spinal column of the *Bruta*: in no other Mammalian order are found such complex vertebral articulations; and here alone are manifested the exceptional instances of affinity to certain *Ovipara* in the lower cervicals with free ribs of the Three-toed, and in the twenty-three costigerous dorsals of the Two-toed, Sloths.

B. *Skull.*—The skull in the insectivorous *Bruta* is long and slender; these proportions reaching their extreme in true Anteaters (*Myrmecophaga*, fig. 269). The occipital condyles, 2, are large and terminal: the superoccipital, 3, inclines forward as it rises to join the parietals, 7, which retain their 'sagittal' suture. The frontals are elongate, and continue the smooth transversely convex cranial roof forward to the nasals, 15; these are still longer, being coextensive with the maxillaries, 21, which, with the mandible, 32, form the walls of the bony tubular sheath of the very long tongue: the premaxillaries are minute. The orbits are feebly defined: the lacrymal, 73, is large and chiefly antorbital. A beginning of the malar is appended to the maxillary: the small squamosal forms the flat surface for the mandibular condyle, but developes no zygoma. The tympanic, 28, retains its separate condition. The petrosal is excavated by a hemispheric cavity for the condyle of the stylohyal, the framework of the tongue in the Anteaters almost equalling the mandible in its amount of bone.

269

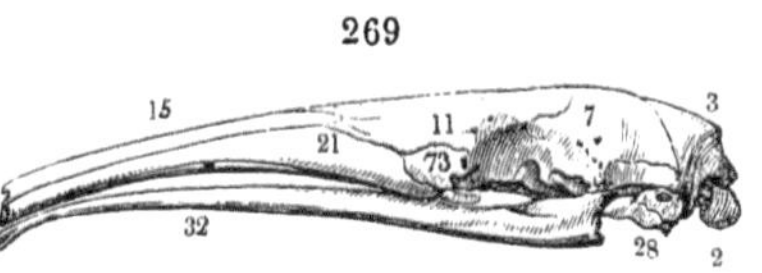

Skull of *Myrmecophaga jubata*. CLI.

In the Pangolins (*Manis*, fig. 270) the cranium also shows the inclination of the occipital surface from below upward and forward, the plane of the foramen magnum being slightly inclined in the same direction. The exoccipitals, 2, meet above the foramen. The superoccipital, 3, is rhomboidal, and its aspect is almost wholly upward. The mastoids, 8, form two hemispheric protuberances at the sides of the occipital region, and the petrosals two smaller protuberances at the sides of the base of

270

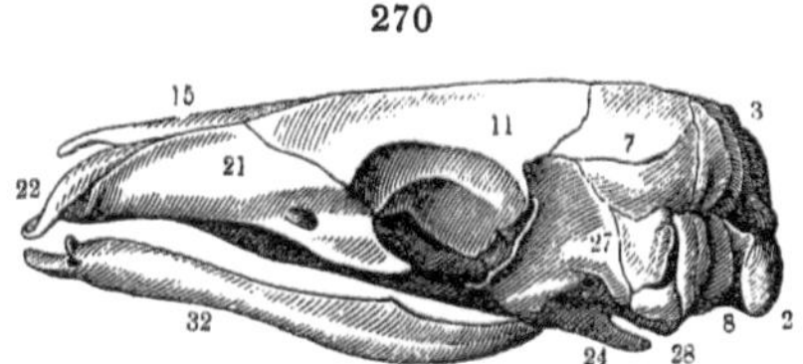

Skull of *Manis*. CLI.

the skull: the pterygoids, 24, extend backward beyond them, and form the sides of a deep and wide postnasal groove. The tympanic bone forms the lower boundary of a hemispheric bulla, which communicates with an equal-sized cavity in the squamosal; a narrow strip of the petrosal intervenes between the tympanic and the broad basioccipital. The zygomatic process of the squamosal, 27, extends but little beyond the joint for the lower jaw: there is no separate malar. The premaxillaries, 22, join the nasals, 15. The small lacrymal, in *Manis longicaudata*, is wedged in between the frontal, 11, and maxillary, 21, at the anterior angle of the orbit. Two tooth-like processes project from the fore-part of the alveolar border of the slender under-jaw.

In the hairy Anteater of the Cape (*Orycteropus*) the petromastoid and tympanic are distinct from each other, and retain their primitive separation from the squamosals. The occipital condyles are bilobed, the inferior and smaller lobe being developed from the basioccipital. The zygomatic arch is slender, but entire. A well-marked venous fossa depresses the inner border of the foramen magnum; the sella is large and moderately deep, with anterior and posterior clinoid processes, bounded on each side by the carotid channels, external to which are the deeper Gasserian fossæ. There are few mammalian skulls in which the cranial cavity is more equally divided, than the present, into the epencephalic, mesencephalic, prosencephalic, and rhinencephalic chambers; but the mesencephalic chamber contains not only the proper mesencephalon, but also, as in other Mammals, part of the backwardly developed prosencephalon, and especially those inferior protuberances called 'natiform.' The petrosals show very narrow cerebellar fossæ. The maxillary has seven alveoli, the mandibular ramus has six: but the small anterior ones are obliterated with the loss of their teeth in old animals. The hyoid arch is large and consists of the stylohyal, ceratohyal, epihyal, and basihyal elements, with the appended thyrohyals, or 'cornua majora.'

271

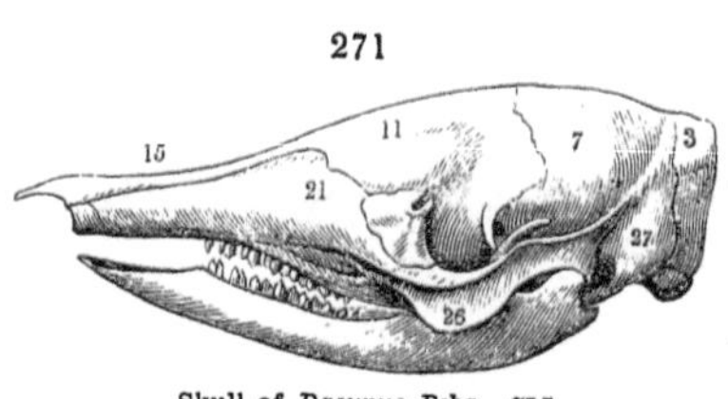

Skull of *Dasypus Peba*. CLI.

In the Armadillos (*Dasypus*, fig. 271) the occipital plane is vertical. There is no paroccipital: the superoccipital, 3, is bent at a right angle to join the parietals, 7, which obliterate, as in *Orycteropus*, the sagittal suture by their union. The mastoid is perforated by a vein from the lateral sinus, and terminates below in an obtuse process. The tympanic is a separate semicircular plate of bone. The alisphenoids join the parietals. The lacrymal is large and chiefly antorbital. The petrosal presents a wide and shallow cerebellar fossa: the canal between the petrosal and the angle of the superoccipital gives exit to a vein and entry to an artery. The rhinencephalic almost equals the epencephalic divi-

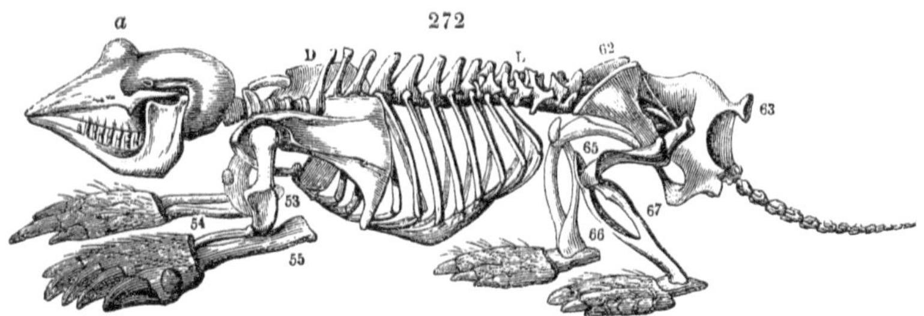

Dwarf Armadillo (*Chlamyphorus truncatus*). LXXXVIII'.

sion of the cranial cavity. The frontals are large, but chiefly occupied by the nasal chamber: in the *Chlamyphorus* the outer table rises into a pair of domes, fig. 272, *a*, augmenting the olfactory cavity. In most Armadillos there are two small prenasal ossicles. The premaxillaries are small and lodge the first tooth in one or two species. The zygomatic arch is complete and strong: the malar part, fig. 271, 27, curves down outside the mandible, and there, in *Glyptodon*, developes a long process for the service of the masseter.

The zygomatic arch culminates in regard to complexity in the Sloths, albeit in the small existing species, fig. 273, the squamosal element, 27, fails, as in the Anteaters, to reach the malar one, 26, *b*. In the Megatherioids this union is effected, fig. 274, and an unusually massive arch is the result. The malar, 26, still sends upward its temporal process, *b*, and downward its masseteric one, *a*. This cranial developement relates, as in the recent and extinct Kangaroos, to the share of the masseter in the business of mastication; and the molars are transversely ridged in *Megatherium* as in *Diprotodon*.

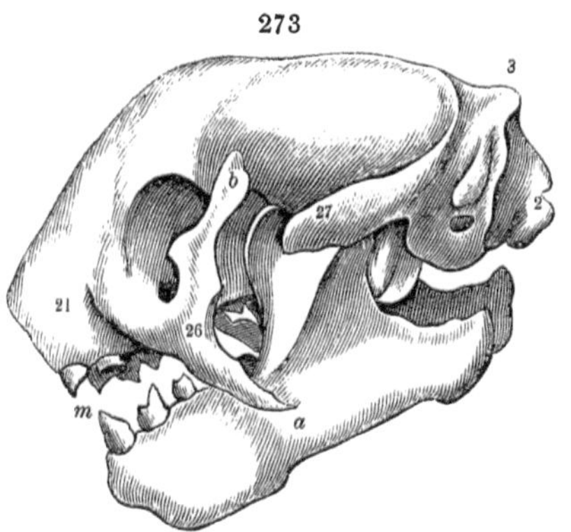

Skull of Ai (*Brad. tridactylus*), half nat. size.

The facial part of the skull in the Sloth is as remarkable for its shortness as in the Anteater for its length. In the Ai, fig. 273, the interparietal coalesces with the superoccipital, 3, before the exoccipitals unite with the super- and basi-occipitals. The malar bone, 26, is freely suspended by its anterior attachment to the maxillary and frontal, and bifurcates behind; one division, *a*, extending downward, outside the lower jaw, the other, *b*, ascending above the free termination of the zygomatic process of the squamosal, 27. The premaxillary is single and edentulous, being represented only by its palatal portion completing the maxillary arch, but not sending any processes upward to the nasals. Within the cranium there is no bony tentorium: the two divisions of the meatus internus commence separately upon the exterior of the petrosal, which is not impressed by a cerebellar fossa. The depression receiving the natiform protuberance of the cerebellum is formed chiefly by the squamosal. The walls of the rhinencephalic fossa are entirely surrounded by the olfactory chamber, which extends above into the frontal and beneath into the sphenoidal sinuses. A well-marked vascular foramen leads downward from the partition between the rhinencephalic and prosencephalic chambers. The rough exterior part of the petrosal forms, as it were, the border of a capsule to the tympanic: the fossa for the stylohyal is well marked at the back part of the border. The pterygoid forms a large quadrate vertical plate. The bony septum narium terminates half-way from the large vertical external nostril. There is a small imperforate lacrymal: the antorbital hole is wanting. In the Unau (*Bradypus didactylus*) the lacrymal is pierced external to the orbit. In the Megathere the foramen is at the orbital margin, and there is a large antorbital foramen: the premaxillaries, though edentulous, are more produced than in existing Sloths, and there is a corresponding production of the symphysial part of the lower jaw, which is grooved above, as in the Anteaters, for the support during its pro- and re-tractile movements of a long tongue, prehensile in the Megathere as in the Giraffe, in reference to the smaller branches of the trees yielding food to the extinct giant. Behind the symphysis the

274

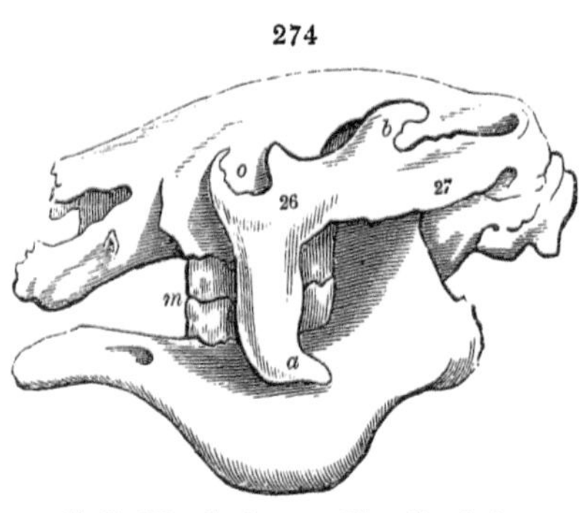

Skull of Megatherium, one-fifteenth nat. size.

mandible is deepened for the long roots and matrices of the ever-growing molars. The extension of the air-sinuses, great in the climbing Sloths, was still more so in the colossal species, whose strength enabled them to uproot and prostrate the trees they browsed on. The skull of the *Mylodon robustus* in the Hunterian Museum shows two extensive fractures of the outer table, one wholly, the other partially, healed: the latter extending to near the occiput, but having broken only into the air-chamber, not into the cranial cavity, the inner, proper or 'vitreous' table of which is everywhere divided by sinuses and sinuous bony plates from the outer table.

Notwithstanding the extreme diversity—singular contrast indeed in several particulars—which the skull presents in the order *Bruta*, the marks of inferiority of position in the Mammalian series, according to the cerebral character, are constant throughout. The terminal position of the occipital condyles and the aspect of the occipital surface, the degree in which the parts of the complex 'temporal bone' of higher Mammals retain their primitive separation, the position of veins conducting from the cerebral sinuses, the low facial angle and small proportional size of the cranial cavity, the small share in which the squamosal contributes to its walls—all exemplify the inferiority of the present unguiculate group of animals to the Gyrencephalous Ungulates.

C. *Bones of the Limbs.*—In all Armadillos the clavicles, fig. 275, 52′, are complete. The scapula, 51, is broad, convex externally, and presents two spines, the normal one of which is produced into an acromion, long in all the species and unusually so in the Chlamyphore, ib. *n*; in most it sends down a process from its base. The coracoid curves downward: there is a well-marked tubercle behind the neck of the scapula. The suprascapular element is represented by a coarsely ossified cartilage attached to the base of the scapula, fig. 260, 51. The humerus is remarkable for its strength and for the great developement of the deltoid ridges. It is perforated above the inner condyle, but not between the condyles. The ulna, fig. 276, 55, is considerably longer and stronger than the radius: the olecranon, fig. 275, 54, is remarkably developed. In *Dasypus* the radius, fig. 276,

275

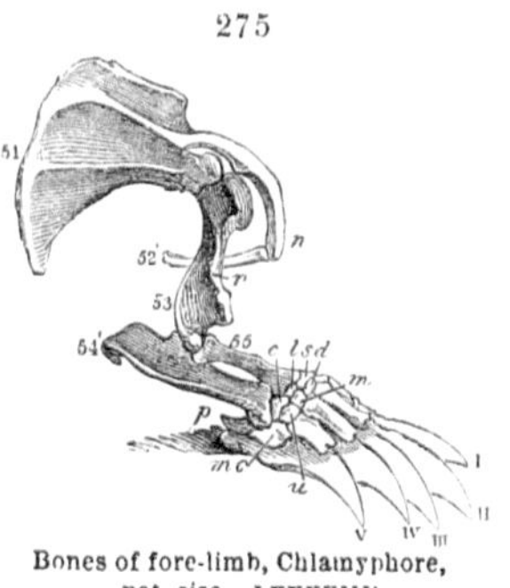

Bones of fore-limb, Chlamyphore, nat. size. LXXXVIII′.

54, is but half the length of the ulna: in all *Bruta* it is free, and rotates on the ulna. The four carpal bones of the proximal row are distinct from one another.

276

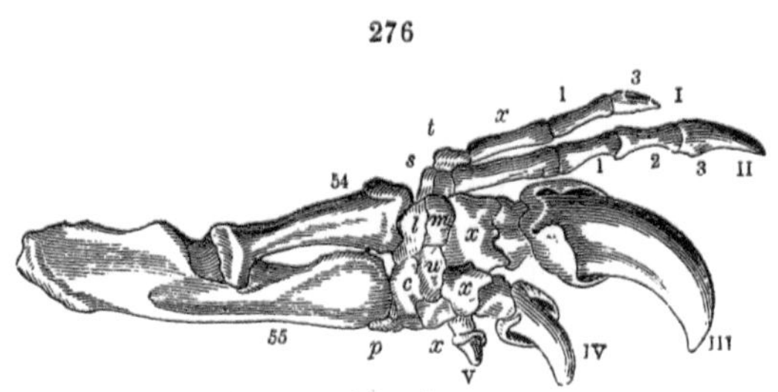

Bones of fore-arm and foot, *Dasypus gigas*. CLI.

The scaphoid, ib. *s*, is the smallest of the four bones of the proximal row. The pisiform, *p*, articulates to the cuneiform, *c*, and ulna, and extends, palmad, to the lunare, with which it forms a large articular cavity, upon which the palmar patella plays. In some the trapezium, *t*, is distinct; in others it is connate with the trapezoides. The magnum, *m*, in most coalesces with the base of the cubical metacarpal of the digitus medius, III. The outer part of the base of that metacarpal rests upon the unciforme, *u*, which also supports the small but thick cubical metacarpus of the annularis, IV, and rudiment of the metacarpal of the minimus, V. The medius and annularis have each but two phalanges; the long and slender index retains the normal number of three phalanges; the base of its metacarpal is wedged between that of the third, the trapezoides, and the trapezium. The chief peculiarity is the very large sesamoid bone developed in the flexor tendons, and filling the palmar aspect of the fore-foot: a second sesamoid is attached by ligament to the apex of the large palmar one. An accessory ossicle, *x*, is wedged into the outer side of the carpus in *Das. gigas*. In *Das. Peba*, fig. 260, there are four digits on the fore-foot, the two middle much exceeding in length and strength the outer and inner ones: the pollex, I, is obsolete. The femur, figs. 260, 277, 65, presents a third trochanter. The proximal and distal extremities of the tibia, 66, and fibula, 67, are connate: their shafts subsequently coalesce therewith, so that a single epiphysis answers to the shafts of both bones at each of their extremities, in the immature Armadillos. The naviculare is remarkable for its two inferior tuberosities, the interspace between which receives the under part of the entocuneiform bone. In *Das. sex-cinctus* it sends downward a process, like that in some Rodents. The calcaneum is less produced in the Chlamyphore, fig. 277, *cl*, than in the Nine-banded Armadillo, fig. 260. The hind-foot is pentadactyle in all. The chief modification of the limb-bones in the extinct gigantic Armadillos (*Glyptodon*) relates to the modification of unguiculate feet to the support and terrestrial progression of species too huge for burrowing, and as heavy as the bulky Pachyderms. The ungual phalanges are accordingly obtuse, short, broad, and thick, for being incased in hoof-like nails, and their phalanges are flat bones, presenting the maximum of breadth in proportion to length. In the third trochanter and the anchylosed tibia and fibula the Dasypodoid characteristics are preserved.

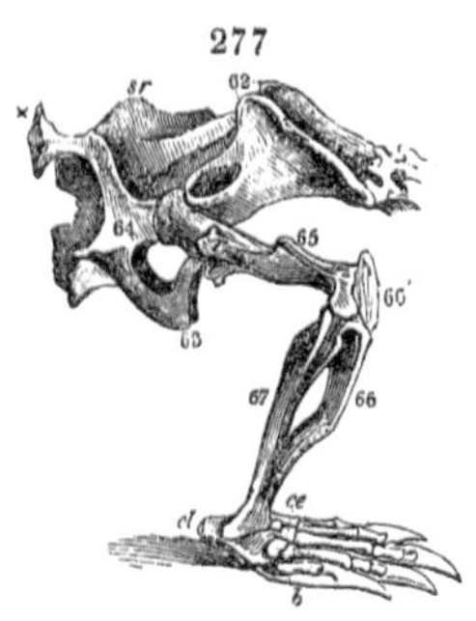

Bones of the hind-limb, Chlamyphore. LXXXVIII'.

The limb-bones of the *Orycteropus* more resemble those of the Armadillos than of the toothless Anteaters. The acromion scapulæ is less elongate: the entocondyloid process of the humerus is recurved, and widely perforated. The wrist-bones are as in *Dasypus*, but the pisiform is long and slender. The pollex is reduced to a stunted metacarpal and phalanx, and the hand has but four claws, of which that of the medius is largest, but less disproportionately so than in *Das. gigas*. In the hind-limb the femur shows the third trochanter. There is a fabella behind the outer condyle. The tibia and fibula coalesce at their upper ends, but not below: the hind-foot is pentadactyle.

In the *Manis*, fig. 153, the spine of the scapula is single, is not prolonged into an acromion, and there are no clavicles: the coracoid is represented by a small distinct tubercle, forming the anterior extremity of the elliptical glenoid cavity for the humerus. The humerus is perforated at the internal condyle. There is an articular sesamoid developed on the outer side of the capsule uniting the radius with the humerus. The scaphoid and lunare coalesce. The digitus medius is disproportionately large, and its ungual phalanx is deeply cleft: that of the index and annularis show slightly the same character. These phalanges are so articulated as to admit of flexion, but not of extension, or retraction, beyond the line of the supporting digit. The femur has no third trochanter. There is a fabella behind the outer condyle of the femur. The bones of the leg retain their distinctness: the extremity of the fibula beyond the outer malleolus bends inward, and terminates in a tuberosity playing in a cavity upon the outer side of the astragalus. There is an accessory tarsal ossicle on the inner side of the entocuneiform and scaphoid. The ungual phalanx of the hallux is simple; those of the three middle toes are cleft at the apex.

In the *Myrmecophaga*, fig. 263, the scapula, 51, is very broad, with a sub-circular contour, and is traversed by two spines, the upper one prolonged as an 'acromion' toward the coracoid, and supporting a small clavicular bone (*Myrm. jubata*) or joined to a complete clavicle (*Myrm. didactyla*). The humerus, 53, is greatly expanded at its distal end, especially by the entocondyloid crest, which is recurved and perforated. The radius, 54, and ulna are of nearly equal length, the acromion, 55, has the lower angle produced. The carpus consists of the usual eight bones, fig. 278, viz. scaphoides *s*, lunare *l*, cuneiforme *c*, pisiforme *p*, which is

278

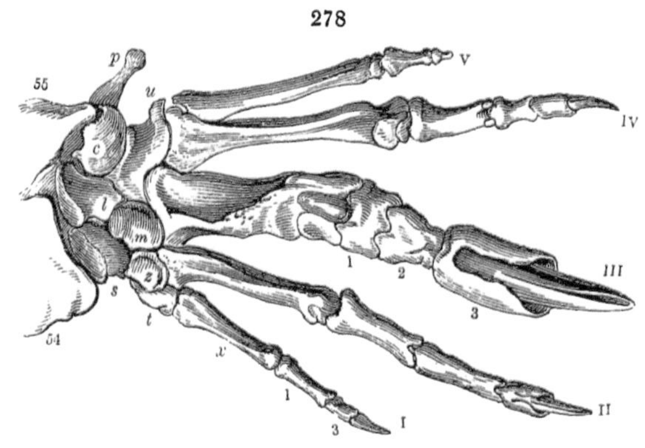

Bones of the hand, *Myrmecophaga jubata*. CLI.

produced like a carpal calcaneum. The trapezium, *t*, supports a slender pollex, I; the trapezoides, *z*, a longer and larger index, II; the strong, four-sided, outwardly-ridged metacarpal of the medius, III, rests its base upon the magnum, *m*, and unciforme, *u*; and is wedged between that of the index, II, and annularis, IV. The minimus, V, which is articulated more to the annularis than to the unciforme, *u*, has but two phalanges, and is clawless. The ungual phalanges are dorsally grooved, not notched. In the *Myrmecophaga didactyla* the metacarpal rudiments of the pollex and minimus are hidden beneath the shin: the annularis metacarpal supports a clawless phalanx: the two conspicuous digits are the index and medius, the latter the largest. The femur, fig. 263, 65, has a crest along its outer margin: the tibia and fibula, 67, are distinct: the foot is plantigrade and pentadactyle, with a calcaneum long and compressed in the great terrestrial species, but short in the scansorial didactyle Anteater, in which a supplementary bone on the inner (tibial) side of the tarsus is produced backward to increase the power of the heel in grasping.

In the Ai, fig. 264, the scapula, 51, is broad, and its outer surface is equally divided by the spine, which is short, but continued into an acromion arching to join the coracoid. The supra-spinal notch is converted into a foramen by the extension of ossification from the superior costa to the base of the coracoid. The same characteristics are reproduced in the scapula of the Megathere, fig. 279, 52, under more massive proportions of these growths for muscular attachments, and with the superaddition of

279

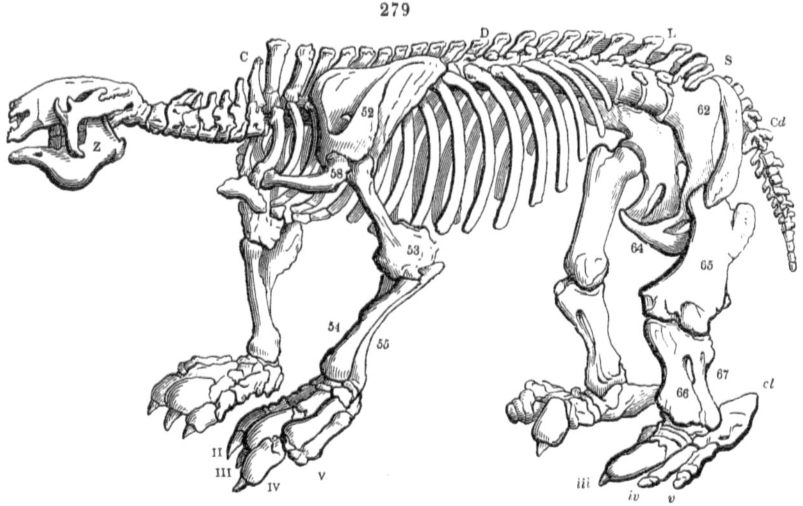

Megatherium americanum.

an inferior spine, as in the Anteater. The clavicle is complete in the Megathere, ib. 58, as in the Two-toed Sloth (*Cholœpus*): in the Ai it exists as a short appendage to the acromion. In the small climbing Sloths the length of the prehensile fore-limbs is attained by that proportion of the humerus, fig. 264, 53, and antibrachial bones, 54, 55: both the latter are bent, leaving a wide interosseous space, and are so articulated as to allow of pronation and supination. In the Megathere the humerus is relatively shorter, but thicker, and is enormously expanded at its distal end, fig. 279, 53: the inner condyle is imperforate, as in *Bradypus tridactylus*: in the *Megalonyx* it is perforated as in the *Brad.* (*Cholœpus*) *didactylus.* The ulna of the Megathere, fig. 279, 55, is equally remarkable for the vast expanse of its proximal end, including the olecranon which is twice as broad as long, and projects backward rather than upward. The proximal end of the radius, 54, is circular, and its articular modifications are as well adapted for rotatory and flexile movements of the antibrachial bones as in the human arm. The interosseous space is shorter and much narrower relatively than in the Sloths. Of these the Ai, fig. 280, has the carpus reduced to six bones, the scaphoid being connate with the trapezium, *s* .... *t*, and the magnum with the trapezoides, *m*. A rudiment of the metacarpal of the pollex, I, has coalesced with that of the index, II, and a rudiment of the metacarpus of the minimus with that of the annularis, *x*, IV. In the three functional digits the proximal and middle phalanges are confluent, 1...2: the ungual ones are of great length, and restricted in their movements, by the production of the back part of their base, to degrees of flexion. The joints of all the digits are deeply trochlear.

280

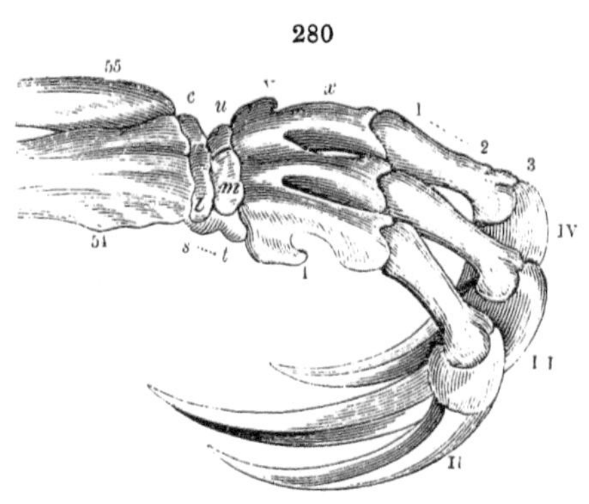

Bones of fore-foot, Ai.

The bones of the hand of the Unau (*Cholœpus*) are described at p. 306, fig. 191, 'Sloth.' In the Megathere the wrist has a scapho-trapezium, fig. 191, *st*, but the trapezoid and magnum are distinct. The pollex is represented by a stunted metacarpal, I: that of the minimus, figs. 191, 279, V, is long and supports one or two short thick phalanges: the second, II, third, III, and fourth, IV, digits are powerfully unguiculate, but the first and second phalanges coalesce only in the medius, 1, 2, III. The massive metacarpal is squared, firmly attached to the contiguous ones, with the outer angle of the base produced and wedged between that of the annularis, the magnum, and unciforme.[1]

In the Sloths the femur, fig. 264, 65, is straight, like the humerus, but is thicker and shorter; the head shows no impression for a ligamentum teres. The tibia, 66, and fibula, 67, are oppositely bent, leaving a wide interosseous space, as in the forearm, but are still shorter than their homotypes. The inner malleolus projects backward and supports a grooved process: the lower end of the fibula fits, like a pivot, into a socket in the

[1] xci. p. 53, pl. xxi.

astragalus. In *Brad. tridactylus* the calcaneum is remarkably long and compressed. The scaphoid, cuboid, and cuneiform bones have become confluent with each other and the metatarsals, of which the first, I, and fifth, *u*, exist only in rudiment. The other three have likewise coalesced with the proximal phalanges of the toes which they support. In the *Brad. didactylus* the ento- and meso-cuneiform bones, the rudimental metatarsal of the hallux, and the metatarsal of the second toe are confluent into one bone: the rudimental metatarsal of the fifth toe has not become united with that of the fourth toe. The functional toes have long prehensile claws like those of the fingers; by the peculiar ankle-joint the foot is turned inward, and the advantage in grasping is obtained at the cost of the power of stepping on flat ground.

The Megathere has a pivoted articulation of the foot with the leg, but the process and the cavity are on reverse parts of the ankle-joint, and the astragalus sends a process to fit a cavity in the tibia. The result, in the inflection of the hind foot, is nearly the same: but an enormous calcaneum, fig. 279, *cl*, and metatarsal of the fifth toe, ib. *v*, rest broadly on the ground. The lower surface of the astragalus transmits the superincumbent weight in two directions; backward upon the heel-bone and forward upon the metatarse. By the naviculare it is transmitted through the ectocuneiforme and the produced angle of the base of the mid-metatarsal to the fourth and thence to the fifth metatarsal. The cuboides receiving the weight from both astragalus and naviculare transmits it by its produced fore-part to the base of the fourth metatarsal, and partly by that medium, but chiefly by direct articulation, to the side of the fifth metatarsal. The tendency of the cuboides to yield under this pressure and slip back is resisted by the abutment of the calcaneum against its back part. The digitus medius, ib. *iii*, was alone developed to sustain and wield a claw; but this was of enormous size, and must have had the power of a pick when worked by the lever of the long heel-bone. The first and second toes were not present, nor was the ento-cuneiform bone. The two outer toes, *iv* and *v*, terminated in tuberous phalanges, evidently imbedded, in the living animal, in a hoof-like thickening of the outer border of the foot. The outer side of the fore-foot presents a similar modification for quasi-ungulate progression on the ground. Thus the Megathere, Mylodon, and allied great terrestrial Sloths seem to have combined ungulate and unguiculate characteristics in the same extremity.

The principle of viewing structures and instruments, in reference to the work that they may do, is shown to be good in gaining insight into the mode of life of extinct animals, in a striking degree through its application to the skeletons of the Megatherioids.[1] The teeth of these conform so closely in all characters with those of the Sloths as to suggest leaves rather than roots to have been their food. In the light slender Sloths the modifications of structure for climbing, clinging, and living altogether in trees, are carried out to an extreme. In the colossal extinct kinds the foliage was obtained in a different way. The huge single claw on the hind foot would be applicable as a pickaxe to clear away the soil from between the ramifications of the roots: a second claw would have interfered with such work. The foot is organised to give great strength to that claw; dislocation of its toe is specially guarded against: the rest of the tarso-metatarsal structure relates to the power of the foot to sustain superincumbent pressure, with a position of the claw bringing its side instead of its point in contact with the ground. The bones of the thigh and leg are remarkable for their massive proportions, for their thickness, and especially their breadth in proportion to their length: the femur in both *Mylodon* and *Megatherium* would rank rather with the 'flat' than the 'long' bones. These osseous columns were needed to support the huge, heavy, expanded pelvis, fig. 267. The iliac expansions are the chief conditions of the other characteristics of this part: and they are unintelligible save in relation to adequate extent of origin of powerful muscles, especially those arising from the crista ilii, 62, the chief of which muscles concentrate their force upon the fore-limbs. This indicates that these limbs were put to some unusual work; and the inferences from the teeth and the hind claw lead to its recognition as the pulling down trees and wrenching off their branches: but, for these operations, the pelvis must have adequate fixity; and to the weight and strength of itself and its supporting limbs there is added a tail so developed as to serve as a third support and give the pelvis the basis of a tripod. Without this view of the function of the hind-parts of the skeleton we can only see that the pelvis is so great and, with its caudal appendage, so weighty as to require the massive proportions and structure of the hind-limbs; and, reciprocally, that these bespeak a proportionate size and weight of the parts to be sustained: but why such developement of sustaining limbs and parts to be supported in reference to any other action and way of life is inconceivable. The excess of bone in the hind-part of the skeleton once recognised as relating to the fixed point of attachment of muscular forces working the fore-limbs,—to the exertion of power adequate to prostrate a

[1] xc·.

tree,—and the rest of the bony organisation becomes intelligible. That of the hind-foot has been explained: the concomitant extent of muscular origin afforded by the broad scapular plate with its many ridges, crests, and processes, is thereby accounted for. The necessity of the firmness imparted to the shoulder joints by the perfect clavicles abutting at one end against a large 'manubrium,' at the other end against the conjoined acromion and coracoid, becomes obvious. The fore-foot retained three huge claws to effect an adequate grasp of the trunk or bough: for their due and varied application the fore-arm enjoys all the variety and freedom of movements which an arm terminated by a hand possesses. A tree being prostrated and its foliage thus brought within reach, every indication in the skull of the size, strength, flexibility, and prehensile power of the tongue harmonises with the foregoing teleological conclusions. The Megatherioids, like the Giraffe, thus plucked off the foliage on which they fed. In the ridged crowns of the grinders of the Giant Ground-Sloth we discern the power of crushing coarser parts—a greater proportion of twigs and stems, e.g.—of the foliage than the diminutive Tree-Sloths take. It needed only evidence of the occasional occurrence of what might happen to a beast in the fall of a tree which it had uprooted, to seal the foregoing physiological inferences with the stamp of truth: and the skeleton of the *Mylodon* in the Hunterian Museum[1] shows that evidence above the right orbit, and at the back part of the cranium, marked *f*, fig. 267.

§ 184. *Skeleton of Cetacea.* — This is characterised by the coarseness and greasiness of the osseous texture, by the shortness of the cervical and the length of the caudal regions, by the loose and diminutive pelvic bones, by the absence of pelvic limbs, and by the large size of the skull, due in most to that of the jaws, which in some Whales (*Balænidæ*, fig. 159, *Physeter macrocephalus*) is excessive.

A. *Vertebral Column.*—Although there is as little outward sign of a neck in a whale as in a fish, the same number of cervical vertebræ are present as in the giraffe. The atlas, fig. 283, 1, is the largest, is characterised by its huge and approximate articular cups, *c*, for the occipital condyles, and by the substitution of a hypapophysis for the true centrum, which coalesces as an odontoid process with that of the axis: both these vertebræ are antero-posteriorly compressed and transversely extended, and the five succeeding cervicals are still shorter in proportion to their height and breadth: they are, in fact, lamelliform, without reciprocal movement, and usually exhibit a greater or less extent of confluence, the whole

[1] xcii'. p. 63, no. 377.

forming one mass like a 'cervical sacrum' in the true Whales (*Balæna*, fig. 159, C), small Cachalots (*Euphysetes*),[1] the Grampus,[2] the Porpoise: the neural arches of the axis and following cervicals are confluent in most. The cervicals thus give a firm support to the large head which has to overcome the resistance of the water when the swift swimmer is cleaving its course through that element. The characters defining the succeeding vertebræ, applicable to comparisons of species and recognition of range of variation, appear to be:—the support of free ribs; the presence of transverse processes formed chiefly by coalesced pleurapophyses, fig. 141, *d*, the articulation of hæmapophyses, fig. 282, *h*, to the centrum, ib. *c*. Thus, in a large British Dolphin (*Delphinus tursio*), the skeleton of which I prepared (and to take the bones from the carcase is almost essential to certainty as to number of ribs and hæmal arches), there are sixty vertebræ. Of the seven cervical the first two only are anchylosed: thirteen vertebræ support free ribs suspended to terminally expanded diapophyses, fig. 281, *d*; then follow twenty-nine with transverse processes only, as in fig. 141, *d*: the thirty-third vertebra from the skull first supports a hæmal arch, but in that and the two following vertebræ the piers or 'hæmapophyses' are small and ununited: the complete arch, as in fig. 282, *h*, is continued, diminishing, to the last six vertebræ, which consist of the centrum only, much depressed. Thus, between the thirteenth dorsal vertebra and the first with hæmapophyses, there are thirteen which might be termed 'lumbar,' fig. 159, D, Cd, but hold the place of lumbar and sacral in other Mammals (*Megatherium*, e.g., fig. 279, L, S). A sacrum is never indicated by vertebral confluence in *Cetacea*, and only obscurely by the position of the pelvic rudiments, fig. 159, 63, 64, loosely suspended below. In the *Delph. tursio* a metapo-

281

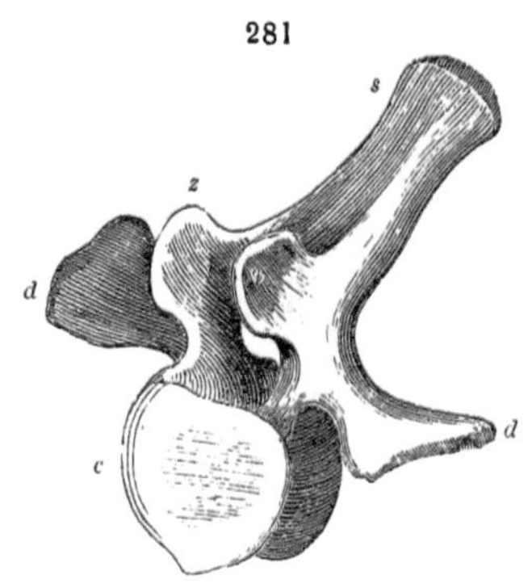

Dorsal vertebra, Whale.

282

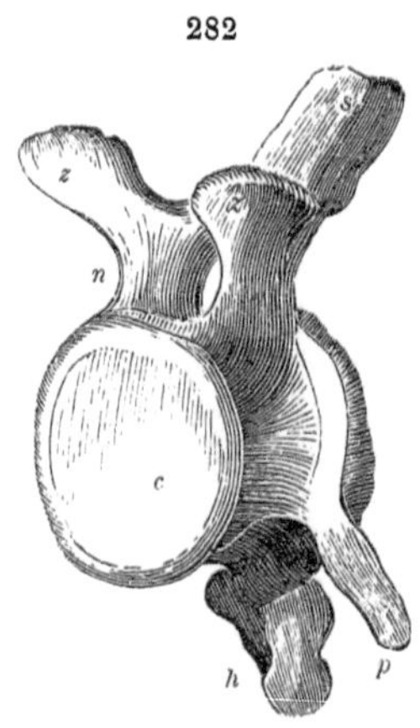

Caudal vertebra, Whale.

[1] XCIII'. [2] XVIII'. p. 520, fig. 214.

physis begins to project from the fore part of the diapophysis of the third dorsal, increases in length to the fourth, and is gradually transferred in the sixth and seventh dorsals to the outer side of the prozygapophyses: in the following vertebræ it seems to take their place, and to occasion a reversing of the usual relative position of the zygapophyses; for whereas in the cervical and anterior dorsal vertebræ the anterior ones are overlapped, as in other Mammals, by the posterior zygapophyses of antecedent vertebræ,—in the succeeding dorsals, beginning with the seventh, the posterior zygapophyses seem to be overlapped and concealed by the anterior ones; but the appearance is due to the place of the zygapophyses being taken by the metapophyses.[1] These latter processes, in fact, continue after the articular surface has ceased to be developed, and after the entire disappearance of the posterior zygapophyses, to project forward from the thirteenth dorsal to the sixth lumbar vertebræ inclusive; beyond which the neural arch is devoid of all exogenous processes, save the spine, *s*, until the middle caudal vertebræ, where rudiments of the metapophyses again reappear. There are no anapophyses in the *Cetacea.*

The four anterior ribs have a head and neck: the rest are suspended by the homologue of the tubercle to the end of the transverse process. The costal cartilages are partially ossified: the first four pairs articulate with the sternum: the original separations of the parts of that bone have disappeared. The first piece or manubrium has an anterior median notch and two broad lateral processes.

In *Delphinus delphis,* of the seven cervical vertebræ the first two have become anchylosed together: there are sixty-three other vertebræ, of which the first fifteen bear moveable ribs; thirty-three vertebræ have transverse processes without ribs: the forty-second vertebra from the skull begins to support hæmapophyses: the eight terminal vertebræ consist of the centrum only, and are much flattened. The metapophysis begins abruptly, as a long well-marked process, from the fore part of the diapophysis of the fourth dorsal, progressively approximates and attains the outside of the prozygapophysis in the eighth dorsal, performs the function of an articular process as far as the sixth lumbar, clamping, as it were, the sides of the back part of the base of the spine of the antecedent vertebra, disappears in the next dozen lumbar vertebræ, and reappears in the caudal vertebræ at the fore part of the base of the spine. The six anterior pairs of ribs support hæmapophyses which unite directly with the sternum.

[1] XLIV. Note, vol. ii. p. 452.

In the common Porpoise (*Phocæna communis*) all the cervicals are anchylosed, and the head of the first free rib rests upon their coalesced bodies: there are fifty-six other vertebræ, thirteen of which are 'dorsal,' or have moveable ribs. The diapophyses and spines of the lumbo-caudal vertebræ incline forward. In the Narwhal all the seven cervicals are free: the wielding of the horn-like tusk of the male is the condition of their greater freedom of movement in the neck. Beyond the cervicals are fifty-six vertebræ, twelve of which have moveable ribs, and of these six pairs join the sternum. In the Bident Dolphin, or Bottle-nosed Whale (*Hyperoodon bidens*), the cervical vertebræ have coalesced with one another: beyond these there are thirty-eight free vertebræ, of which only the nine anterior bear moveable ribs: the twenty-second vertebra first bears hæmapophyses attached to the under part of the centrum. The five anterior pairs of ribs articulate with the sternum, which consists of three bones.

In the *Balæna mysticetus*, fig. 159, there are thirteen dorsals; as many vertebræ without ribs intervene between these and the first vertebra with hæmapophyses; the rest of the column, this inclusive, consists of twenty-two vertebræ, the last dozen being reduced to the centrum, which is much depressed, the last two or three coalesce. The seven cervical, ib. C, are blended into one bone.

In a young or fœtal Whale (*Balæna australis*)[1] the cervical neurapophyses of one side are disunited above from those of the other side, as they are from the centrum below: a compressed diapophysis is sent off from the outer side of each; it is shortest and thickest in the atlas. The third and fourth neurapophyses have coalesced at their upper part on the left side, and those of the last five vertebræ have coalesced on the right side. The cortical portion of the centrum of the atlas is ossified, and forms a wedge-shaped piece of bone, like the corresponding part in the Ichthyosaurus. The centrums of all the cervicals are connate. In the adult true Whales (*Balæna*), the cervicals, fig. 283, 1–7, are distinguishable mainly by the intervals for the passage of the nerves between the neural arches. In *Balænoptera rostrata*, anchylosis does not proceed farther than to unite the atlas with the dentata, and the sixth with the seventh vertebra. In *Bal. boops*, *Bal. patachonicha*, and some other Fin-whales, the atlas retains its individuality. The interval between the lower (pleurapophysial), fig. 283, *p*, and upper (diapophysial), ib. *d*, part of the transverse process is wide, often open, and when circumscribed usually leaves a large foramen. In the seventh cervical the diapophysis, *d*, alone

[1] XLIV'. vol. ii. p. 440, no. 2437.

is developed, as a rule, and the head of the first dorsal pleurapophysis abuts against its centrum. The diapophyses progressively elongate in the dorsals, and support the correspondingly lengthening ribs. The discoid terminal epiphyses long retain their individuality in Cetacean vertebræ, longest in *Balæna.*[1]

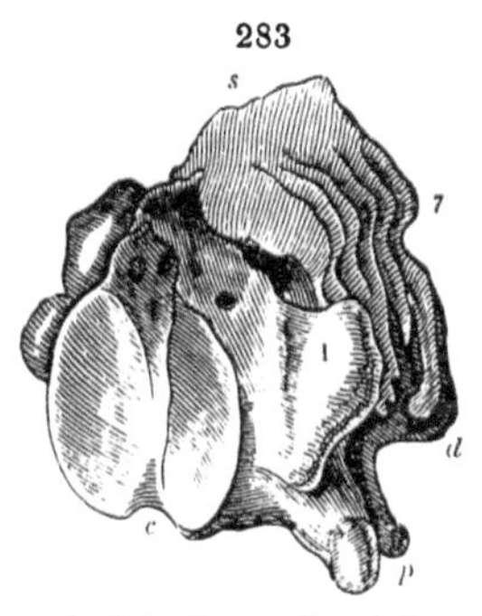

Anchylosed mass of cervicals, *Balæna mysticetus.*

The 'breast-bone' in *Physeteridæ* consists usually of three sternebers, each ossified from a pair of centres which tardily coalesce, save in the first and largest. Four pairs of ribs directly unite with this sternum, as in *Delphinus Tursio,* in which the sternebers ultimately coalesce into a single bone. In *Hyperoodon* and *Ziphius* there are four sternebers, with a vacuity at the middle of each articulation, and five pairs of ribs articulate with the sternum. The sternum in Whales consists of but one bone, to which is usually connected a single pair of ribs. The articular surfaces for these mark its sides: in the more active *Balænoptera* the bone is deeply notched in front, produced behind, fig. 284, where it is ridged below. The sternum is short and broad, shield-shaped, in *Balæna*: rhomboid, sometimes with a central perforation, in *Kyphobalæna,* Esch. One or two of the posterior pleurapophyses are loosely suspended by ligament to the diapophyses of their vertebra in many *Delphinidæ.*[2]

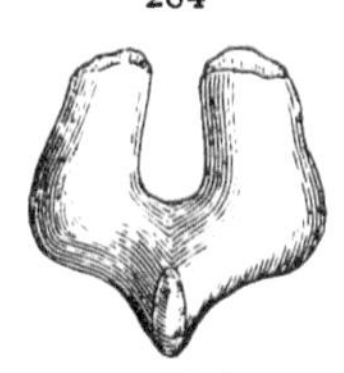

Sternum of *Balænoptera.*

B. *Skull.*—The cranial neural arches continue to manifest the peculiar proportions which are shown in an exaggerated degree in the cervical series. In an advanced fœtal Cachalot (*Physeter macrocephalus*) I find the elements of the epencephalic arch unanchylosed. The lateral margins of the anterior half of the basioccipital are produced and bent obliquely downward. The exoccipitals are much produced and expanded laterally, like the neurapophyses of the atlas in fig. 283, 1: they are deeply notched below. The superoccipital contributes the upper ends of both condyles; it is in the form of a vertical plate, convex from side to side, and developes internally a falciform crest. The superoccipital is overlapped at its lower and lateral

[1] XLIV., vol. ii. p. 440. In Fin-Whales the anchylosis is noted in certain vertebræ of no. 2444, p. 441.
[2] XCVIII'. p. 8. XCXX'., p. 72, taf. 4 and 5.

angles by the exoccipitals, anterior to which it attains to the alisphenoids, and is notched externally for the reception of the upper angle of the squamosals. The basisphenoid, a thick hexagonal bone, concave from side to side below, nearly flat above, is anchylosed to the alisphenoids: it is perforated or grooved by the entocarotids, but has no clinoid processes nor sella turcica. The alisphenoids, perforated near the middle of their base by the foramina ovalia and rotunda, have a thick quadrate plate on their inner side, forming their cranial surface: they extend into a point anteriorly, and articulate with both the frontal and with the parietal angle of the superoccipital. The neural spine of the parietal vertebra is a thin plate partly detached and partly anchylosed to that of the occipital vertebra: the lower angles are confluent with the diapophyses, called 'mastoids,' which here, as in other *Cetacea*, are distinct from the petrosals, and chiefly support the squamosals: these enter a groove of the superoccipital posteriorly, and receive the alisphenoid in a groove anteriorly. The presphenoid and the anchylosed orbitosphenoids form the anterior wall of the cranial cavity, and are perforated by the optic foramina: they articulate with the frontals, sending up a small process into the interspace at the beginning of the frontal suture, which process is impressed by a blind fossa like a small foramen olfactorium on each of its sides: the presphenoid unites with the basisphenoid: the posterior and lateral parts of the orbitosphenoids unite with the alisphenoids: the fore part of the presphenoid is underlapped by the vomer. There is no cribriform plate. The frontal bones are large triangular plates, concave externally, with the outer and fore angle produced into a superorbital process, the channel on the under part of which contracts as it approaches the cranium into a long, deep and narrow groove, which lodged the muscles of the eyeball. The straight median margins of the frontals are thinned off and joined by a squamous frontal suture, the right overlapping the left. The whole posterior and lateral border of the frontal, as far as the junction with the squamosal, presents a broad, oblique, sutural surface, which joins, by overlapping, the contiguous border of the superoccipital. The smooth cerebral surface of the frontals is flat at the middle, arched at the sides, and not impressed by any convolutions. The vomer expands into two aliform processes at its base, which is applied against the presphenoid and orbitosphenoids; it then becomes subcompressed and smoothly excavated, but much more deeply at the left side, where it forms the inner and posterior boundary of the single nasal meatus: it again slightly expands, and afterwards is continued, gradually decreasing, to near the anterior end of the premaxillaries: it is canaliculate above, and occupied by cartilage continued from the coalesced prefrontals. There is no trace of nasal bones. The bone, formed by the coalesced prefrontals, penetrates the posterior part of the groove of the vomer, above which it expands, unequally, into an obtuse prominence rising and inclining to the left side: it is grooved on both sides, and forms the septum of the vertical nasal passage: it is not complicated with turbinal or rhinal capsules, as in the so-called 'ethmoid' of other Mammals. The palatine and pterygoid bones articulate with the sides of the expanded base of the vomer: the margins of the canal excavated in the upper surface of the rostral production of the vomer are overlapped by the premaxillaries.

The palatal is a small, triangular bone, thickest anteriorly, thin, produced and bent posteriorly and above: it commences here by its attachment to the anterior and outer angle of the vomer, bends forward, downward, and inward to circumscribe the nasal meatus, and receives in a groove on its upper and anterior border the palatine prominence of the upper maxillary bone. The whole posterior border of the palatine fits into a groove of the contiguous border of the pterygoid. The pterygoid, which is double the size of the palatine, extends backward to the basioccipital, articulating in its progress by its expanded upper border with the pre-, basi- and ali-sphenoids: from this border the bone descends, arching inward toward its fellow, which it joins along the anterior half of its extent: the remaining free border is divided from this by a deep notch, and circumscribes the large posterior bony aperture of the nostril.

The maxillary expands from its palatine prominence — the essential point of its suspension—backward, outward, but chiefly forward, where it gradually diminishes to an obtuse point. It contracts a union posteriorly with the orbitosphenoid and alisphenoid, and very extensively with the frontal. The nasal process of the maxillary is traversed by a large vertical canal. The premaxillaries are applied against the whole inner surface of the maxillaries between them and the vomer. The right extends much farther back than the left. The capacious basin on the upper surface of the skull, which lodges the valuable product called 'spermaceti,' is formed by the expanded and concave nasal processes of the premaxillaries and maxillaries, which overlap the frontals: a stout ridge divides the inner concave from the outer sloping surface of this part of the maxillary. The malar bone is a moderately long and slender piece, bent upon itself at an acute angle. The upper portion, wedged between the maxillary and frontal, is the thickest: the lower and more slender branch is bent downward and backward, circumscribing the orbit anteriorly and below, and continued by ligament or fibro-cartilage to the short obtuse zygomatic process of the temporal. There are no lacrymal bones. The anterior two-thirds of the middle and under surface of the maxillary is traversed by a vascular and dental groove: rudiments of teeth hidden and buried in the gum are usually found in this groove. The squamosal is a comparatively small, but strong and thick triangular bone: the upper angle represents the expanded squamous part in land Mammals, and is articulated by broad dentated sutural margins to the frontal and exoccipital: its anterior border is grooved for the reception of the alisphenoid: the lower angle is, as it were, truncated, and presents a rough surface for the attachment of the petrotympanic: a short obtuse anterior angle bends forward as the zygomatic process: the under surface presents a smooth shallow cavity for the condyle of the lower jaw; the inner border of the glenoid surface being produced downward into a slender styliform process. The tympanic, here, as in other *Cetacea*, presents a peculiar conchoidal shape, and is extremely dense in texture. An outer plate bends over the thicker, seemingly involuted part, like the outer lip of the univalve *Pyrula*. The 'Eustachian' outlet is at the fore part; and, besides this, may be noted, in *Physeter*, the 'involute convexity,' with its 'outer' and 'inner' lobe, the 'overarching plate,' and the 'rough tympanic process,' by which it joins and coalesces with the 'petrosal:' this is characterised by a deep fossa.[1] The condyle of the mandible projects from the posterior part of the base or ascending ramus, which is compressed and produced into a low obtuse coronoid process above, and into a similar angle below: a wide excavation, beginning on the inner side of the ascending ramus, deepens and contracts into the dental canal, which enters the substance of the horizontal ramus: a fissure is continued along the inner side of the ramus from this canal, and is the sole indication of a compound structure of the jaw. The vessels and nerves emerge from several foramina at the outer side of the ramus, where it is attached by its long symphysis to its fellow: the upper border of the symphysial part of the ramus is excavated by a continuous dental canal or groove, now somewhat resembling that in the upper jaw. The length of the symphysis in the fœtal Cachalot is three-fourths that of the rest of the

[1] XVIII'. p. 526, figs. 220, 225.

ramus. In the adult male the disproportionate growth of this part of the jaw leads to an excess of length of the symphysial part beyond the rest of the ramus. It is coextensive with the dental series, which consists, in each ramus, of twenty-seven teeth, conical or ovoid, according to their state of developement and usage: the smallest teeth are at the two extremes of the series. In the young Cachalot they are conical and pointed, but become obtuse by use, whilst progressive growth expands and elongates the base into a fang, which then contracts, and is finally solidified and terminated obtusely. The teeth are separated by intervals as broad as themselves. In respect to their mode of implantation they offer a condition intermediate between that of the teeth of the Ichthyosaurus and Grampus, being lodged in a wide and moderately deep groove, imperfectly divided into sockets, the

285

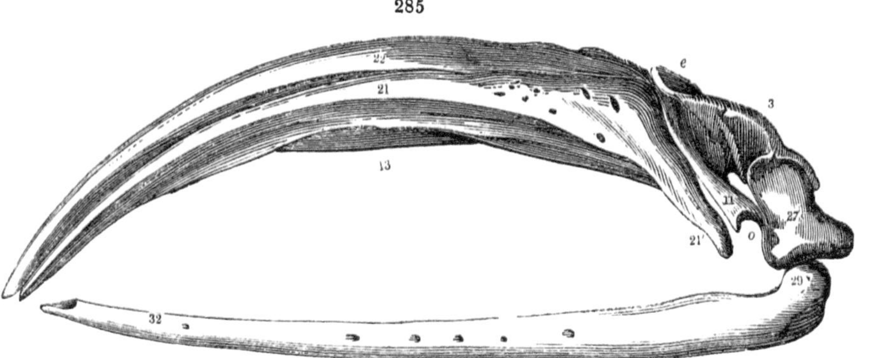

Skull, *Balæna mysticetus.*

septa of which reach only about half-way from the bottom of the groove.

In a fœtal Southern Whale (*Balæna australis*), each frontal is a transversely elongated slender triangle, with its base at the frontal suture, which is a thick vertical symphysis, and its apex at the superorbital ridge: the inferior angle of the base rests upon the prefrontals and upon the sides of the expanded base of the vomer. The frontals take a very small share in the formation of the cerebral cavity. Their cranial surface forms a small concavity at the back part of the base: a half-canal is continued forward from the lower angle of this surface into the nasal cavity. Almost the whole of the upper and outer surface of the frontals is overlapped by the parietals and occipitals, leaving a very narrow exposed transverse strip across the upper part of the skull. The anterior border of each frontal is joined mesially with the nasal, next with the upper end of the premaxillary, and for the rest of its extent with the maxillary bone, which is continued onward to form the antorbital process.

In the mature Mysticete Whale, of which a side view of the skull (now in the British Museum) is given in fig. 285, the maxillaries, 21, are disposed each like an expanded arch along the outside of the coextended premaxillaries, 22; their inferior surface has two facets separated by a longitudinal ridge, to the sides of which the plates of baleen are attached. The premaxillaries are compressed and diverge from each other posteriorly to form the long oval outlet of the nostrils, completed behind by the nasals, which are elongate, as in the *Zeuglodon cetoïdes*: the frontals, *e*, extend outward to form the roof, 11, of the small orbit, *o*; and therewith is coextended the back part of the maxillary, 21′: a small malar, fig. 159, 26, is articulated to the lacrymal, 73, the maxillary, 21, and the squamosal, 27; the most expanded part of which, fig. 285, 27, forms the articulation for the mandible, 29–32. The superoccipital, 3, inclines forward, as it rises, and forms almost the whole upper part of the cranium. The coalesced prefrontals are perforated by the olfactory nerves. The presphenoid is sheathed in the hind part of the canaliculate vomer, 13, which extends far forward along the middle of the roof of the mouth. Each mandibular ramus arches outward and forward from the slightly-raised condyle, 29, to the short, ligamentous symphysis, 32: it is compressed and subtrenchant at both upper and lower margins: a coronoid ridge is feebly marked; there is no ascending ramus. The skull of the whale is more symmetrical than that of toothed *Cetacea*.

286

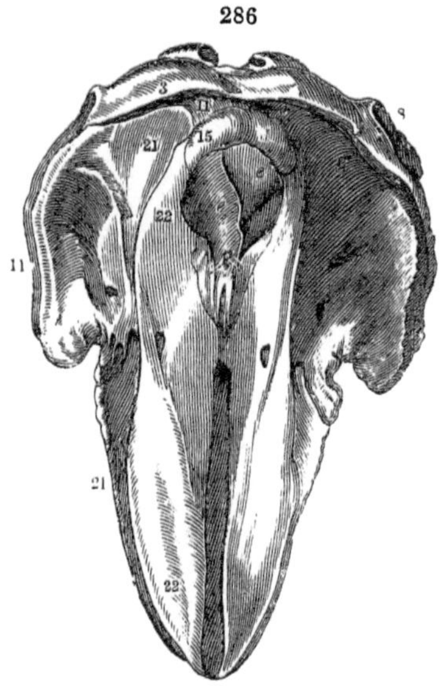

Skull of a Grampus (*Phocæna globiceps*).

In the section of *Delphinidæ* to which the Grampuses and Porpoises belong (*Phocæna*, Cuv.), the facial bears a less proportion to the cranial part of the skull: the latter is broad, elevated, and convex posteriorly. The superoccipital, fig. 286, 3, forms the transverse crest dividing the hinder from the upper surface, where it is met by the frontals, 11, the overlapped parietals coming into view only at the sides, where they expand into the mastoids, 8. The maxillaries, as they extend backward, rise and expand, 21, covering so much of the frontals, 11, as to allow but a narrow strip of these bones to be seen, except where they dilate to form the roof of the orbit. The nasals, 15, are oblong tubercles set deeply in depressions of the frontal, at the back part of the nostrils, *e*, *e*. The premaxillaries, 22, form the front and sides of these apertures, save at the small portion contributed by the maxillaries at *g*. The nasal passages descend almost vertically. The malar is flattened where it helps to form the orbit, and is covered by the maxillary: it sends backward a long and slender process, which articulates with the zygomatic process of the squamosal, and forms the only lower boundary of the orbit. The bony palate has a deep longitudinal channel on each side in some Dolphins.

In the vertical section of the cranium of the Porpoise, fig. 287, is shown the plane of the occipital foramen, 2, inclined from below

287

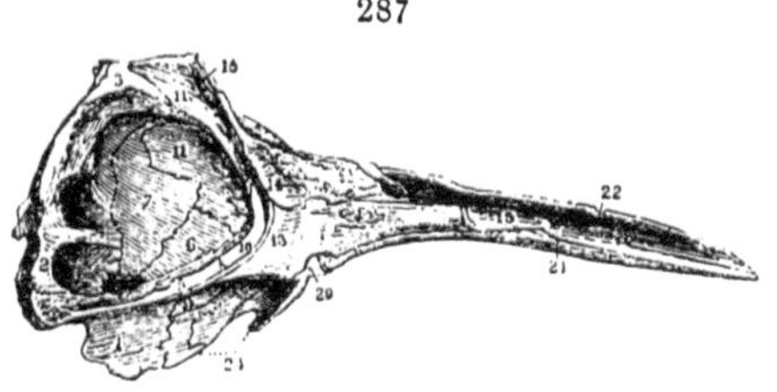

Section of skull, Porpoise.

forward: the proportions of the inner wall of the cranial cavity contributed by the ex- 2, and super- 3, occipitals, by the alisphenoid, 6, the parietal, 7, the orbitosphenoid, 10, and the frontal, 11: the small vacuity between the alisphenoid and exoccipital is blocked up by the loosely attached petrosal. The right nasal passage is exposed, showing the proportion of the septum formed by the vomer, 13, and the coalesced prefrontals, 14: the vomer extends forward to the middle of the upper jaw, which is chiefly composed of maxillary, 21, and premaxillary, 22. The small palatines, 20, articulate with the vomer and maxillary, and send backward the larger pterygoids, 24, which form with the vomer the internal or lower nostril, whilst the canal for the long conical larynx is contributed by the pterygoids, 24, and a corresponding descending plate of the basisphenoid, 6, and basioccipital, 1. The squamosal is excluded, as in Birds and lower Vertebrates, from the cranial cavity. The prefrontals in the Beluga (*Delphin-*

apterus) are large, and ascend into view at the back part of the nostrils, where they coalesce with the frontals. The small nasal bones are wedged into an interspace between them and the frontals at the summit of the nasal apertures.

In *Hyperoodon* the skull is remarkable for the developement of the outer border of each maxillary bone into a broad and lofty vertical crest, and for the backward prolongation of the posterior border of the same bones to the occipital region, where it is developed into what seems to be an occipital crest. In *Platanista*, the corresponding borders of the maxillary, after rising to the vertex, are reflected forward, converging, so as to overarch like a domed roof the circumnarial part of the skull. In *Euphysetes* this concave space is divided behind the nostrils by a vertical ridge. *Euphysetes simus*[1] shows the opposite extreme to *Balæna* and *Physeter*, in the disproportionate shortness of the rostral or 'prenarial' to the cranial or 'postnarial' part of the skull. In *Paraziphius* the vomer is singularly tumid and dense.

The hyoid arch consists, in *Balænidæ*, of a pair of stylohyals, fig. 288, 38, ligamentously connected with the mastoids, and similarly attached by fibrous representatives of ceratohyals, 39, to the pair of processes at the fore part of the basihyal, 41. This large, broad bone is produced outwardly into a pair of compressed bars, thicker than the stylohyals, and representing the thyrohyals.

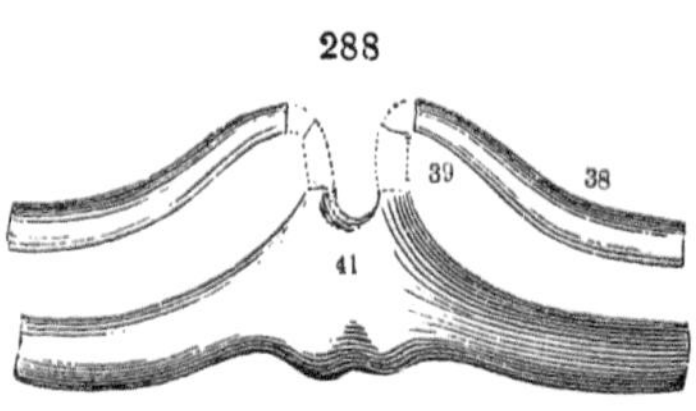

Hyoid arch, *Balænoptera*.

C. *Bones of Limbs.*—The clavicle is absent. The scapula is a flat triangular plate, with one angle truncate to form the glenoid cavity for the humerus, and without the 'spine' along the outer surface. In *Balæna*, figs. 159, 289, 51, the triangle is almost equilateral, with the side forming the base rather convex, and the part supporting the truncate angle, *b*, somewhat produced, forming a 'neck.' The acromion, *a*, projects forward from the outer part of the neck near the anterior border. In *Balænoptera* the base is proportionally longer than the other two sides, and forms a more convex border: in *Bal. longimana*, Rud., the acromion is obsolete, and the coracoid is merely an obtuse production of the fore part of the glenoid cavity. In the Cachalot (*Physeter*), the convex base is the shorter side of the triangle, the vertical exceeding the antero-posterior diameter of the scapula: the acromion is longer and larger than in *Balænidæ*, and there is

[1] xcix.

a long and slender coracoid. In *Euphysetes* the triangle is more equilateral, as it is in *Ziphius* and *Hyperoodon*. In *Delphinidæ* the convex base of the scapula is usually the longest of the three sides: the extension of the bone in the axis of the trunk is remarkable in the Gangetic Dolphin (*Platanista*), in which the acromion projects mid-way between the anterior basal angle and the glenoid cavity.

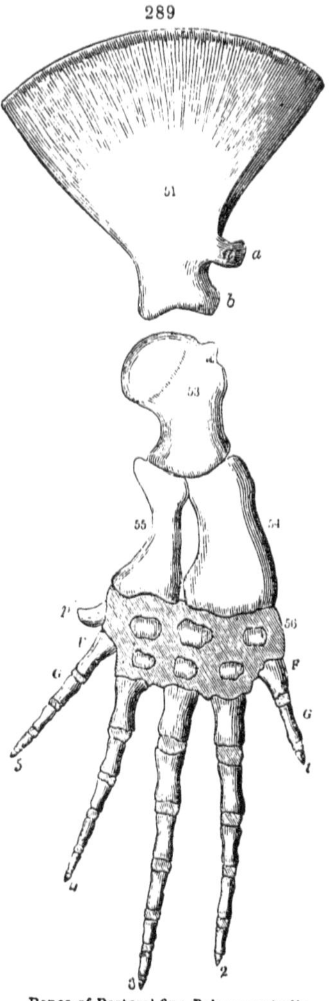

Bones of Pectoral fin: *Balæna australis.*

The humerus, figs. 159, 289, 53, 290, *a*, is remarkably short in proportion to its thickness: the head is large, hemispherical; bent very slightly out of the axis of the bone; with the outer or radial tuberosity feebly marked in most, rather more strongly in the Cachalots, and forming a deltoid tuberosity: the shaft becomes compressed and expanded toward the distal end, which has two ill-defined, flattened surfaces for syndesmotic junction with the radius and ulna. The latter, ib. 55, usually sends backward an olecranon; but this is not developed in Platanista, where the ulna is broader and longer than the radius: usually the radius is the larger bone, as in fig. 289, 54: both bones are flattened, shorter than the humerus in Cachalots and Platanists, longer in Whale-bone Whales, Bottle-nose Whales (*Hyperoodon Ziphius*), Grampuses, Porpoises, Dolphins. The contiguous epiphyses of the humerus and antibrachials first unite with their respective shafts: in an old Cachalot and *Delphinus Tursio*,[1] the radius and ulna are anchylosed with the humerus, fig. 290, A. In a Southern Whale the carpus, fig. 289, 56, consists of seven ossicles: the first on the radial side answers to the scaphoid and trapezium: the second, in *D. Tursio*, is wedged into a distal cleft between the radius and ulna, and corresponds with the lunare in the Chelonian carpus and

[1] XLIV, no. 2483.

that of the Orang: the third is very small, and represents the cuneiforme: the pisiforme is separated from it by the junction of the unciforme with the ulna: the unciforme supports the rudiment of the fifth digit and part of that of the fourth: the magnum supports part of the fourth finger and a great part of the third: the trapezoides is moved to the interspace between the third and second digits, but principally supports the latter. The metacarpal of the first digit, in *D. Tursio,* fig. 290, I, supports one small phalanx: the larger metacarpal of the second digit, II, supports seven phalanges: that of the third, III, supports five phalanges; the metacarpal of the fourth, IV, two phalanges: the fifth, V, is represented only by a rudimental metacarpal bone.

In the Grampuses, Porpoises, and other *Delphinidæ,* the second and third digits are also the longest, with the excessive number of phalanges. The fifth metacarpal articulates nearer to the antibrachium than the others do. In both the Porpoise and Grampus, e. g., it is attached to the ulnar border of the carpus, is broader than long, and supports one or two stumpy phalanges: the first metacarpal is short and slender, but its base is on the distal border of the carpus. In the *Hyperoodon* there are three carpals in the proximal row, and a second row of four small ossicles in the fibro-cartilaginous matrix. The metacarpal of the pollex supports one phalanx: those of the second and third digits have each five phalanges: the fourth metacarpal has three; the fifth, which is the shortest of all, has two phalanges. In the *Cachalots* and *Ziphius,* the fourth digit more nearly equals the third in length: in *Balæna mysticetus,* fig. 159, IV, it rather exceeds the second, ib. II, and, like it, the metacarpal supports three phalanges; the third metacarpal, ib. III, having four phalanges, and the fifth, V, two: the first digit is the shortest, and consists of metacarpal only. In *Platanista* the first metacarpal has two phalanges, the other four each support four phalanges, the fingers being of nearly equal length, and more divergent than usual, supporting a fin correspondingly expanded to its free truncate end.

290

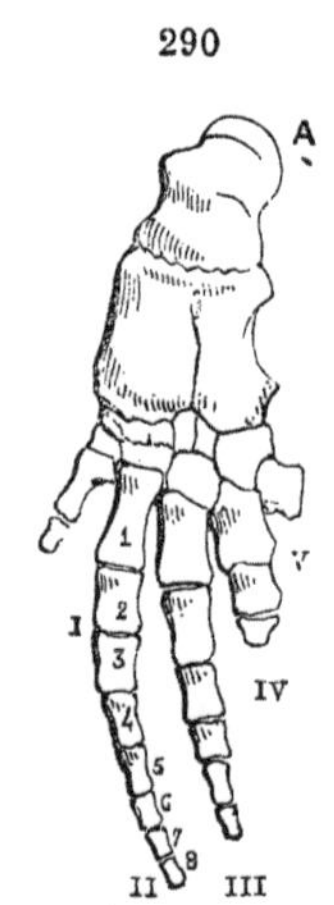

Bones of pectoral fin, *Delphinus.*

In some Piked Whales (*Balænoptera*) the first digit is obsolete, the third and fourth much longer than the second and fifth. In *Bal. longimana* (*Kyphobalæna,* Esch.), the third and fourth each support a metacarpal and six phalanges. All the digits in the entire Cetacean are enveloped in a common fold of integument. The increase of the phalanges of certain digits beyond the number three is a remarkable instance of departure from the mammalian type and of affinity with the extinct enaliosaurs and fishes.

In the *Delphinidæ* there are a pair of pelvic bones larger in males than females, chiefly subserving the origins of the 'erectores penis' and 'clitoridis'; and which, therefore, I regard as ischial bones. In a female *Hyperoodon* 28 feet long, each ischium was 4½ inches long, straight, subtriedal, 8 lines in diameter. In *Balæna mysticetus* there is, besides the ischium, fig. 192, 63, a smaller, more slender and curved ossicle, which, being anterior to it, seems to represent a pubis, ib. 64: the junction of the two bones expands into a surface, representing the acetabulum, to which is ligamentously suspended a bone of similar length to the pubis, but thicker, and expanding, with some flattening, to a transversely extended convex surface, like that at the distal end of a chelonian femur; ib. 65: to which is suspended a smaller rudiment of a tibia, 66. This is the simplest condition of the limb, or appendage, of the pelvic arch known in the Mammalian class.[1] There is no outward indication of it in the Whale. The little bones, of the relative size to the rest of the skeleton as shown in fig. 159, p. 280, are suspended beneath the last two lumbar vertebræ, which may thus be regarded as answering to the sacral in quadrupeds.

§ 185. *Skeleton of Sirenia.*—In this class of marine Mammals the hind limbs are absent, as in *Cetacea*, and the pelvic bones, where best developed, as in fig. 292, *s*, *h*, retain the size and shape of the small contiguous costal arches. The texture of the bones is denser, the neck, though short, is longer than in the *Cetacea*, and the vertebræ are distinct: but the chief differences are found in the relative size and structure of the skull, and in the better developement of the bones of the pectoral limb, the digits of which are not composed of more than the normal mammalian number of phalanges (compare fig. 292 with fig. 159, p. 280). The known existing representatives of the Sirenian order are the Dugongs (*Halicore*) and the Manatees (*Manatus*): the latest extinct form is the edentulous Sirenian, called 'Steller's

[1] The bones described and figured in CLI. t. v. p. 236, pl. xxvi., figs. 24 and 25, were not seen *in situ* by Cuvier, but are described as pelvic bones, on the authority of M. Delalande, the Articulator. The discoverers of the rudimental hind limbs, and authors of LXV′, observed the pelvic bones of the whale *in situ* (p. 151, tab. II.).

Sea-cow' (*Rhytina borealis*, Ill.), last observed in the arctic seas off the shores of Bering's Island: the miocene extinct genus (*Halitherium*) has left its remains in southern Europe.

A. *Vertebral Column.*—In the Dugong, fig. 292, there are 7 cervical, C, 19 dorsal, D, and about 26 lumbo-caudal vertebræ, L, S, CD. To the 29th vertebra, counting from the skull, the pelvic arch is suspended, characterising it as a sacral one, S, and leaving two lumbars in advance, the transverse processes of which long retain the suture indicative of their pleurapophysial part. The second vertebra, beyond the sacral, first supports a hæmal arch, and this is continued to the fourteenth and fifteenth of the caudal series, which, if counted from the sacrum, would include about twenty-four vertebræ. Fig. 291 gives the characters of the transitional vertebræ between the trunk and tail, especially as afforded by modifications of the hæmal arch. In the posterior dorsal vertebræ, the pleurapophysis, *pl*, is the sole ossified element of the hæmal arch; it progressively shortens in the

291

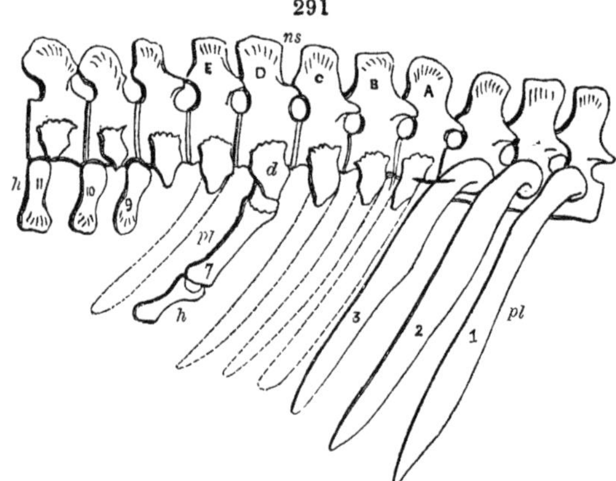

Dorsal, lumbar, sacral, caudal, vertebræ, *Halicore*.

16th, 1, 17th, 2, and 18th, 3, vertebræ, retaining its mobility; and, in the 19th, A, it shortens suddenly, but usually with more extended ossification in the sclerous basis of the rib than is shown in the figure. In the vertebræ, B and C, the pleurapophysis, besides being short, becomes confluent with the centrum, as a 'transverse process,' and characterises them as 'lumbar.' The sclerous or tendinous continuations of the pleurapophyses into the abdominal muscles are indicated by dotted lines. In the vertebra D, the ossification which extends the pleurapophysis, *pl*, 7, beyond the part, *d*, representing the transverse process, retains a ligamentous union therewith, and represents the 'ilium:' a lower ossification in the hæmal arch establishes a bony 'hæmapophysis,' *h*, and represents the 'ischium.' It is ligamentously connected at its lower end to its fellow, completing by such 'symphysis ischii' the pelvic hæmal arch. In the vertebra, E, the proximal part only of the pleurapophysis is ossified, as in the lumbar series; and this is the case with the succeeding vertebræ: but the centrums exhibit, at their under surface, articulations for parts answering to the lower portion, *h*, of the hæmal arch of the vertebra D. The parts in question, 9, 10, 11, *h*, are severally united together by their lower or distal ends, at first ligamentously, but afterwards by coossification, constituting inverted bony arches of the chevron shape, and which are serially homologous with the bony hæmapophyses, *h*, of the pelvis, and the sclerous or cartilaginous hæmapophyses of the trunk: they are dislocated from their pleurapophyses and approximated to their centrums, with a slight horizontal displacement leading to their partial articulation with that of the vertebra succeeding their own (see fig. 188, p. 299). These hæmapophyses, fig. 292, *h*, *Cd*, are not developed in the terminal vertebræ, the last six of which are represented by horizontally flattened centrums, *o*, sustaining, as in *Cetacea*, a horizontal tail-fin.

The ribs of the dorsal vertebræ, fig. 292, *pl*, are massive,

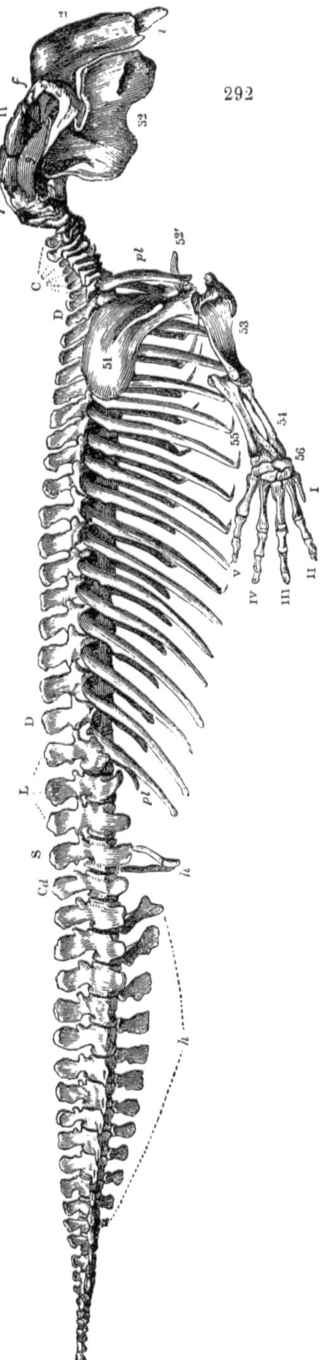

Dugong (*Halicore*).

peculiarly so in the Manatees: the first in the Dugong has a long oblique process from the under part of the neck, and a shorter process, terminated by a rough surface, from the inner border, two inches from the lower end of the rib. The first three or four pairs of ribs join, by cartilaginous hæmapophyses, the sternum, which consists of two bones and a xiphoid cartilage: the two sternebers coalesce into a single bone, of the borders of which the costal articulation occupy the middle third. From the third to the sixteenth dorsal, the ribs are of nearly equal length. Many of the succeeding ribs have a process from the posterior margin, simulating the costal appendages in Birds. The metapophyses are not developed so as to supersede the prozygapophyses, as in *Cetacea*. The neural spines are of equal length and similar inclination slightly backward. In the Dugong, fig. 292 C, the atlas has short par- and di- apophyses, and the neural arch is perforated on each side near its anterior border: in the axis the transverse processes are chiefly by diapophyses. In the fourth cervical, the right process was pierced by the vertebral artery; in the seventh, the left process: the other diapophyses were notched. The centrum of the seventh cervical has a facet on each side for the first pair of dorsal ribs. The side of the centrum of the first dorsal vertebra bears two articular facets; one of which is smaller than the other, looks forward, and receives a part of the head of that rib which articulates with the preceding vertebra; the other looks backward, and receives a large share of the head of the second dorsal rib. The transverse processes are long and strong, and present on their extremity an articular facet which receives the tubercle of the first free or dorsal rib.

In *Manatus Americanus* the cervicals are also very short, but only four of these compressed vertebræ intervene between the axis and the first dorsal: seventeen vertebræ support the moveable ribs, and are followed by about twenty-two lumbo-caudal vertebræ: the hæmal arch commencing at the lower interspace between the fourth and fifth of their series. The pelvic bones are reduced, as in most *Cetacea*, to an ischium giving origin to an 'ischio-cavernosal,' and insertion to an 'ischio-coccygeal' muscle. In a half-grown Manatee I have seen the neurapophyses of the first twenty-nine vertebræ still suturally joined to their centrum. But two pairs of ribs join the sternum, which soon becomes a single bone, with a costal process on each side of the middle part.

The vertebral characters of *Rhytina* agree in the main with those of existing *Sirenia*.[1] Steller assigns to it six cervicals, as in *Manatus*. Nine pairs of ribs are said to have joined the sternum.

B. *Skull*.—The facial or rostral part of the skull, anterior to the orbits, is short, especially so in *Manatus*, fig. 239, in which it slightly descends: in *Halichore*, figs. 292, 294, 22, it is bent down more abruptly: in *Rhytina* the angle of the upper contour of the rostrum is greater than in *Manatus*, that of the lower contour less than in *Halichore*, exemplifying, as in other parts of the skeleton, an intermediate character. All the skull-bones are massive in *Sirenia*, and, save in the instances of anchylosis, are somewhat loosely connected together. In the Dugong the basioccipital, fig. 294, 1, is a triradiate bone, the two short rays diverging posteriorly to join the exoccipitals, and forming the lower end of each condyle. The exoccipitals, 2, almost meet above the foramen magnum: they have a short rough paroccipital process. The superoccipital, 3, is early anchylosed to the parietals, which have equally coalesced into a single subquadrate massive bone, fig. 293, 7, with the sides bent down at nearly a right angle with the almost flat upper part, which is perforated by a 'foramen parietale.' A falciform ridge descends from the inner surface. The basisphenoid, fig. 294, 5, has coalesced with the alisphenoids, which are grooved both behind and before, not perforated, by the trigeminal nerve. The massive pterygoids are anchylosed to the base of the alisphenoids: the posterior ends of the palatines, which are wedged into the interspace between the ento- and ecto- pterygoid processes, send upward a part which appears in the temporal fossa behind the maxillary. The presphenoid, as a compressed 'rostrum,' is wedged between the laminæ of the vomer, and has coalesced with the confluent orbitosphenoids which it supports. There is no 'sella turcica.' The orbitosphenoids are perforated

293

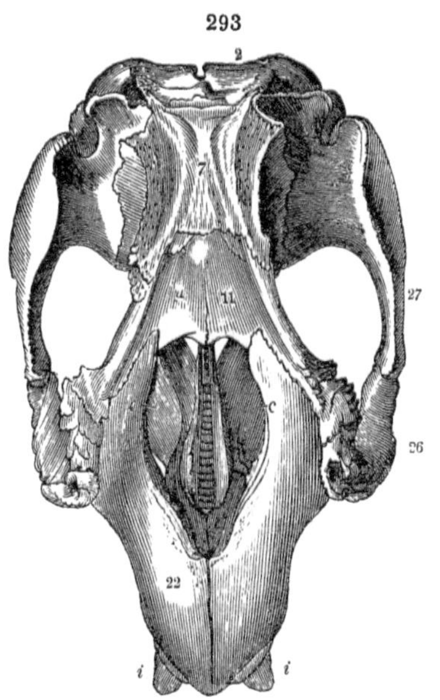

Skull of Dugong (*Halichore indicus*).

[1] xciv. p. 95.

by widely separated optic foramina: they are anchylosed with the coalesced prefrontals, fig. 294, 14. The cribriform plates are lodged in deep fossæ, between which is a crista galli. The frontals, 11, are not confluent; their orbital processes extend far forward and outward from the anterior angles: they are excavated below almost to the posterior margin by the rhinal cavity: the median angles of the nasal border are slightly produced, but there is no trace of a suture there marking out the proper nasals. The cranial plate of the frontal forms a small concave surface, not exceeding the depth and thickness of the posterior part of the bone to which it is confined. A small part marked off by a fissure from the fore end of the orbital process represents an imperforate lacrymal, fig. 292, *f*.

294

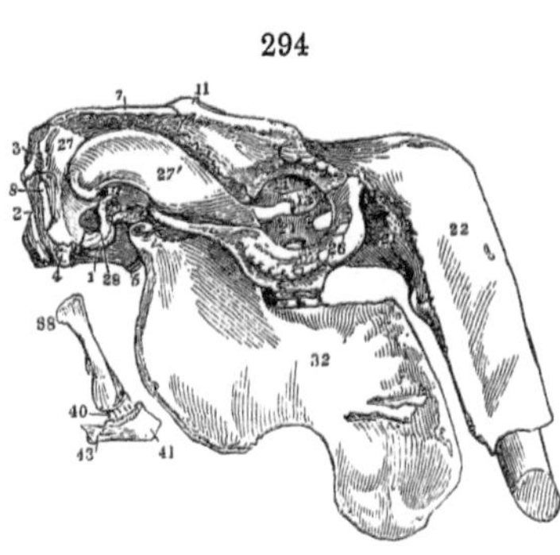

Skull of Dugong (*Halichore*).

The maxillary, fig. 294, 21, is deflected anteriorly; its nasal and malar processes do not meet and circumscribe the great antorbital foramen, but this is closed by the upper end of the malar bone, 26. The premaxillary, 22, is remarkable for its very large and long deflected alveolar portion, and for its slender nasal portion, fig. 293, *c*: it is excavated by the deep alveolus of the incisive tusk, *i*. The squamo-mastoid forms no part of the inner surface of the cerebral cavity, but is deeply and smoothly excavated for the lodgement of the dense petro-tympanic. The mastoid part forms a thick rugged process, 8, wedged between the tympanic, 28, and paroccipital, 4. The zygomatic parts of both squamosal, 27′, and malar, 26, form a strong arch. The petrotympanic fits closely the cavity in the squamo-mastoid, and partially closes the vacuity between it, the occipital and sphenoid bones; the tympanic, 28, describes two-thirds of a circle for the support of the ear-drum, and is less than the dense otic capsule with which it is confluent at both ends. The stapes is an elongate, subcompressed pyramid, with a minute perforation near the base, and an epiphysis at the apex: the incus is also long and narrow; the malleus broad and bilobate.

The mandible, figs. 292, 294, 32, is deep in proportion to its length: the coronoid rises with a slight backward curve: the condyle is small and convex: the ascending ramus has a convex hind border, curving to an advanced feebly-marked angle: between this and the deflected symphysial part the lower border is deeply concave: the sockets for the molar teeth, originally five or six in number, like those in the maxillary, are reduced to two or to one in the old animal: the deflected symphysis forms a flat oval surface anteriorly, with four or five pairs of small alveoli, in one or more of which may be an abortive incisor, fig. 160, *a*, *d i* 3, covered by the thick horny plate attached to the flat rough surface; the dental canal, beginning in advance of the ascending ramus, ends by a wide oblique opening from which channels diverge on the outside of the deflected symphysis.

In the Manatee, a large otocrane is also smoothly excavated in the mastoid, squamosal, and exoccipital bones, to which the petrosal closely fits without coalescence, its posterior surface appearing in the space left between the mastoid, super- and exoccipitals. The basi-sphenoid coalesces with the alisphenoids, prior to confluence with the basioccipital and presphenoid: the latter similarly coalesces with the orbitosphenoids, and is continued, like a rostrum, into the vomerine fissure. I find no distinct nasals anterior to the frontal suture in the new-born Manatee; nor other representatives of them than the small amygdaloid bones, fig. 239, 13, 13, articulated to the frontals at the posterior angles of the nasal aperture: this is large, subrhomboid, horizontal. The wide antorbital foramen is entirely surrounded by the maxillary, ib. 26. The suborbital plate of the malar rests upon the platform extending horizontally outward from above the anterior molars, and extends the floor of the orbit an inch beyond the roof, the eyeball resting upon the concavity of the malar, as on a shelf. The zygoma, ib. 27, is unusually massive. The premaxillaries, ib. 22, in the young Manatee, show a pair of alveoli for abortive incisors: a similar pair impresses the fore part of the mandibular symphysis, and a slight groove extends downward from each. The symphysis is deeply hollowed out behind. The coronoid is produced obliquely upward and forward: the angle of the jaw is not marked.

The ossified parts of the hyoid arch are the basihyal, fig. 294, 41, stylohyals, 38, and the thyrohyals, 43: the ceratohyal, 40, is cartilaginous: the arch is suspended to the angles between the mastoid and paroccipital.

C. *Bones of the Limbs.*—These are limited to the pectoral pair, and their supporting arch is reduced to the scapula, with a short coracoid as a tuberous process. The scapula, fig. 292, 51, is sublongate, recurved, with the convex anterior costa continued into the base, with an angle feebly marked in the Manatee. The posterior costa is concave, deepest in the Dugong. The outer surface has a spine about half the length of the bone, marking off a broad pre-spinal, from a narrow post-spinal fossa: the spine is produced into a slender acromion in the Manatee, not in the Dugong.

The humerus, ib. 53, has the normal mammalian character, though of small size, with the head, tuberosities, and deltoid crest, the twisted shaft, the epicondyloid processes and intermediate trochlear articular surface, for synovial articulation with the coalesced proximal ends of radius, 54, and ulna, 55. The latter developes an obtuse olecranon: the distal ends of the antibrachial bones are extensively united and ultimately by bone. In the Manatee there are six carpals, three in each row: the outermost and largest represents a cuneo-pisiform, and articulates with both the ulna and the fifth metacarpal. The trapezium and trapezoides are represented by one bone articulating with the first and second metacarpal: the magnum supports the third, and the unciforme the fourth and part of the fifth metacarpals. In the Dugong there are but three carpals: the scapho-lunar and cuneo-pisiform in the first row, 56, and a single transversely oblong bone representing the second row, but leaving the major part of the base of the fifth metacarpal to articulate with the cuneiform. The pollex, I, is represented by a styliform metacarpal: the other digits have each three phalanges; and most of the ungual ones, in *Manatus*, support nails. All the limb-bones, like those of the rest of the skeleton in *Sirenia*, are solid.

The herbivorous *Sirenia* have not to move far from their favourite localities for food; they contrast, in that respect, with the *Cetacea* that pursue a living prey: hence the difference in the specific gravity of the bones, which in *Sirenia* is such as to require an effort on the part of the animal to reach the surface of the water for breathing, but enables them to browse, at ease, the vegetation clothing the bottoms of their seas, estuaries, or rivers. The massiveness of the zygomatic arches in the skull contrasts singularly with the slenderness of those parts in Whales: the pterygoid productions offer a similar difference: the external bony nostril is as remarkable for expanse in *Sirenia* as for contraction in *Cetacea*. The movements of the head and jaws, in browsing, call for a flexibility of the short neck in *Sirenia*, incompatible with the fixation of that part which prevails in most *Cetacea*: the dorso-lumbar vertebræ are articulated by true zygapophyses, not metapophyses. The pleurapophyses are as remarkable for thickness and density in *Sirenia*, as in similar-sized *Cetacea* for slenderness and oily porosity of texture. Although the bones of the pectoral limbs are swathed in skin, the fins project more freely from the trunk, the elbow is better marked; the limb-joints are synovial, not syndesmotic merely, as in *Cetacea*; and although there are clearer indications of the digits in the fin of *Sirenia*, none of the digits have more than three phalanges.

§ 186. *Skeleton of Proboscidia.*—With the exception of a very small cavity in the femur and tibia, a light cancello-reticulate structure occupies the centre of the long bones, which have thick and compact osseous walls. The skull-bones are extensively pneumatic.

A. *Vertebral Column.*—In the giant mammal of the land, as in that of the sea, the neck is short, and through loss of length, not of number, of the cervical vertebræ. In the Indian Elephant, fig. 162, the vertebral formula is:—7 cervical, C, 20 dorsal, D, 3 lumbar, 3 sacral, and 31 caudal. Anapophyses are developed from the sixteenth dorsal, and articulate with metapophyses from the seventeenth. The same joints are superadded to the ordinary articular processes, as far as the last lumbar. Five pairs of ribs directly join the sternum, which consists of four bones. The epiphyses continue detached from the bodies of the vertebræ to nearly full growth.

In a half-grown Elephant, the neurapophyses of the atlas are distinct from the hypapophysis, and united to each other above by suture: the centrum is also distinct, but attached to that of the axis, of which it forms the 'odontoid' process. The neurapophyses develope both upper and lower transverse processes, which circumscribe the vertebrarterial foramen. The same is the case with the neurapophyses of the axis, which blend together above and develope a thick bifurcate spine before coalescing with the centrum. The removal of the terminal epiphyses of the short flat bodies of the other cervicals shows that the upper fourth of the body is contributed by the neurapophyses, the rest by the centrum. In the fifth cervical vertebra, a short and slender spine is developed from the summit of the neural arch. The antroverted costal part of the transverse process is connate with the parapophysis, and afterwards coalesces with the diapophysis. In the seventh cervical vertebra, the transverse processes consist of diapophyses only. The articular surface for the head of the first free or dorsal rib is formed, half by the neurapophysis, and half by the centrum. The neural spine has much increased in length, but is slender.

The first dorsal vertebra is remarkable for the strength as well as the height of the neural spine. The diapophyses are shorter and thicker than in the neck. The surfaces for the first and second ribs meet at an acute margin below; they are formed as in the preceding vertebra. In the fourth dorsal vertebra, the spine is still more remarkable for its height and strength: the vertebral body has a greater antero-posterior thickness, but the anterior and posterior costal surfaces still meet below. A larger proportion of these surfaces is contributed by the neurapophyses. In the ninth dorsal vertebra, the posterior costal surfaces, which are almost exclusively formed by the neurapophyses, are separated by a non-articular tract from the anterior ones.

The sixteenth dorsal vertebra shows only a single pair of costal surfaces, which are wholly formed by the neurapophyses: the metapophyses are well developed. In the remaining dorsals the costal surfaces decrease in size. The first and second ribs are almost straight, and expand to join their short sternal parts: as the ribs lengthen, they preserve their slenderness, and are straighter at their lower halves than usual; the vertebral third is bent, subcylindrical, and grooved anteriorly. The lumbar diapophyses are short and depressed. The neural spines of the dorso-lumbar series incline backward, gradually decreasing in height, and indicate no centre of inflexion in the capacious well-ribbed trunk. The thick sides of the three sacrals which join the ilia consist of pleurapophyses which coalesce with both centrum and neural arch. The neural spine subsides after the seventh or eighth caudal: diapophyses continue to the twelfth, and zygapophyses to the fifteenth: the rest are reduced to the centrum.

B. *Skull.*—The cranial much exceeds the facial part in size: its upper part forms an expanded dome: but a section, as in fig. 296, shows that the cavity for the brain occupies but a small proportion of the back part of the dome's base: the rest being formed by air-sinuses, bounded by plates of bone, extending between the remote outer and inner 'tables' in the form of sinuous plates so disposed as to give greatest strength with least material. The occipital condyles are small, approximate below, and project backward from the upper half of the posterior surface of the skull. The occipital slopes as it rises to curve forward to the vertex, and more so in the African than in the Indian species. The position of the epencephalic compartment of the cranium fig. 296, *e*, the suspension of the malar bone, fig. 295, 26, in the middle of the zygomatic arch, the size and connections of the premaxillaries, 22, and their deep and large alveoli for the single pair of incisors, recall characters of Rodentia. The cranial sutures become obliterated; but examination of the skull of a very young Elephant (Indian) has enabled me to give the following details:—The basioccipital is notched behind, and contributes there the lower ends of the occipital condyles: it increases in thickness as it advances to form the flat rough surface for junction with the centrum in advance (basisphenoid). There is a rough depression on each side of the under surface for the insertion of the 'recti capitis antici.' The exoccipitals form small, inferiorly approximate condyles, fig. 295, 2, have no precondyloid foramina, and do not develope paroccipital processes: they meet above to complete the foramen magnum. The superoccipital is much expanded, and supports two supplementary bones (interparietals): it is deeply impressed by the insertion of the liga-

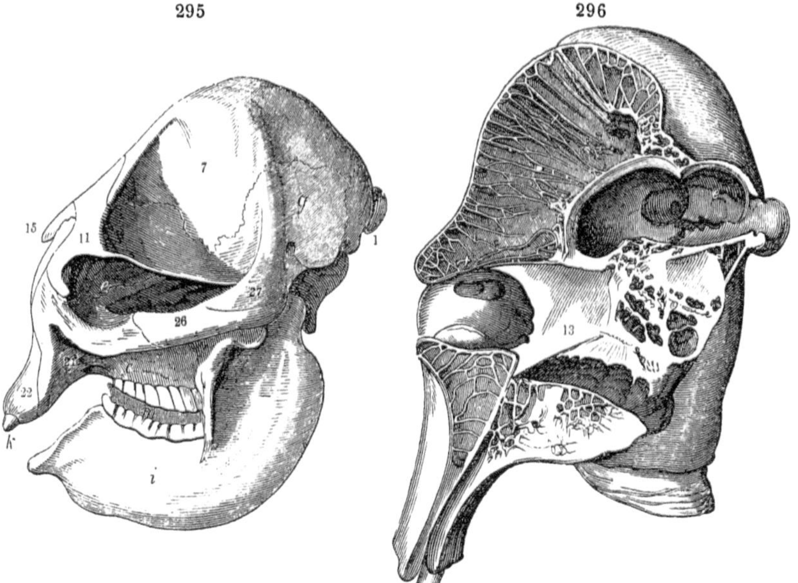

mentum nuchæ. The basisphenoid has coalesced with the alisphenoids, which are separated from their neural spine (parietals, 7) by the intercalated squamosals, *g*. The pterygoid processes are long, much expanded and excavated anteriorly, and are perforated at their base. The alisphenoids are perforated by a wide 'foramen ovale.' The basisphenoid when united with the presphenoid receives air into the cells with which the bone, as it acquires vertical extent, is excavated. The vomer retains its character as a vertical plate, fig. 296, 13. The orbitosphenoids have coalesced with each other at their base, and also with the prefrontals (laminæ mediæ æthmoidei): they are perforated by the optic foramina, and notched posteriorly for the foramen rotundum.

The portions of the olfactory capsule closing the anterior orifice of the cranial cavity form extensive 'cribriform' plates. The frontal, fig. 295, 11, is excessively expanded by the air-cells, fig. 296; its hind border is convex, its front one concave, and extended outward to form the superorbital ridge. The nasals, 15, are short, triangular, and pneumatic: they ultimately coalesce with the frontals. The mastoid is confluent with the squamosal, and, bending forward to near the back part of the zygomatic process, circumscribes the meatus auditorius externus. The tympanic completes the inner part of the meatus, contributes to the back part of the glenoid cavity, and expands into a broad horizontal plate supporting the large ear-drum: it early coalesces with the petrosal. The apex of this bone is grooved by the ento-carotid.

The epencephalic compartment of the cerebral cavity, fig. 296, *e*, as in *Lissencephala*, is wholly behind the pros- and mesencephalic ones: the rhinencephalic compartment is well defined. The 'sella' has slightly-marked clinoid processes. The orbits are continuous with the large temporal fossæ. The palatines form the posterior half of the intermolar part of the roof of the mouth, and bound the hinder nostril; they soon coalesce with the pterygoids and maxillaries, 21: these are remarkable for the large proportional size of their alveolar part, in advance of which the bone extends upward to be wedged between the frontal and premaxillary, downward and forward to strengthen the socket of the tusk, and backward to form the anterior pier of the zygomatic arch and the lower part of the orbit. The maxillary is perforated by a large antorbital foramen. The premaxillary, figs. 152, 295, 22, mainly consists of the part which lodges the base of the great tusk: but its ascending portion reaches the frontal, 11, and excludes, as in Rodents, the maxillary from the nasal: the alveolar part is grooved mesially by the long incisive canal. Both maxillary and premaxillary are pneumatic, fig. 296. The mandible, fig. 295, *i*, is short, the ascending being as extensive as the horizontal ramus, and being also excavated for the formative alveolus of the succeeding molar. The condyle is small, convex, rising above the coronoid process, which is low and projects obliquely forward. The dental canal is wide in reference to the unceasing supply of material for the growth of the great molars. The symphysis is short, small, pointed: in some extinct Proboscidians it was excavated for the alveoli of a pair of tusks; and in one aberrant form (*Deinotherium*) the symphysial tusk-bearing part of the mandible was enlarged, lengthened, and deflected.

The bony nostril, formed by the nasals and premaxillaries, is small, transversely sub-bilobed, and elevated in position. The rhinal cavity expands as it extends backward to be divided by the vomerine septum, fig. 296, 13. The inferior turbinals are slightly-curved laminæ, one on each side the lower border of the 'lamina perpendicularis,' where is the aperture admitting air to the singularly extensive pneumatic structure of the skull. The lacrymal is a small protuberant imperforate bone, serving chiefly to give attachment to the tendon of the 'orbicularis palpebrarum.'

From the middle of the stylohyal a slender pointed process is sent off at an acute angle. There is no bony ceratohyal. The basihyal is transversely extended; and articulates at each end to a gristly epihyal, and a long bony thyrohyal. The base of the stapes is an oval convex plate, with a marginal groove: one crus is thinner than the other, and it is very slender.

297

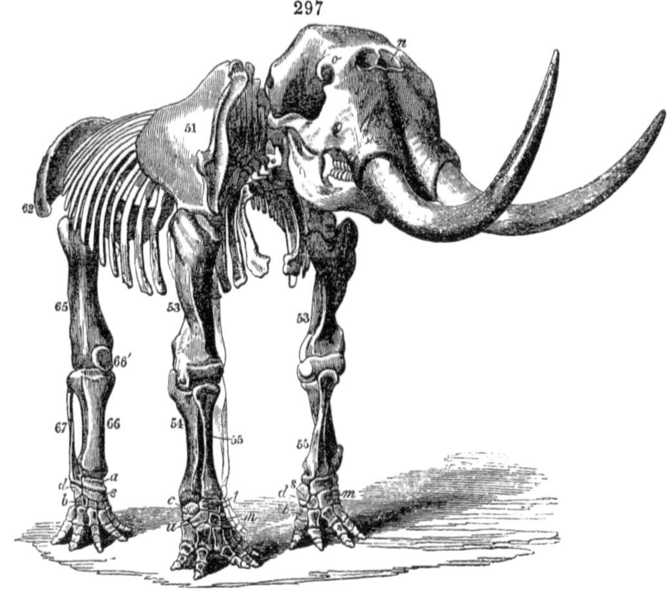

Skeleton of the Mastodon (*Mast. giganteus*).

§ C. *Bones of the Limbs.*—The *scapula* of the elephant, figs. 162, 297, 51, is second only to that of the Megatherioids, fig. 279, 51, in the proportion of breadth to length (dorso-humeral diameter): but the margin answering to the 'inferior costa' of anthropotomy, instead of being the longest, as in the Megathere, is the shortest: it is very concave: the 'base' is convex or bent at almost a right angle: a thick epiphysis is attached to its border: the spine extends into a short pointed acromion, and, as in some Rodents, sends down a process: the coracoid is a mere tuberosity: the glenoid cavity is shallow, twice as long as broad: it looks downward, the scapula rising vertically above the humerus.

The *humerus*, ib. 53, has the great tuberosity extended antero-posteriorly, and rising above the sessile hemispheric head of the bone: the deltoid ridge descends below the middle of the bone: the occipital groove is deep: the ectocondylar ridge rises straight for one-third the length of the humerus, and forms a low angle before subsiding upon the shaft. The distal articular surface is a simple shallow trochlea. The proximal epiphysis is in two parts, one capping the head, the other the great tuberosity: the distal epiphysis is single. The centre of the shaft is almost wholly occupied by a delicate cancello-reticulate structure.

The antibrachial bones are distinct and cross obliquely, the *radius* passing in front of the *ulna* to the inner side of the carpus, as in the Megathere: but the prone position of the fore foot cannot here be changed; for the head of the radius, fig. 297, 55, is wedged between two processes of the ulna, ib. 54, and the expanded distal half has a rough ligamentary union with that bone. The proximal articulation with the humerus is transversely elongate, partly convex and partly concave. The ulna is the larger bone; its olecranon is thick and convex: the proximal epiphysis covers only this process: the distal one forms the articulation for both radius and carpus. This segment includes a small scaphoid, fig. 297, *s*, a larger lunare, *l*, cuneiforme, *c*, and pisiforme, with the usual four bones of the distal row. In the scaphoid, the small surface for the radius is remote from that which joins the trapezium, *t*, and trapezoides, *d*. The single surface of the pisiforme has two facets, the smaller of which joins the ulna. The trapezium extends along half the metacarpal of the index. The phalanges, two in the first and three in each of the other four digits, are broad and short, especially the last, which is firmly encased in the corresponding division of the hoof.

The hind limbs and pelvic arch present opposite proportions to those in the Megatherioids: the skeleton of the extinct Proboscidian leaf-eater, fig. 297, contrasts singularly in this respect with that of the extinct Megatherioid one, fig. 267. To both these giants among land quadrupeds the forests of the primeval world afforded sustenance; but their ways of obtaining it were different, and called for preponderance of developement in the hind part of the skeleton in the one, and of the fore part in the other. The pelvis descends vertically at almost a right angle with the trunk, fig. 162, 62, the ilia forming with the lumbar series an angle of 120°: the ischium and pubis are short, and form a symphysis, fig. 298, *g*, the axis of which runs at an angle of 100° with that of the ilium. This bone arches out from its sacral joint almost transversely, the thick rough crista descending with its angle, *a*, produced to a level with the acetabulum: the anterior or abdominal surface is concave. The ischium, *f*, has the tuberosity, *e*, directed dorsad: the pubis shows a pectineal ridge, *h*. The sciatic notches are widely open: the obturator foramina are smaller than the acetabula, the planes of which incline from the perpendicular about 70°,—a favourable position for transmitting the weight upon the heads of the femora: these, as in the Megathere, have no round ligament, and the acetabulum is simplified accordingly. In a young Elephant I have observed an accessory pelvic ossicle wedged between the ischium and pubis behind the acetabulum.

298

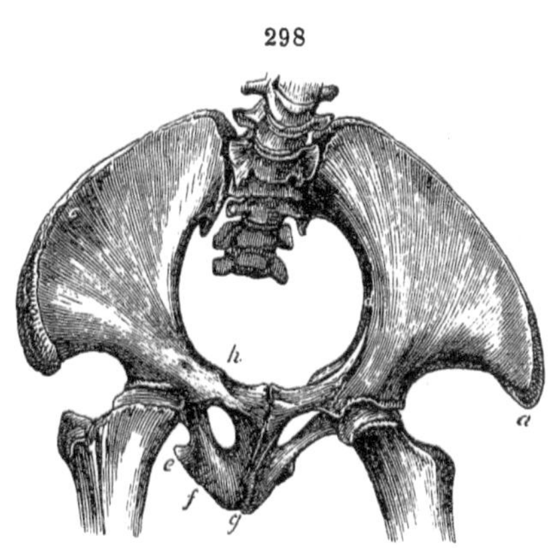

Pelvis of the Elephant, front view.

The great trochanter does not rise so high as the head of the *femur*: the small one is almost obsolete: the post-trochanterian fossa is shallow: the shaft, figs. 162, 297, 65, is straight, simple, and compressed from before backward. The rotular trochlea is subsymmetrical, occupying one-third of the breadth of the distal end: the condyles are divided by a deep popliteal cavity. The proximal epiphysis consists of the part forming the articular ball and that forming the trochanter. The medullary artery enters the back part of the lower third of the shaft, and ascends to a very small medullary cavity.

The two proximal articular surfaces of the *tibia*, ib. 66, are transversely oval, separated by a conical prominence: there is a large rough depression in front of the head of the bone: the middle of the shaft is triedral, the hinder surface is very concave superiorly. The distal articular surface is semicircular, convex behind, and rising externally on the shaft to give articulation to the fibula. The medullary artery passes transversely from the back of the shaft forward to a small medullary cavity. The fibula, 67, retains its distinctness from end to end in the Proboscidian Ungulates. The patella, 66′, is slightly convex lengthwise, and concave transversely at its articular surface. The bones of the foot are described at p. 309, fig. 193.

§ 187. *Skeleton of Perissodactyla.*—A. *Vertebral Column.* All the existing, and so far as is known the extinct, species of this order have more than nineteen dorso-lumbar vertebræ. The Tapir (*Tapirus americanus*, fig. 299) has 7 cervical, 18 dorsal, 5 lumbar, 6 sacral, and 13 caudal vertebræ. The pleurapophysial part of the transverse process extends forward in the third cervical, and underlaps that of the second: the corresponding part of the transverse process progressively expands in the succeeding

299

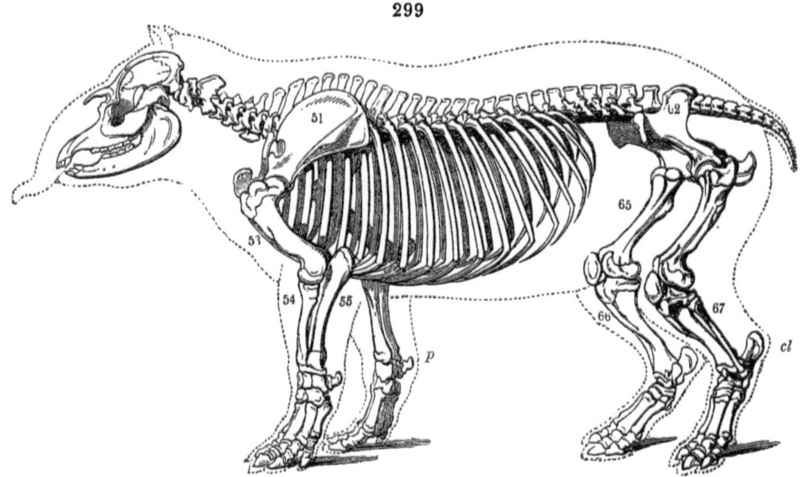

American Tapir (*Tapirus americanus*).

vertebræ to the sixth, where it forms a broad hatchet-shaped plate of bone directed downward and a little outward. In the seventh cervical the transverse process consists of a diapophysis only, and is therefore imperforate. In the anterior dorsal vertebræ the base of the neural arch is perforated on each side by the spinal nerve. In both these and the cervical vertebræ the fore part of the centrum is convex, the hind part concave.

The neural spines gain rapidly in height to the third dorsal, and gradually shorten to the eleventh; after which, they increase in fore-and-aft extent, and, from slightly inclining backward, become vertical. Eight pairs of ribs directly join the sternum, which consists of seven bones, with the xiphoid cartilage. The transverse processes of the last two lumbar and first sacral vertebræ are articulated to one another. Only the first two of the anchylosed sacrals afford articular surfaces to the ilia. Sometimes a coalesced caudal adds a seventh vertebra to the sacrum. The atlas has a recurved hypapophysis: its articular cups are deep: the base of the transverse process is twice perforated by the vertebral artery, the anterior hole opening upon the groove which leads to the foramen in the neural arch common to the vertebral artery and the first spinal nerve.

In the Rhinoceros, fig. 165, the vertebral formula is—7 cervical C, 19 dorsal D, 3 lumbar, 4 sacral, and 22 caudal. In the atlas, the hypapophysis developes a process from the lower part of the anterior surface. The neural arch is perforated transversely by the vertebral artery. In the axis the centrum supports a simple diapophysis, inclining downward and backward. The neural spine is thick, short, tuberculated, and divided by a deep and broad groove into two: the upper part of the spine is prolonged obliquely upward, giving the whole a trifid character. The pleurapophyses, from the fourth to the sixth cervical vertebræ inclusive, have the form of broad subquadrate plates: in the seventh the diapophysis only is developed, and the transverse process is consequently imperforate. The spine of this vertebra suddenly acquires great increase of length, which continues more gradually to the second and third dorsals, beyond which the spines shorten, but gain in antero-posterior extent to the eleventh dorsal, beyond which they continue of the same size, shape, and inclination to the lumbar region. A metapophysis rises in the fourth dorsal from the back of the diapophysis, from which it becomes distinct in the sixteenth dorsal. The diapophysis, which gradually subsides in the dorsal, reappears suddenly in the first lumbar: it becomes shorter in the second; and still more so in the third, in which it is very broad. The lower edge of the diapophysis of the second lumbar articulates with the upper edge of the diapophysis of the third, and the third articulates in the same manner with the first vertebra of the sacrum. The metapophyses are distinct, and are situated on the anterior zygapophyses in the first two lumbars: in the last they have become rudimental, and almost obsolete. The centrum is strongly convex anteriorly, and concave behind, in the cervical vertebræ; the dorsals are opisthocœlian in a less degree.

The ribs are slender in proportion to their length, and more curved than in the Elephant. In the first rib the tubercle is large, with a corresponding articular surface: both this and the second are almost straight, become expanded distally, and have no groove on the posterior margin. The twelfth rib is the longest, measuring three feet. In the nineteenth rib the articular surfaces of the head and tubercle are almost confluent, and the shaft decreases in thickness distally: its length is one foot four inches.

In the sacrum the articular surface for the ilium is formed by the first three of the four coalesced vertebræ. The metapophyses are distinct in the first two. The neural spines are long, strong, and tubercular at the end; the last curves very much backward. The three interarticular cartilages between these four vertebræ long remain unossified. In the caudal series the neural canal does not extend beyond the seventh vertebra.

The little *Hyrax* has not fewer than twenty-nine, or even thirty, dorso-lumbar vertebræ. In the twenty-two dorsal vertebræ of the skeleton of *H. capensis*,[1] the spines incline toward the thirteenth, which is vertical, and indicates the centre of motion of that part of the trunk. In their forms and proportions they resemble those of the Rhinoceros. Seven or eight pairs of ribs directly join the sternum, which consists of six bones. The metapophysis commences on the third dorsal, and attains the outside of the zygapophysis on the fifteenth: it exceeds the diapophysis in length in all the posterior dorsals. In the eighth lumbar vertebræ the diapophyses suddenly acquire great breadth, and gradually increase in length to the last lumbar; the metapophyses are continued throughout the series. No anapophyses are developed.

The transverse process of the atlas is perforated vertically at its fore part by the vertebral artery, which afterwards perforates the neural arch. The hypapophysis developes a short process. The simple transverse process of the dentata is perforated at its base for the vertebral artery, and the neural arch is perforated on each side by the second cervical nerve. The pleurapophysial part of the transverse process is much expanded in the third to the sixth cervical vertebræ inclusive: I have found it wanting on the left side of the seventh vertebræ, but present as a distinct element, or rudimental cervical rib, on the right side, where it completes the foramen for the vertebral artery. The sacro-caudal vertebræ are fourteen in number, of which the first three articulate with the ilia, and the four succeeding have transverse processes.

In an extinct S. American Perissodactyle (*Macrauchenia*),[2] the cervical vertebræ are remarkable for their length, as the name implies; also for the flattening of the terminal articular surfaces, and the imperforate character of the transverse processes,

[1] XLIV. p. 524, no. 3097.

[2] XCV'. p. 35, pls. v. vi. vii.

the vertebral artery entering the neural canal and perforating the neurapophysis, lengthwise. The transverse processes of the last lumbar present each a concave articular surface to corresponding convexities of those of the sacrum.[1]

In the Horse, fig. 300, the vertebral formula is—7 cervical, 19 dorsal D, 5 lumbar, *a*, *b*, *c*, *d*, *e*, 5 sacral, *f*—*l*, and 17 caudal, *p*—*r*. Eight pairs of ribs directly join the sternum, which consists of seven bones and an ensiform cartilage. The neural

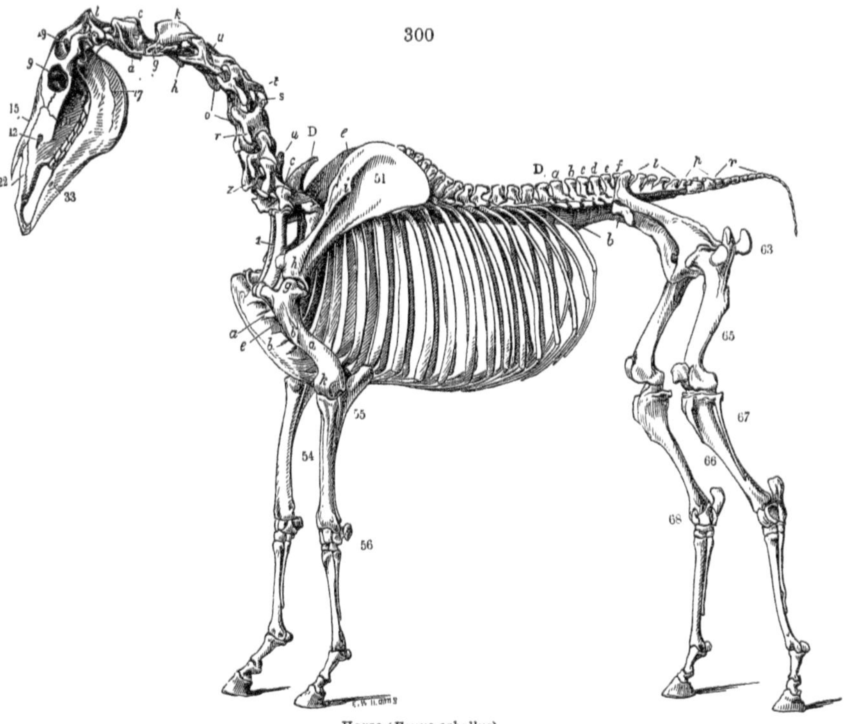

Horse (*Equus caballus*).

arches of the last five cervical vertebræ expand above into flattened, subquadrate, horizontal plates of bone, with a rough tubercle in place of a spine: the zygapophyses are unusually large. The perforated transverse process sends a pleurapophysis, *z*, downward and forward, and a diapophysis, *r*, backward and outward, in the third to the sixth cervicals inclusive: in the seventh the diapophysial part alone is developed, and is imperforate. The neural spines suddenly and considerably increase in length in the first three dorsals, and attain their greatest length

[1] xcv'. pl. viii. fig. 1, *b*.

in the fifth and sixth, after which they gradually shorten to the thirteenth, and continue of the same length to the last lumbar. The metapophysis, commencing as a tuberosity above the diapophysis, passes gradually from that part to the outer side of the prozygapophysis, which it finally attains in the seventeenth dorsal vertebra, and continues in the same place throughout the lumbar series. There are no anapophyses. The lumbar diapophyses are long, broad, and in close juxtaposition; the last presents an articular concavity adapted to a corresponding convexity on the fore part of the diapophysis of the first sacral.

The cervical vertebræ, though shorter than in *Macrauchenia*, are longer than in other Perissodactyles, and rise with an arch to support the head: the joints of the centrums are opisthocœlian. In the third cervical the pleurapophysis is developed below the arterial canal, and extends forward, outward, and downward. The neural spine, *u*, has subsided to a low rough ridge. The hypapophysial ridge and tubercle, *o*, are well marked, as are also the anterior convexity and posterior concavity of the centrum. The inner surface of each neurapophysis is pierced by a small canal in the same place and direction as that which transmits the vertebral artery in *Macrauchenia*; but the artery traverses the base of the transverse process in the Horse, as in most other mammals. In the axis the neural spine, *k*, is a strong but low rugged ridge, which bifurcates posteriorly, and subsides upon the zygapophyses. The diapophyses are short and triangular, with their bases perforated by the vertebral artery. A strong ridge on the under part of the centrum leads to the hypapophysis, *h*. The posterior articular surface of the centrum is deeply excavated. In the atlas, *c*, the anterior articular cavities do not meet below: the diapophysial ridges, *a*, bend down, forming large concavities: the vertebral artery twice pierces their base, which is also traversed by a canal leading to the neural canal, anterior to which the neural arch is perforated on each side. The hypapophysis developes a strong tubercle.

In the skeleton of a Quagga (*Equus Quagga*) I have observed 19 dorsal, 6 lumbar, 5 sacral, and 18 caudal vertebræ; in that of a Zebra (*Equus Zebra*), 18 dorsal, 6 lumbar, 5 sacral, and 17 caudals; whilst in the skeleton of an Ass (*Equus Asinus*), there were 18 dorsal, 5 lumbar, 5 sacral, and 17 caudal vertebræ. The sixth lumbar, fig. 299, *f*, becomes the first sacral by coalescence.

B. *Skull.*—Some common characters of this part of the skeleton in Perissodactyles are given at pp. 283, 284. In the Malayan Tapir (*Tapirus indicus*), the paroccipitals are compressed and slightly incurved: they are strengthened by a long post-tympanic process, developed from the squamosal and articulated to the fore part of the base of the paroccipital, so as to circumscribe a space occupied by the true mastoid which is confluent with the petrosal. One or two vacuities are left in this space for the exit of veins. The post-glenoid process is much developed. The base of the pterygoid process is perforated lengthwise by the ectocarotid; the apex is slightly recurved, and unites with the palatine by a squamous suture. The entopterygoids are thin, small, curved lamellæ applied to the inner side of the base of the pterygoid processes, and uniting with each other below, and clear of, the presphenoid. The major part of the palatine enters into the formation of the large oblique hinder aperture of the nasal passages: the smaller anterior division completes the bony palate which terminates behind between the first and second true molar. The lacrymal canal commences by two distinct orifices. The bases of the nasal bones are deeply grooved, and articulate with the frontals parallel with the back part of the orbit. There is no superorbital foramen or canal. The premaxillaries terminate behind at a considerable distance from the elevated nasals. In the American Tapir (*Tapirus Americanus*), fig. 301, the superoccipital is narrower and more deeply excavated than in the Malayan Tapir: a smaller proportion of the petromastoid is visible between the exoccipital and squamosal, *g*: the frontals, 11, are less expanded and less elevated above the nasals, 15. The petromastoids fit, but not closely, the vacuities on each side the basioccipital. In the cranial cavity the rhinencephalic fossa is well defined.

301

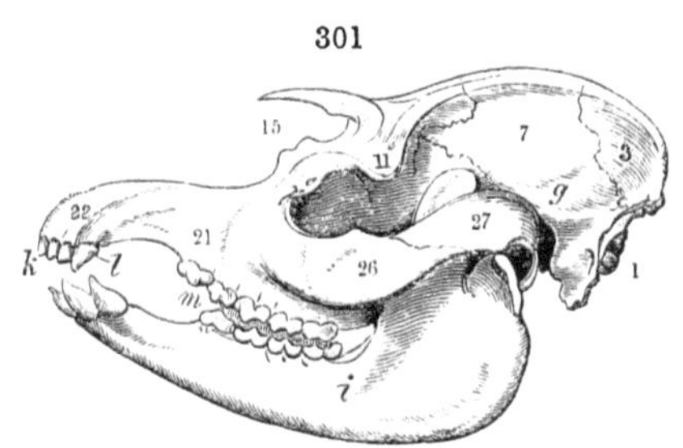

Skull of American Tapir.

In the (Sumatran) Rhinoceros, a smaller proportion of the palatine bones enters into the formation of the bony palate than in the Tapir; they chiefly form the sides of the hinder nasal aperture, the anterior boundary of which is opposite the first true molars. The pterygoid processes are perforated at their base, lengthwise, by the ectocarotid arteries. The nasofrontal suture is in advance of the orbits. The postglenoid process is long, subtrihedal, and obtuse: the post-tympanic process takes the place of the mastoid and is here a strong quadrate process applied to the base of the paroccipital. The orbits are very obscurely marked off from the temporal fossæ: there is no postorbital process. The lacrymal canal commences by two apertures defended by a rough protuberance of the lacrymal bone. There is a well-developed pit for the origin of the inferior oblique. The premaxillaries are small and do not join the nasals. The air-sinuses extend from the frontals to the superoccipital ridge.

In the Indian Rhinoceros (*Rhinoceros Indicus*, fig. 302), the bones, 3, 7, 11, 15, forming the expanded neural spines of the cranial vertebræ, are so curved, that the summit of the superoccipital, 3, and the centre of the nasals, 15, form the two pillars from which are suspended the parietals, 7, and frontals, 11, forming an inverted arch. The highest part of the nasals shows the rough flattened surface for the attachment of the horn: from which part they curve downward, ending pointedly. The premaxillaries, 22, are small, support a pair of incisors, articulate with each other and the maxillaries, and terminate remotely from the nasals. In the African two-horned Rhinoceroses, the premaxillaries are almost obsolete, and usually edentulous in the adult.

302

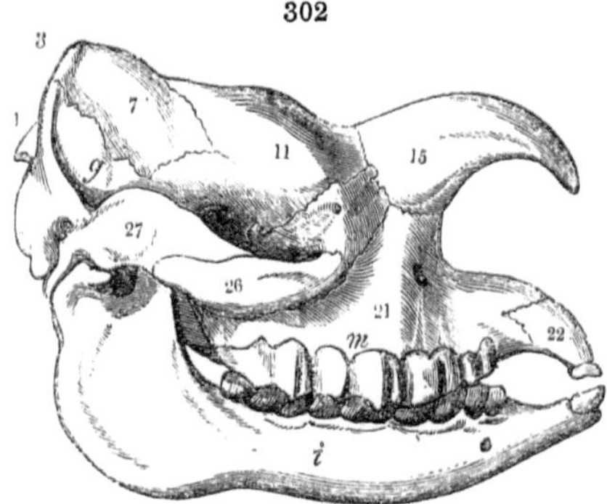

Skull of Indian Rhinoceros.

In certain extinct Rhinoceroses the septum narium was partially (*Rh. leptorhinus*) [1] or wholly (*Rh. tichorrhinus*) ossified. The articulation between the basi- and pre-sphenoids long remains. There is no interparietal. The entopterygoid swells into a tuberosity, and overlaps the palato-pterygoid suture.

In the *Hyrax* the elements of the occipital bone are late in coalescing. I have seen an interparietal wedged into the back part of the sagittal suture, and also the upper half of the superoccipital detached from the rest. The ascending process of the malar articulates with the postorbital process which is formed by both the parietal and frontal bones. The tympanic, which forms the bulla ossea at the basis cranii, has not coalesced with the petrosal. The hinder halves of the palatines enter into

[1] XVIII'. p. 356, figs. 131, 138.

the formation of the palato-nares. The lacrymal canal commences by one or two foramina, defended by a process. The maxillary forms the floor of the orbit, as in the Rhinoceros and Tapir; but the premaxillaries join the nasals. The lower jaw, 7, is remarkable for the backward expanse of the ascending ramus. The coronoid process is perforated lengthwise at its base.

If the equine skull, fig. 303, be compared with that of the Rhinoceros, the basioccipital will be seen to be narrower and more convex. The mastoid, 8, intervenes, as a tuberous process, between the post-tympanic and paroccipital processes, clearly indicating the true nature of the post-tympanic in the Rhinoceros; the Tapir shows an intermediate condition of the mastoid

303

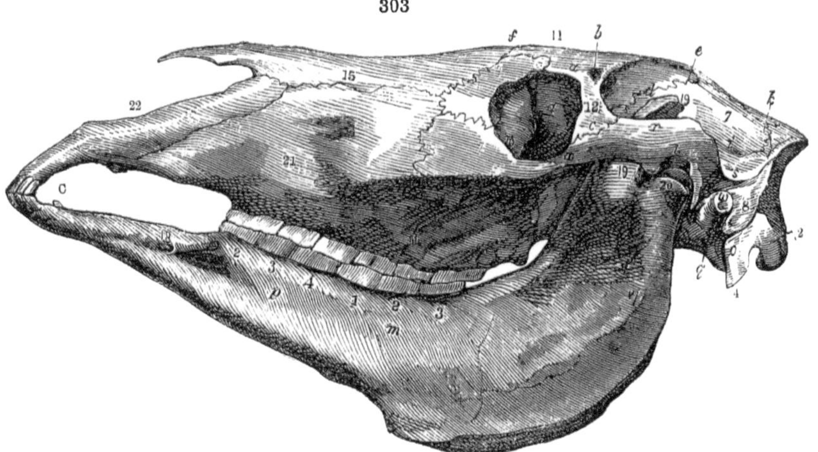

Skull of Horse, *Equus Caballus*.

between the Rhinoceros and Horse. The latter differs from both the Tapir and Rhinoceros in the outward production of the roof of the orbit and the completion of the bony frame of that cavity behind by the junction of the postorbital process, 1, 2, *c*, with the zygoma, *r*: *Equus* resembles *Macrauchenia* in this particular. The temporal fossa, 7, is small in proportion to the length of the skull: the base of the postorbital process is perforated by a superorbital foramen, *b*: the lacrymal canal begins by a single foramen. The premaxillaries, 22, extend to the nasals, 15, and shut out the maxillaries, 21, from the anterior aperture of the nostrils. The chief marks of affinity to other Perissodactyles are seen in the shape, size, and formation of the posterior aperture of the nostrils, the major part of which is bounded by the palatine bones, of which only a small portion enters into the formation of the bony palate, which terminates behind opposite the interspace between the penultimate and last molars. A narrow groove divides the palato-pterygoid process from the socket of the last molar, as in the Tapir and Rhinoceros. The pterygoid process has but little antero-posterior extent: its base is perforated by the ectocarotid. The entopterygoids are thin plates applied like splints over the inner side of the squamous suture between the pterygoid processes of the palatines and alisphenoids. The postglenoid process is less developed than in the Tapir. The Eustachian process of the petro-tympanic is long and styliform. There is an anterior condyloid foramen, and a wide 'fissura lacera.' The broad and convex bases of the nasals, fig. 304, 8, 8, articulate with the frontals, *f*, a little behind the anterior boundary of the orbits. The space between the incisors and molars is of greater extent than in the Tapir, fig. 301: a long diastema is not, however, peculiar to the Horse, and, although it allows the application of the bit, that application depends rather upon the general nature of the Horse, and its consequent susceptibility to be broken in, than upon a particular structure which it possesses in common with many other Herbivora. The air-cells do not extend farther back than the fore part of the frontals above the cranial cavity, and of the basisphenoid beneath. Ossification extends into the base of the tentorium and its continuation into the falx. The upper boundary of the rhinencephalic fossa is much developed.

304

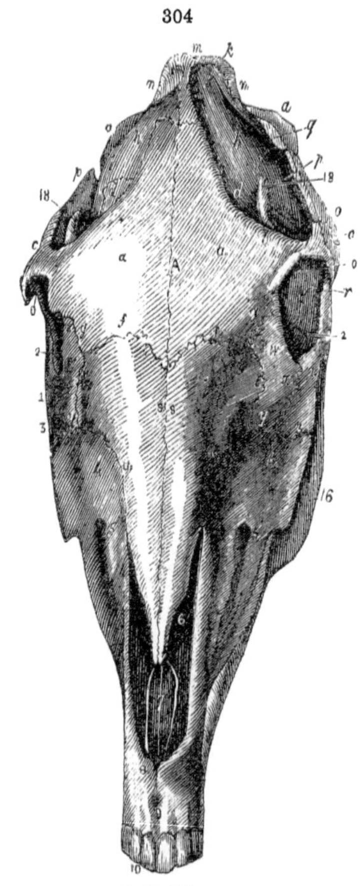

Skull of Horse.

There is a foramen, fig. 303, 9, in the premaxillary suture.

The zygoma, fig. 303, *r*, is chiefly formed by the squamosal, *s*.

The malar extends upon the face, beneath the lacrymal, in advance of the orbit. The ascending ramus of the mandible has a convex hind border curving from the condyle, 20, to beneath the last alveolus, where it ends by a slight projection below the inferior border of the horizontal ramus. The coronoid process, 19, is long, narrow, and recurved.

305

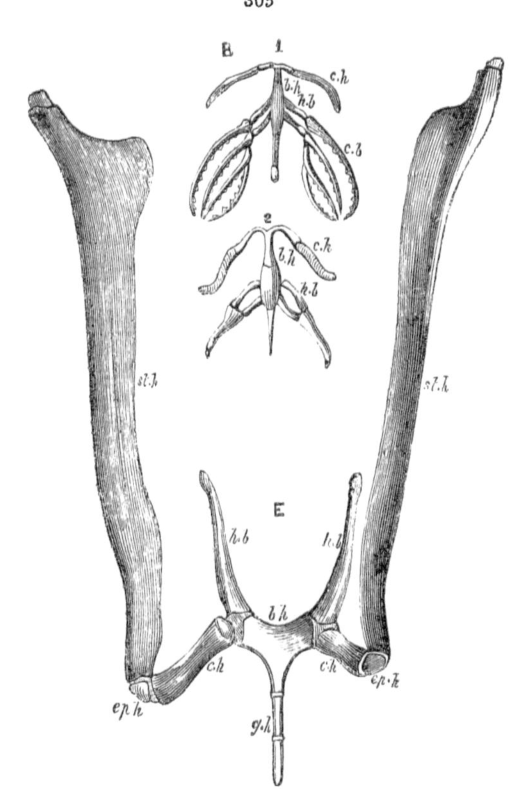

Hyoid arch, Horse.

The stylohyals, fig. 305, *st. h*, long and rib-like, articulate by a rounded 'head' to the petromastoid; expanding beyond this, to form a sort of 'tubercle,' and continued, slightly contracting, to the short epihyal, *e, p, h*; by means of which they articulate with the ceratohyals, *c, h*; which unite with the basihyal, *b, h*, where this is joined to the thyrohyals, *h, b*. The basihyal supports a glossohyal, *g, h*. The homology of the thyrohyals with the ceratobranchials in Fishes and Batrachia is illustrated by the figures B, *h b*, introduced (reversed) in the same cut.

C. *Bones of the Limbs.*—These are chiefly modified in their proportions with reference to degree of swiftness of course in the different species: which have diverged, in this respect, from the old tertiary type as exemplified by the Palæothere, in two directions, the extremes of which are now shown in the Rhinoceros, fig. 165, and the Horse, fig. 300. The segments farthest from the trunk are the seats of chief variety, and here the elongation and attenuation of the bone is attended with suppression of certain of the digits.

The scapula is long in proportion to its breadth, and most so in the Horse, fig. 300, 51: the anterior angle is largely rounded off: the spine developes no acromion, but gradually subsides as it approaches the neck of the scapula: it is situated nearer the hind border in the Tapir (fig. 299, 51), nearer the front border in the Horse, with concomitant differences in the areas of the supra- and infra-spinal fossæ: in the Rhinoceros it equally bisects the blade-bone, and is most prominent at its upper third. The coracoid is a mere tuberosity in all. The front border or 'costa,' in the Tapir, has a wide and deep notch. *Macrauchenia* differs most from other Perissodactyles in the continuation of the spine, without loss of height, to the neck of the scapula, above which it forms a slightly produced angle and is perforated.

In the Rhinoceros, fig. 165, the humerus is remarkable for the strength of the tuberosities and deltoid ridge, and for the smooth basal surfaces between the tuberosities and on the outside of the external one. The medullary artery enters the back part of the bone, and proceeds obliquely forward and downward. In the radius, the surface for the ulna extends along the back part of the ridge bounding that for the humerus. The two antibrachial bones interlock at their distal ends by reciprocally adapted cavities and tuberosities. The usual eight bones are present in the carpus: but the trapezium does not support a digit, and the unciforme is small and has only the digit answering to the fourth: this, with the medius and index, being alone developed in the Rhinoceros. The ilia are massive, short, and less expanded than in the Elephant, subvertical in position, concave anteriorly, and also behind in the transverse direction. The terminal angle of the rough thick crest is bifurcate. The ischia are relatively longer than in the Elephant, with thick outwardly-bent tuberosities. The ischio-pubic symphysis is prominent. The lumbo-iliac angle is 125°.

The head of the femur is impressed by a deep semicircular pit at its margin. The third trochanter, fig. 165, 65, is a remarkable feature, from its great size and forward curvature. Ossification sometimes extends from the great trochanter to the third trochanter. The rotular surface is distinct from those on the condyles. The inner wall of the trochlear surface for the patella is thicker, more prominent, and is prolonged farther up the shaft of the femur than the outer wall is; the condyles are nearly of the same length. The medullary canal commences at the back part in the upper half of the shaft, and inclines forward and downward. The bones of the hind-foot are explained at p. 309, fig. 193.

In the Tapir, the intercondyloid part of the humerus, fig. 299, 53, is perforated, as it is likewise in the Hyrax. I have found the radius, 54, and ulna, 55, partially anchylosed at their distal ends in the Malayan Tapir, and have observed their distal epiphyses to coalesce with each other before uniting with their respective shafts. The carpus resembles that in *Rhinoceros*; but the unciforme is rather larger, and supports the metacarpal of a fifth, as well as of a fourth digit. The first or trapezial digit is absent, and the one articulated to the magnum, answering to the third, is the largest and of symmetrical shape, the whole fore-foot plainly showing the perissodactyle type, though with four toes. The little *Hyrax* and an extinct hornless Rhinoceros (*Acerotherium*) have a similar unsymmetrically tetradactyle fore-foot. That of the *Macrauchenia* was tridactyle. The expanded part of the ilium of the Tapir, ib. 62, is an oblong quadrate plate with the upper and hinder angle articulating with the sacrum. The canal for the medullary artery of the femur, which begins near the small trochanter, extends downward to a small medullary cavity at the middle of the shaft, 65; which is longer than that of the tibia, 66. The bones of the hind-foot closely resemble those of the Rhinoceros, forming the same number of toes: the heel-bone, *d*, is more prominent.

In fig. 190, 'bones of the fore-limb of the Horse,' the suprascapular cartilage is ossified and confluent with the base of the scapula, *g*: *o* is the infraspinal fossa, *p* the supraspinal fossa, *i* the prominent and thickened part of the spine, *h* the neck, *m* the anterior border or 'costa,' *l* the posterior 'costa;' the line from *n* to *n* marks the base of the scapula supporting the suprascapula; *k* is the coracoid protuberance. In the humerus, *a* is the shaft of the bone, *b* the lower part of the deltoid ridge where the 'teres major' is inserted, *e* is the great tuberosity which is grooved by the tendon of the biceps, *f* is the 'neck.' The proximal epiphysis of the young bone forms both the head and the tuberosity. At the distal end, K marks the trochlear surface for the radius, the fore part of which bone, *o*, passes into the depression *l*, when fully flexed: *k* is the inner condyle, *i* the outer condyle; *m* the posterior fossa for the olecranon, when the antibrachium is extended. The ulna, represented by its olecranon, *s*, and upper part of the shaft, *u*, coalesces by the latter, in aged Horses, with the radius: it presents a small articular surface, *t*, for the humerus. The radius and ulna coalesce in *Macrauchenia*. The equine carpus includes, in the proximal row, the scaphoid *w*, lunare *x*, cuneiforme *y*, and pisiforme *z*, which latter is large and prominent. The os magnum, 2, in the second series of carpal bones is remarkable for its great breadth, corresponding to the enormous developement of the metacarpal bone of the middle toe, 4, 5, which forms the chief part of the foot. Splint-shaped rudiments of the metacarpals, answering to the second and fourth, 6, of the pentadactyle foot, are articulated respectively to the trapezoides and the reduced homologue of the unciforme, 3. The miocene Hipparion retained stunted hoofs supported by the second and fourth digits of the fore-foot, as in the hind-foot, fig. 194: but all modern and existing representatives of the genus *Equus* have the digital developement concentrated on the medius: of which, in fig. 190, 12–13 shows the proximal phalanx, called in Hippotomy the 'great pastern'; 14–15, the middle phalanx, called the 'small pastern'; 16, the ungual phalanx, called the 'coffin-bone': 11 and 17 are 'sesamoids,' the latter being called the 'nut-bone.'

The ilium of the Horse, fig. 300, 62, is longer and less expanded superiorly than in the Tapir; but it articulates by the corresponding part to the sacrum, which renders it hammer-shaped. The femur is characterized by the partial division of the great trochanter, and, as in other Perissodactyles, has the third trochanter. The medullary artery enters the middle of the shaft at its postero-internal side, and inclines slightly upward. In fig. 195, *a* is the shaft, *b* the 'neck,' *c* the head; *d d*, the great trochanter, of which the upper division is called 'the spoke;' *f* is the 'third trochanter,' *g* marks the place of a deep fossa giving origin to the gastrocnemius externus, *h* is the outer condyle. In the tibia, *s*–*w* is the protuberance and ridge for the rotular ligament, *v* the articular head of the bone, *u* the outer concavity. The distal end is excavated by a deep oblique double trochlear cavity for the astragalus, 5. The fibula is represented by its head, 1, and a slender styliform portion of the shaft, ending in a point, at 2. There is no representative of the distal end, as in *Macrauchenia* and the Ruminants. The bones of the foot are described at p. 308, fig. 193 (Horse).

The astragalus shows the extreme perissodactyle modification by the depth and obliquity of the superior trochlea, and by the extensive and undivided anterior surface, which is almost entirely appropriated by the naviculare: the ectocuneiforme, which is the homotype of the magnum in the carpus, is equally remarkable for its large size, since it supports that metatarsal, answering to the middle one in pentadactyle quadrupeds, which constitutes the chief part of the hind-foot in the Horse.

§ 188. *Skeleton of Artiodactyla.*—Some of the common osteological characters of this order, with the genera representing it, are given at pp. 285, 286.

A. *Vertebral Column.*—In the Hippopotamus, fig. 306, the

306

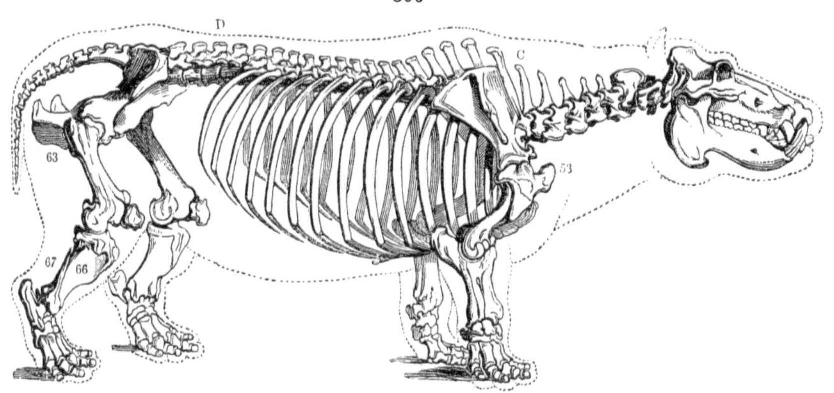

Hippopotamus amphibius.

vertebral formula is:—7 cervical, 15 dorsal, 4 lumbar, 6 sacral, 16 caudal. The pleurapophysial parts of the transverse processes of the third to the sixth cervical inclusive develope hatchet-shaped plates, progressively increasing in size, which overlap each other. The second and third cervicals have bituberculate hypapophyses. The transverse processes of all the cervicals are perforated by the vertebral arteries. The neural spines elongate from the third to the seventh cervical, C. Six pairs of ribs directly join the sternum, which consists of five bones and a broad ensiform cartilage. A metapophysial ridge is developed above the diapophyses of the eighth dorsal, changes its position and shape with increase of size in the two succeeding vertebræ, in the eleventh projects forward from above the prozygapophysis, and so continues throughout the rest of the dorsal and the lumbar series. There are no anapophyses, but a broad plate is developed from the back part of each transverse process of the last lumbar, which presents an articular convexity for a corresponding concavity on the fore part of each transverse process of the first sacral vertebra.

In the Peccari (*Dicotyles*), the vertebral formula is:—7 cervical, 14 dorsal, 5 lumbar, 5 sacral, and 6 caudal. The axis vertebra has a short pointed diapôphysis: the third vertebra has a pleurapophysial lamella coextensive with the centrum. The corresponding lamella increases in the fourth, the fifth, and very remarkably in the sixth cervical, and they overlap each other. The bony plate between the anterior zygapophysis and diapophysis is perforated by the spinal nerve in the last four cervical vertebræ: the third and fourth terminate above in a large platform of bone supported by vertical neurapophysial walls, without a neural spine; in the fifth a neural spine is developed, and the spine progressively increases in length and inclines forward in the sixth and seventh cervicals. The neural spines of the first and second dorsals are vertical, and as long as the pleurapophyses of the same vertebræ. The succeeding dorsal spines gradually diminish in length and incline backward to the twelfth, which is short and vertical. The metapophyses begin to be developed at the third dorsal, and increase in length to the eleventh, after which they rise upon the zygapophyses. The neural arches of all the dorsal vertebræ are directly perforated by the spinal nerves, and the base of the diapophysis is vertically perforated. The diapophysis of the fourteenth dorsal vertebra begins to show the increase of size which characterizes the lumbar series. Seven pairs of ribs directly articulate with the sternum, which consists of six bones.

In the Hog (*Sus Scrofa*), the vertebral formula is:—7 cervical, 13 dorsal, 6 lumbar, 4 sacral, and 23 caudal, with varieties, chiefly depending on the number of moveable ribs developed in the domestic breeds. The fifth and sixth cervical vertebræ are remarkable for the great expanse of the lamelliform, overlapping, and downwardly directed costal parts of the transverse processes, and the seventh cervical for the absence of the pleurapophysis and the sudden increase in the length of the neural spine. This is far surpassed by the spines of the anterior dorsal vertebræ; after which those processes progressively decrease in height to the last three dorsals, where they gain in antero-posterior extent: the verticality of the spine of the eleventh dorsal indicates the centre of motion of the trunk. The dorsal neurapophyses are directly perforated by the spinal nerves, and a bar of bone connects the end of the diapophysis with the hind part of the centrum, circumscribing a vertical perforation on each side. The metapophysis commences as a tuberosity upon the diapophysis of the middle dorsal vertebræ, projects forward midway between the di- and prozyg-apophyses in the tenth, passes upon the prozygapophyses of the eleventh dorsal, and is continued in that position throughout the lumbar series. There are no anapophyses.

307

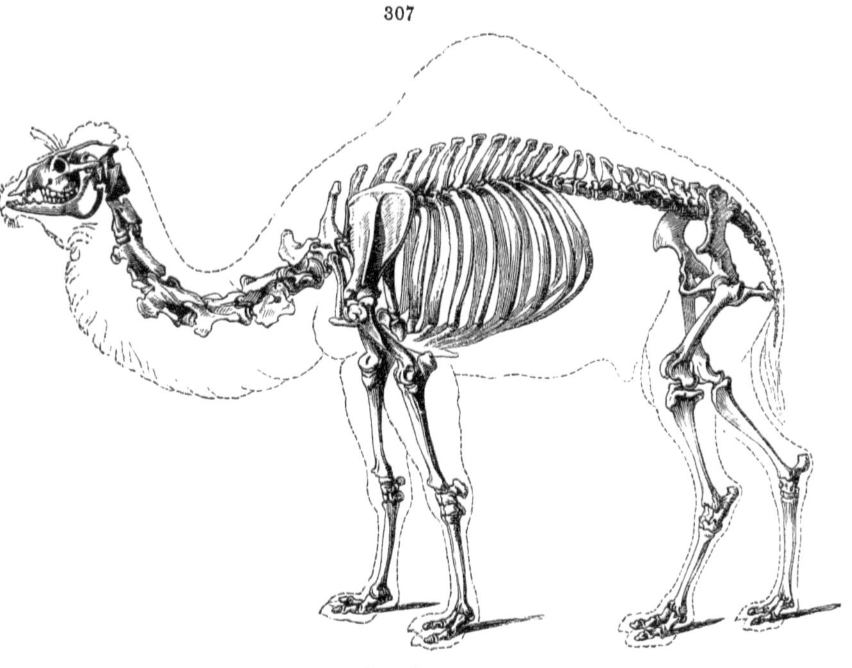

Dromedary.

In the Dromedary, fig. 307, and Camel (*Camelus bactrianus*), the vertebral formula is:—7 cervical, 12 dorsal, 7 lumbar, 4 sacral, and 18 caudal. Seven pairs of ribs articulate directly with the sternum, which consists of six bones, the last being greatly expanded and protuberant below, where it supports the pectoral callosity in the living animal. The cervical region is remarkable for its length and flexuosity; the vertebræ are opisthocœlian, but resemble those of *Macrauchenia* in the absence of the perforation for the vertebral artery in the transverse process, with the exception of the atlas; that artery, in the succeeding cervicals, enters the back part of the neural canal, and perforates obliquely the fore part of the base of the neurapophysis, as shown in the longitudinal section, fig. 308.[1] The costal part of the transverse process is large and lamelliform in the fourth to the sixth cervical vertebræ inclusive: in the seventh it is a short protuberance: in this cervical the neural spine becomes conspicuous. The metapophysial tubercle is developed from the diapophysis in the eleven anterior dorsal vertebræ, and passes upon the zygapophysis in the twelfth, continuing in that position throughout the lumbar series. There are no anapophyses. The spinous process in the first dorsal suddenly exceeds in length that of the last cervical, and increases in length to the third dorsal; from this to the twelfth dorsal the summits of the spines are on almost the same horizontal line, and are expanded and obtuse above, sustaining the substance of the hump (Dromedary) or humps (Camel); the spines of the lumbar vertebræ progressively decrease in length. The diapophyses of the last six lumbar vertebræ are very long: those of the last lumbar do not articulate, in the *Camelidæ* or in any Ruminant, with the sacrum.

308

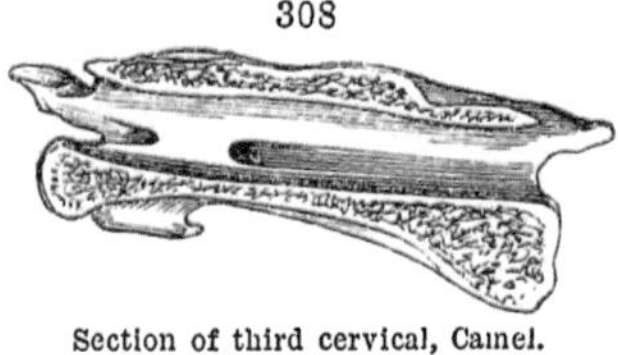

Section of third cervical, Camel.

In the Llama (*Auchenia*) the last sterneber is not so expanded as in the Camel: the vertebral formula is the same: the fifth lumbar has the largest spine: the cervicals, besides having imperforate transverse processes, resemble those of *Macrauchenia* in the flatness of the terminal articular surfaces, and the neck is habitually less bent down than in the Camels.

In the Musk-deer (*Moschus moschiferus*), the vertebral formula is:—7 cervical, 14 dorsal, 5 lumbar, 5 sacral, and 6 caudal. The atlas has a hypapophysis, but no neural spine. The transverse process is a broad thin plate coextensive with the length of the vertebra: it is perforated transversely from the neural canal outward to beneath its base, for the exit of the nerve, and then vertically, by the vertebral artery, which also perforates the neural arch. The axis has a sharp hypapophysial ridge extending from below the base of the odontoid process to beyond the posterior surface of the centrum, where it underlaps the next vertebra. A similar ridge and backwardly produced process are developed from the two succeeding cervicals, beyond which the ridge gradually subsides to the seventh vertebra. From the third to the sixth cervical inclusive, the pleurapophysial part of the transverse

[1] XCII'. p. 218, no. 925, A., figured in XCV'. pl. VI. fig. 2, also in XCVI, fig. 344.

process equals or exceeds the length of the vertebra, and those parts are arranged so as to overlap each other. There is a distinct, but less extensive diapophysial portion projecting external to the vertebrarterial canal: this part alone represents the transverse process in the seventh cervical. The spines of the third and seventh cervical vertebræ are vertical, those of the intermediate ones incline forward. The spines of the anterior dorsal vertebræ are remarkable for their height, those of the posterior dorsal and of the lumbar vertebræ for their antero-posterior extent, the anterior angle being produced forward and overlapping the spine in advance. A distinct metapophysis begins to be developed from the second dorsal, and attains its greatest length on the twelfth. There are no anapophyses. The notches for the nerves increase in depth as the vertebræ recede in position,

309

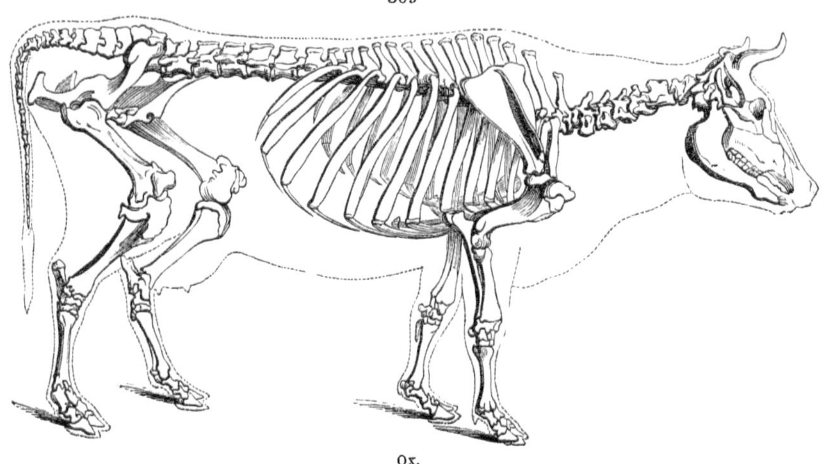

Ox.

and in the last dorsal the neural arch is completely perforated by these, which is likewise the case in most of the lumbar vertebræ. Eight pairs of ribs directly articulate with the sternum, which consists of seven bones. The tubercle disappears from the penultimate pair of ribs, and the diapophysis is reduced to a short rough tuberosity; but in the last pair the costal tubercle with its articular surface reappears, and the diapophysis resumes its normal size and articulation with the rib. In the first lumbar vertebra the diapophysis suddenly increases in length and breadth, and is probably augmented by the ossified and coalesced beginning of a rib.

In the common Ox (*Bos Taurus*, fig. 309), the vertebral formula is:—7 cervical, 13 dorsal, 6 lumbar, 5 sacral, and 21 caudal. The spines of the cervical are short, save in the last, and they incline to that of the third cervical, as the centre of the movements of the neck: these are facilitated by the ball-and-socket articulations of the bodies, common to the true Ruminants with most other Ungulates. The neural spine is longest in the third and fourth dorsals, whence the spines gradually shorten to the tenth: the metapophysis passes from the diapophysis to the zygapophysis in the tenth, eleventh, and twelfth dorsals. In the first lumbar the diapophysis exchanges its short, thickened, obtuse shape for a long, broad, vertically compressed plate: these processes increase in length to the fourth lumbar. The foramina for the spinal nerves directly perforate the neurapophyses of the dorsal vertebræ; they escape by conjugational foramina at the interspaces of the lumbar vertebræ.

The European Bison has 14 dorsal and 5 lumbar vertebræ, the American Bison has 15 dorsal and 4 lumbar, and this is the extreme reached, in the Ruminant order, of moveable pairs of ribs, equalling in number those of the Hippopotamus. The ribs are more slender in *Bison* than in *Bos*.

In the Roan Antelope (*Antilope equina*), the vertebral formula is:—7 cervical, 14 dorsal, 6 lumbar, 4 sacral, and 14 caudal. The atlas and dentata send out strong diapophyses: from that of the third cervical a broad pleurapophysial ridge extends forward and underlaps the diapophysis of the axis: a similar structure is presented by the fourth and fifth cervicals, and in the sixth the pleurapophysis forms a broad subquadrate plate extending downward and a little outward. This element is absent in the transverse process of the seventh vertebra, which is imperforate. The dorsal spines begin progressively to shorten from the fifth; that of the thirteenth is vertical, and indicates the centre of motion of the trunk. A metapophysis is developed from the front of the diapophysis of the second to the ninth dorsal vertebræ inclusive, where it begins to be transferred to the anterior zygapophysis, from which it extends in the last four dorsals and in all the lumbar vertebræ. There is a short anapophysis in the last two dorsals, but not in any of the lumbar vertebræ. Nine pairs of ribs directly join the sternum, which consists of eight bones and the xiphoid cartilage.

These characters are found in the vertebral column of most Antelopes.

In the Wild Sheep of Thibet (*Ovis Nahura*), as in the English domestic *Ovis Aries*, the vertebral formula is:—7 cervical, 13 dorsal, 7 lumbar, 4 sacral, 10 caudal, the latter being subject to variety. The pleurapophysial parts of the transverse processes of the third, fourth, and fifth cervicals underlap the diapophysial parts of those in advance: the pleurapophysis of the sixth cervical is an oblong quadrate plate; the seventh is imperforate, as in Ruminants generally. The neural spines increase in height from the third to the seventh cervical, and are suddenly and greatly

310

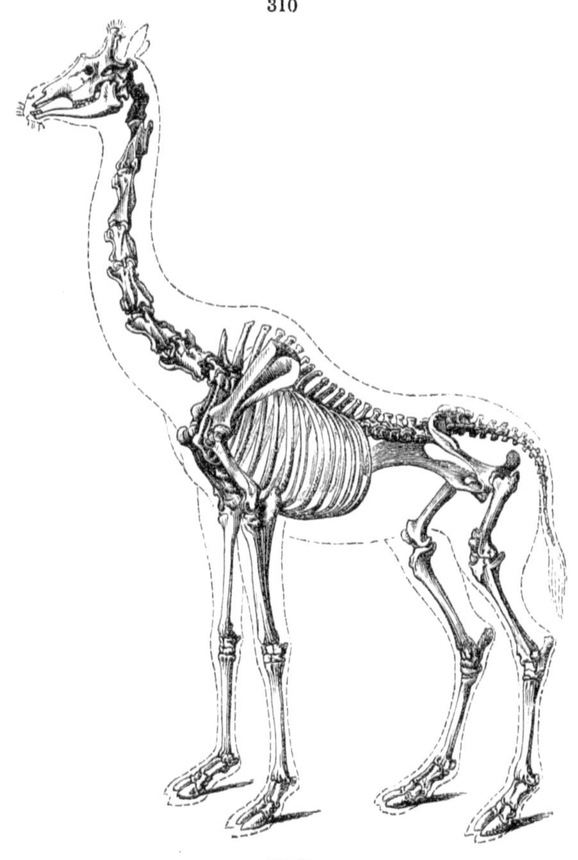

Giraffe.

surpassed in height by those of the anterior dorsals. The metapophysis is developed on the second and succeeding dorsals; attains the anterior zygapophysis in the eleventh; and projects from that part in all the lumbar vertebræ. The last pair of ribs are joined by the head, only, to the vertebra: the seven anterior pairs directly join the sternum, which consists of six bones.

The Giraffe is, in some respects, intermediate between the

'hollow-horned' and 'solid-horned' Ruminants, though partaking more of the nature of the Deer.

In the Nubian Giraffe (*Camelopardalis Giraffa*, fig. 310), the vertebral formula is:—7 cervical, 14 dorsal, 5 lumbar, 4 sacral, and 20 caudal. The vertebral artery perforates the fore part of the neurapophysis of the atlas twice, vertically and transversely: the atlas has a hypapophysis: this process in the dentata is a long thin ridge: the upper and fore part of the transverse process is perforated by the vertebral artery in this and the succeeding cervicals: a pair of exogenous processes is developed from the

311

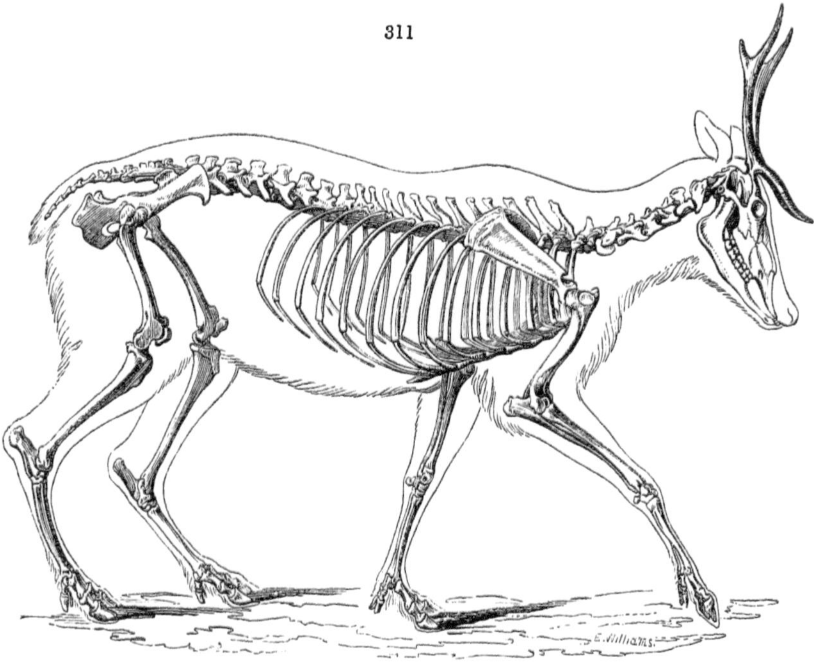

Female Rein-deer (*Cervus tarandus*).

under and fore part of the body in the third to the seventh cervical inclusive: the second to the sixth are remarkable for their length and almost want of neural spines: the short one of the seventh cervical is antroverted: those of the dorsals rapidly increase to the third, which, with those of the fourth and fifth, raise the outline of the back, like a hump: they then gradually diminish to the last dorsal. The ribs are long, corresponding with the great depth of the chest. Seven pairs directly join the sternum, which consists of six bones.

In the Rein-deer (*Cervus tarandus*, fig. 311), the vertebral formula is:—7 cervical, 14 dorsal, 5 lumbar, 4 sacral, and 11 caudal. The pleurapophyses of the third, fourth, and fifth cervicals are developed forward as well as backward; those of the sixth are also of great breadth, and are more produced downward. The metapophysis is distinctly developed upon the second and succeeding dorsal vertebræ, and attains the outside of the zygapophysis in the eleventh. All the dorsal ribs are biarticulate, retaining both head and tubercle. Eight pairs of ribs directly join the sternum, which consists of seven bones. In the Megaceros, fig. 166, as in the Fallow and most other Deer, there are thirteen dorsal and six lumbar vertebræ.

The opisthocœlian ball-and-socket joints of the cervical vertebræ facilitate the habitual inflections of the neck in the grazing and browzing actions in all Ruminants, while the long spines of the anterior dorsals afford adequate surface of attachment to the elastic and muscular structures sustaining the head—heavy in most of them with horns or antlers.

B. *Skull.*—This presents great diversity of shape in the *Artiodactyla*, with some common characters, already noted, which distinguish it from that of *Perissodactyla*.

312

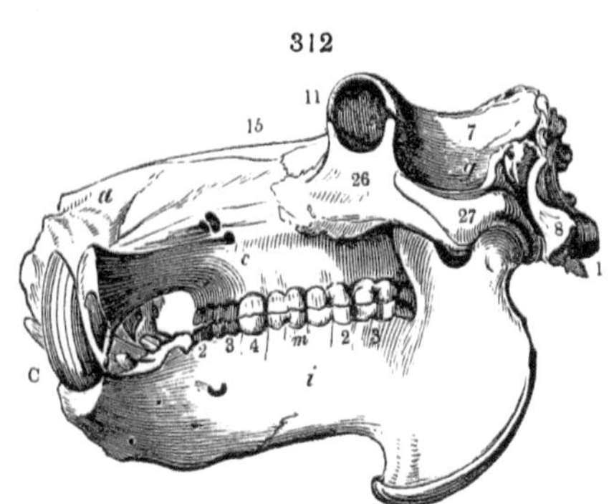

Hippopotamus amphibius.

In the *Hippopotamus*, fig. 312, the occiput is subvertical: from the upper part of its crest the contour of the skull runs nearly straight to the fore ends of the nasals, 15. The orbits, small and with an entire, or almost entire, rim of bone, singularly project both upward and outward, the frontals, 11, rising toward them, and arching lengthwise across their upper half. The upper jaw, *c*, which is almost cylindrical in advance of the molar series, suddenly expands to form the alveoli of the upper tusks, the mandible similarly expanding for those of the lower tusks, *c*; in the upper jaw a second terminal expansion, divided by a deep groove from the first, increases the space for the large tusk-shaped incisors. The depth of the temporal fossæ renders that part of the cranium, 7, narrower across than any part of the face: the fossæ meet above to form a parietal crest in old males. The facial part of the lacrymal is extensive, but the small deep-seated orbital part is perforated by the lacrymal foramen. The malar, 26, sends up a process to the postfrontal, which it rarely reaches: it extends backward to the glenoid cavity, and forms the under part of the zygoma: the upper part is due to the squamosal, 27. The external nostril is terminal, vertical, and formed by the nasals and premaxillaries: the maxillaries are perforated by a moderately large antorbital foramen far in advance of that cavity: the lateral series of molar alveoli slightly diverge anteriorly—a disposition which Cuvier regarded as peculiar to the Hippopotamus among Mammals. The bony palate is deeply notched anteriorly between the premaxillaries: there are two pairs of 'foramina incisiva.' The ascending ramus of the mandible, *i*, has a posterior convexly-curved outline descending to an antroverted angular process; the horizontal rami, divided by a deep notch from the angle, run forward almost parallel with each other, and expand at the symphysis, along whose upper and anterior broad truncated border the incisor sockets, four in existing, six in some extinct, *Hippopotami*, form a straight transverse line, between the tusks, *c*.

In a very young *Hippopotamus* may be observed the following evidences of cranial structure. The basioccipital has partially coalesced with the basisphenoid, but not with the exoccipitals; it forms no part of the occipital condyles, and developes no processes from its under surface: its lateral synchondrosal surfaces are divided into two facets, one for the part of the exoccipital behind the precondyloid foramen, the other for the smaller part in front. These parts of the exoccipital have not coalesced on the inner side of that foramen, which is single: the exoccipital developes, besides the condyloid process, the paroccipital and a broad process to join the mastoid. The superoccipital is a thick, rhomboid, vertical plate. The alisphenoids have coalesced with the basisphenoid: they are short, and are grooved behind by the boundary which they contribute to the foramen common to the foramen ovale and the basicranial foramen lacerum, and more deeply in front by the part they contribute to the foramen common to the foramen rotundum and foramen lacerum anterius: they develope long pterygoid processes, which are imperforate, and articulate along their inner sides with the entopterygoids. The presphenoid has coalesced with the orbitosphenoids and with the rudimental prefrontals, which are connate, compressed, and form the median septum of the great anterior outlet of the cranial cavity. The vomer is a long, slender, pointed bone, deeply grooved above. The parietals articulate with the alisphenoids, orbitosphenoids, squamosals, mastoids, frontals, superoccipital, and each other. The under part of the frontal is divided into a cranial, orbital, and olfactory surface; the orbital surface being the largest, and the superorbital ridge broad and much produced. The petrosal, mastoid,

313

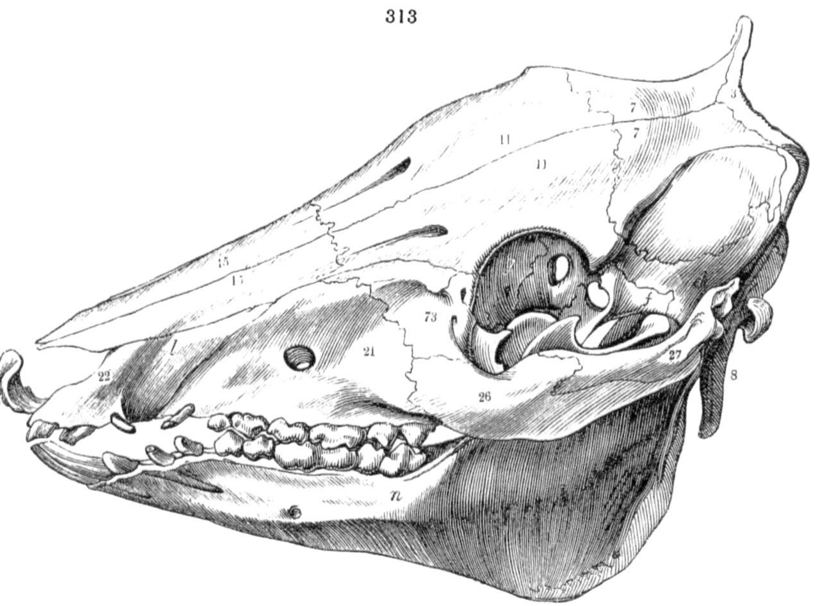

Skull of Young Pig (*Sus*).

tympanic and squamosal elements of the temporal have coalesced. The meatus internus is a deep fossa divided into a cribriform surface below and a canal above: the tympanic swells into a large three-sided conical protuberance below. The palatines prolong the bony palate beyond the series of grinding teeth in use.

314

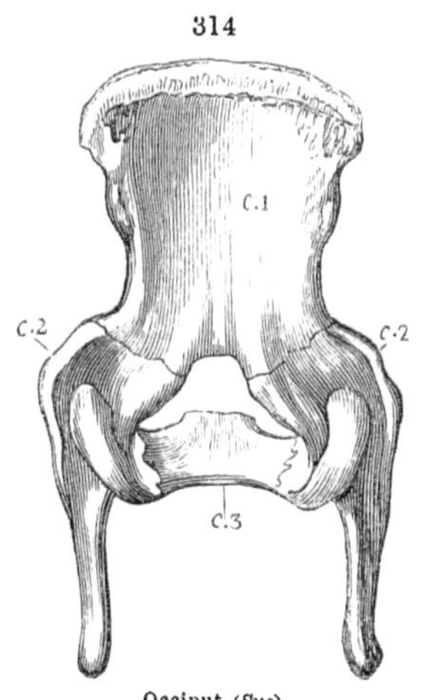

Occiput (*Sus*).

The composition of the Mammalian skull has been more fully exemplified in the young of the genus *Sus* (p. 300, fig. 189). In fig. 314, is given a back view of the neural arch of the occipital vertebra, showing the flattened centrum (basioccipital), *c* 3, the neurapophyses (exoccipitals), *c* 2, with their convex post-zygapophyses or 'condyles,' and long descending diapophyses (paroccipitals). The neural spine (superoccipital), *c* 1, is a vertical, quadrate, expanded plate, which completes the upper part of the neural canal (foramen magnum).

The superoccipital enters into the formation of the upper surface of the skull, as at 3, fig. 313. The parietals present flat supracranial, ib. 7, 7, temporal, and intra-cranial surfaces, fig. 315, 7. The frontals present, also, a flat supracranial surface, fig. 313, 11, an orbital, and an intra-cranial surface, fig. 315, 11. The postorbital process is not joined by a malar one: the superorbital canals are large, as in Ruminants. The nasals, 15, are long and pointed: the premaxillaries, 22, unite and circumscribe with them the external nostril. There is a prenasal bone, *o*, which strengthens the uprooting snout in most of the hog-tribe. The maxillary, 21,

315

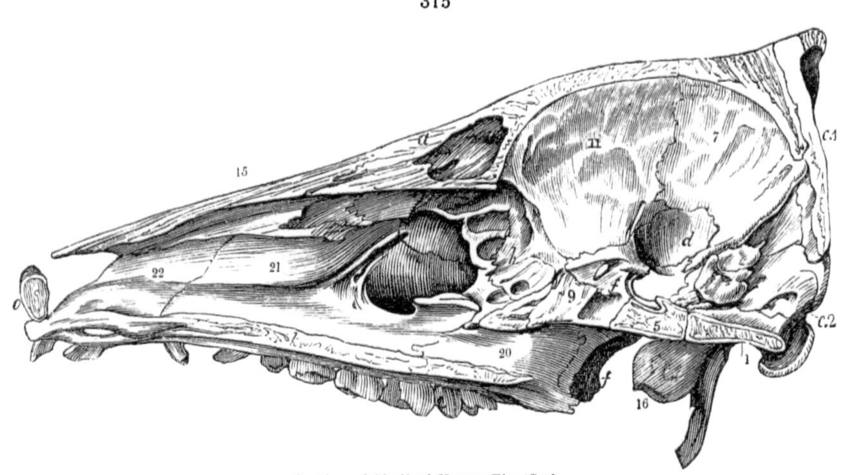

Section of Skull of Young Pig (*Sus*).

in the adult Boar, developes a large outwardly-curved alveolus for the tusk; strengthened above, in the Indian wild Boar, by a longitudinal ridge: the antorbital perforation is of moderate size: the maxillary unites posteriorly with the large facial plate of the lacrymal, fig. 313, 73, and with the malar, 26. This has no postorbital process. It is united with the zygomatic part of the squamosal, 27, by a double notch. The small cranial plate of the squamosal is shown at *d*, fig. 315. The articular surface for the mandible is convex from before backward, concave transversely, in which direction it is most extended. The alisphenoid is marked *f*, in figs. 313 and 315. The floor and sides of the long nasal canal are formed by the premaxillaries, fig. 315, 22, the maxillaries, 21, and the palatines, 20: to the latter succeed the pterygoids, *f*: the depth of the canal is gained by depressing the backwardly-extended bony palate below the level of the basis cranii, 1–9. The petrotympanic bulla, 16, is large, prominent, and subcompressed. In the interior of the cranium the rhinencephalic compartment, *h*, is large and well defined.

The skull of the Babyroussa (*Sus Babyrussa*), as compared with that of *Sus Scrofa* and *Sus larvatus*, shows a broader and lower occiput; the mastoids are larger; the temporal fossæ more approximated on the upper part of the cranium; the bony palate is more produced beyond the last molars. The mastoids show a

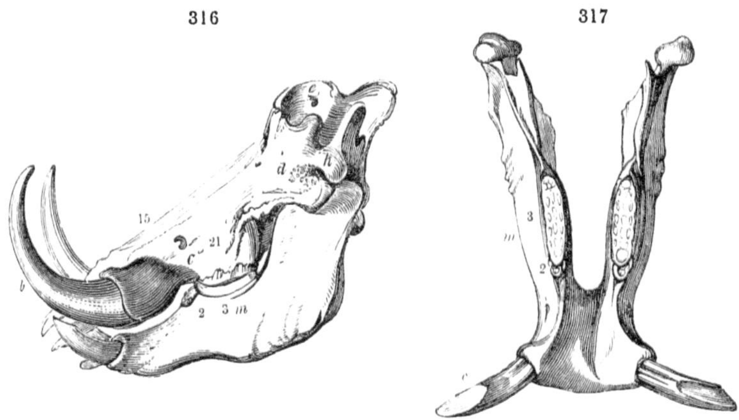

Skull of Wart-hog (*Phacochœrus*). Mandible of Wart-hog.

pneumatic cellular structure, and become confluent with the tympanic and squamosal, not with the petrosal. This bony capsule of the acoustic organ retains its primitive individuality, as such, and may be detached from the surrounding bones forming the otocrane: neither paroccipital nor mastoid are dismemberments thereof, as misinterpreters of developmental phenomena allege. There is no ossified prenasal. In the maxillary the long sockets of the canine tusks bend upward; the naso-maxillary part of the cranium being slightly compressed between them. A remarkable peculiarity is also presented by the fossæ at the inner side of the base of the pterygoids, which lead to sinuses communicating on one or both sides with the sphenoidal sinus. The air-cells extend from the nose to the occiput.

In the Wart-hog (*Phacochœrus Æliani*, fig. 316), the fronto-parietal region is broad and flat, except transversely, where it is rendered concave, as in the Hippopotamus, by the orbits being raised above its level: those cavities, *e*, are placed farther back than in the other *Suidæ*, and are partly defended by a post-orbital process of the malar. The paroccipital processes are long and slender. The mastoids are compressed and pointed, and are much less developed than in the Wild Boar, the Masked Boar, or the Babyroussa. The pterygoid fossæ are simple; not divided into an external and internal compartment, as in the Babyroussa, but they are more extended backward. The sockets of the canines, *c*, have not the process from the upper part, as in the *Sus larvatus*. The maxillo-premaxillary suture is early obliterated, except at the apex of the premaxillaries which extend beyond the sockets of the tusks. The nasals, 15, are of great length. The fore part of the lower jaw, fig. 317, is expanded for the sockets of the tusks, *c*, and truncate, as in *Hippopotamus*; but the sockets of the incisors are soon obliterated. In the interior of the skull a tentorial ridge is developed.

In the Peccary a strong ridge extends from the lower border of the malar. The pterygoids have not the fossæ shown in the Babyroussa and Wart-hog, and are less laterally expanded. The paroccipitals rise more to the outside than in *Sus*. The articular surface for the mandible is concave from before backward.

In the skull of a Camel (*Camelus bactrianus*, fig. 318), the occipital condyles are divided into two surfaces meeting at an acute angle, and they come in contact with each other beneath the basi-occipital, which contributes an equal share with the exoccipitals to their formation. The paroccipitals are small, and shorter than the mastoids. The occipital, 11, and parietal, 9, crests are sharp: the zygomatic arches, in relation to the laniariform teeth, *s*, *a*, *o*, are longer and overspan a wider temporal fossa, 10, than in true Ruminants. The orbit has an entire bony rim. The premaxillaries, 1, do not reach the nasals, 7, and the maxillaries, 2, contribute to form the external bony nostril. In the Llama and Vicugna (*Auchenia*), the premaxillaries exclude the maxillaries from the nostril. A vacuity between the maxillary, lacrymal, frontal, and nasal remains large in Llamas, but is reduced in old Camels to a small size, between the frontal, 8, and maxillary, 2; or it may be obliterated, as is usual in the Vicugna. The antorbital foramen, *b*, opens above the last premolar. The orbital plate of the lacrymal

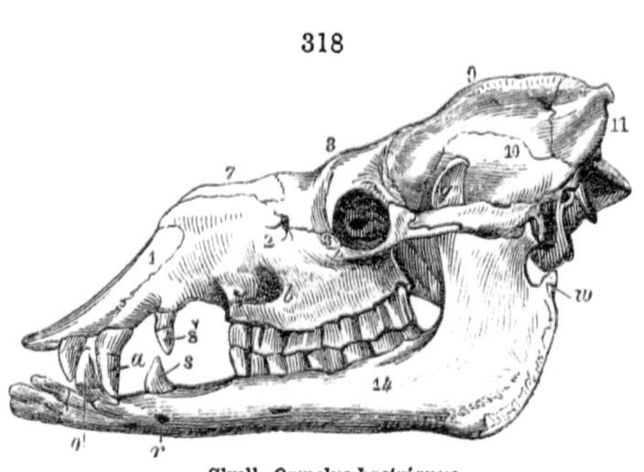

Skull, *Camelus bactrianus*.

shows two perforations. The external pterygoid process is formed by the alisphenoid, the internal one by the true pterygoid; both are far behind the bony palate, which is divided from the last molar alveolus by a notch. The cranial wall in the Camel is unusually thick, with a close cancellous diploë, save where the air-cells penetrate the frontal and presphenoid. There is no bony tentorium. The lateral sinus bifurcates above the petrosal into two wide venous canals. The hinder one again divides, one branch terminating on the superoccipital surface, above the mastoid, the other descending to terminate at the ordinary 'foramen jugulare:' the anterior canal descends to the base of the zygoma, where it also divides, one division opening on the inner and the other on the outer side of the post-glenoid process. In the Llama the venous opening above the root of the zygoma is large: and there is a smaller one at the fore part of the root. The foramen rotundum is blended with the foramen lacerum anterius. The rhinencephalic fossa is narrow but deep. The osseous septum is coextensive with the nasal bones in old Camels. The angle of the mandible, *w*, is singularly elevated, and the contour of the ascending ramus makes a convex sweep to the lower border of the horizontal one. The outlet of the dental canal, *r*, is below the laniariform premolar, *s*. The fore part of the symphysis expands horizontally for the incisor alveoli.

In the true Ruminants the skull is characterised by the small size and edentulous condition of the premaxillaries, the slender zygomatic arches, the entire bony rim of the orbit, the large facial plate of the lacrymal, and by the processes of the frontal bone for the formation of horns or antlers. These latter, however, are wanting in both sexes of the Musk-deer (*Moschus*, *Tragulus*), as in the Camel tribe. The occipital condyles, fig. 319, closely approximate below: the paroccipital is longer than the mastoid. The temporal fossæ, in the formation of which the parietals, 9, take a large share, with the squamosals, 10, are divided above by a parietal crest, and resemble those of the Camel. There is a small vacuity between the frontal, 8, lacrymal, 3, maxillary, 2, and nasal, 7, in *Moschus moschiferus*, which does not exist in *Tragulus*. The malar, 4, is

319

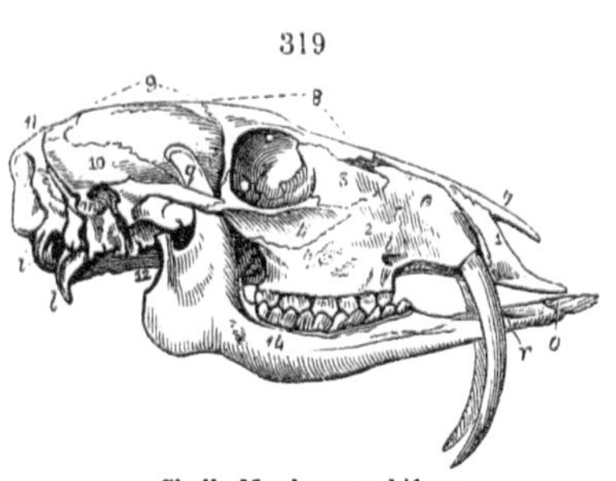

Skull, *Moschus moschiferus*.

marked by a ridge continued from the lower border of the orbit. The petrotympanic forms, in the smaller Musk-deer (*Tragulus*) and Antelopes (*Cephalophus*), a large 'bulla ossea:' and the orbits gain in proportional size as the bulk of the species decreases. The lateral emarginations of the bony palate are usually deeper than the median one, in true Ruminants, the reverse being the case in the Llama and Vicugna. In *Microtherium* the premaxillaries do not reach the nasals, nor yet quite in *Hyæmoschus*.

320

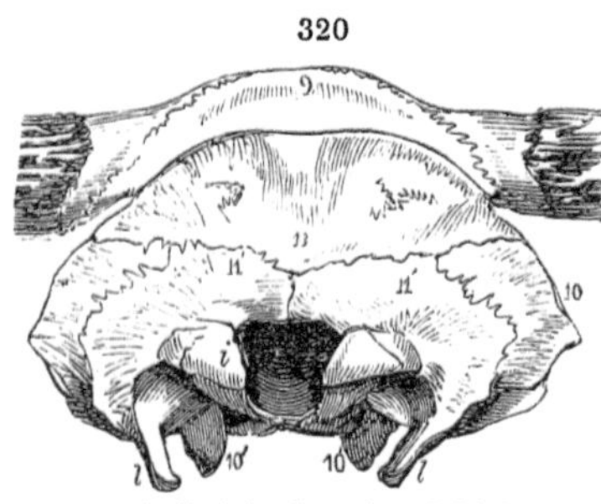

Skull of the Gaur, from behind.

In the skull of the *Bovidæ* I have usually seen that, although the full size and mature dentition have been acquired, the suture between the exoccipitals, fig. 320, 11′, 11′, and that between these and the superoccipital, ib. 11, remain distinct. The occipital condyles, *i*, are wide apart, as in Antelopes and Deer. The paroccipital, l, and fig. 321, *a″*, descends much below the mastoid, 10; the exoccipitals complete the foramen magnum, above: the basioccipital has a pair of tubercles. In the Ox (*Bos taurus*) the whole of the upper surface of the cranium

321

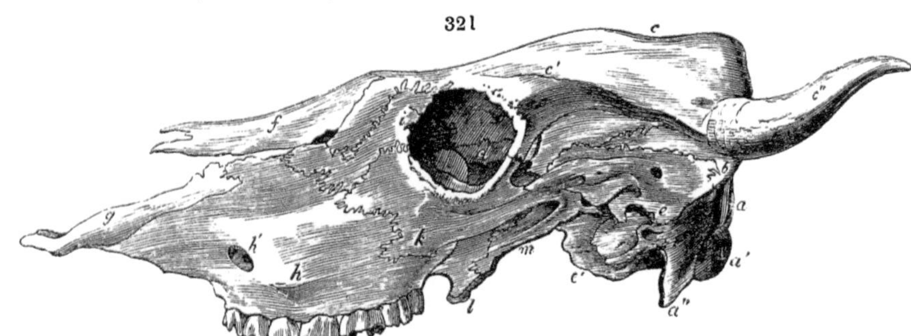

Skull of the Ox (*Bos taurus*).

is formed by the frontals, fig. 321, *c*: the parietals, which, at an earlier period, encroach upon the back part of the upper surface, are now pushed quite to the posterior or occipital aspect. This deposition does not take place in the *Bison*, fig. 320, but the frontals, at the interspace between the horns, are, with the conjoined parietals, 9, developed into a ridge rising above that formed by the superoccipital, 11. The petro-tympanic, fig. 320, 10, 321, *e′*, is prominent and rough. The squamosal, *e*, has a venous outlet above the base of the zygoma. The malar forms the lower part of the orbit and extends largely upon the face, at *k*, to join the maxillary, *h*. The corresponding plate of the lacrymal is still more extensive and here joins the nasal, *f*, leaving a small fissure between those bones and the frontal, *c*. This very extensive bone has a large superorbital fissure. The nasals are cleft at their fore end. The premaxillary, *g*, has but a small or loose junction with the nasals. The maxillary, *h*, is extensive, the antorbital foramen, *h'*, perforates it above the first premolar. In *Bison europæus* the horns arise in advance of the ridge formed by the superoccipital bone, the parietals advancing to the upper surface of the skull and being interposed between the frontal and superoccipital. The Bison differs from the Buffalo (*Bubalus*) in the greater breadth and convexity of the frontal, and in the much greater extent of the orbital processes of that bone, which, with the coextensive processes of the lacrymal and malar, form a prominent cylinder. The nasals are relatively shorter and broader than in the Ox (*Bos*); but the chief distinction between the Bison and the Ox is seen in the shorter premaxillaries, which do not rise to join the nasals: here, therefore, six bones enter into the formation of the external nasal aperture, instead of four, as in *Bos* and *Bubalus*.

The frontal sinuses extend into the horn-cores in all Bovines, but not so in the majority of Antelopes. In this ruminant group, with some exceptions, e. g. *Aigoceros*, *Lyroceros*, *Strepsiceros*, *Dicranoceros*, the facial plate of the lacrymal is impressed, often deeply, by the antorbital cutaneous sac, commonly called 'lacrymal.' In the Duykerbok (*Cephalophus mergens*), the parietals are produced in an angular form between the bases of the horn-cores, which spring as usual from the frontals. In the Chickara (*Tetraceros*) the frontal developes two pairs of horn-cores: and this peculiarity was also manifested by some gigantic Antelopes (*Bramatherium* and *Sivatherium*), now extinct, of the same continent (India): in which, also, the posterior horn-cores were ramified, as in the Prong-horn (*Dicranoceros furcifer*). The Sivathere was also remarkable for the shortness of the facial part of the skull and the termination of the nasals in a down-bent point, fig. 322. In the Duykerbok, Cha-

322

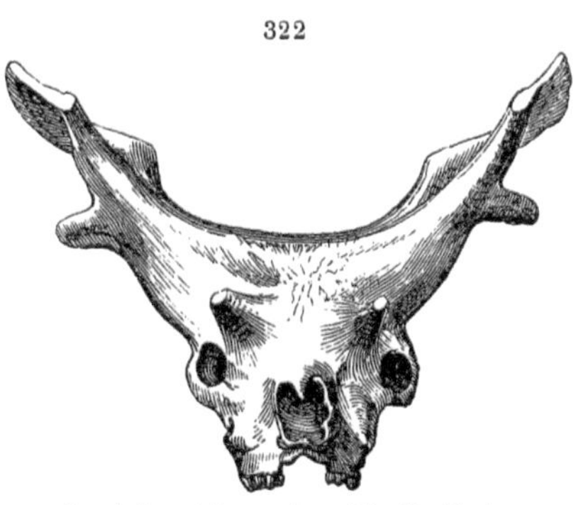

Front view of the cranium of the Sivatherium.

mois, Goral, Saiga, Chiru, and one or two others, the premaxillaries do not join the nasals; but this junction is seen in most Antelopes.

In all, as in the Sheep and its allies, both the superoccipital, fig. 323, 11, and fig. 324, *a*, and the parietals, ib. 9 and *b*, maintain their position at the back part of the vertex: the frontals, ib. 8, 9, and *c*, still form the chief part and alone develope the horn-cores: the nasals, 7, are not expanded posteriorly, as in *Camelidæ*. Both frontals, *c*, and malars, *k*, fig. 324, extend far in advance of the orbit, *d*, but are exceeded in this extension by the lacrymals, *i*, which articulate with the nasals, *f*, for an equal extent with the maxillary, *h*. In the wild *Ovis Ammon* there is a lacrymal pit, and this, in *Ovis Vignei*, deeply impresses the facial plate of the bone. The premaxillaries in the same wild Thibetan sheep join the nasals suturally, but in the domestic *Ovis Aries*, the premaxillaries, *g*, barely touch the nasals. In the Nahura Argali (*Ovis Nahura*), the premaxillaries do not reach the nasals: nor is the lacrymal impressed with the pit. The 'incisive' fissures in the palatal plates of the premaxillaries, figs.

323

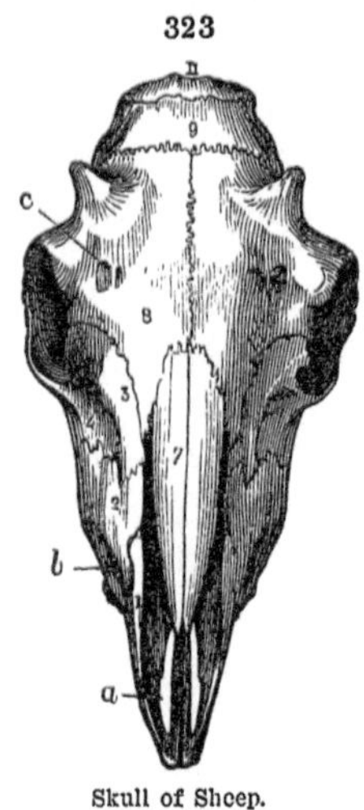

Skull of Sheep.

324

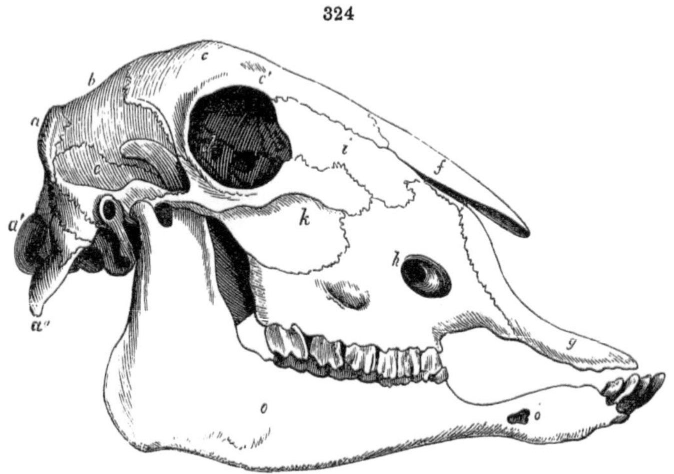

Skull of Sheep, hornless var.

168 and 323, *a*, are long and narrow. The maxillo-palatal sutures, fig. 168, *d*, turn obliquely outward and backward to the inner wall of the socket of the last molar, opposite the hinder half of

which are three posterior palatal notches. The pterygoids form no part of the bony palate.

The following differences may be noticed in comparing the skull of the Goat (*Capra Hircus*) with that of the Sheep (*Ovis Aries*). In the Sheep the postorbital process or plate is broader and more bent outward, forming a deep depression between it and the origin of the horn; it also turns the plane of the orbit more obliquely forward: in the Goat the aspect of this plane is more directly outward. The occiput is higher in proportion to its breadth in the Goat than in the Sheep. The petrosal is relatively longer and deeper in the Goat than in the Sheep. The nasals are relatively smaller in the Goat, where they are shorter than the premaxillaries; their upper surface is concave lengthwise, except at the free points, where they are slightly bent down. In the Sheep the nasals are relatively larger, are longer than the premaxillaries, and their whole upper surface is convex lengthwise. There are also differences in the connections of these bones; in the Sheep the nasals join the lacrymals, rarely the premaxillaries, whilst in the Goat they join the premaxillaries but not the lacrymals,—a vacuity, which is not present or is rudimental in the Sheep, separating them from the lacrymals. The upper border of the maxillary bone is relatively shorter in the Goat, and the anterior border is not notched to receive the upper end of the premaxillary, as it is in the Sheep. The premaxillary is narrower at its alveolar end in the Goat, and its upper end rises so as to overlap the side of the nasal: in the Sheep the premaxillary is relatively broader, and rarely rises to touch the nasal. The lacrymal bone of the Goat is shorter in proportion to its breadth, and is not impressed on its facial surface by a lacrymal fossa; it does not touch the nasal: in the Sheep the lacrymal is longer in proportion to its breadth, and is more regularly quadrate in form; it joins the nasal, and thus obliterates that vacuity which is present in the skull of the Goat; its facial plate is usually impressed by a concavity for the cutaneous lacrymal pit. In comparing the upper contour of the skull, from the occipital ridge to the free extremity of the nasal bones, it forms, in the Goat, nearly a right angle, with the two sides equal: in the Sheep it forms a more open angle, with the anterior side twice as long as the posterior one.

In the skull of the Giraffe (*Camelopardalis Giraffa*, figs. 325, 326), the exoccipitals form a marked protuberance above the foramen magnum and below a deep fossa for the implantation of the ligamentum nuchæ. The parietals are chiefly situated on the upper surface of the skull; the osseous horn-cores are originally distinct, with their bases crossing the coronal suture, and resting equally upon the parietals and frontals: they, however, coalesce therewith in old males, and the frontal and parietal sinuses extend into the lower fourth, the rest of the horn-core being a solid and dense bone. The protuberance upon the frontal and contiguous parts of the nasal bones is due to an enlargement of those bones (as obvious in the section, fig. 326), and not to any distinct osseous part: its surface is roughened by vascular impressions, undermining the basal periphery and simulating a suture. The lacrymal is separated from the nasal by a large vacuity intervening between those bones, the frontal and the maxillary. The premaxillaries, which are of unusual length, articulate with the nasals. The petro-tympanic is a separate bone. The symphysis of the lower jaw is unusually long and slender. The articular surface of the prominent occipital condyles is so extended vertically as to admit of the head being raised into a line with the neck, and

325

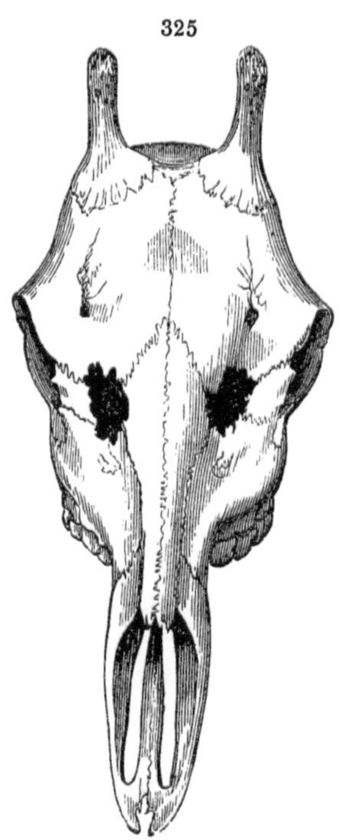

Skull of female Giraffe.

326

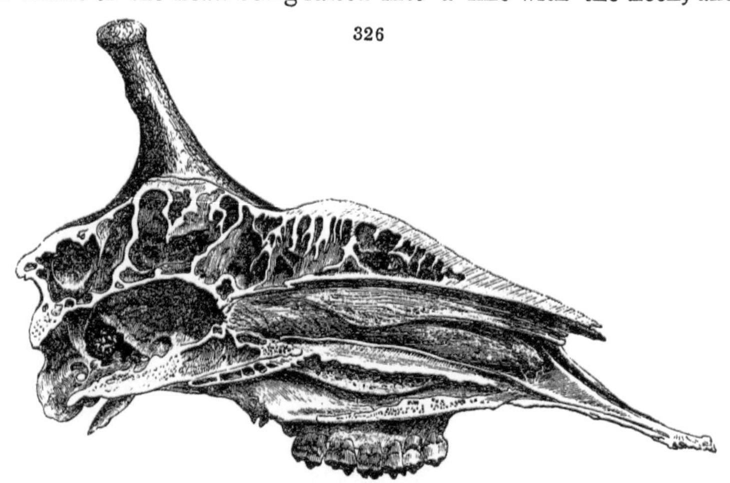

Section of Skull of male Giraffe.

even slightly bent back beyond that line. The great freedom given to the movements of the head relate, like the length of neck and general altitude of the body, to the culling of leaves from the trees browzed on by the Giraffe. The part of the skull to which the elastic ligament is attached is raised considerably above the roof of the cranial cavity by the extension backward of large sinuses, or air-cells, as far as the occiput, fig. 326. The sinuses commence above the middle of the nasal cavity, and increase in

327

Skull of Barbary Deer (*Cervus elaphus*).

depth and width to beneath the base of the horns, where their vertical extent equals that of the cerebral cavity itself. The exterior table of the skull, thus widely separated from the vitreous table, is supported by stout bony partitions, extended chiefly in the transverse direction, and with an oblique and wavy course.

Two of the most remarkable of these bony walls are placed at the front and back part of the base of the horns, intercepting a large sinus immediately over the middle of the cranial cavity, and from a third and larger one behind. The pre-sphenoidal sinuses are of a large size.[1]

The chief peculiarity in the skull of the Deer-tribe is the annual development, from the frontals, of the solid deciduous exostoses which serve as weapons (fig. 326, *d*, *b*) during a portion of the year, in the males of all kinds and in both sexes of the Rein-deer. Most species likewise show vacuities between the frontal, 8, nasal, 7, maxillary, 2, and lacrymal, as in figs. 327 and 328. The base of the zygoma is perforated by a vein from the lateral sinus.

328

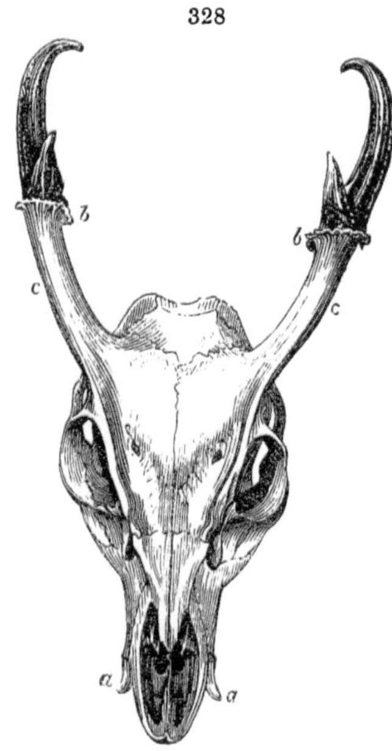

Skull, *Cervus Muntjak.*

The chief peculiarity of the skull of the Elk (*Alces*) is seen in the great length of the premaxillaries and of the edentulous portion of the maxillaries, and in the shortness and breadth of the nasal bones, which do not join the premaxillaries. The vomer is carinate beneath.

In the Rein-deer (*Tarandus*) the antlers spring from within an inch of the superoccipital crest, and the frontal bones are proportionably extended backward on each side of the parietal, in which the sagittal suture becomes obliterated: the frontal suture is persistent, and is complex in its dentations at its posterior half. The large lacrymal presents two canals upon its orbital border and a deep oblong depression on its facial surface, above which is the vacuity leading to the olfactory chamber. The premaxillaries do not join the nasals. In the Fallow-deer (*Dama*) the frontal bones do not extend so far back as in the Rein-deer, and the antlers, in consequence, rise at a greater distance from the occipital crest. The lacrymal bone has two perforations at its outer border, and its facial plate is nearly equally divided into an upper convex and a lower concave surface. The antorbital depressions show but a small perforation, if any.

[1] xcvii'. p. 235.

The skull of the Barking-deer (*Cervus Muntjak*) is remarkable for the great length of the persistent pedicles (fig. 328, *c, c*) which support the antlers, and which are continued from two strong ridges that traverse the outer side of the frontal bone from its junction with the nasals. The lacrymal presents a deep and well-marked fossa, anterior to which is the antorbital vacuity. The sockets of the upper canines are largely developed in the maxillaries.

In all Ruminants, and especially the horned kinds, the temporal fossæ are small, the zygomatic arches weak, the coronoid processes of the mandible fig. 319, *g*, narrow, the base of the ascending ramus expanded; in short, the attachments of the biting muscle are restricted, those of the chewing muscle expanded. That for the masseter is shown by the ridge and fossa continued forward from the zygoma below the orbit: that for the 'pterygoidei' by the backwardly produced and rounded angle of the lower jaw. The exceptions to the edentulous premaxillaries have been noted. The articular surface for the mandible is broad, slightly convex, with a posterior semicircular channel bounded by a ridge.

329

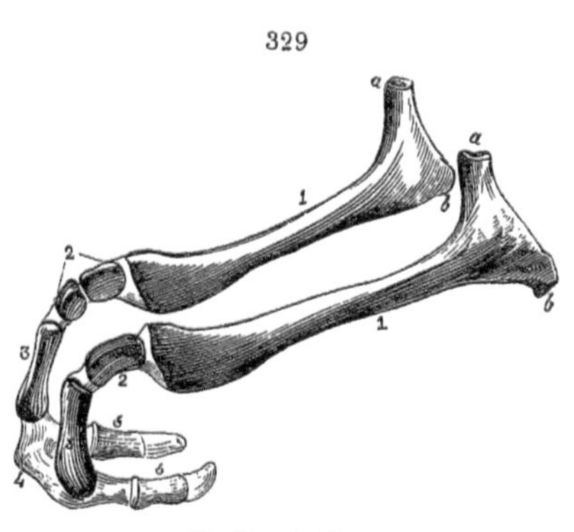

Hyoid arch, Sheep.

The hyoid arch includes long, compressed, hammer-shaped 'stylohyals,' fig. 329, 1, having at their promixal end the articular, *a*, and muscular, *b*, processes, the short 'epihyals,' 2, the ceratohyals, 3, the basihyal, 4, and thyrohyals, 5; attached to posterior angular processes of the basihyal.

C. *Bones of the Limbs.*—Artiodactyles have the limbs terminated by feet of 4 or 2 toes, in symmetrical pairs: but, as in other Ungulates, almost restricted to locomotive functions. The Hippopotamus and the Gazelle manifest in the even-toed series analogous extremes in the proportions of the limbs, as do the Rhinoceros and Horse in the Perissodactyles. The blade-bone is long and narrow; but the spine is more commonly produced into an acromial angle in the Artiodactyles. In the Hippopotamus, fig. 305, this angle is slightly produced: the coracoid is recurved. The greater tuberosity of the humerus is divided into two subequal processes, the inner one separated by a deep and wide bicipital fossa from the lower inner tuberosity. The ulna and radius have coalesced at their extremities and at the middle of their shaft, the interosseous space being indicated by a deep groove and two foramina. The trapezium does not support any digit: of the other four, the two middle ones, answering to the third and fourth, are most developed.

In the pelvis the ilia expand and bend outward from their sacral attachments almost into the same plane with the broad and flat sacrum: the lumbo-iliac angle is about 150°. The ischia, fig. 305, 63, are long and with the dorsal angles of the broad and thick tuberosities produced toward the caudal vertebræ, as in other Artiodactyles, figs. 308 and 310. The ischio-pubic symphysis is long and more backward than in the Rhinoceros; the obturator vacuities are large; the acetabula look downward and outward, their planes being about 50° from the perpendicular. The femur has a straight subcylindrical shaft. The canal for the medullary artery commences at the upper and fore part of the shaft. The fibula is distinct from the tibia, and extends from its proximal end to the calcaneum. The internal cuneiforme is present in the tarsus, but there is no rudiment of the innermost toe: the proportions of the other four resemble those of the fore-foot: the bones of the hind foot are noted at pp. 308, 309, and figured in cut 193, 'Hippopotamus.'

In the Wild Boar (*Sus scrofa*) the spine of the scapula is most developed at its middle, where it is bent back: there is no acromion. The coracoid is a low tubercle: the glenoid cavity is nearly circular. The humerus has an intercondyloid vacuity, as in the Peccari; in which the inner division of the great tuberosity rises above the head of the bone, higher than in *Sus*. The radius and ulna are distinct in *Sus*, but invariably connected by a rough longitudinally grooved surface. The olecranon is large and compressed: the distal end of the ulna presents a small trochlear surface for the carpus and a narrow strip for the radius. In the Peccari the radius and ulna coalesce throughout nearly their whole extent. The trapezium and pollex are not present: the 'index' and 'miminus' digits are small; the 'medius' and 'annularis' large, and chiefly serviceable in progression.

The pelvis is longer and narrower, relatively, in *Suidæ* than in the Hippopotamus: the lumbo-iliac angle is 145°, the ilio-pubic angle 120°. The medullary artery of the femur enters the fore part of its upper third and the canal slopes downward. The tibia and fibila are distinct, and the latter fully developed in both *Sus* and *Dicotyles*. In both, the symmetrical pair, which are most developed and chiefly serviceable in progression, answer to the third and fourth digits of the pentadactyle foot: but in *Sus* the homologues of the fifth and second are present; whilst in *Dicotyles* the fifth as well as the first toe are wanting in the hind foot: in this the second toe is small; the third and fourth are very large, and form a symmetrical pair, showing that the Artiodactyle structure essentially prevails, although the toes, by the non-development of the fifth, are, exceptionally, reduced to three in number in the hind foot of the Peccari.

In the *Camelidæ*, fig. 306, the scapula though longer than in the non-ruminant Artiodactyles, is broader, relatively, than in horned Ruminants: its spine is produced into a short pointed acromion: the coracoid is grooved below, or sub-bifid. The humerus is weaker than in the Ox, stronger than in the Deer, longer relatively to the rest of the limb than in the Giraffe: the great tuberosity does not rise above the head: the ridge upon the outer condyle is less marked. The ulna has coalesced with the radius, and appears to be represented only by its proximal and distal extremities. The carpal bones have the same number and arrangement as in ordinary ruminants, but the pisiforme is proportionally larger. There is no trace of the digits answering to the first, second, and fifth in the pentadactyle foot: the metacarpals of those answering to the third and fourth have coalesced to near their distal extremities, which diverge more than in the ruminants, giving a greater spread to the foot, which is supported by the three phalanges of each of those digits. The last phalanx deviates most from the ordinary form, by its smaller proportional size, rougher surface, and less regular shape: it supports, in fact, a modified claw rather than a hoof. The ilium, in proportion to the ischium, is longer than in the Hippopotamus. In the femur, the chief deviation from the ordinary Ruminant type is seen in the position of the orifice of the canal for the medullary artery, which enters the back part of the middle of the shaft, and inclines obliquely upward. The fibula is represented by the irregularly-shaped ossicle interlocked between the outer side of the distal end of the tibia and the calcaneum. The scaphoid is not confluent with the cuboid as in the normal Ruminant: the rest of the hind-foot deviates in the same manner and degree from that type, as does the fore-foot. In both metacarpals and metatarsals, notwithstanding the intimate blending of the two bones apparent externally, their medullary cavities are distinct: the canal of the medullary artery enters the back part of each, above the middle, and ascends obliquely to its respective cavity.

In true Ruminants the spine of the scapula is not produced, as in *Camelidæ*, but terminates in an acute (*Bos*) or a right angle (*Cervus*): the Musks and Chevrotains agree with the horned families in this character, but the coracoid is a better defined process in the latter: in all, the scapula is a long slender triangle, with two equal or subequal sides, the infraspinal division chiefly expanding to the base, which is truncate in *Bos*, fig. 309, *Antilope* and *Cervus*, fig. 311; but rounded off at the hinder angle in *Camelopardalis*, fig. 310: in this Ruminant the cervix scapulæ is unusually long. The *humerus*, fig. 330, A, is short, but strong, with slightly expanded ends: the outer tuberosity, at the proximal one, rises above the head of the bone, and bounds, with the inner tuberosity, a deep bicipital groove: the deltoid crest, 1, is less prominent than in the Horse. The distal articular end presents three prominences answering to the hollows of the head of the radius, the internal one being the broadest and lowest. The supracondylar ridges are but little produced: the olecranal fossa is deep, and perforated in Musk-deer, Chevrotains, and Microtheres, as in the Hog-tribe. In the Gnu (*Antilope Gnu*) the humerus is as long as the metacarpus: in the Ox, fig. 309, it is longer; in the Giraffe, fig. 310, and Gazelle, fig. 330, A, it is shorter. The *radius*, fig. 330, A, 2, is the chief bone of the antibrachium: its proximal trochlear surface offers three eminences and as many depressions to the humerus, restricting the movements of the fore-leg to one plane. The shaft is slightly bent forward: the distal end is moulded to the irregularities of the carpus, and is most impressed by the scaphoid, especially

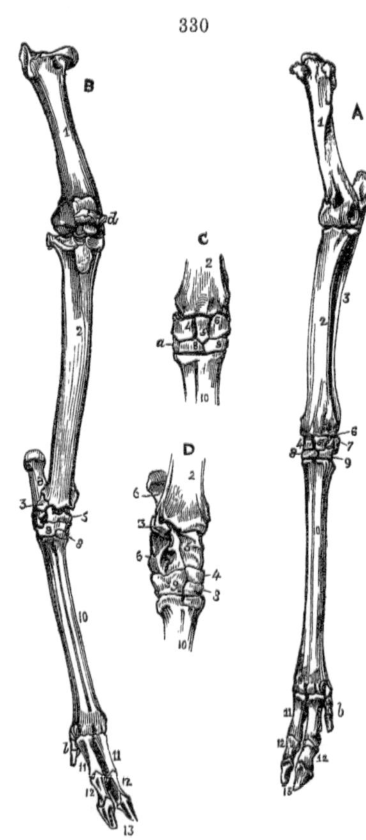

Bones of the fore and hind limbs of the Indian Antelope. XCVI'.[1]

[1] XLIV. p. 605, nos. 3672 and 3673.

in the Chevrotains. The chief part of the ulna, ib. 3, is its compressed olecranon: the slender shaft may be continued to the carpus, as in *Moschidæ*, most Antelopes, Sheep, the Elk, the Rein-deer, fig. 311, the Fallow-deer, and the common Ox. In the Chevrotains it longest maintains its individuality: in the Musk-deer and Elk the distal extremity coalesces with that of the radius; in the Rein-deer the shaft, also; in the Ox this is so confluent as to be hardly traceable from the olecranon to the styloid extremity. In the Giraffe the ulnar shaft is interrupted at its lower third, but the distal end reappears, as the 'styloid process,' but is connate with the distal epiphysis of the radius. The radius and ulna are so interlocked that the fore-foot is kept 'prone,' or with the surface answering to 'palm' turned back and downward: there is a narrow cleft at the upper part of their line of union, and sometimes a second lower down. In the carpus the usual four bones of the proximal row remain distinct: in fig. 330, A, C, 4 is 'scaphoides,' 5 lunare, 6 cuneiforme, 7 pisiforme: the distal row consists of the 'trapezoides,' *a*, in some, and in all of the 'magnum,' 8, supporting the moiety of the metacarpal answering to the 'third' one of the pentadactyle foot, and the unciforme, 9, supporting the moiety answering to the 'fourth' metacarpal. These metacarpals early coalesce into a single 'cannon-bone:' but a longitudinal section, as in fig. 331,[1] shows the medullary canal of each distinct, in *Megaceros*, as in most Ruminants; in a few, e.g. the Yak (*Bos grunniens*) the septum becomes partially absorbed.[2] Longitudinal grooves at the fore (fig. 330, A, 10) and back parts of the cannon-bone, with antero-posterior perforations, are the outward signs of the original separation: they are most strongly marked in the Chevrotains (*Tragulus*); and the severance persists in the Water-Musk (*Hyæmoschus*) as in the extinct Dichodons, Anoplotheres, and Microtheres. Each moiety of the cannon-bone has its distinct distal trochlea, fig. 331, *a*, *b*, which is traversed by a median ridge, *c*, from before backward. To each trochlea articulates a proximal phalanx, fig. 330, A, 11, supporting a middle, 12, and this an unequal phalanx, 13, of a triedral conical shape, modified to be sheathed in a hoof; the unsymmetry of each hoof being such as to form a symmetrical pair. They resemble the single hoof of the horse cleft in twain: whence the

331

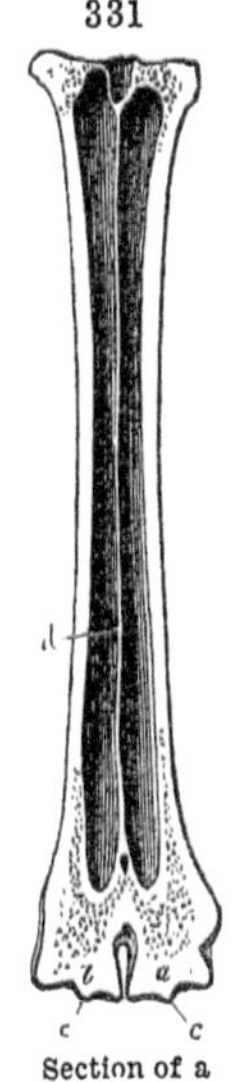

Section of a metacarpal of *Megaceros*. XCVI'.

[1] XCII'. p. 260, no. 1162. [2] Ib. vol. ii. p. 628, no. 3852.

Ruminants are said to 'divide the hoof.' In the Giraffe, fig. 310, Antelopes, fig. 330, and Deer, fig. 311, the proximal phalanx is longer than the next: in the Ox and Musk-deer the difference is small: in the Chevrotains, they are more nearly of the same length. In the Giraffe, as in the Camel-tribe, there is no trace of other toes: in most true Ruminants stunted portions of them are suspended to the back part of the distal end of the cannon-bone, whence dangle the pair of 'spurious hoofs,' fig. 330, *b*. In the Bison the bones of these consist in each of the middle and distal phalanges: and there is a styliform representative of the proximal end of their respective metacarpals articulated, in the fore-foot, one to the connate trapezoid, the other to the unciform and cuneiform bones.[1] In Deer the spurious hoofs are supported by the three phalanges proper to the second and fifth digits, and by a styliform distal end of their respective metacarpals with the point upward: these hooflets are large enough in the Rein-deer, fig.311, to usefully increase the base of the 'snow-shoe,' formed by its broad hairy and horny foot, with the advantage of their collapse as the foot is withdrawn. The *Moschus moschiferus* has a similar bony structure of the second and fifth digits; while the still smaller Chevrotains, like the embryos of larger Ruminants, show so much more of the generalised foot-structure as is exemplified by the extension of the slender metacarpals of those 'spurious hoofs' from them to the carpus.

The os innominatum is elongate with the iliac portion concave lengthwise, convex across, externally, with the expanded anterior end divided by a ridge into the portions *b* and *c*, fig. 332, articulating with the sacrum, *a*, and rising as high as, or above, the sacral spine: the portion, *c*, is thickest and broadest in the heavier Ruminants: the ilium joins the spine at an angle of about 145°. The ischium extends back from the acetabulum, two-thirds or three-fourths the length of the ilium, with the tuberosity, *e*, bending upward: the tuberosity is strengthened in Deer, Antelopes, and Oxen by a ridge, *g*. In the male Chevrotains the ischia join the elongated sacrum by ossifications of the sacrosciatic ligaments, but in the females these retain their normal extensile texture. The tendons and aponeuroses of the dorso-spinal muscles become more or less ossified by age, and a thin roof of bone may thus overarch the pelvis, as e.g. in *Tragulus javanicus*, *Tr. Kanehil*,[2] &c. The pubics, *f*, are slender: they converge to the symphysis at an iliopubic angle of about 135°. The iliopectineal spine is well marked in some

[1] CXL. p. 31, fig. 5. [2] LXXII'. p. 581, no. 3498.

Deer. In the Ox, fig. 332, the symphysis pubis is placed obliquely, so as to cause the anterior pelvic opening to be longer than the posterior one: in the Deer, fig. 333, the symphysis runs more parallel with the sacrum. The acetabula are carried by the length of the ilia opposite the last sacral vertebra, at the apex of the wide arch of the os innominatum: the plane of the aperture is inclined about 40° from the perpendicular. In the Cow, near the period of parturition, the ischial tuberosities, by the relaxation of the ligaments and sinking of the sacrum, become more protuberant than at other times. In the Giraffe the posterior concavity between the ilium and ischium, as in the Deer, fig. 333, is scarcely interrupted by the prominence of the conjoined bones above the acetabulum. The Harderian groove of the acetabulum is wide and deep, and breaks through the border of that cup.

332

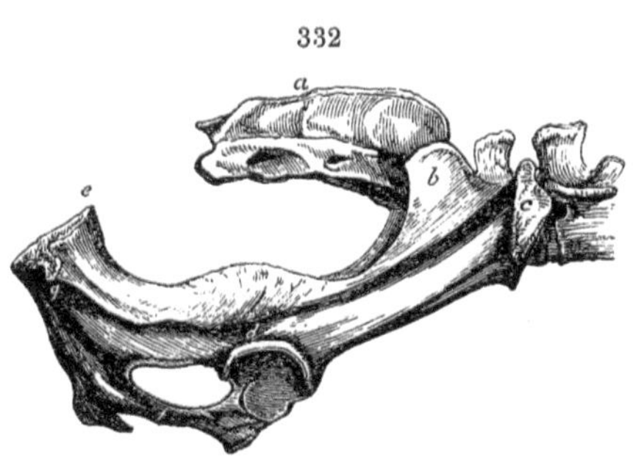

Pelvis of Ox (*Bos*).

The hind-limb, fig. 330, B, exceeds the fore-limb, ib. A, in length in all Ruminants: least so in the *Camelidæ* and Giraffe, most so in the bounding Deer and Antelopes.

333

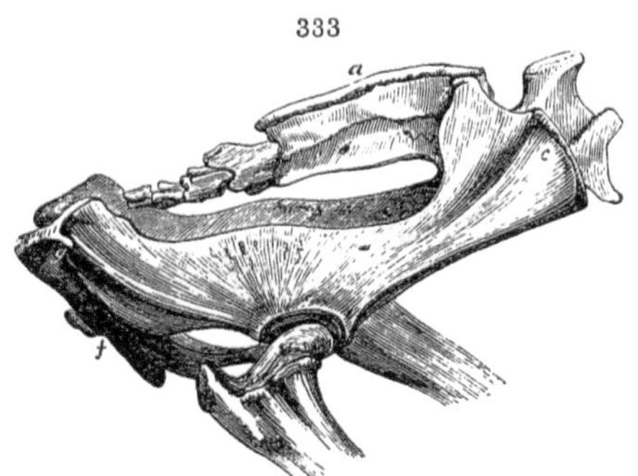

Pelvis of Stag (*Cervus*).

The femur of Ruminants, ib. B, 1, as of other Artiodactyles, has no third trochanter; and the medullary artery enters the fore part of the shaft, usually at the upper third, and goes downward and backward to the medullary cavity: the antero-posterior expanse of the distal end is great, especially in the Giraffe: and the inner border of the rotular channel is more produced than in the Hog-tribe, without developing an irregular prominence as in the Perissodactyles. The *Camelidæ* offer the exception in the position of the foramen and canal of the medullary artery, and in the subequal development of the borders of the rotular channel. The

Moschidæ agree in both characters with the horned Ruminants; but the rotular channel is unusually long and narrow in the Chevrotains. The head of the femur turns but little out of the longitudinal axis of the shaft, of which the great trochanter is the highest part: the articular surface extends from the head toward the base of the trochanter, in a subtrochlear form: the ligamentous pit is in the middle of the hemispheric part of the head. The back part of the proximal end is flattened, especially in *Camelidæ*: the trochanterian fossa is deep, that above the outer condyle is shallow; its outer border is continued from the 'linea aspera,' and is rough: there is a rough eminence at a similar distance above the inner condyle.

The *tibia*, ib. B, 2, like its homotype the radius, is the chief bone of its segment; it is also the longest bone of the hind limb. The proximal end is most expanded: the two articular surfaces rise at the middle of the head, and that of the inner surface higher in *Bovidæ* than in *Cervidæ*: there is a tuberosity at their posterior interspace: the broader base of the anterior spine fills up the anterior one, but is separated from the outer condylar surface by a deep groove: the spine projects from the fore part of the upper fourth of the shaft, inclining outward: it subsides toward the inner side of the shaft, the back part of which is flattened and marked by longitudinal intermuscular ridges. The distal articular surface is subquadrate, with a notch and articular fossa on its outer side for the ossicle representing the distal or 'malleolar' extremity of the fibula,[1] fig. 193, 67, (Ox); this coalesces with the tibia in Chevrotains. In the Rein-deer, as in *Moschus moschiferus* and *M. aquaticus*, the proximal end of the fibula projects downward as a short styliform process from the outer part of the head of the fibula: in *Tragulus Napu* it extends more than half-way down the tibia,[2] and the shaft exists as a sclerous tissue in all Ruminants. The distal articular surface of the tibia presents a pair of deep antero-posterior channels divided by a transversely convex tract of equal extent: they are less oblique than in the Horse-tribe: the inner malleolus descends the lowest.

The tarsus, fig. 193, 'Ox,' consists of astragalus, *a*, calcaneum, *cl*, scapho-cuboid, *s-b*, and ectocuneiform, *ec*: the confluence of the naviculare or scaphoid with the cuboid does not take place in *Camelidæ*: it does so in *Moschidæ*, and the fusion extends to the ectocuneiform in *Tragulus*.[3] A mesocuneiform is present in

[1] CLI. tom. IV. p. 18. [2] XLIV. p. 580, no. 3495.
[3] Ib. and LXXII'. p. 59.

Chevrotains for the support of the slender metatarsal of the toe answering to the second of pentadactyle feet. The corresponding metatarsal, fig. 334, *c*, of the fifth or outermost toe, ib. *b*, articulates with the cuboid, 9. In *Moschus aquaticus*, the second and fifth metatarsals coexist with an almost complete severance of the third and fourth, which, in a state of confluence, represent the metatarsal segment in other Ruminants. The true Musk-deer and most horned Ruminants have the distal ends of the second and fifth metatarsals ossified, and supporting the small digits terminated by the 'spurious hoofs,' fig. 193, II, V: in the Giraffe and Camel tribe these are wholly absent, as are their homotypes in the fore-foot. The digital phalanges of the hind-foot, fig. 330, B, 11, 12, 13, *b*, closely correspond with those of the fore-foot.

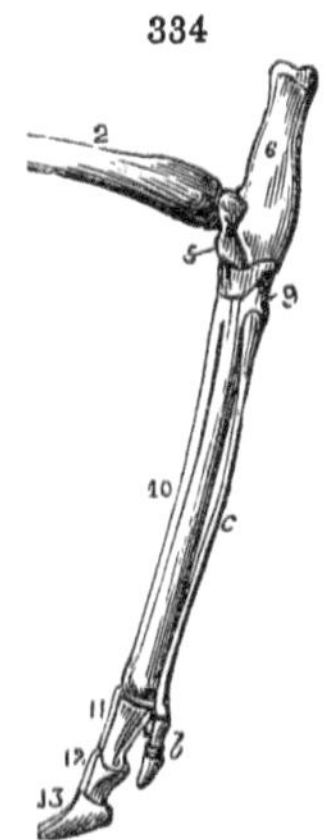

Bones of hind-foot of Chevrotain (*Tragulus*). XCVI'.

In all Ungulates the encasing of the end of the digit in a hoof is accompanied by a junction of the radius and ulna such as to prevent reciprocal rotation of those bones on each other, and by a joint with the humerus restricting the movements of the antibrachium to flexion and extension in one plane. The expansions of the humerus for attachment of pronator and supinator muscles are uncalled for; while the proximal processes giving leverage to the permitted motors of the limb may project in a degree which would impede its more varied and freer motions. The length of the blade bone and of muscles arising from it is increased at the expense of their breadth; the acromion is stunted, and clavicles are absent. The concomitant modifications of the skull and jaws in relation to masticating vegetable food are best exemplified in the Ruminant Ungulates, and have been specified at p. 471.

§ 189. *Skeleton of Carnivora.*—In the Unguiculate *Gyrencephala* the whole frame is modified, in degrees corresponding with the perfection of the claws as prehensile weapons, for mastery and destruction of other animals. The mandible, fig. 341, 32, is short and strong, it is articulated by a close-fitting joint to the skull almost restricting its movements to one plane, as in opening and closing the mouth, for biting, and for dividing, not pounding the food. The coronoid process giving insertion to the temporal muscles is broad and high; the fossæ from which they rise are large and deep, and augmented by peripheral ridges of bone. The zygomatic arch spans across the muscle, bending outward to give space for its passage, and arching upward for the more extensive and favourable attachment of accessory fascicles. Each toe has its distinct metacarpal or metatarsal. The digits, especially in the fore-limb, enjoy freedom of motion and power of reciprocal approximation and divarication; the terminal phalanx is compressed and deep, with a plate of bone reflected forward from the basal periphery, beyond which the apex of the phalanx projects like a peg from a sheath: the claw is fixed upon the peg, its base being firmly wedged into the interspace between the peg and the sheath. In the Felines, which are the most perfect carnivorous Unguiculates, the claw phalanx is retractile. The fore-paw, so armed, is attached to the radius and ulna, which are entire, distinct, and strong bones: these articulate with the humerus by a joint, which, although well knit, allows both freedom of motion in bending and extending, and also a reciprocal play of the two bones, the radius rotating on the ulna, and carrying with it, by the greater expanse of its lower end, the paw, which can thus be turned 'prone' or 'supine,' whereby its efficacy as an instrument for seizing and tearing is enhanced. The humerus has strong ridges from the outer and inner sides above the condyles for extending the origins of the muscles of the paw; and, to defend the main nerve and artery of the fore-leg from compression during the action of these muscles, a bridge of bone spans across them in the feline, and some other *Carnivora*. The upper end of the humerus has a long and strong deltoid ridge; but the tuberosities do not project beyond the round head of the bone so as to impede its movements in the socket. The scapula is of great breadth, with well-developed spine, acromion, and coracoid. A small clavicular bone is interposed in most *Carnivora* between a muscle of the head and one of the arm, giving additional force and determination of action to them. Such are the chief modifications of the framework of the unguiculate as contrasted with the ungulate *Gyrencephala*.

A. *Vertebral Column.*—This part of the skeleton of *Carnivora* is modified in relation to the medium of life, degree of carnivority, and modes of motion of the species. In no Carnivore do cervical vertebræ articulate by ball-and-socket joints; and in all, the seventh has the transverse processes imperforate, consisting only of diapophyses. The Harp Seal (*Phoca grœnlandica*, fig. 332) has 15 dorsal, D, 5 lumbar, L, 4 sacral, S, and 8 caudal. Ten pairs of ribs directly join the sternum, which consists of eight bones: the manubrium, 52′, is much produced, for extending the fore-and-aft origins of the pectoral muscles. The neural arches of the middle dorsal vertebræ are slender, leaving wide intervals of the neural canal. The bones of the neck are modified to allow of great extent and freedom of inflection. The perforated transverse processes of the third to the sixth cervicals inclusive are remarkable for the distinctness of their di- and pleurapophysial parts. Metapophyses are developed on the last five dorsal vertebræ: the strong hypapophysial ridge of the lumbar vertebræ divides into two tuberous processes. These processes indicate the great developement of the anterior vertebral muscles, e.g. the 'longi colli' and 'psoæ,' and relate to the important share

335

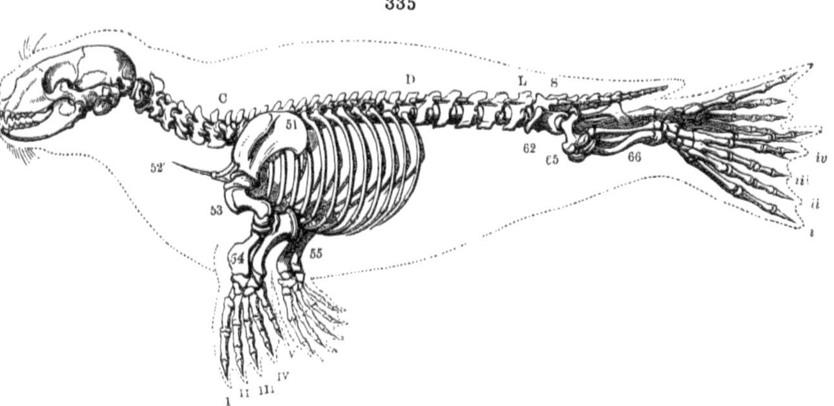

Harp Seal (*Phoca grœnlandica*). LXXIII'.

which the vertebræ and muscles of the trunk take in the locomotion of the Seal-tribe, especially when on dry land, where they shuffle along on their belly.

In the Sterrink or Saw-toothed Seal (*Stenorhynchus serridens*), with 15 dorsal, 5 lumbar, 3 sacral, and 11 caudal, the metapophyses commence as tubercles outside the prezygapophysis on the second dorsal, are distinct on the third dorsal, pass on the fore part of the diapophysis in the fourth, and continue rudimental as far as the tenth dorsal, on which they are well and distinctly developed; they again pass upon the outside of the prezygapophysis in the eleventh and twelfth dorsals, and so continue throughout the lumbar, sacral, and anterior caudal vertebræ. The anapophyses are mere rudimental projections from the back part of the diapophysis. The transverse processes of the axis are more developed than in the *Phoca grœnlandica*; they show as distinctly as in the other cervicals, but on a smaller scale, the pleur- and di-apophysial parts of the process. The cervical and anterior dorsal vertebræ have a hypapophysial ridge, which, in the latter, is produced into a tuberosity: the lumbar vertebræ are characterized by a pair of hypapophyses from near the hinder end of the centrum. The eared Seals have the same vertebral formula: the anterior sacral vertebræ are narrow.

The Walrus (*Trichecus Rosmarus*), like the Otter, has 14 dorsal and 6 lumbar vertebræ. Nine pairs of ribs directly join the sternum, which consists of eight bones. The anterior sacrals have greater relative breadth than in *Phoca* or *Otaria*: as in true Seals the tail is short, with 9 or 10 vertebræ.

In the Bear-tribe, as in the Seal-tribe, the number of true vertebræ is 27, as a rule, and 14 of these usually bear movable

336

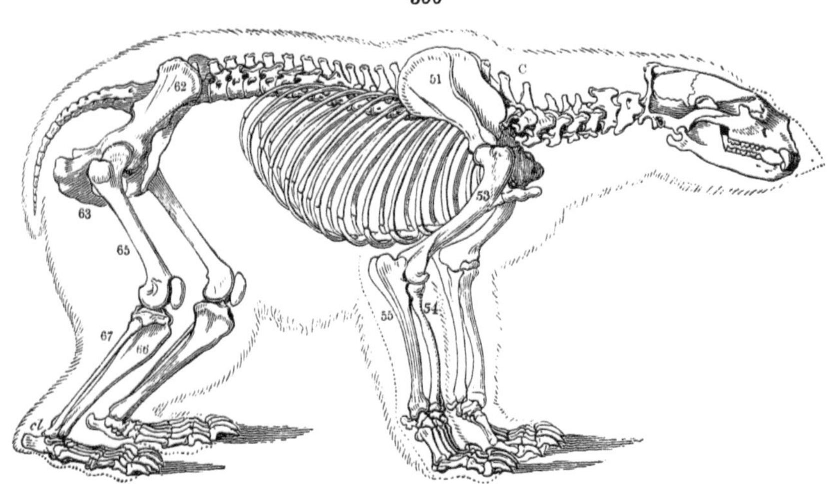

Polar Bear (*Ursus maritimus*). LXXIII'.

ribs; but I have seen 15 'dorsal' vertebræ in *Ursus maritimus*, fig. 333, and in *U. labiatus*, the latter having 5 lumbar instead of 6, which is the common number. Nine pairs of ribs articulate directly with the sternum, which consists of eight bones, with a xiphoid appendage. The manubrium is truncate anteriorly. The number of anchylosed sacral vertebræ may vary from 5 to 7, that of the caudal vertebræ rarely exceeds 10. The met- and an-apophyses are distinct on the twelfth dorsal, diverge and increase on the succeeding dorsals, the metapophyses continuing throughout the lumbar series; the anapophyses, after underlapping the pro-zygapophyses of the first and second lumbar, rapidly subside. The neural spines are better developed than in the Seal-tribe, but do not indicate a centre of motion, save in the smaller and more active subursines. In the Racoon they converge to the twelfth dorsal: the caudal vertebræ, 16 in this plantigrade and 18 in the Badger, increase to 31 in number in the Kinkajou, where the tail is prehensile, whence the name *Cercoleptes caudivolvulus.* The Benturong (*Arctictis*) has a similar caudal developement, with hæmal arches on the ten anterior vertebræ. In the last five cervical vertebræ of the Ratel the neural arches are longer than the centrums and overlap each other in an imbricated manner, giving great strength to the articulations of this part of the vertebral column. The number of dorsal vertebræ is 15, as in *Mydaus* and *Meles,* with 5 lumbar ones.

In the more digitigrade Mustelines, the Sable (*Mustela zibellina*) has 14 dorsal, 6 lumbar, 3 sacral, and 18 caudal. The eleventh dorsal vertebra is that toward which the spines of the other trunk-vertebræ converge. The anapophyses begin to be developed upon the ninth dorsal, and are continued to the penultimate lumbar vertebra. Ten pairs of ribs directly join the sternum, which consists of nine bones, with a xiphoid cartilage. In the Ermine (*Putorius ermineus*), with a similar vertebral formula, the spines of the tenth and eleventh dorsal vertebræ converge towards each other and almost meet, indicating the centre of motion of the trunk. Ten pairs of ribs directly join the sternum. The neck is strengthened by the overlapping of the costal parts of the transverse processes of the third to the sixth cervical vertebræ. Some of the anterior caudal vertebræ have hæmapophyses. The Otter (*Lutra vulgaris*) has 25 or 26 caudal vertebræ. Ten pairs of ribs directly join the sternum, which consists of nine bones and an ensiform cartilage. The spine of the eleventh dorsal vertebra is vertical, and those before and behind it converge towards it. The metapophyses begin to be developed on the twelfth dorsal vertebra, and are continued throughout the lumbar series; they are low and obtuse. The anapophyses commence at the eleventh vertebra, and are continued to the penultimate lumbar. The spines of the three sacral vertebræ have coalesced to form a vertical crista. Hæmapophyses are developed beneath several of the anterior caudal vertebræ; they are articulated, and some of them become anchylosed to short hypapophyses or exogenous processes from the under and fore part of the centrum, and then are continued in several of the succeeding vertebræ, which have not the hæmal arch complete. The neural arch is incomplete beyond the eighth caudal vertebra. The entire tail is longer and much stronger than in the terrestrial

Mustelidæ: it is the chief organ in regulating the course of the Otter through the water.

In the Civet (*Viverra civetta*) the transverse processes of the atlas have a more extensive origin than in the Otter, and are perforated both horizontally and vertically by the vertebral artery before it pierces the neural arch. In the axis the median inferior ridge, and the two lateral ones continued upon the transverse processes, are longer, deeper, and sharper than in the Otter. Certain Viverrines, e.g. the Palm-cats (*Paradoxurus*) have the tail organised for prehension, including upwards of 30 joints, with hæmal arches beneath the interspaces of the first eight or ten. The prozygapophysis of one lumbar vertebra is received into the interspace between the postzygapophysis and the anapophysis of the antecedent vertebra: this interlocking which commenced in the plantigrade and musteline *Carnivora*, is continued in the present (viverrine) and subsequent families.

Among the *Canidæ* the Wolf (*Canis lupus*), has 13 dorsal, 7 lumbar, 3 sacral, and 15 caudal. Eight pairs of ribs articulate directly with the sternum, which consists of eight bones. The eleventh dorsal vertebra is that towards which the spines of the other trunk-vertebræ converge. Metapophyses begin to be developed on the eighth dorsal, and are continued to the fourth lumbar vertebra.

The Dog agrees with the Wolf in vertebral characters.

In a Fox (*Canis rufus*) the vertebral formula is the same, save that the tail-joints are more slender and numerous, being 22. A few of the anterior caudal vertebræ have hæmapophyses: the supporting processes or 'hypapophyses' are developed from a greater number. The sacrum is remarkable for its sudden diminution of size, as compared with the lumbar vertebræ, and only the first sacral vertebra articulates directly with the iliac bones.

The Hyænas have 15 dorsal and 5 lumbar vertebræ: the striped kind (*H. vulgaris*) has 3 sacral and 23 caudal: the spotted kind (*H. crocuta*) has 4 sacral and but 16 or 18 caudal. Eight pairs of ribs articulate directly with the sternum, which consists of eight bones. The transverse processes of the atlas are perforated longitudinally and vertically by the vertebral artery before this perforates the neural arch. The strong spine of the axis is bifid posteriorly. The convergence of the dorso-lumbar spines towards that of the thirteenth dorsal is feeble compared with other Carnivora. Anapophyses begin to be developed on the thirteenth dorsal and subside on the penultimate lumbar vertebræ.

The Lion (*Felis leo*, fig. 337) has 13 dorsal, 7 lumbar, 3 sacral, and 23–25 caudal vertebræ. The spine of the axis has great height, length, and posterior breadth, arching forward and backward, overlapping the third, of which the spine is obsolete; that of the fourth is short and vertical, indicating a centre of the motions of the neck. The anterior dorsal spines are lofty and strong, for the origin of muscles implanted in the ridged and pitted back part of the skull, whereby the head can be raised together with the prey which the jaws have seized: a Lion thus draws along the carcase of a Buffalo, and can with ease raise and bear off the body of a man. The eleventh dorsal is that toward which the

337

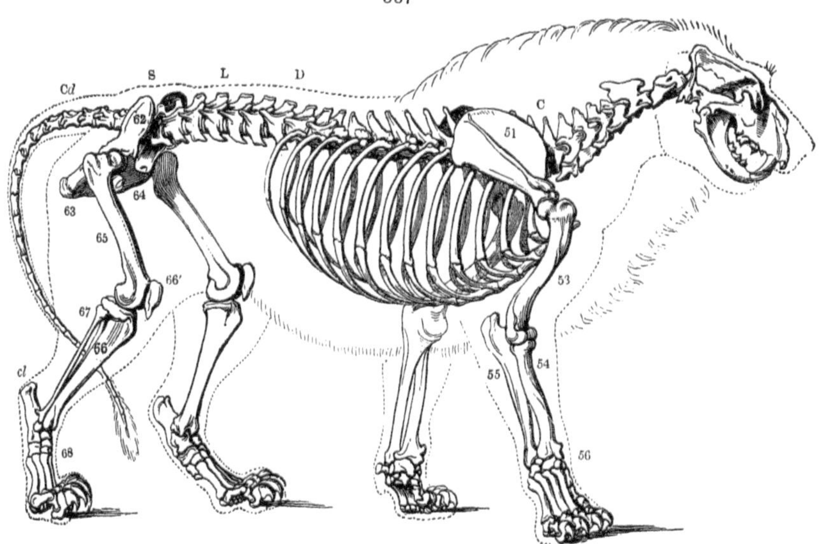

Lion (*Felis Leo*). LXXIII'.

spines of the other trunk-vertebræ converge: the anapophyses begin to project backward from this vertebra, and are continued to the penultimate lumbar. Eight pairs of ribs directly join the sternum, which consists of eight bones. The lumbar diapophyses are long and antroverted. The tail is the chief seat of variety in the vertebral column of the feline group. The Lynx (*F. Lynx*) e.g. has the number of caudal vertebræ reduced to 15. In a tailless variety of domestic cat a stunted mass of 4 or 5 coalesced caudals has become hereditary.

The carnivorous unguiculate do not, like the herbivorous ungulate Gyrencephala, show two series by numerical characters of trunk-vertebræ; the constancy of twenty dorso-lumbars is remarkable and significant: the exceptions are not only rare, but abnormal. Where the trunk is lithe and subject to varied and agile turns and bends, the number of pairs of free elongate pleurapophyses is small, and that of the vertebræ wanting them great; thus the springing Cats, the swift-footed Dogs and Foxes, the climbing Benturong, have 13 dorsal and 7 lumbar: the Viverrines and Mustelines commonly show 14 dorsal and 6 lumbar; the stiffer-trunked Hyænas, Bears, and Seals have 15 dorsals and 5 lumbar: and mostly, where an exceptional excess occurs in any of these groups in one series of vertebræ, it is balanced by as exceptional a deficiency in the other series.

B. *Skull.*—In the Harp Seal (*Phoca grœnlandica*) the basioccipital is a thin plate, and shows a vacuity in front of the foramen magnum: it early coalesces with the basisphenoid: the paroccipital is small, subretroverted: the mastoid large, swollen, not prominent. The frontal, fig. 338, 11, gives its larger proportion to the orbital and olfactory chambers. In the latter the confluent prefrontals and vomer form an extensive bony septum between the meatuses which are blocked up anteriorly by the complex turbinals. Both the tentorium and posterior part of the falx are ossified. The shallow 'sella' has overhanging posterior clinoid processes. The petrosal is perforated by the entocarotid and impressed by a deep transverse cerebellar fossa. The tympanic forms a 'bulla.' The meatal portion of the tympanic is slightly bent and directs the external auditory aperture obliquely forward and upward. The squamosal has a small cranial plate, *g*, and a large thick zygomatic process, 27, with rises at its junction with the malar, 26, to partially define the orbit posteriorly.

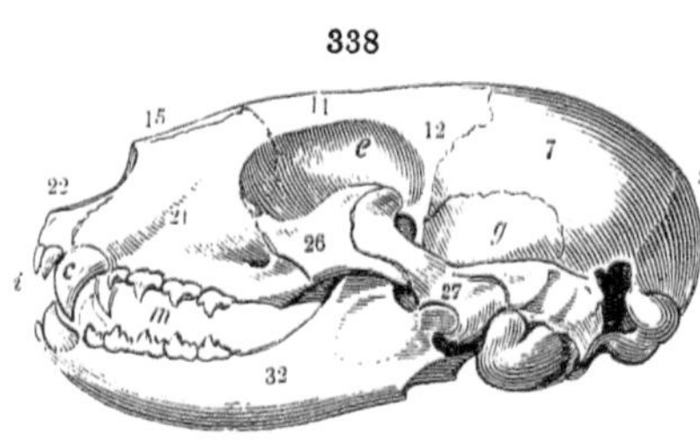

Skull of a Seal (*Phoca*).

The seals, like other carnivora, have the orbit, *e*, incomplete behind, and continuous with a large temporal fossa; the nasals, 15, are short, and the nostril looks more or less upward, in reference to their common sojourn in water and the necessity of rising to the surface to breathe. The condyle is the hindmost part of the mandible.

In the Grey Seal (*Halichœrus griseus*) the skull is remarkable for the straightness of its upper contour and the sudden bending down of the equally straight line formed by the deep and narrow premaxillaries. There is a deep depression in the superoccipital, overarched by a thickly-developed occipital ridge; the squamosal and malar rise abruptly at their junction at the middle of the zygoma. The acoustic bulla receives the meatus auditorius by an expanded and oblique opening. The olfactory fossæ contain, as in all Seals, large and complex turbinal bones. The bony palate is terminated behind by a semicircular notch.

In the Monk Seal (*Pelagius monachus*) the upper contour of the skull presents a sigmoid curve. The temporal ridges meet, and form a low sagittal crest over the posterior half of the frontals and parietals. The upper jaw is much less deep than in the *Halichærus,* the canines are relatively larger and the nasal bones are much shorter. The entocarotid canal perforates the back part of the petrosal as in the *Phoca grœnlandica*: the ecto-carotid does not pierce the pterygoid process.

In the Sterrinks (*Stenorhynchus*) the skull is longer, more 'canine' in the proportions of jaws to cranium, than in other Seals. The malar is long and slender, defining the orbit below: a lacrymal process of the maxillary projects from the anterior rim. The basis cranii is long and narrow in *Stenorhyncus leptonyx.* In the saw-toothed Sterrink (*Stenorhynchus serridens*) the facial part tapers more gradually than in the *Stenorhynchus leptonyx.* The paroccipitals are small, but distinct. The petrosals are perforated posteriorly for the entocarotids; the pterygoid processes are imperforate. The temporal ridges meet upon the sagittal suture, but do not develope a crista. The malar bones are slender, strongly curved, bifurcate posteriorly, the upper prong rising to form, with the zygomatic process of the squamosal, the postorbital boundary. There is no corresponding process from the frontal. The antorbital process of the maxillary is small, but distinct. The premaxillaries are narrow and slender, but do not reach the nasals. The posterior border of the bony palate is terminated by a deep semi-elliptic notch. A single superoccipital venous canal opens, in *Sten. leptonyx,* within the border of the foramen magnum. The basioccipital shows two depressions. The sella turcica is very shallow, and is defined only by a posterior clinoid ridge, between which and the platform for the optic chiasma there is a long tract. The petrosals terminate by obtuse subdepressed apices. The foramina lacera anteriora are of unusual size, and appear to include the foramina rotunda: there is no ridge indicating the division between the anterior and middle lobes of the cerebrum. The rhinencephalic fossæ are small, but deep and well defined, and completely divided by a broad and thick crista galli.

In the Ursine Seal (*Arctocephalus australis*) the border of the superoccipital, forming the upper part of the foramen magnum, shows the orifices of two venous sinuses. The posterior border of the bony palate has an angular notch. The pterygoid processes are pierced for the ectocarotids. The frontal developes a superorbital plate. The mastoid projects free of the tympanic. The olfactory chambers extend backward exterior to the rhinencephalic fossa.

In the Hooded Seal (*Cystophora cristata*), the thin basioccipital shows a small vacuity. The superoccipital inclines from below upward and forward. The temporal cristæ have not met above the parietals. The premaxillaries do not reach the nasals: they form with the maxillaries an antorbital prominence. In the great Proboscis-Seal, or Sea-Elephant (*Cystophora proboscidea*), the occipital condyles meet upon the basioccipital: the paroccipitals are less prominent than in the *Cystophora cristata*. Traces of the suture between the basisphenoid and the basioccipital and between the basisphenoid and presphenoid long remain. The entocarotid canals at the back part of the petrosals are very conspicuous: there are no ectocarotid canals. The sagittal crista is feebly indicated, but the occipital crest is conspicuous for its great height and thickness; the lower border of the superoccipital presents two vertical venous perforations, which are likewise present in the *Cystophora cristata*. The tentorium is less ossified than in the *Otaria leonina*. The walls of the cranium formed by the parietals are thick with a coarse diploë: a very small proportion of the squamosals enters into the formation of those walls: the mastoid has a dense structure where it coalesces with the base of the zygomatic process. Two vertical venous sinuses terminate above the foramen magnum: the basioccipital is also perforated by a similar venous sinus near its middle part. The petrosal is excavated by a deep but narrow cerebellar fossa; a long groove or notch upon its upper surface leads to the meatus auditorius internus: the petrosal is, as it were, bent upwards upon this groove. The tympanic bulla supports the under part of the petrosal like a capsule. The tympanic cavity is divided into two chambers, one above the termination of the meatus externus, the other beneath and internal to it. The carotid canal perforates the tympanic internal to this part of the chamber. The Eustachian groove commences from the angle between the supra- and infra-meatal divisions, and grows deeper and wider until it forms the canal at the fore part of the tympanic bone. The rhinencephalic fossa is divided by a strong and sharp crista galli. The frontal bones form an unusually small proportion of the cranial cavity: they are extensively overlapped posteriorly by the parietals. Besides its superior size, the skull of *Cystophora proboscidea* differs from that of the *Cystophora cristata* in the form and proportions of the palatine bones, the posterior borders of which present three notches; in the relatively shorter extent of the nasal processes of the premaxillaries; in the greater prominence of the antorbital processes of the maxillaries; and the absence of the depression beneath the antorbital foramen. In the skull of a young Proboscis-Seal I have seen traces of a suture partially dividing the orbital from the rostral part of the maxillary, extending from the side of the nasal aperture into the antorbital foramen: this incompletely separated part might be compared with a large lacrymal, but there is no trace of a distinct bone or of any lacrymal perforation.

In the 'Sea-Lion' (*Otaria jubata*) the superoccipital is broader and more nearly vertical than in the preceding species of Seal: the basioccipital is carinate below; the paroccipitals form an obtuse angle, but are less prominent than the large mastoids. The petrosals and tympanics are not expanded into a bulla ossea, but send down a subcompressed smooth tuberosity: the entocarotid pierces the petrosal. The pterygoids are pierced by the ectocarotids. The bony palate is very long, and remarkably concave, from the bending down of its sides: its posterior border is transversely truncate. The sagittal and occipital cristæ are singularly elevated. Each frontal sends out an obtuse process near its junction with the parietal, into the middle of the extensive temporal fossa, and each developes large, horizontal, triangular, postorbital processes. In old males, the parietal also sends out a ridge, and the great temporal muscle seems thus to have been divided into three masses: there is a ridge from the inner side of the parietal, dividing the middle from the anterior lobe of the cerebrum, parallel with the external ridge projecting into the temporal fossa. The maxillaries develope antorbital processes. The nasals are short and broad, and articulate with the premaxillaries as well as the maxillaries.

The posterior part of the falx and the whole of the tentorium are ossified. The superoccipital sinus, commencing by a common aperture at the hinder extremity of the longitudinal sinus, diverges on each side into the substance of the exoccipitals, and terminates in a deep and wide fossa on the inner side of the condyle, from which fossa one canal leads backward to open external to the condyle, and another downward and inward to terminate in the foramen jugulare. The bony tentorium terminates anterior to the petrosal, which has an obtuse expanded inner apex, and shows no petrosal pit. There is no Gasserian fossa. A ridge divides the foramen ovale from the foramen rotundum. The sella turcica is broad and shallow: it is defined by posterior clinoid processes: there are no anterior ones. The rhinencephalic fossa is narrow, but of unusual longitudinal extent: the optic nerves traverse a common canal of nearly an inch in extent before it divides. The ascending plates from the palatine processes of the maxillary form a deep groove for the reception of the vomer. The superior turbinals occupy that part of the olfactory fossa which overarches the rhinencephalic chamber: this is divided by a broad crista galli. A large oblong vacuity at the outer and posterior side of the nasal passages between the frontal, presphenoid, palatine and maxillary bones, is closed by membrane in the recent animal. There is a smaller vacuity in the corresponding part of the skulls of some other species of Seals.

In all Seals, the convex mandibular condyle is transversely extended, terminal, the border of the jaw extending from below the condyle forward, and rarely developing an angle: this is best marked in *Phoca groenlandica*: in *Otaria* it seems to project just below the condyle.

The hyoid arch consists of stylo- epi- and cerato-hyals, and of a basi-hyal in form of a transverse bar, with a pair of thyro-hyals: the stylo-hyals are attached by ligament to the outer side of the petrosals.

In the Walrus (*Trichecus rosmarus*, fig. 339), the basioccipital is subcarinate below. The superoccipital, 3, inclines a little upward and forward, is divided by a median crista, and is bounded above by a broad rugged tract. The venous fossa on the inner side of the condyles is divided by a bony bar. There is a wide sphenopalatine vacuity. The paroccipitals are broad, but not very prominent: the hinder surface of the skull is much extended laterally by the great development of the mastoids, 8. The alisphenoid is excluded from the parietal, 7, by the junction of a small part of the frontal, 11, with the squamosal, 27. There is no trace of a lacrymal bone, but a small elliptical canal perforates the base of the antorbital process of the frontal slightly upwards. The zygomatic process of the squamosal is remarkably thick. The malar sends up a lofty postorbital process, but there is none on the frontal: the maxillary, 21, developes a large but low sub-bifid antorbital process: it is perforated by a large antorbital foramen, and excavated by a large and deep socket for the canine tusk. The premaxillaries, 22, are minute. There is a large oval vacuity in the lateral wall of the posterior nares. The skull is singularly expanded, short, obtuse, and, as it were, truncated anteriorly; and, being constricted between the orbits, the upper surface presents an hour-glass form. The parietes of the cranium are thick and dense, with a diploë, gradually degenerating into a coarse cellular texture, in the enormous mastoids. The tentorium is bony, the sella turcica large and shallow, with anterior and posterior clinoid processes, and the crista galli is prominent. The petrosal terminates below in three obtuse processes, but there is no bulla ossea. The pterygoid process is perforated by the ectocarotid. The bony roof of the palate is very concave towards the mouth, and terminates behind by a broad biangular notch. The tympanic cavities are smooth, and almost hemispheric: the antorbital canal is large: the nasal fossæ contract as they pass forward to the vertical external nostril. The osseous part of the septum narium is formed by the canaliculate vomer and the coalesced plates of the prefrontals, dividing the posterior halves of the olfactory chamber. The lateral sinuses are completely surrounded by bone. A vein perforates the back part of the parietals and terminates in the longitudinal sinus. The bony tentorium terminates above the base of the petrosal; a thick, smooth ridge enters the lower half of the fissure between the anterior and posterior cerebral lobes. A similar but shorter ridge from the inner side of the frontal more completely defines the rhinencephalic chamber: an elliptic foramen leads from the lower and outer corner of this fossa into the back part of the orbit between the orbitosphenoids and frontals. The mandible, 32, articulates by a thicker condyle than in true Seals: it is terminal: the feeble angle slopes forward from it: the coronoid is oblique and rounded.

339

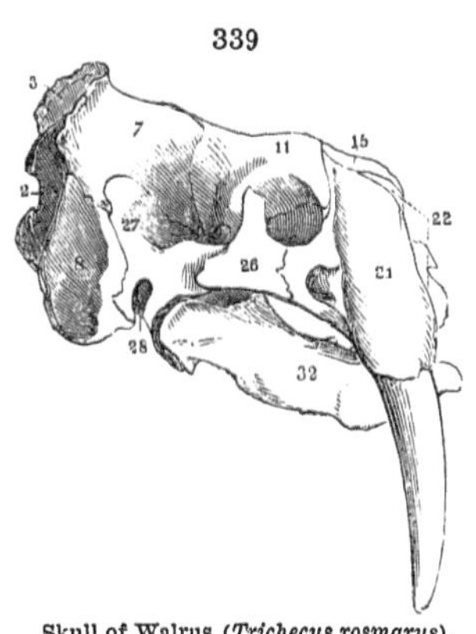

Skull of Walrus (*Trichecus rosmarus*).

In the Bear-tribe, as in the Seals, the tentorium is ossified: the interparietal unites with and forms a triangular process of the superoccipital: the alisphenoid articulates with the parietal: the ectocarotid pierces the pterygoid process. There is no pterygoid fossa.

In the European Black and Brown Bears (*Ursus arctos*) the frontal region of the skull is raised and convex. In the American

Grisly Bear (*Ursus ferox*), the facial part of the skull is relatively longer than in the *Ursus arctos*, and the nasal processes of the premaxillaries are much longer, are more slender, and articulate directly with the anterior processes of the frontals. In the Brown Bear, the maxillaries articulate with the small part of the nasals and separate the premaxillaries from the frontals. In the Polar Bear (*Ursus maritimus*, fig. 340), the lower extremities of the occipital condyles are united by a ridge, which, however, is less prominent than in the *Ursus ferox*. The precondyloid foramen is exposed. The superoccipital, 3, terminates above in a strong ridge overhanging the condyles. Both paroccipitals, 4, and mastoids, 8, are well developed, but the latter are the larger

340

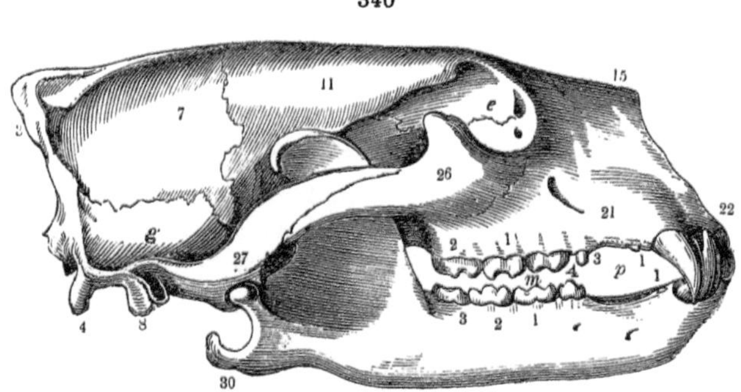

processes. The temporal ridges, commencing at the postorbital processes, converge at a right angle and meet at about two inches behind the orbits, and form a long and prominent sagittal crest, the upper border of which is straight; the frontal region is low and flattened. Within the cranium the cerebellar fossa is formed by the bony tentorium above, and by a shorter osseous ridge below, separating the cerebellum from the upper part of the medulla oblongata. The commencement of the entocarotid canal may be seen distinct from the fore part of the fossa jugularis; the petrosal fossa is divided into two cells for the reception of the cerebellar appendages. The mastoid is occupied by a close diploë, which receives no air-cells from the tympanic cavity. The meatus auditorius terminates obliquely within the tympanic cavity. A triangular constriction separates the prosencephalic from the rhinencephalic chamber. The malar, 26, alone forms the postorbital rising: the squamosal, 27, does not reach so far; it developes a low subquadrate cranial plate, *g*. The mandible developes a long angular process, 30, which rises toward the condyle.

In the Racoon (*Procyon*), and Coati (*Nasua*), the entocarotid pierces the inner border of the tympanic bulla: there is no ectocarotid canal. The mastoid is thicker than the paroccipital. The bony tentorium terminates upon the petrosal above the shallow depression of the cerebellar appendage. The upper cranial parietes are moderately thick and with a diploë. In the Coati the olfactory chamber, with the superior turbinals, extends above the whole rhinencephalic fossa, and forms in part the frontal elevation of the cranial contour. In the Benturong (*Ailurus*), the ectocarotid perforates the pterygoid, as in Bears.

The skull of the Badger (*Meles taxus*) is chiefly remarkable for the closeness with which the transverse condyles of the lower jaw are grasped by the borders of the articular grooves at the base of the zygomatic processes, so that the mandible cannot be disarticulated without some violence. The lateral sinus terminates behind the glenoid cavity, as in other *Ursidæ*, and the subpetrosal sinus terminates at the entocondyloid foramen. There is no ectocarotid canal.

In *Ratelus* the transversely extended base of the paroccipital is applied to the back part of the bulla. In the Glutton (*Gulo*) the cranial cavity is less expanded posteriorly, and less constricted anteriorly, than in the Ratel. There is a smooth articular surface in the basioccipital, but it is less distinctly continuous with the occipital condyles than in the Ratel. The zygomatic arches are larger, stronger, and more curved: the palate is relatively broader: both the paroccipital and the mastoid processes are feebly developed.

In the Stoats and Weasels (*Putòrius*), the meatus auditorius is an oblique perforation in the lateral and inferior parietes of the skull, directed from within outward and forward, and not produced upon an auditory process. The bulla tympanica is very extensive. The bony tentorium, which projects rather from the upper than the back wall of the cranium, terminates upon the back part of the petrosal, above the deep circular pit for the cerebellar appendage. The rhinencephalic fossa is less distinctly defined than in Plantigrades from the rest of the cranial cavity; the olfactory chamber extends backward both above and beneath that fossa, causing the cranium to appear dilated at that part: the air must be filtered, as it were, through the complex turbinals before passing into the canal of the posterior nares.

In the Otter (*Lutra*), a narrow articular surface upon the basioccipital connects together the two condyles: the temporal ridges, commencing from the postorbital processes, meet at an open angle, and extend backward, as a low and straight sagittal crest, as far as the broader occipital crest. The zygomatic arches are strong and boldly curved; they bifurcate anteriorly to surround the large antorbital foramen. The cranial walls are thin, without diploë: the impressions of the convolutions are strongly marked: there are no frontal sinuses. The cranial cavity is remarkable for its great posterior expanse and its extreme contraction between the prosencephalic and rhinencephalic divisions. The bony tentorium terminates upon the petrosal above the small pit for the cerebellar appendage. The sella turcica is shallow. The tract for the optic chiasma is long and narrow. The crista galli extends backward through nearly the whole of the rhinencephalic fossa. The longitudinal sinus communicates behind with two small venous foramina in the superoccipital bone. The olfactory chamber commences directly in front of the rhinencephalic fossa, the cribriform plate, or back part of the olfactory capsule with the coalesced prefrontals, separating them. The entry to the nasal passages is almost blocked up by the large and complex turbinals.

In the Civet (*Viverra Civetta*), the occipital condyles are separate from each other at their lower extremities. The paroccipitals and mastoids have coalesced and form a triangular plate of bone, applied to the posterior part of the tympanic bulla, like the capsule of the acorn to the seed. This bulla is more circumscribed and much more developed than in the Otter: the bony meatus auditorius is much shorter, and opens directly into the tympanic cavity. The nasal processes of the upper maxillaries extend backward much further than the nasal bones, the reverse being the case in the Otter. The pterygoid processes are perforated by the ectocarotids. The cranial cavity is longer and narrower, and the postorbital constriction much less, than in the *Mustelidæ*. The bony tentorium is continued forward beyond the petrosal, and terminates above the foramen rotundum. The petrosal is impressed by a deep pit for the cerebellar appendage. A vertical inverted tract of the cranial walls divides the prosencephalic from the rhinencephalic compartments. The olfactory fossa is continued backward above as well as beneath the rhinencephalic compartment. The crista galli is rudimental. When the squamosal is removed, the extensive surface of the parietal and alisphenoid is exposed to which it was applied, and the small vacuity in the suture between those bones which was left for it to cover in completing the cranial walls. In some Viverrines (*Ichneumon, Mangusta*), the orbital processes of the frontal and malar meet and circumscribe the rim of the orbit.

In the Common Fox (*Canis Vulpes*), the paroccipital is triangular, and applied to the back part of the acoustic bulla, but is smaller and thicker than in the Viverrines[1], and stands off more from the bulla. The alisphenoid articulates with the parietal. The interparietal, which has anchylosed with the superoccipital, penetrates the posterior interspace of the parietals. The nasal processes of the maxillaries are truncate, and terminate on the same transverse line as the nasals. The maxillaries directly articulate with the middle part of the nasals below the bony tentorium, which appears to be developed from the superoccipital.

The skull of the Wolf (*Canis Lupus*), as of the Jackal (*Canis aureus*), differs from that of the Fox in the median depression and transverse convexity of the frontal region produced by the bending down of the postorbital processes; in the greater posterior extension of the nasals, as compared with the maxillaries; and in the encroachment of the lacrymal on the face. The frontal bones of the Wolf preserve a more uniform breadth than in the Jackal, being less expanded posteriorly where they join the parietals. The short and wide meatus auditorius terminates obliquely in the tympanic bulla. The base of the zygomatic process is pierced by a vertical venous canal.

Like the Jackal and Wolf, the Dingo (*Canis Australis*) differs from the Fox in the greater transverse convexity of the frontals, especially opposite the postorbital processes, and in the greater longitudinal depression between the frontals; in the greater posterior extension of the nasals, as compared with the maxillaries; and in the encroachment of the lacrymal bone upon the face. In a comparison of the skull with that of the Marsupial Carnivore (*Thylacinus Harrisii*) from the same part of the world, which equals the Dingo in size, the most striking difference is the comparative superiority of the cerebral cavity in the wild Dog, and of the olfactory cavity in the Thylacine, the proportions being reversed in the two specimens. The superoccipital overhangs the foramen magnum in the Dog, but is on the same vertical plane with it in the Thylacine. The paroccipitals are more compressed in the Thylacine, and their base is not applied to the acoustic bulla, which is of much smaller size and formed exclusively by the alisphenoid,

[1] The extinct genus *Galecynus* of the Œningen miocene indicates the transition from *Viverra* to *Canis*, CIV'. p. 55.

not by the petrosal and tympanic, as in the Dog. The tympanic has preserved its distinctness in the Thylacine, but has coalesced with other elements of the temporal bone in the Dog. A wide and deep groove divides the bulla from the basisphenoid in the Thylacine, but the sides of the basisphenoid in the Dingo are swollen and abut against the large tympanic bullæ. The articular cavities for the lower jaw are much nearer the occiput in the Thylacine than in the Dingo, and the malar bones enter partially into their formation. There are two large vacuities in the back part of the bony palate in the Thylacine, but this part is entire in the Dingo. The antorbital foramina are larger in the Thylacine, and much nearer the orbits than in the Dingo; they are also formed partly by the malar, and are not wholly perforated in the maxillary bone, as in the Dingo: the lacrymal bone is much larger in the Dingo, and encroaches much more upon the face: the nasal bones are broader posteriorly in the Dingo, and extend further back, as compared with the maxillaries. The petrosals are much larger in the Dingo, and send bony plates into the tentorium, which plates are not present in the Thylacine. The chief bony part of the tentorium projects from near the middle of the occiput, and does not reach the petrosal in the wild Dog. The sella turcica is defined by the posterior clinoid processes in the Dingo, but not in the Thylacine. The foramina optica and lacera anteriora are blended together in the Thylacine, but are distinct in the Dog. Although the olfactory chamber is so much larger in the Thylacine, the rhinencephalic fossa is smaller than in the Dog. The lower jaw, besides its greater length and slenderness in the Thylacine, differs by the bending in of the angle, which is the characteristic of the Marsupials. In most of these distinctions the Thylacine manifests its nearer affinity to the oviparous type of skeleton.

The chief distinction between the wild and domestic Dogs is the greater proportional size of the cranium to the face in the latter, and this increases as the size of the variety diminishes.

The affinity of the Hyæna to the *Viverridæ* is shown, in the skull, by the broad, triangular, rough plate formed by the paroccipital and mastoid, and applied to the back part of the acoustic bulla: but the pterygoid processes are not pierced by the ectocarotids. The strength of the muscles which work the jaw is shown by the extent of the temporal fossæ, the height of the sagittal crest, the thickness and the expanse of the zygomatic arches, the height of the coronoid processes, and the depth of the strongly-defined fossæ into which the great muscles of mastication are inserted. The antorbital foramina are small semilunar slits. The nasal processes of the maxillaries extend further back than the nasals.

The specific character of the Lion (*Felis Leo*), as compared with the Tiger (*Felis Tigris*), is shown by the obtusely-pointed termination of the nasal process of the maxillary, and its extension backward to the same transverse line as that which the hinder ends of the nasals reach. In the Tiger the nasal bones are relatively longer and extend further back than the nasal processes of the maxillaries, which are, as it were, truncated. The concavity of the frontal platform between the deflected postorbital processes is narrower than in the Lion; the suborbital foramina are smaller.

The carnivorous character of the skull, fig. 341, as exemplified by the sagittal, 7, and occipital, 3, crests, by the strength and expanse of the zygomatic arches, 27, by the depth and shortness of the jaws, by the height and breadth of the coronoid processes, and by the extent of the muscular fossæ of the lower jaw, reaches its maximum in the skull of the old males of both these large Felines. The triangular occipital region is remarkable for the depth and boldness of the sculpturing of its outer surface. The conjoined paroccipital, 4, and mastoid, 8, form a broad and

341

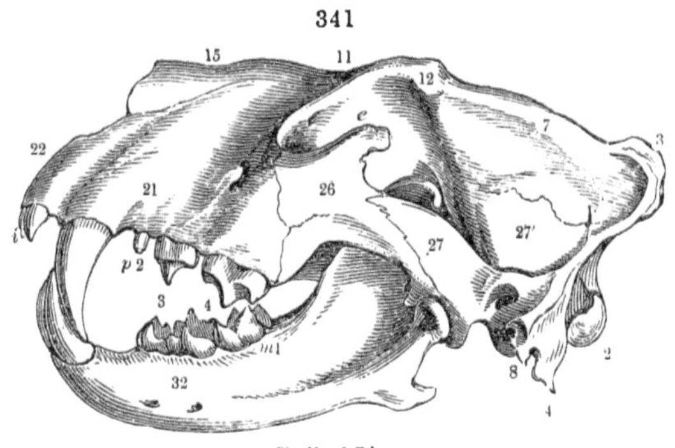

Skull of Lion.

thick capsular support for the back part of the acoustic bulla. The pterygoid processes are imperforate. A well-marked groove extends on each side of the bony palate from the posterior to the anterior palatine foramina. The premaxillaries, 22, are comparatively short, and one half of the lateral border of the nasals, 15, directly articulates with the maxillaries, 21.

The bony tentorium extends above the petrosal to the ridge over-hanging the Gasserian fossa: the petrosal is short, its apex is neither notched nor perforated: the cerebellar pit is very shallow.

The sella turcica is deep, and well defined by both the anterior and posterior clinoids. The rhinencephalic fossa is relatively larger than in most *Carnivora*, and is defined by a well-marked angle of the inner table of the skull from the prosencephalic compartment: the olfactory chamber extends backward both above and below the rhinencephalic fossa: the upper part of the chamber is divided into two sinuses on each side: the superior turbinals extend into the anterior sinus, and below into the presphenoidal sinus. The inner surface of the squamosal is tripartite; the upper facet rough for the broad squamous suture, the anterior and inferior one smooth and deep for the natiform protuberance of the hemisphere, and the posterior facet smooth and undulated where it is applied to the petrosal capsule, its juncture with which is effected by the medium of the mastoid, which is anchylosed to both.

The strengthening of the cranium in *Carnivora*, in reference to the forcible action of the muscles attached thereto, is gained by the growth of bone in the form of ridges both from the outer and the inner surfaces of the cavity. This is so completely filled by the brain, its blood-vessels and membranes, that were any concussion conceivable of cerebrum against cerebellum through an active bound or leap, an interposed membrane so elastic as to yield and recover would best meet the contingency: to suppose that a hard plate between the two soft masses had any such relation to the spring of the stealthy feline implies both dull physiological reasoning and limited knowledge of the comparative osteology of the *Carnivora*: the commonly ascribed final cause of the bony tentorium of the Cat is refuted by the presence of that part in the plantigrade Bears that do not move by bounds, and in the pinnigrade Seals that can only shuffle along the ground, and are pillowed by the waves during their swiftest and most habitual movements.

The hyoid arch of Felines consists of stylo-cerato- and basihyals, with the appended thyro-hyals. The stylo-hyals, as a rule, connect the arch to the base of the skull: but in the Lion a long ligament intervenes between the stylo- and cerato-hyals, allowing more freedom of motion to the base of the tongue and larynx, in relation to the characteristic vibratory roar of the king of beasts.[1]

C. *Bones of the Limbs.*—The general characters of these in the *Carnivora* have been defined, and the principal modifications determining the pinni- planti-, and digiti-grade modes of locomotion are illustrated in figs. 172–175, pp. 288, 289.

The pinnigrades are pentadactyle, and without trace of clavicle.

[1] CCXXXVI, ol. vii. p. 38.

The scapula is broad and curved backward, the anterior and basal borders are continued in *Phoca* by a bold convex line to the angle terminating the posterior costa, which is as strongly concave, fig. 335. In *Otaria* the breadth is increased by the production of the fore part of the scapula, causing a disproportionate extent of the prespinal surface, on which is a low accessory ridge, anterior to the true 'spine,' not posterior to it as in *Megatherium*. The spine is farther from the posterior costa in *Trichecus*. In all *Phocidæ* it terminates in a short acromion. The humerus is shorter than the scapula in *Phoca*, longer in *Otaria*; it is remarkable for the great development of the inner tuberosity and of the deltoid ridge, which is deeply excavated on its outer side. The inner condyle is perforated in *Phoca*, not in *Otaria*, *Monachus*, and *Trichecus*. The middle of the distal end is excavated by the articular trochlea; an olecranal fossa is feebly or not at all marked. The antibrachial bones are compressed, and firmly united, the interosseous space being widest in *Otaria*: the olecranon is large and hatchet-shaped. The fore-part of the lower half of the radius is produced. The scaphoid and lunar bones are connate: the fifth metacarpal articulates with the cuneiform as well as with the unciform: the magnum is the least of the carpals. Although the pollex or the first digit exceeds the third, fourth, and fifth in length, it presents its characteristic inferior number of phalanges, by which the radial border of the fin is rendered more resisting. The pelvic arch is remarkable for the stunted development of the ilia, and the great length of the ischia and pubes: the symphysis is short, and divaricable in parturition, as in the Guinea-pig (p. 380). The femur is equally peculiar for its shortness and breadth: its head has no pit for a 'round ligament.' The tibia and fibula present the more usual proportions, but are anchylosed at their proximal ends. The astragalus, fig. 173, *a*, has its proximal articular surface in two facets, one for the tibia, *b*, the other for the fibula, *m*: a part of the bone projects 'proximad' of these surfaces; and it is produced 'distad' to articulate with the naviculare, *s*: the co-extended calcaneum, *cl*, is applied to the tibial side of the astragalus. In *Otaria* the calcaneal process is longer. The entocuneiform, *i*, mesocuneiform, ectocuneiform, *c*, and cuboides, *b*, have the usual connections. The bones of the foot are much developed, and are modified to form the basis of a large and powerful fin: in *Phoca*, the middle toe is the shortest, and the rest increase in length to the margins of the foot: in *Otaria* and *Trichecus* the toes are subequal in length. The long-bones of Seals have no medullary cavity.

In the plantigrade *Carnivora* the clavicle is wholly wanting. In the Bear-tribe, the scapula, fig. 336, 51, is remarkable for its almost quadrate form, and for the strong development of the ridge between the infraspinatus and teres major, constituting almost a second spine. The inner condyle of the humerus is not perforated, save in *Ursus ornatus.* The antibrachials little, if at all, exceed the humerus in length; their shafts are of equal strength. The scaphoid and lunar bones of the carpus have coalesced: the pisiforme is elongated and expanded at its free end like a calcaneum. The fore-foot is 5-dactyle, the pollex being a little shorter than the other toes, which are subequal in length; the basal sheath of the ungual phalanges is thickened and tuberculate below: the claw-bearing part is long, subcompressed, and slightly arched. The ilia are shorter, thicker, and broader than in Digitigrades: the ischia are short and expanded, forming with the strong pubics a long symphysis. The acetabula are large and deep; the ilio-pubic angle is 125°. The femur is remarkable, in Bears, for its great length, and superficial resemblance to that in man; but its shaft is relatively thicker, straighter, and rather flattened from before backward; it differs also in the more shallow pit for the round ligament, in the great trochanter being longer though less prominent above, in the less projection of the small trochanter, in the minor expansion of the distal condyles, and in the smaller size of the rotular channel. The medullary cavity is confined to the middle third of the bone. The medullary artery, which enters at the posterior and inner side, below the middle of the shaft, takes an oblique course upward. The tibia is one-fourth shorter than the femur: the fibula is much smaller and compressed: but the medullary cavity extends through nearly the whole of the shaft of this slender bone. In fig. 174, *cl* marks the calcaneum, *s* the naviculare, *e* the ectocuneiform, *b* the cuboides; the astragalus is almost as broad as long, without a calcaneal process. The hallux is rather shorter than the other toes, which are of subequal length, and form the basis of a broad flat foot.

In the scapula of the Racoon (*Procyon*), the pre- and postspinal fossæ are of equal extent. The inner condyle of the humerus is perforated as in all Subursines. In *Nasua* and *Arctictis*, a supplementary carpal ossicle is wedged between the scapholunar and the metacarpal of the pollex, external to the trapezium: the tarsus shows a corresponding ossicle wedged between the naviculare and entocuneiform. In the Racoon, the fibula is characterised by three processes behind its distal end: the malleolar process is very short, but plays upon a well-marked articular surface of the astragalus. In a Kinkajou (*Cercoleptes*), I have seen the condyle notched in the right and perforated in the left humerus. In the Badger (*Meles taxus*), the scapula presents a subquadrate form, crossed diagonally by the spine, and with one angle produced to form the glenoid cavity: the coracoid is represented by a low tubercle: there is no inferior ridge or spine. In a Ratel (*Ratelus mellivorus*), with a similar shaped scapula, the coracoid is sub-bifid, and the acromial tubercle is slightly produced. I have seen both humeri perforated between the condyles, only the right one above the inner condyle. There is no medullary cavity in the tibia. The humerus of *Mydaus* shows both the intercondylar and entocondylar holes. In the Glutton (*Gulo*), the scapula is of a trapezoidal form, equally and obliquely bisected by the spine, which developes a bifid acromion: there is a distinct coracoid tubercle. The inner condyle of the humerus is perforated. The deltoid ridge terminates on the middle of the shaft. Both pollex and hallux are relatively shorter to the other toes, in most Subursines, than in the true Bears. Besides the patella, the fabellæ are commonly present at the knee-joint.

In *Mustelidæ* the acromion is more distinctly bifurcate than in *Subursidæ*: the posteriorly produced plate is broad in the Polecat (*Putorius*), in which the glenoid surface is continued upon the coracoid tubercle. In that of the Otter may be noticed the greater expanse of the prespinal portion and the well-marked division of the acromion, the broader and posterior part bending down, and the narrow and anterior one extending forward: the coracoid tubercle is rudimentary. The humerus is remarkable for the compression of the shaft, which is strongly bent forwards, and for the continuation of a ridge from the deltoid as far as the distal condyles. The inner one is perforated. The ulna is much longer, and is stronger than the radius. The supplementary ossicle answering to that marked *i* in fig. 361, is present in the carpus of both *Lutra* and *Putorius*. The diminution of the pollex proceeds: that of the hallux in a less degree: the third and fourth digits are the longest in both fore and hind feet.

In the *Viverridæ* the scapula is longer, more quadrate, and more equally bisected by the spine than in *Mustelidæ*: the acromion is bifid, but the divisions are less distinct. There are detached clavicular styles. The innermost digit is relatively shorter than the rest in both fore and hind feet, taking no share in the support of the body. In *Mangusta tetradactyla* the pollex is absent. In the Civet, and Cynogale, the humerus is pierced between the condyles, but not, or rarely, above the inner condyle. In the Genet, the humerus shows the entocondylar, but not the intercondylar hole. In the femur a ridge extends from the great trochanter more than half-way down the shaft of the bone. In the Ichneumon (*Mangusta*), the upper contour of the scapula is slightly sigmoid, very convex anteriorly, and the prespinal is larger than the postspinal fossa. The acromion is bifid. The humerus is pierced both between the condyles and above the inner condyle. The supplementary ossicle at the radial side of the carpus is present in most *Viverridæ*. Its homotype exists in the tarsus of *Cynogale* and *Bassaris*. The hallux is wanting in both *Mangusta penicillata* and *M. tetradactyla*.

In the *Canidæ* the scapulæ, and especially the limb-bones, are longer and more slender, relatively, than in the foregoing carnivorous families. There are clavicular styles. The humerus has the intercondylar vacuity, not the entocondylar perforation. The pollex is reduced to the 'dew-claw' appendage; and, in *Canis pictus*, to a metacarpal style, which is concealed. The ulna and radius are closely and extensively united: swift course is the characteristic of the present digitigrade family. The slender fibula closely adheres to the lower half of the tibia. The hallux is reduced to a minute beginning of the metatarsal. The accessory carpal ossicle and the fabellæ are present.

In the *Hyæna*, the humerus is usually pierced between the condyles: it is thicker in proportion to its length than in the Dog, but is more bent and twisted: the same characters mark the radius and ulna, which are still shorter in proportion than in the Dog. The pollex is reduced to a rudiment of its metacarpal. In fig. 191 (*Hyæna*, p. 306), *sl* marks the 'scapholunar' common to the carpus of all *Carnivora*, *c* is the cuneiforme, *p* the pisiforme; *t* trapezium, *d* trapezoides, *m* magnum, and *u* unciforme. The femur is more compressed antero-posteriorly than in the Dog, and the small trochanter is more posterior in position. The neck is longer, and the head of the bone larger: there is a fabella behind each condyle. The tibia is shorter than the femur: the rotular ridge is less produced than in the Dog. The fibula is less flattened at its lower half, and more independent of the tibia than in the Dog. The entocuneiform supports a rudiment of the metatarsus of the hallux, as in the Dog: the calcaneal process is shorter and thicker.

All Felines have the clavicular bone *s*. The humerus perforated above the inner condyle, but not between the condyles. In the Lion, fig. 337, the supraspinal fossa of the scapula, 51, is less deep than the infraspinal one, and its border is almost uniformly convex: the acromion is bifid, the recurved point being little larger than the extremity or anterior point. A supplementary ossicle is wedged in the interspace between the prominent end of the scapho-lunar bone and the proximal end of the metacarpal of the pollex. The ilia, fig. 342, *c*, are long and narrow, but thick; placed so obliquely upon the vertebræ, *a*, as to form an angle of about 155° with the lumbar series: the ischia, *e*, are also long and directed on the same antero-posterior plane: the length, ridged strength, and great obliquity of the 'innominate' bone, afford powerful attachment and advantageous leverage to the muscles acting upon the hind limbs. The boundary of the ischiatic notch is feebly marked at *g*. The pubis is short, but the ischio-pubic symphysis, *f*, is long: the ilio-pubic angle is 120° in the Lion. The posterior exceeds the anterior pelvic outlet in size. The os penis exists in all *Carnivora*, and is remarkably developed in Bears and Seals.

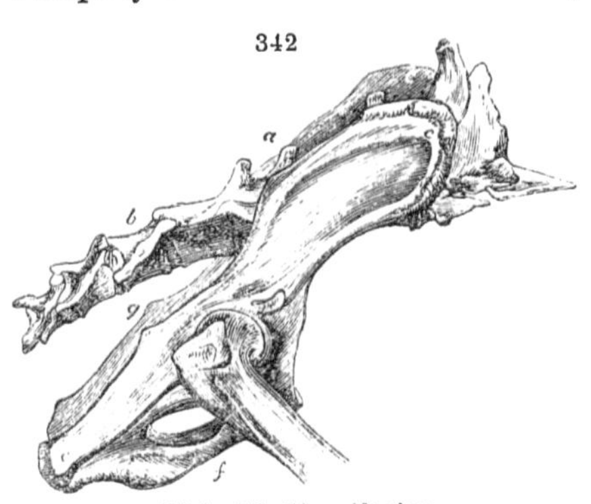

Pelvis of the Lion, side view.

The pollex, in the Felines, is retained on the fore-foot, and, like the other toes, is terminated by a large, compressed, retractile ungual phalanx, forming a deep sheath for the firm attachment of the large curved and sharp-pointed claws. This highly-developed unguiculate structure, with the dental system and concomitant modifications of the skull, completes the predatory character of the typical *Carnivora*.

§ 190. *Skeleton of Quadrumana.*—The *Quadrumana* combine the opposable thumb in the hind limb with complete clavicles, and a greater relative capacity of cranium than in foregoing *Gyrencephala*. The orbits are turned more forward, have the bony rim entire, and in most of the order are partitioned off by bone from the temporal fossa. In no Quadrumane is the hyoid arch complete, or articulated by bone to the basis cranii.

To the *Quadrumana* the transition is, not from the *Gyr-*, but from the *Liss- encephala*. For promoting the Colugos to the Lemurs the grounds are almost as good as for degrading them to the Bats. It has required a thorough knowledge of the structure of the Aye-aye to gain a majority against keeping *Cheiromys* amongst the mice. The singular group of Lemuridæ, which, from the superior brain-development, especially the posterior extension of the cerebrum over the cerebellum, I associate with the *Gyrencephala*,[1] and by the hinder thumb with the *Quadrumana*, offers great diversity in dentition and minor characters.

A. *Vertebral Column.*—All *Quadrumana* have the seven cervical vertebræ. In the Lemurine or Strepsirrhime group the following are the numbers of the other vertebræ:—

	D	L	S	C
Galeopithecus	13 or 14	6 or 7	3 or 4	18 or 19
Tarsius spectrum	13	6	2 or 3	29 or 30
Cheiromys madagascariensis	13	6	2 or 3	22 or 23
Perodicticus Potto	15 or 16	6 or 7	2 or 3	19 or 20
Stenops tardigradus	16	8	3	7 or 8
„ *gracilis*	14 or 15	9	3	5 or 6
Otolicnus Peli	13	7	3	23
„ *crassicaudatus*	13	6	3	27
Lichanotus Indri	12 or 13	8 or 9	4	10 or 11
Tarsius spectrum	13	6	3	29 or 30
Lemur	13	6	2 or 3	28 or 29

The majority, including the type-form, of the *Lemuridæ* thus have 19 dorso-lumbar vertebræ: the slow nocturnal species have longer and less flexible trunks, approaching in the number of dorso-lumbars—24—to the vertebral characters of the lissencephalous Sloths. The tail is, as usual, the seat of the greatest diversity; the slow lemurs, again, in the shortness of this terminal appendage, recal a bradypodal character. In *Stenops gracilis* a metapophysial tubercle is developed on each of the twelve anterior dorsals: on the thirteenth it takes the place of the diapophysis, and in the fourteenth extends forward, and offers an articular surface for the outer side of the postzygapophysis: it has the same disposition in the lumbar series, where the diapophyses are serial repetitions of the base supporting the anchylosed rib in the first lumbar vertebra. The succeeding lumbars slightly decrease in size as they approach the sacrum. No centre of motion of the trunk is indicated by the direction of the dorso-lumbar neural spines. In the more active and flexible-bodied *Lemuridæ* the trunk-vertebræ resemble in proportions, connections, and direction of neural spines those of the agile *Carnivora*. In *Lemur*

[1] See CI', p. 8, pl. xi. figs. 6, 7. The extension of the cerebellum over more or less of the cerebrum is the primary and more constant character of the group called, from the secondary character of convolutions, 'Gyrencephala.' The smooth brain of the small Monkey (*Midas rufimanus*) is figured in LXIV'. to illustrate such primary character. To be consistent, Mr. Murray would have to remove the Marmosets as well as the Galagos to the Insectivora; CI.' pp. 9 and 10.

nigrifons the metapophysis begins to be developed in the middle dorsal vertebræ, and, in the tenth, projects above, but distinct from, the diapophysis. In the eleventh the diapophyses have disappeared, and the metapophysis is on the outside of the prozygapophysis. From this vertebra a well-marked anapophysis is developed, which is continued from all the succeeding vertebræ. The diapophysis reappears upon the first lumbar, and increases in length and breadth as the other lumbar vertebræ approach the sacrum. The centre of motion of the back is indicated by the vertical spine of the tenth dorsal vertebra, towards which those of the other dorsal and of the lumbar vertebræ incline.

343

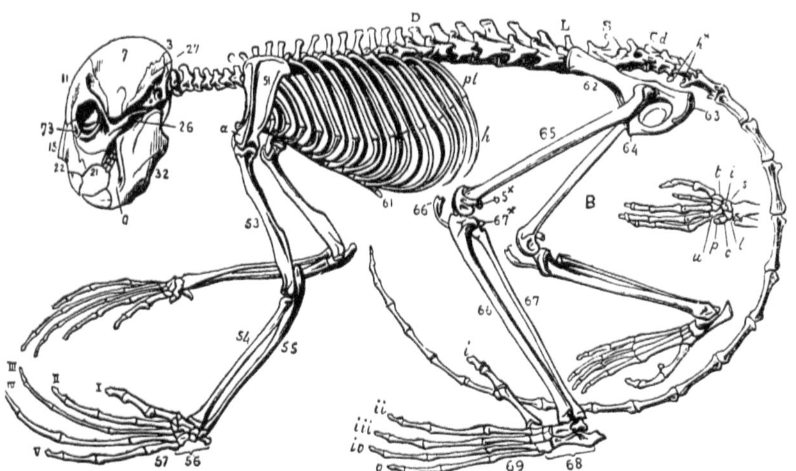

Skeleton of the Aye-aye. (*Cheiromys madagascariensis.*) B. Hand of Potto.

In the Aye-aye (*Cheiromys*), fig. 343, the true vertebræ describe one slight curve convex backward from the middle dorsal to the penultimate lumbar, beyond which there is a slight bend in the opposite direction to and including the sacrum. The bodies of the dorsal vertebræ gradually lengthen and deepen as they approach the loins, with a narrower and at last almost carinate under surface. The diapophysis, longest on the first dorsal, very gradually shortens to the eleventh, where the beginnings of the metapophysis and anapophysis are manifest. These processes become widely separated in the twelfth and thirteenth dorsals, and the diapophysis is lost. The neural spines are of equal length throughout the dorsal series. The vertical one is on the eleventh dorsal, towards which the rest of the dorso-lumbar series slightly incline.

The vertebræ go on increasing in size to the fifth of the lumbar series,—the diapophyses more especially, which recommence in the first lumbar; these processes are directed forward and downward, as well as outward, are truncate, with the anterior angle a little produced; that of the last lumbar is similar in shape and direction, but is smaller than the two preceding. The anapophysis overlaps the front margin of the following vertebra to the fifth lumbar, in which it becomes too short; it disappears in the sixth. The metapophysis overhangs the back part of the neural arch of the preceding vertebra. The neural spine decreases from the third to the last lumbar, where it has 3 lines of length. The last two ribs join their own centrum close to the front intervertebral space; the rest have the usual intervertebral articulation of the head. The first rib is the shortest (9 lines) and thickest; the others increase in length to the ninth, and then gradually shorten to the thirteenth, which is 1 inch 3 lines in length. The tubercle and diapophysial articulation exist to the eleventh rib; the twelfth and thirteenth articulate only by the head. The first cartilage articulates with the manubrium, the second to the seventh inclusive with the joints of seven sternebers, the eighth with the seventh, and the ninth to the joint between the seventh and eighth sterneber.

The bodies of the cervical vertebræ are broad, short, and flattened below in the last five. The last three have no neural spines: there are tubercular beginnings of these in the fourth and third; in the second it is 2 lines long, thick, and produced anteriorly; in the atlas it is as a small tubercle. The seventh cervical has a simple slender diapophysis, 2 lines in length; in the sixth it coalesces with the tubercle of a short pleurapophysis, also confluent by the head with the centrum, and projecting outward, backward, and downward, with an obtuse end. The vertebral artery, in its forward course, enters the canal between the pleur- and di-apophyses. The pleurapophysis simply completes that bony canal in the fifth cervical, making a short angular projection outward and forward in the fifth, fourth, and third cervicals. The low flat neural arch is narrowest in the fifth. The shape and disposition of the zygapophyses give an imbricate character to the union of those arches in the last six cervicals. The body of the axis is carinate below; that of the atlas has the usual state of an 'odontoid process;' the hypapophysial bar uniting with the neurapophysial pillars or crura of the atlas is carinate. Besides the wide canals for the vertebral arteries in the 'transverse processes' of the atlas, the neural arch is perforated above the base of that process on each side for the passage of a nerve.

In the short-tailed Indri (*Lichanotus Indri*), the atlas has a short hypapophysis, but no neural spine: the transverse process is moderately long and broad, and is perforated lengthwise and vertically by the vertebral artery, which afterwards pierces the neural arch. The transverse process is perforated in all the other cervical vertebræ: the pleurapophysial portion of that of the sixth forms a broad lamella directed downward and outward. Each of these cervicals has its hypapophysial ridge and neural spine, the latter moderately long and slightly increasing to the seventh. The broad neural arch is fissured behind. The spines of the dorsal vertebræ are continued of equal length throughout that region, and have the same direction. The dorsal diapophyses support each a metapophysial tubercle, which augments as they diminish, and seems to take their place in the eleventh and twelfth vertebræ, the ribs of which have no tubercle. In the twelfth dorsal the metapophysis projects from above the prozygapophysis, and is continued backward upon a well-developed anapophysis, which commences at once in that vertebra, and continues to be developed, although decreasing in length, to the penultimate lumbar inclusive. The metapophyses, which are prominent in the anterior lumbar vertebræ, gradually subside as these approach the sacrum. The diapophysis has a low rough tubercle on the first lumbar, which is developed into a depressed plate increasing in length and breadth as the succeeding lumbars approach the sacrum. As in the true Lemurs, eight pairs of ribs directly join the sternum, which consists of seven bones and an ensiform cartilage.

Nineteen is the usual number of dorso-lumbar vertebræ in the Platyrrhine group, the Spider-monkeys (*Ateles*) offering the exception of eighteen, viz. D 14, L 4: the varieties which have been formulised in the type-genus *Cebus* are due to freedom or confluence of pleurapophyses, as e.g. D 12, L 7, *Cebus hypoleucus*; D 13, L 6, *C. capucinus*; D 14, L 5, in most Capuchins. The tail is long in all, and prehensile in most Platyrrhines; it rarely has so few as 18 (*Callithrix sciureus* and *C. Spixii*), usually 30 vertebræ, or upwards, as in *Ateles paniscus*, which has 33 caudals.

In the little Ouistiti (*Hapale Jacchus*), the accessory tubercle appears upon the middle dorsal vertebra; it divides into met- and anapophyses on the tenth dorsal, where a diapophysial prominence still articulates with the tubercle of the rib. The diapophysis disappears in the succeeding dorsals in which the met- and anapophyses become distinct and remote, with progressive increase of size. The diapophysis reappears in the first lumbar as a short depressed process, and increases in length and breadth to the penultimate lumbar. In this vertebra the anapophysis becomes much shorter, and almost disappears in the last lumbar. The transverse process of the atlas is perforated lengthwise and vertically by the vertebral artery, and the neural arch is perforated. The bodies of the succeeding cervicals are produced posteriorly into a convex prominence which fits into a concavity on the fore part of the centrum behind. Eight pairs of ribs directly articulate with the sternum, which consists of seven bones.

In the Capuchin (*Cebus capucinus*), the tubercles representing met- and an-apophyses project distinctly, the one from the fore part, the other from the back part, of the diapophysis of the fifth dorsal: they progressively increase in size, and become quite distinct in the thirteenth dorsal, in which the metapophysis has risen upon the anterior zygapophysis. The anapophyses continue to be developed to the penultimate lumbar. The diapophyses progressively increase in length from the first to the last lumbar vertebræ. Hæmal arches are articulated to the inferior interspaces of the six anterior caudals, and are supported by distinct hypapophyses from the fourth caudal, which processes continue to be developed after the hæmapophyses have ceased to be so. Nine pairs of ribs articulate directly with the sternum, which consists of seven bones and an ensiform cartilage.

In the black Spider-monkey (*Ateles niger*), the tuberosity above the dorsal diapophyses becomes a ridge in the eleventh dorsal, and is produced forward into an angular metapophysis: in the thirteenth dorsal it is produced to the same extent backward into an anapophysis: in the fourteenth dorsal these processes are distinct and well-developed but the diapophysis has disappeared. The anapophysis is developed from the first and second lumbar vertebræ, and the diapophysis from all the lumbars, progressively increasing to the penultimate one. A pair of hypapophyses begin to be developed from the fifth caudal, and increase in size in the sixth and seventh. The hæmal arch is anchylosed to these processes in the eighth and ninth caudals, but the hypapophyses continue to be developed, without the addition of that arch, throughout the succeeding caudal vertebræ. The anterior zygapophyses disappear in the ninth caudal, but the metapophyses which support them in the preceding caudals continue to be developed to near the end of the tail. The diapophyses are single on each side in the seven anterior caudals, but are divided into an anterior and posterior portion on each side of the vertebræ throughout the rest of the tail. The third to the sixth cervical vertebræ inclusive show an anterior concavity and a posterior convexity of the articular ends of the centrums in the transverse direction, an anterior convexity and posterior concavity in the vertical direction, producing an interlocking joint, combining strength with freedom of motion, and analogous to that in the neck of birds. The pleurapophysial part of each transverse process is a broad depressed plate, with its anterior margin produced, and progressively increasing in size from the third to the sixth vertebra. A similar increase is presented by the neural spines, especially in the sixth vertebra. As in the Capuchins, the transverse process of the atlas of the Spider-monkeys is perforated lengthwise only by the vertebral artery, which afterwards perforates the neural arch. The atlas has a hypapophysial ridge, and the axis shows a corresponding tubercle. Nine pairs of ribs articulate directly with the sternum, which consists of eight bones and an ensiform cartilage.

The vertebral column of the Platyrrhine *Quadrumana* is the seat of greater and more important varieties: the caudal portion is reduced to a stunted 'coccyx,' the lumbar region is shortened and strengthened, and the sternum is composed of fewer and broader bones in the Apes properly so called. In the Monkeys and Baboons, the dorso-lumbar vertebræ are nineteen in number as a rule, either D 13, L 6, or D 12, L 7. The latter is the common formula in the Macacques: the caudals varying from upwards of 20 in *Macacus radiatus* to 15 in *M. rhesus*, and being reduced to 3 or 4 in *M. inuus*. In the Baboons (*Cynocephalus*), the caudals also vary from 25 in *C. porcarius* to 10 very small and stunted vertebræ in the Mandril (*C. papio*, fig. 344). In this, as in the Black-faced Drill (*C. porcarius*) and Thoth (*C. Thoth*), the dorso-lumbar vertebræ are reduced to D 12, L 6. An anapophysial tubercle is developed from the diapophysis of each dorsal vertebra, increasing in length to the two last, in which it has an independent origin. The metapophysis is suddenly developed from the tenth dorsal, and presents an articular surface to a second facet on the outer side of the hinder zygapophysis of the vertebra in front. The anapophyses continue to be developed from all the lumbar vertebræ, progressively decreasing as these approach the sacrum, and appearing in the last as a mere ridge on the upper part of the base of the diapophysis. The homotypal ridge may be recognised on the first sacral vertebra. There are rudiments of hypapophyses on the middle caudal vertebræ. Seven pairs of ribs articulate with the sternum, which consists of seven bones and an ensiform cartilage. The transverse process of the atlas is perforated lengthwise and vertically by the vertebral artery; which afterwards pierces the neural arch: the neural spine is represented by a small tubercle, and there is a hypapo-

344

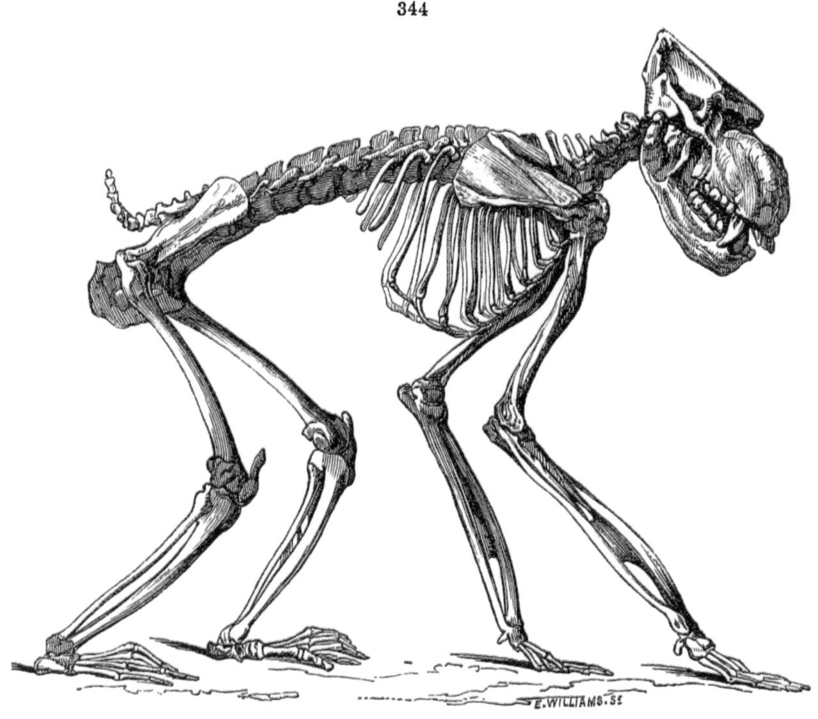

Mandrill. (*Papio Maimon.*) LXIX'.

physial ridge. The centrum of the axis is much produced backward, underlapping that of the third vertebra: this character is gradually lost in the succeeding vertebræ: the transverse process of the axis ends in two tubercles. The lower (pleurapophysial) division of the process is compressed in the third cervical, and becomes developed into a plate, progressively increasing, and disproportionately so in the sixth cervical: it is absent in the seventh cervical, the transverse process of which is, however, still perforated by the vertebral artery. The neural spines are simple, and increase in length from the third to the seventh cervicals. Those of the dorsal vertebræ are longer and stronger, but diminish in length as they approach the loins: that of the tenth indicates the centre of motion of the trunk.

In the Pig-tailed Macacque (*Macacus nemestrinus*), the atlas has a strong hypapophysis, but no neural spine or tubercle: the transverse process is perforated obliquely. The back part of the centrum of the axis is much produced; that of the third cervical is less produced. The spine of the axis is long and bent backward. A pleurapophysial plate extends obliquely from the transverse processes of the third, fourth, and fifth cervicals, and projects downward and outward as a distinct broad plate from that of the sixth vertebra. The long and simple transverse process of the seventh is not perforated by the vertebral artery. Metapophysial tubercles are developed upon the diapophyses of the second and succeeding dorsal vertebræ, increasing in distinctness and size to the tenth: in the eleventh the anapophyses become separate processes, and the metapophyses develope a facet for the accessory articular surface of the posterior zygapophysis of the tenth vertebra. This additional interlocking is continued to the antepenultimate lumbar, the joint being further strengthened by the underlapping of the long anapophyses: these disappear in the last lumbar. The diapophysis is a rudimental ridge in the last dorsal, but becomes a distinct depressed sharp plate in the first lumbar, and progressively increases in size with an antroverted direction in the succeeding lumbar vertebræ. Eight pairs of ribs articulate directly with the sternum, which consists of eight bones and an ensiform cartilage.

The Doucs (*Colobus, Nasalis, Semnopithecus*) have commonly D 12, L 7: but sometimes D 13, L 6 (*S. melalophis*). In *Semnopithecus Entellus*, the cervical transverse processes incline downward: their pleurapophysial divisions from the second to the sixth increase; but this part is wanting in the seventh, and the transverse process is imperforate. The accessory tubercle is well developed on the diapophysis of the ninth and tenth dorsals; the diapophysial part disappears on the eleventh and twelfth dorsals, in which the accessory tubercle becomes divided into well-marked met- and an-apophyses. The diapophysis reappears on the first lumbar, and progressively increases to the antepenultimate one. The metapophysis exists as an elongated tubercle outside the prozygapophysis from the eleventh dorsal to the last lumbar, and the anapophysis is present from the tenth dorsal to the sixth lumbar. The hæmal arch is present in a few of the anterior caudal vertebræ. Seven pairs of ribs directly articulate with the sternum, which consists of six bones, slender as in all previous *Quadrumana*.

In the Gibbons, with D 13, the lumbar vertebræ are 5, save in *Hylobates syndactylus*, fig. 189, where they are reduced to 4. In the Silvery Gibbon (*H. leuciscus*), the transverse process of the atlas is only perforated lengthwise and the neural arch grooved by the vertebral artery. A pleurapophysial part of the transverse process begins to project forward on the fifth cervical, and becomes a distinct and larger depressed plate on the sixth: the transverse process of the seventh is a simple diapophysis, and is imperforate. The metapophysis and anapophysis become distinct in the twelfth dorsal, and diverge from each other with increase of size in the thirteenth. The anapophysis disappears in the lumbar vertebræ, whilst the diapophysis reappears and the metapophysis is retained. The interlocking joints, common to the preceding *Quadrumana* with *Carnivora*, here and henceforth cease. Seven pairs of ribs directly join the sternum, which consists of the manubrium, the body, which consists of two or more anchylosed broad and flat bones, and a slender bony base of the 'ensiform cartilage.' Two pairs of ribs, and part of a third pair, articulate with the manubrium.

In the Siamang (*H. syndactylus*, fig. 189), the last dorsal shows well the separate diapophyses, metapophyses, anapophyses, and zygapophyses, more particularly the distinction between the anterior zygapophysis and the now superadded metapophysis. The diapophyses are broad depressed plates, progressively increasing in the first three lumbar, whilst the anapophyses diminish and disappear on the third lumbar. The metapophysis recedes from the anterior zygapophysis in the last lumbar, and becomes quite distinct from it in the first sacral, in which, nevertheless, the articular surface of the zygapophysis has a nearly vertical position. The sacrum, by its greater breadth and the number of vertebræ forming it, indicates the nearer affinity of the Siamang, than of other Gibbons, to the Orangs.

In the Orang-utan (*Pithecus Satyrus*), the vertebral formula is:—7 cervical, 12 dorsal, 4 or 5 lumbar, 5 or 6 sacral, 2 or 3 caudal. The transverse process of the atlas is bituberculate, and is perforated lengthwise by the vertebral artery, which afterwards grooves the neural arch: there is a low hypapophysial tubercle, but no neural spine. The transverse process of the axis is deeply grooved, but not perforated; consisting almost entirely of the pleurapophysial portion. In the third vertebra the two portions of the transverse process are united, external to the perforation by the vertebral artery. In the fourth cervical the pleurapophysial part projects distinctly below the diapophysial part, and progressively diverges in the fifth and sixth, increasing in size, especially in the latter, without, however, acquiring that antero-posterior breadth which gives it the lamelliform character in the inferior apes. The transverse process of the seventh cervical consists of the diapophysis only, and is grooved below, not perforated, by the vertebral artery. The distinct nature of the equally simple transverse process in the second and seventh cervical vertebræ of this Orang is well shown by their different relative positions to the groove with which the vertebral artery has impressed them. The neural spine of the axis is bifurcate; that of the third cervical is simple, long, and slender; those of the succeeding cervicals are still longer, and progressively increase in thickness as well as length. The metapophysis appears as a tubercle near the base of the anterior zygapophysis of the twelfth dorsal: it is equally distinct on the first lumbar, but subsides to a slight eminence on the succeeding lumbar vertebræ. The anapophysis is only distinguishable from the diapophysis upon the first lumbar vertebra; it is not so developed as to interlock, but serves to illustrate the relation of the diapophysis of that vertebra to those of the antecedent dorsals and the succeeding lumbars. The spine of the third dorsal has an anterior and posterior prominence: the succeeding spines gradually diminish in length, but increase in breadth and antero-posterior extent to the penultimate lumbar. Seven pairs of ribs directly articulate with the broad sternum, which consists of the manubrium and four pairs of ossicles, the two lower pairs of which have coalesced. The manubrium is relatively shorter than in the Gibbons, and receives only the first and part of the second pairs of ribs. As a rule, the number of dorso-lumbar vertebræ, in *Pithecus*, is 16: that of the sacro-caudal vertebræ 8. The first rib is less curved, and describes a smaller portion of a circle than in Man: its head is relatively larger, and is supported on a shorter neck: it has an epiphysis, as in Man. The distal portion is relatively less expanded than in Man. The other ribs chiefly differ in their more compressed form and their more gradual and equable curvature.

In the Chimpanzee (*Troglodytes niger*, fig. 345), the vertebral formula is:—7 cervical, 13 dorsal, 3 or 4 lumbar, 5 or 6 sacral, and 2 or 3 caudal. The pleurapophysial portion of the transverse process of the atlas is shorter than in the Orang, and has not united with the longer diapophysial division: the canal for the vertebral artery is thus not quite circumscribed by bone: the artery afterwards pierces the neural arch on the left side, and deeply grooves it on the right side. The two portions of the transverse process of the axis have coalesced, and form a thick tubercle externally, surrounding the vertebral artery: this tubercle increases in breadth in the third, and in length in the fourth; in the fifth it sends a distinct tubercle from its lower part, and the answerable part forms an antroverted, obtuse, broad process in the sixth. The pleurapophysial element is wanting in the seventh, in which the diapophysis is deeply grooved below for the vertebral artery. The spines of the 4—7 cervicals are long and simple. A metapophysis may be distinguished in the eleventh and twelfth dorsals, which becomes distinct from the diapophysis in the thirteenth, and projects from the outside of the prozygapophysis in all the lumbar vertebræ. The diapophyses are longest in the first and second lumbars, are shortest in the third, and are augmented in the fourth by the developement of a thick anapophysis at their back part, which here articulates with the first sacral vertebra. In old males the fourth lumbar becomes the first sacral by a more complete coalescence. Seven pairs of ribs directly join the sternum, which consists of five flat bones and an ensiform part: the fourth and fifth bones have coalesced: the manubrium, as in the Orangs, is the broadest, and receives the first pair and part of the second pair of ribs. These are shorter, and their neck relatively longer than in the Orang, and they are more curved. The thirteenth rib retains a distinct articular tubercle and neck.

345

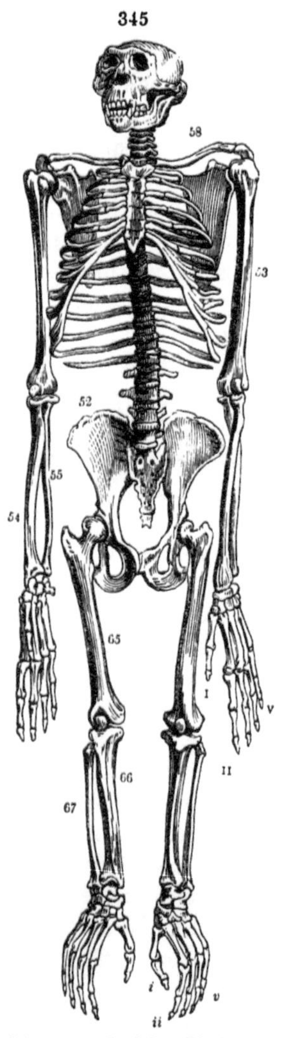

Chimpanzee (*Troglodytes Niger*). CIII'.

In the Gorilla (*Troglodytes Gorilla*, fig. 346), the dorso-lumbar vertebræ, as in the Chimpanzee, are 17 in number, the thirteenth dorsal answering to the first lumbar in Man, with the pleurapophyses retained as distinct elements. The bodies of the middle dorsal vertebræ are shorter in proportion to their breadth; the diapophyses are thicker, stand more directly outward, and the costal surfaces are more concave and oblong than in Man; the metapophysis, which projects distinctly in the eleventh vertebra in Man, does not so appear until the twelfth in the Gorilla. In the first dorsal the diapophysis projects directly outward; the proportionate increase of the centrum is greater than in Man; the neural spine is less obliquely bent backward, and is thicker antero-posteriorly, though not longer; the anterior zygapophyses are more produced; the diapophyses are broader and somewhat shorter. In the eleventh dorsal the neural spine is much expanded at its extremity. In the twelfth, there are distinct and well-developed metapophyses, projecting from the fore part of the diapophyses, and overhanging the anterior zygapophyses: this vertebra corresponds in this character

346

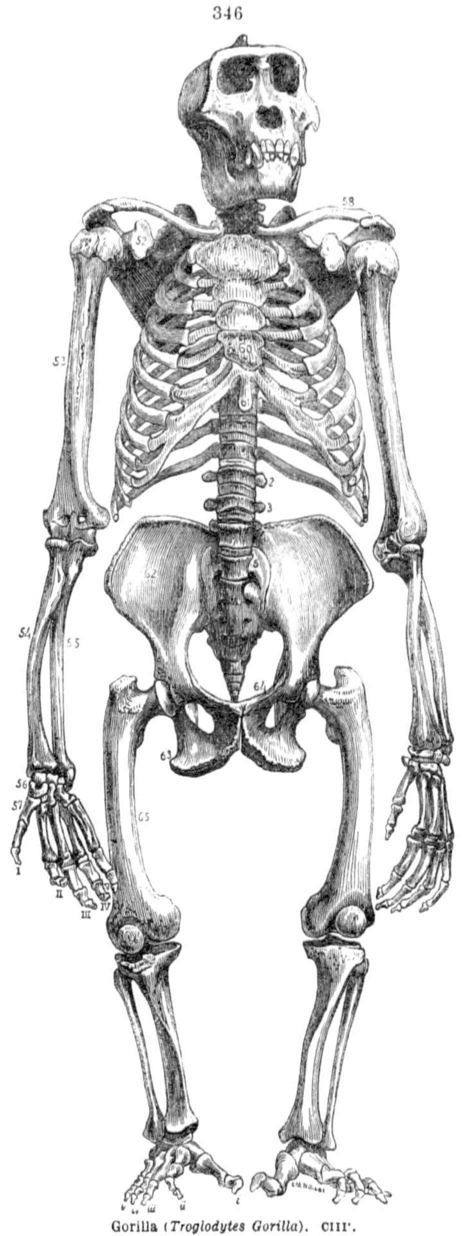

Gorilla (*Troglodytes Gorilla*). CIII'.

with the eleventh of the Human subject. The neural spine is broader and thicker, especially superiorly; there is but one costal surface on each side; the diapophyses are reduced in size, the metapophyses equalling them, the body and neural spine increasing. The thoracic ribs are longer and thicker, more convex on their inner side, with the subcostal groove not defined, except in two or three of the longest ribs near their vertebral end; the neck is shorter and thicker than in Man; the longest rib is one foot four inches in length,—that of the longest rib in an average-sized man being thirteen inches. The manubrium sterni is much broader than in Man (fig. 183), and less deeply excavated for the clavicles; the three or four sternebers which coalesce to form the 'body' of the breast-bone have a like character. The cervical vertebræ differ most from the Human in the extraordinary length of the spines of the last five vertebræ; that of the fourth cervical is not less than three inches and a half; the spines of the sixth and seventh cervicals gradually decrease in length and increase in thickness: the spine of the dentata is trihedral, the surfaces being divided by produced sharp ridges: the canal for the vertebral artery decreases in diameter from the sixth forward to the atlas. The bodies of these vertebræ are longer in proportion to their breadth than in Man, and the lower (pleurapophysial) part of the transverse process of the sixth is more suddenly increased in length and breadth, and diverges more from the upper division of the same process. The atlas is narrower than in Man, with a wider neural canal, especially between the condyles, which are smaller than in Man. An obtuse process is developed backward from the part representing the body, which is broader than in Man; the perforation of the transverse process is smaller, and that process is narrower, especially vertically; the groove behind the upper articular processes is deeper and narrower. The axis or dentata differs chiefly in the greater size of the neural canal, and in the greater length and less breadth of the neural spine; the zygapophyses are smaller, the transverse processes are more directly perforated by the arterial foramina, and the diapophyses are more produced.

In the first lumbar vertebra, fig. 346, 2, the metapophysis is still large and distinct; the anterior zygapophysis becomes more convex and oblique in position; the diapophysis is suddenly elongated, as compared with that of the corresponding (second) Human lumbar vertebra; the chief difference is seen in the smaller size of the neural canal which relates to the inferior developement of the lower extremities. The same difference obtains in the second (3, answering to the third Human) lumbar vertebra; the diapophyses are broader and more depressed in the Gorilla; a fossa divides the anterior zygapophysis from the metapophysis; the centrum is as broad as in Man, but is deeper and longer; the neural spine extends more obliquely backward, and its expanded apex is bifid. In the third lumbar vertebra, 4, the difference is very striking in the minor expanse of the centrum in the Gorilla, especially behind, in the much smaller and more depressed form of the neural canal, in the shorter and broader diapophysis, the more distinct metapophysis, in the convex anterior and more approximated posterior zygapophyses, and in the greater length of the centrum. In old males this vertebra is included by the ilia. The whole series of true vertebræ in the Gorilla form but one curvature, which is slightly concave forward, especially in the dorsal region.

The sacrum departs in a greater and more instructive degree from the Human type; it consists of five or six anchylosed vertebræ, but they are longer and narrower than in Man, and present a very slight curve, with the concavity forward; the neural foramina are much smaller, the neural spines much more developed, and coalesce to form a single strong bony ridge, extended over and gradually subsiding on the last sacral vertebra, the neural arch of which is entire; the first sacral vertebra, ib. 5, answers to the fifth lumbar in Man; the zygapophyses are smaller, but the metapophyses are present and well developed. The posterior outlets for the sacral nerves are very small, and the whole neural canal is much more contracted than in Man.

B. *The Skull.*—The skull of the Aye-aye, fig. 343, in comparison with that of lower mammals of similar size, is remarkable for the large proportion of the cranium to the face, and the extreme shortness of the latter in advance of the orbits. Its profile contour, from the upper border of the foramen magnum, curves rapidly from the occipital to the parietal region, and is continued with a bold convexity to the root of the nose, whence it slopes straight to the nostril. The cranium is still more convex transversely; it expands a little in advance of the lambdoid ridge, and gradually, but very slightly, contracts to the post-orbital processes; these, meeting with the malars, complete the rim of the orbit, which opens widely beneath that part of the frame into the temporal fossa.

In the complete circumscription of the rim of the bony orbit *Chiromys* exemplifies its quadrumanous affinity; whilst it shows the special family to which in that order it belongs by the deficiency of the wall partitioning the orbital from the temporal cavity. The Lemurs, in this defect, indicate the transition to the lower unguiculate *Mammalia*, the *Galeopithecus*, fig. 253, offering the last step by the incompleteness of the orbital frame-ring behind. The outlook of the orbits, in the Aye-aye, obliquely forward, upward, and outward, but least so in the last direction, differs significantly from the direct outward aspect of those ill-defined cavities in most Rodents.

The basioccipital extends to the fore part of the large tympanic bullæ, to abut against which its margins are slightly produced. The occipital condyles are long and narrow. The plane of the foramen magnum forms with the basioccipital an angle of 125°, its aspect being downward and backward. The paroccipital is a low eminence, and the mastoid in front of it is hardly more prominent; neither process extends freely downward. The superoccipital, ib. 3, is a thin plate moulded on the middle and lateral lobes of the cerebellum, and showing outwardly their respective prominences. The petrosal is impressed by the pit for the cerebellar appendage.

There is a small triangular interparietal. The basisphenoid is expanded by a large sinus, and coalesces with the presphenoid. The alisphenoid developes the ectopterygoid ridge, extending from between the squamosal and tympanic to the outer side of the entopterygoid; both plates are imperforate. The natiform protuberances form deep depressions in the alisphenoid, on each side the flat square platform of the cranial surface of the basisphenoid, in the middle of which is the subcircular pituitary pit. There are no clinoid processes. The alisphenoids join the parietals, which contribute the greatest share to the formation of the calvarium. The tympanic, coalescing with the petrosal, is, together with that element, expanded into an oval bulla on each side the basisphenoid. The parietals, 7, impressed from within to transparent thinness by the longitudinal convolutions of the cerebrum, do not exceed half a line in thickness elsewhere.

The coronal suture crosses the cranium transversely three lines behind the postorbitals: the frontal suture remains, as in other *Lemuridæ*, and, like the sagittal, it is a harmonia. The fore part of the frontals, 11, project a little between the origin of the nasals, and also between the nasals and maxillaries; they then join the lacrymals, form the upper half of the inner wall of the orbit, and unite behind with the orbitosphenoid, alisphenoid, and parietal. The rhinencephalic fossa is subcircular and large: the median septum is produced into a 'crista galli.' The frontal sinuses give no outward indication, but are extensive; they are divided from each other by a median bony septum; each division communicates with the nasal chamber by a median orifice and by a lateral one with the antrum. The nasals, 15, join above with the frontals and at the sides with the premaxillaries, 22. The presphenoid is short, smooth on the under surface, and concave there transversely. The vomer quickly assumes the form of a vertical plate, with the free hind border concave. The palatines form the hinder third of the bony palate; the suture of each with the maxillary is slightly convex forward: they are divided from the inner alveolar wall of the last two molars by a groove which deepens into a fissure, bounded beyond the last molar by the pterygoid. The maxillary forms more than the middle third of the palate, leaving the smallest share of the roof of the mouth to the premaxillary. The facial plate of the maxillary, 21, extends by a narrow produced apex to the lacrymal, 73, but is excluded from the frontal by the junction of the lacrymal with the premaxillary; it is perforated by a small antorbital foramen. The premaxillaries constitute a larger share of the facial wall, rising as high as the nasals, between which and the maxillaries they interpose a broad plate, circumscribing, with the nasals, the external nostril. The socket of the incisor curves upward and backward to the maxillary, in which it is continued to beneath the orbit. The malar bone, 26, is long and deep, especially below the orbit, of which it forms the lower half; and where it bends outward to expand that cavity, it unites with the lacrymal and extensively with the maxillary anteriorly, and bifurcates behind,—the narrower branch mounting to the postorbital, the broader one continuing backward to the squamosal, 27. This essentially facial or maxillary element is anchylosed not only with the mastoid and petrosal, but also with the tympanic; its cranial plate terminates by a convex border overlapping the contiguous borders of the alisphenoid and parietal. The articular surface for the mandible is broad and flat, save where its inner border bends down upon the side of the petro-tympanic bulla. There is no ridge behind it to prevent the free movements of the mandible backward and forward, accompanying the rodent action of the great scalpriform incisors: in this the Aye-aye differs from other *Lemuridæ*.

The mandible, 32, is short and deep: each ramus is compressed and straight; they converge at an acute angle to a short ligamentous symphysis. The condyle is sessile, narrow, rather long, convex both across and lengthwise, and the latter most so, looking backward and upward, and placed on the level of the grinding-teeth.

The thin borders of the ascending ramus diverge from the condyle as they pass, the one downward and inward to the low angle, and the other forward and upward to the better-marked and more advanced coronoid, the obtuse end of which is nearer the last molar than the condyle. A slight ridge above the angle bounds the surface for muscular insertion behind; and here the angle is a little inflected.

In the Woolly Lemur (*Lichanotus laniger*, fig. 177), the cranium has a short paroccipital and a shorter mastoid process which coalesces with the base of a large petro-tympanic bulla. The squamosal is perforated by a venous foramen anterior to the auditory meatus. The malar extends backward almost to the glenoid cavity, which, as in following *Lemuridæ*, is defended by a posterior ridge. The large orbits reduce the intervening part of the frontal to a narrow channel. The premaxillaries are divided anteriorly by an angular cleft separating in the same degree the anterior or mid-incisors from each other. The lower jaw is remarkable for the great production of its broad and rounded angle: the back part of its symphysis is also produced.

In *Stenops gracilis*, and especially in *Tarsius spectrum*, the most remarkable feature in the cranium is the expanded frame of the orbits, which are closely approximated above the nasal bones. These overhang the premaxillaries, the most produced part of which forms the lower boundary of the external nostril, from which, in the Slender Lemur, the premaxillaries slope downward and backward to the incisive alveoli. The temporal ridges are widely separated along the upper part of the globular cranium, where the coronal and fronto-sagittal sutures intersect each other at right angles. As in the Aye-aye and most *Lemuridæ*, the cranial sutures are 'harmoniæ.'

347

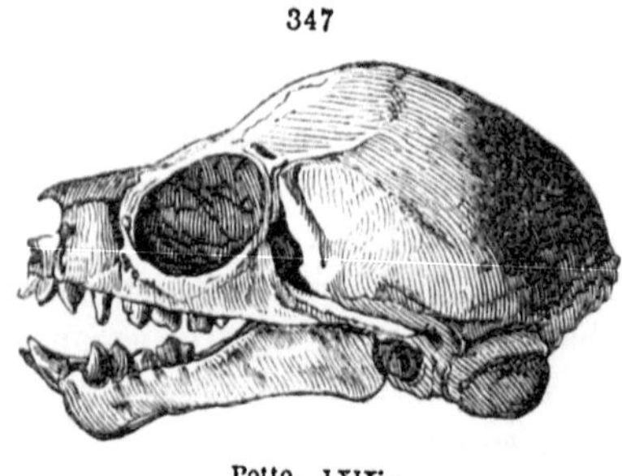

Potto. LXIX'.

In the Slow Lemur (*Stenops tardigradus*), the orbits are less closely approximate than in the *Stenops gracilis*, and the anterior surface of the small premaxillaries is more nearly vertical. The vomer divides the nostrils to their posterior apertures.

In the Potto (*Perodicticus*, fig. 347), as in other Slow Lemurs (*Stenops*), the cranial expansion behind the meatus auditorius forms one-third the length of the skull, owing to the great proportional size of the occipital and mastoid. The interorbital space is broader than in *Stenops tardigradus.*

In the true Lemurs the facial part of the skull is more produced; it is formed by the lacrymals, nasals, and maxillaries; the premaxillaries continuing very minute. In *Lemur Macaco* the petrosal has a large and deep cerebellar fossa: a short tentorial ridge projects anterior to this. There is a low postclinoid ridge. The lateral sinus pierces the petrosal where it joins the parietal and meets a second venous channel traversing the middle fossa of the cranium to terminate at the postglenoid foramen. The foramen ovale is a small fissure between the petrosal and alisphenoid, less than the foramen rotundum, which is close to the foramen lacerum anterius: the outlet of the foramen ovale is in the Eustachian fossa.

The grey Lemur (*Chirogaleus griseus,* fig. 348) has the more common abbreviation of the antorbital part of the skull, in which the lacrymal foramen is conspicuous. The malar is perforated by the 'nervus subcutaneus malæ.' The coronoid process of the mandible, well developed in all *Lemuridæ,* is here very high.

348

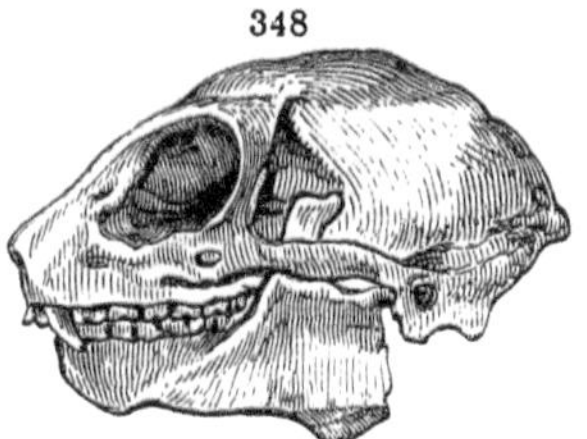

Grey Lemur. LXIX'.

The anterior cornua of the hyoid, in *Cheiromys* and other *Lemuridæ,* are longer than the posterior, and include epi- and cerato-hyals, supporting a cartilaginous stylo-hyal.

In the Platyrrhines the cranium is proportionally larger and the jaws less, as the species are smaller in size: they thus exemplify the immature characters of the larger species. The cranium is more globular, the occiput more protuberant, the 'foramen magnum' more advanced in position, and with a more downward aspect, in the Marmosets (*Jacchus*), and Ouistitis or Ti-tis (*Callithrix*), than in the Howlers (*Mycetes*). The frontal suture is obliterated in all, and the single bone, thence resulting, is triangular with the apex extending back, between the parietals, in some Capucins (*Cebus*) as far as the superoccipital (fig. 239, *Cebus*): thus repeating a piscine collocation of supra-cranial bones. The entocarotid perforates the back part of the petrosal.

In all Platyrrhines a division of the lateral cerebral venous sinus excavates the base of the petrosal to terminate at the postglenoid fossa, as in most Lemurs: the malar is similarly perforated by a facial nerve: the plate which divides the orbital from the temporal fossa exhibits a small unossified vacuity in most Platyrrhines.[1] The petrosal has a deep cerebellar depression. The postclinoid plate is more developed, the rhinencephalic fossa is smaller, and the orbital walls project more into the cranial cavity than in the Strepsirrhines. The lacrymal is not extended upon the face, and the foramen is within the orbit.

In *Callithrix sciureus* the orbits do not communicate with the temporal fossæ. There are no paroccipitals, and only a feeble mastoid ridge. The petrosals are slightly swollen at the basis cranii. The parietals articulate with the malars. There is a vacuity in the interorbital septum.

In *Cebus*, also, there are neither paroccipitals nor mastoids, and the petrotympanics form slightly swollen convexities. Besides the postglenoid venous foramen, there is a second at the end of the squamosal suture. The foramen ovale is between the petrosal and alisphenoid. The superoccipital plate has two large depressions, as in *Callithrix*. The orbital plate of the malar shows a small hole near its junction with the alisphenoid.[2] The basi-hyal is excavated behind; not so in *Callithrix*: the anterior cornua are long, and formed by epi- and cerato-hyals; the thyro-hyals are broader, not longer.

349

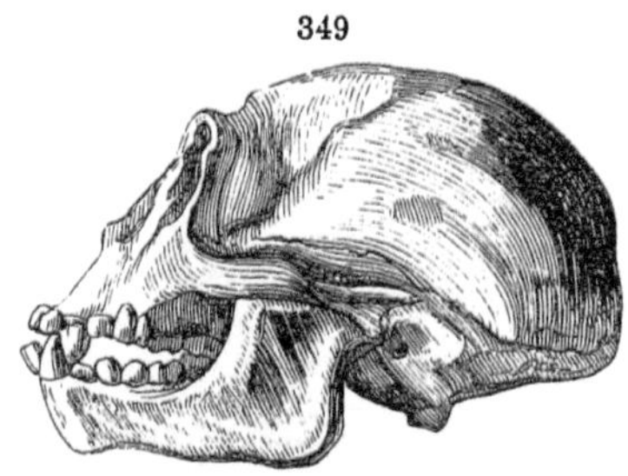

Capucin (*Cebus Apella*). LXIX'.

In the Spider Monkeys (*Ateles*) the paroccipitals and mastoids form rough tubercles. There is the same venous foramen as in *Cebus*, formed by the meeting of two converging sinuses between the squamosal and alisphenoid. Ossification has extended into one half of the tentorium. The cerebellar fossa in the petrosal is of great depth. The foramen ovale is formed by the petrosal and alisphenoid. The vomer extends to the posterior nares. The incisive foramen is large and single.

The symphysis of the lower jaw is completely anchylosed, and the angle of the jaw rounded off, as in most Platyrrhines. The condyles and small coronoid processes are of equal height: in Marmosets the coronoid is higher, and in *Hapale Jacchus* the mandibular angle is slightly produced. The basi-hyal is a convex plate: the cerato-hyals are shorter than in *Cebus*: the thyrohyals are longer.

[1] CIV'. pl. vi. (*Cebus*, Douroucouli, Chamek.)

[2] These relics of the orbito-temporal vacuity were first noticed as such by Prof. Filippi.

In the Red Howler (*Mycetes seniculus*, fig. 350) the superoccipital region is almost flat and vertical, at right angles with the parietal surface, from which it is separated by a well-defined ridge: the foramen magnum looks almost directly backward. The maxillo-premaxillary sutures demonstrate the junction of the premaxillaries with the nasals. The ectopterygoids much exceed the entopterygoid plates in size. The large malar foramen communicates with the orbit: the suborbital foramina of the maxillary are two in number, and small. The chief feature of peculiarity in the skull of the Howler is the extraordinary depth of the mandibular rami, especially of their angular and ascending portions. This development relates to the protection and support of the still more extraordinarily developed hyoidean and laryngeal apparatus—the organs of the loud and dissonant cries which have procured for these South American Monkeys their common name. The superior length of the postglenoid process, in relation to the larger and heavier lower jaw, is worthy of notice. An obtuse paroccipital ridge extends from the condyle to the mastoid ridge. The precondyloid, jugular, and carotid foramina all open into an irregular fossa between the petrosal and paroccipital ridge. There is a small venous foramen outside the mastoid, and a second at the anterior border of the squamosal. The hyoid arch is reduced to the basi- and thyro-hyals; but the former is enormously developed, and expanded into a capacious sac with thin walls, and a posterior opening, admitting a laryngeal pouch. A narrow transverse plate descends from the roof of the bony sac. The cerato-hyals are obsolete. The thyro-hyals long, for suspending the sac to the upper angles of the large thyroid cartilage.

350

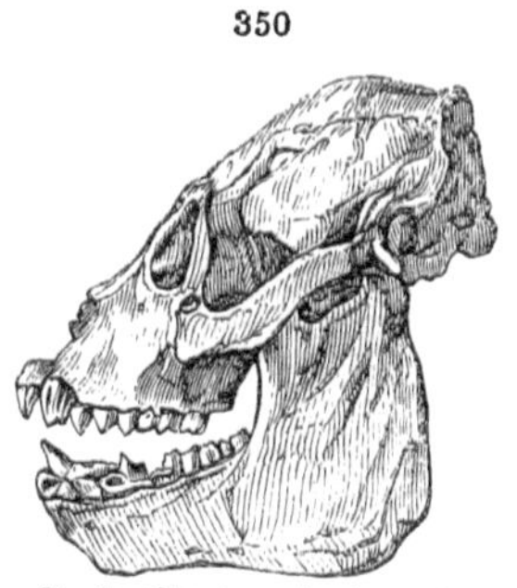

Howler (*Mycetes ursinus*). LXIX'.

There is much greater diversity, and more marked ascending steps of structure, in the skull of the 'Old World' than of the 'New World' Monkeys. No Catarrhine shows ossification of the tentorium; and in all the preclinoid, as well as postclinoid, processes defend the sella. The same remark, as to concurrence of immature proportions of cranium and jaws with infantile stature, applies to the Catarrhine as to the Platyrrhine Quadrumana. But the larger species of the lower groups (*Cynocephalus, Papio,* e.g.) show more carnivorous or brutish proportions of skull than do those (Orang, Gorilla) of the higher group.

In the Black-faced Drill (*Cynocephalus porcarius*, fig. 351) the facial much exceeds the cranial part of the skull. The superoccipital is almost flat: but, sloping upward and backward, forms an acute angle with the parietal, from which it is divided by a strong ridge, where the diploë is obliterated. The mastoid is more developed than the paroccipital prominence; but both are low. The jugular fossa is distinct from the precondyloid and carotid foramina; outside the latter is a short 'vaginal' process. The petrosal bifurcates anteriorly into a 'eustachian' and an 'apical' process: the latter underlaps the base of the pterygoid process: the inner surface of the petrosal is closely applied to the basisphenoid and basioccipital as far as the 'foramen jugulare:' there is, thus, no 'foramen lacerum basis cranii.' The squamosal is perforated near its middle by one or two small foramina, but there are no 'post-glenoid' outlets of the lateral sinuses. The foramen ovale is between the petrosal and alisphenoid, and the nerve which it transmits pierces the base of the broad ectopterygoid: the entopterygoid plate is comparatively small, but ends in a hamular process. The glenoid articular surface projects from the under part of the base of the zygoma, and is slightly convex: it is defended by a postglenoid process. The vomer divides the posterior nostrils, and there is a venous sinus or foramen between its base and the presphenoid. The coalesced nasals are prominent and gradually expand as they advance forward: they unite with a small proportion of the premaxillaries. The fossæ between the nasals and maxillary tuberosities are short and wide. The pterygoid fossæ are large and deep. The alisphenoid is separated by the squamosal from the parietal. The upper angle of the mastoid is wedged between the superoccipital and parietal. The limits of the interparietal may be traced upon the inner surface of the calvarium. There is a shallow cerebellar fossa above the meatus internus. The optic foramina are approximated. The entry to the rhinencephalic fossa is much contracted by the bulging prominence of the roofs of the orbits.

351

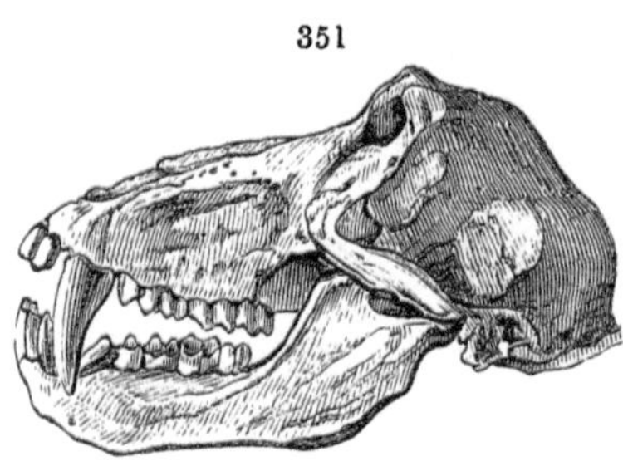

Skull (*Cynocephalus porcarius*). LXIX'.

In the Magot (*Macacus Inuus*, fig. 352) and other Macacques, with a general reduction of the size of the animal, the jaws are concomitantly reduced, so that the cranial cavity forms one half of the length of the skull. The general characters are those noted in the Drill. In *Macacus nemestrinus*, a process of styliform shape is developed from the lower end of each mastoid. The posterior clinoid plate is largely developed and is perforated. The cerebellar fossa is moderately deep; the foramen ovale is between the alisphenoid and petrosal. The entry to the rhinencephalic fossa is contracted by a pair of lateral processes.

352

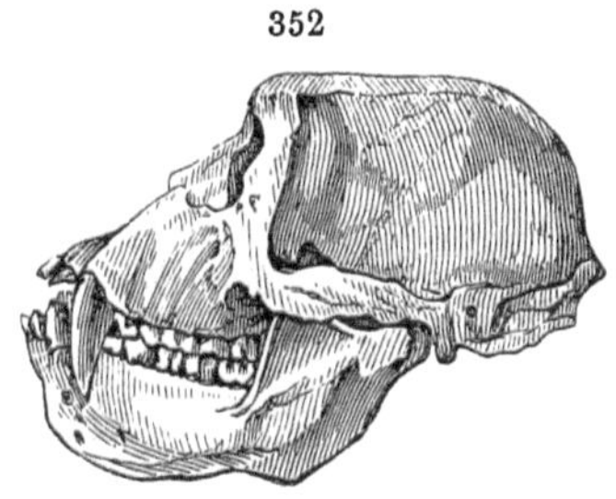

Skull of Macacus Inuus. LXIX'.

In the still smaller Monkeys (*Cercopithecus*, fig. 353) the cranial cavity forms a larger portion of the skull. In *C. ruber*, the alisphenoid joins the parietal on the left side, not on the right. In all the premaxillaries rise high between the maxillaries and nasals: the interior of the cranium shows the cerebellar pit of the petrosal, and the well-developed crista galli dividing the deep rhinencephalic fossa. The postglenoid process is pointed, and in some (*Cerc. albogularis*) the mastoid also: the entocarotids pierce the petrosals. The Doucs (*Semnopithecus*) have a similar proportion of cranial cavity: in which the cerebellar fossa of the petrosal is both large and deep. The entry to the rhinencephalic fossa is constricted by the approximation of its lateral margins, which almost touch at the middle. The foramen ovale is between the petrosal and the alisphenoid. The tympanic air-cells extend into the mastoid and squamosal. The bony septum between the orbital and temporal fossæ is entire in all Catarrhines.

353

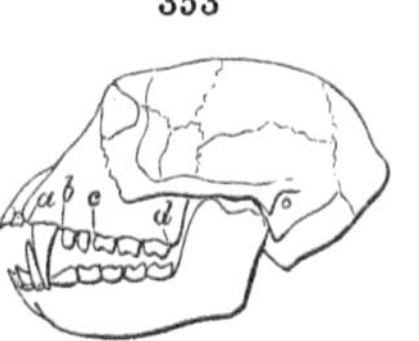

Skull of Monkey (*Ceropithecus ruber*).

In the Gibbons (*Hylobates*, fig. 354) the jaws are more shortened, the cranium more expanded. The alisphenoid is perforated by the foramen ovale, and joins the parietal. The premaxillaries do not reach the nasals.

354

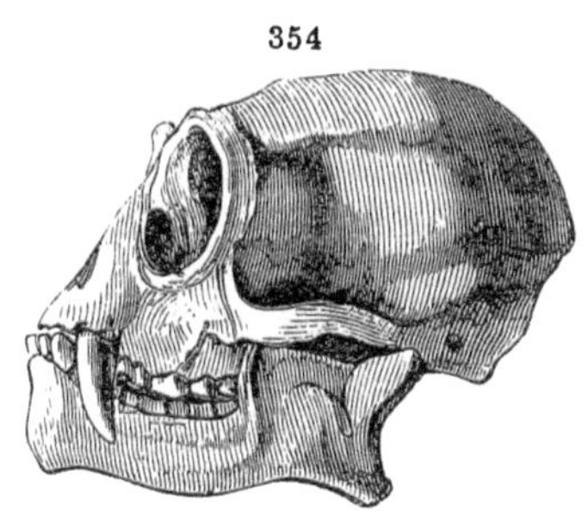

Siamang (*Hylobates syndactylus*). LXIX'.

The petrosal still shows the cerebellar fossa: its exterior surface is no longer swollen into a cellular bulla, but exhibits a well-marked eustachian process. The ento- and post-glenoid processes are well developed. The orbital border is thick and prominent, but the superciliary portions do not meet above the nasal. In the skull of a young Gibbon I have seen the exceptional extension of the frontal backward to the occipital, as in fig. 239 (*Cebus*). The mandibular symphysis is more nearly vertical and the angle more produced in the Siamang than in other Gibbons.

In the Orangs and Chimpanzees the foramen ovale is pierced in the alisphenoid, and the entocarotid traverses the petrosal, which has no cerebellar pit. The cranial and facial parts of the skull are about equal in the adult males, with fully developed laniary canines: in the females, with smaller canines, the jaws are less; and the cranial cavity predominates still more in the immature individuals. In some varieties of Orang (*Pithecus Satyrus*, fig. 355) the cranium rises higher than in others: and this feature is increased, in old males, by the growth of the parietal crest, which bifurcates anteriorly, defining a flat triangular space upon the frontal, and posteriorly to form the lambdoid crests,—a provision, as in *Carnivora,* for the large and powerful temporal muscles. The superorbital ridge does not project above the nasal bone: this, usually single and small, is flat. The premaxillaries coalesce with the maxillaries when the sockets of the permanent laniaries are developed: and about the same time the basisphenoid coalesces with the basioccipital. The sphenoidal sinus is almost wholly formed by the presphenoid, and it is divided by a longitudinal septum. The lower border of the basi-occipito-sphenoidal floor of the cranium is parallel with the bony palate or floor of the nostrils. The plane of the occipital foramen forms an open angle with the straight basi-occipito-sphenoidal line. The alisphenoid, 6, joins the parietal, 7; the precondyloid foramina are usually double on each side. The mastoid is not a

355

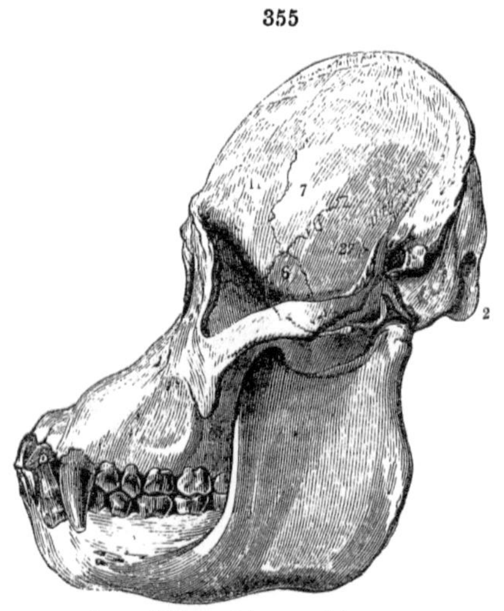

Orang (*Pithecus Satyrus*, male). CIII'.

prominent process; the tympanic air-cells are continued therefrom into the squamosal. The interorbital sinuses do not ascend to within half an inch of the upper level of the orbits, and there is consequently no proper frontal sinus: a cancellous structure occupies the usual place of this, below which the part of the interorbital septum formed by the hinder crista of the nasal bone and the frontal presents a very compact dense structure. The small venous canal continued from the foramen cæcum traverses the base of this septum to terminate at the lower end of the short nasal bone. The 'lamina perpendicularis æthmoidei,' or coalesced prefrontals, presents a quadrate form. The floor of the nasal cavity is long and thick, as compared with that in Man, and a larger proportion of it is contributed by the premaxillary. The orbits are directed forward and have a full oval shape. The area of the nasal cavity equals more than one third of that of the cranial cavity. The most anterior part of this cavity is formed by the deep, narrow, and well-defined rhinencephalic fossa: the 'crista galli' is rudimental. The division of the prosencephalic compartment, for the anterior and middle lobes of the cerebrum, is very slightly defined by the orbitosphenoid. The tentorial ridge is not continued backward beyond the petrosal. The nasal end of the incisive canal is divided by the process extending from the premaxillary to the maxillary; but this is the only part of the premaxillary which does not coalesce with the maxillary. The turbinal plates are less developed than in the Gorilla; the lower one is shorter than the one above; and there is not any plate answering to the small superior turbinal in the Gorilla and in Man. There is, in some Orangs' skulls, a process, formed by the anchylosed base of the stylohyal, which is defended in front by a low and obtuse vaginal process. The compact wall of the mandibular symphysis is thick and dense. The symphysis slopes from above downward and backward.

356

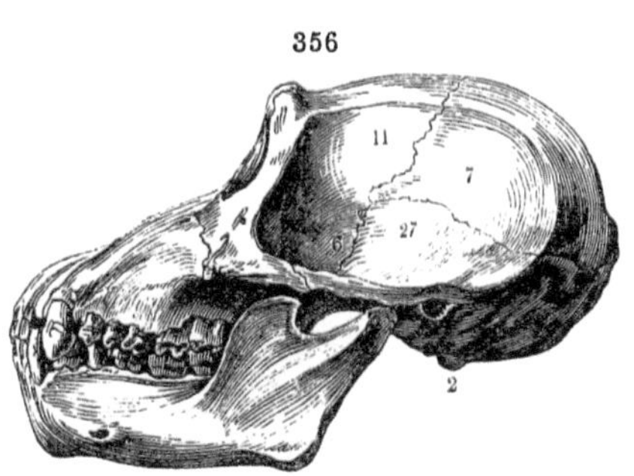

Chimpanzee (*Troglodytes niger*, male). CIII'.

In the genus *Troglodytes*, the squamosal, fig. 356, 27, usually articulates with the frontal, 11; the premaxillaries coalesce with the maxillaries earlier than in *Pithecus*, the alveolar part of the suture being obliterated before the nasal portion; the palatal part long remains. In the smaller species of Chimpanzee (*Tr. niger*) the temporal ridges meet, in old males, upon the sagittal line, but rarely develope a crest: the lambdoidal boundary-ridges are better marked.

Independently of the superiority of size of the *Tr. Gorilla* over the *Tr. niger*, the skull of the former, figs. 357—359, presents well-marked differences of form, differences in the developement and proportions of the intermuscular ridges, in the disposition of certain sutures, and in the structure and proportions of certain teeth. Compared in profile, the skull of both species, figs. 356 and 357, presents a striking difference from that of the Orang, fig. 355, in the prominence of the superorbital ridge; but the temporal ridges, after their junction upon the frontal, rise, in the *Tr. Gorilla*, into a strong and lofty sagittal crest, which is continued to the lambdoidal crest, the great extent of which masks the posterior convexity of the occiput. The zygomatic arch is proportionally much stronger in the Gorilla, and also differs from that in *Tr. niger* by the squamosal part being of equal depth with the malar part, and by its having its upper border convex, or produced into an angle instead of being straight or slightly concave. The alisphenoid is longer and narrower in *Tr. Gorilla*, and contributes less to the back wall of the orbit than in *Tr. niger*, in which it forms a much smaller proportion of that part than in Man. The spheno-maxillary fissure is not only larger in *Tr. Gorilla*, but is narrower and more vertical, not angularly bent as in *Tr. niger*. The extent of the premaxillary bones below the nostril is not only relatively but absolutely less in *Tr. Gorilla*, and the profile of the skull less convex at that part, or less 'prognathic,' than in *Tr. niger*. The breadth of the premaxillaries and of the incisor teeth is the same in both, whilst in all other dimensions the *Tr. Gorilla* greatly surpasses the *Tr. niger*: this is seen in the height of the sagittal crest, the thickness of the great superorbital bar of bone, the prominence of the ectorbital walls, and of the inferior tumid malar boundaries of the orbits, fig. 358, 26. The nasal bones have united together

357

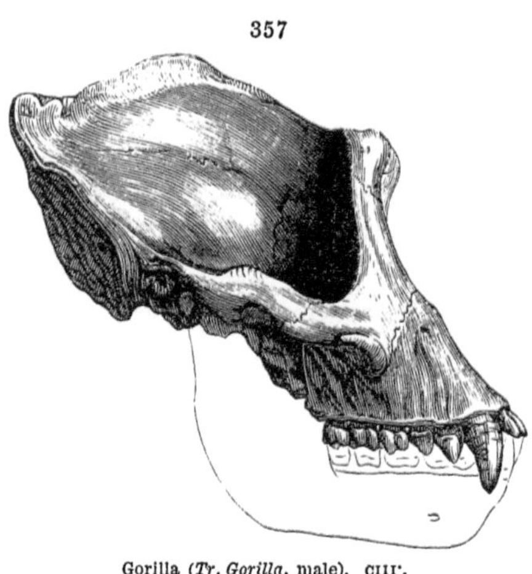

Gorilla (*Tr. Gorilla*, male). CIII'.

in *Tr. Gorilla* as in *Tr. niger*, but less completely, a linear indication of the median suture remaining along the exterior surface: the coalesced upper portions of the nasals ascend higher above the nasal processes of the maxillary than in *Tr. niger*, become contracted between those processes and there project slightly, their median coalesced margins being produced forward: they expand at their lower halves, and articulate not only with the maxillaries, 21, but with an expanded superior portion or dismemberment of the premaxillaries, 22. In the immature *Tr. niger*, the maxillo-premaxillary sutures show that each premaxillary bone terminates above in a point which does not reach the nasals. The orbits have a more subquadrate form, with the angles rounded off, in *Tr. Gorilla* than in *Tr. niger*; but their periphery is less sharply defined, especially below, than in *Tr. niger*. The ethmoidal cells are more swollen out, giving the interorbital space a greater breadth below and the lachrymal fossæ a more anterior aspect in *Tr. Gorilla*. The infraorbital canal issues upon the face relatively lower and further from the orbit. The whole nasal bone is relatively longer, and the distance from the orbits to the external nostril greater in the *Tr. Gorilla*. The malar bone, 26, is more convex outwardly, and is more remarkable for its vertical extent: it is flatter and developed more transversely in the *Tr. niger*. The larger proportional size of the canines in *Tr. Gorilla* impresses a corresponding difference upon the alveolar part of the maxillary bone in that species. Fig. 357 contrasts the broad flattened superoccipital surface of the Gorilla with the convexity of the same part in the *Tr. niger*, fig. 356: the difference is due to the much thicker and broader lambdoidal ridge in the larger species, which prolongs the surface beyond the cerebellar fossa, and gives the condyles and foramen magnum a rather more advanced position as compared with the *Tr. niger*. The next character, explicable in relation to the greater weight of the skull to be poised upon the atlas, is the greater prominence of the mastoid processes in the *Tr. Gorilla*, which are represented by only a rough ridge in the *Tr. niger*. These protuberances are cellular, and with a very thin outer layer of bone in the *Tr. Gorilla*. The lower surface of the long tympanic or auditory

358

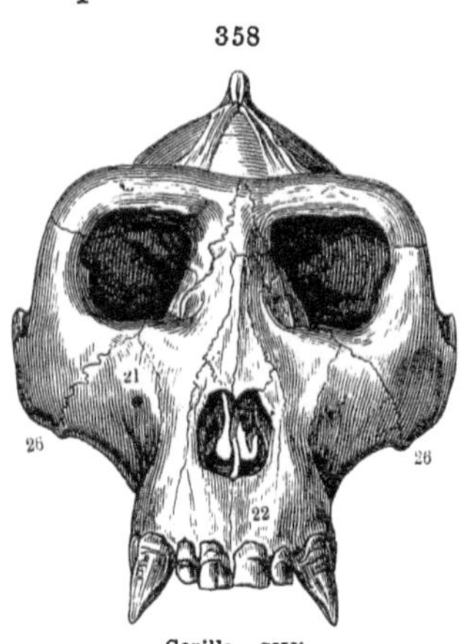

Gorilla. CIII'.

process is smooth and flat, or slightly concave, in *Tr. niger*, and developes a slight tubercle anterior to the stylo-hyal pit: in the *Tr. Gorilla* the same process is more or less convex below, and developes a ridge, answering to the vaginal process, on the outer side of the carotid canal. The processes posterior and internal to the glenoid articular surface are better developed, especially the internal one in the *Tr. Gorilla*, than in the *Tr. niger*: the ridge which extends from the ecto-pterygoid along the inner border of the foramen ovale terminates in *Tr. Gorilla* by an angle or process answering to that called 'styliform' or 'spinous' in Man, but of which there is no trace in the *Tr. niger*.

The palate is narrower in proportion to its length in the *Tr. Gorilla*, but the premaxillary portion is relatively longer in *Tr. niger*. Two anterior palatine foramina, one on each side the almost confluent incisive foramina, are more constant and conspicuous in *Tr. Gorilla*: the posterior palatine foramina are nearer the posterior border of the bony palate in *Tr. niger*. The pterygoid fossæ are relatively deeper and longer in *Tr. niger*. The stronger zygomatic arches, with the more developed sagittal and lambdoidal crests, are adaptive developments concomitant on the presence of larger canines and stronger mandible in the Gorilla; but the larger proportional molars and the smaller proportional incisors, the prominence of the nasal bones at their median line of coalescence, and the reappearance of the premaxillaries upon the face above the nostril with their longer enduring sutures, constitute a series of differential characters of more importance than such as are due to greater bulk or activity of muscles, and are inexplicable by the operation of external influences. The basi-hyal, as in the Chimpanzee, is deeply excavated behind: the cerato-hyals are obsolete: the thyro-hyals long and nearly straight. Further characteristics of the skull of the highest known Quadrumanous species will be shown in comparison with the cranial characters of the lowest races of Man.

359

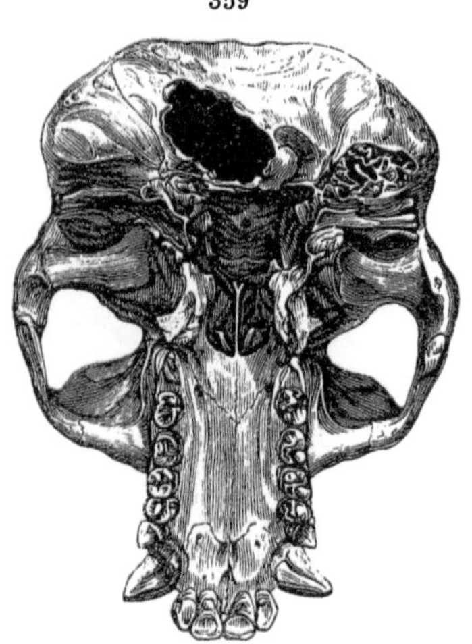

Base of skull, Gorilla. CIII.

C. *Bones of the Limbs.*—In Quadrumanes, as in Quadrupeds,

both pairs of limbs are concerned in support and locomotion; but are made prehensile in relation to an arboreal sphere of life by an opposable thumb-like condition of the innermost of the five digits, always conspicuous upon the hind-limbs, and, in most, upon the fore-limbs. Complete clavicles, and an elbow-joint allowing both rotatory and flexile movements of hand and fore-arm, are present in all.

The scapula of *Chiromys*, fig. 343, 51, differs from that of Rodents, and resembles that of Lemurs, in the proportions of the pre- and post-spinal fossæ. The subscapular surface does not show the inter-muscular cristæ which are usually so well marked in Rodents. The length of the acromion, *a*, is 6 lines; that of the coracoid is 7 lines: it is a simple compressed process. The glenoid cavity is a long oval, with the apex above and rather produced. The clavicle has a double bend upward and outward, and a half twist on itself.

The head of the humerus, 53, has a long-oval form, regularly convex, and surpassing in both breadth and length those dimensions of the glenoid cavity. The great tuberosity projects on one side to the same height; the small tuberosity is somewhat lower. A sharp deltoid ridge extends from the fore part of the great tuberosity halfway down the shaft. The supinator crest begins below the middle of the shaft, near its back part, standing well out, and thence passes in an almost straight line to the ectocondyloid tuberosity. The internal ridge projects from nearly the fore part of the distal fourth of the shaft, bridging over the humeral artery and median nerve on its way to the entocondyloid tuberosity where it coalesces with a shorter and sharper ridge, completing the epicondyloid foramen. The inner tuberosity is much more prominent than the outer one. The anconeal fossa is oblong, of moderate depth, and imperforate. The tubercle for the radius forms nearly half of the fore part of the elbow-joint; the back part is exclusively formed by the well-defined trochlear cavity for the ulna. The humerus reaches to the tenth rib, when bent upon the chest: and this proportion of length is characteristic of most *Lemuridæ*.

The radius, 54, is of equal length with the humerus; the head is nearly circular. The ulna, 55, is the longest bone of the fore-limb: it is compressed below the humeral joint, and gradually narrows to the lower fifth of the shaft.

The wrist-bones, 56, are ten in number, including a supplemental sesamoid on the outer side of the scapho-trapezial joint. The scaphoid is the longest, presenting its convex articular surface to the outer two thirds of the radial concavity, and articulating with the lunare, which completes the wrist-ball; at its distal surface it joins the 'intermedium,' the trapezium, and the trapezial sesamoid: the cuneiform offers a cup for the hemispheric end of the styloid process of the ulna and a flatter surface for the pisiform; this wrist-bone is long, and its articular surface is divided between the ulnar process and the cuneiform. The intermedium and cuneiform combine to form the cup for the ball common to the magnum and unciform, of which the latter bone contributes the larger share. The intermedium articulates with the trapezoid. The distal series of carpal bones have the usual relations to the metacarpals. The first, second, fifth, and fourth metacarpals, 57, progressively increase in length, with similar proportions as to thickness; but the middle metacarpal is double the length of the second, and suddenly contracts into a shaft more slender by half than the contiguous metacarpals. The phalanges of the same digit, III., are filamentary, and support the hooked probe-like finger adapted for the extraction of the xylophagous larvæ—the favourite food of the Aye-aye—from the canal in the wood which has been exposed by the scalpriform incisors.

The ilium, 62, is long and narrow, slightly expanded at both ends: it articulates with the two first sacral vertebræ, just touching the second by a projection above its middle. The iliac bones incline to the acetabula at an angle of 140° with the lumbo-sacral axis. There is an elongate tuberosity above the acetabulum for the origin of the rectus femoris. The ischia, 63, are continued almost in a line with the ilia, the posterior contour describing a very feeble curve concave backward; the tuberosities are slightly everted: a small projection behind the lower part of the acetabulum divides the great from the small ischiadic notches, both of which are very shallow. The obturator vacuities are large. The pubic oones, 64, pass from the acetabula at almost a right angle with the ilio-ischial axis; they converge to a short symphysis. There is a slightly marked ilio-pectineal prominence. The femur, 65, has a straight shaft, one third longer than that of the humerus. The neck is short: the great trochanter rises to the height of the head, and at the outer and lower part is developed into a small tubercle. Opposite to this the lesser trochanter projects from the inner side to a greater degree. The orifice for the medullary artery is at the back part, one fourth of the length from the head; the canal ascends. The inner condyle is rather the larger. The outer border of the rotular groove projects most. There is a sesamoid bone ('fabella') in each origin of the gastrocnemius. The tibia, 66, is about two lines shorter than the femur, and soon contracts below the head to a compressed shaft, giving a long and narrow subelliptic section; at the upper half it is very slightly bent, with the convexity forward. A roughish surface is continued from the tuberosity nearly one third of the way down the fore and outer part of the shaft. The orifice of the medullary canal is one fourth of the way down, just within the posterior border; the canal slopes downward. The tibia is one fifth longer than the ulna. The fibula, 67, touches the tibia only by the two extremities articulating with that bone, leaving an interosseous space co-extensive with their shafts. The outer malleolus is shorter and thicker than the inner one. There is a sesamoid in the external lateral ligament of the knee-joint, at its insertion into the head of the fibula.

The tarsal bones, 68, are seven in number. The naviculare has its shallow concavity for the astragalus supplemented by the strong ligament arising from its posterior and inferior margin, and inserted into the fore part of the inner malleolus; anteriorly it articulates with the three cuneiform bones, and externally at its fore part with the os cuboides: its depth exceeds its length. The calcaneum offers two articular surfaces to the astragalus, rather far apart; the lever projects moderately beyond the hinder surface, and is curved a little upward and inward. The ento-cuneiform offers at the anterior half of its outer part a trochlear surface, concave in one direction, convex in the opposite, to the powerful hallux. The meso- and ecto-cuneiform bones are narrower, the outer one is of nearly equal length with the inner, the middle one being the shortest. The cuboid is large and long, with the lower half of its calcaneal surface convex, the upper half concave, for an interlocking joint with that bone; it is grooved externally and beneath for the peroneus longus, and, as usual, supports the two outer toes. The base of the metatarsal of the hallux is broad, and its under border is produced into contact with that of the second metatarsal. The third metatarsal is a little longer than the second; the fourth has nearly the same length, and so has the fifth, 69, by reason of the backward production of the outer angle of its base. The proximal phalanx of the fourth toe is the longest. Fig. 343 shows how little the phalanges of *ii–v* differ in length or breadth.

With the exception of the attenuated state of the third digit of the fore-foot, the characters of the limb-bones of *Chiromys* are closely repeated in other *Lemuridæ*. The Pottos (*Perodicticus*), however, offer an anomaly, in the fore-hand, by the stunted phalanges of the index digit; and the pollex is large and opposable, fig. 343, B. The Galagos (*Otolicnus*) and the Spectres are, also, exceptional, by the excessive length of the calcaneum and naviculare in the hind-hand, whence the generic name *Tarsius* given to the latter Lemurs. In the relative length of the tarsus to the leg and to the rest of the foot *Chiromys* most resembles *Lichanotus* and *Propithecus*: it is rather shorter than in *Lemur* proper, being less than one third the length of the tibia, and only about one fourth the length of the whole foot. The scaphoid and calcaneum are proportionately rather shorter than in *Lemur perodicticus*[1].

In the Indri (*Lichanotus*) the scapula is remarkable for the length and strength of its coracoid process. The humerus, as in *Chiromys*, is perforated above the inner condyle, but not between the condyles. The interosseous space is considerable between the long and slender radius and the more slender ulna. The ilium has a strong tubercular process above the acetabulum. The femur is so long as to equal in length seventeen vertebræ of the trunk, measured from the first dorsal backwards. The fore part of both the astragalus and calcaneum is unusually produced. In the slender Lemur (*Stenops gracilis*) the humerus is perforated above the inner condyle, and has a wide intercondyloid vacuity. The iliac bones, fig. 360, *a*, are long, slender, and extended almost in the same line with the sacrum. The pubic bones, *b*, *c*, join the ilia at a right angle, and are inclined to each other at an angle of 40°; they form a very short symphysis. There is a small ossified patella. The feeble development of the vertebræ in the long lumbar region, the small sacrum, and contracted pelvis are points of resemblance with the Bat-tribe; and, together with the long and slender bones of the extremities, relate to the slow movements of this climbing quadruped.

360

Pelvis of the slender Lemur.

In *Lemur Catta* I have found the pubic symphysis ossified; the ilium has an epicotyloid ridge. The coracoid and acromial processes of the scapula are subequal. The humerus is perforated above the inner condyle. Two fabellæ are usually attached to the capsule of the knee-joint. The fourth digit is the longest on both limbs of all *Lemuridæ*.

In *Hapale Jacchus* the coracoid sends a short process backward.

[1] The tarsal bones figured as those of *Chiromys* in CIV. 'Lemurs,' pl. v., belong to the *Otolicnus crassicaudatus*, CII'. p. 35.

The humerus is not perforated either above or between the condyles. The ungual phalanges are compressed and falcate, and the pollex is on a line with the rest of the digits of the fore-limb, not opposed to them. In the hind-limb the ungual phalanx of the hallux is depressed for the support of a nail, and it is opposed as a thumb to the other digits which have falcated ungual phalanges. The ilium is long and narrow, with a supracotyloid ridge. In the Marmoset (*Callithrix sciureus*), the pollex is partially opposable; as it is also in *Cebus*. In a young *C. capucinus* I have found the humerus perforated both between the condyles and above the inner condyle. There are fabellæ behind the knee-joint. A sesamoid is wedged between the entocuneiform and metatarsal of the hallux. A pair of sesamoids are developed beneath the proximal joints of each of the toes, and a single sesamoid beneath the last joint of the hallux.

In the Spider-monkeys (*Ateles*) the long and large coracoid has an angular tuberosity, which sometimes joins the anterior costa, so as to circumscribe the prescapular notch. The humerus is not perforated either above or between the condyles. This bone, and, still more, the radius and ulna, are remarkable for their length and slenderness; as are also the bones of the digits, with the exception of the pollex, which is reduced to a rudiment of its metacarpus, and is concealed beneath the skin in the recent animal. The femur, tibia, and fibula are longer than in the other Platyrrhines, but the tibia is not attenuated in the same proportion. The inner border of the naviculare is much produced. The thumb of the hind-foot is complete and well-developed. The prehensile tail compensates for the loss of that quality in the hand.

In the Catarrhine group the African Doucs (*Colobus*) repeat the abortive condition of the pollex; but in all the rest it is a true thumb, though smaller and weaker than that of the hind-hand.

In the Baboons the coracoid shows an angular ridge, but less developed than in the Capucins. In *Macacus nemestrinus* the short and broad coracoid has an angular tuberosity. I have observed an intercondyloid vacuity in this species; but, as a rule, the humerus is imperforate at its distal end. The 'intermedium' is present in the carpus of all Baboons, Macacques, and Doucs, as in the Gibbons and Orangs, fig. 361, *h*; and there are fabellæ behind the knee-joint. In most there is an ossicle, ib. *i*, wedged between the scaphoid, *a*, and trapezium, *d*, in the wrist, and between the cuboid and fifth metatarsal in the ankle. The ischia expand into rough flattened tuberosities in all those Catarrhines that have the corresponding dermal callosities.

In the Long-armed Apes (*Hylobates,* fig. 180) both acromion and coracoid are large, and much produced. The clavicles are of unusual length, equalling the extent of the eleven anterior dorsal vertebræ. The bones of the arm and fore-arm are still more remarkable for their length and slenderness, as well as those of the fingers of the hand, the thumb of which is comparatively short and slender. The femur is long and nearly straight. The tibia is slightly bent. The thumb of the hind-foot is strong and well-developed, with two phalanges.

The great length of the pectoral limbs, and the provision made for the extensive origin of some of their muscles by the breadth of the thorax and the size of the scapulæ and clavicles, relate to the chief share which these limbs take in the rapid and characteristic locomotion of this species, which swings itself thereby from branch to branch, with a force that propels the body through considerable distances.

The Siamang offers the peculiarity of a common tegumentary sheath of the proximal phalanges of the second and third digits of the hind-hand, whence the name (*Hyl. syndactylus*).

In the Orangs (*Pithecus*) the clavicle is less curved than in

361

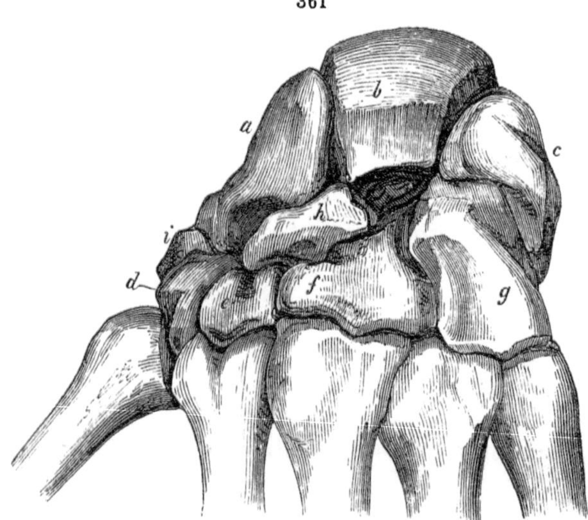

Carpus of the Orang. LXIX'.

Man, and the distal end is less expanded. The scapula approaches, by its breadth, to the form of that of Man, but the acromion is narrower, longer, and more antroverted. The humerus, in some Orangs, shows a small perforation between the condyles. The radius and ulna are remarkable for their length, and the extent of the interosseous space. The wrist, fig. 361, consists of nine bones, as in the inferior Apes,—resulting, as in them, from the presence of the intermedium, *h*: the scaphoid, *a*, and lunare, *b*, articulate with the radius; the cuneiform, *c*, is attached by ligament to the styloid process of the ulna; the 'sesamoid,' *i*, is imbedded in the tendon of the abductor longus pollicis. The metacarpals have only half the breadth of the proximal phalanges at their middle part. The phalanges are long, bent towards the palm, and expanded at their middle. The bones of the thigh and leg are disproportionately short: the articulation of the latter with the tarsus is adjusted to turn the sole obliquely inward. The hallux is disproportionately short, and, in some Orangs, has but one phalanx. The bones of the other toes have great length, especially the metatarsals and proximal phalanges, which are bent toward the sole, indicating the habitual application of the foot in the act of grasping and climbing. The joint of the hind-limb is made as free as that of the fore-limb, by the absence of the inter-articular 'ligamentum teres.' The calcaneum projects but little beyond the astragalus, the tibial surface of which is inclined obliquely inward, so that the foot presents its outer edge to the ground,—a mode of articulation favouring its prehensile power.

In the Chimpanzee, fig. 345, the clavicle, 58, is relatively shorter than in the Orang; the sigmoid curvature is more marked, the sternal end is thicker, and the acromial end broader: the scapula is longer in proportion to its breadth, and the acromion is broader than in the Orang. The bones of the anterior extremity, especially those of the fore-arm, are shorter than in the Orang. The humerus, 53, is imperforate at its distal end; it is shorter and stronger than the Orang; both tuberosities are more developed, especially the inner one, and the bicipital groove is deeper: the antero-internal surface, bounded outwardly by the deltoidal ridge, is flatter than in the Orang: the supinator ridge commences above the middle of the shaft. The trochlear prominence of the distal articulation is more developed, and the canal which separates it from the ball for the radius is both deeper and wider. The radius, 54, is shorter in proportion to its breadth, and presents a more marked sigmoid curvature; the borders of the circular proximal end are more produced; the trihedral character of the distal half is better marked. The distal end is more suddenly expanded, and the grooves for the extensor tendons are deeper and better defined. The ulna, 55, differs from that in the Orang in the proportion of its length and thickness. The outer or ulnar division of the great sigmoid cavity is less developed than in the Orang, and its margin is more extensively interrupted at its middle part: the radial division of the same cavity extends more nearly to the back part of the olecranon. The lesser sigmoid cavity is more nearly semicircular than in the Orang. The ridge continued a short way downward from the inner and ulnar angle of the great sigmoid cavity is sharply defined, but the fossa which it bounds is much less deep than in the Orang. The interosseous ridge is not marked, the bone being there rounded off in the Chimpanzee. The styloid process is better developed than in the Orang. The carpus consists of eight bones, as in Man. The thumb, 1, is relatively longer and stronger than in the Orang. The pelvis is longer in proportion to its breadth than in the Orang. The tuberosities of the ischia are expanded, flattened, and bent outward, as in the Orang. The expanded part of the ilium, 62, is slightly concave anteriorly, but in the Orang it is plane. The spine of the ischium is parallel transversely with the middle of the obturator foramen, but in the Orang it is parallel with the upper border of that foramen. The ilio-ischial angle is 165°. The ischio-pubic symphysis is longer than in the Orang: but retains its longitudinal parallelism with the sacro-lumbar series of vertebræ. The posterior wall of the acetabulum is still the deepest. The bones of the hind extremity are relatively longer and stronger, especially the femur, than in the Orang; but the most marked distinction between the two great anthropoid Apes is seen in the length and strength of the hallux, *i*, in the Chimpanzee. The articulation of the tarsus with the leg still, however, favours the oblique position of the foot, and adapts it for grasping. The femur, 65, shows the pit upon its head for the ligamentum teres: both trochanters are relatively larger than in the Orang: the neck is longer, thicker in proportion to the head, and passes off at a less obtuse angle with the shaft. The shaft is slightly bent forward; it is not straight: the condyles are more expanded, especially the inner one.

In the Gorilla, fig. 346, the scapula is broader than in the Chimpanzee, but differs from that of Man in the more oblique course of the spine, which gives greater extent to the superior costa; in the greater length and breadth of the coracoid, 52; in the straightness of the inferior costa; and in the greater convexity of the base, especially as it approaches the lower angle: the plane of the glenoid cavity is less parallel with the base than in Man, it looks more obliquely upward; the suprascapular notch is not defined.

The clavicle, 58, is thicker than that in Man, with a subtrihedral shaft and the sigmoid flexure less marked; the sternal articular surface is less oblong; the acromial end is broader and flatter below.

The humerus, 53, though surpassing in length that of Man, fig. 183, 53, is thicker and stronger in all its ridges and processes; especially at the lower extremity, the transverse diameter of which surpasses that of the upper extremity of the bone in a greater degree than in Man: both tuberosities are relatively greater, the 'lesser' one more especially. Immediately above the distal articular surface are two depressions divided by a ridge continued to the prominence between the radial and ulnar articulations; the outer or radial depression is the smaller and shallower, the inner or ulnar one is larger: it answers to the 'coronoid fossa' in Man, but becomes a foramen in full-grown Gorillas, by absorption of the thin plate of bone dividing it from the anconeal fossa behind. The ectocondyloid prominence is more marked than in Man: the entocondyloid one is more produced, is angular, and compressed. The back part of the humerus shows, as in Man, the musculo-spiral tract dividing the ridges for the external and internal heads of the 'triceps extensor.' The configuration of the lower articular surface is closely similar to that in Man; the whole surface extends a little further below the condyloid prominences, allowing to that extent a more free sweep of the fore-arm in flexion and extension, and adding power to the leverage of the tendons inserted into the antibrachial bones.

The medullary artery enters the fore part of the shaft, but nearer the middle of the bone in the Gorilla than in Man: in both the course of the canal is towards the elbow-joint. The head of the radius, 54, has an elliptical contour: the shaft bends outward so as to leave a wider interosseous space than in Man. The neck expands to the tuberosity which shows an oblong rough prominence for the insertion of the tendon of the biceps, behind or 'ulnad' of the smoother prominence supporting the bursa interposed between it and the tendon. Below the tuberosity the shaft assumes a pyriform transverse section through the developement of the interosseous ridge, which extends to near the 'sigmoid cavity.' The styloid process is represented by a prominence which gives a larger surface than in Man for the insertion of the tendon of the 'supinator longus.' It is not impressed, behind or externally, so deeply by the two grooves for the extensor muscles of the metacarpal and first phalangeal bones of the thumb: a still stronger tuberosity divides the fossa for the radial extensors of the wrist, from the wider and deeper one for the strong tendons of the extensor communis digitorum. The semicircular depression for the lower end of the ulna is well marked: the distal articular surface is divided, as in Man, by two concave facets, the larger one for the os scaphoides, the lesser for the os lunare: the anterior border of the latter is much produced, giving a greater proportional antero-posterior extent to the 'lunar' surface than in Man. The orifice for the 'arteria medullaris' is situated as in Man, and the direction of the canal is 'proximad' or towards the elbow-joint.[1] The shaft of the ulna, 55, presents two slight opposite curves, the upper one concave, the lower one convex, on the ulnar or inner aspect. Viewed sideways the whole bone has a slight bend convex backward. The lower half of the shaft becomes subcylindrical as it descends. The ridge, commencing below the lesser sigmoid cavity, is strongly marked and more vertical than in Man. The distal end of the ulna suddenly expands into a convex reniform articular surface, thickest at the middle, where it plays upon the lateral concavity of the radius. The difference from the Chimpanzee, most significant of their relative position in the Quadrumanous series, presented by the antibrachium of the Gorilla, is its inferiority of length compared with the humerus; fig. 346, with figs. 345 and 180.

The bones of the wrist, 56, agree in number and relative position with those of Man; but the differences of shape and proportion give a greater breadth to the carpal segment in proportion to its length, in the Gorilla. The radial surface is nearly circular in shape, instead of being oval and oblong as in Man. The pisiforme of the Gorilla is much longer in proportion to its breadth than in Man; whilst the articular surface for the cuneiforme is but little larger: its superior length gives stronger leverage to the great ulnar flexor of the wrist. The trapezium of the Gorilla differs most from its homologue in Man, by the production of its outer unarticular surface into two diverging tuberous processes: the articular surface, moreover, for the metacarpal of the thumb is relatively much smaller than in Man. This metacarpal, 57, is a little longer than in Man, but not quite so broad: the proximal trochlea is more concave vertically and more convex transversely, and the distal surface is more convex. The proximal phalanx is one fifth longer, and is more slender than in Man. The metacarpals of the other fingers are more than one third larger and longer than in Man, their shaft is more bent; the

[1] It may be noted that the hair covering the arm and fore-arm has a direction corresponding with that of the medullary arteries of the brachial and antibrachial bones.

tuberosities beneath the proximal articular surfaces are better developed. The proximal phalanges differ not only in their greatly superior size, but in the deep excavation of their under or anterior surface, which is bounded by rough lateral ridges; they are also more flattened and rather more bent. The distal phalanges of the anterior extremity are longer, more slender, and less expanded at their rough terminations.

Each os innominatum in the adult male Gorilla, 62, is one foot three inches in length, that of Man being seven inches and a half: the breadth of the ilium is eight inches and a half, that of Man being six inches. The ilium is less concave, of a more triangular figure, the anterior border being much longer and straighter. The more elongated and narrower form of the sacral surface corresponds with what has been noticed in the sacrum: the posterior angle or spine of the ilium is above that surface, not behind it as in Man: the distance between the antero-superior and antero-inferior spines is much greater in the Gorilla: the antero-inferior spine is situated, as in Man, just above the acetabulum. The upper ischiatic notch is much less deep than in Man, and there is a very feeble rudiment of the tuberosity dividing it from the lower notch. The acetabulum is not much larger than that of Man: the posterior is deeper than the upper wall, providing for resistance to the femur in a semi-flexed rather than in an erect position. The ischium extends, as in the Chimpanzee, far below the acetabulum, where it forms a strong subtrihedral column, terminating in a large flattened outwardly bent tuberosity, the aspect of which is wholly downward, not backward, as in Man: the united plates of the ischium and pubes, bounding the obturator foramen internally, are considerably broader than in Man. The plane of the ilium is twisted almost at right angles with that of the ischium and pubes.

The femur, 65, is shorter than in Man, and much shorter in proportion to the breadth of the shaft; the head is more relieved from the neck, and shows a less deep depression for the ligamentum teres; the neck is less oblique than in Man; the great trochanter rises to a level with the upper border of the head; the small trochanter is less prominent, but has a larger base than in Man, and is more remote from the great trochanter. The linea aspera is less developed, and the back part of the lower half of the shaft is flat and smooth: the inner angle of the popliteal space presents a well-marked rough depression, which is not present in the Human femur, and the shaft more gradually expands to the condyles. The outer articular condyle is narrower than the inner one, the reverse being the case in Man: the inner condyle is not longer than the outer one, as in Man. The rotular surface is shallower, the lateral borders are better defined: the medullary artery enters the middle of the back part of the shaft, and the course of the canal is proximad or upward.

The length of the tibia is one foot six lines, and its shaft is as thick as in Man, and expands more gradually to the distal end: the conformation of the proximal surface is similar to that in Man; the spine is rather stronger, and an anterior spine or tuberosity is more distinctly developed. The internal tuberosity in front of the fibular one is better defined; the interosseous ridge is very feebly marked in the Gorilla, and the anterior ridge of the shaft is much less marked than in Man. The astragalar surface is more undulating, less concave, and more directly continued upon the internal malleolus: the side of the distal end next the fibula, instead of being concave, forms an angular projection. The fibula is stronger in proportion to its length than in Man; the lower articular surface of the fibula is flatter, and divided into two facets more distinctly, than in Man.

The astragalus of the Gorilla equals in size that of Man, but is broader in proportion to its length: the surface for the tibia is less defined, especially from the inner facet, which in the Gorilla is almost horizontal and appears as a concave inner termination of the upper surface. The anterior surface is more convex, especially vertically, and more directly continued into the anterior calcaneal surface. The inner tuberosity is larger and more advanced: the Gorilla differs from the Chimpanzee in the greater size of this process, and in the greater proportional size of the scaphoid convexity, in which respect its astragalus more resembles that of Man. The calcaneum of the Gorilla is a longer and more slender bone than in Man, which is chiefly due to the greater length and slenderness of the posterior or calcaneal process. The lower surface of the bone is smoother, narrower, and more concave longitudinally: the groove for the flexor tendons beneath the inner astragalar surface is wider and better defined: that astragalar surface is broader in proportion to its length, and there is a deep longitudinal groove on the outer side below the outer astragalar surface, which does not exist in Man. The anterior cuboidal surface is placed further from the outer side of the bone than in Man; the outer side forming a rough convex protuberance at its anterior half. The naviculare is one third larger than in Man, the increase being in its transverse extent, and due to the greater development of the rough convex protuberance at the inner end of the bone. The entocuneiform has an equal vertical, but a minor longitudinal, extent than in Man, and chiefly differs in the convexity of the articulation for the hallux, which articnlar surface in Man is nearly flat: this difference is very significative of the different function of the hallux in the two species; the chief fulcrum of the foot requiring a firm articulation in Man, but in the Gorilla great extent of motion for the functions of an opposable grasping thumb. The metatarsal of the hallux is fully as large as that in Man; it differs in the deeper concavity of the proximal articular surface, and in the more prominent convexity of the distal one. The proximal phalanx of the hallux also equals that of Man in size; the borders of its proximal concavity are less neatly defined. The ungual phalanx is somewhat less than that of Man, especially in its terminal rough tuberosity; it is concave below instead of being convex. The remaining metatarsals of the foot are much longer and stronger than in Man; the upper border is more bent. The first and second phalanges are larger and more bent. The ungual phalanges are longer and narrower in proportion than in Man.

In all the characters by which the bones of the foot of the Gorilla depart from the Human type, those of the Chimpanzee recede in a greater degree, the foot being in that smaller ape better adapted for grasping and climbing, and less for occasional upright posture and motion upon the lower limbs. The lever of the heel is relatively shorter and more slender; the hallux has still more slender proportions, and the whole foot is narrower in proportion to its length, more curved towards the planta, and more inverted in the Chimpanzee.

On a retrospect of the skeletons of the latisternal tailless Catarrhines, it may be observed that no Orang, Chimpanzee, or Gibbon has mastoid processes; they are present in the Gorilla, but smaller than in Man. In the Chimpanzee, as in the Orangs, Gibbons, and inferior *Simiæ*, the lower surface of the long tympanic or auditory process is more or less flat and smooth, developing in the Chimpanzee only a slight tubercle, anterior to the stylo-hyal pit. In the Gorilla the auditory process is more or less convex below, and developes a ridge, answering to the vaginal process, on the outer side of the carotid canal. The processes posterior and internal to the glenoid articular surface, especially the internal one, are better developed in the Gorilla than in the Chimpanzee; the ridge which extends from the ectopterygoid along the inner border of the foramen ovale terminates in the Gorilla by an angle or process answering to that called 'styliform' or 'spinous' in Man, but of which there is no trace in the Chimpanzee, Orang, or Gibbon.

The orbits have a full oval form in the Orang; they are almost circular in the Chimpanzee and Siamang, more nearly circular, and with a more prominent rim, in the smaller Gibbons; in the Gorilla alone do they present the form which used to be deemed peculiar to Man. The occipital foramen is nearer the back part of the cranium, and its plane is more sloping, less horizontal in the Siamang than in the Chimpanzee and Gorilla. Considering the less relative prominence of the fore part of the jaws in the Siamang, as compared with the Chimpanzee, the occipital character of that Gibbon and of other species of *Hylobates* marks well their inferior position in the quadrumanous scale. The characteristics of the limbs in Man are their near equality of length, but the lower limbs are the longest. The arms in Man reach to below the middle of the thigh; in the Gorilla, fig. 346, they nearly attain the knee; in the Chimpanzee, fig. 345, they reach below the knee; in the Orang they reach the ankle; in the Siamang, fig. 180, they reach the sole: in most Gibbons the whole palm can be applied to the ground without the trunk being bent forward beyond its naturally inclined position on the legs. These gradational differences coincide with other characters determining the relative proximity of the Apes compared with Man.

In the Gorilla, the humerus, though less long compared with the ulna than in Man, is longer than in the Chimpanzee; in the Orang it is shorter than the ulna; in the Siamang and other Gibbons it is much shorter. The peculiar length of arm in those 'long-armed apes' is chiefly due to the excessive length of the antebrachial bones.

The difference in the length of the upper limbs, as compared with the trunk, is but little between Man and the Gorilla. The elbow-joint in the Gorilla, as the arm hangs down, is opposite the 'labrum ilii,' the wrist opposite the 'tuber ischii;' it is rather lower down in the Chimpanzee; is opposite the knee-joint in the Orang; and opposite the ankle-joint in the Siamang. The iliac bones are not so broad in proportion to their length in any ape as in the Gorilla. In the Orang they are flat, or present a concavity rather at the back than at the fore part. In the Siamang they are not only flat, but are narrower and longer, resembling the iliac bones of tailed monkeys and ordinary quadrupeds.

The lower limbs, though characteristically short in the Gorilla, are longer in proportion to the upper limbs, and also to the entire trunk, than in the Chimpanzee: they are much longer in both proportions and more robust than in the Orangs or Gibbons. But the guiding points of comparison here are the heel and the hallux.

The heel in the Gorilla makes a more decided backward projection than in the Chimpanzee; the heelbone is relatively thicker, deeper, more expanded vertically at its hind end, besides being fully as long as in the Chimpanzee. Among all the tailless Apes the calcaneum in the Siamang and other Gibbons least resembles in its shape or proportional size that of Man. Although the foot be articulated to the leg with a slight inversion of the sole it is more nearly plantigrade in the Gorilla than in the Chimpanzee. The Orang departs far, and the Gibbons farther, from the Human type in the inverted position of the foot. The great toe which forms the fulcrum in standing or walking is perhaps the most characteristic peculiarity in the Human structure; it is that modification which differentiates the foot from the hand, and gives the character to his order (*Bimana*). In the degree of its approach to this developement of the hallux the quadrumanous animal makes a true step in affinity to Man. The Orang-utan and the Siamang, tried by this test, descend far and abruptly below the Chimpanzee and Gorilla in the scale. In the Orang the hallux does not reach to the end of the metacarpal of the second toe; in the Chimpanzee and Gorilla it reaches to the end of the first phalanx of the second toe: but in the Gorilla the hallux is thicker and stronger than in the Chimpanzee. In both, however, it is a true thumb by position, diverging from the other toes, in the Gorilla, at an angle of 60 degrees from the axis of the foot.

362

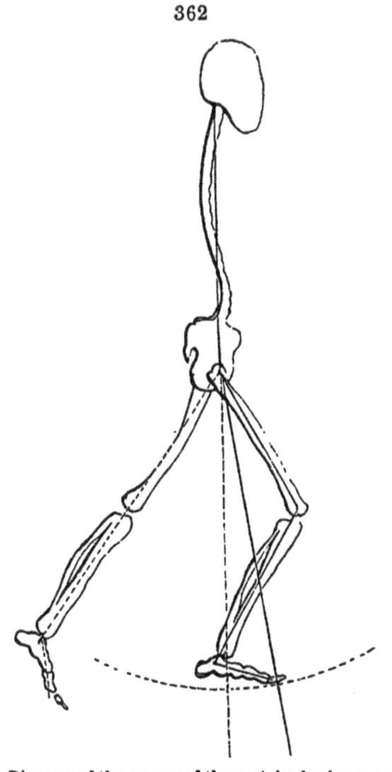

Diagram of the curves of the vertebral column, as supported on one leg, in the act of walking.

§ 191. *Skeleton of Bimana.*—The parts of the bony frame of Man, fig. 183, are co-ordinated for station and locomotion on and by the pelvic limbs, which sustain the trunk erect, and liberate the pectoral, now the upper limbs, for other uses.

A, *Vertebral Column.*—This is disposed in an undulating series of opposite curves, fig. 362; backward in the chest and sacrum, forward in the loins and neck. The vertebræ which rest on the base of the broad sacral wedge gradually decrease in size to the third cervical. In regard to breadth, they decrease to the fourth dorsal, then increase to the first dorsal, and again decrease to the second cervical. A soft elastic cushion of 'intervertebral' substance lies between the bodies of the vertebræ. The distribution and libration of the trunk, with the superadded weight of the head and arms, are favoured by these gentle curves, and the shock in leaping is broken and diffused by the numerous elastic intervertebral joints. The expansion of the cranium behind, and the shortening of the face in front, give a globe-like form to the skull, which is poised by a pair of condyles, advanced to near the middle of its base upon the cups of the atlas; so that there is but a slight tendency to incline forward when the balancing action of the muscles ceases, as when the head nods during sleep in an upright posture. The free or 'true' vertebræ are C 7, D 12, L 5. The metapophysis becomes distinct in the eleventh and well-developed in the twelfth dorsal, in which the anapophyses are recognisable, and the diapophyses reduced to tubercles without an articular facet. The neural spines increase in length and inclination 'sacrad' (downward in Man, backward in brutes) from the fourth to the twelfth dorsal, and in length and direction 'dorsad' from the fourth dorsal to the last cervical: they all have tuberous terminations.

363

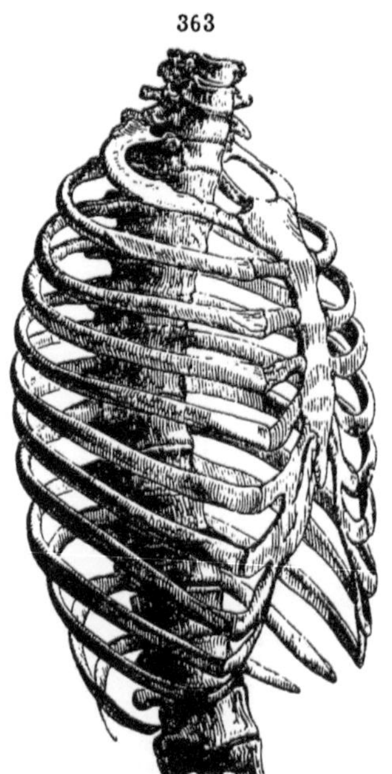

Bones of human thorax. CXXIII'.

The dorsal series of twelve vertebræ, with all their elements, constitute the 'thorax' in anthropotomy, fig. 363, the parts of the much developed hæmal arches, not anchylosed like those of the neural arches with the centrum, being reckoned as distinct bones. The pleurapophyses are termed 'costæ' or ribs, the hæmapophyses 'costal cartilages' being rarely ossified: as many of the hæmal spines as may be ossified are called 'sternum.' The first and largest, which longest retains its individuality, is the 'manubrium,' fig. 364, *b* 1; it receives the cartilages of the first pair, and part of those of the second pair of ribs. The four succeeding 'sternebers,' ib. *e*, *s*, *s*, coalesce to form the 'gladiolus' or 'body' of the sternum: the sixth piece, which commonly remains distinct, is the 'xiphoid appendage, *x*.' The parts of the sternum are usually developed each from a single centre, as at *b*, 1, 2, 3, 4: but occasionally, and usually the lower ones, are developed from a pair of centres, *c*, 1, 3, 4, as in the broader breast-bone of the Gorilla and Orang: a fissure or foramen may persist as an anomaly, fig. 364, *d*, *s*: but the union of the pairs transversely usually precedes that in a longitudinal direction. In fig. 364, *a* represents the primordial cartilages of the dorsal hæmapophyses and spines; ib., *b* and *c*, varieties in the ossific nuclei. Occasionally a pair of tubercles, ib. **d* *, indicate episternal rudiments.[1] The second to the seventh pairs of costal cartilages articulate with the sternal body: the second between it and the manubrium: the seventh between it and the xiphoid appendage: the first to the manubrium exclusively. The ribs which thus join the sternum are called 'true;' the five remaining pairs 'false,' and of these the last two are 'floating' ribs. The proportions of the dorsal pleurapophyses, or bony ribs, and their degrees of curvature, are shown in fig. 363, in their position after deep expiration. In each is distinguished the 'head,' fig. 365, *c*, the 'neck,' *f*, the 'tubercle,' *g*, and the 'shaft,' the part of the curve marked *h* being called the 'angle:' *d* is the 'sternal,' or rather hæmapophysial end to which the shaft usually slightly expands. The first and last three ribs have a single articular surface on the head: in the rest it is divided into two facets. The tubercular articular surface is wanting in the last two pairs. When the pleurapophyses

364

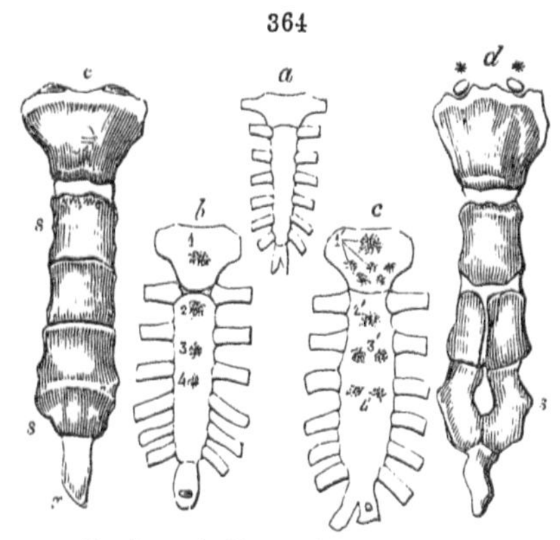

Development of human sternum. cxxiii'.

365

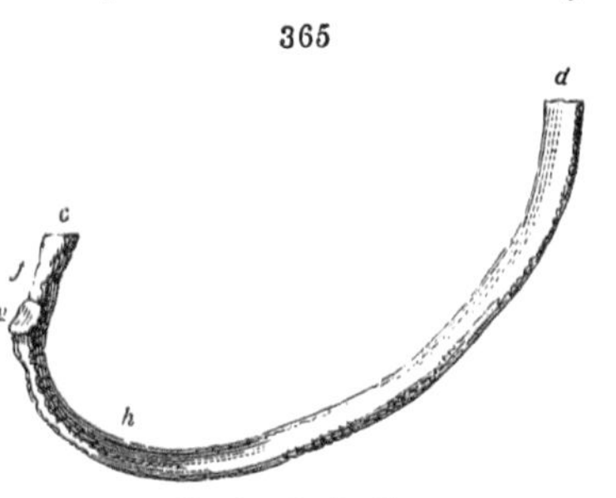

The fourth rib: Human.

[1] cviii'. p. 91.

of the last cervical or first lumbar vertebræ happen to be elongated and free, they are added to the numbers of 'pairs of ribs' in anthropotomical computation.

As a rule, the coalesced pleurapophyses make long 'transverse processes' in the lumbar vertebræ; in the first of which, as compared with the last dorsal, the centrum is much increased in size, and the neural spine in extent. The metapophysial tubercles are also enlarged, but do not project so freely, by reason of the extension of the articular surfaces of the anterior zygapophyses upon the inner sides of their base. The anapophysial tubercles are still distinct. The second lumbar vertebra chiefly differs from the first by a slight increase in the size of the centrum and in the length of the diapophysis. The anterior zygapophyses are larger and look more directly inwards. Both metapophysial and anapophysial tubercles are distinct. The backward production of the posterior zygapophyses occasioning the deep posterior emargination of the neural arch is a characteristic distinction of the Human lumbar vertebræ. Both metapophysial and anapophysial tubercles continue distinct on the third lumbar vertebra. The body of the fourth lumbar vertebra, though much broader, is not longer than that of its homologue, the third lumbar, in the Chimpanzee. It likewise shows a marked diminution in the antero-posterior extent of the neural arch, occasioned principally by a diminished length and increased breadth of the posterior zygapophysis. The anapophysial tubercles are distinctly developed. The fifth lumbar vertebra is characterized not only by its superior size, but by the great transverse expansion of the hinder part of the neural arch concomitant upon the superior development and outward expansion of the posterior zygapophyses. The diapophyses and neural spine are shortened: the anapophyses appear like a part of the upper border of the base of the diapophysis pinched up and produced backwards. The metapophysial tubercles are separated by a groove from the anterior zygapophyses.

The sacrum, fig. 366, consists of five anchylosed vertebræ. They differ from those of the Gorilla by their greater breadth and by their anterior concavity both lengthwise and transversely. The anterior nervous foramina, *b*, are relatively much larger: the spinous processes are shorter and thicker. The coalesced pleurapophyses, *pl*, *b*, of the two anterior sacrals chiefly form the sacro-iliac joint. The neural arch of the last two sacral vertebræ, *d*, *c*, is incomplete.

The first coccygeal vertebra, ib. C, *e*, is less flattened and is shorter than in the Chimpanzee: the neurapophyses, *b*, are longer, the diapophyses, *c*, are shorter: the terminal coalesced vertebræ, *c*, *d*, are reduced to their 'centrums.' Each of the three upper

366

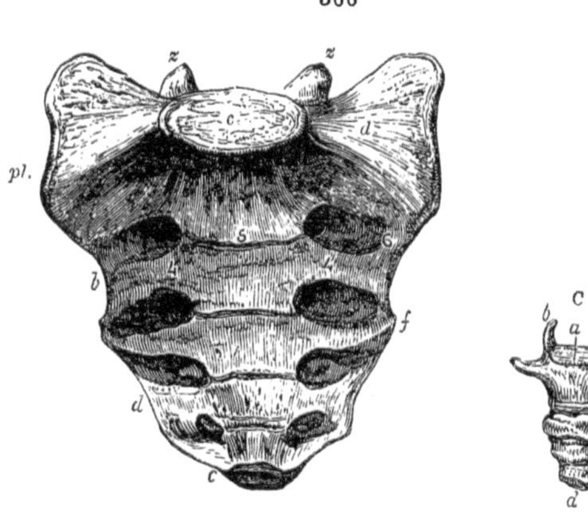

Anterior surface and base of human sacrum. Coccyx. CV'.

sacral vertebræ are developed from five primary nuclei, one, fig. 367, *a*, for the centrum, a pair for the neurapophyses, and a second pair, *b*, for the pleurapophyses: the accessory ossifications form, as epiphyses, the articular surfaces of the centrums. In the fourth and fifth sacrals the transverse processes are exogenously developed.

367

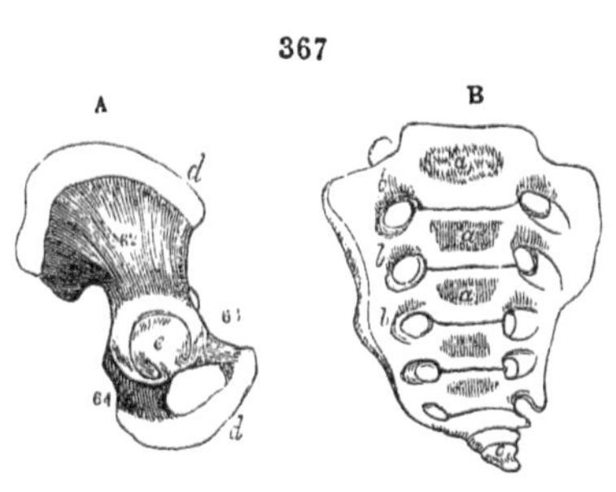

Development of the human pelvis. CV'.
A, innominatum; B, sacrum.

The spines of the six lower cervicals are short and bifurcate. As a rule, the vertebrarterial canal is completed in the seventh as in the other cervicals.

In the atlas there is a tubercle from the hypapophysis representing the body, and a rough surface on the neural arch in place of a spine. The vertebral artery perforates the transverse process lengthwise, and afterwards grooves the neural arch behind the produced angles of the anterior zygapophysis. The body is longer and deeper in proportion to its breadth than in the Gorilla. The surface for the odontoid is more nearly circular and better defined. The cavities for the condyles are relatively larger, deeper, with their margins more produced. The arterial foramina are relatively larger and the posterior zygapophyses are relatively much larger than in the Chimpanzee and Gorilla.

These differences chiefly relate to the more secure articulation and support of the vertically sustained head in the Human species, and to the larger size of the cerebral organ in part nourished by the vertebral arteries. The development of the zygapophyses gives a greater antero-posterior extent to those parts of the atlas, and the transverse processes are thicker in proportion to their length

The foregoing observations on individual vertebræ are drawn from an examination of the vertebræ of a male Australian.

B, *Skull.* Taking the lowest form of Human skull that has come under my observation, figs. 368–370, the difference is great and abrupt from that of the highest Ape, in the superior capacity of the cranium and small size of the face. On a comparison of a front view, fig. 368, with fig. 358, the cranial dome

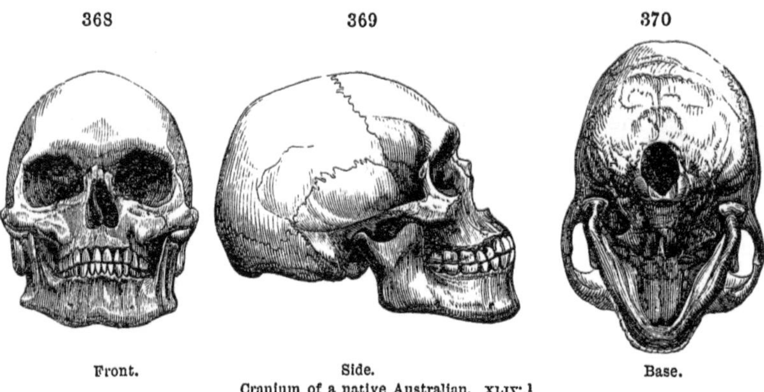

368 Front. 369 Side. 370 Base.

Cranium of a native Australian. XLIV'.[1]

forms the background for all the parts of the skull above the zygomata. The frame of the orbits is not produced clear of the dome, so as to project beyond it. The malars and contiguous parts of the maxillaries have, relatively, much less depth. The nasals are arched transversely. The bony nostril is larger and higher in position, rising to between the orbits. The alveolar border is arched transversely, and the curve is not interrupted by excessive expansion of the socket of any single tooth. On comparing the side views, fig. 369, with fig. 357, the larger cranium of the savage is still more conspicuous; the expansion is not only upward and lateral, but backward and downward, bringing the

[1] Vol. II. p. 823, no. 5304.

floor of the cavity to the level of the lower teeth, when the mandibular ramus rests on a horizontal plane. The supranasal ridge is not so produced as in the Gorilla. The zygomatic arch is shorter and more slender; the mandible shows both the angle and the 'mentum,' instead of being rounded off at both ends as in the Gorilla. But the most important differences are brought out in the base-views, figs. 370 and 359. In the lowest as in the highest Human race, the foramen magnum is placed nearer the centre of the base of the skull, the anterior end of the condyles reaching the transverse line which equally bisects the base. The condyles are relatively larger. The mastoids are developed into processes of the size and form which gave rise to the name. The zygomatic arches are in the anterior half of this view of the skull, but are opposite the middle third in the Gorilla. The stylo-hyals are anchylosed, and are supported anteriorly by a ridge from the tympanic called the 'vaginal process.' The eustachian process of the petrosal is shorter. A short styliform process is developed from the lower and outer angle of the alisphenoid. The glenoid cavity for the mandibular condyle is deeper, and is formed behind by the tympanic. There is also a low postglenoid prominence. The bony palate is much shorter, but is proportionally deeper and broader, and the teeth are arranged in a full semielliptic contour without any natural interspace, the crowns being of equal length. In the Gorilla the alveoli of the molars and canine of one side are in a straight line, parallel with those of the other side.

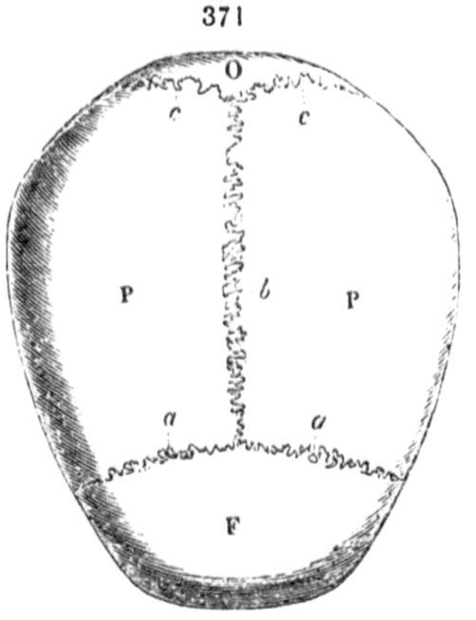

'Calvarium,' or upper view of human cranium.

As a general rule the form of the Human cranium, seen from above, fig. 371, is a full oval, with the small end forward, and the largest diameter across the 'parietal eminences,' fig. 375, *c*. The bones seen in fig. 371 are the superoccipital, O, the parietals, P, and the frontal, F; the sutures are the 'lambdoid,' *c*, the 'sagittal,' *b*, and 'coronal,' *a*.

In fig. 372, O marks the expanded and outwardly convex superoccipital, articulated by the 'lambdoid' suture, *c*, with the parietal, and by the 'mast-occipital' suture with the mastoid: *g* marks the poster-inferior angle of the parietal, P, which unites by the 'mast-occipital' suture with the mastoid: this angle is impressed on the inner side by the lateral sinus; *e* is the 'squamous' and *a* the 'coronal' suture. The frontal, F, is joined by the 'external angular process,' *i*, to the malar, which, with the alisphenoid, divides the orbital from the temporal fossa. The alisphenoid, *s*, and co-articulated portion of parietal, *f*, divide by a broader tract the frontal, F, from the temporal, T, as compared with the Australian. In more intellectual races the cranial cavity is relatively larger, especially loftier and wider. The fore-parts of the upper and lower jaws, concomitantly with earlier weaning, are less produced, and the contour descends more vertically from the longer and more prominent nasals. The ascending ramus of the mandible, K, is loftier. The malar, *t*, is less protuberant, and the mastoid, *m*, more so.

372

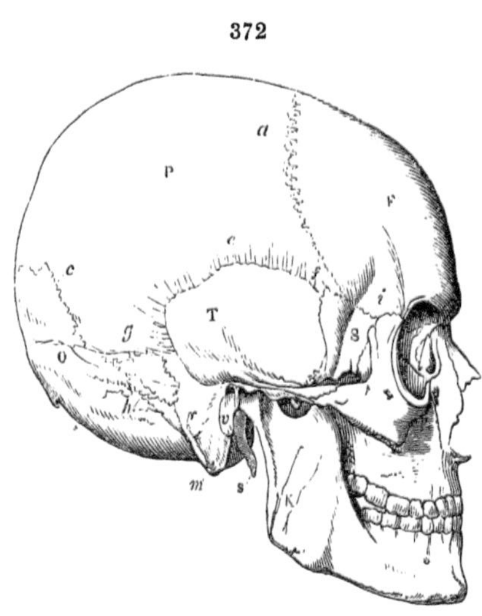

Skull of a well-formed European.

The vertical longitudinal section, fig. 373, of a well-formed European skull, best exemplifies, in comparison with fig. 395, the characteristic proportion of the human cranial cavity. The basioccipital, 1, coalesces with the basisphenoid, 5, and this with the presphenoid, such base of the cranium rising as it advances. The chief part of the foramen magnum is formed by the exoccipitals, 2; the plane of the foramen looks downward, with a slight inclination forward. The superoccipital, 3, is expanded and bulged outward by the cerebellum and posterior cerebral lobes. The petrosal, perforated by the foramen auditorium internum, is 16; between this and the alisphenoid, 6, is the squamosal; they contribute but small proportions to the cranial walls, which are chiefly due to the expanded neural spines called 'parietal,' 7, and 'frontal,' 11, with the above-mentioned superoccipital, 3: 14 is between the 'orbito-sphenoid' ('lesser ala of the sphenoid,' in anthropotomy) and the coalesced 'prefrontals' ('perpendicular plate of the ethmoid' with the 'crista galli,' ib.). The rhinencephalic fossa is shallow, ill-defined, relatively small, and floored by the 'cribriform plates.' In the nasal cavity the inferior 'turbinal,' *d*, and the 'middle turbinal,' *c*, are shown. The bony palate arched both lengthwise and transversely is formed by the palatines, 20, maxillaries, 21, and small confluent premaxillaries, 22, supporting the incisor teeth. The pterygoid appendage is marked 25.

373

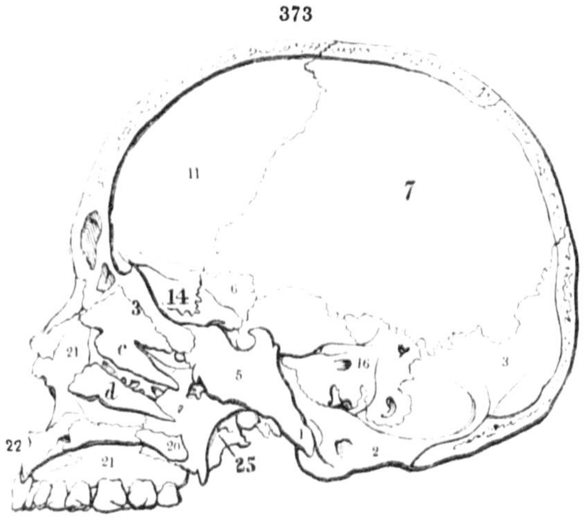

Longitudinal vertical section of European skull.

The 'hyoid bone,' fig. 374, consists of a 'body' (basi-hyal), two 'lesser cornua' (stunted cerato-hyals), and two 'greater cornua' (thyro-hyals). The body, B, 41, is compressed antero-posteriorly, curved and extended transversely, with a prominent tubercle from the fore part, answering to that which supports the 'glosso-hyal,' fig. 305, *gh*, in the horse; it is not expanded and excavated behind, as in Apes. The cerato-hyals, 40, are reduced to mere pisiform nodules of bone projecting from the line of union of the basi- and thyro-hyals. The ligaments which pass from the 'lesser cornua' to the 'styloid processes' represent the rest of the 'cerato-hyals' with the 'epihyals,' in primitive sclerous tissue. The greater cornua, 46, are attached to the body by an expanded end; a layer of cartilage intervenes to a late period; the opposite end is slightly expanded and sometimes bears an

374

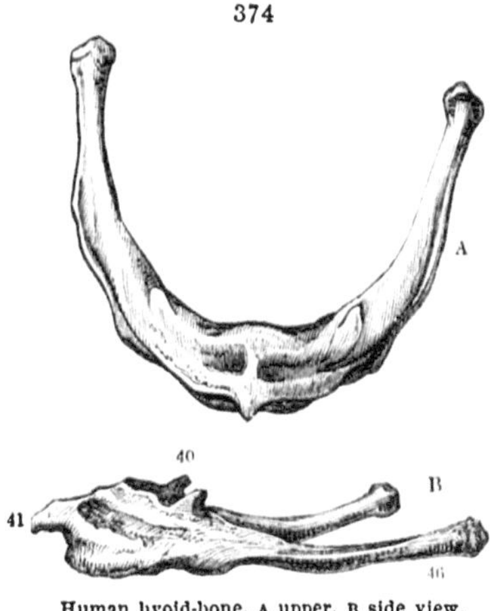

Human hyoid-bone, A upper, B side view, nat. size.

epiphysis: they are joined by ligament to the thyroid cartilage; on which account, although homologous with a pair of the 'cerato-branchials' of fishes and batrachians, they are termed 'thyro-hyals.'

The Human skull presents varieties related to sex, age, and race. Those of sex are exemplified in the smaller size of the female skull, the more delicate proportions of the facial bones, the minor prominence of the malars, mentum, and angles of the jaw.

In the skull of the child at birth, fig. 375, the jaws, through the non-developement of the teeth and their sockets, are relatively smaller than in the adult; but the facial angle, owing to the rapid growth of the brain, is, perhaps, nearer to the Greek ideal, at the period when the deciduous teeth are in place: both the cerebral cavity and the orbits are then relatively greater. The bones of the face are shorter vertically, through the non-developement of the ethmoidal and maxillary sinuses; the regular convexity of the forehead is not broken by the prominences of the frontal sinuses. The sutures of the cranium are more linear, less dentated, and more numerous, through the non-coalescence of the elements of the adult cranial bones. The angle is more open between the ascending and horizontal ramus of the mandible: the mentum is vertical or recedes.

375

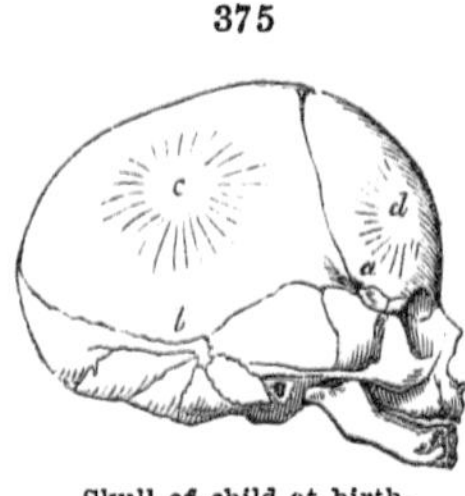

Skull of child at birth.

In the adult, fig. 371, the vertical extent of the jaws is increased by the growth of the teeth and their sockets, while the whole face is expanded by the developement of the maxillary sinuses and olfactory cavity, through the full growth of the nasal and turbinal bones and of the ethmoidal sinuses. The palatine arch has extended backward, and the posterior nares have become more vertical. The ascending mandibular ramus forms almost a right angle with the horizontal one or 'body' of the bone.

376

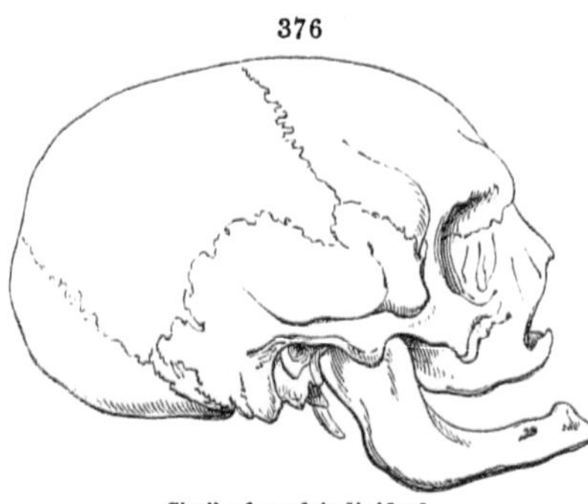

Skull of aged individual.

In extreme age, fig. 376, the teeth are lost, the alveoli become absorbed, and the jaws are reduced in vertical extent to infantile proportions. The mandibular angle again becomes more open; but the chin projects and is brought nearer to the nose when the mouth is shut. Some of the ordinary cranial sutures of the adult become obliterated.

The observed range of ethnic variety in the configuration of the Human skull and proportions of its parts is much more limited than in domesticated breeds of lower Mammals, e. g. the canine races. There is no osteological or dental difference of specific value. Assuming the skull of the Australian, figs. 368—370,[1] to be the lowest known form, the extent of variation will be exemplified by comparing the figures given with corresponding ones of the European skull, figs. 389—391. Besides the increased capacity of cranium concomitant with increased size of the intellectual organ in the educated Man, the orbital rim is more sharply defined, though thinned and less protuberant; the malars are less prominent; the nasals more prominent and longer: the alveolar parts of both jaws are more vertical anteriorly, and their entire extent is less, owing to the relatively smaller size and less complex implantation of the molar series of teeth; the ascending ramus of the mandible is deeper, and the angle less everted or squared. The profile views, figs. 369 and 390, show, in the Australian, the greater longitudinal and less vertical extent of the face, the produced jaws and receding forehead, the deep depression between the superorbital ridge and the shorter nasals: the base views, figs. 370 and 391, whilst exhibiting the same position of condyles and great foramen in relation to the erect posture, alike differentiating both extremes of Humanity from the nearest allied Ape, fig. 359, show the vacuities resulting from the stronger zygomatic arches and the narrower intertemporal part of the cranium in the Australian. The vertical longitudinal section, fig. 396, also shows, as compared with fig. 373, the thicker cranial walls of the Australian and the absence of frontal sinuses. But, whilst the characters brought out by this comparison are pretty constant in the Australian race, they are far from being so in the European: and this difference depends on the comparatively uniform low intelligence and sameness in the mode of life of the savage as compared with the state of civilized man. The cranium of the Australian may vary somewhat in the degree of compression, of shelving of the roof from the mid-line of the vertex, of the convexity of the arch from before backward; and in the presence or absence of the suture between alisphenoid and parietal: but besides the narrow cranium, with its contracted and retreating forehead and the prognathic jaws common to the Melanian races, the Australian skull is characterized by the thick

[1] XLIV. p. 823, no. 5304.

and prominent superorbital ridge, which is continued across the glabella and overhangs the deep-set, small, and slightly prominent nasals: another well-marked characteristic is seen in the large proportional size of the molars, premolars and canines, but more especially of *m* 1 and *m* 2, and in the almost constant distinction of the two external fangs of these teeth, in both jaws. In most skulls the vertex is raised, and the sides of the calvarium slope away from the sagittal elevation. The sutures are less dentated. The malar bones are small, but moderately prominent and rugged. The alisphenoid is narrow, and the squamosal is unusually closely approximated to the frontal, if it does not directly articulate therewith. The frontal sinuses are seldom developed.[1] Between the extremes brought out by the above comparison lie subjects for ethnological notice of cranial diversity, seemingly inexhaustible, of which the following are selected examples.

In the diminutive Boschisman race of South Africa, by some reckoned amongst the lowest of the aborigines of that continent, the cranium, figs. 377–379, is flatter at the vertex and relatively broader at the parietal protuberances than in the Australian race, and the forehead, though low and narrow, is more prominent. A larger proportion of the alisphenoid joins the parietal. The border of the orbit is thick and relieved, but the superorbital ridge is not carried so strongly across the glabella as in the Australian race, and the origin of the nasals is less sunk: the nasals are narrower and flatter and the malar protuberances are more regularly convex and prominent. (The prognathic character of the jaws is affected by the absorption of the alveoli due to age, in the specimen figured.)[2]

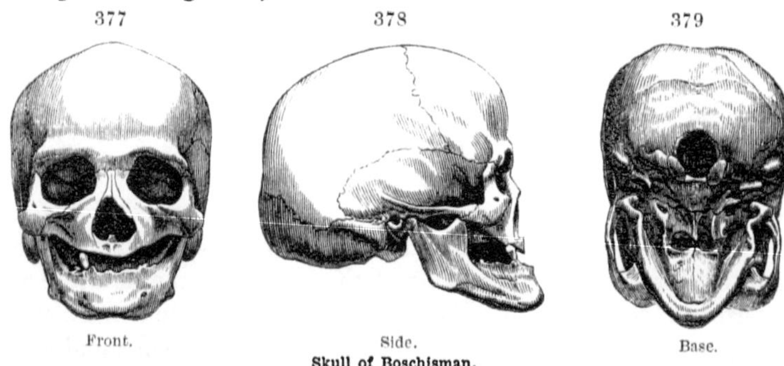

Front. Side. Base.

Skull of Boschisman.

The cranium of the Hottentot[3] resembles that of the Boschis-

[1] Minor characters, such as the suborbital depression, supra-mastoid ridge, &c., are cited in XLIV. pp. 805–830.

[2] XLIV. no. 5357.

[3] Ib. no. 5359.

man, in the contracted but almost vertical forehead, continued, with a very slight prominence of the glabella, to the narrow flattened nasals, and in general shape: the malar bones are equally prominent, and the facial parts of the maxillaries are similarly depressed, but the superorbital ridges are less thickened and less produced. The alisphenoid joins the parietal on both sides of the head. The molars are small or average-sized. The upper border of the squamosals is on a level with the fronto-malar suture. The superoccipital region rises immediately from the hinder margin of the foramen magnum.

In a Negro from the Gold Coast, Africa,[1] the cranium is large at the parietal protuberances, though narrow at the forehead. The nasal bones are broad and flat, but are continued from the same vertical line as the glabella. The alisphenoids articulate largely with the parietals. The jaws are produced. The molars are not larger than in the White races. The cranial walls are thick in most West-Coast Negroes. The uneducated African, like the uneducated European, has a minor cranial capacity than

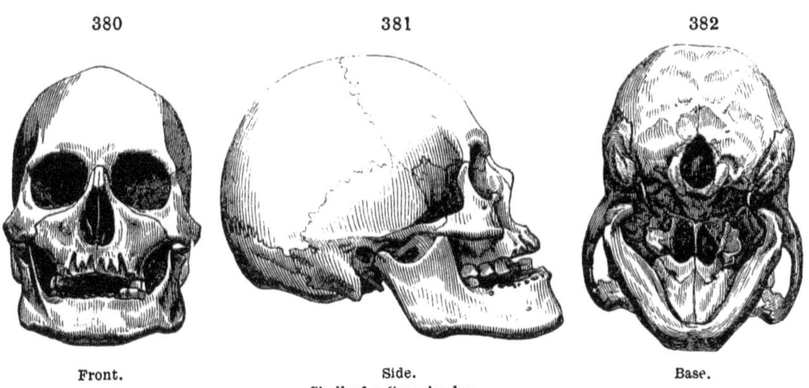

Front. Side. Base.

Skull of a Greenlander.

the educated African or European, but this becomes a race-character only when, as in the Australians and Tasmanians, all are sunk in barbarism, or none risen above that oldest known state of man.

In the skull of a Greenlander, figs. 380-382,[2] the cranium presents an elongate form, with the sides sloping from a median sagittal eminence. The parietal protuberances are feebly developed. The glabella is not very prominent, scarcely produced above the root of the nose: the superorbital ridge is thin and well defined. The nasals are prominent: the upper jaw is pro-

[1] XLIV. no. 5364. [2] Ib. no. 2479.

duced; but the chief characteristic of the skull is presented by the large and prominent cheek-bones, the lower border of which terminates a plane extending from the ectorbital process downward, outward, and forward. The zygomata are long and strong. The lower jaw is large, with a well-marked chin. These characters are repeated, with slight modifications, in the Esquimaux, but with varying proportions of length to breadth in the cranium. Among the Laplanders, with similar characters of zygomata and jaws, and the sloping of the calvarium from the sagittal line, the cranium is short, averaging 6·90 inches, with a breadth of 5·78 inches. But the so-called 'pyramidal type,' as exemplified in fig. 380, and in most races inhabiting high northern latitudes, and extending southward in Asia, is associated with both long (dolichocephalic) and short (brachycephalic) crania. Blumenbach's 'Mongolian' characters are, in the main, those of Pritchard's 'pyramidal type.' Where much uniformity of manner of life and of degree of mental power prevails, as, e.g., in the Lapps and the Esquimaux, a certain constancy of cranial character is associated therewith: where difference of work and of social grade creeps in, then cranial characters become inconstant. This is, now, manifested instructively by extended comparison of the skulls of the wide-spread Polynesian peoples. Prognathism is still the most constant feature in them, concomitant probably with late weaning of the infant. It is a conspicuous character in the skull of the native of Tahiti, fig. 383,[1] in which the forehead is narrow and sloping: the parietal protuberances moderately developed and the cranium of moderate length; it is narrower and flatter at the sides than in the White races generally. The nasal bones are prominent. Of the varieties exhibited by the aborigines of the two American continents, the works of Dr. Morton [2] give ample evidence.

383

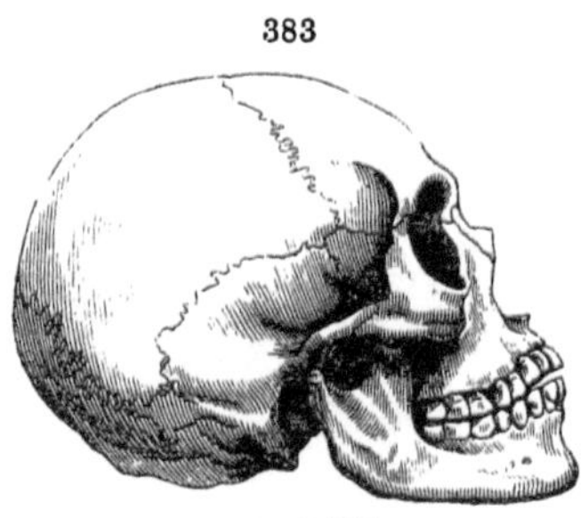

Skull of a Tahitian.

In the skull of a Macusi Indian, from Guiana, figs. 384, 385,[3] the cranium is symmetrically formed, narrow at the forehead, expanded at the parietal bosses, with the broad and rather low nasals coming off in a line with the glabella; the upper jaw is produced, but the zygomata and the mandible have European characters.

[1] XLIV. no. 5386. [2] CIX'. [3] XLIV. no. 5405.

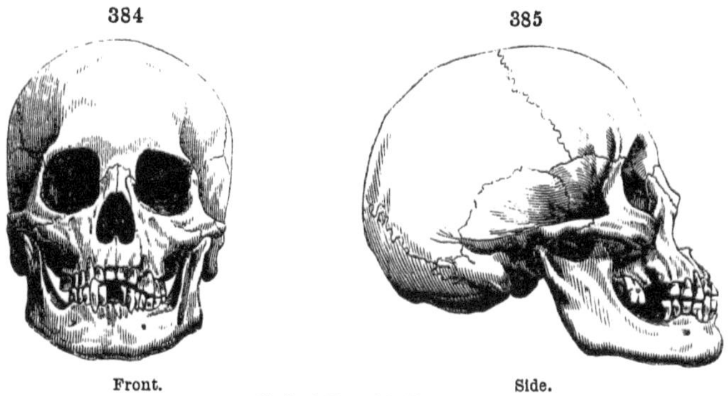

Front. Side.

Skull of Macusi Indian.

The cranium of a Peruvian of the modern or Inca race is short, broad, and high, especially behind, owing to the habit of carrying the infant with the back of the head resting upon a flat board, the pressure usually producing a slight unsymmetrical distortion of the occipital part of the skull. The forehead is narrow and receding. The glabella slightly prominent. In the older race the cranium was singularly and artificially distorted to the form, e.g., shown in figs. 387 and 388;[1] in which, with a sudden slope and slight

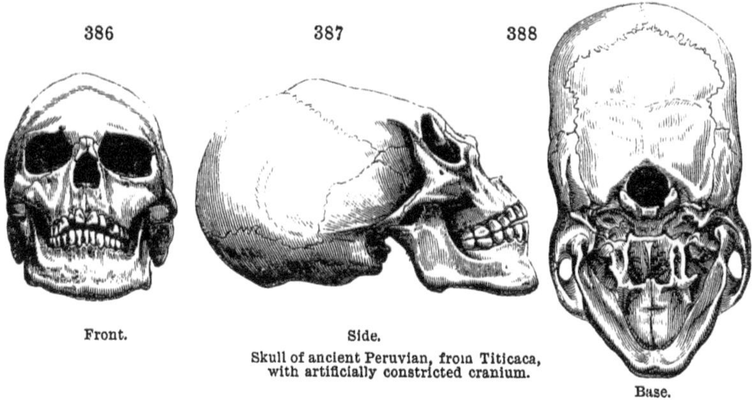

Front. Side.

Skull of ancient Peruvian, from Titicaca, with artificially constricted cranium.

Base.

convexity of the frontal, there is an annular constriction of the cranium behind the coronal suture; the flattening, constriction, and elongation having been produced by ligature at that part during infancy. The nasal bones are large, moderately prominent, and continued forward from the same sloping line with the glabella. The jaws are much produced, but the chin is well developed. Notwithstanding the deformity and the low character imparted artificially to this skull, the cranial cavity is as capacious

[1] XLIV. p. 844.

as in other American races: the brain was as large, but was differently placed. The transverse line equally bisecting the lower surface of the skull here crosses the middle, instead of the forepart, of the foramen magnum.

In the Indians of the Columbia River, called 'Flat-heads,' the cranium is deformed by the application of flattened boards to the frontal and superoccipital regions, occasioning a singularly depressed, broad or side-bulging, subelongate figure. But they resemble most other Indians in the large and almost flattened nasals being continued forward in a line with the glabella. The upper jaw is produced, and the chin moderately prominent.

The skull of the Patagonian agrees in general shape with that of the modern Peruvian, the occiput presenting the same height, breadth, and slight unsymmetrical flattening, but it is distinguished by its superior size, obviously belonging to a larger race of men. The frontal sinuses are well developed. The nasal bones are narrow, but prominent. The malars are large and prominent. The upper jaw is moderately produced. In a Fuegian I found the cranium subelongate, moderately expanded at the parietal bosses, with a narrow and protuberant superoccipital; the forehead narrow and low. The glabella was prominent, and the nasals produced. The malars were moderately prominent; the jaws prognathic; the chin well developed. The base of the skull presented paroccipital protuberances, large styliform processes of the sphenoid, and small but distinct eustachian processes of the petrosal. Traces of the maxillo-premaxillary suture remained on the palate. The molar teeth were of moderate size, worn on the inner border in the upper jaw and on the outer border in the lower jaw.

In the Indians of the Pampas the head is generally rounded, nearly ellipsoid, contracted in length and but little compressed laterally, with a forehead moderately prominent and not falling back. In the Chiquitos the same character is exaggerated and the head is nearly circular, while in the Moxos it is more oblong: this last form is very nearly that of the Guarani, or Paraguay Indians. The heads of the Caribs, as well of the Antilles as of Terra Firma, are naturally rounded. The skulls of the individuals of the continental Caribs are ovate, viewed from above: the occiput is not flattened as in the Peruvian and Californian Indians, but is moderately prominent, rounded and rather narrow. The forehead is narrow and slopes with a gentle curve directly from the interorbital space, which is more prominent than the supraciliary ridges and has no median vertical impression. The alisphenoid presents a margin of half an inch in length to join the parietal on both sides of the head. The cheek-bones and lower border of the orbit are moderately prominent: the nasal bones are continued with a very slight depression from the glabellar prominence: the superior maxillary bones are produced: the lower border of the malar process of the maxillary bone is slightly concave. The lower border of the orbit is a little more concave than the upper one: the spheno-orbital fissure is widely open anteriorly. The cranium of the Macusi Indian, fig. 384, is more oblong and ellipsoid, viewed from above: the forehead is broader, the parietal region narrower, or at least not broader, than it is in the shorter crania of the Carib tribe. The frontal sinuses cause the superorbital ridges to project beyond the interorbital space: the malar bones are equally prominent: the outer angle of the malar processes of the maxillary bones overhangs the concave line leading thence to the alveolar processes. The general character of the facial part of the skull resembles that of the Patagonian Indian, but the prominent convex occiput and general form of the cranium approach nearer to the Carib form. The Carib, Guianian, and Columbian skulls all agree in the roundness or convexity of the occipital region, and differ in this respect, as well as their more symmetrical figure, from the skulls of the Peruvians, Chilians, and Patagonians. All the American skulls manifest the same inferiority in the size of the true molar teeth as compared with the skulls of the Australians: the incisors, canines, and premolars, or bicuspides, are not smaller than in the Black races.

In the average skull of the Chinese the cranium presents the moderate or medium proportions of length, height, and breadth. The sagittal region is not unusually elevated. The plane of the glabella is slightly affected by the frontal sinuses, and the large and prominent nasals are continued therefrom with a very slight depression. The malars are large and slightly prominent. The upper jaw is not produced. The chin is well developed. The paroccipital tubercles are well marked. The chief distinction which such skull presents from the average form of those of European races is in the size and prominence of the malar bones.

Most well-formed skulls of educated Whites present the characteristics ascribed by Blumenbach to his Caucasian race. The contour of the cranium, as well as that of the face, is oval: the forehead is moderately vertical, high, and broad: the nasal bones are prominent and well developed: the malars are vertical, and the orbital boundaries are neatly defined. The upper jaw is not produced: the lower jaw has the chin well marked.

The range of variety is, however, considerable. From an old and well-filled European graveyard may be selected specimens of 'klinocephalic' (slope- or saddle-skull), 'conocephalic' (cone-skull), 'brachycephalic' (short-skull), 'dolichocephalic' (long-

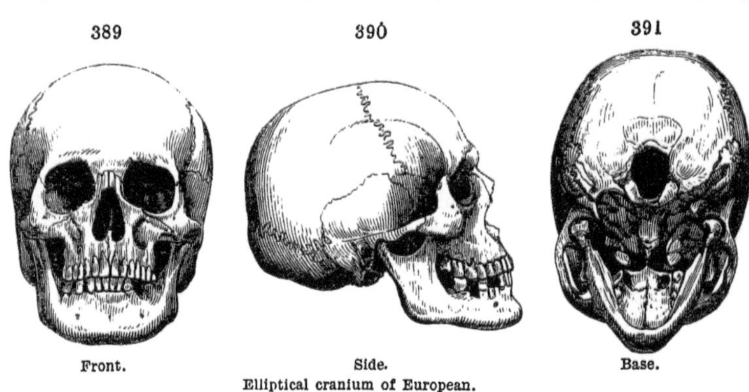

Front. Side. Base.
Elliptical cranium of European.

skull), 'platycephalic' (flat-skull), 'leptocephalic' (slim-skull), and other forms of cranium equally worthy of penta- or hexa-syllabic Greek epithets. There are varieties in the degree of projection of the supranasal and superorbital ridge, but never attaining that exhibited as a constant and specific character in the Gorilla, fig. 395. There are varieties in the sutures, in the time and degree of their obliteration,[1] and in their intercalated 'wormian' bones, &c. &c.

[1] Rokitanski [a] appears first to have conceived, in relation to the skull of a young person in which the lower ends, for rather more than an inch, of the coronal suture were obliterated,[b] that it was the cause of a transverse contraction of the cranium at that part.

What this skull actually shows is the coincidence of partial confluence of parietals and frontals with a least transverse diameter at the temporal fossæ, a high and rather short cranium, with a general inferior capacity of the brain-case. But the relation of cause and effect in this instance is not reasoned out by the great pathologist. The ultimate or adult size of the cerebrum is due to inherent, or inherited, capacity of brain-developement, with the accident of such culture, or of the absence thereof, through which that developement might be influenced. The growth of the brain governs the capacity of the cranium, and, in a general way, is anterior in the order of the phenomena: it influences its bony case, moreover, not by mechanical expansion, but by exciting the modelling action of the absorbents in co-operation with the arterial depositors of the bony matter. The coronal, sagittal, and lambdoid sutures are, as a rule, and in the cranium in question, too intricately interwoven to admit of any forcible drawing asunder. On what facts it is assumed that the obliteration of

[a] cxi'. Bd. ii. p. 148:—'Durch seitliche Synostose der Scheitel- und Stirn-beine, d. h. Verknöcherung des seitlichen unteren Theiles der Kranznaht, wird eine quere Verengerung des Schädels bedingt.'

[b] Figured in Lucae, Tafel VIII.

The progressive superiority of the cranial over the facial division of the skull is best illustrated in the Mammalian class; but, to show the full gradational extent of diversity, two examples, in this retrospective summary, are borrowed from lower vertebrate classes.

In the cold-blooded Crocodile, fig. 392, the cavity for the brain, in a skull three feet long, will scarcely contain a man's thumb. Almost all the skull is made up of the instruments for gratifying an insatiable propensity to slay and devour; it is the material symbol of the lowest animal passion.

392

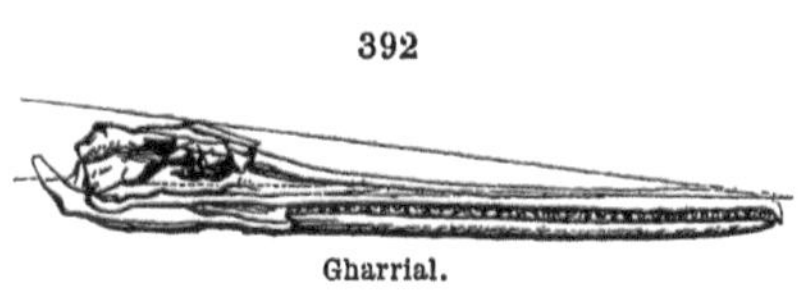

Gharrial.

In the Bird, fig. 393, the brain-case has expanded vertically and laterally, but is confined to the back part of the skull. In the small singing-birds, with shorter beaks, the proportion of the cranial cavity becomes much greater.

393

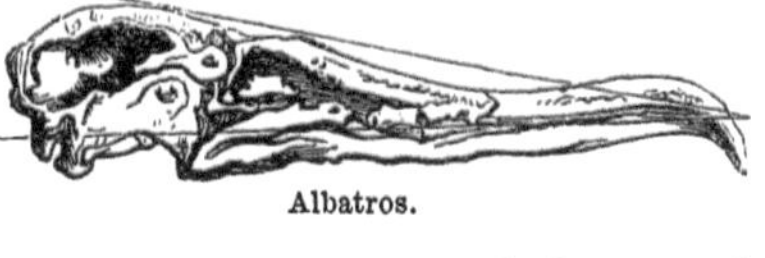

Albatros.

In the Dog, fig. 394, the brain-case, with more capacity, begins to advance further forward. In the Gorilla, fig. 395, the capacities of the cranium and face are about equal. In Man, fig. 396, the cranial area vastly surpasses that of the face.

394

Dog.

A difference in this respect is noticeable between the savage and civilised races of mankind; but it is immaterial as compared with the contrast in this respect presented by the lowest form of the human head, fig. 396, and the highest of the brute species. Such as it is, however, the more contracted cranium is commonly accompanied by more produced premaxillaries and thicker walls of the cranial cavity, as is exemplified in the negro or Papuan skull.

the parts named of the coronal suture caused or conditioned ('bedingt') the transverse contraction of the cranial cavity is not stated. If the mechanical idea prevailed that obliteration of a suture prevented the previously distinct bones being pulled apart, so as to allow, or stimulate, disproportionate growth at the margins of the stretched bones, then we should have expected that the elongation of the cranial box would have been prevented in the direction at right angles to the obliterated suture, producing contraction in the longitudinal instead of in the transverse direction.

If a line be drawn from the occipital condyle along the floor of the nostrils, and be intersected by a second touching the most prominent parts of the forehead and upper jaw, the intercepted angle gives, in a general way, the proportions of the cranial cavity and the grade of intelligence; it is called the 'facial

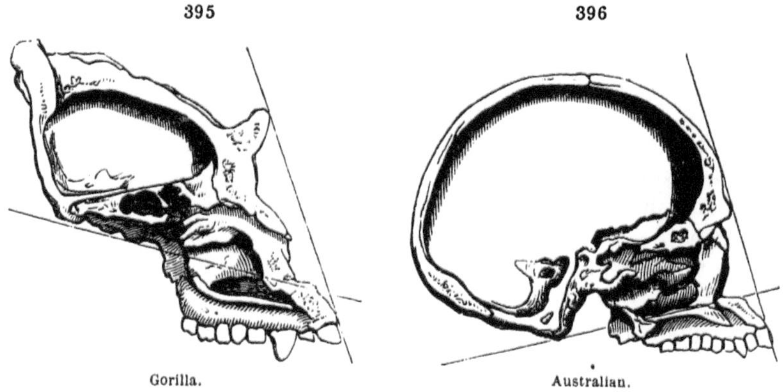

395 Gorilla. 396 Australian.

angle.'[1] In the Dog this angle is 20°; in the Gorilla it is 40°, but the prominent superorbital ridge occasions some exaggeration; in the Australian it is 85°; in the educated White it averages 95°. The ancient Greek artists adopted, in their beau ideal of the beautiful and intellectual, an angle of 100.

C. *Bones of the Limbs.*—The Human clavicle, fig. 183, 58, is more slender in proportion to its length, and its curves are always better marked than in the great Apes: the tubercle for the conoid ligament is usually more developed. The peculiarities of the Human scapula, as brought out by the same comparisons, are its great breadth in proportion to its length, the more transverse direction of the spine and acromion, and the disproportionate extent of the subspinal as compared with the supraspinal tract. The upper angle is less rounded; the extent of the upper border between that angle and the superscapular notch is relatively greater, and is more nearly straight; the notch itself is smaller and deeper. The smooth triangular surface near the origin of the spine, upon which the trapezius muscle glides, is relatively greater. The surface for the teres minor muscle, on the outer side of the bone, near the lower border, is broader; as is that for the teres major, nearer the lower angle. The deep part of the subscapular bed, being parallel with the attachment of

[1] For illustration of other 'angles,' e.g. the 'palato-facial' and 'basi-facial,' reference may be made to CIII'. and CX'. p. 21, pls. x. xi. and xii.

the spine of the scapula, is situated nearer the upper border than in the Gorilla or Chimpanzee. The surface for the upper origin of the serratus magnus is relatively less than in the Gorilla. The long narrow surface between the obtuse lower boundary of the subscapular fossa and the lower border of the scapula is flat, or is less concave than in either the Gorilla or Chimpanzee.

The humerus of the male Australian, ib. 53, is more slender than that of the average-sized male European; both show the inferior developement of the condyloid processes as compared with the Gorilla; and the same difference in relation to muscular attachments is exemplified by the lower tuberosities at the upper end of the bone. The intercondyloid perforation is occasionally seen in the Human humerus. The characteristics of the Human radius, ib. 54, are its greater relative shortness to the humerus (seldom noted in anthropotomical descriptions of the bone); its more slender and less bent shaft; the better definition and greater depth of the grooves for the three tendons acting on the thumb at the back part of the distal expansion, and the more produced styloid process; whilst the tuberosity above it for the attachment of the supinator longus is much less developed than in either the Gorilla or Chimpanzee. The chief distinctions presented, in the same comparison, by the ulna, ib. 55, are its minor length compared with the humerus; its greater relative slenderness; the less proportional expansion of the proximal end; the somewhat minor production of the coronoid process; and the greater straightness of the shaft, especially on the side view.

In the Gorilla the hand is an instrument of great power of grasp, capable of easily sustaining the weight of the body suspended by the fingers: the length and strength of the whole pectoral limb accord with the mechanical adjustments of the hand as a hook, and as a crutch in moving along the ground. In Man the framework of the hand, ib. 56, 57, bespeaks an organ of varied and delicate prehension; and the form and proportions of the whole upper limb relate to the free motions and complex functions of the instrument. In Man the length of the three bones of the thumb, I, nearly equals one third the length of the humerus: in the Gorilla it is little more than a fifth of that length. The metacarpal of the index digit in the Gorilla is twice the length of that of the pollex: in Man it is little more than one fourth larger. The shafts of the proximal and middle phalanges of the fingers are less expanded than in the Gorilla; their distal ends are broader than the shaft instead of being narrower:

the terminal portions of the ungual phalanges are longer, broader, and flatter than in the Gorilla, considerably so in relation to the size of the whole hand,[1] having reference to sustaining the developed surface for a refined sense of touch.

The ilium, fig. 367, A, 62, ischium, ib. 64, and pubis, ib. 63, coalescing, the two latter at the sixth year, and both with the ilium at about the twenty-fifth year, have been described, according to the usage of anthropotomy in such instances, as a single bone, under the designation of 'os innominatum.' The Human characteristics are strongly marked in this part of the skeleton.[2] The ilium is broader than it is long, and is more concave anteriorly, fig. 398, 4, than in the Gorilla; it is also more concave posteriorly, fig. 397, especially in the vertical direction, in which it is slightly convex in the Chimpanzee. The sacro-iliac symphysis, fig. 398, 2, 3, *n*, *b*, is subquadrate, instead of being long and narrow as in the Chimpanzee. The 'crest,' *a*, *b*, *c*, is much thicker and much more curved; and both angles or 'spines,' but especially the posterior one, *b*, are more produced. These modifications, and especially the developement of the 'external labrum,' fig. 397, *c*, relate chiefly to the needful increased surface of attachment for the large muscles which sustain the trunk upright upon the hinder, now become the lower, limbs. The anterior border of the innominatum, figs. 397, 398, *a*, *e*, *f*, especially that part formed by the ilium, *a*, *u*, *d*, is much shorter and thicker, and the 'anterior inferior spine,' *d*, is better developed. The acetabulum, fig. 397, 4, is turned more toward the back of the os innominatum. The great ischiatic notch, *m*, is shorter, but much deeper; the spine of the ischium, *l*, is more produced; the lesser ischiatic notch, *k*, is deeper, more concave, but of the same length. The tuberosity of the ischium, *i*, is convex, and is continued upon the outer part of the bone to near the acetabulum; in the Gorilla and Chimpanzee it is more flattened, is carried further down from the acetabulum, and its outer margin is produced or everted. The pubis, *q*, *s*, is shorter and much thicker than in the Chimpanzee. The symphysial boundary of the obturator foramen, *o*, is much narrower and less curved. The oblique groove, *t*, beneath the pubic boundary of the foramen in Man is not present in either the Chimpanzee or Orang-utan. The cotyloid notch, 5, is broader, and the symphysis pubis, fig. 398, *g*, is much shorter than in the Anthropoid apes.

The backward developement of the ilium, *b*, *n*, for the ectoglutæus, and the ant. inf.-spine, *d*, for the rectus femoris, relate to

[1] CIII'. vol. v. pl. 10. [2] Ib. vol. v. pl. 6.

the important share taken by both muscles in maintaining the erect position.

With the outer surface of the ilium turned to the observer, as in fig. 397, is seen the same surface of both ischium and pubis, together with the acetabulum; but, in the Gorilla, the twist of the innominatum is such as to present only the outer margin of the ischium, with a side view of the acetabulum; and, in the Chimpanzee, the greater twist brings the inner surface of the pubis into view and almost excludes the acetabulum.

397

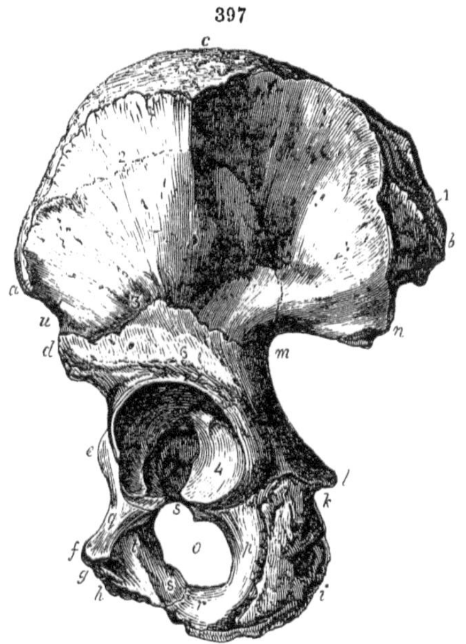

External view of the innominate bone.

The sacro-iliac surface is divided into a 'syndesmotic,' fig. 398, 1, 7, and a 'synchondrosal,' ib. *n*, 2, part: the latter is more especially termed the 'articular,' and sometimes, from its shape, the 'auricular' part; it is united by 'fibro-cartilage' to the first and second, and a small part of the third, sacral vertebræ. The concavity, 4, is the 'internal iliac fossa.' The ridge, 5, transmits, like a buttress, the weight sustained by the articular surface, 2, to the back wall of the acetabulum: the ridge, 6, thence continued to the spine of the pubis, *f*, is termed 'ilio-pectineal.'

The human pelvis, formed by the sacrum, coccyx, and ossa innominata, offers

398

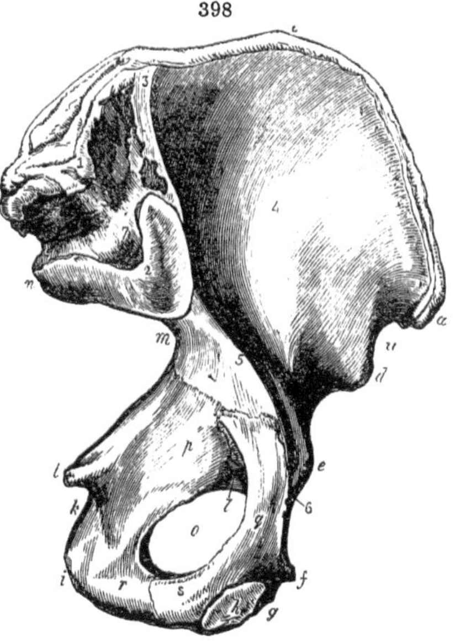

Internal view of the os innominatum.

characters of sex and race. The male pelvis is shown in fig. 183, 62, 63, 64: the female pelvis in fig. 399. In the latter the sacrum is relatively wider, and, anteriorly, it is less concave transversely above, *a*, more concave vertically below, *b*. The ilia are broader and shorter, with more capacious fossæ: the ' obturator foramen ' is triangular; the ischial tuberosities are wider apart, and the symphysis pubis less deep. Anthropotomists call the part which is above the linea ilio-pectinea, *o*, *f*, and promontory of the sacrum, *a*, the ' false pelvis;' that beneath, the ' true pelvis.' Of this the ' brim,' or ' superior circumference ' *e*, *f*, *b*, incloses the ' inlet '; the ' inferior circumference,' bounded by the ischial tuberosities, pubic symphysis, and tip of the coccyx, incloses the ' outlet ' of the ' true pelvis.' The diameter from the sacral promontory, *a*, to the pubic symphysis, *b*, is called the ' conjugate ' or ' antero-posterior ' one; that between the ilia taken at *e*, *f*, or half way between the sacro-iliac joint and the pectineal eminence, is the ' transverse ' diameter; the ' oblique ' diameter is between the point of the brim nearest the pectineal eminence, *c*, and the sacro-iliac joint of the opposite side, *d*. Of the pelvic outlet two diameters are noted—the ' antero-posterior ' from the tip of the coccyx to the lower part of the pubic symphysis, and the ' transverse ' taken between posterior parts of the ischial tuberosities. The following may be regarded as the normal extent of the above diameters in the two sexes:—

399

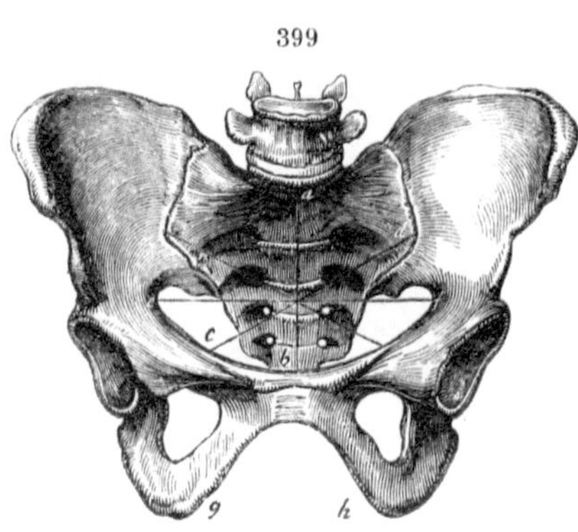

Anterior view of the female pelvis.

' BRIM.'	MALE.		FEMALE.	
	In.	Lines.	In.	Lines.
Transverse	4	6	5	1
Oblique	4	6	5	0
Antero-posterior	4	0	4	5
' OUTLET.'				
Transverse	3	3	4	5
Antero-posterior	3	4	4	2

In Man alone are the boundaries of the superior outlet on one plane: the section through the ilium, in fig. 400, shows this to be due to the direction of the body of the pubis, which is on the same plane with that of the cotylo-sacral tract of the ilium. In this section, *a a′* is the line of *fulcrum* falling in the transverse vertical plane of the trunk; *c c′* is the line of *weight* passing through the centre of the sacro-iliac joint; *b b′* is the line of *power*, or pubic projection, giving attachment to the extensor muscles of the thigh; *d d′* is the line of sacral projection; *e f* gives the ‘cotylo-sacral

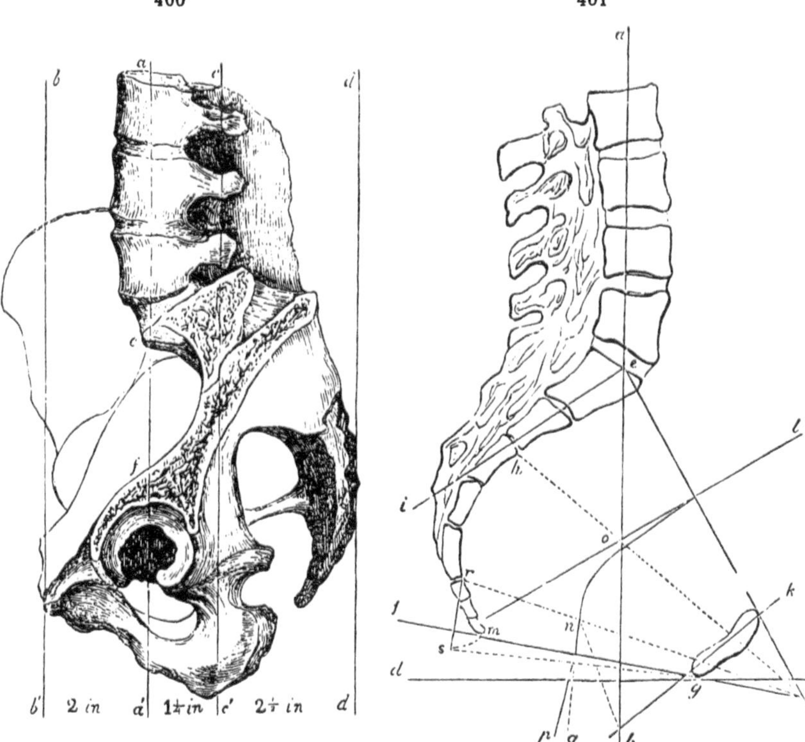

Section through cotylo-sacral arch: Human pelvis. CV′.

Angles of inclination and axes: Female pelvis. CV′.

curve;’ *a′ b′* is the pubic arm of the lever; *a′ c′* is the cotylo-sacral arm; *a′ d′* gives the length of the gluteal arm; *c′ d′* that of the posterior spinal arm. In the diagram, fig. 401, the lines *a e*, *e c*, mark the angle of inclination of the pelvis, or the ‘pelvi-vertebral angle;’ *f g d* gives the inferior angle of inclination of the pelvis, which is about 10° with the horizon, *g d*. The two lines of the superior and inferior planes, when prolonged anteriorly, cut each other at *c*, and include an angle of about 50°, *e c f*. The ‘sacro-vertebral angle’ is shown at *a e i*, which is about 117° in the male and 130° in the female. The angle *a b k*, taken through the long diameter of the pubic symphysis, is the complementary angle of the sacro-vertebral one in the female, owing to the general parallelism of the pubic with the sacral wall of the 'true pelvis.' The 'axis of the brim' is the line *l m*, drawn from the centre of the superior plane at right angles thereto. The 'axis of the inferior outlet,' *n p*, is drawn at right angles to the centre of the inferior plane. The axis of the pelvic cavity, *l o r p*, is an irregular parabolic curve, passing from the fixed axis of the brim and moveable forward, through the flexibility of the coccyx at its inferior extremity, with the moveable axis of the inferior outlet, with which it coincides below.[1]

The observed range of variety in the Human pelvis is restricted to some slight difference in the breadth and curve of the sacrum, in the contour of the iliac crest (fig. 397, *a*, *c*, *b*), in the interspace between the ant.-superior, *a*, and ant.-inferior, *d*, spines,[2] and in proportions that modify the shape of the upper aperture of the 'true pelvis,' whereby it might be approximately defined as 'oval,' 'oblong,' 'round,' or even approaching to 'quadrate.' According to my experience, these are not characteristic of race, nor uniformly concomitant with cranial varieties, as, e. g. the 'round' pelvis with the 'brachycephalic,' the 'oblong' with the 'dolichocephalic,' or the 'square' with the 'pyramidal' form of skull.[3] Vrolik[4] has noted a more vertical direction of the ilia, and the proximity of the highest part of the crest to the posterior superior spine, in the pelvis of a Negro: I have noted the smaller and narrower iliac bones of an Australian female as compared with an European;[5] but the size accorded with a general dwarfishness of stature, and the difference of proportion was too slight to affect the characteristic human configuration of the bone. Save in regard to Europeans, the requisite number of observations of the pelvis in the same races or tribes of mankind is yet a desideratum.

In the typical Mammalian foot the digits decrease from the middle to the two extremes of the series of five toes; and in the modifications of this type, as traced through the gradations (p. 308, fig. 193), the innermost, *i*, is the first to disappear. In Man it is the seat of excessive developement, and receives the name of 'hallux,' or 'great toe;' it retains, however, its characteristic inferior number of phalanges. The tendons of a powerful muscle, which in the Orangs and Chimpanzees are inserted into the three middle toes, are blended in Man into one, and this is inserted into the hallux, upon which the force of the muscle now called 'flexor longus hallucis' is exclusively concentrated.

[1] CV'. pp. 133–135. [2] XLIV. p. 839 (Polynesian). [3] CVI'.
[4] CVII'. [5] XLIV. p. 806.

The arrangement of other muscles, in subordination to the peculiar developement of this toe, makes it the chief fulcrum when the weight of the body is raised by the power acting upon the heel, the whole foot of Man exemplifying the lever of the second kind. The strength and backward production of the heel-bone relate to the augmentation of the power. The tarsal and metatarsal bones are coadjusted so as to form arches both lengthwise and across, and receive the superincumbent weight from the tibia on the summit of a bony vault, which has the advantage of a certain elasticity combined with adequate strength. In proportion to the trunk, the pelvic limbs, fig. 183, 65-68, are longer than in any other animal; they even exceed those of the Kangaroo, fig. 211, and are peculiar for the superior length of the femur, fig. 183, 65, and for the capacity of this bone to be brought, when the leg is extended, into the same line with the tibia, ib. 66. The inner condyle of the femur is longer than the outer one, so that the shaft inclines a little outward to its upper end, and joins a neck longer than in other animals, and set on at a very open angle. The weight of the body, received by the round heads of the thigh-bones, is thus transferred to a broader base, and its support in the upright posture facilitated. The pelvis is modified so as to receive and sustain better the abdominal viscera, and to give increased attachment to the muscles, especially the 'glutei,' which, comparatively small in other Mammals, are in Man vastly developed, to balance the trunk upon the legs, and reciprocally to move these upon the trunk. In comparison with that of the Apes the Human femur is distinguished by its greater length, both absolute and relative to the trunk, by the more angular and less cylindrical shape of the shaft; by its forward bend, and the buttress-like developement of the 'linea aspera;' by the greater proportional expanse of the distal end, especially at and above the inner condyle, and by the greater backward production of both condyles, especially of the inner one. Only in the Chimpanzee and Gorilla is the 'cervix femoris' relatively as long as in Man; but it stands out at a different angle, and the femora are parallel, fig. 340, 65, not converging to the knee-joints, through the double obliquity of 'neck' and 'shaft' which characterises the human femur, fig. 183, 65. The great trochanter does not rise so high as the head of the bone in Man: the small trochanter is more prominent and circumscribed. The terminal expansion of the shaft is chiefly toward the inner or tibial side. The major part of the rotular surface is on the outer condyle.

After the femur, the tibia, 66, is the longest bone of the skeleton in Man: in the Gorilla it is the shortest of all the long bones of the limbs: the Human tibia also differs from that of the Gorilla in the more equable diameter of the shaft and more parallel contour of the outer and inner sides, with a considerable reduction of the interosseous space between it and the fibula. The crest descends in Man near the middle of the anterior surface of the shaft, with a slightly sigmoid or wavy course. The lower articular surface is uniformly concave from before backward, and is continued at a less open angle and to a greater extent upon the articular surface of the inner malleolus, the articulation with the astragalus being deeper and firmer than in the Gorilla and other apes. The outer or fibular malleolus descends in Man lower and more vertically than the inner malleolus, interposing a greater obstacle to lateral inflections of the foot upon the leg than in the Gorilla. The foot is shorter in proportion to the leg, in Man, than in any Quadrumane, and is so articulated that the sole is directed downward: the tarsus is longer and narrower. The entocuneiform presents a flat, reniform surface, anteriorly, to the base of the hallux. The four outer toes are very slender compared with the innermost; and their proximal and middle phalanges are very feeble compared with those of the Gorilla: all the five toes have the same direction, forward.[1]

The osseous texture of the Human bones is remarkable for its delicacy and finish: it is exemplified in fig. 402 by longitudinal sections of parts of the three chief bones of the pelvic limb. In the section of the upper end of the femur, A, the outer, compact tissue, *a*, is extremely thin upon the head of the bone, begins to gain thickness at its under part, and at the corresponding part of the great trochanter, and increases until it forms the wall of the medullary cavity. In the cancellous or reticular tissue forming the substance of the head and neck, a tendency to a radiating disposition, diverging from the under part of the neck, and favourable to strength, may be discerned in the principal laminæ. In the head of the tibia, B, the compact tissue is also very thin, where it encloses the reticular structure occupying the proximal end, and becomes thicker as that structure is absorbed. In both the tibia and fibula is shown the line indicative of the union of the upper 'epiphysis' with the 'shaft:' and in the femur there is a similar indication of the epiphysial condition of the great trochanter.

The more constant sesamoid bones of the Human skeleton,

[1] For the details of a comparison of the limb-bones of Man with those of the Apes, see CIII'. vol. v. p. 1, pls. i.–xiii.

which have an articular facet playing upon a joint, are the 'patella,' fig. 183, 66′, the pair beneath the metatarso-phalangial joint of the great toe, and the pair at the corresponding part of the thumb. Of those playing on surfaces of bone, there is one in the tendon of the peroneus longus which glides on the groove of

402

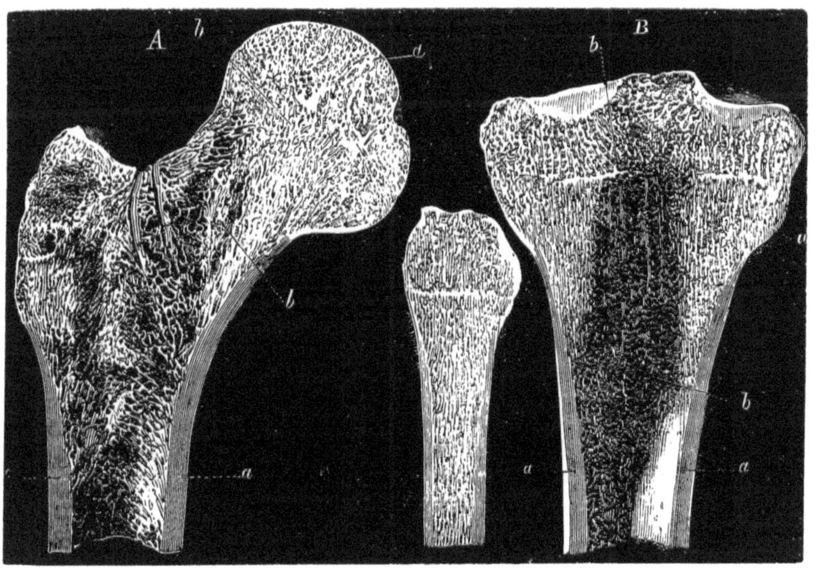

Sections of A femur, B tibia, and fibula.

the cuboid; one in the tendon of the tibialis anticus opposite the smooth facet on the entocuneiform; one in the tendon of the tibialis posticus opposite the inner side of the astragalus; and one (fabella) in the outer head of the gastrocnemius behind the outer condyle of the femur. The os penis, common in *Quadrumana,* is never developed in Man.

D. *Relations to Archetype.*—Finally, in regard to the skeleton of *Bimana,* there remain a few observations on its relations to the general vertebrate archetype (vol. i. fig. 21), from which it departs so widely.

The skull shows the following extreme modifications. In the occipital segment the hæmal arch is detached and displaced, as in all Vertebrates above fish; its pleurapophysis (scapula, fig. 403, *pl*, 51) has exchanged the long and slender for the broad and flat form; the hæmapophysis (coracoid, 52) is rudimental, and coalesces with 51: the diverging appendage, 53–57, of this arch becomes the 'pectoral limb.' The neurapophyses (exoccipitals, 2) coalesce with the neural spine (superoccipital, 3), and next with the centrum (basioccipital). This afterwards coalesces with the centrum (basisphenoid, 5, *c*) of the parietal segment. With this centrum also the

403

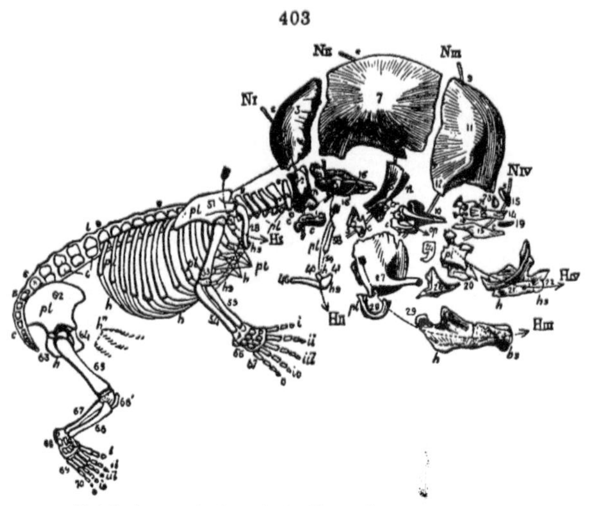

Vertebral segments shown in the Human fœtal skeleton. CXL.

neurapophyses, called 'alisphenoids,' *n*, the centrum of the frontal vertebra, called 'presphenoid,' and its neurapophyses (orbitosphenoids, 10), become anchylosed. The neural spine (parietal, 7) retains its primitive distinctness, but is enormously expanded, and is bifid, in relation to the vast size of the brain in Man. The parapophysis (mastoid, fig. 404, *c*) becomes confluent with the otic capsule (petrosal), the tympanic, *d*, squamosal, *a*, and with the pleurapophysis, called 'stylo-hyal,' fig. 403, 38, of the hæmal (hyoidian) arch. The hæmapophysis is ligamentous, save at its junction with the hæmal spine when it forms the ossicle called 'lesser cornu of the hyoid bone,' ib. 40, the spine itself being the basi-hyal, 41. The whole of this inverted arch is much reduced in size, its functions being limited to those of the tongue and larynx, in regard to taste, speech, and deglutition. The neurapophyses (orbitosphenoids, 10) becoming confluent with the centrum (presphenoid, 9) of the frontal vertebra, and the latter coalescing with that of the parietal vertebra, the compound bone called 'sphenoid' in Anthropotomy results, which combines the centrums and neurapophyses of two cranial vertebræ, together with a diverging appendage (pterygoid) of the maxillary arch.

The knowledge of the essential nature or 'general homology' of such a compound bone gives a clue to the phenomena of its developement from so many separate points, which neither embryology nor teleology could have afforded. As the centrum, 5, becomes confluent with 1, a still more complex whole results, which has accordingly been described as a single bone, under the name of 'os spheno-occipitale' in some anthropotomies. Such a bone has not fewer than twelve distinct centres of ossification, corresponding with as many distinct bones in the cold-blooded animals that depart less from the vertebrate archetype. The spine of the frontal vertebra (frontal bone) is much expanded and bifid, fig. 405, *d*, *d*, like the parietal bone; but the two halves more frequently coalesce into a single bone, with which the parapophysis (postfrontal, *b*) is connate. Much of the hæmal arch is consumed by the rapidly-growing 'ossicles of the ear,' and the proper pleurapophysis (tympanic bone, fig. 404, *d*) is reduced to the function of supporting the ear-drum, *b*; and becomes anchylosed to the squamosal, *a*, and mastoid, *c*. The hæmapophysis, fig. 403, 29, *hs*, is modified to form the dentigerous lower jaw, but articulates, as in other Mammals, with a diverging appendage (squamosal, 27), of the antecedent hæmal arch, now interposed between it and its proper pleurapophysis; the two hæmapophyses, originally separate, as in fig. 405, become confluent at their distal ends, forming the symphysis mandibulæ.

404

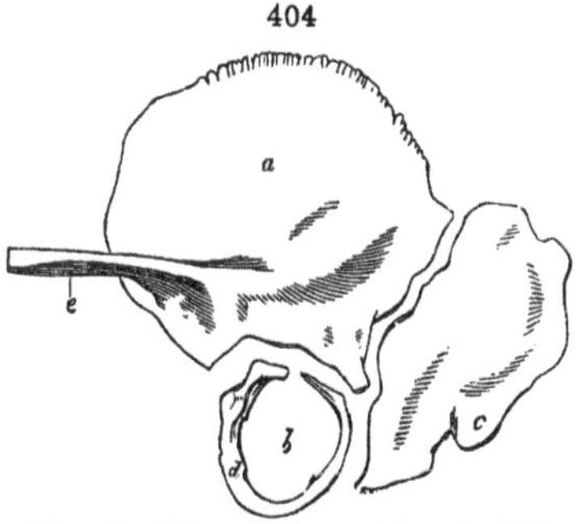

Elements of Human 'temporal bone,' outside view: the confluent ear-capsule, called 'petrous portion,' is not here seen.

The centrum of the first or nasal vertebra, like that of the last vertebra in Birds, is shaped like a ploughshare, and is called 'vomer,' fig. 403, 13; the neurapophyses have been subject to similar compression, and are reduced to a pair of vertical plates, which coalesce together, ib. 14, and with parts of the olfactory capsules (upper and middle turbinals), forming the compound bone called 'ethmoid.' The prefrontals assume this confluence and concealed position even in some fishes—*Xiphias*, e. g.—and repeat the character in all Mammalia and in most Birds; but they become partially exposed in the Ostrich and Batrachia. The spine of the nasal vertebra (nasal bones, ib. 15) is usually bifid, like those of the two succeeding segments; but it is much less expanded. The hæmal arch, called 'maxillary,' is formed by the pleurapophyses (palatines, 20) and by the hæmapophyses (maxillaries, 21), with which the halves of the bifid hæmal spine (premaxillaries, 22) are partly connate, and become completely confluent. Each moiety, or premaxillary, is reduced to the size required for the lodgment of two vertical incisors. As the canines in Man do not exceed the adjoining teeth in length, and the premolars are reduced to two in number, the alveolar extent of the maxillary is short, and the whole upper jaw is very slightly prominent.

Of the diverging appendages of the maxillary arch, the more constant one, called 'pterygoid,' 24, articulates with the palatine, but coalesces with the sphenoid; the second pair, formed by the malar, 26, and squamosal, 27, has been subject to a greater degree of modification: this appendage still performs the function assigned to it in Lizards and Birds, where it has its typical, ray-like figure, of connecting the maxillary with the tympanic, or one rib with the next; but the second division of the appendage (squamosal), which began to expand in the lower Mammalia, and to strengthen, without actually forming part of, the walls of the brain-case, as in fig. 140, 27, now attains its maximum of developement, and forms an integral constituent of the cranial parietes, filling up a very large cavity between the neural arches of the occipital and parietal segments. It coalesces, moreover, with the tympanic, mastoid, and petrosal, and forms, with the subsequently anchylosed stylo-hyal, a compound bone called 'temporal' in human anatomy. Embryology shows, empirically, the facts of developement: the key to the complex beginning of this 'cranial bone' is given by the discovery of the general pattern on which the skulls of the vertebrate animals have been constructed. In relation to that pattern, or to the archetype vertebrate skeleton, the Human temporal bone includes two pleurapophyses, 38 and 28, a parapophysis, 8, part of a diverging appendage, 27, and a sense-capsule, 16.

405

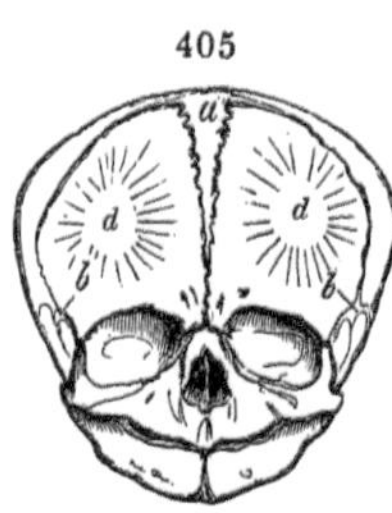

Front view of Human fœtal skull.

In the Human embryo the cartilaginous follows the fibrous stage of the brain-case in all the neurapophyses, viz.: exoccipitals, alisphenoids, orbitosphenoids, prefrontals. The latter already show their lateral confluence, closing the cranium anterior to the multiperforate part for the divisions of the olfactory nerve, called 'cribriform plate of the ethmoid,' and forming the 'crista galli' above, and the 'lamina perpendicularis' below, that plate; in connection with which are the 'turbinal' capsules, or supporters of the 'sense-organ,' which are also cartilaginous. The chondrified bases of the alisphenoids descend into the basisphenoid. The exoccipital cartilage ascends into the lower half of the superoccipital. The cartilaginous capsule of the ear-organ also sends a thin plate to the superoccipital, and a thicker process behind which becomes the basis of the mastoid. All the above cartilaginous parts are more or less continuous or confluent, and when separate from the un-chondrified extensions of the brain-capsule have been, illogically, termed 'primordial cranium' (Primordialschädel). But the actual embryonal or primordial skull is originally wholly membranous, and at the stage above described includes parts unchondrified, as well as those showing the intermediate histological conversion into cartilage. The bones, ossification of which begins in membrane, are the basioccipital, vomer, upper half of superoccipital, parietals, frontals, nasals, lacrymals, malars, squamosals, palatines, pterygoids. A pair of cartilaginous buds from the prefrontals form the piers of the yet unclosed anterior hæmal, or 'maxillary,' arch. A pair of cylindrical cartilages, called 'Meckel's,' are developed in the blastemal basis of the tympano-mandibular arch. The body and the stylo-hyal parts of the cornua of the hyoid are gristly before they ossify: much of the cerato-hyal parts of this thin hæmal arch retain their primitive fibrous condition. The capsule of the essential parts of the organ of vision is in the same 'sclerous' predicament in Man and Mammals: that of the organ of hearing becomes cartilaginous before it ossifies: the perfection of this organ in the well-brained Mammals calls for accessory parts, which show their true nature by their rapid growth.[1] The true comprehension of the developemental phenomena of the Human and Mammalian skull is afforded by that of its vertebral archetype: the artificial nature of the classification of the ossified parts into 'primordial skull-bones' and 'lid-bones' ('Deckknochen') is hereby plainly manifested: it is akin to that which divides them into 'eight bones of the cranium' and 'fourteen bones of the face.'

The first seven segments of the trunk consist each of 'centrum,' 'neurapophyses,' fig. 403, *n*, and 'pleurapophyses,' *pl*, the ultimate confluence of which forms the bone called 'cervical vertebra:' the centrum of the first of these coalesces with that of the second, and forms the 'odontoid process:' its place in the 'atlas' is taken by a 'hypaphophysis.' The pleurapophyses of the seventh cervical are occasionally elongated as 'ribs,' fig. 185, A, *b*. In the seven segments which succeed the cervicals, the pleurapophyses, *pl*, are elongated, and retain their freedom; and after the first they are shifted to the interspace between their own centrum and the

[1] The precocious developement of the ear-organ and its complex appendages in Mammals sorely perplex the devotees of developemental phenomena: the superadded bones of the ear-drum, growing straightway to full size, and appropriating much of the blastema of the pleurapophysial or tympanic part of the hæmal arch, have been veritable 'will-o'-the-wisps' to hunters of homologies on embryological ground.

next in advance (or above). The hæmapophyses, *h*, are gristly and interposed between the pleurapophyses and the hæmal spines, the conversion of which into the 'sternum' has been already explained. The fact of this short and slender bone in Man being ossified from a longitudinal series of centres (fig. 364, *b*) is learnt from embryology, the reason from general homology. The hæmal spine here repeats the variability of its homotype, the neural one, being sometimes entire, sometimes bifid (ib. *c*, *d*). In the three succeeding segments the pleurapophyses become shorter and the hæmapophyses are attached by their attenuated ends each to that in advance. In the next two segments the still shorter pleurapophyses resume the exclusive articulation with their proper centrum and terminate freely. The centrum and neurapophyses of each of the segments, with free and elongate pleurapophyses, constitute by their coalescence the 'dorsal vertebræ,' which are 'twelve' in number. Each of the five succeeding segments is represented by the centrum, neural arch, and short confluent pleurapophyses, forming the 'lumbar vertebræ:' the hæmapophyses of these segments are represented by the 'inscriptiones tendineæ musculi recti,' *h'*, which are the homologues of the gristly or bony 'abdominal ribs' of reptiles. The constitution of the Human 'os sacrum' has already been given. Part of a sacral pleurapophysis expands to form the 'ilium,' fig. 403, 62, *pl*. Two hæmapophyses called 'ischium,' 63, and pubis, 64, coalesce with 62 to constitute the 'innominatum:' the inverted arch, supporting the appendage which becomes developed into 'pelvic limb,' is completed by the ischio-pubic symphysis.

406

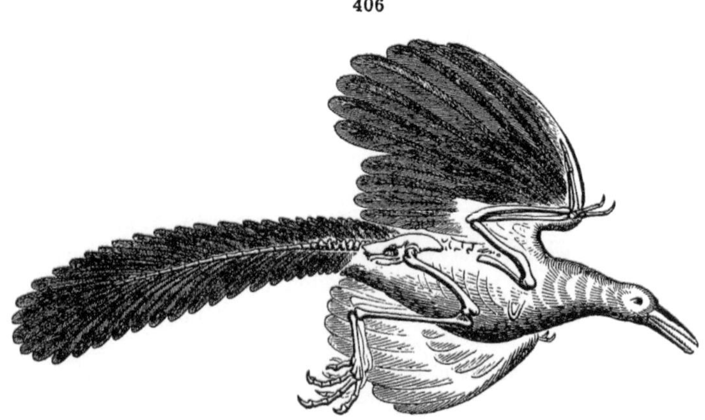

Restoration of *Archeopteryx*, a mesozoic bird.

TABLE I.—SYNONYMS OF THE BONES OF THE HEAD, ACCORDING TO THEIR GENERAL HOMOLOGIES. [To face page 586, Vol. II.

The material originally positioned here is too large for reproduction in this reissue. A PDF can be downloaded from the web address given on page iv of this book, by clicking on 'Resources Available'.

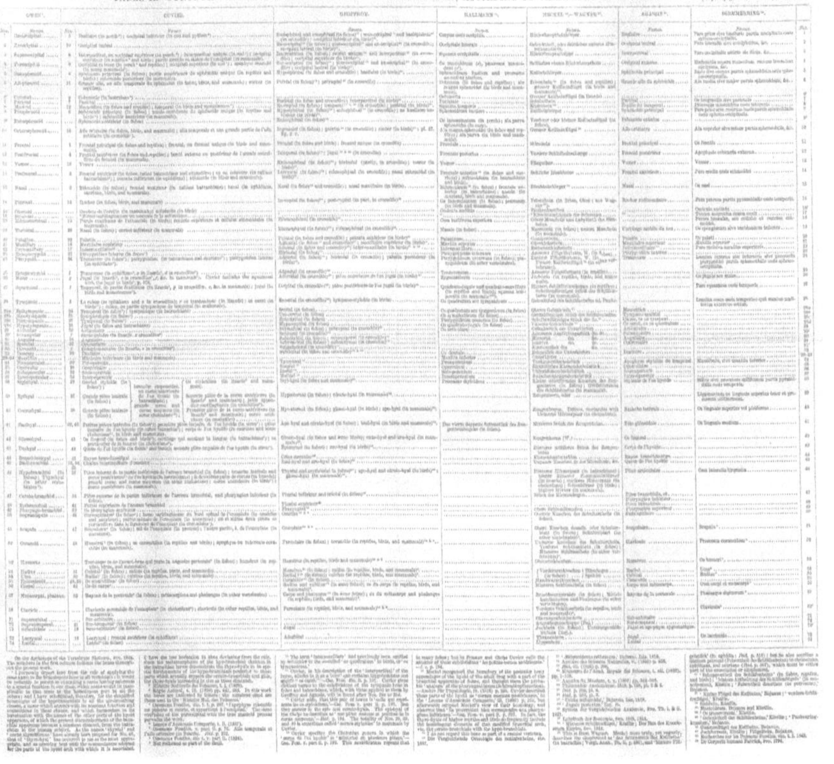
TABLE II.—SYNONYMS OF THE BONES OF THE HEAD OF VERTEBRATES, ACCORDING TO THEIR SPECIAL HOMOLOGIES.

The material originally positioned here is too large for reproduction in this reissue. A PDF can be downloaded from the web address given on page iv of this book, by clicking on 'Resources Available'.

Table III.—SYNONYMS OF THE ELEMENTS OF THE TYPICAL VERTEBRA.

OWEN.[1]	GEOFFROY.[2]	CARUS.[4]	MULLER.[5]	CUVIER.[6]	SOEMMERRING.[7]
Autogenous Elements.					
Centrum (κέντρον, centre)	Cycléal	Tertiar-wirbel	Wirbel-körper	Corps de vertèbre	Corpus vertebræ.
Neurapophysis (νεῦρον, nerve, and ἀπόφυσις, a process of bone).	Périal	Deckplatten and Grundplatten	Oberer Wirbelbogen	Partie annulaire, ou lames vertébrales.	Arcus posterior vertebræ, seu radices arcus posterioris.
Pleurapophysis (πλευρὰ, a rib, and ἀπόφυσις).	Paraal	Rückentheil and Ober-sternal-theil des Urwirbelbogens.	...	Côtes vertébrales	Processus transversus vertebræ cervicalis. Costa, seu pars vertebralis, seu ossea, costæ.
Hæmapophysis; by syncope for hæmato-apophysis (from Gr. αἶμα, blood, and ἀπόφυσις).	Cataal	Unter-sternal-theil des Urwirbelbogens.	Unterer Wirbelbogen	Côtes sternales (in thorax); côtes abdominales, ou cartilages ventraux (in abdomen); os ployé en chevron (in tail).	Cartilago costæ seu pars sternalis costæ; (in the abdomen) inscriptiones tendineæ musculi recti.
Neural spine	Périal (in fishes); épial (in other vertebrates).	(Its base is the) Oberer Tertiar-wirbel, (its apex is the) Oberer Dornfortsatz.	Oberer Dornfortsatz	Apophyse épineuse supérieure	Processus spinosus vertebræ.
Hæmal spine	Paraal (in fishes); cataal (in other vertebrates).[3]	Sternal-wirbel Körper	Unterer Dornfortsatz	Apophyse épineuse inférieure	Ossa sterni et processus ensiformis; (in the abdomen) linea alba.
Exogenous Parts.					
Parapophysis (παρὰ, across, and ἀπόφυσις).	Paraal (in the tail of fishes).	Querfortsatz	Unterer Querfortsatz	Apophyse transverse	Radix prior seu antica processus transversi vertebræ.
Diapophysis (διὰ, across, and ἀπόφυσις)	Paraal (in reptiles and mammals).	Querfortsatz	Oberer Querfortsatz	Apophyse transverse	Radix postica processus transversi vertebræ, (and) processus transversus.
Zygapophysis (ζυγὸς, junction, and ἀπόφυσις).	...	Seitlicher Tertiar-wirbel	Gelenk-fortsatz	Apophyse articulaire	Processus obliquus vertebræ.
Anapophysis (ἀνὰ, upward or backward, and ἀπόφυσις).	...	...	...	Prolongement postérieur d'apophyse transverse; seconde apophyse articulaire; apophyse styloïde.	
Metapophysis (μετὰ, between, and ἀπόφυσις).	...	...	...	Apophyse articulaire prolongée en une pointe.	Processus accessorius processui transverso et articulari superiori interpositus.
Hypapophysis (ὑπὸ, below, and ἀπόφυσις).	...	...	...	Apophyse épineuse inférieure	

[1] Description of the Plesiosaurus macrocephalus (April 1838), in 'Geological Transactions,' 2nd series, vol. v. p. 518; and CXL.

[2] Geoffroy St.-Hilaire (Étienne), Mémoires du Muséum, 4to. t. ix. 1822, p. 89.

[3] The dermal spines which sustain the dorsal and anal fins of fishes are called respectively 'épiaux' and 'cataaux' by Geoffroy.

[4] Carus (Carl Gustav), Urtheile des Knochen- und Schalen-gertistes, fol. 1828.

[5] Müller (Johannes), Vergleichende Anatomie der Myxinoiden: Abhand. Akad. der Wissenschaften zu Berlin, 1834. The terms adopted in most of the recent works of the German zootomists correspond with those of John Müller.

[6] Leçons d'Anatomie Comparée, t. i. edit. 1835.

[7] De Corporis Humani Fabricâ, 8vo. 1794.

WORKS

REFERRED TO BY ROMAN NUMERALS AND DOT IN THE SECOND VOLUME.

(Those referred to by Roman Numerals without Dot are in the First Volume.)

I·. Becquerel and Breschet. Recherches sur la Chaleur animale au moyen des appareils thermo-électriques, in Archives du Muséum, t. i.

II·. Martin. Description des effets du Sommeil sur la Chaleur du corps humain, in Journal de Physique, 1773, t. ii.

III·. Brown-Séquard. Experimental Researches (on the Normal Degree of the Temperature of Man).

IV·. Martins. Sur la Température des Oiseaux palmipèdes du Nord de l'Europe, in Mémoires de l'Académie des Sciences et Lettres de Montpellier, 1856, t. iii.

V·. Chossat. Influence du Système nerveux sur la Chaleur animale. 8vo. 1820.

VI·. Budge. De l'Influence de la Moëlle épinière sur la Chaleur de la Tête, in Comptes Rendus de l'Académie des Sciences. 1853.

VII·. Owen. Art. Aves, Cyclopædia of Anatomy, vol. i. 1836.

VIII·. Nitzsch. Osteographische Beiträge zur Naturgeschichte d. Vögel. 8vo. 1811.

IX·. Vigors. Observations on the Natural Affinities that connect the Orders and Families of Birds. Transactions of the Linnæan Society, vol. xiv. 1827.

X·. Owen. On the Megatherium, Philosophical Transactions, 1851.

XI·. Owen. On the Anatomy of the Southern Apteryx (*Apteryx australis*), Transactions of the Zoological Society, vol. ii. 1838, and vol. iii. 1842.

XII·. Owen. On the Osteology of the Great Awk (*Alca impennis*), Transactions of the Zoological Society, vol. v. 1864, and
Cuvier, Leçons d'Anatomie Comparée.

XIII·. Blythe, in Orr's edition of Cuvier's Animal Kingdom. 8vo. 1840.

XIV·. Eyton. Osteologia Avium. 4to. 1861–4.

XV·. Owen. On the Archeopteryx, Philosophical Transactions, 1863.

XVI·. Owen. On Dinornis, Transactions of the Zoological Society of London, 1839–1864.

XVII·. Owen. Palæontology. 8vo. 2nd ed. 1861.

XVIII·. Owen. On Notornis, Transactions of the Zoological Society, vol. iii. 1848.

XIX·. Owen. On the Affinities of *Gastornis parisiensis*. Quarterly Journal of the Geological Society. 1856.

XX·. Owen. Catalogue of the Physiological Series in the Museum of the Royal College of Surgeons.

XXI·. Owen. On the Anatomy of the Flamingo (*Phœnicopterus*), Proceedings of the Zoological Society, vol. ii. 1835.

XXII·. Owen. On the Anatomy of the Gannet (*Sula*), Proceedings of the Zoological Society, vol. i. 1833.

XXIII·. Owen. On the Anatomy of the Red-backed Pelican (*Pelecanus rufus*), Proceedings of the Zoological Society, vol. iii. 1836.

XXIV·. Owen. Memoir on the Apteryx, Proceedings of the Zoological Society, vol. x. 1842; and
Yarrell, Wm., On the Xiphoid Bone, &c., of the Cormorant.

XXV·. Owen, R. On the Anatomy of the Hornbill (*Buceros cavatus*), Transactions of the Zoological Society, vol. i. 1836.

XXVI·. Owen, R. On the Dodo, Transactions of the Zoological Society, vol. iii.

XXVII·. Strickland and Melville. The Dodo and its Kindred. 4to. 1848.

XXVIII·. Yarrell, Wm. On the Structure of the Beak and its Muscles in *Loxia curvirostra*, Zoological Journal, vol. iv.

XXIX·. Mayer. Analekten für Vergleichende Anatomie. 4to. 1839.

XXX·. Müller, J. Physiologie des Gesicht-Sinnes, 8vo. 1826; and
Bishop. Art. Voice, in Cyclopædia of Anatomy, vol. iv. (CCCXX. vol. i.).

XXXI·. Owen, R. On the Olfactory and Trigeminal Nerves of the Vulture, Turkey, and Goose, Proceedings of the Zoological Society, Part v. 1837.

XXXII·. Audubon, J. J. Ornithological Biography. 8vo. 5 vols. 1831–1839.

XXXIII·. Edwards, G. A Natural History of Uncommon Birds, &c. 4to. 1743–1751.

XXXIV·. Crampton, Sir P. Description of an Organ by which the eyes of Birds are accommodated to the different distances of objects, Annals of Philosophy, 1813. 8vo.

XXXV·. Valentin. Fortgesetzte Untersuchungen über die Flimmerbewegung, in Repertorium d'Anatomie; and
Purkinje. Commentatio Physiologica de Phenomeno motus vibratorii continui. 4to. 1835.

XXXVI·. Geoffroy St.-Hilaire, Ét. Système Dentaire des Mammifères et des Oiseaux. 8vo. 1824.

XXXVII·. Nitzsch, in Meckel's Archiv für Physiologie, 1826.

XXXVIII·. Newton, Alfred. On the supposed Gular Pouch of the Male Bustard, in the 'Ibis,' April 1862.

XXXIX·. Lund. De Genere Euphones. Hafn. 1829.

XL·. Pérault. Mémoires de l'Académie Royale des Sciences, Paris, t. iii.

XLI·. Lauth. Mémoire sur les Vaisseaux Lymphatiques des Oiseaux, Annales des Sciences Nat., vol. iii. 1825.

XLII·. Panizza. Osservazione Antropo-zootomico-fisiologiche. Fol. 1830.

XLIII·. Macartney. Art. Birds, in Rees' Cyclopædia.

XLIV·. Owen. Catalogue of the Osteology, Royal College of Surgeons; and
Barkow. Untersuchungen über das Schlagadersystem d. Vögel, Meckel's Archiv für Physiologie. 1828.

XLV·. Rainey. Memoir on the Minute Structure of the Lungs, and on the Minute Anatomy of the Lung of the Bird, Medico-Chirurgical Transactions, vols. xxii. and xxiii.

XLVI·. Yarrell, Wm. On the Organs of Voice in Birds, Transactions of the Linnæan Society, vol. xvi.; and
Meckel. Vergleich. Anatomie.

XLVII·. Yarrell, Wm. Observations on the Tracheæ of Birds, Transactions of the Linnæan Society, vol. xv.

XLVIII·. Yarrell, Wm. On a new species of Wild Swan, &c., in Transactions of the Linnæan Society, vol. xvi.

XLIX·. Müller, J. Ueber Compensation der physischen Kräfte am menschlichen Stimmorgane; mit Bemerkungen über die Stimme der Säugethiere, der Vögel und der Amphibien, 8vo. 1835. Vergleichende Anatomie der Stimmorgane bei Vögeln, Berlin. Akad. 1845.

L·. Geoffroy St.-Hilaire. Mémoire sur les Appareils sexuels et urinaires de l'Ornithorhynque, in Mémoires du Muséum, t. xv.; and Composition des Appareils génitaux, urinaires et intestinaux dans l'Autruche, etc., in Mémoires du Muséum, t. xx.

LI·. Cuvier, F. Sur le Développement des Plumes, in Mémoires du Muséum, t. xiii.

LII·. Yarrell, W. On the Laws that regulate the Changes of Plumage in Birds, Transactions of the Zoological Society, vol. i.

LIII·. Jaeger, Gustav. Das Os humeroscapulare der Vögel, Sitzungsberichte der Mathem.-naturwissensch. Classe, Kaiserl. Akad. der Wissensch. in Wien. 2tes Heft. 1857.

LIV·. Nitzsch, C. L. System der Pterylographie. 4to. 1840.

LV·. Owen. John Hunter's Observations on Animal Development edited, and his Illustrations of that Process in the Bird described. Fol. 1841.

LVI·. Gould, John, F.R.S. On the Brush Turkey (*Talegalla Lathami*), and On the Habits and Characters of *Leipoa ocellata*, Proceedings of the Zoological Society, vol. viii.

LVII·. Gould, John, F.R.S. On the 'Run' or Playing-house constructed by the Satin-bird (*Ptilonorhynchus holosericeus*), &c., of Australia, Proceedings of the Zoological Society, vol. viii.

LVIII·. Gould, John, F.R.S. The Birds of Australia. 7 vols. fol. 1840–48.

LIX·. Schwann. De necessitate Aeris Atmospherici ad evolutionem Pulli in Ovo incubito. 4to. 1834.

LX·. Soemmerring, S. M. De Sectione Oculorum.

LXI·. Sappey. Recherches sur l'Appareil respiratoire des Oiseaux. 4to. 1847.

LXII·. Owen, R. On *Thylacotherium* and *Phascolotherium*, Transactions of the Geological Society, 2nd series, vol. vi. 1839.

LXIII·. Owen, R. Description of the Fœtal Membranes and Placenta of the Elephant, Philosophical Transactions, 1857.

LXIV·. Owen, R. On the Characters, Principles of Division, and Primary Groups of the Class Mammalia, Proceedings of the Linnæan Society, February and April, 1857. 'Rede's Lecture' on the same subject. 8vo. 1858.

LXV·. Eschricht and Reinhardt. Om Nordhvalen. 4to. 1861.

LXVI·. Paley. Natural Theology. 10th ed. 8vo. 1805.

LXVII·. Müller, J. Ueber zwei verschiedene Typen in dem Bau der erectilen männlichen Geschlechtsorgane bei den straussartigen Vögeln. 4to. 1838.

LXVIII·. Owen, R. Hunterian Lectures on the Nervous System, in Medical Times, 1840.

LXIX·. Vrolik. Art. Quadrumana, Cyclopædia of Anatomy, vol. iv.

LXX·. Owen. On the Structure of the Brain in Marsupial Animals, Philosophical Transactions, 1837.

LXXI·. Turner, H. N. On the Evidences of Affinity afforded by the Skull in the Ungulate Mammalia. Ann. and Mag. of Nat. History, Dec. 1850, ser. 2, vol. iii.

LXXII·. Milne-Edwards, Alf. Recherches sur la Famille des Chevrotains, in Annales des Sciences, 5th ser. vol. ii. 1864.

LXXIII·. Pander and D'Alton. Die Vergleichende Osteologie. Fol. 1821–7.

LXXIV·. Owen, R. Outlines of a Classification of the Marsupialia, Zoological Transactions, vol. ii. 1839.

LXXV·. Owen, R. On the Generation of the Marsupial Animals, with a description of the Impregnated Uterus of the Kangaroo, Philosophical Transactions, 1834; and on the Osteology of the Marsupialia, Zoological Transactions, Part 1, vol. ii. 1841; Part 2, vol. iii. 1845.

LXXVI·. Owen, R. On the Mammary Glands of the *Ornithorhynchus paradoxus*, Philosophical Transactions, 1832, p. 517.

LXXVII·. Owen, R. On the Ova of the *Ornithorhynchus paradoxus*, Philosophical Transactions, 1834.

LXXVIII·. Owen, R. On the Young of the *Ornithorhynchus paradoxus*, Zoological Transactions, vol. i. 1834.

LXXIX·. Owen, R. Art. Monotremata, Cyclopædia of Anatomy, vol. iii. 1841.

LXXX·. Owen, R. Art. Marsupialia, Cyclopædia of Anatomy, vol. iii. 1841.

LXXXI·. Meckel. *Ornithorhynchi paradoxi* Descriptio Anatomica. Fol. 1826.

LXXXII·. Owen, R. Anatomy of *Capromys Fournieri*, Proceedings of the Committee of Science of the Zoological Society of London. 8vo. 1832.

LXXXIII·. Owen, R. On some Outline-drawings and Photographs of the Skull of the *Zygomaturus trilobus*, Macl., *Nototherium*, Owen, Quarterly Journal of the Geological Society, vol. xv. 1858.

LXXXIV'. PETERS. Naturw. Reise nach Mosambique: Säugthiere. 4to.
LXXXV'. PETERS. Ueber das normale Vorkommen von nur sechs Halswirbeln bei *Choloepus Hoffmanni*, Monatsbericht der Königl. Akad. der Wissenschaften zu Berlin, Dec. 1864.
LXXXVI'. OWEN. On the Cervical and Lumbar Vertebræ of the Mole (*Talpa europæa*, L.), Transactions of the British Association, 1858.
LXXXVII'. MIRAIN, Ed. Ueber einen eigenthümlichen Bau des Gehörorganes bei den Nagern. 8vo. 1840.
LXXXVIII'. YARRELL, Wm. On the Osteology of the *Chlamyphorus truncatus*, in the Zoological Journal, vol. iii. 1830.
LXXXIX'. OWEN. On the Glyptodon. Transactions of the Geological Society, 2nd ser. vol. vi., and Catalogue of the Fossil Mammalia in the Royal College of Surgeons. 4to. 1845.
XC'. OWEN, R. Description of the Skeleton of the Mylodon. 4to. 1842.
XCI'. OWEN, R. Memoir on the Megatherium. 4to. 1860.
XCII'. OWEN, R. Catalogue of Fossil Organic Remains (Mammalia and Aves) in the Museum of the Royal College of Surgeons of England. 4to. 1845.
XCIII'. MACLEAY, W. S. History and Description of the Skeleton of a New Sperm-whale, &c. 8vo. Sydney: 1851.
XCIV'. BRANDT, J. F. *Symbolæ Sirenologicæ.* 4to. 1846.
XCV'. OWEN, R. Fossil Mammalia of the Voyage of the Beagle. 4to. 1840.
XCVI'. COBBOLD, T. S. Art. *Ruminantia*, Cyclopædia of Anatomy, Supplement, 1859.
XCVII'. OWEN, R. Memoir on the Anatomy of the Nubian Giraffe, Zool. Trans., vol. ii.
CXVIII'. STANNIUS, H. Beiträge zur Kenntniss der Amerikanischen Manati. 4to. 1845.
XCIX'. OWEN, R. On Indian Cetacea, Transactions of the Zoological Society, vol. v. (1866).
C'. RAPP, W. v. Die Cetaceen zoologisch-anatomisch dargestellt. 8vo. 1837.
CI'. MURRAY, A. On the genus Galago, Edinb. New Philosophical Journal. New ser., Oct. 1859.
CII'. OWEN, R. Monograph on the Aye-aye (*Cheiromys madagascariensis*, Cuv.). 4to. 1863.
CIII'. OWEN, R. Osteological Contributions to the Natural History of the Orangs (*Pithecus*) and Chimpanzees (*Troglodytes*), Transactions of Zoological Society, vols. ii., iii., and (*Troglodytes Gorilla*) vols. iv. and v. Parts I.–VIII.
CIV'. DE BLAINVILLE, H. D. Ostéographie, &c. 4to. and fol. (plates). 1841.
CV'. WOOD, Jno. Art. Pelvis, Cyclopædia of Anatomy, Supplement.
CVI'. WEBER, M. J. Die Lehre von dem Ur- und Racen-formen der Schädel und Becken des Menschen. 1830.
CVII'. VROLIK, W. Platen behoorende tot de Beschowing van het Verschel der Bekkens in onderscheidene Volkstammen. 1826.
CVIII'. BRESCHET. Recherches sur différentes Pièces du Squelette des Animaux Vertébrés, etc., in Annales des Sciences Naturelles. 2e série, t. x. (Zoologie).
CIX'. MORTON, Dr. Crania Americana. 1839.
CX'. OWEN, R. Memoir on the Gorilla (*Troglodytes Gorilla*, S.). 4to. 1865.
CXI'. ROKITANSKI. Lehrbuch der Pathologische Anatomie.
CXII'. ZAWARYKIN, Th. Ueber die Aufbewahrung von Blutkrystallen mittelst Aether, in Kais. Akad. der Wissenschaften in Wien, Sitzungsberichte, 8vo. Nr. IV. p. 14.
CXIII'. TSCHERINOFF, M. Ueber die Abhängigkeit des Glykogengehaltes der Leber von der Ernährung, in Kais. Akad. der Wissenschaften in Wien, Nr. XI. pp. 64–65.
CXIV'. TÖRÖK, Aurel. Beiträge zur Kenntniss der ersten Anlagen der Sinnesorgane und der primären Schädelformation bei den Batrachiern, in Kais. Akad. der Wissenschaften in Wien, *Ib.* Nr. XXIX. pp. 206 bis 207.

CXV·. STRICKER, S. Ueber den Bau und das Leben der capillaren Blutgefässe, in Kais. Akad. der Wissenschaften in Wien, *Ib.* Nr. XXII., pp. 164–165. Ueber die Histologie der Gehirnentzündung, in Kais. Akad. der Wissenschaften in Wien, Sitzungsberichte, 8vo. Nr. XXVI. pp. 186–187.

CXVI·. HYRTL, J. Ueber einen freien Körper im Herzbeutel, in Kais. Akad. der Wissenschaften in Wien, *Ib.* Nr. IX. p. 48.

CXVII·. HYRTL, J. Ein *Pancreas accessorium* und *Pancreas divisum*, in Kais. Akad. der Wissenschaften in Wien, *Ib.* Nr. XXI. p. 148.

CXVIII·. HYRTL, J. Eine quere Schleimhautfalte in der Kehlkopfhöhle, in Kais. Akad. der Wissenschaften in Wien, *Ib.* Nr. XXI. p. 148.

CXIX·. GOETHE. Zur Naturwissenschaft überhaupt, besonders zur Morphologie. 4 Bde. M. Kupf. 8vo. 1817–23.

CXX·. HARTWIG, G. Gott in der Natur, oder die Einheit der Schöpfung. Mit Abbild. 8vo. 1864.

CXXI·. RUDOLPHI, K. A. Zur Anthropologie u. allgem. Naturgesch. 8vo. 1812.

CXXII·. BUFFON et DAUBENTON. Histoire naturelle générale et particulière. 14 vols. 4to. 1766.

CXXIII·. HUTCHINSON. Art. Thorax, Cyclopædia of Anatomy, vol. iv.

CXXIV·. L'HERMINIER. Recherches sur l'Appareil sternal des Oiseaux. Mémoires de la Soc. Linn. de Paris, tom. iii. 1827.

CXXV·. BLANCHARD, Em. Ostéologie des Oiseaux. Annales des Sciences Nat., sér. iv. tom. ix. 1859.

END OF THE SECOND VOLUME.

LONDON

PRINTED BY SPOTTISWOODE AND CO.

NEW-STREET SQUARE

For EU product safety concerns, contact us at Calle de José Abascal, 56–1°, 28003 Madrid, Spain or eugpsr@cambridge.org.

www.ingramcontent.com/pod-product-compliance
Ingram Content Group UK Ltd.
Pitfield, Milton Keynes, MK11 3LW, UK
UKHW040735110726
473066UK00021B/277

* 9 7 8 1 1 0 8 0 3 8 2 6 3 *